**Corrosion Resistance of Copper
and Copper Alloys**
Corrosive Agents and Their Interaction
with Copper and Copper Alloys

Corrosion Resistance of Copper and Copper Alloys

Corrosive Agents and Their Interaction
with Copper and Copper Alloys

Editors

Prof. Dr.-Ing. Michael Schütze
Executive Director
Karl-Winnacker-Institut of DECHEMA e. V.
Society for Chemical Engineering and Biotechnology
Theodor-Heuss-Allee 25
60486 Frankfurt
Germany

Prof. Dr. Ralf Feser
FB Informatik und Naturwissenschaften
University of Applied Sciences South Westphalia
Frauenstuhlweg 31
58644 Iserlohn
Germany

Dr. rer. nat. Roman Bender
Chief Executive of GfKORR e. V.
Society for Corrosion Protection
Theodor-Heuss-Allee 25
60486 Frankfurt
Germany

Cover Illustration
Source: Karl Winnacker Institute of DECHEMA e. V., Frankfurt (Main), Germany

Warranty Disclaimer

This book has been compiled from literature data with the greatest possible care and attention. The statements made only provide general descriptions and information.

Even for the correct selection of materials and correct processing, corrosive attack cannot be excluded in a corrosion system as it may be caused by previously unknown critical conditions and influencing factors or subsequently modified operating conditions.

No guarantee can be given for the chemical stability of the plant or equipment. Therefore, the given information and recommendations do not include any statements, from which warranty claims can be derived with respect to DECHEMA e. V. or its employees or the authors.

The DECHEMA e. V. is liable to the customer, irrespective of the legal grounds, for intentional or grossly negligent damage caused by their legal representatives or vicarious agents.

For a case of slight negligence, liability is limited to the infringement of essential contractual obligations (cardinal obligations). DECHEMA e. V. is not liable in the case of slight negligence for collateral damage or consequential damage as well as for damage that results from interruptions in the operations or delays which may arise from the deployment of this book.

■ This book was carefully produced. Nevertheless, editors, authors and publisher do not warrant the information contained therein to be free of errors. Readers are advised to keep in mind that statements, data, illustrations, procedural details or other items may inadvertently be inaccurate.

Library of Congress Card No.: Applied for.

British Library Cataloguing-in-Publication Data:
A catalogue record for this book is available from the British Library.

Bibliographic information published by Die Deutsche Bibliothek
Die Deutsche Bibliothek lists this publication in the Deutsche Nationalbibliografie; detailed bibliographic data is available in the Internet at <http://dnb.ddb.de>.

© 2011 DECHEMA e. V., Society for Chemical Engineering and Biotechnology, 60486 Frankfurt (Main), Germany

All rights reserved (including those of translation into other languages). No part of this book may be reproduced in any form – nor transmitted or translated into machine language without written permission from the publishers. Registered names, trademarks, etc. used in this book, even when not specifically marked as such, are not to be considered unprotected by law.

Printed in the Federal Republic of Germany
Printed on acid-free paper

Typesetting Kühn & Weyh, Satz und Medien, Freiburg
Printing Strauss GmbH, Mörlenbach
Binding Strauss GmbH, Mörlenbach
Cover Design Graphik-Design Schulz, Fußgönheim

ISBN: 978-3-527-33224-3

Contents

Preface *VII*

How to use the Handbook *IX*

Warranty disclaimer *1*

Acetates *3*
L. Hasenberg

Acetic Acid *5*
G. Elsner

Acid Halides *27*
G. Elsner

Aliphatic Aldehydes *31*
G. Elsner

Aliphatic Amines *37*
L. Hasenberg

Aliphatic Ketones *41*
H. Barkholt

Alkaline Earth Chlorides *43*
R. Weidemann

Alkaline Earth Hydroxides *45*
A. Weser

Alkanecarboxylic Acids *47*
L. Hasenberg

Alkanols *57*
K. Hauffe

Aluminium Chloride *61*
L. Hasenberg

Amine Salts *63*
K. Hauffe

Ammonia and Ammonium Hydroxide *69*
P. Drodten

Ammonium Salts *97*
K. Hauffe

Atmosphere *119*
K. Baumann

Benzene and Benzene Homologues *153*
K. Hauffe

Bromides *157*
K. John

Bromine *163*
K. John

Carbonic Acid *167*
P. Drodten

Carboxylic Acid Esters *175*
L. Hasenberg

Chloroethanes *183*
H. G. Spilker

Chloromethanes *197*
H. G. Spilker

Chlorine and Chlorinated Water *217*
K. Hauffe

Chlorine Dioxide *227*
L. Hasenberg

Drinking Water *243*
G. Heim, K. Reeh

Ferrous Chlorides (FeCl$_2$, FeCl$_3$) *249*
A. Werner

Fluorides *261*
K. Hauffe

Fluorine, Hydrogen Fluoride, Hydrofluoric Acid *263*
K. Hauffe

Formic Acid *271*
H. Leyerzapf

Hot Oxidizing Gases *275*
K. Hauffe

Hydrochloric Acid *293*
A. Bäumel, P. Drodten

Hydrogen Chloride *305*
H. Barkholt

Hypochlorites *307*
L. Hasenberg

Lithium Hydroxide *319*
K. John

Methanol *323*
H. G. Spilker

Mixed Acids *351*
M. B. Rockel

Nitric Acid *355*
K. Hauffe

Phosphoric Acid *369*
L. Hasenberg

Polyols *389*
G. Elsner

Potassium Chloride *395*
L. Hasenberg

Potassium Hydroxide *417*
P. Drodten

Seawater *423*
P. Drodten

Sodium Chloride *465*
M. B. Rockel

Sodium Hydroxide *521*
P. Drodten

Sodium Sulfate *529*
J. Küpper-Feser

Soil (Underground Corrosion) *553*
G. Elsner

Steam *559*
H. Leyerzapf

Sulfonic Acid *567*
K. Hauffe

Sulfur Dioxide *569*
L. Hasenberg

Sulfuric Acid *585*
L. Hasenberg

Waste Water (Industrial) *631*
E. Heitz, G. Subat

Bibliography *633*

Index of materials *715*

Subject Index *729*

Preface

Practically all industries face the problem of corrosion – from the micro-scale of components for the electronics industries to the macro-scale of those for the chemical and construction industries. This explains why the overall costs of corrosion still amount to about 2 to 4% of the gross national product of industrialized countries despite the fact that zillions of dollars have been spent on corrosion research during the last few decades.

Much of this research was necessary due to the development of new technologies, materials and products, but it is no secret that a considerable number of failures in technology nowadays could, to a significant extent, be avoided if existing knowledge were used properly. This fact is particularly true in the field of corrosion and corrosion protection. Here, a wealth of information exists, but unfortunately in most cases it is scattered over many different information sources. However, as far back as 1953, an initiative was launched in Germany to compile an information system from the existing knowledge of corrosion and to complement this information with commentaries and interpretations by corrosion experts. The information system, entitled "DECHEMA-WERKSTOFF-TABELLE" (DECHEMA Corrosion Data Sheets), grew rapidly in size and content during the following years and soon became an indispensable tool for all engineers and scientists dealing with corrosion problems. This tool is still a living system today: it is continuously revised and updated by corrosion experts and thus represents a unique source of information. Currently, it comprises more than 12,000 pages with approximately 110,000 corrosion systems (i.e., all relevant commercial materials and media), based on the evaluation of over 100,000 scientific and technical articles which are referenced in the database.

Last century, an increasing demand for an English version of the DECHEMA-WERKSTOFF-TABELLE arose in the 80s; accordingly the first volume of the DECHEMA Corrosion Handbook was published in 1987. This was a slightly condensed version of the German edition and comprised 12 volumes. Before long, this handbook had spread all over the world and become a standard tool in countless laboratories outside Germany. The second edition of the DECHEMA Corrosion Handbook was published in 2004. Together the two editions covered 24 volumes.

The present book compiles all information on the corrosion behaviour of Copper alloys that was compiled in the volumes of the corrosion handbook. This compilation is an indispensable tool for all engineers and scientists dealing with corrosion problems of Copper and its alloys.

Copper and its alloys have been utilized for more than 10,000 years. Today, copper is one of the most commonly used metals in the world; 24 million tons are consumed worldwide.

A wide variety of copper alloys are used in a range of applications. As well as good mechanical properties, the excellent electrical conductivity and thermal conduction are reasons copper alloys are deployed in many industrial fields. Copper plays a role in electronic and electrical applications and all forms of heat transfer. In automobiles as well as in houses copper could not be replaced. In the sanitary industry copper and brass are well established, for example, drinking water pipes have been used for decades without problems.

While the corrosion resistance of copper and its alloys is excellent in unpolluted air and drinking water, corrosion rates in impure environments can be much higher and lead to severe material damage. Corrosion is a system property, so it is important to find the right copper material with regard to the environmental conditions it will be exposed to.

This handbook highlights the limitations of the use of copper and its alloys in various corrosive solutions and provides vital information on corrosion protection measures.

Corrosion is a complex phenomenon that depends on a number of parameters, related to both the environment and the metal. In this handbook the behaviour of aluminium and of the commonly used aluminium alloys in different corrosive environments is shown.

The chapters are arranged by the agents leading to individual corrosion reactions, and a vast number of aluminium alloys are presented in terms of their behaviour in these agents. The key information consists of quantitative data on corrosion rates coupled with commentaries on the background and mechanisms of corrosion behind these data, together with the dependencies on secondary parameters, such as flow-rate, pH, temperature, etc. This information is complemented by more detailed annotations where necessary, and by an immense number of references listed at the end of the handbook.

An important feature of this handbook is that the data was compiled for industrial use. Therefore, particularly for those working in industrial laboratories or for industrial clients, the book will be an invaluable source of rapid information for day-to-day problem solving. The handbook will have fulfilled its task if it helps to avoid the failures and problems caused by corrosion simply by providing a comprehensive source of information summarizing the present state-of-the-art. Last but not least, in cases where this knowledge is applied, there is a good chance of decreasing the costs of corrosion significantly.

Finally the editors would like to express their appreciation to Lieselotte Wolf and Dr. Horst Massong for their admirable commitment and meticulous editing of a work that is encyclopedic in scope.

They are also indebted to Gudrun Walter of Wiley-VCH for their valuable assistance during all stages of the preparation of this book.

Michael Schütze, Ralf Feser and Roman Bender

How to use the Handbook

The Handbook provides information on the chemical resistance and the corrosion behavior of aluminium in different attacking chemical media and mixtures.

The user is given information on the range of applications and corrosion protection measures.

Research results and operating experience reported by experts allow recommendations to be made for the selection of aluminium and to provide assistance in the assessment of damage.

The objective is to offer a comprehensive and concise description of the behavior of Aluminium in contact with a particular medium.

The information on resistance is given as text, tables, and figures. The literature used by the author is cited at the corresponding point. There is an index of materials as well as a subject index at the end of the book so that the user can quickly find the information given for a particular keyword.

The Handbook is thus a guide that leads the reader to aluminium materials that have already been used in certain cases, that can be used or that are not suitable owing to their lack of resistance.

The resistance is coded with three evaluation symbols in order to compress the information. Uniform corrosion is evaluated according to the following criteria:

Symbol	Meaning	Area-related mass loss rate		Corrosion rate
		$g/(m^2 h)$	$g/(m^2 d)$	mm/a
+	resistant	≤ 0.3	≤ 0.8	≤ 0.1
$\oplus$	fairly resistant	< 0.3	< 8.0	< 1.0
−	not resistant	> 0.3	> 8.0	> 1.0

The evaluation of the corrosion resistance of metallic materials is given

- for uniform corrosion or local penetration rate, in: mm/a
- or if the density of the material is not known, in: $g/(m^2 h)$ or $g/(m^2 d)$.

Pitting corrosion, crevice corrosion, and stress corrosion cracking or non-uniform attack are particularly highlighted.

The following equations are used to convert mass loss rates, x, into the corrosion rate, y:

from x_1 into $g/(m^2 h)$ from x_2 into $g/(m^2 d)$ where

$$\frac{x_1 \cdot 365 \cdot 24}{\rho \cdot 1000} = y\ (mm/a) \qquad \frac{x_2 \cdot 365}{\rho \cdot 1000} = y\ (mm/a)$$

x_1: value in $g/(m^2 h)$
x_2: value in $g/(m^2 d)$
ρ: density of material in g/cm^3
y: value in (mm/a)
d: days
h: hours

In those media in which uniform corrosion can be expected, if possible, isocorrosion curves (corrosion rate = 0.1 mm/a) are given.

Unless stated otherwise, the data was measured at atmospheric pressure and room temperature.

The resistance data should not be accepted by the user without question, and the materials for a particular purpose should not be regarded as the only ones that are suitable. To avoid wrong conclusions being drawn, it must be always taken into account that the expected material behavior depends on a variety of factors that are often difficult to recognize individually and which may not have been taken deliberately into account in the investigations upon which the data is based. Under certain circumstances, even slight deviations in the chemical composition of the medium, in the pressure, in the temperature or, for example, in the flow rate are sufficient to have a significant effect on the behavior of the materials. Furthermore, impurities in the medium or mixed media can result in a considerable increase in corrosion.

The composition or the pretreatment of the material itself can also be of decisive importance for its behavior. In this respect, welding should be mentioned. The suitability of the component's design with respect to corrosion is a further point which must be taken into account. In case of doubt, the corrosion resistance should be investigated under operating conditions to decide on the suitability of the selected materials.

Warranty disclaimer

This book has been compiled from literature data with the greatest possible care and attention. The statements made in this book only provide general descriptions and information.

Even for the correct selection of materials and correct processing, corrosive attack cannot be excluded in a corrosion system as it may be caused by previously unknown critical conditions and influencing factors or subsequently modified operating conditions.

No guarantee can be given for the chemical stability of the plant or equipment. Therefore, the given information and recommendations do not include any statements, from which warranty claims can be derived with respect to DECHEMA e.V. or its employees or the authors.

The DECHEMA e.V. is liable to the customer, irrespective of the legal grounds, for intentional or grossly negligent damage caused by their legal representatives or vicarious agents.

For a case of slight negligence, liability is limited to the infringement of essential contractual obligations (cardinal obligations). DECHEMA e.V. is not liable in the case of slight negligence for collateral damage or consequential damage as well as for damage that results from interruptions in the operations or delays which may arise from the deployment of this book.

Acetates

Author: L. Hasenberg / Editor: R. Bender

Copper and copper alloys

Copper and its alloys are relatively resistant to aqueous aluminium acetate solutions. For example, copper, CuAl-, CuSi- and CuSn-alloys are resistant to basic aluminium acetate [1]. Corrosion rates in aqueous solutions of low concentration are below 0.5 mm/a (19.7 mpy) at 295 K (22 °C) [2–4].

CuAl- and CuSi-alloys are resistant to ammonium acetate solutions [1]. Highly concentrated and moist solid ammonium acetate attacks copper and its alloys (above 1.25 mm/a (49.2 mpy)) [2–4].

Copper and its alloys lose up to 1.25 mm/a (49.2 mpy) in aqueous 10 % lead acetate solutions even at 295 K (22 °C); this value is exceeded at higher temperatures (373 K (100 °C)) [3, 4].

Aqueous potassium acetate solutions even at their boiling temperatures do not attack copper and its alloys. Drying pansdrying pans of copper, in which completely anhydrous potassium acetate used for rectifying alcohols is produced at 613 K (340 °C), show only slight corrosion attack [1]. The corrosion rates of 99.9 % copper and of alloys having copper contents above 85 % are below 0.5 mm/a (19.7 mpy) in solutions of all concentrations at temperatures up to the respective boiling point; higher rates (up to 1.25 mm/a (49.2 mpy)) are found in the melt above 590 K (317 °C) [2–4].

Corrosion rates for copper and its alloys in calcium acetate solutions are around 0.05 mm/a (1.97 mpy) at virtually all concentrations [2–4].

Neutral and alkaline solutions or suspensions of copper acetates virtually do not attack copper, bronzes and copper-nickel alloys in the absence of air, whereas acidic solutions corrode these materials in the presence of air or other oxidizing agents.

Paris Green, $Cu(C_2H_3O_2)_2 \cdot 3Cu(AlO_2)_2$, can be precipitated in copper kettles. Reaction vessels made of copper or aluminium bronze can be used in organic halogenations using cuprous acetate as catalyst [1].

Brass having a high zinc content is unsuitable as a material for handling copper acetates [1]. The corrosion rates exceed 1.25 mm/a (49.2 mpy) [1–4,]. In the case of alloys containing at least 85 % copper and of pure copper, corrosion rates in 10 % and 90 % copper acetate solutions are about 0.05 mm/a (1.97 mpy) and do not exceed 0.5 mm/a (19.7 mpy) [2–4].

Pure copper is not susceptible to stress corrosion cracking in copper acetate solutions [5].

Copper and its alloys having Cu contents above 85 % are very resistant to sodium acetate solutions (corrosion rates below 0.05 mm/a (1.97 mpy)) in solutions of any concentration up to their boiling points; at temperatures above 422 K (149 °C) (in the salt melt), however, the rates are between 0.5 and 1.25 mm/a (19.7 and 49.2 mpy)

[1, 4, 6, 7]. The CuNi-alloys, however, are highly resistant even at 613 K (340 °C) (less than 0.05 mm/a (1.97 mpy)) [4, 7].

Altogether, copper and its alloys are relatively resistant to sodium acetate solutions, with rates of corrosion usually around 0.05 mm/a (1.97 mpy).

Containers, piping and drying pans made of copper and its alloys, or valves, pumps and fittings made of bronze and brass, are used for handling sodium acetate solutions [1]. Sodium acetate solutions having a pH value of 8 can cause stress corrosion cracking in 70/30 brass [8].

Acetic Acid

Author: G. Elsner / Editor: R. Bender

Copper

In the production of acetic acid from calcium carbide, copper is used for acetaldehyde condensers and for distillation vessels, fractionating columns etc. [9].

For the catalytic purification of crude acetic acid between 474 and 573 K (201 and 300 °C) and also for the distillation to terpentine oil or benzene which contain acetic acid, copper equipment is used. Autoclaves for the preparation of furfural (393 K (120 °C), 20 % sulfuric acid, 5 % acetic acid, 0.2 % formic acid) are lined with copper or manufactured from aluminium bronze [9, 10].

Crude acetic acid (70 % acetic acid, 0,5 % formic acid) is in practice concentrated in columns made of high-purity copper or silicon bronze (Cu-1.5Si0.25 to 1.0 Mn). However, copper heating coils in these columns corrode rapidly in some cases; therefore, it is better to replace them by those made of austenitic CrNiMo-steel [9, 11].

In the processing of pyroligneous acid according to Suida and Othmer, copper, bronze, Everdur® (CuSi4Mn1, cf. UNS C87300, C65500), Everdur® 1015 (Cu-1.5Si0.25Mn) and Herculoy® (Cu-1.75 to 3.0 Si-0.25 to 0.5Sn) are also of particular importance [9].

Rectifying columns and heat exchangers in plants for wood chemicals remained resistant for 27 years if made of copper produced in America and only 5 to 7 years if made of copper produced in Russia. The reason for this is the difference in oxygen content of the copper. Of 7 Russian copper grades containing 99.38 to 99.92 % copper, 5 had a higher corrosion rate (test duration 30 h in 65 % acetic acid) than American copper grades. The corrosion rate depends directly on the oxygen content of the copper and increases with increasing acetic acid concentration and temperature, if the oxygen content is the same [12].

Copper and Cu-alloys and also nickel and Monel® have higher service lives in acetic acid-formic acid mixtures than in glacial acetic acid, while for austenitic steels the opposite is the case. For hot (393 K (120 °C)) mixtures of acetic acid with more than 2 % of formic acid, stainless steels are not as suitable as Cu-alloys or Monel®; however, for dilute acetic acid-formic acid solutions CrNiMo-steels are preferable to Cu- and Ni-alloys [13].

An acetic acid rectifying column for a mixture of 55 % acetic acid, 8 to 10 % formic acid and small amounts of propionic acid and esters consists of copper-plated cast iron [14].

In the presence of peracetic acid or acetaldehyde monoperacetate, copper and its alloys must not be used because of explosion risk. Even the brass sleeves for the thermal elements of the Dowtherm heating of the reaction boiler were replaced by those made of stainless steel for safety reasons [15–17]. Therefore, when acetic acid is prepared, copper is replaced as far as possible by CrNi- or CrNiMo-steel, which under operating conditions, even in welded joints, has a maximum material consumption rate of only 0.0006 g/m^{2}h, even if the reaction is not carried out under nitrogen but under air [18].

In contrast to the high-alloy steels, copper and Cu-alloys are resistant to acetic acid if it contains reducing components.

Copper and Cu-alloys (see also data of the American Copper Development Association) are only slightly attacked by acetic acid in the absence of oxidizing agents, for example air, oxygen, chromic acid, ferric chloride, potassium permanganate, and are among the most frequently used materials for handling acetic acid of all concentrations and temperatures [9, 11, 13, 19–24]. The exceptions are completely anhydrous acetic acid at the boiling point and Cu-alloys with more than 30% zinc [9, 11] and the peroxide containing systems already mentioned.

However, in the food industry copper is not considered resistant in acetic acid media [25]. Copper is also unsuitable for the processing (distillation) of technical grade acetic acid (from wood): It loses 2.3 mm/a (90.6 mpy), while titanium is completely resistant under these conditions and CrNiTi-steel 21 5 has a corrosion rate of 0.6 mm/a (23.6 mpy) [26].

The losses of copper, aluminium and iron in 1 to 15% acetic acid increase abruptly on increasing the temperature to 373 K (100 °C) and are independent of the acetic acid concentration. In boiling solutions, too, the attack on copper depends only slightly on the acetic acid concentration.

0.2 to 0.6% phenols increase the corrosivity of the acid on copper [27].

Investigations on the corrosion behavior of Cu in methanol, ethanol or water with addition of acetic acid were made. The corrosion rates are higher in alcoholic than in aqueous solutions, especially in methanolic solutions. Corrosion rate/time curves indicate a remarkable increase in the corrosion of copper in methanolic acid solutions [28].

It is true that Cu-alloys and steel form a galvanic element in dilute acetic acid, but this combination is hardly important in practice because steel is in most cases not taken into consideration as construction material anyway [29].

Copper is attacked in 10% aerated acetic acid in conductive contact with titanium due to contact corrosion [30].

The corrosion losses of copper welded seams in 50% acetic acid are as high as those of unwelded specimens, nor is there a difference between joints welded in the presence or absence of an inert gas [31].

Mixtures of acetic acid and acetic anhydride, for example at ratios of 90:10 or 40:60, severely attack copper at the boiling point at a material consumption rate of 200 to 300 g/m^2 d (about 8.1 to 12.1 mm/a) and slightly at room temperature at a material consumption rate of 0.6 to 1 g/m^2 d (about 0.025 to 0.04 mm/a) [9].

Mixtures of acetic acid + ethyl alcohol and acetic acid + formic acid cause corrosion rates of only less than 0.1 mm/a (3.94 mpy) [9].

However, according to other sources, oxygen and oxidizing agents such as peroxides and peracids make acetic acid solutions highly corrosive towards copper and Cu-alloys [20, 32, 33] and impurities such as cupric acetate and basic ferric acetate also increase the corrosion rate in, for example, 25 to 100% acetic acid (1,000 ppm metal ions, 391 K (118 °C), 120 h) [20]. The strong influence of oxygen can be considerably reduced by the addition of reducing substances such as sodium bisulfite or of precipitants for cupric salts such as sulfuric acid, oxalic acid or hydrogen sulfide. Thus, in boiling glacial acetic acid through which oxygen is bubbled, the material consumption rate of copper is reduced from 1680 $g/m^2 d$ to 96 $g/m^2 d$ by means of 0.05% concentrated sulfuric acid. However, sulfuric acid is consumed, thus reducing the protective effect with time. Oxalic acid is said to be more effective [9].

In air-saturated acetic acid solutions, the maximum of copper corrosion is between 344 and 350 K (71 and 77 °C) [34].

The corrosion not only in cold but also in hot acetic acid takes place in combination with oxygen depolarization [12, 35]. The attack also depends on the particle size and the type of heat treatment of copper [12].

A disadvantage of the use of copper and Cu-alloys is the fact that acetic acid turns blue if air is admitted [9].

In mixtures of 83% ethyl acetate, 9% acetic acid and 8% water at 293 to 373 K (20 to 100 °C) and a pH of 3.24, Cu is only resistant if oxygen is largely excluded [18].

Anodic dissolution of copper in anhydrous acetic acid solutions oxidizes copper only to the cuprous ion [33].

In the presence of $LiClO_4$, $NaClO_4$ or $HClO_4$ at room temperature in anhydrous acetic acid the kinetics of the anodic dissolution of Cu change. A study of the mechanisms were performed in [36] with rotating Cu-disk electrodes. Similar experiments were made with Cu and glacial acetic acid with the same additions [37].

o-Toluidine is mentioned as corrosion inhibitor for copper in dilute acetic acid [38].

As for the corrosion behavior of copper in acetic acid, see Table 1 [9, 11, 24, 39, 40].

Investigations on the suspectibility to intercrystalline corrosion (ICC) in dependence on the purity of the copper were performed in 20 and 30% acetic acid solutions at 353 K (80 °C) with welded and unwelded specimens [41, 42]. In [42] it was shown that especially Fe-contaminated Cu tended to ICC, which was attributed to Fe-accumulation at the grain boundaries. A decrease in the Fe-amount from 0.05 to 0.005% improved the resistance to ICC.

– *Acetic acid vapors* –

Mixtures of acetic acid, acetic anhydride and water in vapor form do not attack copper even at 973 K (700 °C), nor are they changed catalytically [9]. Small amounts of formic acid in acetic acid are decomposed in the gas phase by passing the vapors over copper catalysts [19].

The observation that copper is attacked by evaporating acetic acid more than by condensing acetic acid mainly based on the fact that the condensate contains much less oxygen [9].

As for the effect of the temperatures of the metal on the corrosion of copper heating surfaces in acetic acid under heat transfer conditions, see Figure 1 and Figure 2 [43].

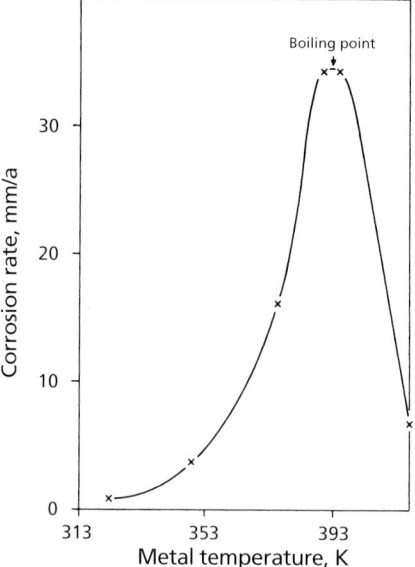

Figure 1: Effect of the metal temperature on the corrosion of copper heating surfaces in glacial acetic acid [43]

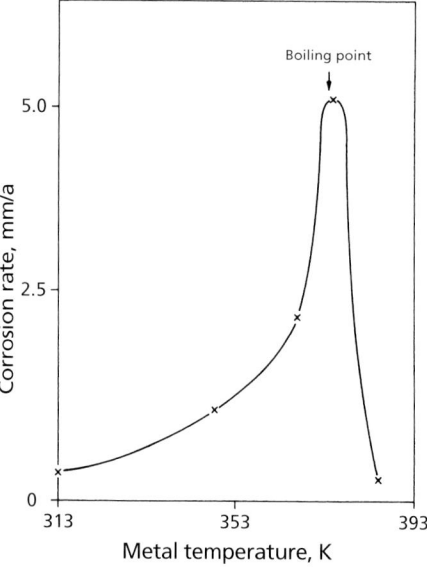

Figure 2: Effect of the metal temperature on the corrosion of copper heating surfaces in 30% acetic acid [43]

Under heat transfer conditions, laboratory tests using copper in glacial acetic acid or 30% acetic acid in the temperature range 348 to 413 K (75 to 140 °C) showed a strong increase in the corrosion rate up to a "peak" with respect to higher and lower temperatures, if the temperature of the heat transfer surface was kept approximately at the boiling point of the heated liquid [43, 44].

– Acetic anhydride –

In the preparation of acetic anhydride by thermal cracking of acetone [45], copper is not used, because the strength of copper is too low at the temperatures employed. Moreover reducing gases are formed in the reaction which have a negative effect on the structure of copper [46].

Copper and Al-bronzes are used in the ketene process for the preparation of acetic anhydride for condensers, pipelines and stages of the refining process in which air can be excluded [47]. Nitrogen is used as sweep gas in this reaction to minimize damage to the copper parts of the apparatus [46].

Metal	Acetic acid %	Temperature K (°C)	Corrosion rate mm/a (mpy)	Remarks	Literature
Copper	6	293 (20)	0.03 (1.18)	H_2 bubbling through	[9, 48]
Copper	6	293 (20)	0.7 (27.6)	O_2 bubbling through	[9, 48]
Copper	50	293 (20)	0.09 (3.54)	H_2 bubbling through	[9, 48]
Copper	50	293 (20)	2.1 (82.7)	O_2 bubbling through	[9, 48]
Copper	glacial acetic acid	293 (20)	0.05 (1.97)	H_2 bubbling through	[9, 48]
Copper	glacial acetic acid	293 (20)	$0.05^{1)}$ (1.97)	O_2 bubbling through	[9, 48]
Copper	1 to 100	293 (20)	about 0.2 (7.87)	air-free	[9]
Copper	1 to 100	293 (20)	about 0.6 (23.6)	moderate access of air	[9]

1) This value is probably due to an experimental error, for in hot glacial acetic acid the ratio between the corrosion rates in unaerated and aerated acetic acid is 1:50 [9].
2) This value refers to pure acetic acid; in 80% crude acetic acid the corrosion rate is only 0.14 mm at the bottom of the tank [9].
3) The nitrogen oxygen mixtures are passed through acetic acid (1 h before the samples are suspended). During the experiment, the acetic acid is not agitated [9].

Table 1: Corrosion behavior of copper in acetic acid of various concentrations and temperatures

Table 1: Continued

Metal	Acetic acid %	Temperature K (°C)	Corrosion rate mm/a (mpy)	Remarks	Literature
Copper	1 to 100	293 (20)	about 1.6 (63.0)	strong access of air	[9]
Copper	100	boiling point	about 0.6 (23.6)	slight access of air	[9]
Copper	5	boiling point	0.1 (3.94)	liquid phase	[9, 49]
Copper	5	boiling point	0.09 (3.54)	vapor phase	[9, 49]
Copper	50	boiling point	0.2 (7.87)	liquid phase	[9, 49]
Copper	50	boiling point	0.3 (11.8)	vapor phase	[9, 49]
Copper	98	boiling point	0.8 (31.5)	liquid phase	[9, 49]
Copper	98	boiling point	0.3 (11.8)	vapor phase	[9, 49]
Copper	99.9	boiling point	0.6 (23.6)	liquid phase	[9, 49]
Copper	99.9	boiling point	0.07 (2.76)	vapor phase	[9, 49]
Copper	6	293 (20)	0.14 (5.51)	intermittent immersion	[9, 50]
Copper	80	room temperature	0.28[2] (11.0)	immersed at tank bottom	[9, 49, 51]
Copper	80	room temperature	0.5 (19.7)	immersed in upper part of tank	[9, 49, 51]
Copper	7.2	295 (22)	0.09[3] (3.54)	nitrogen with 10% oxygen	[9, 48]
Copper	7.2	295 (22)	0.16[3] (6.30)	nitrogen with 16% oxygen	[9, 48]
Copper	7.2	295 (22)	0.16[3] (6.30)	nitrogen with 21% oxygen	[9, 48]
Copper, 85–99.9%	100	373 (100)	> 1.25 (> 49.2)	acetic acid vapor	[39, 40]

1) This value is probably due to an experimental error, for in hot glacial acetic acid the ratio between the corrosion rates in unaerated and aerated acetic acid is 1:50 [9].
2) This value refers to pure acetic acid; in 80% crude acetic acid the corrosion rate is only 0.14 mm at the bottom of the tank [9].
3) The nitrogen oxygen mixtures are passed through acetic acid (1 h before the samples are suspended). During the experiment, the acetic acid is not agitated [9].

Table 1: Corrosion behavior of copper in acetic acid of various concentrations and temperatures

Table 1: Continued

Metal	Acetic acid %	Temperature K (°C)	Corrosion rate mm/a (mpy)	Remarks	Literature
Copper	80	297 (24)	> 1.25 (> 49.2)	acetic acid, aerated	[39]
	100	297 (24)	< 0.5 (< 19.7)	acetic acid, aerated	[39]
Copper	100	297 (24)	< 0.05 (< 1.97)	acetic acid, air-free	[39]
	50	up to 373 (100)	< 0.5 (< 19.7)	acetic acid, air-free	[39]
	100	325 (52)	< 0.5 (< 19.7)	acetic acid, air-free	[39]
	100	373 (100)	> 1.25 (> 49.2)	acetic acid, air-free	[39]
Copper, 85–99.9 %	all conc. up to 100	298 (25)	> 1.25 (> 49.2)	acetic acid, aerated	[40]
Copper, 85–99.9 %	all conc. up to 100	298 (25)	< 0.05 (< 1.97)	acetic acid, air-free	[40]
	50	up to 373 (100)	< 0.5 (< 19.7)	acetic acid, air-free	[40]
	100	323 (50)	< 0.5 (< 19.7)	acetic acid, air-free	[40]
	100	373 (100)	> 1.25 (> 49.2)	acetic acid, air-free	[40]
	10	423 (150)	transgranular attack	acetic acid, air-free	[40]
Copper	100	up to 373 (100)	< 0.05 (< 1.97)	acetic acid + ethyl alcohol	[39]
Copper, 85–99.9 %	100	up to 373 (100)	< 0.05 (< 1.97)	acetic acid + formic acid	[39, 40]

1) This value is probably due to an experimental error, for in hot glacial acetic acid the ratio between the corrosion rates in unaerated and aerated acetic acid is 1:50 [9].
2) This value refers to pure acetic acid; in 80 % crude acetic acid the corrosion rate is only 0.14 mm at the bottom of the tank [9].
3) The nitrogen oxygen mixtures are passed through acetic acid (1 h before the samples are suspended). During the experiment, the acetic acid is not agitated [9].

Table 1: Corrosion behavior of copper in acetic acid of various concentrations and temperatures

Table 1: Continued

Metal	Acetic acid %	Temperature K (°C)	Corrosion rate mm/a (mpy)	Remarks	Literature
Copper, 85–99.9%	100	298 (25)	> 1.25 (> 49.2)	acetic acid + hydrobromic acid	[39, 40]
Copper, 85–99.9%	100	298 (25)	> 1.25 (> 49.2)	acetic acid + hydrochloric acid	[39, 40]
Copper	100	up to 297 (24)	> 1.25 (> 49.2)	acetic acid + mercury salts	[39]
Copper	100	297 (24)	< 0.5 (< 19.7)	acetic acid + sulfur dioxide	[39]
Copper	20	373 (100)	< 0.5 (< 19.7)	acetic acid + acetates + chlorides	[40]

1) This value is probably due to an experimental error, for in hot glacial acetic acid the ratio between the corrosion rates in unaerated and aerated acetic acid is 1:50 [9].
2) This value refers to pure acetic acid; in 80% crude acetic acid the corrosion rate is only 0.14 mm at the bottom of the tank [9].
3) The nitrogen oxygen mixtures are passed through acetic acid (1 h before the samples are suspended). During the experiment, the acetic acid is not agitated [9].

Table 1: Corrosion behavior of copper in acetic acid of various concentrations and temperatures

Copper and Cu-alloys are very resistant to acetic anhydride and its mixtures with acetic acid in the absence of air and oxidizing substances with the exception of those containing more than 20% zinc. However, since air cannot be excluded entirely, the products can turn green due to the contact with copper [46].

In the presence of oxygen, copper is attacked by acetic anhydride in any case [52, 53]. However, in practice distillation boilers made of copper are used successfully for mixtures of 90% acetic anhydride + 10% acetic acid [46].

As for the behavior of copper towards acetic anhydride, see Table 2 [11, 39, 40, 54].

– *Acetic anhydride vapor* –

Even at 973 K (700 °C), copper is resistant when exposed to acetic anhydride vapor and does not have an unfavorable catalytic effect on the thermal decomposition of acetic acid to acetic anhydride and water [46].

Metal	Acetic anhydride %	Temperature K (°C)	Corrosion rate mm/a (mpy)	Remarks	Literature
Copper	pure	298 (25)	0.06 (2.36)		[11]
	pure	348 (75)	1.16 (45.7)		
	pure + 40% glacial acetic acid	298 (25)	0.02 (0.79)		
Copper	not specified	293 (20)	0.057 (2.24)		[54]
	not specified	348 (75)	0.81 to 1.21 (31.8 to 47.6)	dependent on the presence of air	
Copper, 85–99.9%	10	298 (25)	< 0.05 (< 1.97)	acetic anhydride in acetic acid	[40]
	40	298 (25)	< 0.05 (< 1.97)		
	60	298 (25)	< 0.05 (< 1.97)		
	80	298 (25)	< 0.05 (< 1.97)		
	90	298 (25)	< 0.5 (< 19.7)		
	20	323 (50)	< 0.5 (< 19.7)		
	20	348 (75)	< 0.5 (< 19.7)		
	40	323 (50)	< 0.05 (< 1.97)		
	10	398 (125)	1.25 (49.2)		
	60	398 (125)	> 1.25 (> 49.2)		
	80	398 (125)	> 1.25 (> 49.2)		
	90	398 (125)	< 0.05 (< 1.97)	this value seems questionable	
Copper	100	297 (24)	< 0.5 (< 19.7)		[39]
	100	325 to 352 (52 to 79)	0.5 to 1.25 (19.7 to 49.2)		
	100	373	> 1.25 (> 49.2)		

Table 2: Behavior of copper towards acetic anhydride of various concentrations and temperatures

– Vinegar –

Copper and Cu-alloys are not suitable for the rapid vinegar process, since copper ions have a toxic effect on the vinegar bacteria [9, 24]. In a fermentation acetic acid manufacturing plant, copper is used for pipelines, extraction, stripping and concentration columns and is not attacked by hot concentrated acetic acid, if air is excluded [55].

However, vinegar must not be stored in copper containers, because of the formation of poisonous basic copper acetate (verdigris) [56].

As for the behavior of copper and Cu-alloys towards wine vinegar (4% acetic acid), see Table 3 [9, 57, 58].

Material composition %	Corrosion rate mm/a (mpy)		
	Fully immersed	Aerated	Spray test
Cu	0.032 (1.26)	0.206 (8.11)	0.809 (31.9)
CuSn20	0.071 (2.80)	0.503 (19.8)	0.587 (23.1)
CuSn10Pb15	0.241 (9.49)	0.435 (17.1)	0.593 (23.3)
CuSn5Pb5PbZn5	0.114 (4.49)	0.395 (15.6)	0.569 (22.4)
Cu Sn8Pb6	0.100 (3.94)	0.322 (12.7)	0.443 (17.4)
CuSn10Fe4	0.066 (2.60)	0.269 (10.6)	0.703 (27.7)
Cu-4.5Al4Mn2.5Fe	0.068 (2.68)	0.305 (12.0)	0.890 (35.0)
CuAl8	0.003 (0.12)		

Table 3: Behavior of copper and copper alloys towards wine vinegar at room temperature, test duration 100 h [9, 57, 58]

The different effects of container materials on fermentation processes are illustrated, inter alia, by the following experience:

When Emmental cheese is allowed to ripen (postfermentation) in steel tubs, more acetic acid and propionic acid but less lactic acid are formed than in copper tubs [59].

Copper-aluminium alloys

– Acetic acid –

CuAl-alloys behave similar to copper. In acetic acid, the resistance of Al-bronze (Cu-7Al2.5Fe, cf. CW303G) is worse than that of chemically resistant steels.

In the preparation of acetic acid from calcium carbide, Al-bronze is used for acetic acid distillation and fractionating columns or condensers [9]. Pumps for dilute acetic acid may also be manufactured from Al-bronze [55].

The KMC alloy (Cu-7Al1Ni0.3Fe0.05Cr) is superior to the materials Cu 99.9%, Cu-6.15 Al and Cu-5.12 Sn in acetic acid and is recommended for condenser tubes [60].

Under heat transfer conditions, Al-bronze (Cu-5.66Al0.02Ni0.01Fe) [61, 62] suffers severe corrosion in 56% acetic acid with 20 ppm of chloride + 26 ppm of sulfate (specimen temperature 373 K (100°C), solution temperature 351 K (78°C), test duration 96 h); the corrosion rate is higher than 1.25 mm/a (49.2 mpy) [13, 61, 62]. In a mixture of 56% acetic acid + 5% sulfuric acid, the corrosion rate for the same Al-bronze (specimen temperature 383 K (110°C), solution temperature 373 K (100°C), test duration 73 h) is also higher than 1.25 mm/a (49.2 mpy) [61, 62].

When Al-bronze is tested for stress corrosion cracking in 20% acetic acid at 453 K (180°C) (behavior of U-band specimens after a test duration of 150 h), no cracks developed in the cold forming up to 60% [63].

Further corrosion rates of Al-bronze can be seen in Table 4.

– *Acetic acid vapors* –

Aluminium bronze (Corrix® metal) ((%) 88 Cu-8.7Al3.1Fe) is not attacked at 731 K (458°C) by mixtures of acetic acid + acetic anhydride + water in vapor form, nor do the mixtures change catalytically [9]. As for the behavior of Corrix® metal in 50% acetic acid at 353 K (80°C), see [9].

– *Acetic anhydride* –

A contact apparatus made of aluminium bronze (Corrix®) has proved to be very suitable in the thermal cracking of glacial acetic acid and in the older ethylidene diacetate process to give acetic anhydride at 773 K (500°C) [46, 54]. The corrosion rate of Ampco® aluminium bronze by acetic anhydride is stated to be smaller than 0.4 mm/a (15.7 mpy) [46, 64].

For the behavior of aluminium bronze towards acetic anhydride, virtually the same applies as for copper. The corrosion rates in 100% acetic anhydride are less than 0.5 mm/a (19.7 mpy) at 297 K (24°C), between 0.5 to 1.25 mm/a (19.7 to 49.2 mpy) at 325 to 352 K (52 to 79°C) and higher than 1.25 mm/a (49.2 mpy) at 373 K (100°C) [39]. Aluminium bronze can be used for pipelines, pumps and for hoods of copper distillation plants and equipments [54] as well as for fittings and valves [46].

Copper-nickel alloys

For CuNi-alloys, virtually the same applies as for copper. Impurities in acetic acid such as cupric acetate in small amounts have a corrosion-accelerating effect [20].

As for the behavior of CuNi-alloys towards acetic acid, see Table 5 [9, 11, 40].

Copper-tin alloys (bronzes)

CuSn-alloys behave very similar to copper. Due to the insufficient resistance to organic acids, phosphor bronzes are also severely attacked by acetic acid. This applies, for example, to the phosphor bronzes:

- CDA 505 (CuSn1.25)
- CDA 510 (CuSn5)
- CDA 521 (CuSn8)
- CDA 524 (CuSn10)
- CDA 544 (88Cu-4Sn4Zn)

Therefore, bronze spindles in gate valves are severely attacked by mixtures of air and hot acetic acid, as expected [65].

Material composition %	Acetic acid %	Temperature K (°C)	Corrosion rate mm/a (mpy)	Remarks	Literature
Al-bronze (Cu-8.7Al3.1Fe)	100	room temp.	< 0.15 (< 5.91)		[9, 66]
Al-bronze (Cu-10Al3.5Fe)	10	295 (22)	0.05 (1.97)		[11]
	35	295 (22)	0.075 (2.95)		[11]
	35	357 (84)	0.50 (19.7)		[11]
	50	295 (22)	0.100 (3.94)		[11]
Al-bronze (without further details)	100	373 (100)	> 1.25 (> 49.2)	acetic acid vapor	[39]
	≤ 80	297 (24)	> 1.25 (> 49.2)	acetic acid, aerated	[39]
	≤ 100	297 (24)	< 0.5 (< 19.7)	acetic acid, aerated	[39]
	100	297 (24)	< 0.05 (< 1.97)	acetic acid, air-free	[39]
	≤ 50	373 (100)	< 0.5 (< 19.7)	acetic acid, air-free	[39]
	100	325 (52)	< 0.5 (< 19.7)	acetic acid, air-free	[39]
	100	373 (100)	> 1.25 (> 49.2)	acetic acid, air-free	[39]

Table 4: Corrosion behavior of aluminium bronze in acetic acid of various concentrations and temperatures

Table 4: Continued

Material composition %	Acetic acid %	Temperature K (°C)	Corrosion rate mm/a (mpy)	Remarks	Literature
	100	373 (100)	< 0.05 (< 1.97)	acetic acid + ethyl alcohol	[39]
	100	373 (100)	< 0.05 (< 1.97)	acetic acid + formic acid	[39]
	100	297 (24)	> 1.25 (> 49.2)	acetic acid + hydrobromic acid	[39]
	100	297 (24)	> 1.25 (> 49.2)	acetic acid + hydrochloric acid	[39]
	100	297 (24)	> 1.25 (> 49.2)	acetic acid + mercury salt	[39]
	100	297 (24)	< 0.5 (< 19.7)	acetic acid + sulfur dioxide	[39]

Table 4: Corrosion behavior of aluminium bronze in acetic acid of various concentrations and temperatures

Material composition %	Acetic acid %	Temperature K (°C)	Corrosion rate mm/a (mpy)	Remarks	Literature
Copper-nickel alloy Cu-(11–33)Ni	all conc. up to 100	298 (25)	> 2.35 (> 92.5)	acetic acid, aerated	[40]
Nickel silver Cu-22Ni18Zn	0.1 to 70	293 (20)	0.2 (7.87)		[9]
Nickel silver	0 to 40	293 (20)	< 0.1 (3.94)		[11]
	40 to 70	293 (20)	< 0.2 (< 7.87)		[11]

Table 5: Corrosion behavior of CuNi-alloys in acetic acid of various concentrations and temperatures

In the preparation of acetic acid from calcium carbide, bronze is used for the crude acetic acid pump [9]. Bronze containing 10 to 20 % tin is said to be suitable for the casing of stop valves for 4 to 30 % acetic acid at 293 to 353 K (20 to 80 °C) [67].

Extruders for acetic acid acetylcellulose solutions are manufactured from bronze [10].

As for the behavior of CuSn-alloys towards acetic acid, see Table 6 [9, 11, 24, 39].

– Acetic anhydride –

For the behavior of Sn-bronze towards acetic anhydride, virtually the same applies as for copper. The corrosion rates in 100 % acetic anhydride are less than 0.5 mm/a (19.7 mpy) at 297 K (24 °C), between 0.5 and 1.25 mm/a (19.7 and 49.2 mpy) at 325 to 352 K (52 to 79 °C) and higher than 1.25 mm/a (49.2 mpy) at 373 K (100 °C) [39].

Large acetylation vessels made of bronze containing 4 % tin are used successfully in practice [46].

Bronze (without zinc) is also used for fittings, valves, pumps, pipelines, condensers etc. for handling crude acetic acid [46, 54].

– Vinegar –

Phosphor bronze loses 0.46 g/m^2 d (0.019 mm/a) in about 6 % vinegar at room temperature [9].

Copper-tin-zinc alloys (red brass)

– Acetic acid –

For CuSnZn-alloys, virtually the same applies as for copper. Their behavior in acetic acid is shown in Table 7 [9, 24, 39].

– Acetic anhydride –

The Cu-alloys which contain zinc reach at best limited resistance to acetic anhydride. Thus, red brass loses between 0.5 and 1.25 mm/a (19.7 and 49.2 mpy) in 100 % acetic anhydride at 297 to 352 K (24 to 79 °C) and more than 1.25 mm/a (49.2 mpy) at 373 K (100 °C) [39].

Material	Acetic acid %	Temperature K (°C)	Corrosion rate mm/a (mpy)	Remarks	Literature
Bronze (CuSn5)	12	room temp.	0.09 (3.54)	fully immersed, not agitated, attack at the waterline 0.10 mm, duration 386 d	[9, 48]
Bronze (Cu-3.4Sn)	33	boiling point	0.06 (2.36)		[9, 11]
Bronze (Cu-3.5Mn0.13Sn)	33	boiling point	0.07 (2.76)		[9, 11]
Bronze (Cu-4.9Sn0.2Fe)	33	boiling point	0.09 (3.54)		[9, 11]

Table 6: Corrosion behavior of copper-tin alloys towards acetic acid of various concentrations and temperatures

Table 6: Continued

Material	Acetic acid %	Temperature K (°C)	Corrosion rate mm/a (mpy)	Remarks	Literature
Sn-bronze (no further data given)	100	373 (100)	> 1.25 (> 49.2)	acetic acid vapor	[39]
	≤ 80	297 (24)	> 1.25 (> 49.2)	acetic acid, aerated	[39]
	≤ 100	297 (24)	< 0.5 (< 19.7)	acetic acid, aerated	[39]
	100	297 (24)	< 0.05 (< 1.97)	acetic acid, air-free	[39]
	≤ 50	373 (100)	< 0.5 (< 19.7)	acetic acid, air-free	[39]
	100	325 (52)	< 0.5 (< 19.7)	acetic acid, air-free	[39]
	100	373 (100)	> 1.25 (> 49.2)	acetic acid, air-free	[39]
	100	373 (100)	< 0.05 (< 1.97)	acetic acid + ethyl alcohol	[39]
	100	373 (100)	< 0.05 (< 1.97)	acetic acid + formic acid	[39]
	100	297 (24)	> 1.25 (> 49.2)	acetic acid + hydrobromic acid	[39]
	100	297 (24)	> 1.25 (> 49.2)	acetic acid + hydrochloric acid	[39]
	100	297 (24)	> 1.25 (> 49.2)	acetic acid + mercury salts	[39]
	100	297 (24)	< 0.5 (< 19.7)	acetic acid + sulfur dioxide	[39]
	10	293 (20)	0.075 (2.95)		[11]
	10	323 (50)	0.225 (8.86)		[11]
	10	boiling	2.0 (78.7)		[11]
	50	293 (20)	0.125 (4.92)		[11]
	50	323 (50)	0.55 (21.6)		[11]
	50	boiling	0.9 (35.4)		[11]
	75	293 (20)	0.200 (7.87)		[11]
	75	323 (50)	0.625 (24.6)		[11]
	75	boiling	1.0 (39.4)		[11]

Table 6: Corrosion behavior of copper-tin alloys towards acetic acid of various concentrations and temperatures

Table 6: Continued

Material	Acetic acid %	Temperature K (°C)	Corrosion rate mm/a (mpy)	Remarks	Literature
	100	293 (20)	0.650 (25.6)		[11]
	100	323 (50)	0.950 (37.4)		[11]
	100	boiling	35.0 (1378)		[11]
	33	297 (24)	0.079 (3.11)		[68]

Table 6: Corrosion behavior of copper-tin alloys towards acetic acid of various concentrations and temperatures

Material	Acetic acid %	Temperature K (°C)	Corrosion rate mm/a (mpy)	Remarks	Literature
Red brass	100	373 (100)	> 1.25 (> 49.2)	acetic acid vapor	[39]
	≤ 100	297 (24)	> 1.25 (> 49.2)	acetic acid, aerated	[39]
	10	373 (100)	< 0.5 (< 19.7)	acetic acid, air-free	[39]
	80	297 (24)	0.5 (< 19.7)	acetic acid, air-free	[39]
	100	297 (24)	< 0.5 (< 19.7)	acetic acid, air-free	[39]
	100	297 (24)	< 0.5 (< 19.7)	acetic acid + ethyl alcohol	[39]
	100	297 (24)	> 1.25 (> 49.2)	acetic acid + hydrobromic acid	[39]
	100	297 (24)	> 1.25 (> 49.2)	acetic acid + hydrochloric acid	[39]
	100	297 (24)	> 1.25 (> 49.2)	acetic acid + mercury salts	[39]

Table 7: Corrosion behavior of CuSnZn-alloys towards acetic acid of various concentrations and temperatures

Copper-zinc alloys (brass)

– Acetic acid –

CuZn-alloys behave similar to copper. The material consumption per unit area of 80:20 (α) and 59:41 (α + β) brass in 6 % acetic acid at 308 K (35 °C), test duration 3 d, are almost of the same order as those in 3.65 % hydrochloric acid, that is 91 mg/dm^2 (0.384 mm/a), 85 mg/dm^2 (0.358 mm/a) for 80:20 brass and 102 mg/dm^2 (0.45 mm/a), 93 mg/dm^2 (0.41 mm/a) for 59:41 brass [69]. The corrosion of 80:20, 70:30 and 60:40 brass in acetic acid is supported by hydrogen peroxide, while dezincification simultaneously declines [70, 71].

The silicon-brass diecasting copper alloy No. 879 ((%) 63.0 to 67.0 Cu-0.8 to 1.2Si max. 0.25Sn, max. 0.25Pb, max. 0.15Fe, max. 0.50 other elements, balance Zn) has good resistance to acetic acid [72], while a corrosion rate of 2.19 mm/a (86.2 mpy) is given for the Russian silicon-brass type CuM3S in acetic acid containing a small amount of sulfuric acid at 353 K (80 °C) [73].

As for the behavior of CuZn-alloys towards acetic acid, see also Table 8 [9, 11, 24, 39, 40].

Thiocresol exhibits good inhibition effect for brass (CuZn37, cf. CW508L) in acetic acid [74].

– Acetic acid vapors –

Nickel-plated brass handles which were kept in wooden boxes before their actual use were severely attacked by the wood components during the winter months. Corrosion products are in particular nickel acetate und maybe a basic copper acetate, which can be formed by the acetic acid which is eliminated from the wood or from vinyl acetate (component of adhesive) [75].

Cu-based alloys can be attacked by acetic acid used as catalyst in some silicones [76].

Material	Acetic acid %	Temperature K (°C)	Corrosion rate mm/a (mpy)	Remarks	Literature
Cu-28Zn1Sn	12	298 (25)	0.08 (3.15)	fully immersed, not agitated	[9, 48]
Cu-15Zn	12	298 (25)	0.13 (5.12)	fully immersed, not agitated attack at waterline 0.55 mm, duration 386 d	[9, 48]
Cu-40Zn2.8Pb	33	293 (20)	2.4 (94.5)		[9, 11]

Table 8: Corrosion behavior of the CuZn-alloys (brass) towards acetic acid of various concentrations and temperatures

Table 8: Continued

Material	Acetic acid %	Temperature K (°C)	Corrosion rate mm/a (mpy)	Remarks	Literature
Cu-29Zn1Sn	6	room temp.	0.08 (3.15)	fully immersed, not agitated	[9, 50]
Cu-29Zn1Sn	6	room temp.	0.3 (11.8)	stirred in the presence of air	[9, 50]
Cu-29Zn1Sn	6	room temp.	0.2 (7.87)	alternating immersion test	[9, 50]
Cu-29Zn1Sn	6	room temp.	0.44 (17.3)	spray test (30 d)	[9, 50]
Brass	100	297 (24)	> 1.25 (> 49.2)	aerated (dezincification)	[39]
70 to 80 Cu + Zn, Sn or Pb	all conc. up to 100	298 (25)	> 1.25 (> 49.2)	aerated	[40]
59 to 93 Cu + Al, Zn or As	all conc. up to 100	298 (25)	> 1.25 (> 49.2)	aerated (partitioning into elements)	[40]
Yelloy brass	up to 100	297 (24)	> 1.25 (> 49.2)	air-free (dezincification)	[39]
Brass	10	373 (100)	< 0.5 (< 19.7)	air-free	[40]
70 to 80 Cu + Zn	80	298 (25)	< 0.5 (< 19.7)	air-free	[40]
Sn or Pb	100	298 (25)	< 0.5 (< 19.7)	air-free	[40]
59 to 83 Cu + Al, Zn or As	all conc. up to 100	298 (25)	> 1.25 (> 49.2)	air-free	[40]
Yellow brass	100	373 (100)	> 1.25 (> 49.2)	acetic acid vapor	[39]
70 to 80 Cu + Zn, Sn or Pb	100	373 (100)	> 1.25 (> 49.2)	acetic acid vapor	[40]
59 to 93 Cu + Al, Zn or As	100	373 (100)	> 1.25 (> 49.2)	acetic acid vapor	[40]
70 to 80 Cu + Zn, Sn or Pb	100	298 (25)	> 1.25 (> 49.2)	acetic acid + hydrobromic acid	[39, 40]
59 to 93 Cu + Al, Zn or As	100	298 (25)	> 1.25 (> 49.2)	acetic acid + hydrobromic acid	[39, 40]

Table 8: Corrosion behavior of the CuZn-alloys (brass) towards acetic acid of various concentrations and temperatures

Table 8: Continued

Material	Acetic acid %	Temperature K (°C)	Corrosion rate mm/a (mpy)	Remarks	Literature
70 to 80 Cu + Zn, Sn or Pb	100	298 (25)	> 1.25 (> 49.2)	acetic acid + hydrochloric acid	[39, 40]
59 to 93 Cu + Al, Zn or Pb	100	298 (25)	> 1.25 (> 49.2)	acetic acid + hydrochloric acid	[39, 40]
Yellow brass	100	297 (24)	> 1.25 (> 49.2)	acetic acid + mercury salts	[39]

Table 8: Corrosion behavior of the CuZn-alloys (brass) towards acetic acid of various concentrations and temperatures

In direct contact with plastics (for example phenol-novolak resin, phenol-resol resin, phenol-melamine resin, urea-formaldehyde resin) or rubber, copper and its alloys can also be attacked, since at 308 K (35 °C) or under the influence of electric voltages these materials give off vapors containing acetic acid, formaldehyde, formic acid, ammonia and phenol, which are corrosive when the air humidity is high [77, 78].

– *Acetic anhydride* –

Cast brass is not resistant to acetic anhydride. It loses more than 1.25 mm/a (49.2 mpy) in 100 % acetic anhydride at 297 K (24 °C) [39]; the same value is observed for brass with (%) 59 to 93 Cu + Al, Zn or As in mixtures of 100 % acetic anhydride + acetic acid [40].

– *Vinegar* –

The gravity diecasting alloy U-Z40Y30 according to AFNOR® A 53-703, Cu-35.5Zn1.4Pb0.6Sn0.5Al-0.5Fe0.4Ni, is attacked by wine vinegar over a period of 5,000 h at a corrosion rate of 0.106 mm/a (4.17 mpy) [79].

Other copper alloys

– *Acetic acid* –

For copper-silicon alloys, virtually the same applies as for copper. As for the behavior of these alloys towards acetic acid, see Table 9 [9, 11, 24, 39].

Material	Acetic acid %	Temperature K (°C)	Corrosion rate mm/a (mpy)	Remarks	Literature
Silicon bronze Cu-3Si	10	297 (24)	0.005 (0.20)	1)	[9, 11]
	25	297 (24)	0.04 (1.57)		
	50	297 (24)	0.05 (1.97)		
	75	297 (24)	0.10 (3.94)		
	99.5	297 (24)	0.32 (12.6)		
Silicon bronze (no further data given)	up to 100	297 (24)	> 1.25 (> 49.2)	acetic acid, aerated	[39]
	up to 70	297 (24)	< 0.05 (1.97)	acetic acid, air-free	[39]
	80	297 (24)	< 0.5 (< 19.7)	acetic acid, air-free	[39]
	90	297 (24)	< 0.5 (< 19.7)	acetic acid, air-free	[39]
	100	297 (24)	< 0.5 (< 19.7)	acetic acid, air-free	[39]
	up to 90	373 (100)	< 0.5 (< 19.7)	acetic acid, air-free	[39]
	100	325 (52)	> 1.25 (> 49.2)	acetic acid, air-free	[39]
	100	373 (100)	> 1.25 (> 49.2)	acetic acid vapor	[39]
	100	373 (100)	< 0.5 (< 19.7)	acetic acid + ethyl alcohol	[39]
	100	373 (100)	< 0.05 (< 1.97)	acetic acid + formic acid	[39]
	100	297 (24)	> 1.25 (> 49.2)	acetic acid + hydrobromic acid	[39]
	100	297 (24)	> 1.25 (> 49.2)	acetic acid + hydrochloric acid	[39]
	100	297 (24)	> 1.25 (> 49.2)	acetic acid + mercury salts	[39]

1) Fully immersed, not agitated, duration 7 d

Table 9: Corrosion behavior of copper-silicon alloys towards acetic acid of various concentrations and temperatures

Copper-beryllium alloys are not as resistant as copper and are attacked by warm acetic acid [9, 80].

Beryllium-copper which contains about 2% beryllium loses 0.03 to 0.25 mm/a (1.18 to 9.84 mpy) in 2.5 to 10% acetic acid at room temperature (fully immersed, without stirring) [9, 81].

– Acetic anhydride –

For the behavior of silicon bronze towards acetic anhydride, virtually the same applies as for copper. The corrosion rates in 100 % acetic anhydride are smaller than 0.5 mm/a (19.7 mpy) at 297 K (24 °C), between 0.5 and 1.25 mm/a (19.7 and 49.2 mpy) at 325 to 352 K (52 to 79 °C), and higher than 1.25 (49.2 mpy) at 373 K (100 °C) [39]. Silicon bronze is therefore used at moderately high temperatures for fittings, valves, pumps etc. [46, 54].

Acid Halides

Author: G. Elsner / Editor: R. Bender

Copper

Copper is often used for handling the aqueous solutions of fluorophosphoric acid, and it is also used for short-term handling of concentrated or anhydrous fluorophosphoric acids [82]. It is not resistant to 10 to 90 % chlorosulfonic acid at temperatures as low as 297 K (24 °C) and is no longer suitable for handling 100 %, dry acid at 422 K (149 °C) [83].

Copper and its alloys are not suitable for the sulfochlorination of paraffin hydrocarbons with sulfur dioxide and chlorine with exposure to light (manufacture of mersols) [84]; when oxidizing media are excluded, however, copper is said not to be attacked by "mersols" [84].

Copper is resistant at room temperature to mixtures of o- and p-toluene sulfonyl chlorides containing hydrochloric acid, sulfuric acid and water and is therefore used for centrifugal drums in which the impurities are removed [85]. Copper is stated to be severely attacked, however, by 100 % toluene sulfonyl chloride at 297 K (24 °C) [83].

Fluorosulfonic acid does not react in the cold with copper [86], which is also suitable as material for piping and other equipment for handling dry phosgene (in the absence of even atmospheric moisture) at 298 K (25 °C); the corrosion rates are between 0.05 and 0.5 mm/a (1.97 and 19.7 mpy) [87, 88].

Fluorination of acetyl fluoride with the equivalent quantity of fluorine in the vapor phase (using dry nitrogen as the diluting agent), is carried out in a copper pipe packed with copper gauze when monoacetyl fluoride, diacetyl fluoride and a little trifluoroacetyl fluoride are produced [89].

Copper is not resistant to 10 % acetyl chloride at 297 K (24 °C), although it is resistant to 100 % acetyl chloride up to 373 K (100 °C) [83]; according to another source [90], however, copper should never be brought into contact with acetyl chloride. It is nevertheless used as material for seals [91].

With corrosion rates between 0.05 and 0.5 mm/a (1.97 and 19.7 mpy) at up to 373 K (100 °C), copper is resistant to pure butyryl chloride when air is excluded [83]. In the absence of oxidizing media (air, etc.) copper is also not attacked by butyryl chloride which still contains small quantities of hydrochloric acid, so that butyryl chloride can be distilled in copper equipment [92]. In mixtures of butyric acid and phosphorus trichloride such as occur in the manufacture of butyryl chloride, properly welded copper loses 19.3 g/m^2 d (corresponding to 0.78 mm/a (30.7 mpy)) in the liquid phase at 353 K (80 °C), and loses 14.9 g/m^2 d (equivalent to 0.60 mm/a (23.6 mpy)) in the vapor phase [93].

Copper can therefore be used as material for seals and packings for use in contact with butyryl chloride [92].

Copper is adequately resistant to benzoyl chloride up to 373 K (100 °C) [83], with a corrosion rate of 0.05 to 0.5 mm/a (1.97 to 19.7 mpy); it is therefore used as material for seals and packings, including combinations with asbestos, for use in contact with benzoyl chloride [94].

The behavior in dry, air-free chlorobenzoyl chloride is identical.

Copper-aluminium alloys

These alloys have virtually the same properties as copper. Aluminium bronze is not suitable for handling chlorosulfonic acid (10 % to 90 %) at 297 K (24 °C); the corrosion rate at temperatures up to 422 K (149 °C) is 0.5 to 1.25 mm/a (19.7 to 49.2 mpy) in 100 %, air-free chlorosulfonic acid [83]. CuAl-alloys therefore cannot be used for sulfochlorination processes. The same considerations also apply to toluene sulfonyl chloride at 297 K (24 °C) [83] or 10 % acetyl chloride at 297 K (24 °C); on the other hand, the alloy is resistant to 100 % acetyl chloride up to 373 K (100 °C).

Aluminium bronze is to be regarded as resistant to butyryl chloride and benzoyl chloride up to 373 K (100 °C) (corrosion rate 0.05 to 0.5 mm/a (1.97 to 19.7 mpy)), and it is also resistant to corrosion by chlorobenzoyl chloride [83].

Copper-nickel alloys

These alloys show virtually the same properties as copper. They can often be used for aqueous solutions and in the short term for concentrated or anhydrous fluorophosphoric acids [82]. They are, however, unsuitable for use in contact with sulfochlorination processes [84].

CuNi-alloys are said not be attacked by alkane sulfonyl chlorides ("mersols") when oxidizing media (air) are excluded [84].

CuNi-alloys behave like copper towards toluene sulfonyl chloride [85].

Copper-tin alloys (bronzes)

These alloys closely resemble copper in corrosion resistance properties because of the corrosion rate in excess of 1.25 mm/a (49.2 mpy), tin bronze is not suitable for handling 10 to 90 % chlorosulfonic acid at 297 K (24 °C); in 100 % acid the corrosion rates are not between 0.5 and 1.25 mm/a (19.7 and 49.2 mpy) until higher temperatures are reached (422 K (149 °C)) [83, 95]. CuSn-alloys are also not suitable for use in contact with sulfochlorination processes. They are said, however, not to be attacked by mersols when oxidizing media (including air) are excluded [84]. But since, however, according to another source bronze containing 10 to 20 % Sn should not be used for the casings of isolating valves in contact with 10 % solutions of "Mersol D" at 313 K (40 °C), bronze should never be used in these media. Bronze is also unsuitable for use in contact with toluene sulfonyl chloride and 10 % acetyl chloride at 297 K (24 °C), whereas it is said to be resistant to 100 % acetyl chloride up to 373 K (100 °C) with a corrosion rate less than 0.05 mm/a (1.97 mpy). Corrosion losses of between 0.05 and 0.5 mm/a (1.97 and 19.7 mpy) are to be expected in dry, air-free benzoyl chloride and chlorobenzoyl chloride at this temperature.

Copper-tin-zinc alloys (red brass)

These alloys closely resemble copper in corrosion resistance properties; accordingly, red brass is not resistant to 10 and 90 % chlorosulfonic acid at 297 K (24 °C) [83] and is unsuitable for use in contact with sulfochlorination processes [84]. In the absence of water, red brass is resistant to corrosion by mixtures of o- and p-toluene sulfonyl chloride and is therefore used as material for valves [85]; according to other sources, however, red brass is said to lose 0.5 to 1.25 mm/a (19.7 to 49.2 mpy) in pure toluene sulfonyl chloride at 297 K (24 °C) [83]. According to the same source, red brass is not resistant to 10 % acetyl chloride at 297 K (24 °C) (corrosion rate in excess of 1.25 mm/a (49.2 mpy)), although it is resistant to 100 % acetyl chloride up to 373 K (100 °C) (less than 0.05 mm/a (1.97 mpy)) [83]. Red brass loses less than 0.5 mm/a (19.7 mpy) in air-free butyryl chloride and benzoyl chloride at 373 K (100 °C) [83].

Copper-zinc alloys (brass)

These alloys closely resemble copper in corrosion resistance properties. They are not suitable for use in contact with sulfochlorination processes [84, 87, 88]. With corrosion rates of less than 0.5 mm/a (19.7 mpy), brass is moderately resistant to dry phosgene at 298 K (25 °C) and can therefore be used as material for piping and other plant equipment. Brass is to be regarded as not suitable for handling acetyl chloride [83, 90]; in butyryl chloride (air-free) it loses less than 0.5 mm/a (19.7 mpy) at 373 K (100 °C), whereas it loses more than 1.25 mm/a (49.2 mpy) in chlorosulfonic acid (10 and 90 %) at 297 K (24 °C) [83].

Other copper alloys

These alloys closely resemble copper in corrosion resistance properties. For example, silicon bronze is unsuitable for handling chlorosulfonic acid (10 and 90 %) at 297 K (24 °C), but is moderately resistant to the 100 % acid (dry) at 297 K (24 °C), the corrosion rate varying between 0.05 and 0.5 mm/a (1.97 and 19.7 mpy) [83]. Silicon bronzes also behave like copper in contact with toluene sulfonyl chloride [85]; the corrosion rate is less than 0.5 mm/a (19.7 mpy) at 297 K (24 °C) [83]. The alloy is not resistant to 10 % solutions of acetyl chloride at 297 K (24 °C), whereas it is said to lose less than 0.05 mm/a (1.97 mpy) in pure acetyl chloride at 373 K (100 °C). The loss in air-free butyryl chloride at 373 K (100 °C) is less than 0.5 mm/a (19.7 mpy), the same figure as in benzoyl chloride and chlorobenzoyl chloride [83].

Aliphatic Aldehydes

Author: G. Elsner / Editor: R. Bender

Copper

Copper is resistant to formaldehyde, acetaldehyde, butyraldehyde, metaldehyde, paraldehyde and paraformaldehyde at 297 K (24 °C); depending on concentration and temperature (up to 373 K (100 °C)), copper is generally usable in contact with the aqueous solutions of these aldehydes as well as with acrolein, caproic aldehyde, crotonaldehyde and chloroacetaldehyde; in contrast, copper is said to be unsuitable for propionaldehyde at temperatures as low as 297 K (24 °C). Formaldehyde solutions occasionally become colored, whereas acetaldehyde may react explosively with copper [96–98].

Copper is resistant to mixtures of formaldehyde, water and methanol; even in the ;presence of formic acid, the corrosion rate is not greater than 0.05 mm/a (1.97 mpy). e;Copper apparatus is as suitable as stainless steel apparatus for the distillation of formaldehyde [99]. 40 % formaldehyde solutions strongly etch polished copper [100].

The initial acid content (0,034 %) of an aqueous formaldehyde solution has increased after 60 days' contact with copper to 0.083 % and the corrosion rate was 0.0050 mm/a (0.20 mpy); after this time the color of the formaldehyde solution was blue-green and a precipitate had formed [101, 102].

The following examples illustrate the behavior of copper in formaldehyde manufacturing plants:

a) copper in a mixing vessel containing 40 % formaldehyde, 10 % methanol, 0.1 % formic acid, balance water, moderate aeration and turbulence, test duration 28 days, probably at room temperature; corrosion rate 0.0508 mm/a (2.00 mpy);

b) copper in a storage vessel containing 37 % formaldehyde, up to 1 % methanol, balance water, acid content not stated, temperature about 309 K (36 °C), agitation once daily for 15 to 30 minutes to equalize temperatures, test duration 87 days; corrosion rate 0.0965 mm/a (3.80 mpy);

c) copper in the lower and middle sections of a fractionation column, 10 % formaldehyde solution with traces of formic acid, temperature 390 K and 377 K (117 and 104 °C), test duration 87 days, turbulence; corrosion rate 0.012 mm/a (0.47 mpy);

d) copper in the head of a fractionation column, 2 % formaldehyde solution, 0.2 % formic acid, small quantities of methanol, temperature about 393 K (120 °C), turbulence, test duration 138 days; corrosion rate 0.0076 mm/a (0.30 mpy);

e) copper on the tray of the fractionation column from the top, 20 % formaldehyde solution, 10 to 15 % other volatile substances such as acetone and 0.1 %

formic acid, temperature about 408 K (135 °C), turbulence, test duration 71 days; corrosion rate 0.0050 mm/a (0.20 mpy);

f) copper on the inlet tray of a fractionation column, 12 to 15 % formaldehyde solution, 2 % formic acid, temperature 408 K (135 °C), turbulence, test duration 27 days; corrosion rate 0.0203 mm/a (0.80 mpy) [101, 102].

Copper and its alloys are very resistant to commercial formaldehyde solutions; the corrosion rates are below 0.15 mm/a (5.91 mpy) even at boiling temperature, and are mostly less than 0.05 mm/a (1.97 mpy). However, the small quantities of copper going into solution strongly produce a blue color in the formaldehyde solutions and therefore make them unsuitable for some purposes [101, 102].

Formaldehyde in 0.1 mol/l formic, acetic or propionic acid increases the corrosion of copper [101, 103].

Vessels of copper and its alloys without surface protection must not be used for storage and transportation of formaldehyde and its solutions because of the danger of coloration [104]. The same considerations apply to the storage and shipment of other aliphatic aldehydes [105]. Copper, in addition to aluminium and the stainless steel AISI 304, is more suitable than mild steel and cast iron (which are attacked) for vessels in which a solution of 40 % formaldehyde, 10 % methanol, 0.01 % formic acid and 50 % water is mixed by moderate aeration/agitation at 323 K (50 °C) [106].

Copper catalysts, though still used for converting alcohols to aldehydes are inferior to silver catalysts in respect of yield, throughput rate and life [107].

Copper catalysts are particularly suitable for the manufacture of butyraldehyde [108]. The reaction products consisting of butyraldehyde, butanol and water, are distilled in copper columns with Raschig rings [109]. Seals suitable for butyraldehydes include soft seals clad with copper [109].

A catalyst used in the catalytic dehydrogenation of primary alcohols is copper obtained by reduction of copper oxide at 533 to 563 K (260 to 290 °C) [105, 110].

Copper plate 5 to 6 mm thick is used for the reaction chambers in which a mixture of methanol and air is catalytically converted to formaldehyde [111].

In the ICI formaldehyde process, the reactor is protected by a copper-tin alloy flame trap in the connecting pipe to the evaporator flame trap [112].

The apparatus for the catalytic oxidation of methanol to formaldehyde may be made from copper pipes. The silver catalyst is usually spread on fine-mesh copper wire networks. Copper stills are suitable for distillation of formaldehyde under pressure, since copper does not disturb the equilibrium between liquid and vapor phases [99, 113]. The contact chambers for the conversion of formaldehyde + acetaldehyde to acrolein at 473 to 673 K (200 to 400 °C) can be made from copper, bronze or aluminium bronze [113]. Sealing material which may be used for acetaldehyde and crotonaldehyde include soft seals clad with copper [114, 115].

Dehydrogenation of caproic alcohol to caproic aldehyde is performed in the presence of copper catalysts at 573 to 623 K (300 to 350 °C) in pipes of unalloyed or low-alloy steel (3 % chromium) [116]. Distillation is performed in copper columns with porcelain Raschig rings [116]. Copper and its combinations with asbestos may be used for seals and packings for caproic aldehyde [116].

Equipment for the manufacture of crotonaldehyde, e.g. acid vessels, columns, separators, storage vessels and piping – are best made from copper, especially when nitrogen is used as the circulating gas. Valves and fittings are of red brass or brass [115].

Olefins, e.g. propylene, are converted to acrolein by gas phase oxidation with oxygen at 643 K (370 °C) and 0.5 MPa in a copper coil [117].

Corrosion of the copper, aluminium and iron base materials inside encapsulated electrical devices depends on the microclimate, which is determined by moisture and certain evaporation products from the insulating materials, e.g. formaldehyde from phenolic or amino resins [118].

It must always be borne in mind that vapors given off from phenol-formaldehyde and amino resins at 308 K (35 °C) and 100 % relative humidity are highly corrosive to copper, brass, aluminium, zinc and cadmium because of their formaldehyde content (and the formic acid produced therefrom). These metals are corroded in an atmosphere containing, for example, 0.002 mg/l formaldehyde at 308 K (35 °C) and a relative humidity of 100 % [119]. In contrast to the surface layers formed on copper, aluminium, cadmium and iron, the corrosion products produced on zinc have no protective effect against corrosion by formaldehyde [120, 121]. The formaldehyde vapors generated from adhesives (e.g. based on phenol-formaldehyde resin which are used for plywood in moist air (90 % relative humidity at 298 to 303 K (25 to 30 °C)) may severely corrode copper, brass, zinc, cadmium, mild steel, lead and solder in sealed packages during long-term storage [122].

Copper and its alloys should also not be used in the presence of mixtures of formaldehyde and ammonia or amines (e.g. hexamethylenetetramine) [113].

Bimetallic pipes, copper or brass with mild steel can be used for formaldehyde coolers and condensers [123].

Corrosion of copper, steel, tin and other structural materials in 0.1 to 72.5 % perchloric acid can be inhibited by the addition of 0.05 to 0.10 % chloral hydrate [124].

Copper-aluminium alloys

Virtually the same as has been said for copper also applies to these alloys. Some examples showing the behavior of CuAl-alloys in industrial formaldehyde manufacturing plants are listed below:

a) in the still head of a fractionation column, 2 % formaldehyde solution, 0.2 % formic acid, small quantities of methanol, about 393 K (120 °C), turbulence, test duration 138 days, corrosion rate 0.0025 mm/a (0.10 mpy);
b) on the second tray from the top in the fractionation column, 20 % formaldehyde solution, 10 to 15 % other volatile substances (e.g. acetone) and 0.1 % formic acid, about 408 K (135 °C), turbulence, test duration 71 days, corrosion rate 0.0127 mm/a (0.5 mpy);

c) on the inlet tray in a fractionation column, 12 to 15 % formaldehyde solution, 2 % formic acid, 408 K (135 °C), turbulence, test duration 27 days, corrosion rate 0.0177 mm/a (0.70 mpy) [101, 125].

Copper-aluminium alloys are regarded as "resistant to formaldehyde under most operating conditions" [126]. Ampco® aluminium bronze loses < 0.15 mm/a (< 5.91 mpy) in formaldehyde solutions [127].

Copper-nickel alloys

Virtually the same as has been said for copper applies to these alloys. Some examples showing the behavior of CuNi-alloys in industrial formaldehyde manufacturing plants are listed below:

a) in the still head of a fractionation column, 2 % formaldehyde solution, 0.2 % formic acid, small quantities of methanol, about 393 K (120 °C), turbulence, test duration 138 days, corrosion rate 0.0025 mm/a (0.10 mpy);
b) on the second tray from the top in the fractionation, 20 % formaldehyde solution, 10 to 15 % other volatile substances (e.g. acetone) and 0.1 % formic acid, about 408 K (135 °C), turbulence, test duration 71 days, corrosion rate 0.0101 mm/a (0.40 mpy);
c) on the inlet tray in a fractionation column, 12 to 15 % formaldehyde solution, 2 % formic acid, temperature 408 K (135 °C), turbulence, test duration 27 days, corrosion rate 0.0152 mm/a (0.60 mpy) [101, 102].

CuNi-alloys containing 10 to 30 % nickel as well as CuZnNi-alloys containing 15 to 30 % zinc and 10 to 20 % nickel are described as being resistant to formaldehyde under most operating conditions [126].

Copper-tin alloys (bronzes)

Virtually the same as has been said for copper applies to these alloys. CuSn-alloys are described as being resistant to formaldehyde under most operating conditions [126].

Copper-tin-zinc alloys (red brass)

Virtually the same as has been said for copper applies to these alloys. Cast-iron cocks with red brass plugs are in use for neutral acetaldehyde [114]. Valves and fittings of red brass or brass are used for crotonaldehyde [115].

Copper-zinc alloys (brass)

Virtually the same as has been said for copper applies to these alloys. A few examples of the behavior of admiralty brass (CuZn28Sn1Sb) in industrial formaldehyde manufacturing plants are listed below:

a) Admiralty brass in the still head of a fractionation column, 2 % formaldehyde solution, 0.2 % formic acid, small quantities of methanol, temperature about 393 K (120 °C), turbulence, test duration 138 days, corrosion rate 0.0076 mm/a (0.30 mpy);
b) Admiralty brass on the second tray from the top in the fractionation column, 20 % formaldehyde solution, 10 to 15 % other volatile substances (e.g. acetone) and 0.1 % formic acid, about 408 K (135 °C), turbulence, test duration 71 days, corrosion rate 0.0101 mm/a (0.40 mpy);
c) Admiralty brass on the inlet tray of a fractionation column, 12 to 15 % formaldehyde solution, 2 % formic acid, temperature 408 K (135 °C), turbulence, test duration 27 days, corrosion rate 0.0304 mm/a (1.20 mpy), [101, 102, 128].

Brass is not recommended for handling formaldehyde, since the solutions become colored by the dissolved copper ions which again, promote oxidation to formic acid [101, 129]. Brass valves and fittings are, however, used for crotonaldehyde [115].

Butyraldehyde is manufactured by dehydrogenation of butanol at 673 K (400 °C) in a contact chamber of brass with loose brass wire mesh coils (67.9 Cu, 32 Zn, 0.1 Fe and traces of Al) as catalyst. Propionaldehyde, isobutyraldehyde and higher aliphatic aldehydes can be manufactured in the same way [130]. When evaporators in sugar beet factories are cleaned with boiling dilute hydrochloric acid, 40 % formaldehyde solution inhibits the corrosion of brass, iron or steel [131].

Formaldehyde and urea inhibit the corrosion of brass in nitric acid. The effect of formaldehyde increases with time or with its concentration. Accelerated corrosion was observed after 1 to 2 hours' exposure to low formaldehyde concentrations [132].

Corrosion of brass, copper, aluminium and ferrous materials inside encapsulated electrical devices depends on the microclimate, which is determined by moisture and certain decomposition products (e.g. formaldehyde from phenolic resins) from the insulating materials [118].

Because of their formaldehyde content, the vapors given off by various plastics (phenol-formaldehyde and urea-formaldehyde resins), at 308 K (35 °C) and 100 % relative humidity are always highly corrosive to brass, copper, aluminium, zinc and cadmium [120]. The formaldehyde vapors emanating from synthetic adhesives (for plywood) based on phenol-formaldehyde resin in moist air (90 % relative humidity at 298 to 303 K (25 to 30 °C)) may also substantially corrode brass, copper, zinc, etc. in sealed packages during long-term storage [122].

Paraldehyde is an excellent inhibitor of the corrosion of 60/40 brass in 0.2 mol/l potassium persulfate solution; it influences the cathodic reaction [133].

Aliphatic Amines

Author: L. Hasenberg / Editor: R. Bender

Copper

At a standard potential of +0.345 V (Cu/Cu^{2+}), copper is a relatively noble metal. However, it has a marked tendency for complex formation with inorganic and organic substances and is dissolved, for example, by ammonium salt solutions with complex formation. It shows the same behavior in aniline (phenylamine), where it is strongly corroded [134, 135].

Copper is more or less attacked by amines, such as methylamines, butylamines, amylamines and chloramines and also hexamethylenetetramine, ethylenediamine, diethylamine, butylamine, octadecylamine, benzylamine, cyclohexylamine and diphenylamine. Discoloration of the products occurs almost in each case [136–147]. The corrosion rate of copper and copper alloys in amines is more than 1.0 mm/a (39.4 mpy) at room temperature [148].

The corrosion rate caused by saturated methylamine solutions is more than 1.27 mm/a (50 mpy) at 298 K (25 °C) [149]. According to [136], copper cannot be used in methylamine solutions because it has a corrosion rate of more than 2.5 mm/a (98.4 mpy).

The corrosive attack of copper caused by amines is reinforced by the presence of oxygen. The use of copper in amines can lead to discoloration of the substances. The corrosion rate to be expected in any particular case should appropriately be determined individually.

Apart from uniform corrosion, stress corrosion cracking of copper materials by amines (such as in the case of ammonia) must be expected [150]. The corrosion behavior of copper in basic (pH 12) sodium sulfate solution did not deteriorate by addition of cyclohexylamine, which instead stabilized the oxide layer on the copper surface [151]. In a current of air having a humidity of 80 and 100%, corrosion tests were carried out in the laboratory using copper samples. These were wrapped in paper which had been impregnated with amine solutions as volatile inhibitors. No signs of corrosion could be detected on any of the copper samples after a test duration of three years. The solutions consisted of benzotriazole, nitromethane, triethylamine, and dimethyl phthalate in a ratio of 1:0.3:0.3:5 and benzotriazole, nitromethane, triethylamine, and acetone in a ratio of 1:0.3:0.3:5 [152].

Copper-aluminium alloys

CuAl-alloys hardly differ from other copper materials in their corrosion resistance to amines. The corrosion rate in methylamines at 293 K (20 °C) is expected to be more than 1.0 mm/a (39.4 mpy) (see Table 10) [148]. However, even in the case of a minor corrosive attack, discoloration of the amines is certainly to be expected.

Copper-nickel alloys

Copper-nickel alloys have only moderate corrosion resistance to amines. Some of the corrosion rates in lower alkylamines are above 1.0 mm/a (39.4 mpy) (Table 10). These alloys can be used only if a relatively large corrosive attack can be tolerated. They are attacked even by the relatively weak bases among the amines [148]. Corrosion rates are not mentioned.

Copper-tin alloys (bronzes)

As all copper materials, copper-tin alloys have only moderate corrosion resistance to amines. In lower alkylamines, corrosion rates of more than 1.0 mm/a (39.4 mpy) are expected due to complex formation. This attack already starts at 293 K (20 °C) and is reinforced by the presence of air or oxygen. Stress corrosion cracking must be expected with these alloys. Corrosion rates for amines are summarized in Table 10 [148].

Copper-tin-zinc alloys (red brass)

CuSnZn-alloys are only moderately corrosion-resistant to amines and in particular to the lower alkylamines. Depending on the medium and other conditions, such as, for example, aeration, which promotes corrosion, corrosion rates of up to and higher than 1.0 mm/a (39.4 mpy) are to be expected (Table 10) [148].

Copper-zinc alloys (brass)

As with all copper materials, the corrosion resistance of these materials to amines is quite low.

In aerated methylamine solutions, corrosion rates of more than 1.27 mm/a (50 mpy) and stress corrosion cracking probably have to be expected. Surface corrosion at this level is confirmed in [148]. Corrosion rates in amines are summarized in Table 10.

These materials are unsuitable for use with amines in general because of their high corrosion rates.

Apart from this corrosive effect on brass, amines are good corrosion inhibitors in other media. Thus, the corrosion-inhibiting effect of amines in acetic acid increases with increasing concentration [153]. The inhibitors used were methylamine (33%), ethylamine (50%), isopropylamine (70%), and cyclohexylamine. The samples used were cold-rolled brass strips (63 Cu, 37 Zn). The best inhibiting effect of the amines is shown at the point where, for example, acetic acid becomes neutralized by their addition. Further addition of amines decrease the inhibiting effect again and leads to increased corrosion by them. If neutralization of the acid by the amine is not possible, no inhibition is to be expected.

Aliphatic Amines

Medium	Copper	G-CuSn10-14 (cf. CC480K, CC483K)	G-CuSn10Zn	G-CuSn5Znpb (cf. CC491K) / G-CuSn7Znpb (cf. CC493K)	G-CuPb 25	G-CuPb 5-Sn (cf. CC495K, CC496K, CC497K)	G-CuAl9	G-CuAl10Fe (cf. CC331C)	G-CuAl9Ni (cf. CC332C) / G-CuAl10Ni (cf. CC333C)	G-CuAl9Mn	G-Cu 65 Zn	G-Cu 60 Zn	G-Cu 55 Zn (Mn)	G-Cu 55 ZnAl 1-2	G-Cu 55 ZnAl 4	G-Cu Zn 16 Si 4	G-CuNi30 (cf. CC383H)
Amines	4	4	4	4	4	4	4	4	4	4	4	4	4	4	4	4	4
Benzylamine	3	3	4	4	3	4	4	3	3	4	4	4	4	4	4	4	3
Chloramine	2	2	2	2	2	2	2	2	2	2	3	3	3	3	3	3	2
Chloroaniline	3	3	3				3	3	3	3	3	4	4	3	3	3	2
Diethanolamine	4	4	4	4	4	4	4	4	4	4	4	4	4	4	4	4	4
Diethylamine, dry	4	4	4	4	4	4	4	4	4	4	4	4	4	4	4	4	4
Diethylenetriamine	4	4	4	4	4	4	4	4	4	4	4	4	4	4	4	4	4
Dimethylamine	3	3	3	3	3	3	3	3	3	3	3	3	3	3	3	3	3
Ethylenediamine	4	4	4	4	4	4	4	4	4	4	4	4	4	4	4	4	4
Methanolamine, impure	4	4	4	4	4	4	4	4	4	4	4	4	4	4	4	4	4
Methylamine, dry	4	4	4	4	4	4	4	4	4	4	4	4	4	4	4	4	4
Naphthylamine	2	2	2	2	2	2	2	2	2	2	2	2	2	2	2	2	2
Triethanolamine	4	4	4	4	4	4	4	4	4	4	4	4	4	4	4	4	4
Triethylamine, pure	4	4	4	4	4	4	4	4	4	4	4	4	4	4	4	4	4
Triethylamine, moist	3	2	2	2	2	3	2	2	2	2	3	3	3	3	3	3	2
Tributylamine, all conc.	4	4	4	4	4	4	4	4	4	4	4	4	4	4	4	4	4
Trimethylamine	4	4	4	4	4	4	4	4	4	4	4	4	4	4	4	4	4

1: < 0.05 mm/a (< 1.97 mpy), 2: < 0.5 mm/a (< 19.7 mpy), 3: 0.5 to 1.0 mm/a (19.7 to 39.4 mpy), 4: above 1.00 mm/a (39.4 mpy)

Table 10: Corrosion rates of copper materials in amines at 293 K (20 °C) [148]

Other copper alloys

Copper is not corrosion-resistant to amines. The reason for this is its tendency to complex formation (see section copper).
Additional alloying elements do not give any improvement to this insufficient corrosion resistance (Table 10) [148].

Aliphatic Ketones

Author: H. Barkholt / Editor: R. Bender

Copper-zinc alloys (brass)

Copper and its alloys are often used for handling acetone and its aqueous solution; even at the boiling point of acetone, they are still not attacked [154–158].

The corrosion rate of copper and its alloys is generally 0.1 mm/a (3.94 mpy). However, brass with 30 to 40 % zinc loses up to 1.0 mm/a (39.4 mpy) [159].

The corrosion rate of aluminium bronzes (e.g. Ampco® with 8 to 14 % Al) is less than 0.15 mm/a (5.91 mpy).

The material consumption rate of E-copper in anhydrous acetone in the presence of acid is listed in Table 11 [160].

	Temperature, K (°C)		
	294 (21)	313 (40)	329.6 (56.45)
Acid and concentration	Material consumption rate, g/m² d		
0.1 mol/l HCl	2.5	15	17
0.05 mol/l H_2SO_4	0.2	0.2	0.4

Table 11: Corrosion behavior of E-copper in acid-containing acetone [160]

This attack is the reason for the substantial discoloration of weakly acid acetone in copper equipment.

Fractionating columns and condensers for acetone can be manufactured from copper [161]. Bronze with 10 to 20 % tin is suitable as a material for valve casings [162].

When acetone vapors are passed through copper pipes at 923 to 993 K (650 to 720 °C), ketene is produced whose formation is not catalytically affected by copper. Steel pipes made of chromium-nickel steel are also used for this reaction [163].

Copper can be inhibited in aqueous (5 % H_2O) acetone solutions of secondary cellulose (cellulose acetate fiber production) by the addition of 0.3 % benzotriazole with an inhibition efficiency of 62.5 % [166].

Alkaline Earth Chlorides

Author: *R. Weidemann* / Editor: *R. Bender*

Copper and copper alloys

The corrosion behavior of copper in chloride-containing solutions is determined by the cathodic depolarization process

$$O + 2e^- + H_2O \rightarrow 2OH^-$$

The dissolution of copper and of high-copper alloys in alkaline earth chloride solutions is therefore decisively dependent – just as the alkali chloride solutions – on oxygen availability. It initially increases with rising salt concentration, reaches a maximum and decreases with further increase in the concentration due to the lower solubility of oxygen in this case. An analogous behavior is also shown in the temperature dependance [167].

The behavior of copper in calcium chloride and barium chloride solutions is in general designated as "still good" [168] and the material despite slight attack as "in most cases usable" [169].

The copper-zinc alloys containing up to 20% of zinc and the copper-nickel and copper-aluminium alloys behave similarly to copper; copper-zinc alloys are even slightly superior to copper [168]. However, rod and cast alloys having a copper content below 67% cannot be used in calcium chloride and barium chloride solution.

Cast alloys containing 10% of aluminium and 1 to 4% of iron and castings containing about 90% of copper and 3 to 4% of silicon are fairly resistant [170]. Table 12 contains some values as reference points.

Material	Testing medium and conditions	Corrosion rate mm/a (mpy)	Literature
Cu	40% $CaCl_2$, 353 K (80°C), aerated	0.61 (24.0)	[171]
Cu	>~15% $CaCl_2$, 293 K (20°C), aerated 1 d	0.6 (23.6)	[172–174]
Cu85Al10Fe3Mn1		0.01 (0.39)	[175]
Cu56Fe1Mn1PbAl		0.05 (1.97)	[175]
Cu65Fe2Al31SnMnPb	5% $CaCl_2$ + 5% $MgCl_2$, aerated, 293 K (20°C)	0.03 (1.18)	[175]
Cu67Pb4Sn1 (G)		0.07 (2.76)	[175]
Cu88Sn10Pb0.5 (G)		0.03 (1.18)	[175]
Cu85Sn9Zn2.5Pb3.5 (G)		0.05 (1.97)	[175]

Table 12: Quantitative data on the corrosion rate of copper and copper alloys in alkaline earth chloride solutions [170]

Table 12: Continued

Material	Testing medium and conditions	Corrosion rate mm/a (mpy)	Literature
Cu + < 40 % Zn	40 % $CaCl_2$ aerated, 353 K (80 °C)	0.34 (13.4)	[171]
CuNi10	55 % $CaCl_2$, 373 K (100 °C), 29 d	0.15 (5.91)	[176]
CuNi30		0.10 (3.94)	[176]
CuNi10	62 % $CaCl_2$, 423 K (150 °C), 18 d	0.57 (22.4)	[176]
CuNi30		0.42 (16.5)	[176]
CuNi10	73 % $CaCl_2$, 454 K (181 °C), 18 d	1.90 (74.8)	[176]
CuNi30		0.70 (27.6)	[176]
Cu	saturated $BaCl_2$ solution, 459 K (186 °C) pH 7.2	0.06 (2.36)	[171]

Table 12: Quantitative data on the corrosion rate of copper and copper alloys in alkaline earth chloride solutions [170]

Alkaline Earth Hydroxides

Author: A. Weser / Editor: R. Bender

Copper and copper alloys

Copper is resistant to solutions of alkaline earth hydroxides, provided that no ammonium salts or cyanides are present. There is the risk of corrosion in the presence of organic amines.

Care should also be taken that acetylene is absent in an alkaline medium, since copper can form very explosive acetylides.

Pure copper is highly resistant to aqueous calcium hydroxide solutions, if the above-mentioned conditions are taken into account [177]. Pure copper is not resistant in the system calcium hydroxide-calcium hypochlorite (bleaching lime, chlorinated lime) [178]. Compounds based on silica-alkali oxides and alkaline earth oxides (silicates) have a predetermined dissolution rate in aqueous systems and are used as inhibitors for copper lines, for example in cooling water systems [179].

The subject of alkaline earth hydroxides allows to give a summary of the various copper alloys with respect to their behavior.

The copper alloys are distinguished by good formability and generally high corrosion resistance. As in the case of pure copper, the risk of corrosion by ammonium salts and cyanides in alkaline lime solution applies to its alloys as well, in particular to the alloys which are rich in zinc and nickel. Organic amines can also lead to severe corrosion damage by the formation of complexes.

In [180] reference is made to this general possibility of attack of the copper alloys. Copper alloys containing tin and phosphorus have higher resistance in the alkaline range [180].

In the calcium hydroxide – calcium hypochlorite system, phosphor bronze can replace titanium or stainless steel as construction material. Other copper alloys are not resistant [178].

In Table 13 the chemical composition of the alloys tested is given.

Material	Chemical composition, %				
	Cu	Sn	Si	P	Pb
Pure copper-I	97.64	–	–	–	–
Pure copper-II	99.37	–	–	–	–
Silicon bronze (low-alloy)	98.37	–	1.46	–	–
Silicon bronze (high-alloy)	96.8	–	2.97	–	–
Tin bronze	89.93	9.41	–	–	–
Tin bronze (cont. lead)	79.35	5.85	–	–	12.77
Phosphor bronze 6	94.42	4.5	–	0.20	–
Phosphor bronze 7	90.78	7.8	–	0.18	–
Phosphor bronze 8	87.66	11.2	–	0.46	–

Table 13: Chemical composition of copper alloys [178]

In Table 14 the corrosion potentials and corrosion current densities are listed.

The corrosion rates are between 0.0662 and 0.635 mm/a (2.61 and 25 mpy). Phosphor bronze 6 had the lowest values. Good corrosion resistance to weak bases is reported for the Cu-grades Cu-0.4S, Cu-0.15Zr, Cu-1.9Be and Cu-0.95Cr. The same applies to CuAl-, CuAlCo- and CuSiMn-bronzes [177, 180].

Material	Corrosion potential		Current density mA/cm^2
	V_{SCE}*	V_{SHE}**	
Pure copper-I	0.665	0.907	29.85
Pure copper-II	0.66	0.902	28.18
Silicon bronze (low-alloy)	0.65	0.892	28.18
Silicon bronze (high-alloy)	0.645	0.887	47.32
Tin bronze	0.65	0.892	35.48
Tin bronze (high lead content)	0.42	0.662	47.32
Phosphor bronze 6	0.635	0.877	19.95
Phosphor bronze 7	0.66	0.902	70.79
Phosphor bronze 8	0.655	0.897	39.81
Stainless steel	0.58	0.822	1.12
Titanium	0.38	0.622	3.85

* Saturated Calomel Electrode
** Standard Hydrogen Electrode

Table 14: Corrosion potential and corrosion current densities of various materials in the system calcium hydroxide – calcium hypochlorite at 303 K (30 °C) [178]

Alkanecarboxylic Acids

Author: L. Hasenberg / Editor: R. Bender

Copper

Copper is resistant to propionic acid in all concentrations, including the pure acid, up to the boiling point (414 K (141 °C)), provided oxygen and air are not present [181]. The corrosion rate under these conditions is then no more than 0.008 mm/a (0.31 mpy). In the manufacture of propionic acid in copper-lined high-pressure towers (20 MPa, 523–543 K (250–270 °C)) in accordance with the equation $C_2H_4 + CO + H_2O \rightarrow C_3H_5OOH$ the generalized corrosion of copper increases to 0.98 mm/a (38.6 mpy) [181]. It is recommended that silver be used under these conditions.

Figures 3 and 4 show the corrosion behavior of copper in anhydrous monocarboxylic acids (alkanecarboxylic acids).

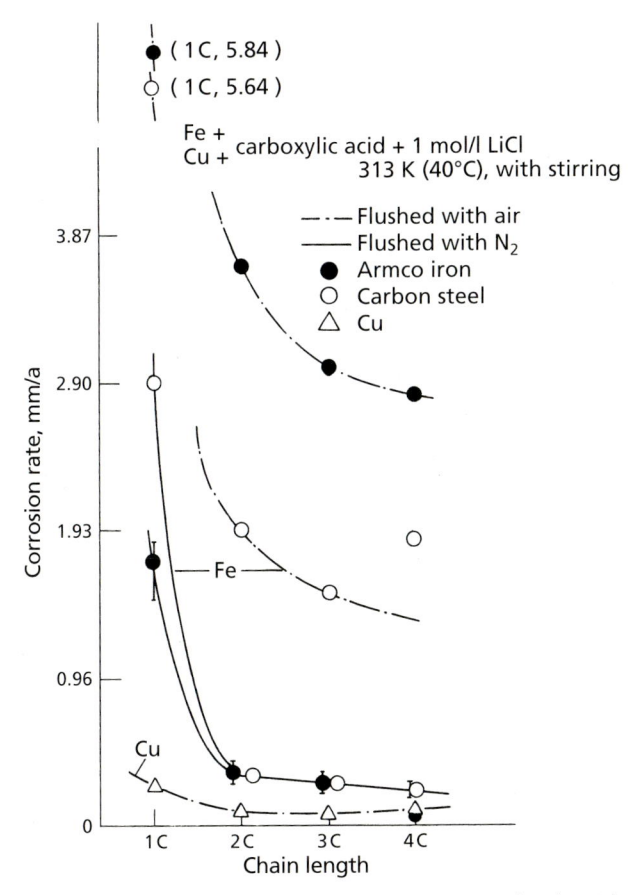

Figure 3: Corrosion rates of iron and copper in monocarboxylic acids with the addition of lithium chloride [182]

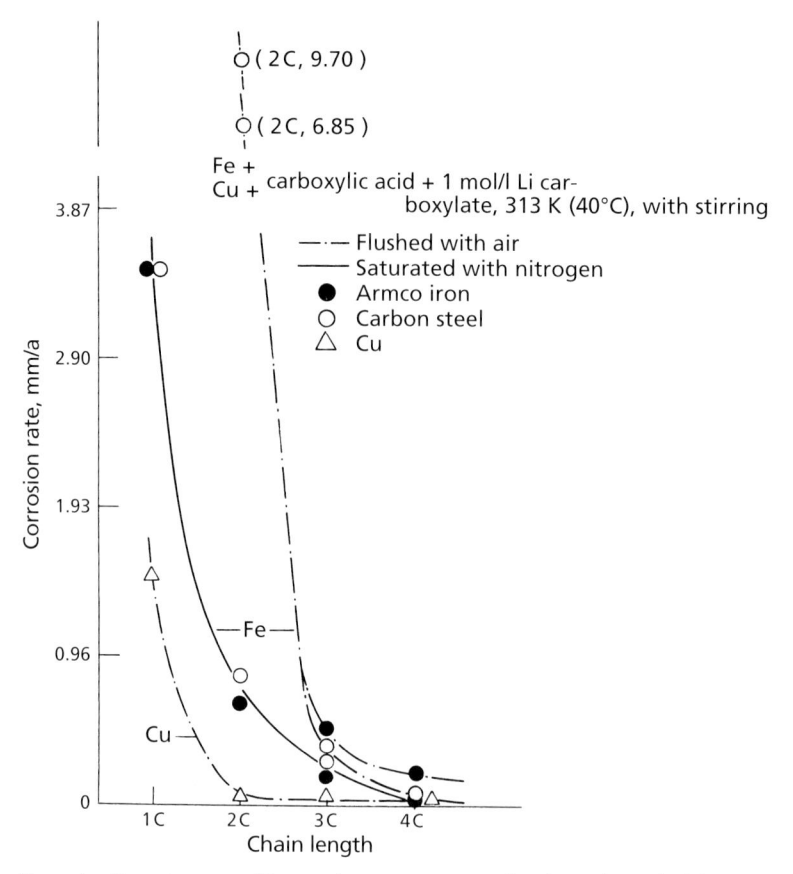

Figure 4: Corrosion rates of iron and copper in monocarboxylic acids to which lithium carboxylate has been added [182]

Salts were added to improve the conductivity for the electrochemical investigations. The illustrations show that corrosive attack on copper by alkanecarboxylic acids decreases significantly with increasing chain length. Corrosion by propionic acid is roughly comparable with that by acetic acid. The curves for these comparable corrosion rates by acetic acid according to F. F. Berg are shown in Figure 5 [183].

As shown in Figures 3 and 4, copper is attacked to roughly the same degree by acetic acid, propionic acid and butyric acid. The corrosive attack by acetic acid on copper shown in Figure 5 may therefore also be assumed for propionic acid and butyric acid.

In Table 15 further corrosion rates are listed.

This table clearly shows that aerated alkanecarboxylic acids, in this case acetic acid, attack copper relatively severely even at 293 K (20 °C). The corrosion rates are between 0.6 and 1.5 mm/a (23.6 and 59.1 mpy) (293–298 K (20–25 °C)) in 5 to 60 % acetic acid [184].

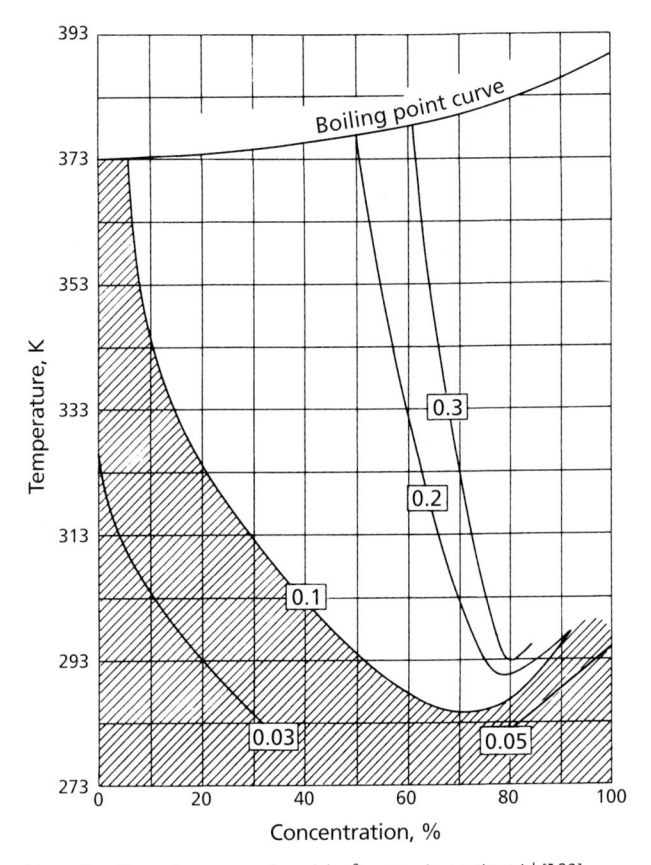

Figure 5: Corrosion curves (mm/a) of copper in acetic acid [183]

These corrosion values are also to be expected in propionic acid. In 1% butyric acid, copper loses 0.07 mm/a (2.76 mpy) at 293 K (20 °C) and 0.44 mm/a (17.3 mpy) at 336 K (63 °C). The corrosion rate in 50% acid at 293 K (20 °C) through which hydrogen was passed was 0.10 mm/a (3.93 mpy). The corrosion rate increased to 0.56 mm/a (22 mpy) when oxygen was passed through butyric acid instead of hydrogen. In boiling, concentrated acid corrosion rates of 0.48 to 1.21 mm/a (18.8 to 47.6 mpy) were found. Corrosive attack is very much dependent on aeration and may increase by a factor of 3 to 4. The addition of 0.1% of sulfuric acid to 99% butyric acid at 437 K (164 °C) decreased the corrosion rate of copper to 0.12 mm/a (4.72 mpy) [181].

Investigations of corrosion of copper M 3 (Russian type) in alkanecarboxylic acids having 7 to 9 carbon atoms in the chain at 373 K (100 °C) showed maximum corrosion rates of 0.03 mm/a (1.18 mpy). Increasing the temperature to 573 K (300 °C) has no influence on the corrosion behavior of copper. Corrosion rate in the gas phase above the liquid at 573 K (300 °C) was 0.02 mm/a (0.79 mpy) [185].

Alkanecarboxylic Acids

To summarize, it can be stated that copper is corroded by alkanecarboxylic acids having three carbon atoms or more, corrosion becoming increasingly severe with increasing air content of the aqueous solution. Copper is resistant to corrosion in oxygen-free solutions [184].

One of the disadvantages of using copper and its alloys when handling alkanecarboxylic acids is that the products take on a greenish discoloration and even small quantities of copper compounds accelerate oxidation of these acids. This leads to the products becoming rancid [186].

Copper-aluminium alloys

What has been said about copper also roughly applies to CuAl-alloys; they are resistant to oxygen-free alkanecarboxylic acids and their aqueous solutions [184]. They are not as sensitive to the presence of oxygen as copper and aluminium-free copper alloys. They are corroded at the rate of 6 mm/a (236 mpy) in concentrated propionic acid [187].

Copper-aluminium alloys are resistant up to the boiling point in aqueous propionic acid solutions and in pure acid. The corrosion rates are less than 0.01 mm/a (0.39 mpy) [181].

Material	Temperature K (°C)	Concentration %	Corrosion rate, mm/a (mpy)			
			aerated		air excluded	
			mobile	static	mobile	static
Copper	293 (20)	0.8	0.3 (11.8)	0.11 (4.33)		
	293 (20)	5.1	0.6 (23.6)	0.27 (10.6)	0.3 (11.8)	0.1 (3.94)
	293 (20)	7	0.6 (23.6)	0.3 (11.8)	0.28 (11)	
	298 (25)	6	0.67 (26.4)			0.04 (1.57)
	298 (25)	20	1 (39.4)		0.3 (11.8)	0.15 (5.9)
	298 (25)	60	1.5 (59)			
	298 (25)	100	0.6 (23.6)			0.05 (1.97)
	298 (25)	conc. anhydride			0.16 (6.3)	0.2 (7.87)

Table 15: Corrosion rates of copper and copper alloys in acetic acid [184]

Table 12: Continued

Material	Temperature K (°C)	Concentration %	Corrosion rate, mm/a (mpy)			
			aerated		air excluded	
			mobile	static	mobile	static
	323 (50)	conc. anhydride		1.12 (44.1)	1.3 (51.2)	
	373 (100)	20		1.76 (69.3)		
	373 (100)	conc.	2.5–8.0 (98.4–315)			
	373 (100)	crude acetic acid			0.4 (15.7)	
	373 (100)	100			0.03 (1.18)	0.01 (0.39)
	373 (100)	conc. anhydride	20.5 (807)		3	
G-CuSn3.4 Phosphor bronze (Not standardized)	293 (20)	10–15	0.18 (7.1)		0.08 (3.15)	
	293 (20)	33	0.3 (11.8)		0.2 (7.87)	
	293 (20)	50		0.125 (4.92)		
	293 (20)	75		0.2 (7.87)		
	293 (20)	100		0.65 (25.6)		
	323 (50)	10		0.22 (8.66)		
	323 (50)	50		0.55 (21.6)		
	323 (50)	75		0.63 (24.8)		
	323 (50)	100		0.95 (37.4)		
	boiling point	10		2.0 (78.7)		

Table 15: Corrosion rates of copper and copper alloys in acetic acid [184]

Table 15: Continued

Material	Temperature K (°C)	Concentration %	Corrosion rate, mm/a (mpy)			
			aerated		air excluded	
			mobile	static	mobile	static
	boiling point	33			0.5 (19.7)	0.3 (11.8)
	boiling point	50			0.9 (35.4)	
	boiling point	75			1.0 (39.4)	
	boiling point	100			35.0	(1,377)
G-CuSn5ZnPb (CW491K)	293 (20)	1.2	0.22 (8.66)		0.14 (5.51)	
	353 (80)	1.2	0.54 (21.3)		0.2 (7.87)	
CuZn30 (CW505L)	293 (20)	0.6	0.2 (7.87)	0.1 (3.94)		
	333 (60)	1.1	0.4 (15.7)	0.1 (3.94)		
CuZn16Si4 (CC761S)	293 (20)	4	0.15 (5.91)	0.08 (3.15)	0.03 (1.18)	
	373 (100)	4	1.1 (43.3)	0.67 (26.4)	0.23 (9.1)	
CuSi3Mn1 (CW116C)	296 (23)	10	0.005 (0.2)			
	296 (23)	25	0.04 (1.57)			
	296 (23)	50	0.05 (1.97)			
	296 (23)	75	0.1 (3.94)			
	296 (23)	99.5	0.32 (12.6)			

Table 15: Corrosion rates of copper and copper alloys in acetic acid [184]

Table 15: Continued

Material	Temperature K (°C)	Concentration %	Corrosion rate, mm/a (mpy)			
			aerated		air excluded	
			mobile	static	mobile	static
CuAl10Fe (CuAl10Fe and CuAl9Fe(Ni))	295 (22)	10		0.05 (1.97)		< 0.01 (< 0.39)
	295 (22)	35		0.075 (2.95)		0.01 (0.39)
	295 (22)	50		0.1 (3.94)		0.01 (0.39)
	357 (84)	35		0.5 (19.7)		0.015 (0.59)

Table 15: Corrosion rates of copper and copper alloys in acetic acid [184]

According to [184], the corrosion rate at 293 K (20 °C) in butyric acid and oxygen-free propionic acid is less than 0.5 mm/a (19.7 mpy). In the case of alloys containing nickel the corrosion rates decrease to below 0.05 mm/a (19.7 mpy) under the stated conditions. Aluminium bronzes are corroded at the rate of 0.16 mm/a (6.30 mpy) in butyric acid (98 %, 383 K (110 °C)). They are used as materials for construction of pumps and condensers [181].

Copper-nickel alloys

These alloys are similar in corrosion behavior to the copper-aluminium alloys and are also not as susceptible as copper and its other alloys in the presence of oxygen. They are rather more resistant than copper-aluminium alloys to corrosion by 98 % butyric acid at 383 K (110 °C). The corrosion rate will be below 0.16 mm/a (6.30 mpy). Loss of material due to corrosion in the liquid is greater than in the vapor above that liquid [181].

The corrosion rate in oxygen-free propionic and butyric acids (no concentrations stated) is below 0.5 mm/a (19.7 mpy) [184].

CuNi-alloys corrode at the rate of 9.35 mm/a (368 mpy) in 100 % boiling, unaerated propionic acid and at the rate of 1.7 mm/a (66.9 mpy) in a mixture of 66 % propionic acid, 17 % butyric acid and 17 % isobutyric acid at 442 K (169 °C). The corrosion loss in butyric acid at 394 K (121 °C) is 2.4 mm/a (94.4 mpy) [187].

These results show that CuNi-alloys cannot be used in hot alkanecarboxylic acid solutions. If their use is being considered, their corrosion behavior should be tested.

Copper-tin alloys (bronzes)

Copper-tin alloys are rather more resistant to corrosion than copper in alkanecarboxylic acids, in particular when oxygen is present, to which they are not as sensitive [188]. According to [184], they are described similar to copper materials with a corrosion rate of less than 0.5 mm/a (19.7 mpy) in oxygen-free propionic acid and butyric acid at 293 K (20 °C).

A copper alloy containing 6 % Sn is used as material for eyeglass frames, which have to be resistant to corrosion by sweat, which largely consists of alkanecarboxylic acids [189].

Investigations of corrosion of bronzes (no detailed data) in alkanecarboxylic acids with 7 to 9 carbon atoms showed corrosion rates of up to 0.03 mm/a (1.18 mpy) at 373 K (100 °C). Temperatures up to 573 K (300 °C) were without influence on the corrosion behavior of these materials [185].

Copper-tin-zinc alloys (red brass)

These alloys, termed red brass, are resistant to corrosion by alkanecarboxylic acids at 293 K (20 °C) when oxygen is excluded. The corrosion rates for propionic acid in aqueous solutions and pure acid when oxygen is excluded are less than 0.1 mm/a (3.93 mpy) up to boiling point [181]. They are used as materials for valves and isolating valves.

Copper-zinc alloys (brass)

The corrosion rates for copper-zinc alloys are approximately the same as for copper and its alloys in alkanecarboxylic acids. The corrosion resistance to oxygen-free solutions of alkanecarboxylic acids of all concentrations at 293 K (20 °C) is stated as being less than 0.2 mm/a (7.87 mpy) for the pure acid up to the boiling point [181]. Brass with a zinc content above 25 % is very sensitive to the presence of even the slightest traces of oxygen [184]. Resistance to corrosion decreases rapidly in the presence of oxygen and may lead to destruction of the material.

According to corrosion investigations in alkanecarboxylic acids with 7 to 9 carbon atoms at 573 K (300 °C), the corrosion rate of brass containing about 23 % zinc is 2 mm/a (78.7 mpy) in the liquid and 1.8 mm/a (70.9 mpy) in the gas phase above the liquid [185]. Copper materials other than brass are more resistant than brass to these acids.

Other copper alloys

These copper alloys include those containing Be and Si. They behave like copper in alkanecarboxylic acids and aqueous solutions thereof and are resistant or moderately resistant in oxygen-free solutions. The corrosion rates for alloys containing silicon up to the boiling point are below 1 mm/a (39.4 mpy) [181]. Specimens fully immersed in 25 to 99% propionic acid are corroded at the rate of only 0.1 to 0.3 mm/a (3.93 to 11.8 mpy). Copper-silicon alloys are corroded at the rate of only 0.16 mm/a (6.3 mpy) at 383 K (110 °C) in 98% butyric acid and are used as materials for pumps.

Silicons bronzes are corroded at the rate of 0.15 mm/a (5.9 mpy) at 299 K (26 °C) in 100% butyric acid, this increasing to as much as 3.4 mm/a (134 mpy) at 383 K (110 °C), while complete destruction occurs at 418 K (145 °C). These alloys also offer very good resistance to corrosion in mixtures of alkanecarboxylic acids in the presence of acetic acid.

Alkanols

Author: *K. Hauffe* / Editor: *R. Bender*

Copper

In pure alkanols and mixtures containing water, copper is resistant at all concentrations up to the boiling temperature [190]. In 2 mol/l hydrogen chloride-alkanol mixtures, the corrosion rate of copper decreases with increasing number of carbon atoms of the alcohol; only from heptanol onwards do the corrosion rates drop below those obtained with water as solvent [191].

In addition to the corrosion current density, the corrosion rate of copper calculated therefrom is listed in Table 16.

Solvent	Current density 10^4, A/cm^2	Corrosion rate mm/a (mpy)
Water (H$_2$O)	0.06	0.07 (2.76)
Methanol (CH$_3$OH)	1.08	1.2 (47.2)
Ethanol (C$_2$H$_5$OH)	0.37	0.43 (16.9)
Propanol (C$_3$H$_7$OH)	0.133	0.15 (5.91)
Butanol (C$_4$H$_9$OH)	0.132	0.15 (5.91)
Pentanol (C$_5$H$_{11}$OH)	0.108	0.12 (4.72)
Nonanol (C$_9$H$_{19}$OH)	0.033	0.038 (1.50)

Table 16: Corrosion current density and corrosion rate of copper in 2 mol/l HCl-alkanol at 293 K (20 °C) [191]

In the production of chloroprene, chloroacetaldehyde and polyethyleneimine, substantial corrosion problems arise on copper in ethanol solutions which contain 1 mol/l HCl from the effect of cupric chloride additives and the flushing with oxygen at room temperature (Table 17). The attack on copper by hydrogen chloride in ethanol at a polarization voltage of –0.050 V_{SCE} is significantly higher than that in water [192].

A comparable result of a different order of magnitude is obtained for the corrosion of copper in 0.01 mol/l sulfuric acid which contains ethanol or water respectively as solvent at a corrosion rate of 3.4 or 0.34 mm/a (134 or 13.4 mpy) at 313 K (40 °C) [193].

The corrosion rate of copper determined between 240 and 700 h in a solution of methylene chloride + ethanol (9:1 parts by weight), which is used for the preparation of a triacetate spinning solution, was about 0.002 to 0.004 mm/a (0.08 to 0.16 mpy) at 300 K (27 °C) [194].

H_2O	$CuCl_2$	Current density	Material consumption rate	Corrosion rate
%	mol/l	A/cm²	g/m² h	mm/a (mpy)
≥ 0.04	–	1.6×10^{-5}	15.4	15 (591)
	0.01	–	11.2	11 (433)
	0.05	–	24.8	24 (945)
	0.10	–	64.4	62 (2,441)
2	–	1.2×10^{-5}	11.6	11 (433)
10	–	5.2×10^{-6}	6.2	6 (236)
20	–	8.3×10^{-6}	5.4	5 (197)
100	–	–	2.0	2 (78.7)
100	0.10	–	56.0	54 (2,126)

Table 17: Effect of water and cupric chloride additions on the corrosion of a copper electrode in ethanol with 1 mol/l HCl (flushed with oxygen) and on the corrosion current density at −0.050 V_{SCE} (Saturated Calomel Electrode) [192]

The corrosion rates of copper SF-Cu (CW024A, Cu-DHP, 2.0090) in solutions with chlorinated hydrocarbons and 33 % by volume 2-propanol at 323 K (50 °C) based on 96 h tests are reproduced in Table 18 [195]. Under the test conditions mentioned, the corrosion rate of copper is about 0.01 mm/a (0.39 mpy) at most.

Medium	Water content	Corrosion rate
	g/l	mm/a (mpy)
Trichloroethylene + 2-propanol	0.40	0.001 (0.04)
Trichloroethylene + 2-propanol	saturated	0.01 (0.39)
Perchloroethylene + 2-propanol	0.35	0.0 (0.0)
Perchloroethylene + 2-propanol	saturated	0.0 (0.0)
1,1,1-Trichloroethane + 2-propanol	0.40	0.01 (0.39)
1,1,1-Trichloroethane + 2-propanol	saturated	0.01 (0.39)

Table 18: Corrosion rate of copper SF-Cu (CW024A, Cu-DHP, 2.0090) in water-containing solutions of chlorinated hydrocarbons + 33 % by volume 2-propanol at 323 K (50 °C), test duration 96 h [195]

These results are confirmed by a further investigation. Thus, according to Figure 6, the material consumption per unit area of copper SF-Cu is lowest in perchloroethylene-isopropanol mixtures at 323 K (50 °C). In a mixture of perchloroethylene

+ 30 % by volume 1-propanol and ethanol or methanol, the corrosion rate is 0.019 and 0.028 or 0.043 mm/a (0.75 and 1.10 or 1.69 mpy) [196].

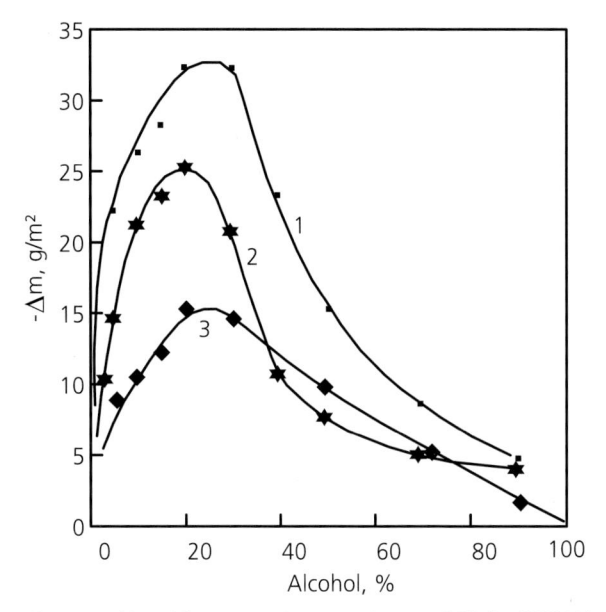

Figure 6: Material consumption per unit area of SF-Cu (CW024A, Cu-DHP, 2.0090) in perchloroethylene-alkanol mixtures at 323 K (50 °C) as a function of the alkanol content in 30 d tests [196] ① methanol; ② ethanol; ③ 1-propanol

For the cleaning of soldered copper conductor plates, that is, those partially provided with lead-tin films, the following two solid mixtures have proved to be suitable for removing rosin residues: 1,1,2,2-tetrachloro-1,2-difluoroethane + 50 % 1-propanol and toluene + 37.5 % 1-propanol + 27.5 % heptane + 5 % butyl acetate [193].

The inhibition efficiency of alkanols on the dissolution of copper in 3 mol/l and 6 mol/l nitric acid was investigated by a thermometric method. In agreement with the results in Table 16, the inhibition efficiency increased in the following order: methanol < ethanol < n-butanol < 1-propanol [197].

Copper-zinc alloys (brass)

Brass is used for handling alkanols.

CuZn-alloys are completely resistant to ethanol up to 343 K (70 °C) and to butanol up to 333 K (60 °C) [198].

The addition of as little as 0.1 mol/l 1-pentanol to 3 mol/l nitric acid was already sufficient to achieve a 92 % inhibition efficiency on the corrosion of brass [197].

Aluminium Chloride

Author: *L. Hasenberg* / Editor: *R. Bender*

Copper and copper alloys

Copper and its alloys with aluminium, silicon, beryllium, and tin are largely resistant to anhydrous aluminium chloride and aluminium chloride solutions at 293 K (20 °C), when air is excluded. Manganese bronze is used, for example, as a material for stirrers. Boiling solutions, however, attack these materials severely [199].

The catalyst for the conversion of normal into branched pentanes, (60 % aluminium chloride, higher hydrocarbons and some hydrochloric acid) also attacks copper and its alloys severely; the attack is intensified in stirred or flowing solutions.

At temperatures above 293 K (20 °C), corrosion rates of about 1 mm/a (39.4 mpy) or higher must be expected for copper and its alloys [200].

Copper is dissolved very rapidly in aluminium chloride melts [201, 202]; the rate of dissolution (on the basis of an 11 hours' test) at 473 K (200 °C) is 26.3 mm/a (1,035 mpy) in the melt and 110.0 mm/a (4,331 mpy) in the vapor space. On addition of 19 % sodium chloride to the aluminium chloride melt, the dissolution increases to 49.1 mm/a (1,933 mpy) in the melt, but decreases to 61.4 mm/a (2,417 mpy) in the vapor space. The most violent attack takes place in the vapor space above an aluminium chloride melt containing 19 % of calcium chloride (585.9 mm/a (23,067 mpy)), while only 15.7 mm/a (618 mpy) are dissolved in the melt.

Copper and its alloys are therefore only suitable to a limited extent in aluminium chloride solutions and completely unsuitable for handling aluminium chloride melts.

Amine Salts

Author: K. Hauffe / Editor: R. Bender

Copper

Triethylamine-2,4-dinitronaphtholate is an effective inhibitor for anodically polarized copper in borate-buffered solution with the addition of 0.001 mol/l sodium sulfate in the presence of nitroaromatic derivatives at 293 K (20 °C). Figure 7 shows the current density vs. potential pattern of copper in this solution in the presence of 0.01 mol/l phenolate (Curve 1), hexamethyleneimine-o-nitrophenolate (Curve 2) and triethylamine-2,4-dinitronaphtholate (Curve 3) [203]. The data are converted from µA/m^2 into mm/a on the right-hand ordinate. Up to a potential of +1.4 V$_{SHE}$ the corrosion current is < 2 < µA/cm^2, i.e. the corrosion rate is < 0.013 mm/a (0.51 mpy). The corresponding p-nitrophenolate exhibits an approximately similar behavior in the corrosion of copper.

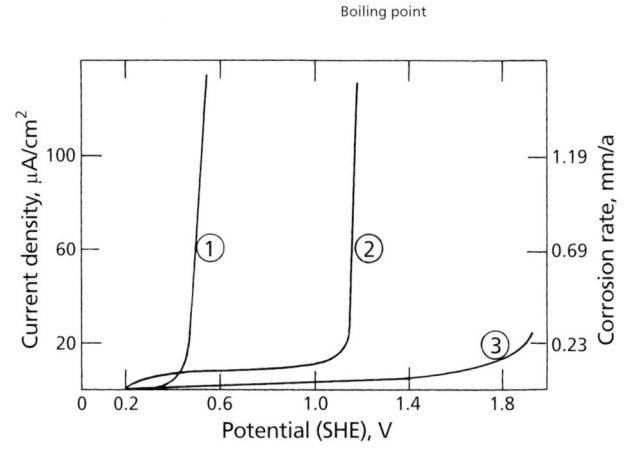

Figure 7: Current density and corrosion rate of Cu in borate-buffered solution (pH 7 at 293 K (20 °C)) with additions of 0.0001 mol/l Na$_2$SO$_4$ with 0.01 mol/l o-nitrophenolate ①, hexamethyleneimine-o-nitrophenolate ② and triethylamine-2,4-dinitronaphtholate ③ [203]

Copper polished in nitric acid was washed in an alcoholic solution containing 14 % hexamethyleneimine-m-nitrobenzoate, with 24 % hexamethyleneimine-o-nitrobenzoate and with 4 % hexamethyleneimine-3,5-dinitrobenzoate and then kept in steel vessels for 21 and 36 months. In no case there was a change in the condition of the surface observed. Only with prior treatment of the copper in 5 % alcoholic piperidine-3,5-dinitrobenzoate solution did black spots occur on one side of the specimen.

Copper specimens wrapped in paper impregnated with piperidine-3,5-dinitrobenzoate which were additionally wrapped in paper impregnated with paraffin remained unchanged in appearance after storage for 15 months under industrial conditions [204].

Since adsorption of the hexamethyleneamine benzoate inhibitor increases with increasing deviation from the stoichiometric composition of the copper oxide surface, the inhibiting effect on the corrosion of copper is improved [205].

The amine salts decyl (C_{10})-, dodecyl (C_{12})- and tetradecyl (C_{14})- ammonium acetate do not exhibit their inhibiting effect on the corrosion of copper in a 4% sodium chloride solution until below -0.4 V_{SCE} and particularly from -1 V_{SCE}. The corrosion current densities at 293 K (20 °C) and a potential of -1.0 V_{SCE} are 0.15, 0.08 and 0.15 mA/cm^2, corresponding to corrosion rates of 1.7, 0.9 and 1.7 mm/a (66.9, 35.4 and 66.9 mpy). A corrosion current density and corrosion rate which is greater by a factor of 6 to 9 occurs in the sodium chloride solution free from amine salts [206].

Copper (99.5% Cu) exposed to a simulated industrial atmosphere (0.01% SO_2 by volume in air at RH 100%) at 295 K (22 °C) corroded at the rate of 0.015 mm/a, (0.59 mpy) and in the simultaneous presence of cyclohexylamine carbonate in the atmosphere at the rate of 0.0072 mm/a (0.28 mpy), corresponding to an inhibition efficiency of 51% [207].

Cyclohexamine chromate, in addition to phenylthiourea has proved to be a good inhibitor of corrosion of copper in the vapor phase [208].

Table 19 summarizes the corrosion behavior of copper in a solution of polyethylene glycol in ethoxylated alcohol (373 K (100 °C)) with amine salt additions of 0.01 mol/l [209].

Inhibitor 0.01 mol/l	Material consumption rate g/m^2 h	Corrosion rate mm/a (mpy)
Cyclohexylamine mercaptobenzothiazole	0.061	0.059 (2.32)
Triethanolamine mercaptobenzothiazole	0.000	0.000 (0.0)
NaNO$_2$ triethanolamine mercaptobenzothiazole	0.19	0.18 (7.09)

Table 19: Inhibiting effect of amine compounds on corrosion of copper in polyethylene glycol ethoxylated alcohol at 373 K (100 °C) [209]

By virtue of the great increase in the angle of wetting on copper immersed for 7 days in hexamethylimine-o-nitrobenzoate solution, good resistance to atmospheric corrosion was achieved due to formation of a hydrophobic film [210]. Treatment with cyclohexylamine chromate was not so effective. The angle of wetting increased only from 100 to 138% compared with the increase from 100 to 192% with the first amine salt solution.

As shown in Table 20, copper is not to be recommended for the hydrolyzer and the receiver of a production plant for caprolactam because of its unsatisfactory resistance, especially so for the hydrolyzer [211].

Reaction medium	Corrosion rate, mm/a (mpy)
Hydrolyzer: 120–150 g/l $(NH_2OH)_2 H_2SO_4$ + 120 g/l H_2SO_4; T = 383 K (110 °C)	1.37 (53.9)
Receiver: hydroxylamine sulfate; T = 353 K (80 °C)	0.29 (11.4)
Receiver: 45 % $(NH_2)_2SO_4$ + 15 g/l H_2SO_4; T = 343 K (70 °C)	0.16 (6.3)

Table 20: Corrosion rate of Cu under the production conditions of caprolactam manufacture (test duration 3,600 hours) [211]

Complete corrosion protection was achieved with papers, impregnated with the following inhibitors, used for wrapping copper items: cyclohexylamine benzoate, cyclohexylamine cinnamate, cyclohexylamine dihydrocinnamate, cyclohexylamine chromate, benzylamine benzoate, benzylamine cinnamate, benzylamine dihydrocinnamate, guanidine carbonate, guanidine chromate, dicyclohexylamine nitrate and hexamethyleneimine benzoate. The surface of the metal remained unchanged after storage for more than 1 year [212].

Tissue paper, woven material and non-woven matting are impregnated with dicyclohexylamine nitrite and used as packing material for preventing tarnishing of bright copper surfaces susceptible to atmospheric influences [213].

Copper-zinc alloys (brass)

Both copper and the brass specimens CuZn10 (CW501L), CuZn20 (CW503L), CuZn30 (CW505L) and CuZn37 (CW508L) are severely attacked at the surface in saturated copper tetramine solutions. Stress corrosion cracking simultaneously occurs due to the formation of CuO in accordance with the reaction $Cu(NH_3)_4^{2+} \cdot H_2O \leftrightarrow 2NH_3 + 2NH_4^+ + CuO$. This, however, can be suppressed by the addition of 2 g/l ammonium chloride. The susceptibility of brass to stress corrosion cracking is however, significantly less than that of the copper-gold alloys. An alternating load with alloys susceptible to stress corrosion cracking decreases the service life by a substantially greater degree than under static load, as a result of corrosion fracture with low mixed crystal concentrations [214].

As shown in Table 22, the corrosion resistance of brass L-68 (Russian type: Cu-(30–32)Zn-0.35Pb) varies in dilute hydrochloric acid containing amine chloride. Corrosion becomes significantly less with increased additions of hydroxylamine (NH_2OH) [215].

Whereas brass L-62 (Russian type: Cu-38Zn) without inhibitor in an atmosphere of 100 % relative humidity with induced condensation already exhibits corrosion after one day, this does not occur until after 8 days when the specimens are washed in an alcoholic solution containing either 4 % hexamethyleneimine-3,5-dinitrobenzoate or 24 % hexamethyleneimine-o-nitrobenzoate. In the absence of induced condensation the brass specimens remain without signs of corrosion even after 60 days [204]. For lengthy storage (one to two years) under industrial conditions the brass components should be wrapped in paper impregnated with piperidine-3,5-dinitrobenzoate and then packed in paper impregnated with paraffin.

Brass (CuZn36, CW507L) exposed to a simulated industrial atmosphere 0.01% SO_2 by volume in air with a relative humidity of 100% at 295 K (22 °C) corroded at the rate of 0.024 mm/a (0.94 mpy), and in the simultaneous presence of cyclohexylamine carbonate as inhibitor in the atmosphere corroded at the rate of 0.015 mm/a (0.59 mpy), corresponding to an inhibition value of 40% [207].

At room temperature copper-zinc alloys in dilute sodium hydroxide solution corroded at a rate between 0.05 and 0.5 mm/a (1.97 and 19.7 mpy) and at higher temperature between 0.5 and 1.7 mm/a (1.97 and 66.9 mpy). According to Table 21, among other inhibitors m-and p-aminophenol exhibited satisfactory inhibition efficiency (100) $(k_0-k)/k_0$, where k_0 and k are the corrosion rates without and with inhibitor respectively) when brass 63/37 was exposed to corrosion in sodium hydroxide solution. The negative data in brackets indicate accelerated corrosion, i.e. no inhibition [216].

NaOH, mol/l	0.2	0.5	1.0
Inhibitor, %		Corrosion rate, mm/a (Inhibition efficiency, %)	
m-aminophenol			
0.001	0.068 (40)	0.127 (1.7)	0.068 (21)
0.01	0.042 (63)	0.095 (27)	0.039 (54)
0.1	0.020 (82)	0.021 (84)	0.022 (75)
0.5	0.014 (87)	0.019 (86)	0.012 (86)
p-aminophenol			
0.001	0.072 (36)	0.136 (–6)	0.105 (–22)
0.01	0.021 (81)	0.102 (21)	0.047 (46)
0.1	0.027 (76)	0.022 (83)	0.019 (77)
0.5	0.050 (56)	0.066 (49)	0.043 (50)
Without inhibitor	0.11	0.13	0.086

Table 21: Inhibition of corrosion of brass 63/37 in NaOH solutions by additions of aminophenol; inhibition efficiency in brackets; 5 day tests [216]

Other copper alloys

Table 23 summarizes the corrosion behavior of copper-zirconium alloys Bz6-ZnT (Cu-6.26Zr-5.69Sn-3.79Pb) and Am-58 (Cu-41.56Zr-0.56Pb) in a solution of polyethylene glycol in ethoxylated alcohol at 373 K (100 °C) with 0.01 mol/l additions of three amine salts [209]. The second inhibitor listed in Table 23 is particularly effective.

Cu^{2+}, g/l	Fe^{3+}, g/l	Inhibitor NH_2OH, g/l	Temperature K (°C)	Material consumption rate $g/m^2 h$	Corrosion rate mm/a (mpy)
–	0.5	3.0	295 (22)	1.15	1.2 (47.2)
–	5.0	30.0	323 (50)	0.61	0.63 (24.8)
–	1.0	6.0	323 (50)	0.50	0.52 (20.5)
0.5	0.5	2.0*	295 (22)	0.26	0.27 (10.6)

* + 2 g/l thiourea (($NH_2)_2CS$)

Table 22: Corrosion resistance of brass L-68 in 1% hydrochloric acid with addition of NH_2OH [215]

Inhibitor	Material consumption rate, $g/m^2 h$ (Corrosion rate, mm/a)	
0.01 mol/l	Am-58	Bz6-ZnT
Cyclohexylamine mercaptobenzothiazole	0.072 (0.071)	0.090 (0.12)
Triethanolamine mercaptobenzothiazole	0.014 (0.014)	0.030 (0.041)
Triethanolamine mercaptobenzothiazole $NaNO_2$	0.135 (0.13)	0.085 (0.12)

Table 23: Inhibiting effect of amine salts on the corrosion of CuZr-alloys in polyethylene glycol ethoxylated alcohol at 373 K (100 °C) [209]

Ammonia and Ammonium Hydroxide

Author: P. Drodten / Editor: R. Bender

Copper

On account of their good corrosion behavior in moist atmosphere, copper and copper based alloys have numerous applications in drinking water, process water and high temperature water systems [217]. Apart from their good corrosion resistance, their further application field in outdoor weather exposure, in fresh water pipelines, fittings, condensers, heat exchangers, in chemical plant construction and many other uses, is based also on their good strength properties as well as their high thermal and electrical conductivity. Furthermore, they can be processed very well.

The properties and compositions of the copper grades and copper alloys are specified in DIN EN 1982 [218], DIN V 17900 [219] and DIN 1787 [220]. The material numbers of the former German standards are no longer valid in the EN standards and have been replaced by a new numbering system comprising a sequence of six characters consisting of letters and numerals [221]. Thereby the first character is always the letter "C" for designating the copper material.

Their good corrosion resistance in approximately neutral to alkaline aqueous media is based on the formation of oxide layers giving good protection, consisting of Cu_2O and/or CuO depending on the kind of medium and the corrosion potential [222]. The chief applications of the copper materials are in the field of drinking water, cooling water and process water.

Dry ammonia gas as well as water-free and oxygen-free liquid ammonia attack copper very little or not at all. Yet, it has been known for long that already traces of moisture in gaseous as well as in liquid ammonia cause strong corrosion. Copper and copper alloys are strongly attacked in moist ammonia containing air with a water content of 300 ppm [223]. Therefore copper and copper alloys are unsuitable for storing and transporting liquid ammonia containing about 0.2 % of water that is utilized for fertilizer purposes.

In the published tables of the stability of metals it is usually stated that copper and the copper materials are resistant to water-free ammonia, but not resistant to moist ammonia or aqueous ammonia solutions [224, 225].

The technical regulations for steam boilers (TRD) the TRD 451 that is applicable to plants for storage of pressure-liquefied ammonia and its annexes 1 and 2 that apply correspondingly to mixtures of ammonia and water, do not permit the utilization of copper and copper alloys for parts in contact with the medium.

Liquid ammonia is also utilized as refrigerant in refrigeration plants. DIN 8975 Part 2 must be taken into consideration for the selection of materials for refrigeration plants [226]. There the instruction is contained under Section 3.2.1 "Copper and copper alloys" that "copper must not be utilized together with ammonia. Copper alloys may be utilized only after testing their compatibility with ammonia". The standard DIN EN 378 that is valid for refrigeration plants and heat pumps contains a similar stipulation in part 2, according to which copper materials containing a large fraction of copper may be utilized for plant components in contact with ammonia only after appropriate tests [227]. The trade association regulations BGV D 4 (formerly VGB 20) "Refrigeration plants, heat pumps and refrigeration facilities" make reference to the specifications in DIN 8975-2 regarding the selection of materials [228]. For condenser pipes in refrigeration engineering with liquid ammonia as refrigerant, the utilization of copper and copper alloys is discouraged. In contrast to these cases of experience, the investigations described in [229] show a remarkably good stability of the utilized copper materials in moist ammonia refrigeration machine oil systems. In these investigations specimens of the materials

Cu-DHP   CuZn37   CuZn39Pb3   CuNi10Fe1Mn   CuNi30Fe2Mn2

were utilized in autoclave tests such that they were separately exposed to the three phases machine oil, liquid ammonia and gaseous ammonia. The test temperature was around 353 K (80 °C) and the test duration was 200 hours. The water contents in the ammonia varied between 370 and 1,900 ppm. As example, Table 24 specifies the material consumptions determined on the specimens with the highest water content in the test medium.

Material	Cu-DHP CW024A (formerly: SF-Cu)	CuZn37 CW508L	CuZn39Pb3 CW614N	CuNi10-Fe1Mn CW352H	CuNi30-Fe2Mn2 CW353H
Medium	liquid	liquid/ gaseous	liquid	liquid/ gaseous	liquid/ gaseous	liquid/ gaseous
Corrosion mg/dm^2	0.23	0.91	1.7	0.67	0.37	0.11

Table 24: Material consumptions of the investigated specimens after 200 h in liquid and liquid/gaseous ammonia [229]

On loop specimens of the brass material CuZn37 (CW508L) most strongly endangered by stress corrosion cracking, after exposure in liquid ammonia and in ammonia gas, no signs of stress corrosion cracking were found in the subsequent metallographic examination.

Copper materials are principally unsuitable in aqueous ammonia solutions, because in the presence of oxygen copper reacts according to Equation 4 with the following reaction steps to produce copper tetrammin complex.

Equation 1 $Cu + 2\,NH_3 \rightarrow [Cu(NH_3)_2]^+ + e^-$ anodic reaction

Equation 2 $Cu + 4\,NH_3 \rightarrow [Cu(NH_3)_4]^{2+} + 2\,e^-$ anodic reaction

Equation 3 $½\,O_2 + H_2O + 2\,e^- \rightarrow 2\,OH^-$ cathodic reaction

As overall reaction

Equation 4 $\quad Cu + \frac{1}{2} O_2 + 4\, NH_3 + H_2O \rightarrow [Cu(NH_3)_4]^{2+} + 2\, OH^-$

With low oxygen concentration the reaction rate is determined by the transport of the oxygen to the metal surface. With increasing concentration of the dissolved copper tetrammin ions these, as strong oxidizing agent, can take over the cathodic reaction at the copper surface according to

Equation 5 $\quad [Cu(NH_3)_4]^{2+} + e^- \rightarrow [Cu(NH_3)_2]^+ + 2\, NH_3$

Depending on the actual content of Cu^+ ions, copper oxide layers may be produced:

Equation 6 $\quad 2\, [Cu(NH_3)_2]^+ + 2\, OH^- \rightarrow Cu_2O + 4\, NH_3 + H_2O$

In electrical plant components in which copper as contact material is utilized, ammonia fractions in the atmosphere can lead to corrosion problems. To increase the corrosion resistance of the contact materials, the influence of various alloying elements in pure copper on the corrosion behavior in moist air containing ammonia was investigated [230]. The specimens were exposed to a gas mixture of air with an ammonia fraction of 20 mg/m³ and a relative humidity of 95 to 98 % at a temperature of 293 to 298 K (20 to 25 °C). The test duration was 96, 204 and 504 hours. The results show that no improvement of the corrosion behavior can be achieved with the alloying elements Cr, Mo, Nb, Ni, Ti and Zr (1 to 10 %).

On copper and copper alloys in moist ammonia/air mixtures the appearance of stress corrosion cracking must always be reckoned with. Thereby the oxygen content of the copper appears to play a role. According to older publications, oxygen-free pure copper is said not to be damaged by stress corrosion cracking on exposure to ammonia [231, 232]. Thus specimens of high purity practically oxygen-free copper (< 0.0007 % O_2, max. 0.01 % total impurities) after exposure for 50 days to an ammonia/water vapor/air mixture, showed no inter-crystalline attack, and tensile tests led only to forced rupture. However, considerable loss of strength due to strong uniform surface corrosion was observed [231].

On the other hand, in investigations of stressed hard drawn wires of high conductivity copper (99.96 % and 99.99 % Cu) in the vapor atmosphere above ammonia solutions, and with changing temperatures (343 to 298 K (70 to 25 °C)), cracks appeared at right angles to the wire axis [233].

Waters that contain ammonia – also in traces – can induce inter-crystalline and/or trans-crystalline stress corrosion cracking of parts that are internally stressed by the construction, production or processing, or that are under operational stress. Although stress corrosion cracking is rarely observed on pure copper, nevertheless on SF copper pipes in domestic installations cracks have sometimes been observed on the exterior of pipes. Unsuitable organic thermal insulation materials that liberate ammonia on the cold pipes when condensation water forms, are a possible cause of the damage [217, 234].

The fact that ammonia unexpectedly brought in from other areas leads to damage on copper components is emphasized by the following examples [235].

- On a thermometer tube made of copper in a central hot water system pitting corrosion took place because hydrazine had been added to the hot softened water to bind the oxygen. The ammonia produced by the decomposition of the hydrazine led to the local attack of the copper tubes.
- In a steam condenser ammonia originating from the chemicals of the boiler feed water processing was dragged over with the steam and accumulated in the condensate on horizontally oriented copper pipes. This led to such a strong material consumption on the underside of the pipes that wall perforations appeared after less than one year of operation.
- Ammonia from the leakage of a nearby plant condensed in the moist atmosphere of a sewage water line on the copper pipes, and the resulting uniform surface corrosion led to failure of the pipes within three years.
- The copper pipes of a drinking water line in a cowshed were severely attacked by the ammonia originating from the cow dung.

Ammonia can also be produced in waters by microbiological processes on the surface of copper materials, leading to local corrosive attack in the form of depressions or pitting corrosion.

Within the scope of damage investigation on heat exchanger pipes made of copper that were exposed to underground water and damaged by pitting corrosion, copper specimens were tested in the same underground water [236, 237]. In one test the water was sterilized and in another test activated bacteria were added to it to produce a biofilm. After an exposure time of two weeks strong pitting corrosion was found under the biofilm, whereas the specimens in sterilized water showed no attack. The biological investigations showed that the bacteria *Staphylococcus* were specifically by fermentation of protein able to liberate ammonia, thus producing local corrosion. Also in [238] damage on heat exchanger pipes made of copper materials by microbiologically induced corrosion (MIC) traceable to the production of ammonia is reported.

Copper-aluminium alloys

The behavior of the copper-aluminium alloys with respect to ammonia and its aqueous solutions corresponds to that of copper. Also the homogeneous α-alloys (Al < 7.8 %), that belong to the more corrosion resistant grades and are superior to pure copper in sea water, are unsuitable like the other copper-aluminium alloys for exposure to moist ammonia gas and aqueous ammonia solutions, particularly in the presence of oxygen [223]. Since Cu-Al materials are often utilized in cooling water lines, it must be ensured that the cooling waters are not contaminated with ammonia.

On condensers with pipes made of aluminium bronze through which sea water was flowing, corrosion took place on the outside in the condensation area due to the appearance of ammonia. Nickel plated Cu-Al pipes were utilized as remedy [239].

In a large power plant a heat exchanger was operated with sea water as cooling liquid on the inside of the pipes for cooling steam on the outside. As material for the pipes, depending on the position, either copper-nickel alloys, CrNiMo steel or aluminium bronze ware utilized. Corrosion problems were encountered exclusively on a small number of the aluminium bronze pipes. The cause was found to be a high concentration of ammonia in condensate drops that formed under the particular cooling conditions in this region on the underside of these pipes [240].

Copper-nickel alloys

Copper-nickel alloys form uninterrupted mixed crystals; this has a positive influence on the corrosion stability. Thus the copper-nickel materials of technical interest

CuNi10Fe1Mn	CW352H	(former Material No. 2.0872)
CuNi30Mn1Fe	CW354H	(former Material No. 2.0882)
CuNi30Fe2Mn2	CW353H	(former Material No. 2.0883)

that also contain iron and manganese, belong to the most corrosion resistant copper-based materials. The addition of iron achieves a considerable improvement of the corrosion stability. The good corrosion protection of these materials is due to the formation of oxidic surface films that consist of Cu_2O on the metal side and of complex corrosion products with high iron and nickel content on the solution side [222].

The CuNiFe materials with nickel contents of at least 10% have the highest resistance of all copper materials to stress corrosion cracking and are also superior to the relatively resistant copper-aluminium materials with aluminium contents higher than 8% (cf. Figure 14). However, at higher temperatures these materials, too, become susceptible to stress corrosion cracking.

Pipes made of copper materials, in particular those made of copper-nickel alloys, are widely utilized in heat exchangers and are there in contact with various waters. On such pipes a special kind of corrosive attack is also observed that is known as "hot spot corrosion" [241]. All copper materials can be affected by this kind of corrosion.

The classical **hot spot corrosion** takes place predominantly in clean uncontaminated waters when the temperature on the product side is higher than 403 K (130 °C). The attack is characterized by pitting corrosion that originates from the cooling water side at places with locally increased temperatures. Reprecipitated metallic copper is often found in these pitting corrosion locations. It is assumed that the cause is galvanic element formation in places with different oxygen concentration.

Contamination of the cooling water with sulfides or ammonia can also initiate this type of corrosion. Slow flow rates or stagnant conditions of the cooling water are prerequisite, e.g. in "dead spaces", with simultaneous heat transfer from the product to the water. From investigations in the laboratory and evaluation of damage incidents it is known that ammonia contents of 2 mg/l in the water and temperature differences of less than 283 K (10 °C) between the cooling water and the outer pipe wall already suffice for initiating hot spot corrosion. Ammonia from organic degradation products is frequently present in contaminated sea water, and damage has therefore frequently been found on oil coolers of ferry ships for which the oil temperatures reach 343 to 353 K (70 to 80 °C) and the flow rates are small.

Of the copper materials listed in Table 25 that are frequently utilized for heat exchangers, copper materials with 30 % of nickel have been found to be most sensitive for classical hot spot corrosion at temperatures of 403 K (130 °C) and higher. Copper-zinc-aluminium alloys and CuNi10Fe2Mn1 show the least sensitivity. In sea water containing 2 mg/l of ammonia all mentioned materials have about the same sensitivity for this type of corrosion.

Alloy type	Cu	Zn	Ni	Fe	Mn	Al	Sn	As	Cr
CuZn20Al2 (2.0460[†])	76	22				2		0.04	
CuZn28Sn1 (2.0470[†])	71	28					1	0.04	
CuNi10Fe1Mn (2.0872[†])	87		10	2	1				
CuNi30Mn1Fe (2.0882[†])	68		30	1	1				
CuNi30Fe2Mn2 (2.0883[†])	66		30	2	2				
Alloy 722	82		16	0.8	1				0.5

[†] old material number

Table 25: Nominal composition (mass%) of copper materials frequently utilized for heat exchanger pipes [241]

Hot spot corrosion induced by ammonia can be prevented reliably by correct design and proper operation to avoid local stagnant conditions or impeded flow rate.

Apart from good corrosion stability, copper materials with high copper content also show high resistance to fouling and are therefore utilized for pipelines conveying sea water as well as for heat exchangers cooled with sea water. When ammonia is present damage can appear in such heat exchangers not only due to stress corrosion cracking but also by local corrosive attack, in particular as crevice corrosion. Since to avoid fouling chlorine as oxidizing agent is often added to the cooling water, the influence of chlorine and ammonia in sea water on the corrosion stability of some of the materials listed in Table 25 was investigated [242]. In addition thereto, the effect of Fe^{2+} ions as inhibitor was investigated. The test facility shown in Figure 8 was utilized for the investigations. This permitted simulation of various corrosion conditions, such as crevice corrosion in cold and warm areas as well as the erosion corrosion in the impact region of the entering water.

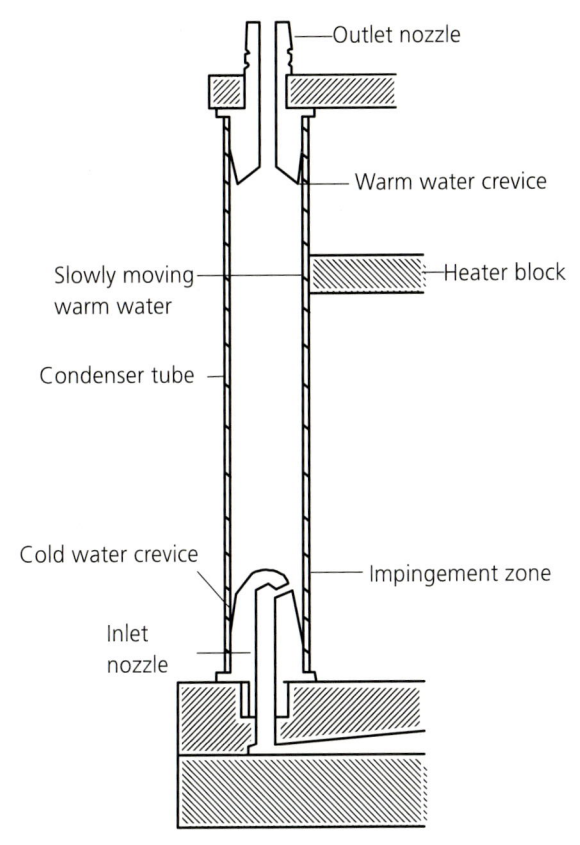

Figure 8: Test facility for cooling water pipes with various corrosion regions [242]

Each test facility contained two specimen pipes made of five materials. In order to test the influence of the ammonia content on the stress corrosion cracking behavior, pipe sections conically widened by 25 % were mounted on the outlet of the test pipes. The tests were carried out in cross-flowing harbor water in the harbor of Portland in England. The temperature of the heated section was set to 363 ± 5 K (90 ± 5 °C). The inflow rate of the water in the impact region was about 7.5 m/s. A typical analysis of the harbor water is shown in Table 26. In the test program three test sequences, each with six test facilities and a test duration of 2 months were carried out. Table 27 shows an overview of the test program.

Chloride, ppm	19,890	sodium, ppm	11,100
Sulfate, ppm	2,400	calcium, ppm	413
Bromide, ppm	73	magnesium, ppm	1,210
pH	7.8–8.2	temperature, K (°C)	282–291 (9–18)

Table 26: Analysis values of the harbor water [242]

		Addition ppm					
Test series		1	2	3	4	5	6
1	NH_3	10	10	5	5	0	0
	Fe^{2+}	0	0.042	0	0.042	0	0.042
2	NH_3	2	2	1	1	0	0
	Fe^{2+}	0	0.042	0	0.042	0	0.042
3	NH_3	2	2	2	2	0	–
	Cl_2	0.5	0.5	0	0	0	–

Table 27: Test program [242]

As the values reported in Table 28 show, for all of the investigated materials, the addition of ammonia up to 2 ppm had no significant influence on the erosion corrosion behavior in the impact zone. Simultaneous addition of 2 ppm of ammonia and 0.5 ppm of chlorine led to increased attack of the two alloys Alloy 722 and CuNi30Fe2Mn2.

Pitting corrosion was observed on all five materials with ammonia contents of 2 ppm in crevices in the region of the heat transfer. Crevice corrosion did not take place when in addition to ammonia either 0.05 ppm of chlorine or 0.042 ppm of iron(II) ions were present. A general corrosive attack and fine stress corrosion cracks with a maximum depth of 50 µm on the alloy CuZn20Al2 were detectable only on the pipe sections of the test series 1 with ammonia contents of more than 2 ppm.

	Without addition of iron				With addition of iron		
Material	NH_3 ppm			$NH_3 + Cl_2$ ppm	NH_3 ppm		
	0	1	2	2 + 0.5	0	1	2
CuZn20Al2	0.19	0.17	0.21	0.19	0.20	0.22	0.26
	0.14	0.12	0.18	0.22	0.26	0.22	0.26
CuNi10Fe1Mn	0.03	0.09	0.07	0.11	0.02	0.02	0.02
	0.06	0.05	0.05	0.11	0.01	0.01	0
CuNi30Mn1Fe	0.11	0.12	0.10	0.15	0.10	0.21	0.05
	0.13	0.12	0.16	0.14	0.11	0.07	0.01
Alloy 722	0.02	0.07	0.04	0.22	0.03	0.03	0.01
	0.03	0.07	0.03	0.13	0.02	0.04	0.02
CuNi30Fe2Mn2	0.03	0.04	0.02	0.08	0.01	0.02	0.02
	0.03	0.04	0.02	0.08	0	0.02	0.01

Table 28: Attack depth (mm) in the impact zone for the test series 2 [242]

Copper-tin alloys (bronzes)

Copper-tin alloys are frequently utilized for pump components, valves and similar parts in water conveying systems. Since these bronze materials, like all copper alloys, are sensitive to corrosion produced by small amounts of ammonia, it must be ensured that no ammonia can enter into such water systems. The following two cases of damage are mentioned as examples of unexpected sources of ammonia [235]:

- On an impeller wheel made of Cu-Zn bronze in a pump for feeding a salt solution containing 20 % sodium chloride, 0.3 % potassium dichromate and 0.1 % sodium hydroxide corrosion damage appeared after 12 months of operation. The investigations revealed that for about 2 months the water had been contaminated with about 2.5 % of ammonia coming from a leaking evaporator of an absorption refrigeration plant. The leak in the evaporator was traced back to corrosion caused by the inadequate inhibitor addition (potassium dichromate) in the salt solution.
- Boiler feed water that contained 6 ppm of ammonium ions from the own spring water supply of the factory led to corrosion damage and thus to failure of a pressure valve made of Cu-Zn bronze.

Copper-zinc alloys (brass)

The copper-zinc alloys can be classified in three groups according to their zinc content:

- single phase α-alloys with up to 37.5 mass% of zinc
- two phase (α+β)-alloys with zinc contents of 37.5 to 46 mass%
- single phase β-alloys with 46 to 50 mass% of zinc.

The corrosion stability of the brass types is generally less favorable than that of copper. The single phase alloys have better corrosion stability than the (α+β)-alloys that show a tendency for preferential attack of the Zn-richer β-phase (zinc depletion) already in aggressive atmosphere as well as in acidic media and in media containing chloride ions. Apart from the reactions described in the copper section (see Equation 1 to Equation 6), zinc here goes into solution according to the

Equation 7 $\quad Zn + 4\,NH_3 \rightarrow Zn(NH_3)_4^{2+} + 2\,e^-$

as resistant complex ion.

According to reactions shown in Equation 5 and Equation 6 the fraction of dissolved Cu^+- and Cu^{2+}-ions plays an important role in the production of cover layers on copper materials. This effect was studied in investigations on brass in aqueous ammonia solution with copper additions [243]. The tests were carried out on sheet specimens of a commercial CuZn30 (Ms70) alloy containing 63.3 % Cu and 36.7 % Zn in a 15.7 molar aqueous ammonia solution (pH 13.4) and with various contents of copper from 2 to 10 g/l at 303 K (30 °C). The copper was added to the solutions in the form of metallic copper powder, copper sulfate ($CuSO_4 \times 5\ H_2O$) and copper chloride ($CuCl_2 \times 2\ H_2O$). As shown in Figure 9, the pH of the solution is not changed by the addition of copper powder, whereas the pH is lowered with increasing concentration of the two salts.

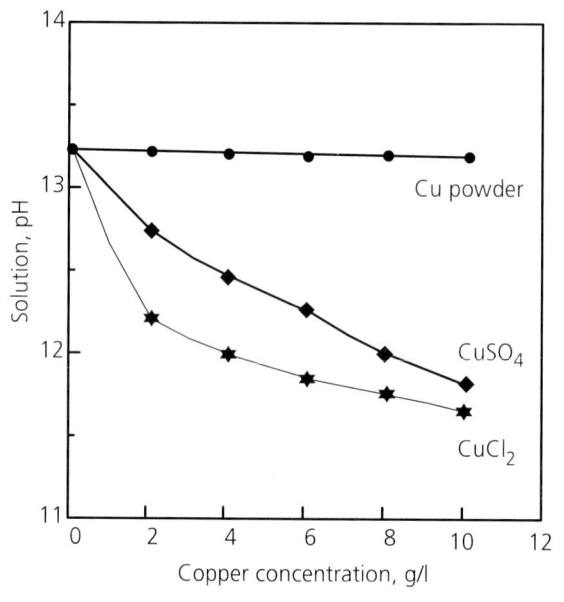

Figure 9: Change of the pH of an aqueous ammonia solution by the addition of copper powder, copper sulfate or copper chloride [243]

Table 29 shows the corrosion rates calculated from the measured material consumptions of the specimens after a test duration of 20 h in the various solutions. In all cases the corrosion rate of the specimens is increased by the copper salt addition in the test solution. When copper is added as metal powder, the corrosion increases strongly until reaching a copper concentration of 4 g/l. As from a concentration of 6 g/l a cover layer is formed and further increase of the material consumption is slowed down significantly, but the values are still higher by a factor of 10 than those in the copper-free solution. The Cu^{2+}-ions produced by the corrosion according to Equation 4 can be reduced by the metallic copper in the solution according to the

Equation 8 $[Cu(NH_3)_4]^{2+} + Cu \rightarrow 2\ [Cu(NH_3)_2]^+$

and then form cover layers of copper oxide according to Equation 6.

Copper content g/l	Added as copper powder	Added as $CuSO_4 \times 5\,H_2O$	Added as $CuCl_2 \times 2\,H_2O$
0	0.23	0.23	0.23
2	12.60	14.66	15.85
4	14.99	18.77	27.28
6	3.38	34.47	31.17
8	2.31	51.84	38.56
10	2.31	60.63	61.29

Table 29: Corrosion rates (mm/a) of CuZn30 (Ms70) specimens in aqueous ammonia solution with various copper contents [243]

In the presence of sulfate and chloride anions the material consumptions increase continuously with increasing copper content of the solution, without formation of cover layers. Copper sulfate and copper chloride react with the solution in ammonia according to Equation 9 and Equation 10 and produce ammonium sulfate and ammonium chloride.

Equation 9 $CuSO_4 + 2\,NH_4OH \rightarrow Cu(OH)_2 + (NH_4)_2SO_4$

Equation 10 $CuCl_2 + 2\,NH_4OH \rightarrow Cu(OH)_2 + 2\,(NH_4)Cl$

It is known that both salts in ammonia solution dissolve copper(I) oxide according to the reactions of Equation 11 and Equation 12.

Equation 11 $Cu_2O + (NH_4)_2SO_4 + 2\,NH_4OH \rightarrow 2\,[Cu(NH_3)_2]^+ + SO_4^{2-} + 3\,H_2O$

Equation 12 $Cu_2O + 2\,(NH_4)Cl + 2\,NH_4OH \rightarrow 2\,[Cu(NH_3)_2]^+ + 2\,Cl^- + 3\,H_2O$

Thus they prevent the formation of cover layers. According to the reactions according to Equation 13 and Equation 14, sulfate and chloride ions also promote the zinc depletion of the brass materials.

Equation 13 $Zn + SO_4^{2-} \rightarrow ZnSO_4 + 2\,e^-$

Equation 14 $Zn + 2\,Cl^- \rightarrow ZnCl_2 + 2\,e^-$

The general surface corrosion of brass in aqueous ammonia solutions can be decreased significantly by the addition of inhibitors [244, 245] as shown by the specifications in Table 30 for some examples.

Inhibitor	Corrosion rate, g/m² d			
	Concentration of ammonia solution			
	0.1 M	1.0 M	2.0 M	3.0 M
None	4.73	11.08	12.62	14.71
Resorcinol	0.67	0.83	1.42	2.47
Hydroquinone	0.82	1.89	1.84	3.06
Phloroglucin	0.65	0.39	2.55	1.86
Pyrogallol	0.68	0.28	1.99	3.66
Pyrocathechol	0.46	1.80	3.64	4.53

Concentration of the inhibitors: in each case 10 mmol

Table 30: Effectiveness of various inhibitors on the corrosion of brass 70/30 in aqueous ammonia solution at 302 K (29 °C) [245]

Brass materials that contain further elements such as aluminium, nickel, iron, manganese, tin, silicon or arsenic up to a few percent are designated as special brass and apart from higher strength also have improved corrosion and erosion stability, because these elements are incorporated in the cover layer and increase its stability. Among these ternary alloys, preferentially CuZn28Sn1As (CW706R, old designation CuZn28Sn1, Material No. 2.0470) and CuZn20Al2As (CW702R, old designation CuZn20Al2 Material No. 2.0460) containing approx. 2 % Al are proven since many years as materials for cooler pipes, condenser pipes and heat exchanger pipes in less strongly contaminated cooling waters. By virtue of their favorable corrosion stability in the atmosphere and water, Cu-Zn alloys are utilized primarily also for fittings and hosepipes. CuZn20Al2As (CW702R) is also distinguished by good resistance to sea water as well as to erosion.

[246] reports investigations in the laboratory and in a pilot plant with regard to the influence of impurities in sea water as sulfide ions and ammonium ions on the corrosion of a copper-zinc-aluminium alloy containing 76.8 % Cu, 20.8 % Zn and 2.4 % Al. The tests in the laboratory were carried out on rotating disks with a rotation speed of 52.3/sec. In the pilot plant pipe sections were tested through which liquid was flowing with a speed of 2.55 m/s. The test solution was a 3.5 % sodium chloride solution with a temperature of 303 K (30 °C), to which various amounts of sulfide and ammonium were added. The test temperatures of the metal specimens were between 323 and 363 K (50 and 90 °C). As the results of the tests in the pilot plant contained in Table 31 show, the corrosion is perceptibly increased by the addition of the contaminants. Thereby the addition of ammonium ions has a stronger effect than that of sulfide ions.

Metal temperature K (°C)	Solution temperature K (°C)	Content of S^{2-} mg/l	Content of NH_4^+ mg/l	Corrosion rate mm/a (mpy)
323 (50)	303 (30)	0	0	0.075 (2.95)
323 (50)	303 (30)	0	5	0.143 (5.63)
323 (50)	303 (30)	10	0	0.704 (27.7)
323 (50)	303 (30)	10	5	1.398 (55.0)
343 (70)	303 (30)	5	2.5	0.472 (18.6)
363 (90)	303 (30)	0	0	0.109 (4.29)
363 (90)	303 (30)	0	5	0.126 (4.96)
363 (90)	303 (30)	10	0	0.789 (31.1)
363 (90)	303 (30)	10	5	1.600 (63.0)

Table 31: Corrosion rate of a CuZnAl alloy in flowing NaCl solution with various additions of sulfide and ammonium ions [246]

The copper-zinc alloys, like copper and the other copper alloys, are corrosion resistant in absolutely water-free and oxygen-free ammonia, but of all copper alloys they have the greatest sensitivity to stress corrosion cracking in the presence of the slightest traces of moisture. Thus in bending tests of a CuZn39 alloy exposed to the vapor space above a 28% aqueous ammonia solution, stress corrosion cracking was observed already with tensile stress in the surface of about 35 MPa after 30 minutes [247] (see also Figure 14).

This high sensitivity to stress corrosion cracking has become known at the beginning of the 20th century as "season cracking", because with the British troops in India cracks and fractures appeared on the brass cartridges and shell sleeves regularly at the time of the monsoon. At that time, the cause was found to be the storage of the ammunition in horse resistants where it came into contact with moist ammonia vapors during the rain season. The tensile stress required for initiating the stress corrosion cracking was the residual inherent stress after the cold forming during manufacturing of the cartridges.

In aqueous ammonia solutions, depending on the composition of the solution, the potential, certain metallurgical factors and especially the extent of cold forming, inter-crystalline as well as trans-crystalline stress corrosion cracking can take place [248, 249].

Inter-crystalline cracking course is usually observed on heat-treated components under environmental conditions that led to the formation of surface films or cover films. Therefore its appearance is attributed to the mechanism of the cracking of cover layers, i.e. a preferred anodic dissolution at the crack tip when the layer breaks up due to plastic deformation.

Trans-crystalline crack course is explained by periodically appearing potential minima at the crack tip. According to the investigations described in [249] and [250] carried out on a CuZn30 (Ms70) alloy in aqueous ammonia solution, these potentials lie significantly above the potential for hydrogen discharge, so that the participation of hydrogen in the crack initiation or crack growth is hardly likely.

The work published in [251] also reports similar investigations of the mechanism of trans-cyrstalline stress corrosion cracking of CuZn30 (CW505L) in aqueous ammonia solutions with and without the addition of oxidizing agents. From these results it is concluded that a selective dissolution of zinc (zinc depletion) is insufficient for initiating cracks, but that instead anodic dissolution of copper is required, for which a corresponding anodic potential is necessary which can be reached only when an oxidizing agent is present.

The question of whether with the stress corrosion cracking of brass in aqueous ammonia solutions hydrogen is liberated with the corrosion at the crack tip and can participate in the crack propagation, was also investigated by modeling the participating chemical and electrochemical reactions [252]. In spite of some simplifications the calculations show that under the stationary conditions of the anodic dissolution of copper and zinc at the crack tip no hydrogen can be discharged.

Even if zinc depletion does not lead to the initiation of cracks, stresses in the vicinity of the propagating cracks can affect the kinetics and extent of a zinc depletion. In this context the influence of a static load below and above the yield point of the material on the dissolution of copper and zinc in 10 molar aqueous ammonia solution was investigated on a commercial CuZn30 (Ms70) alloy [253]. The chemical composition and the mechanical properties of the investigated sheet metal specimens are specified in Table 32.

Cu %	Zn %	Hardness HV	$R_{p0.2}$ MPa	Rm MPa	Elongation %
72.27	27.65	98	156	355	33

Table 32: Chemical composition and mechanical properties of the investigated brass [253]

Before the test the specimens were stress relief annealed for 1 hour at 773 K (500 °C). With a total test duration of 2 hours, the zinc and copper content in the test solution was determined at intervals of 20 minutes. The material consumptions calculated therefrom depending on the test duration for various stress magnitudes are shown in Figure 10 for zinc and in Figure 11 for copper.

For both alloying elements the dissolution rate out of the alloy is increased by applying tensile stress. For zinc the dissolution increases significantly with increasing stress, whereby this effect is more clearly evident in the elastic region below the yield point. Copper shows a corresponding behavior, with the difference that the corrosion rates are less on the whole and that the intensifying effect of the tensile stress is higher above the yield point with plastic deformation. The corrosion rate calculated from these material consumptions is shown in Figure 12.

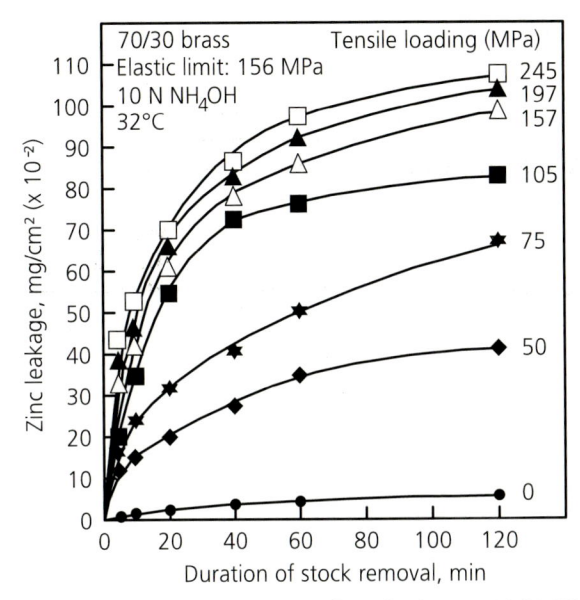

Figure 10: Material consumption of zinc for the CuZn30 (Ms70) specimens depending on the test duration and the mechanical tensile load [253]

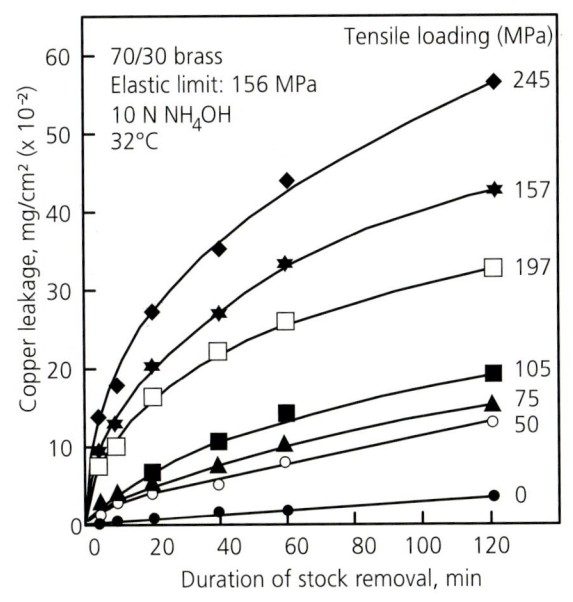

Figure 11: Material consumption of copper for the CuZn30 (Ms70) specimens depending on the test duration and the mechanical tensile load [253]

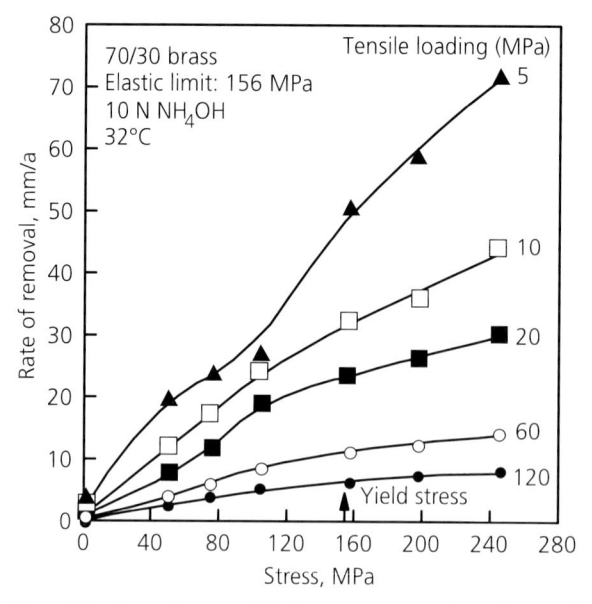

Figure 12: Corrosion rate of the brass specimens depending on the test duration and the mechanical loading of the specimens [253]

From this figure it is evident that the corrosion of the brass in aqueous ammonia solution decreases with increasing test duration, but it is significantly increased by a tensile load, particularly with short test times. From the ratio of the zinc and copper contents in the corrosion solution and in the alloy, a so-called zinc depletion factor "Z" was calculated according to Equation 15.

Equation 15 $Z = (Zn/Cu)_{solution}/(Zn/Cu)_{alloy}$

Figure 13 shows the zinc depletion factors obtained in this way for various test durations and stress magnitudes. A maximum for Z is found with a load of approx. 100 MPa. Within the first five minutes of the exposure the zinc depletion factor increases from 5.5, the value for non-stressed specimens, to the value of 31 under the load of 105 MPa. For loads above the yield point the value for Z becomes smaller again, because in this region the dissolution of the copper increases more strongly than that of the zinc. For every loading magnitude the zinc depletion decreases with increasing test duration.

From these investigation results it is concluded that the zinc depletion assisted by a tensile stress in the immediate vicinity of cracks promotes the trans-crystalline crack propagation of brass in media that do not produce cover layers.

In spite of reports in the literature since decades on cases of damage and fundamental investigations in the system of brass/ammonia (see older literature in [254–262]), new cases of damage still appear again and again.

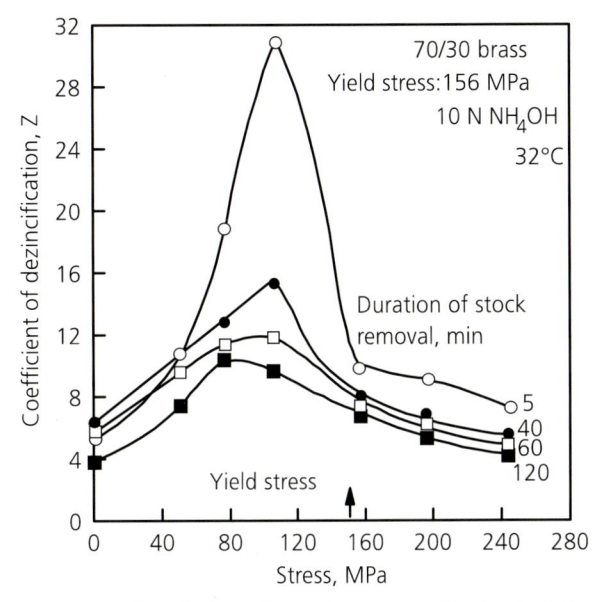

Figure 13: The influence of a tensile stress on the zinc depletion of CuZn30 (Ms70) in aqueous ammonia solution [253]

For example, [263] reports a case of damage on the outside of steam condenser pipes made of CuZn27Sn1As0.1 (UNS C44300). On the pipes, in the region of the holding plates, strong trench-form material loss, as well as pitting corrosion and crack formation were observed. The cause was recognized to be the condensate draining off on the holding plates, in which ammonia and oxygen had become enriched. The ammonia originated from hydrazine and cyclohexamine compounds that had been added to the boiler feed water for controlling the oxygen content and had been dragged over with the steam. Due to an unrecognized leak, oxygen was also able to come in, so that the conditions for material consuming corrosion and stress corrosion cracking were fulfilled.

Ammonia dragged over with the steam from the feed water processing was also the cause for two cases of corrosion damage on condenser pipes made of brass, that are described in [235]. In one case the ammonia content of the steam condensate originating from the hydrazine addition to the feed water led to wall perforation of the pipes after 18 months. In the other case the damage of the pipes appeared only after 14 years of operation. The damage investigation showed that in normal operation the saturated steam at 314 to 327 K (41 to 54 °C) had a pH of 8.7 to 9.0, and that the ammonia content was less than 0.3 ppm and the dissolved oxygen content was around 10 to 20 ppb. During the start-up phase of the plant, that took place once or twice each year, the ammonia content increased to 75 ppm and the oxygen content increased up to 3,500 ppb, in spite of a hydrazine content of 10 to 20 ppb. The condensate that was strongly enriched with ammonia in this phase collected at the bottom end of the pipe and caused the damage there after a correspondingly long time.

In the cases of damage described in [264], cracks and ruptures appeared after some time ranging from a few days up to several months on components made of CuZn alloys (CuZn32, CuZn37, CuZn42) in measuring and control facilities under conditions of outdoor exposure. The surface of the components was still mostly bright metallic, and the metallographic examination revealed a chiefly trans-crystalline crack course with little branching. The cause of this damage was found to be the advection of ammonia from the feces of a nearby animal farm.

Also the small amounts of ammonia contained in cooling waters are often the cause of cases of damage of brass pipes in heat exchangers or condensers [265, 266].

A number of cases of damage in domestic installations and also on exposure to the atmosphere are reported in [267]. In cases of damage to installation components such as fittings, valves, and pipes high stresses are usually present that originate from the production, processing or installation. The appearance of ammonia is often traced back to the thermal insulating material that contains or can produce ammonia or ammonium compounds.

For copper-aluminium alloys and copper-nickel alloys the resistance to stress corrosion cracking increases again with increasing alloying content, after passing through a minimum, but the sensitivity with respect to stress corrosion cracking of the copper-zinc alloys increases continuously with increasing zinc content, as shown in Figure 14 [217].

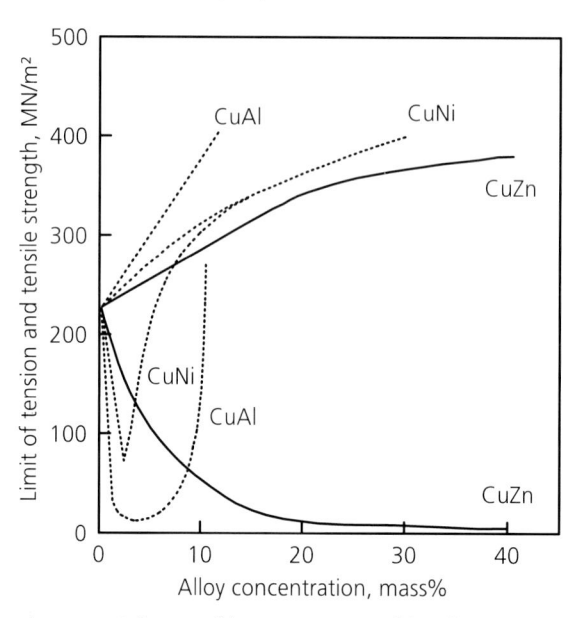

Figure 14: Influence of the concentration of the alloying elements in copper materials on the tensile strength (upper set of curves) and on the sensitivity for ammonia induced stress corrosion cracking (lower set of curves) [217]

With alloys that have higher zinc contents (Zn ≥ 20%) stress corrosion cracking is observed also on components believed to be stress-free, such as stress relief annealed specimens in moist ammonia atmosphere [254]. Even on specimens that stood under external pressure stress, stress corrosion cracking took place, as shown by the investigations described in [268].

For these investigations specimens of a Cu-Zn (α+β) alloy containing about 38 % of Zn were stress relief annealed at 553 K (280 °C) and cooled in the furnace. Two kinds of specimens were utilized, the one kind fracture mechanics specimens (WOL specimens) and the other kind a cross specimen with annular groove according to the sketch in Figure 15.

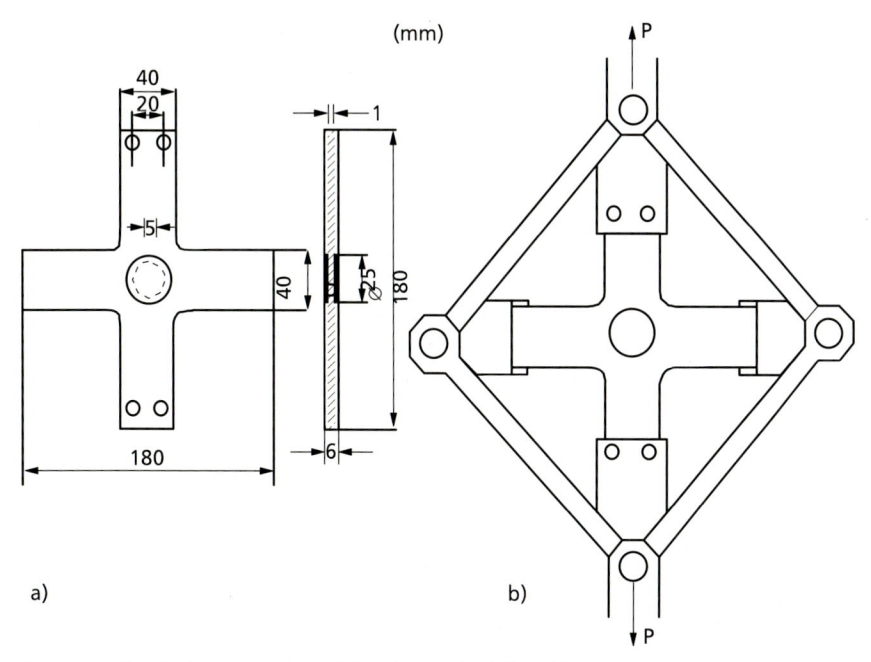

Figure 15: Notched cross specimen (a) and stress kind (b) [268]

A calculation of the stress field in the region of the annular notch using the method of linear elastic fracture mechanics revealed that all stress components were negative, as shown in Figure 16.

The stress corrosion cracking tests on the specimens were carried out at room temperature in a 1 molar aqueous ammonia solution that contained 5 g/l of $CuCl_2$. The obtained results are shown in Figure 17 and in Figure 18.

These results show that the limit values of the stress intensity for producing stress corrosion cracking under compression stress are three to four times higher than those under tensile stress. The incubation times for the appearance of stress corrosion cracking are also very much longer in the presence of compression stress than in the presence of tensile stress. When both kinds of stress are present in a specimen or component it must therefore be assumed that the stress corrosion cracking leads to failure due to the tensile stress, before stress corrosion cracking induced by compression stress can appear.

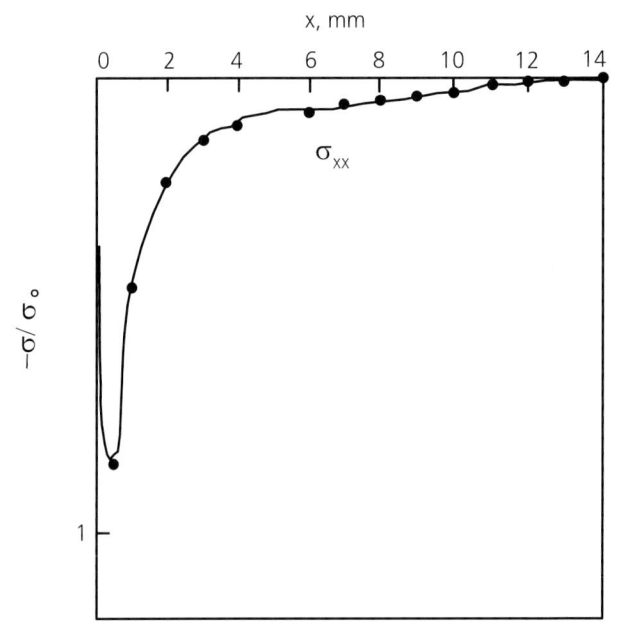

Figure 16: Stress distribution in the cross specimen depending on the distance (x) from the annular notch [268]

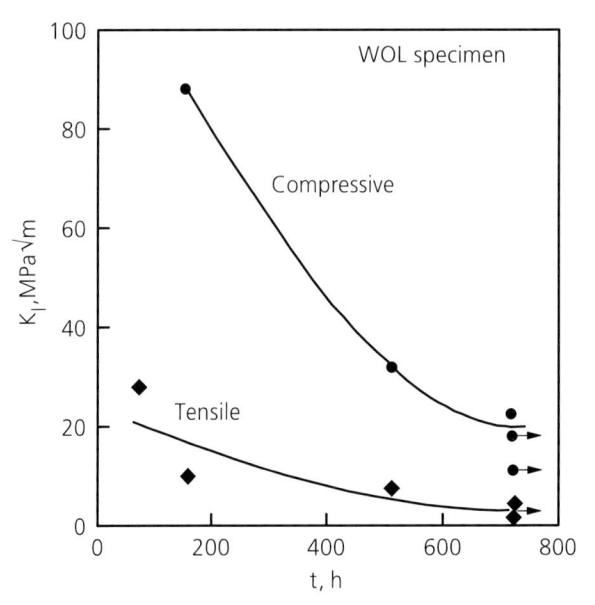

Figure 17: Stress intensity factor K_I plotted against time until stress corrosion cracking appeared under compression stress and tensile stress for WOL specimens [268]

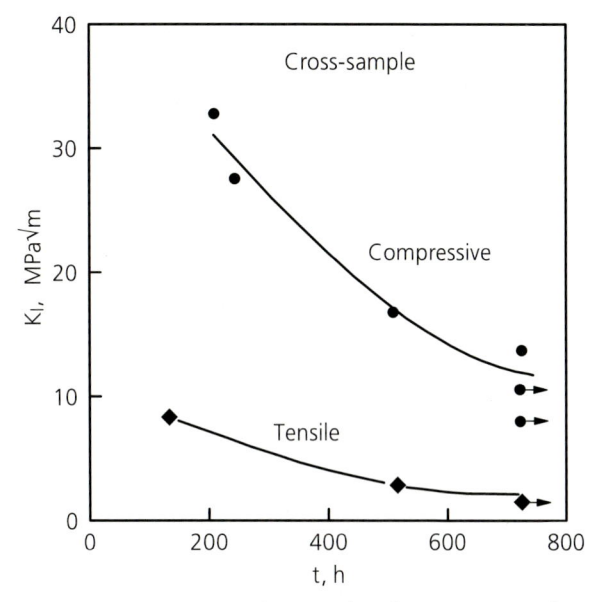

Figure 18: Stress intensity factor K_I plotted against time until stress corrosion cracking appeared under compression stress and tensile stress for cross specimens [268]

From the tests described in [269] that were carried out on a copper-zinc alloy (α-alloy with 35 % Zn) it is concluded that a relationship exists between zinc depletion of the alloy in the surface region and the initiation of stress corrosion cracking.

By cold forming the sensitivity for stress corrosion cracking of brass in aqueous ammonia solutions is significantly increased. For this tensile tests under constant load were carried out on commercial α-brass in aqueous ammonia solutions [270]. The specimens consisting of 1 mm thick sheets were tested in the state as delivered, after stress relief annealing and with subsequent cold forming. The cold forming took place by rolling the specimens in the longitudinal direction to 20 % and 40 % of their original thickness. The chemical composition of the material is specified in Table 33. The mechanical key parameter values determined in the states after the various treatments are collected in Table 34.

Cu %	Zn %	As %	Sn %
69.39	30.54	0.02	traces

Table 33: Chemical composition of the investigated α-brass [270]

State	0.2 % yield strength MPa	Tensile strength MPa	Uniform dimensional extension %	Grain size μm
State as delivered	204.4	420.8	37.6	42
Annealed, 2 h/ 623 K (350 °C)	138.9	371.6	50.2	115
20 % cold formed	396.8	493.3	8.91	20
40 % cold formed	561.2	585.2	9.51	15

Table 34: Mechanical key parameter values of the investigated α-brass [270]

The stress corrosion cracking tests were carried out in 1 to 5 mol/l aqueous ammonia solutions at temperatures between 300 and 335 K (27 and 62 °C). The time until fracture of the specimens was taken as measure for the sensitivity with respect to stress corrosion cracking. Figure 19 shows as example the dependence of the time until fracture of the specimens on the loading in 2 mol/l ammonia solution.

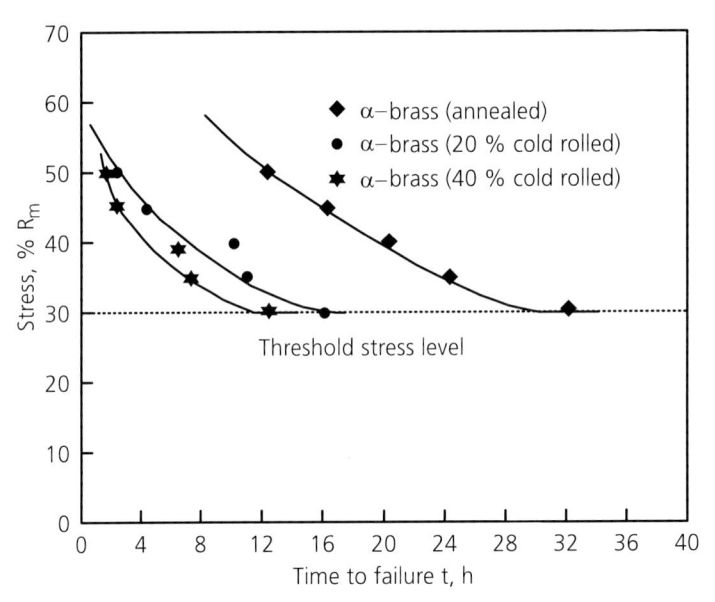

Figure 19: Dependence of the fracture time on the loading of the specimens in 2 mol/l ammonia solution at 300 K (27 °C) [270]

In the loading range between 50 % and 30 % of the tensile strength the fracture times increase with decreasing loading. This suggests that a limiting stress exists below which only the corrosion process, but no crack formation, takes place. With increasing deformation of the specimens the fracture times become significantly shorter with the same loading.

Table 35 shows the influence of the ammonia concentration in the test solution on the fracture time taking as example the test at 300 K (27 °C) and with a specimens loading 45 % of the respective tensile strength.

Table 36 shows accordingly the influence of the temperature of the test solution on the fracture time taking as example the test in an aqueous ammonia solution with 2 mol/l NH_3 with a specimens loading of 45 % of the respective tensile strength. As expected, the sensitivity for stress corrosion cracking increases with increasing ammonia content in the test solution and with increasing temperature.

Ammonia content mol/l	Time until fracture of the specimens min		
	Annealed, 2 h/623 K (350 °C)	20 % cold rolled	40 % cold rolled
1	1,080	645	558
2	930	170	95
3	690	90	72
4	640	55	43
5	555	35	25

Table 35: Influence of the ammonia concentration in the test solution on the time until fracture at 300 K (27 °C) and a specimen loading of 45 % of the tensile strength [270]

Temperature K (°C)	Time until fracture of the specimens min		
	Annealed, 2 h/623 K (350 °C)	20 % cold rolled	40 % cold rolled
300 (27)	930	170	95
315 (42)	330	152	74
325 (52)	266	135	65
335 (62)	225	113	58

Table 36: Influence of the temperature of the test solution on the time until fracture in a 2 mol/l ammonia solution and a specimen loading of 45 % of the tensile strength [270]

The cracking rate increases if for the stress corrosion cracking an alternating stress is superimposed on the tensile stress. For this a pulsating bending stress was superimposed on the tensile stress specimens in the stress corrosion cracking tests of a commercial brass grade having the nominal composition shown in Table 37 during the crack propagation phase [271]. The crack growth was measured according to the separation of the respective markings of the crack front (Δx), i.e. of the vibrating strips, on the fracture surface, depending on the time between the pulse intervals (Δt). The cracking speed is then expressed by the ratio $\Delta x/\Delta t$.

% Cu	% Sn	% As	% P	% Fe	% Zn
70.0–73.0	0.91–1.20	0.02–0.10	0.02–0.10	max. 0.06	balance

Table 37: Analysis values of the investigated CuZn alloy [271]

To build up the internal stresses the specimens were heated in a helium atmosphere at 823 K (550 °C) for 1 hour. The test was carried out in a 15 mol/l aqueous ammonia solution that contained 8 g/l of dissolved copper. The tensional stress specimens were prestressed to a basic load (P) of 668 N corresponding to about 0.2 % of the permanent limit of elongation. Then an additional bending stress (ΔP) was briefly applied at certain time intervals. The diagram in Figure 20 illustrates the stress applied to the specimens.

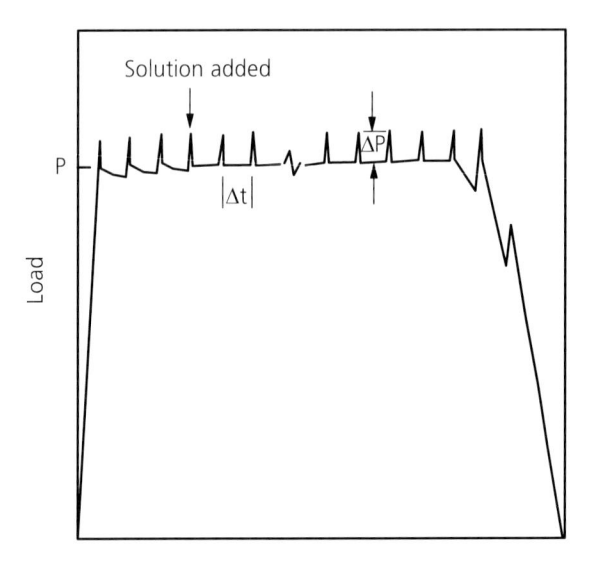

Figure 20: Schematic course of the stress applied to the samples as tensile load and superimposed bending load [271]

Since the specimens creep during the first pulses, the load initially falls under the applied basic load. Therefore the test solution was added only after the creep process had come sufficiently to a standstill. During the final test phases the load again fell under the basic load level due to the formation of cracks.

In a first test series the pulse interval (Δt) was varied between 1 and 500 s with a constant ΔP value of 0.05 P and an average pulse stress time of about 1.2 s. In these tests the time until fracture was between 145 and 220 minutes, but it was evidently independent of Δt. The examination of the fracture surfaces showed that in all cases numerous subsidiary cracks had appeared and that the fracture was traceable to the initiation and growth of various separate stress corrosion cracks. The fractures always started inter-crystalline, changed after about 100 μm to predominantly trans-crystalline course and showed characteristic crevice fracture surfaces. Specimens that were tested only under the basic stress showed the same appearance so that it cannot be attributed to the pulsating stress.

The distances (Δx) between the markings of the crack front on the fracture surfaces were measured and the relationship found between Δx and Δt is shown in Figure 21.

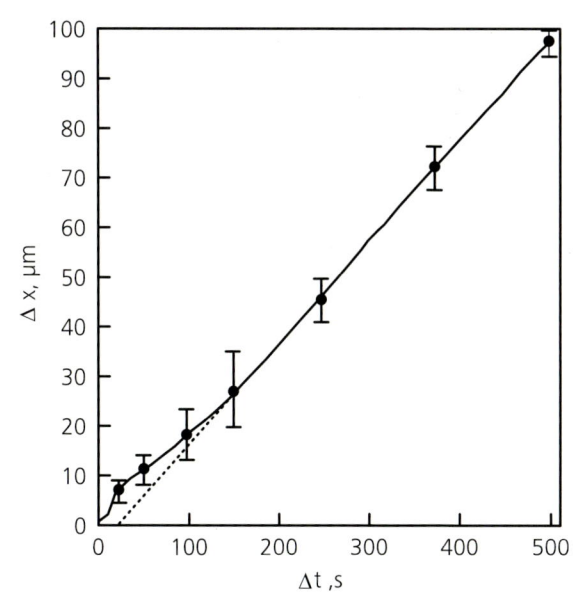

Figure 21: The relationship between the interval time Δt and the distance between the crack front markings Δx in the trans-crystalline fracture region [271]

A linear relationship exists between Δt and Δx for Δt values higher than approx. 150 s. The relationship between the crack growth rate $\Delta x/\Delta t$ and Δt is shown in Figure 22.

The constant value of 1.9×10^{-7} attained for Δt values higher than about 100 s must be assigned to a speed according to type II.

In a second test series the interval between the pulses was held constant at $\Delta t = 100$ s, the lowest Δt value for which with a ΔP value of 0.05 P according to Figure 22 a constant crack propagation rate results, was sustained and the magnitude of the pulse load was varied between 0.05 and 0.2 P. The results shown in Table 38 indicate that the crack growth rate derived from the time until fracture is practically independent of the magnitude of the pulsating load.

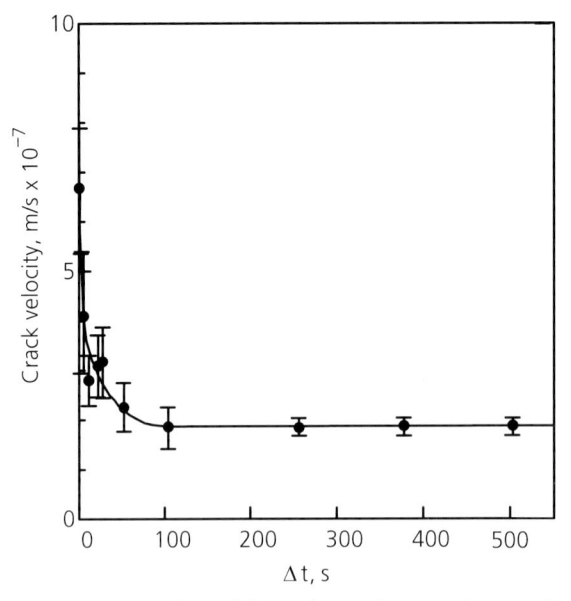

Figure 22: Dependence of the crack growth rate on the interval time (Δt) [271]

ΔP *)	Crack growth rate 10^{-7} m/s
0	0.8
0.05	0.8
0.10	0.8
0.15	1.0
0.20	0.9

*) Basic load P = 668 N

Table 38: Crack growth rates with different magnitudes of the pulsating load [271]

The utilization time from the incubation time until the appearance of micro-cracks was determined for a component endangered by stress corrosion cracking, but without any defects at the time of commissioning. In the investigations described in [272] it was therefore attempted to establish a model using the methods of fracture mechanics, in order to develop a correlation between the incubation time for the crack initiation and the critical stress as well as the corrosion potential, especially for the stress corrosion cracking system CuZn30/aqueous ammonia-containing solution. With the investigations carried out in this context with regard to the transition from local corrosive attack to critical micro-crack initiation it was found that this transition already takes place with a local attack depth of approx. 10 µm, whereby this value increases with decreasing stress.

The slight fractions of ammonia required for initiating stress corrosion cracking of brass can also be produced in a system that was thought to be unproblematic, by microbiological degradation products. In the damage investigation described in [273], in a heat exchanger operating as oil cooler trans-crystalline stress corrosion cracking was found on the pipe plates made of CuZn38AlFeNiPbSn, CW715R (old Material No. 2.0525, CuZn38SnAl) as well as on the pipes made of CuZn20Al2As (CW702R, old Material No. 2.0460). The cracks originated from the water side. Large amounts of manganese and iron as well as a certain content of sulfur were found in the deposits on the pipe surface. From this it was concluded that iron and manganese oxidizing as well as sulfate reducing microorganisms were present. Under these conditions an anaerobic degradation of organic materials with the liberation of ammonia must be reckoned with.

Also in the case of damage investigated in [274] by stress corrosion cracking on condenser pipes made of admiralty brass (70.85 % Cu; 27.96 % Zn; 0.01 % Pb; 0.02 % Fe; 1.11 % Sn; 0.05 % As) must be traced back to fouling with nitrogen reducing microorganisms.

Ammonium Salts

Author: K. Hauffe / Editor: R. Bender

Copper

According to Figure 23, the corrosion of copper in $(NH_4)_2SO_4$ solutions increases with increasing salt concentration (from 5×10^{-5} to 0.5 mol/l). Above 1 mol/l $(NH_4)_2SO_4$, the corrosion decreases again (see Curves ⑤ and ⑧ in Figure 23). The addition of as little as 0.05 mol/l $CuSO_4$ to 0.5 mol/l $(NH_4)_2SO_4$ reduces the material consumption rate of copper at room temperature after 48 hours from 10 to 6.5 g/m² h (6.3 mm/a) [275].

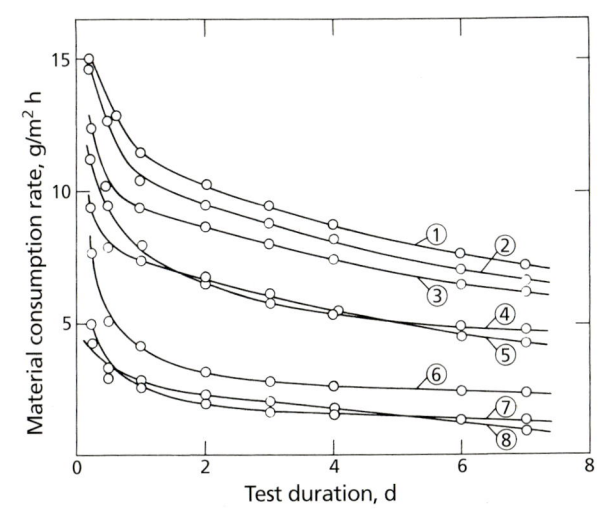

Figure 23: Material consumption rate of copper in $(NH_4)_2SO_4$ solutions as a function of the concentration and the test duration at room temperature [275]
① 0.5, ② 0.25, ③ 0.05, ④ 0.025, ⑤ 1.25, ⑥ 5×10^{-3}, ⑦ 5×10^{-5} and ⑧ 2.5 mol/l salt
(1 g/m² h ≙ 0.97 mm/a)

The corrosion rate of copper increases with increasing sulfate concentration of up to 0.5 mol/l (addition of K_2SO_4) in a 0.5 mol/l $(NH_4)_2SO_4$ solution after a test duration of 7 days by about a factor of 3 and drops to half of the initial value in the presence of 2.5 mol/l sulfate [275]. Increasing copper sulfate content (corrosion product) also reduces the corrosion.

The increased corrosion of copper in the vapor phase above a boiling $(NH_4)_2SO_4$ solution (see Figure 24) is noteworthy. Thus, for example, the corrosion of Cu in 40% $(NH_4)_2SO_4$ is 0.65 mm/a (25.6 mpy) and in the vapor phase 1.6 mm/a (63.0 mpy). The dependence of the corrosion behavior on the salt concentration is complicated and not proportional [276].

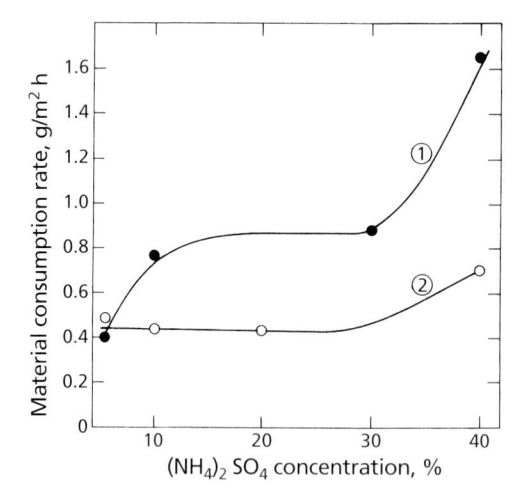

Figure 24: Material consumption rate of copper as a function of the salt concentration in boiling $(NH_4)_2SO_4$ solution [276]
① vapor phase, ② solution
($1\ g/m^2\ h \triangleq 0.97\ mm/a$)

In the case of high $(NH_4)_2SO_4$ concentrations (1 mol/l) the anodic dissolution of copper is determined by the diffusion of the $Cu(NH_3)^{2+}$-ions into the inside of the solutions [277]. This finding is confirmed by the increase in the corrosion rate of copper with increasing rotational speed of Cu [278].

Current density/potential curves on a Cu-anode in 1 mol/l $(NH_4)_2SO_4$-solution at 298 K (25 °C) show an increase in the current density from about 8 to 100 mA/cm^2 at 0.5 V_{SCE} after the addition of 0.1 to 2 mol/l NH_3. This increase is attributed to the formation of the soluble complex $[Cu(NH_3)_4]SO_4 \times H_2O$ [279]. Above +0.8 V, the current density is drastically reduced because of the salt passivation of the Cu-anode in contrast to the electrolyte which is free of or low in NH_3.

Contrary to pure NH_3, which gives a corrosion rate of copper of about 0.01 mm/a (0.39 mpy) at 773 K (500 °C), it increases in mixtures of (%) 8 NH_3 + 10 H_2O + 2 CO_2 up to about 60 mm/a (2,362 mpy) [280].

Data on the effect of temperature, NH_3 content and rotational speed of a Cu-anode on the corrosion rate of Cu is summarized in Table 40. The substantial increase in corrosion with increasing rotational speed of the electrode indicates in this case, too, a rate-determining diffusion of the copper-ammonia complex into the solution [281].

The corrosion rate and stationary corrosion potential of copper in 0.1 mol/l Na_2SO_4 containing increasing amounts of $(NH_4)_2SO_4$ at 298 K (25 °C), using the bimetallic system Cu/Zn in an open current circuit are listed in Table 39 [282].

In the sulfate run-down tank of a caprolactam plant containing 15 g/l of a sulfate solution ($(NH_4)_2SO_4$ + 45 % H_2SO_4) at 343 K (70 °C), the corrosion of copper was 0.16 mm/a (6.30 mpy) [283].

$(NH_4)_2SO_4$ mol/l	Stationary corrosion potential V*	Material consumption rate $g/m^2 h$ (Corrosion rate, mm/a)	
–	–0.426	–	–
0.02	–0.412	0.178	(0.2)
0.04	–0.411	0.178	(0.2)
0.07	–0.414	0.222	(0.25)
0.10	–0.424	0.498	(0.55)

* Saturated Calomel Electrode

Table 39: Stationary corrosion potential $_{SCE}$ and material consumption rate of copper in 0.1 mol/l Na_2SO_4 containing admixtures of $(NH_4)_2SO_4$ (298 K (25 °C)) [282]

Temperature K (°C)	$c(NH_4^+)$	$c(NH_4^+)$*	$c(Cu^{2+})$	Corrosion current density mA/cm^2 at various rps				Corrosion Potential (SHE)** V
				0.93	4.14	8.30	14.82	
288 (15)	4.0	0.7	0.45	38	63	76	85	–0.063
318 (45)				99	170	208	245	–0.050
288 (15)	6.5	3.1	0.48	28	46	56	68	–0.156
318 (45)				85	154	189	234	–0.152

* $c(NH_4) = c(NH_4) - 2(c(Cu + SO_4))$ with $c(SO_4) = 1.2$ mol/l
** Standard Hydrogen Electrode

Table 40: Corrosion current density in mA/cm^2 (1 $mA/cm^2 \triangleq 22\ g/m^2 h \triangleq 21$ mm/a) of a rotating Cu-electrode in a $[Cu(NH_3)_4]SO_4$-containing $(NH_4)_2SO_4$ solution as a function of temperature, electrolyte composition and rotational speed [281]

The corrosion resistance of copper in a 5 % NH_4Cl solution at 293 K (20 °C) having a material consumption rate of 0.263 $g/m^2 h$ (0.29 mm/a) is still satisfactory [284].

According to current density/concentration curves of copper in 0.6 mol/l NH_4Cl at 0.1 V and 1.8 mol/l NH_4I at 0.2 V, the material consumption rate of Cu is calculated in both cases to be < 2 $g/m^2 h$ (< 2 mm/a) [285]. Accordingly, copper should not be used in these solutions without inhibitors.

In a silver plating bath consisting of (g/l) 20 $AgNO_3$ + 50 Na_4Y + 53 $(NH_4)_2SO_4$, copper is severely attacked when plated with silver. The exchange rate between copper and silver is between 4,240 and 36,600 $g/m^2 h$ depending on the composition of the electrolyte [286].

According to Figure 25, the corrosion of copper, after dropping initially, increases again after one day above 0.1 mol/l NH_4Cl, contrary to the curve in Figure 23. Below this concentration, the corrosion decreases gradually and approaches a stationary final value after 1 to 6 days, depending on the NH_4Cl concentration [275]. Increasing additions of up to 0.5 mol/l $CuCl_2$ increase the corrosion rate by up to 2 powers of ten.

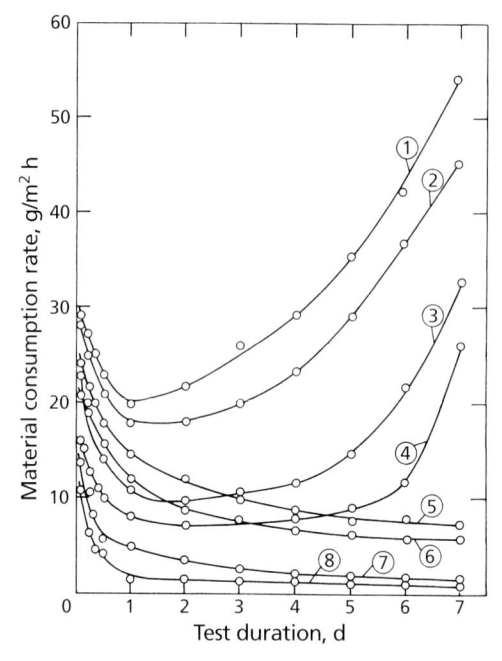

Figure 25: Material consumption rate of copper in NH_4Cl solutions as a function of the concentration and the test duration at room temperature [275]
① 1, ② 0.5, ③ 2.5, ④ 5.0, ⑤ 0.1, ⑥ 0.05, ⑦ 0.01 and ⑧ 0.001 mol/l salt
(1 $g/m^2 h \triangleq$ 0.97 mm/a)

Below 0.05 V_{SHE}, the corrosion current density on copper in 0.2 mol/l NH_4Cl at 313 K (40 °C) becomes $10^{-4} \leq A/cm^2$ ($\leq$ 2.5 mm/a ($\leq$ 98.4 mpy)) [287].

Copper pipes in contact with town gas have fairly good resistance at room temperature. In gas containing 0.05 and 3% NH_4Cl, the corrosion rate is 0.01 and 0.07 mm/a (0.39 and 2.76 mpy) [288].

Copper has proved to be suitable as material for heat exchanger tubes in contact with NH_4^+-containing waste water with 0.01 mm/a (0.39 mpy) at pH 7.5, although substantially higher at pH 6.5 [289].

The addition of 0.001 mol/l thioxine reduces the material consumption rate of copper in a solution of 0.5 mol/l HCl + 0.025 mol/l NH_4Cl at room temperature from 0.6 to about 0.1 $g/m^2 h$. In a solution of 0.5 mol/l H_2SO_4 + 0.0125 mol/l $(NH_4)_2SO_4$, thioxine acts as weak inhibitor only above a concentration of 0.01 mol/l [290].

Triazoles have a good inhibition efficiency on the corrosion of copper in $(NH_4)_2SO_4$ and NH_4Cl solutions [291, 292]. According to Table 41, the efficiencies of

0.001 mol/l benzo- and tolyltriazole are identical [291]. On the other hand, triazole and naphthotriazole are unsuitable for inhibiting the corrosion of copper in a 0.1 mol/l NH$_4$Cl solution [292].

The inhibition efficiency (IE = 1 − Δm_{Cu} (inhibitor)/Δm_{Cu}) · 100%, where Δm is the weight decrease) is only 31% in the presence of 0.001 mol/l naphthotriazole in NH$_4$Cl solutions, but in (NH$_4$)$_2$SO$_4$ solutions it is about 100% comparable with the other triazoles [292].

Referring to the example benzotriazole, it has been shown that the inhibition efficiency on the corrosion of copper in a 0.2 mol/l NH$_4$Cl solution increases and the corrosion decreases with the degree of coverage of the surface (see Table 42) [293].

Additions of 0.02% thiophenol or thiocresol reduce the corrosion of copper in 0.02 mol/l NH$_4$Cl at 304 K (31 °C) from 0.05 to 0.002 (1.97 to 0.08 mpy) and 0.0005 mm/a (0.02 mpy) respectively, which corresponds to an inhibition efficiency of 97 and 99% [294].

Because of the not firmly adhering layer consisting of a 1,2,4-triazole copper oxychloride complex, the inhibition efficiency on the corrosion of Cu in 0.2 mol/l NH$_4$Cl is unsatisfactory [295].

The corrosion of copper in NH$_4$Cl solutions can only be reduced to about 50 to 65% by adding agar-agar and gelatine [296, 284].

2-mercaptobenzoxazole has the greatest inhibition efficiency (test duration 2 days) of the organic inhibitors listed in Table 43 [297].

According to Table 44, sunlight irradiation increases the corrosion of copper in a CuCl$_2$-containing ammonium salt solution.

Solution	Without inhibitor		Triazole		Benzotriazole		Tolyltriazole	
	Corrosion rate mm/a (mpy)	pH value	Corrosion rate mm/a (mpy)/ inhibition efficiency %	pH value	Corrosion rate mm/a (mpy)/ inhibition efficiency %	pH value	Corrosion rate mm/a (mpy)/ inhibition efficiency %	pH value
0.1 mol/l NH$_4$Cl	0.075 (2.95)	5.3–8	0.04 (1.57)/42	4.7–7.9	0.0015 (0.06)/98	5.3–7.9	0.002 (0.08)/97	5.3–7.2
0.05 mol/l (NH$_4$)$_2$SO$_4$	0.09 (3.54)	5.4–8.4	0.006 (0.24)/ 100	4.8–6.3	0.0002 (0.01)/99	5.4–5.5	0.0002 (0.01)/99	

Table 41: Inhibition efficiency (%) of various triazoles (0.001 mol/l) on the corrosion of copper (mm/a) in 0.11 mol/l NH$_4$Cl and 0.05 mol/l (NH$_4$)$_2$SO$_4$ at 295 K (22 °C) [291, 292]

Benzotriazole mol/l	Material consumption rate g/m² h	Corrosion rate mm/a (mpy)	Degree of coverage	Inhibition efficiency %
–	0.0431	0.042 (1.7)	–	–
0.001	0.0169	0.016 (0.6)	0.608	61
0.003	0.0075	0.0073 (0.3)	0.825	83
0.005	0.0016	0.0015 (0.1)	0.963	96
0.010	9×10^{-5}	9×10^{-5} (3×10^{-3})	0.998	100
0.030	9×10^{-5}	9×10^{-5} (3×10^{-3})	0.998	100

Table 42: Material consumption rate of Cu as a function of the degree of coverage of the Cu-surface in the presence of benzotriazole in 0.2 mol/l NH_4Cl at 298 K (25 °C) after 120 hours [293]

Inhibitor	Concentration ppm	Material consumption rate g/m² h	Inhibition efficiency %
2-Mercaptobenzthiazole	–	0.363	–
	10	0.277	23
	40	0.263	27
	100	0.183	49
	200	0.091	75
2-Mercaptobenzimidazole	10	0.348	5
	40	0.132	64
	200	0.064	83
Benztriazole	10	0.277	24
	40	0.124	66
	100	0.071	80
	200	0.0006	100
Benzimidazole	10	0.338	9
	40	0.217	41
	100	0.039	90
	200	0.020	94
Indole	10	0.277	24
	100	0.208	43
	200	0.022	94
2-Mercaptobenzoxazole	10	0.0042	99
	40	0.0021	100
	100	0.0021	100

Table 43: Material consumption rate of Cu in 1 mol/l NH_4Cl at 305 K (32 °C) without and with inhibitors (test duration 2 d, 0.1 g/m² h ≙ 0.097 mm/a) [297]

Solution %		Material consumption rate $g/m^2 h$ (mm/a)	
		Without light	With light
Solution 1	10	2.78 (2.7)	3.89 (3.8)
	40	15.0 (14.5)	17.5 (17.0)
	100	45.3 (44.0)	56.5 (55.0)
Solution 2	10	4.42 (4.3)	5.84 (5.7)
	40	12.0 (11.6)	16.2 (15.7)
	100	33.0 (32.0)	39.5 (38.0)

Table 44: Corrosion behavior of Cu in 25 g/l $CuCl_2 \times 2H_2O$ + 50 g/l NH_4Cl at 305 K (32 °C) in the dark and upon exposure to sunlight, with the addition of 50 ml/l NH_3 (solution 1) and with the addition of 50 ml/l NH_3 + 50 g/l CH_3COONH_4 (solution 2) [298]

A similar effect of light on the corrosion of Cu also occurs in a solution consisting of 50 g/l $(NH_4)_2S_2O_8$ + 5 ppm $HgCl_2$ at 305 K (32 °C). In this case, the material consumption rate in the dark is 15.3 $g/m^2 h$ and in sunlight 23 $g/m^2 h$. In the presence of 50 ml/l of concentrated orthophosphoric acid in a solution of 25 g/l $(NH_4)_2S_2O_8$ + 25 g/l NaCl, the material consumption rate of Cu is 10.3 and 12.5 $g/m^2 h$ [299].

On dislocations of a Cu-specimen, a solution of $(NH_4)_2S_2O_8$ + NH_4OH + NH_4Br in a ratio of 1 to 1.5:6:0.3 produces etch pits of pyramidal structure having (110)-faces. At etching times of 10 seconds, the pit depth is between 8 and 30 µm at 280 K (7 °C) depending on the speed of the stirrer (from 0 to 5 rps) [300]. Since the ammonium persulfate attack on copper is severe and uniform, it is recommended as etchant [301].

In order to achieve the necessary activation of the surface for applying nickel coatings to copper structural components, 120 g/l NH_4Cl are added to the nickel plating bath consisting of (g/l) 40 $NiSO_4 \times 7 H_2O$ + 30 $NaH_2PO_4 \cdot H_2O$; the pH is adjusted within the range of 4.0 to 4.2 [302].

A neutral $CuCl_2$ solution with the following composition is recommended for the chemical etching of copper: (g/l) 50 cuprous chloride dihydrate + 200 ethanolamine + 50 NH_4NO_3 + 50 NH_4Cl up to a pH of 7.6 [303]. In this solution, benzotriazole cannot be used as etching additive, since it precipitates in this pH-range as a blue-green compound not only at room temperature but also at 330 K (57 °C) and impairs the appearance of the surface.

By using rapid potentiodynamic polarization measurements, it was possible to detect formations of CuCl on Cu in a 1 mol/l NH_4Cl solution [304].

The addition of ammonium oxalate as depassivator for the anodic dissolution of copper in a Cu-cladding electrolyte consisting of (g/l) 100 $CuSO_4 \times 5 H_2O$ + 380 $K_2P_2O_7 \times 3 H_2O$ (P_2O_7:Cu = 7:1, pH 8.6) gave the best results for cladding at higher current densities [304].

In NH$_4$HCO$_3$ solutions, copper is said to be substantially attacked, therefore preventing its use as material [305].

In the presence of oxygen and argon as carrier gas (saturated with (NH$_4$)$_2$S), phosphorus that was alloyed with copper for the manufacture of solders, caused an increase in corrosion (see Table 45) [306]. The corrosion increases with increasing phosphorus content, which is further enhanced in the presence of oxygen as carrier gas.

Material	Test duration d	Carrier gas	Temperature K (°C)	Corrosion rate mm/a (mpy)
Cu	9	argon	298 (25)	7×10^{-6} (0.0003)
BCuP-3				9×10^{-6} (0.0004)
Cu-10P				2×10^{-5} (0.001)
Cu	12	oxygen	298 (25)	3×10^{-5} (0.0011)
BCuP-3				4.5×10^{-5} (0.002)
Cu-10P				10^{-4} (0.0004)
Cu	9	argon	343 (70)	4×10^{-4} (0.02)

Table 45: Corrosion behavior of Cu and CuP-solder in (NH$_4$)$_2$S-saturated argon and O$_2$ [306]

In a 20 and 80% (NH$_4$)$_2$S solution, the corrosion rate of the alloy BCuP-3 was 3×10^{-5} and 10^{-4} mm/a (0.001 and 0.004 mpy) at room temperature [306].

In the production of methyl methacrylate, apart from the nickel-molybdenum alloy, copper and Monel® still have, according to Table 46, satisfactory corrosion resistance to NH$_4$HSO$_4$, methacrylic acid and sulfuric acid between 363 and 408 K (90 and 135 °C) [307]. In Table 46, only those production units are listed in which NH$_4$HSO$_4$ occurs. Intergranular corrosion was not observed.

According to tests in sodium hydroxide (200 g/l NaOH) lasting 95 hours, the addition of NH$_4$Cl only caused an increase in the corrosion of copper from about 0.024 to 0.029 g/m^2 h (0.028 mm/a) [308].

In the electrochemical contacts Pd/0.1 mol/l NH$_4$Cl/Cu, the current density of 0.4 µA/cm^2 at 293 K (20 °C) and –0.25 V$_{SCE}$ rises to about 40 µA/cm^2 when the potential is changed to –0.2 V$_{SCE}$, which corresponds to an increase in the corrosion rate from 0.0092 to 0.92 mm/a (0.36 to 36.2 mpy) [309].

Production plants	Composition of the solution, %	Temperature K (°C)	Corrosion rate mm/a (mpy)		
			Cu	Monel®	Ni-27Mo
Esterification plant	methacrylamide + methacrylic acid + NH_4HSO_4 + H_2SO_4 + methyl methacrylate + water (no % given)	363 (90)	3.8 (150)	0.84 (33.1)	0.06 (2.36)
1. Distillation column	25–30 methyl methacrylate + 1.5–2 methacrylic acid + 25–30 H_2SO_4 + 35 NH_4HSO_4 + 8–8.5 water	383 (110)	0.6 (23.6)	0.26 (10.2)	–
2. Distillation column	15–20 methyl methacrylate + 1.5–3 methacrylic acid + 30–35 H_2SO_4 + 38–40 NH_4HSO_4 + 5.5–12 water	393 (120)	0.32 (12.6)	0.1 (3.94)	0.034 (1.34)
3. Distillation column	5 methyl methacrylate + 3–5 methacrylic acid + 35–40 H_2SO_4 + 35–40 NH_4HSO_4 + 15–17 water	393 (120)	0.5 (19.7)	0.7 (27.6)	0.042 (1.65)

Table 46: Corrosion rate of Cu, Monel® and Ni-27Mo in production plants passed through by NH_4HSO_4 [307]

Copper-aluminium alloys

If oxidic surface coatings can be formed in the corroding solution, CuAl-alloys, in particular aluminium bronzes, are also corrosion-resistant in ammonium salt solutions. In solutions containing 0.05 mol/l NH_4^+, the CuAl-alloys containing 4 % Al go through a minimum of fracture strength at 298 K (25 °C) of about 10 N/mm² and then rise to a fracture strength of more than 300 N/mm² above an Al-content of 8 to 9 % and thus increase the resistance to stress corrosion cracking [310]. Conversely, the CuAl-alloys with ≤ 8 % Al are well protected against uniform corrosion if the alloys are homogeneous.

In a 5 % NH_4Cl solutions at 293 K (20 °C), the material consumption rate of the alloy Cu-9.0Al and Cu-10.5Al is 0.0417 and 0.150 g/m²h (0.043 and 0.16 mm/a) [284].

A low NH_4^+-content as a result of ammonia impurities in harbor water (synthetic solution consisting of 3 % NaCl + 3 ml/l NH_4OH) leads to oscillation of the dissolution potential of a Cu-8.3Al alloy between –0.080 and –0.210 V_{SCE}, which can last for days. This phenomenon is dependent on the Al-content of the alloy and the NH_4OH content of the solution [311].

In a Young etching solution (($(NH_4)_2S_2O_8$:NH_4OH:NH_4Br in a ratio of 1 to 1.6:6:0.3) the depth and shape of pits formed on dislocations of CuAl-alloys at 280 K (7 °C) was determined. In Figures 26 and 27, the pit depth is given as a function of the Al-content of the alloy and the $(NH_4)_2S_2O_8$ concentration of the solution [300].

At 1 % Al, the alloy reaches the lowest pit depth of 0.011 mm after an etching time of 10 seconds.

The heat treatment of a CuAl-alloy (AZhN 10 – 4L, Russian type) at 1,023 K (750 °C) in a powder mixture of (%) 35 Al + 15 Mn + 49 Al_2O_3 + 1 NH_4Cl improved the wear resistance by surface alloying [312]. In this case, NH_4Cl acted as a gaseous diffusion activator.

Copper-nickel alloys

The alloy Cu-30Ni-0.7Fe is more corrosion-resistant to $(NH_4)_2CO_3$ solutions than the alloy Cu-10Ni-1.5Fe and also more resistant than the conventional brass types (see section CuZn-alloys (brass)). Quantitative data were not given [313]. This behavior of the copper-nickel alloys is also confirmed for other ammonium salt solutions [314].

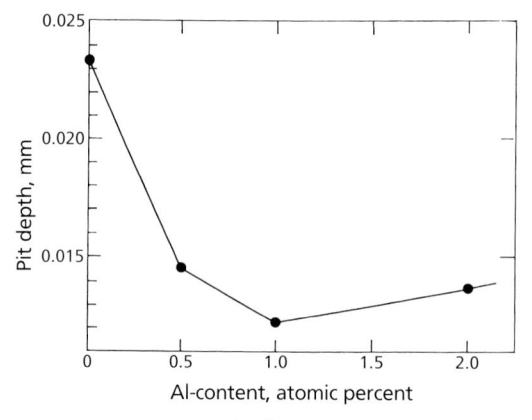

Figure 26: Corrosion pit depth in a CuAl-alloy in (mol/l) 1 $(NH_4)_2S_2O_8$ + 6 NH_4OH + 0.3 NH_4Br at 280 K (7 °C) after 10 seconds as a function of the Al-content [300]

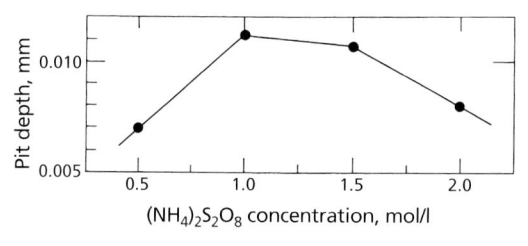

Figure 27: Effect of the $(NH_4)_2S_2O_8$ concentration on the corrosion pit depth of Cu-2Al after 10 seconds in a solution consisting of $(NH_4)_2S_2O_8$ + 6 mol/l NH_4OH + 0.3 mol/l NH_4Br at 280 K (7 °C) [300]

After exposure for 45 days to a solution at 323 K (50 °C) consisting of 3% NaCl + 0.3% NH$_4$Cl, the polarization resistance of the alloys Cu-11.4Ni-0.96Fe-0.56Mn and Cu-31.2Ni-0.48Fe-0.96Mn at a flow rate of 2 m/s remains constant at the initial value of 400 to 500 Ω · cm^2 even after 5 days, which indicates resistance of the alloys [315].

Reflectance photometry is recommended for comparative measurements of the formation of barrier layers on metals, since this technique does not change the structure and extent of the layer formed. Without removing the oxide layer on Cu-30Ni-(Fe), it could be shown by this method that an addition of 3% NH$_4$Cl to seawater increases its aggressivity by more than a factor of 2 [316].

"Nickel silver", a copper alloy (Cu-18Ni-27Zn), shows stress corrosion cracking in seawater containing ≥ 1 g/l NH$_4$Cl in the presence of air at room temperature, but not at 0.5 g/l NH$_4$Cl. The pH value of the solution has a substantial effect on this behavior. Thus, even at 4 g/l NH$_4$Cl, stress corrosion cracking occurs only in the pH-range from 8 to 9.5, while the critical pH-range at 1 g/l NH$_4$Cl is between 8.5 and 9 [317]. In air-free solutions, no stress corrosion cracking could be observed on the alloy even in the critical pH-range. In Table 47, the pH-dependent appearance of stress corrosion cracking on a copper alloy in 4 g/l NH$_4$Cl-containing seawater ((g/l) 24.5 NaCl + 5.2 MgCl$_2$ + 4.09 Na$_2$SO$_4$ + 0.16 CaCl$_2$ + 0.67 KCl + 0.20 NaHCO$_3$ + 0.10 KBr + 0.027 H$_3$BO$_3$ + 0.025 SrCl$_2$ + 0.003 NaF) saturated with pure oxygen, is illustrated.

pH Value	6	6.5	7	7.5	8	8.5	9	9.5	10
Time to stress corrosion cracking, d	–*	–	–	–	36	12	10	4	–

* – no SCC

Table 47: pH-dependence until stress corrosion cracking (SCC) of Cu-18Ni-27Zn looped specimens in seawater containing 4 g/l NH$_4$Cl at room temperature [317]

The addition of 0.5 g/l CuSO$_4$ × 5 H$_2$O to NH$_4$Cl-containing (1 and 4 g/l) seawater prevented the occurrence of stress corrosion cracking even in the critical pH-range (with the exception of pH 9.5) [317].

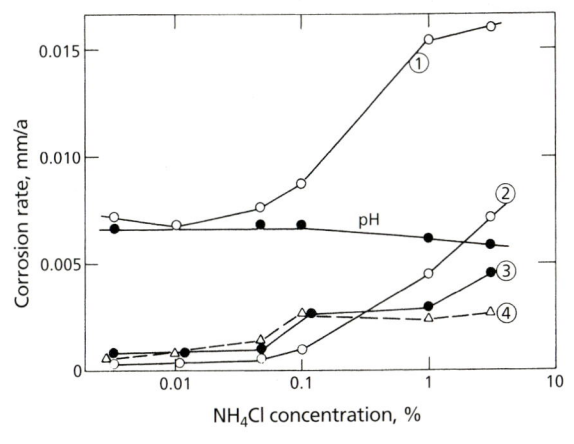

Figure 28: Dependence on NH$_4$Cl concentration of the corrosion rate of ① C-steel, ② oxygen-free Cu, ③ Cu-10Ni and ④ Cu-30Zn and the pH value at room temperature (test duration 1,000 h) [288]

According to Figure 28, copper, 70/30 brass and copper-nickel (90/10) are still sufficiently resistant to 5–10% NH_4Cl solutions at room temperature [288]. The average corrosion rate of Cu and Cu-10Ni in 0.05% NH_4Cl is 0.006 and 0.008 mm/a (0.24 and 0.31 mpy). The pH value in this solution decreases from about 7.2 to 6.3.

Copper-tin alloys (bronzes)

According to Table 48, the corrosion of CuSn-alloys in 5% NH_4Cl solution is still acceptable for technical use [284, 318, 319].

Material	Material consumption rate (Corrosion rate)	
	$g/m^2 h$	mm/a (mpy)
Cu-3.2Sn	0.425	0.41 (16.1)
Cu-4.9Sn	0.358	0.35 (13.8)
Cu-11.3Sn	0.625	0.64 (25.2)
Cu-9.0Sn-3.6Pb	0.642	0.63 (24.8)
Cu-27.2Sn-3.8Pb	0.575	0.606 (23.9)

Table 48: Corrosion behavior of CuSn-alloys in 5% NH_4Cl at 293 K (20 °C) [318, 319]

Copper-tin-zinc alloys (red brass)

No corrosion data are available on the behavior of these alloys in ammonium salt solutions. However, since tin is resistant and zinc preferentially goes into solution from tin-zinc alloys in acidic NH_4Cl solutions, the behavior of CuSnZn-alloys in ammonium salt solutions should be comparable to that of copper. CuSnZn-alloys which have a higher Zn-content should, however, be similar to CuZn-alloys in their corrosion behavior (see Section CuZn-alloys (brass)).

In solutions containing $(NH_4)_2SO_4$, NH_4OH and $CuSO_4$, stress corrosion cracking was observed on naval brass (Cu-34Zn-1Sn) in the pH-range between 5.3 and 7.2. The crack propagation took place preferentially through regions high in zinc [320].

Copper-zinc alloys (brass)

According to Table 49, the corrosion rate of brass increases with increasing Cu-content and with increasing $(NH_4)_2SO_4$ concentration in 0.1 mol/l Na_2SO_4.

The resulting corrosion potential shows only a slight and undefined change with the composition of the alloy and the solution [282].

Copper-zinc alloys are susceptible to stress corrosion cracking. As an example, the cracking rate of α-brass (Cu-30Zn) in the Mattsson solution [321], which consists of 0.48 mol/l $(NH_4)_2SO_4$ + 0.05 mol/l $CuSO_4$ at a pH of 7.2 to 7.4 (by addition of NH_4OH), at room temperature is provided by Figure 29 as a function of the mechanical stress [322]. The crack propagation was determined continuously and by interrupted tests in intervals of 10 days. Despite the substantially divergent test results, they still remain in an acceptable range (shown by the hatched area); for example, at 4 MPa√m they are between approximately 0.0036 and 0.0108 mm/h. The homogeneous corrosion of brass can be recognized by a blue-black tarnish.

$(NH_4)_2SO_4$ concentration mol/l	Material consumption rate, g/m² h (Corrosion rate, mm/a)					
	Cu-37Zn	V_c, mV	Cu-32Zn	V_c, mV	Cu-27.9Zn	V_c, mV
–	–	–480	–	–464	–	–466
0.02	0.15 (0.16)	–447	0.089 (0.092)	–449	0.053 (0.054)	–432
0.04	0.51 (0.53)	–467	0.50 (0.52)	–454	0.59 (0.61)	–445
0.07	1.04 (1.09)	–471	0.89 (0.92)	–456	0.98 (1.01)	–448
0.10	1.47 (1.54)	–477	1.03 (1.07)	–465	0.89 (0.91)	–456

Table 49: Stationary corrosion potential V_c (against Hg/Hg_2SO_4, 0.1 mol/l Na_2SO_4 where E° is +614 mV) and corrosion behavior of CuZn-alloys in 0.1 mol/l Na_2SO_4 at 298 K (25 °C) with additions of $(NH_4)_2SO_4$ [282]

The addition of as little as 0.02 mol/l NaCl to the Mattsson solution increased the time until failure of α-brass from about 1 hour to more than 70 hours [323].

The incubation time before crack formation is independent of the applied mechanical stress [324]. The pH value found in the cracks was less than the 7.4 of the solution.

The crack formation rate of an alloy Cu-29.3Zn-0.02Sn-0.04As in a solution consisting of 1 mol/l $(NH_4)_2SO_4$ + 0.15 mol/l $CuSO_4 \times 5\ H_2O$ + NH_4OH (to adjust the pH to 6.8) was about 0.25 mm/h at a stress intensity factor of 15 $MN/m^{3/2}$ [325].

70/30 brass sheets (20 % cold-rolled and annealed at 900 K (627 °C) for 30 min) rapidly showed tarnishing in combination with stress corrosion cracking under tensile stress in the Mattsson solution (0.05 Cu^{2+} as $CuSO_4 \times 5\ H_2O$ + 1.0 mol/l NH_4^+ as $(NH_4)_2SO_4$ + NH_4OH, pH 7.2).

If the unstressed specimen is subsequently treated in this solution and then in air, transgranular fracture occurs. If the specimen is subjected to stress in the solution, intergranular cracks occur, while the growth of the transgranular cracks ceases [326].

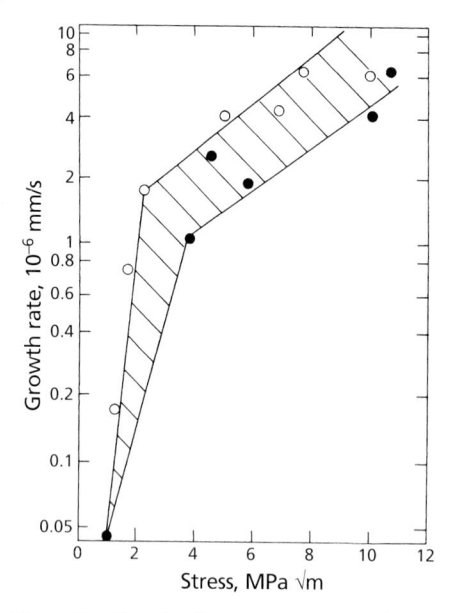

Figure 29: Growth of stress corrosion cracks in α-brass in Mattsson solution (298 K (25 °C)) as a function of mechanical stress MPa√m [322]
○ continuous and ● interrupted tests

According to Figure 30, the susceptibility of a brass specimen (CuZn30, CW505L) to stress corrosion cracking in the Mattsson solution (see above) increases with increasing temperatures at high crosshead rates (for example 1 mm/min) and increases with decreasing temperatures at lower rates (for example 10^{-5} mm/min). Thus, at a crosshead rate of 1 mm/min at 323 K (50 °C), crack formation occurred at 300 MPa, and at 298 K (25 °C) not until 400 MPa.

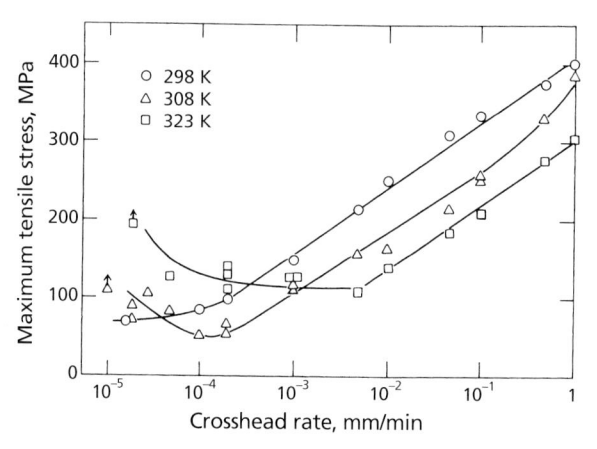

Figure 30: Effect of the crosshead rate on the maximum of the tensile stress leading to fracture in stress corrosion cracking tests using CuZn30 (Ms70) specimens in the Mattsson solution (pH 7.2) at various temperatures [327]

In contrast, at a crosshead rate of 10^{-5} mm/min, crack formation at 298 K (25 °C) occurred at a stress of as low as about 80 MPa, while it did not occur at 323 K (50 °C) until about 250 MPa. Under these test conditions, only intergranular cracks were observed [327, 328]. In Figure 31, the crack propagation behavior of a bright brass specimen (CuZn30) and one provided with a transparent insulator in the Mattsson solution (pH 7.2, 298 K (25 °C)) at a crosshead rate of 10^{-4} mm/min is given as a function of time.

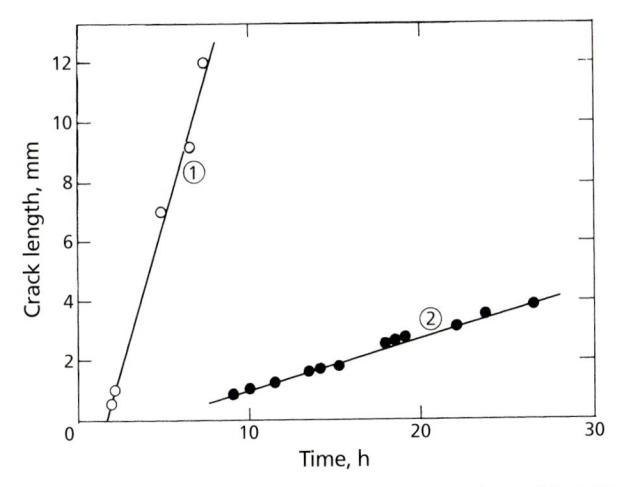

Figure 31: Crack length during stress corrosion cracking of CuZn30 (Ms70) specimens in Mattsson solution (pH 7.2, 298 K (25 °C)) at a crosshead rate of 0.006 mm/h as a function of time [328]
① uncoated area, k = 2.2 mm/h
② coated specimen, k = 0.17 mm/h

Fractographic observations on α-brass specimens in Mattsson test solution (0.04 mol/l $CuSO_4 \times 5\ H_2O$ + 0.75 mol/l $(NH_4)_2SO_4$) show not only at pH 4.3 (no surface film was formed) but also up to pH 7.5 (surface film formation) predominantly intergranular cracking with changes to transgranular cracking at pH values of 8 to 9.8 [329].

In Table 51, the pH-dependence of the time fracture of 65/35 brass in a solution of 65 g/l $(NH_4)_2SO_4$ (309 K (36 °C)) at a potential of +0.01 V_{SCE} is listed [330].

According to Table 50 (test conditions the same as in Table 49), the time fracture is shown as a function of the applied potential at a constant pH value of 8.2 of the solution [330].

Anodic polarization measurements of rotating (130 Hz) α-brass specimens scratched with diamond in solutions of 1.0 mol/l (NH_3 + NH_4^+) and 0.04 mol/l $CuSO_4$ (pH 7.3) showed repassivation at −0.23 V_{SCE} after 2 to 3 ms. During this time, up to 5 atom layers of zinc can go into solution. If stress corrosion cracking occurs, zinc goes into solution predominantly at the grain boundaries [331].

Potential, V_{SCE}	Time, min	Type of fracture	Color of the film
+0.15	80	T + I	dark gray
+0.10	70	T + I	dark gray
+0.05	35	I + T	black
+0.01	32	I	entirely black
−0.01	33	I	entirely black
−0.05	36	I + T	black
−0.10	42	T + I	dark gray

I intergranular/T transgranular/I + T intergranular fracture is predominant/T + I transgranular fracture is predominant

Table 50: Potential dependence of the time to fracture of a 65/35 brass specimen in 65 g/l $(NH_4)_2SO_4$ solution (pH 8.2, 309 K (36 °C)) [332]

To evaluate the aggressivity of salt-containing water on CuZn30 (CW505L) and Cu-21Zn-2Al, tests using aqueous solutions of $(NH_4)_2SO_4$ and/or CuCl were carried out at room temperature [333]. An excess of sulfate or chloride ions was achieved by adding Na_2SO_4 or NaCl. According to Figure 32, the corrosion of the alloy Cu-21Zn-2Al in chloride ion-free solutions of 2.5 mg/l NH_4^+ and 25 mg/l SO_4^{2-} is greater than that in sulfate-free solution. Thus, according to the straight lines 1 and 4 in Figure 32, the material consumption rate is 0.013 and 0.0042 $g/m^2 h$ (0.013 and 0.0043 mm/a). Brass 70/30 shows similar corrosion behavior.

Between pH 6 and 8, the material consumption rate of pure 70/30 brass in 0.5 mol/l $(NH_4)_2SO_4$ solution at room temperature was 0.0715 $g/m^2 h$ (0.074 mm/a) and 7.56×10^{-4} and 4.4×10^{-3} $g/m^2 h$ respectively (7.7×10^{-4} and 4.5×10^{-3} mm/a) in the simultaneous presence of 0.001 mol/l triazole (0.069 g/l) and benzotriazole (0.119 g/l). This means that both inhibitors have a 100 % efficiency [334]. With decreasing and increasing pH value of the solution, the inhibition efficiency decreases [334, 335].

pH-value	Time to fracture min	Type of fracture	Color of the film
6.0	145	T + I	dark gray
7.5	50	I + T	dark gray
8.2	32	I	black
9.0	65	T + I	dark gray
10.0	120	no cracks, homogeneous dissolution	no surface film
11.0	130	no cracks, homogeneous dissolution	no surface film
12.5	120	T + I	blackish white
13.5	125	T + I	dark gray

I intergranular/T transgranular/I + T intergranular fracture is predominant/T + I transgranular fracture is predominant

Table 51: pH-dependence of the time fracture of a 65/35 brass specimen in 65 g/l $(NH_4)_2SO_4$ solution (+0.01 V, 309 K (36 °C)) [322]

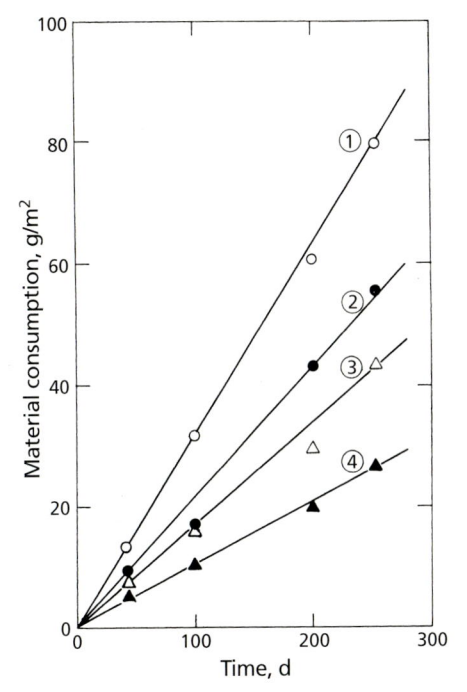

Figure 32: Corrosion behavior of Cu-21Zn-2Al in Cl⁻- and SO_4^{2-}-free solutions containing 2.5 and 25 mg/l NH_4^+ (① and ② Cl⁻-free; ③ and ④ SO_4^{2-}-free) as a function of time [333]
① 25 mg/l NH_4^+ + 670 mg/l SO_4^{2-}
② 2.5 mg/l NH_4^+ + 670 mg/l SO_4^{2-}
③ 25 mg/l NH_4^+ + 490 mg/l Cl⁻
④ 2.5 mg/l NH_4^+ + 490 mg/l Cl⁻

Cold-rolled specimens made of Cu-34.5Zn and Cu-40.0Zn under tensile stress in Mattsson solution (0.05 mol/l $CuSO_4$ + 0.5 mol/l $(NH_4)_2SO_4$, pH 7.5 by means of NH_4OH, 298 K (25 °C)) (according to Weibull) show crack formation after only 1/10 of the time (1 hour) compared with specimens tempered at 793 K (520 °C) for 2 hours [336].

Cold-rolled α-brass specimens of 0.05 mm in thickness were exposed to a 0.4 mol/l NH_4Cl solution at 298 K (25 °C) in the absence of stress for various periods of time (30 min to 50 h), then dried on filter paper (without rinsing or rubbing) and exposed to a defined stress outside the corrosion medium. At R_{air} = 30 kJ/m², the ratio RC/R_{air} after corrosive attack for 30 h was 0.27 for the crack propagation along the direction of rolling and at R_{air} = 20 kJ/m² about 0.8 perpendicular to the direction of rolling [337].

Stress corrosion cracking occured on brass L-62 (Cu-38Zn, Russian type) both in $(NH_4)_2SO_4$ and in Mattsson solution only if, in addition to the shifting of the potential to more noble values, mechanical stresses were also present [338].

If the Mattsson solution is stirred in the region of the copper oxide formation of 70/30 brass, a substantial current increase takes place, which is caused by increased dissolution of the protective oxide film. In the stationary solution, the copper oxide film provides sufficient protection at a uniform material consumption rate. However, ruptures of this oxide film, as already mentioned, heal rapidly in the absence of mechanical stress [310].

Brass which corrodes at room temperature in 0.1 mol/l NH_4Cl (0.096 g/m^2 h) is inhibited to virtually 100 % by the addition of 0.001 mol/l triazole, benzotriazole or naphthotriazole, as a result of which the corrosion rate decreases to about 2.5×10^{-4} to 6.5×10^{-4} mm/a (0.01 to 0.03 mpy) [334].

The inhibition efficiency of 1,5-diphenyl carbazide on the corrosion of α-brass (Cu-36.6.Zn-0.07Sn) in 0.1 mol/l NH_4Cl is 82 % even at 0.08 % carbazide and thus unsatisfactory. Nevertheless, the corrosion rate of 0.035 mm/a (1.38 mpy) is sufficiently low [339].

According to Table 52, the critical passivation current density and thus the corrosion current density of brass in 1 mol/l NH_4Cl + 0.1 mol/l HCl increases with increasing Zn-content. As a result of the simultaneous presence of HCl, the corrosion becomes very extensive. At the beginning of the anodic dissolution, the potential shifts to more negative values with increasing zinc content [340].

Material	Corrosion current density mA/cm^2	Corrosion rate mm/a (mpy)
Cu	9.40	216 (8,504)
CuZn13	11.06	254 (10,000)
CuZn21	16.34	375 (14,764)
CuZn31	19.47	447 (17,598)
CuZn38	24.92	573 (22,559)

Table 52: Average values of the corrosion current density and the corrosion rate of CuZn-alloys in the passive state in 1 mol/l NH_4Cl + 0.1 mol/l HCl at 298 K (25 °C) [340]

Even with 10 ppm NH_4Cl, stress corrosion cracking could be observed on phosphor bronze after as little as 6 weeks [341]. Comparable behavior is also caused by NH_4NO_3 [342].

According to Table 53, 2-mercaptobenzoxazole inhibits the corrosion of brass (Cu-36.7Zn) in 1 mol/l NH_4Cl solution at 305 K (32 °C) most efficiently [297].

In Table 54, the corrosion rates of 70/30 and 60/40 brass specimens in inhibitor-free and inhibitor-containing 0.1 mol/l NH_4Cl solution at room temperature (test duration 10 weeks) are summarized [284]. The protective effect provided by agar-agar and gelatine additions is unsatisfactory.

The corrosion of 63/37 and 70/30 brass in 0.005 to 0.01 mol/l NH_4Cl is said to be inhibited by as little as 0.01 % K_2CrO_4 to almost 100 % (test duration 2 weeks). However, in 1 mol/l NH_4Cl, sodium sulfite and dextrin are said to be better inhibitors [343].

According to Table 55, the corrosion of 63/37 brass in NH_4Cl-containing solutions is increased by exposure to sunlight [298].

Inhibitor	Concentration ppm	Corrosion rate mm/a (mpy)	Inhibition efficiency %
2-Mercaptobenzthiazole	10	0.14 (5.51)	19
	100	0.047 (1.85)	72
	200	0.027 (1.06)	84
2-Mercaptobenzimidazole	10	0.033 (1.30)	80
	40	0.0050 (0.20)	96
	200	0.0037 (0.15)	97
Benztriazole	10	0.012 (0.47)	93
	40	0.0036 (0.14)	98
	200	0.00065 (0.03)	100
Benzimidazole	10	0.16 (6.30)	3
	100	0.047 (1.85)	82
	200	0.013 (0.51)	92
Indole	10	0.15 (5.91)	9
	100	0.079 (3.11)	50
	200	0.028 (1.10)	87
2-Mercaptobenzoxazole	10	0.0044 (0.17)	99
	20	0.0022 (0.09)	100
Without inhibitor	–	0.16 (6.30)	–

Table 53: Corrosion rate of 63/37 brass in 1 mol/l NH_4Cl at 305 K (32 °C) without and with inhibitors (test duration 2 d) [297]

Brass	Addition %	Material consumption rate $g/m^2 h$	Corrosion rate mm/a (mpy)	Inhibition efficiency %
70/30	–	0.094	0.10 (3.9)	–
60/40	–	0.102	0.11 (4.3)	–
70/30	0.2 A	0.020	0.03 (1.2)	35
60/40	0.2 A	0.044	0.05 (2)	56
70/30	0.2 G	0.029	0.03 (1.2)	52
60/40	0.2 G	0.091	0.10 (3.9)	26.5
70/30	0.1 A + 0.1 G	0.031	0.03 (1.2)	37
60/40	0.1 A + 0.1 G	0.083	0.09 (3.5)	35

Table 54: Corrosion behavior of 70/30 and 60/40 brass in 0.1 mol/l NH_4Cl containing agar-agar (A) and gelatine (G) at 293 K (20 °C) [284]

Solution		Material consumption rate g/m²h	
		Dark	Sunlight
Solution 1	10 %	0.97	2.12
	100 %	47.5	56.1
Solution 2	10 %	2.57	3.27
	100 %	50.7	57.3

Solution 1: (g/l) 25 $CuCl_2$ + 50 NH_4Cl + 50 ml/l NH_3
Solution 2: (g/l) 25 $CuCl_2$ + 50 NH_4Cl + 50 CH_3COONH_4 + 50 ml/l NH_3 (1 g/m²h, ≙ 1.1 mm/a (43.3 mpy))

Table 55: Corrosion behavior of 63/37 brass at 305 K (32 °C) after 2 h tests [298]

In a solution consisting of 50 g/l $(NH_4)_2S_2O_8$ + 5 ppm $HgCl_2$, the corrosion of 67/37 brass at 305 K (32 °C) is also increased by sunlight, for example from 16.8 to 26.5 g/m²h (28.6 mm/a (1,126 mpy)) [299].

The brass types Cu-40Zn-2Pb, Cu-38Zn-1Pb, Cu-36Zn-2Pb-0.03As, Cu-36Zn-2Pb-0.15As, Cu-31Zn-2Pb-0.8Sn-0.7Si-0.5Mn-0.9As, Cu-31Zn-2Pb-0.9Sn-0.8Si-0.6Mn-0.07As, Cu-34.5Zn-2Pb-1Sn-1Al-0.03As, Cu-31Zn-2Pb-08.Sn-1.5Al-0.03As, Cu-34Zn-2Pb-0.05As and gunmetal (Cu-6Zn-4Pb-4.7Sn) tested are, with the exception of gunmetal and single-phase α-brass containing an addition of arsenic, susceptible to dezincification in etching solutions of NH_4Cl and $(NH_4)_2S_2O_8$ [344].

High-alloy brass of low copper content (Cu->46Zn-22.5Mn-24Pb-0.31Fe) shows a corrosion behavior similar to that of standard brass types in 3 % NH_4Cl [345].

To reduce the attack on brass structural components caused by NH_4HF_2-containing 5 % hydrochloric acid (1.5 kg per 100 l HCl) in a sugar factory during the removal of the oxide film, of the silicates and the decomposition products of the sugar, the inhibitor Rodine® 213 Special (Agromore Ltd. Bangalore, India) is recommended, the effect of which can be seen from Table 56 [346].

Rodine® 213 Special	Material consumption rate g/m²h	Corrosion rate mm/a (mpy)	Inhibition efficiency %
–	21.2	22.0 (866.1)	–
1.5 l/100 l conc.* HCl	0.417	0.43 (16.9)	93.5
2.0 l/100 l conc.* HCl	0.259	0.27 (10.6)	97.7
2.5 l/100 l conc.* HCl	0.209	0.22 (8.7)	98.2

* concentrated

Table 56: Corrosion behavior of brass during the removal of the oxide film at 348 to 353 K (75 to 80 °C) in 5 % HCl + NH_4HF_2 + Rodine® [346]

Other copper alloys

Contrary to α-brass, Cu-3.71Ti-0.006Fe-0.002Pb is insensitive to stress corrosion cracking in the Mattsson solution (0.5 mol/l $(NH_4)_2SO_4$ + 0.05 mol/l $CuSO_4 \times$ 5 H_2O) in the acidic and neutral pH-range at room temperature. Above 0.3 mol/l NH_3 in the solution, stress corrosion cracking occurs, with 0.6 mol/l NH_3 already after 9 hours. In the neutral pH-range of 6.8 to 7.5, homogeneous corrosion is negligible [347]. The presence of $Cu(NH_3)_4^{2+}$-ions favors the susceptibility to stress corrosion cracking.

CuMn-alloys (Cu-7.1Mn, Cu-15.8Mn, Cu-22.9Mn) are also susceptible to stress corrosion cracking in the Mattsson solution. A 30% cold forming of the alloy Cu-7.1Mn increased the time to fracture occurred under a mechanical stress of 200 N/mm^2 after heat treatment of 1 hour at 773 K (500 °C) from 6 to about 72 hours [348].

The corrosion resistance of the beryllium bronze (Cu-2Be) in 10% NH_4Cl (293 K (20 °C)) is excellent with about 3.9×10^{-6} mm/a (0.0002 mpy) [284].

The first of the two spring alloys Beryvac 200® (CuBe2, CW101C) and Beryvac 60® (CuCoBe) corrodes in 10% NH_4Cl (373 K (100 °C)) at a rate of 0.1 to 1 mm/a (3.94 to 39.4 mpy). The second alloy is comparable to the first one in its corrosion behavior at room temperature, but is no longer resistant at 373 K (100 °C) at a corrosion rate of about 1 to 10 mm/a (39.4 to 394 mpy). In 20% $(NH_4)_2S_2O_8$, both spring alloys are unsuitable even at 293 K (20 °C) (≥ 10 mm/a (≥ 394 mpy)) [349].

Atmosphere

Author: K. Baumann / Editor: R. Bender

Copper

Apart from its utilization as conductor material in electrical engineering, copper is also used for architectural applications. When copper is exposed to the atmosphere an oxidic warm brown to brown-black top layer of oxides and sulfates is formed [350]. The formation of patina (verdigris) indicates the end of the copper corrosion and constitutes a permanent protective coating. The patina changes from green to gray-black and appears in the course of five to fifteen years, often sooner along maritime coasts. In some climates no patina is formed, but only a gray-black discoloration [351].

The components of the corrosion products, apart from CuO, are chiefly Cuprite Cu_2O, the sulfate-containing products Bronchantite $Cu_4SO_4(OH)_6$, Posnjakite $Cu_4SO_4(OH)_6 \times 2\ H_2O$ and Antlerite $Cu_3SO_4(OH)_4$, in chloride-containing atmospheres also Atacamite $Cu_7Cl_4(OH)_{10} \times H_2O$ [352, 353]. The fractions of sulfides, sulfates, carbonates, chlorides and hydroxyl compounds in the corrosion products are small. In industrial atmosphere the corrosion products are colored chiefly green or blue, in coastal atmosphere chiefly brown [353].

The corrosion products adhere well on the surface, so that the quantity of corrosion products agrees with the mass loss of copper [354]. The corrosion products are electrically conducting and hygroscopic [355].

In general pure copper corrodes least and copper containing arsenic corrodes most strongly. In most cases nickel has no effect on the corrosion rate. Copper roofs on buildings in Copenhagen showed corrosion rates of 1 to 2 µm/a in the course of 200 years of exposure. Thereby the main component of the produced patina is basic copper sulfate [356]. 300 year old copper gutters in Europe showed a corrosion loss of less than 100 µm [351].

The corrosion rate of copper is not constant, but decreases with the duration of exposure as shown in Figure 33 [357].

In tests with a duration of 10 to 20 years the mass loss rate was found to be 0.006 to 0.26 $g/m^2 d$. The small corrosion attack is due to the top layer formation. The patina is insoluble in water. The exact composition depends on the environmental components. In particular nitrates, carbonates and chlorides are often trapped in the basic copper sulfate layer of the patina [358].

The following average corrosion rates are reported for [359]:

- Rural atmosphere   < 0.5 µm/a
- Industrial atmosphere   1 up to 5 µm/a
- Coastal atmosphere   1 µm/a.

80% of the corrosion-conditioned mass loss are traceable to acidic rain, i.e. the corrosivity thereof is four times that of the pure atmosphere [360].

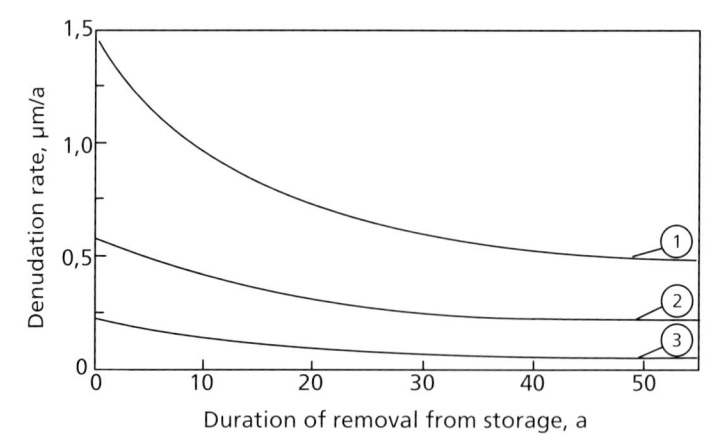

Figure 33: Corrosion rate of copper as a function of the exposure period under various atmospheric conditions [357]
① Coastal atmosphere
② Humid rural atmosphere
③ Dry rural atmosphere

– *Seasonal and constructional influences* –

The specimen orientation affects the corrosion. In general the corrosion is insignificant on the surface of the specimens facing away from the weather, but at least smaller than the corrosion on the front side of the specimen [357, 361].
For exposure in horizontal orientation the corrosion rate of copper is about 5% greater than in vertical orientation [362]. Flat specimens of electrolytic copper Cu 99.0 exposed facing south in 45° orientation at Prague-Letnany with a sulfur dioxide immission rate of 84 mg/m²d and 2,970 h moistening duration per year showed a mass loss of 15.0 g/m² during the first year. Spiral specimens of copper wire with 2 to 2.5 mm diameter and 1,000 mm length exposed parallel thereto showed a mass loss of 18.2 g/m² [363].

– *Behavior in outdoor climate* –

In North Bohemia copper is corroded at a rate which becomes smaller in the course of time. The mass loss in strongly aggressive industrial atmosphere was found to lie between 50 and 170 g/m² after five years of exposure, corresponding to a corrosion rate of 1.1 to 3.8 μm/a (0.04 to 0.15 mpy) [364].

On the other hand, in one year corrosion tests with copper Cu 99.9 in North Bohemia corrosion rates of 2.2 to 7.1 μm/a (0.09 to 0.28 mpy) with an average value of 3.1 μm/a (0.12 mpy) were found. The average corrosion rates in the Friedland region in north Bohemia at four test locations were found to be significantly smaller with a value of 1.7 μm/a (0.07 mpy) having a small scatter width of 0.11 μm/a (0.004 mpy) [365]. The joint investigations carried out in the Czech Republic and Sweden with Cu 99.5 in the course of eight years produced the results shown in Table 57.

After four years of exposure in south Finland only discolorations of the copper surface were found, but no measurable corrosion effects [366].

The mass losses found in the course of exposure to outdoor climates in East Europe, (1) in the industrial urban atmosphere of Prague-Letnany, (2) in the industrial atmosphere of Aussig, (3) and (4) in the rural atmospheres of Orgowany (Hungary) and Swenigorod (Russia), (5) in the urban atmosphere of Moscow, (6) and (7) in the coastal atmospheres of Sarafowo (Bulgaria) and Batumi (Georgia), are shown in Figure 34 in relation to the exposure period [367].

The corrosion rate of copper in the rural atmosphere of the Tajo region in central Spain, as the average of one year tests at 26 exposure sites, lies around 1.3 ± 0.3 µm/a (0.05 ± 0.01 mpy). Related to the average annual moistening duration of 2,750 h, this corresponds to a corrosion rate of 0.47×10^{-3} µm/h [368]. Analogously with Cu 99.92 at 24 exposure test locations in coastal atmosphere, an average corrosion rate of 1.48 µm/a (0.06 mpy) was found with horizontal orientation of the specimens, and 1.47 µm/a (0.06 mpy) with vertical orientation of the specimens. On the other hand, in the midland of Galicia with a corrosion rate of 1.33 µm/a (0.05 mpy) in horizontal exposure compared with a corrosion rate of 1.23 µm/a (0.05 mpy) in vertical orientation, the corrosion attack was slightly less on the vertical test surfaces at nearly all test locations [369].

The corrosion rates reported in Table 58 were measured at various exposure places in the USA [354, 356, 357].

Type of atmosphere	Test place	Mass loss, g/m² after years			
		1	2	4	8
Rural atmosphere	Hurbanovo	15.7	33.3	53.4	88
	Kasperske Hory	13.3	21.9	30.0	47
	Ryda	8.6	12.5	16.3	34
Sub-arctic	Gällivare	5.0	8.4	11.5	21
Urban atmosphere	Prague-Letnany	12.1	27.2	52.6	85
	Stockholm	5.0	9.6	15.7	31
Industrial atmosphere	Ustí (Aussig)	28.9	49.3	78.2	121
	Kopisty	34.7	60.4	94.4	139
Coastal atmosphere	Bohus Malmön, 1 km distant from the coastline	14.7	18.5	23.6	47
	Bohus Malmön, 50 m distant from the coastline	31.6	31.2	30.3	57
Urban	Göteborg	7.9	13.0	21.7	48

Table 57: Mass loss of specimens of electrolytic copper Cu 99.5 in exposure tests in the outdoor climate of Sweden and the Czech Republic [370, 371]

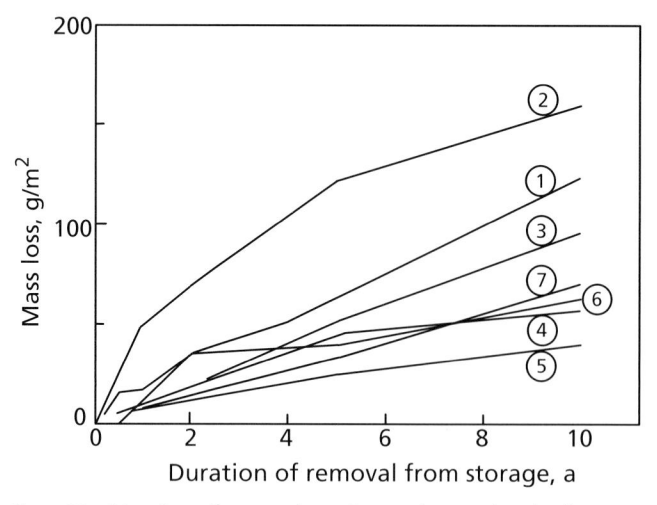

Figure 34: Mass loss of copper depending on the test duration for exposure at selected locations in east Europe [206]
① Prague-Letnany, ② Aussig, ③ Orgowan, ④ Swenigorod, ⑤ Moscow, ⑥ Sarafowo, ⑦ Batumi

The corrosion rates measured at Murray Hill and Norriston (USA) on copper Cu 99.9 with 0.007% phosphorus fraction were on the average around 1.27 µm/a (0.05 mpy) after 20 years and around 0.43 µm/a (0.017 mpy) after 65 years [356].

Atmosphere	Test place	Test period a	Corrosion rate µm/a (mpy)
Rural atmosphere	State College/PA	2	1.25 (0.05)
		7	0.75 (0.03)
		20	0.36 (0.014)
	Washington/DC	1	1.1 (0.04)
Warm/dry	Phoenix/AR	20	0.09 (0.035)
Industrial atmosphere	Newark/NY	2	1.88 (0.07)
		7	1.45 (0.06)
	Altoona/PA	20	1.31 (0.05)
	New York	20	1.19 (0.046)
	New Comb/NY	1	2.1 (0.08)
	Chester/NY	1	2.2 (0.086)
	Research Triangle Park	1	2.4 (0.09))

Table 58: Average corrosion rates of copper Cu 99.9 in outdoor climate of various North American types of atmosphere [354, 356, 357].

Table 58: Continued

Atmosphere	Test place	Test period a	Corrosion rate µm/a (mpy)
	Cleveland/OH	1	1.15 up to 1.68 (0.045 up to 0.066)
	Ottawa/ONT	1	0.86 up to 1.64 (0.034 up to 0.064)
	South Bend/PA	1	1.44 up to 2.09 (0.056 up to 0.082)
	Trail/BC	1	1.60 up to 2.50 (0.063 up to 0.098)
Coastal atmosphere	Key West/FL	20	0.50 (0.02)
	La Jolla/CA	20	1.43 (0.056)
	Kure Beach/NC, 25 m distant from the coastline	1	2.34 up to 3.20 (0.092 up to 0.126)
	Kure Beach/NC, 250 m distant from the coastline	1	3.28 (0.129)

Table 58: Average corrosion rates of copper Cu 99.9 in outdoor climate of various North American types of atmosphere [354, 356, 357].

The results obtained in test series extended up to 7.5 years show the decline of the corrosion rate with the duration of the exposure [357].

Stainless steels cladded with copper on both sides (10/80/10) showed the same corrosion behavior as solid copper after exposure to rural atmosphere for 7.5 years. In coastal atmosphere (Miami, Kure Beach) and in the chloride-containing urban industrial atmosphere of Dallas the cladding on the stainless steels AISI 409 (cf. 1.4512) and AISI 430 (1.4016) corroded only slightly, but contact corrosion occurred on the ferritic steel core. In contrast thereto, the copper cladded austenitic steel AISI 321 (1.4541) behaved as in the rural atmosphere [357].

Figure 35 shows a comparison of the corrosion course of copper in temperate and in hot humid tropical climate [372].

Specimens of copper (0.01 % Ni, 0.005 % Pb, 0.0067 % P) exposed at six outdoor stations in Argentina for three years showed the corrosion rates reported in Table 59 [373].

At Koeberg on the South African west coast corrosion rates of 5 µm/a (0.197 mpy) at a distance of 20 m from the coastline and of 4 µm/a (0.157 mpy) at a distance of 50 m from the coastline, were found [374].

In Egypt the corrosion protection given by the corrosion products layer on copper at the test locations became smaller in the order Cairo > Ismaila = Suez > Port Said > Alexandria [375].

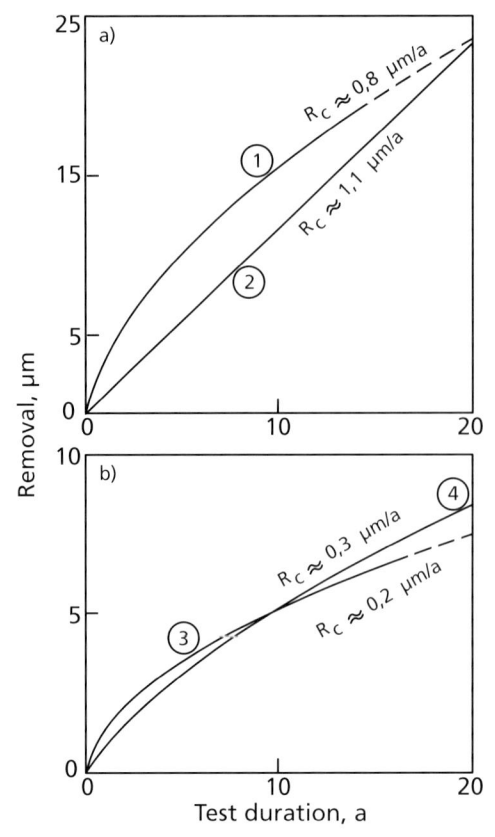

Figure 35: Comparison of the corrosion course of copper in tropical and in temperate climate [372]
Rc = Corrosion rate in the linear range (μm/a)
a) Coastal atmosphere: ① Cristobal (Panama) ② La Jolla (California)
b) Rural atmosphere: ③ Miraflores (Panama) ④ State College (Pennsylvania)

Exposure place	Corrosion rate, μm/a (mpy)		
	After 1 year	After 2 years	After 3 years
Jubany Base	2.06 (0.08)	1.17 (0.046)	
Cidepint	1.32 (0.052)		
San Juan	0.17 (0.007)	0.15 (0.006)	0.15 (0.006)
Iguazu	0.84 (0.033)	0.91 (0.034)	0.86 (0.034)
Camet	2.22 (0.087)	1.65 (0.065)	1.30 (0.051)
M. Martelli	0.80 (0.031)	0.53 (0.021)	0.52 (0.021)

Table 59: Corrosion rates of copper specimens exposed at six locations in Argentina for a period of 1 to 3 years [373]

In the tropical coastal climate of south India (Mandapam Camp) mass losses of 0.215 g/m² d at sea level and of 0.187 g/m² d at a height of 2 m above the sea were found. Thereby the mass loss rate was smallest during the period from October through March and greatest from July through September [376]. At a distance of 400 m from the coastline the annual average mass loss was 70 g/m², 91% of which took place during the 2nd and 3rd quarter of the year [356].

In the coastal atmospheres of Bhavnager and Mithapur the mass loss rate was 35.2 to 78.6 g/m² a. With additional industrial influence this mass loss rate increased to 102.7 to 114.7 g/m² a (Porbandar, Mithapur), but in medium industrial atmosphere (Bhavnagar, Rajkot) it was only 7.7 to 8.8 g/m² a [377].

The tropical humid climate of Vietnam produced the mass losses reported in Table 60 on copper Cu 99.9 and brass CuZn34 in three year and five year tests [378].

The corrosion behavior of copper Cu 99.0 was compared with that of cast bronze CuSn7Zn4Pb6 in an extensive international weather exposure program. The exposure was made outdoors and in exterior room climate at a wide range of different exposure places in Europe and North America. A compact protective layer formed very rapidly on the copper at all locations with rough climate, low sulfur dioxide concentrations and high ozone values. At measuring locations with high sulfur dioxide burden and high chloride concentration no passivation of the surface was found during a period of four years. Porous top layers containing large amounts of sulfate were formed.

Test place	Immission rate, mg/m² d		Mass loss, g/m²					
	Chloride	SO₂	Cu 99.9			CuZn34		
			After years					
			1	3	5	1	3	5
Hanoi	0.1	2.7	8.5	17.5	22	6.0	13.5	17
Nhatrang	5.8	3.9	35	52.5		8	13.5	
Ho-Chi-Minh Town	0.9	4.7	9	19.0		6	16.5	

Table 60: Mass losses on copper Cu 99.9 and brass CuZn34 during outdoor exposure to the tropical humid climate of Vietnam [378]

Within the scope of the ISO-CORRAG program, copper specimens were exposed to the outdoor climate at 53 exposure places on four continents, and after one year of test duration the corrosion values were determined for characteristic atmospheric stress situations. The results are summarized in Table 61 [379].

Climatic region	Type of atmosphere	Corrosion rate μm/a (mpy)
Cold climate	Rural atmosphere	0.08 (0.003)
	Antarctic	2.0 (0.078)
	Coastal atmosphere	1.7 (0.066)
Subtropical climate	Rural atmosphere, dry	0.2 (0.0078)
	Urban atmosphere	2.0 (0.078)
	Coastal atmosphere	2.2 (0.087)
Hot humid climate	Coastal atmosphere	5.5 (0.217)
Temperate climate	Rural atmosphere	1.3 (0.051)
	Urban atmosphere	1.1 (0.043)
	Industrial atmosphere	1.2 (0.047)
	Heavy industrial atmosphere	2.6 (0.102)
	Coastal atmosphere	2.7 (0.106)

Table 61: Average corrosion rates for copper during the first test year in various types of atmosphere, calculated from the ISO-CORRAG program four continents [379]

– *Pollutant influences* –

The moistening duration, the sulfur dioxide immission and the temperature are decisive for the corrosion of copper and its alloys during the first three months. Outdoor exposure tests over a period of at least four years are necessary to obtain a satisfactory correlation between the corrosion rate of the copper and the atmospheric stress factors [380].

The corrosion rate of the copper materials depends chiefly on the thickness of the moisture film on the surface. With increasing thickness the corrosion rate increases. Independent of the sulfur dioxide pollution, the critical relative humidity for clearly evident increase of the corrosion lies at 87 % [356].

Of all copper materials, pure copper is the most sensitive material with respect to sulfur-containing gases in the atmosphere. Thereby hydrogen sulfide and carbon oxysulfide (COS) act more corrosive than sulfur dioxide or carbon disulfide (CS_2) by four orders of magnitude. Ozone and ultraviolet radiation have no effect on the corrosion. But they accelerate the sulfiding reaction, with ozone in the presence of hydrogen sulfide by a factor of three and by a factor of two with ultraviolet radiation [381]. The sulfiding is reduced by implantation of boron in the surface [382].

Sulfur dioxide is the dominant factor for corrosion attack in outdoor climate [356]. In the absence of water sulfur dioxide has no significant effect. In the presence of moisture the mass loss increases with increasing relative humidity and increasing SO$_2$ concentration in the atmosphere, as shown in Figure 36 [356].

For 70 % relative humidity and 298 K (25 °C) a non-linear functional relationship exists between the corrosion rate of the copper and the sulfur dioxide concentration [383].

However, after eight years of outdoor exposure the mass loss of the copper depends linearly on the sulfur dioxide immission rate [371].

In urban atmosphere green patina is formed as a result of the sulfur dioxide influence faster than in rural atmosphere. In rural atmosphere the corrosion rate is less than 1 µm/a (0.039 mpy). In urban atmosphere it is 1 to 3 µm/a (0.039 to 0.118 mpy). However, the correlation between the mass loss and the sulfur dioxide immission rate for copper is not as good as for steel or zinc.

Prediction of the course of the copper corrosion in the course of time is of great importance for electrical engineering. From laboratory tests of relatively short duration (about 30 days) in the condensation water continuous exposure test with 0.1 % sulfur dioxide addition at 303 to 313 K (30 to 40 °C), the sulfur dioxide immission rates found in practice and the known temperature/humidity relationship, the mass loss to be expected in practice can be predicted with the help of a nomogram [364].

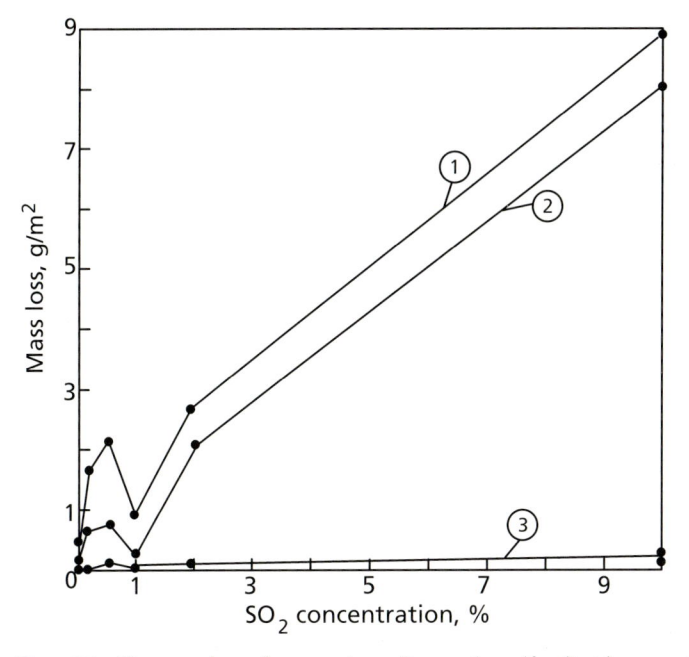

Figure 36: The mass loss of copper depending on the sulfur dioxide concentration in the atmosphere at high relative humidity values [356]
① 99 % relative humidity
② 75 % relative humidity
③ 50 to 63 % relative humidity

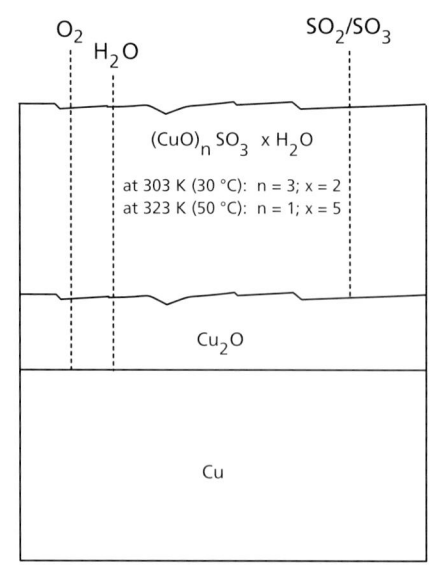

```
O₂              SO₂/SO₃
  H₂O

    (CuO)ₙ SO₃  x H₂O
    at 303 K (30 °C): n = 3; x = 2
    at 323 K (50 °C): n = 1; x = 5


            Cu₂O


            Cu
```

Figure 37: Schematic depiction of the structure of the corrosion products layer on copper after contact with air (303 K and 323 K (30 to 50 °C)) and 98 % relative humidity with addition of 50 ppm sulfur dioxide [384]

The controlling step for the corrosion of copper is the sulfur dioxide take-up on the outside of the oxide layer. Hydrated copper sulfate is here formed on the oxide film. At 303 K (30 °C) with 98 % relative humidity and 50 ppm sulfur dioxide addition the layer of the corrosion products consists of 97 % Antlerite and 3 % Cuprite. The schematic structure of the corrosion products layer is shown in Figure 37 [384].

Increasing fraction of chlorine or hydrogen chloride gas in the atmosphere significantly increases the corrosion rate of copper [383].

According to Spanish investigations, the copper corrosion primarily depends on the chloride content of the atmosphere. An empirical relationship between the corrosion and the chloride immission rate was determined from tests at 69 exposure places [385].

A different quantitative relationship follows from the results for the mass loss of copper in the course of eight years in the coastal atmospheres of south Sweden with various chloride immission rates, so that it appears to be impossible to find a reasonable global equation [386].

Also according to tests in the coastal atmospheres of Murmansk, Wladiwostok and Batumi, the corrosion rate increases approximately linearly with the chloride immission rate [387]. The top layers formed in rural or coastal atmosphere (Sweniorod, Batumi) give poorer protection than those formed in industrial atmosphere (Moscow) [388].

On the basis of the evaluation of results from 33 Spanish exposure stations, it was found that with simultaneous action of sulfur dioxide and chloride, the chloride content must be taken into account additionally in the calculation of the corrosion rate [362, 389].

On exposure to a gas mixture having the composition (in µg/m³) 810 SO$_2$, 940 NO$_2$, 334 O$_3$, 8.6 Cl$_2$ and 21 H$_2$S the increase of the corrosion rate at 298 K (25 °C) with 70 % relative humidity is not so pronounced. The concentration of SO$_2$ or H$_2$S here determines the corrosion rate [383].

Copper is corroded at a constant rate in air containing nitrogen dioxide at 298 K (25 °C) with 95 % relative humidity. The mass increase produced by growing top layers increases with increasing NO$_2$ concentration, as shown by Figure 38 [390].

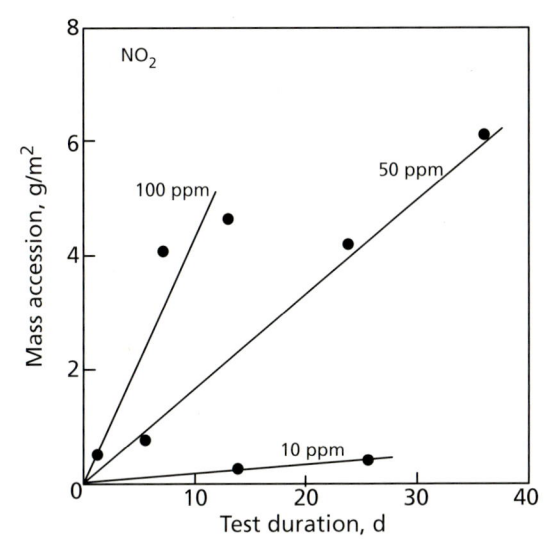

Figure 38: Mass increase of copper in air with contents of 10, 50 and 100 ppm nitrogen dioxide at 298 K (25 °C) and 95 % relative humidity, depending on the duration of the test [390]

With 95 % relative humidity in 30 day laboratory tests, nitrogen dioxide produces significantly greater mass losses than sulfur dioxide, as is shown in Figure 39 [391].

Systematic addition of nitrogen dioxide and sulfur dioxide at 295 K (22 °C) with 90 % relative humidity gave after 1,000 h in laboratory test the mass losses on Cu 99.9 as reported in Table 62 [392].

Pollutant concentration, ppm		Mass loss
NO$_2$	SO$_2$	g/m²
	1.3	0.9
0.2	1.3	5.6
3.0	1.3	20.5
3.0	0.2	10.2
3.0		0.9

Table 62: Mass loss of copper Cu 99.9 in the presence of nitrogen dioxide and sulfur dioxide at 295 K (22 °C) with 90 % relative humidity after 1,000 h test duration [392]

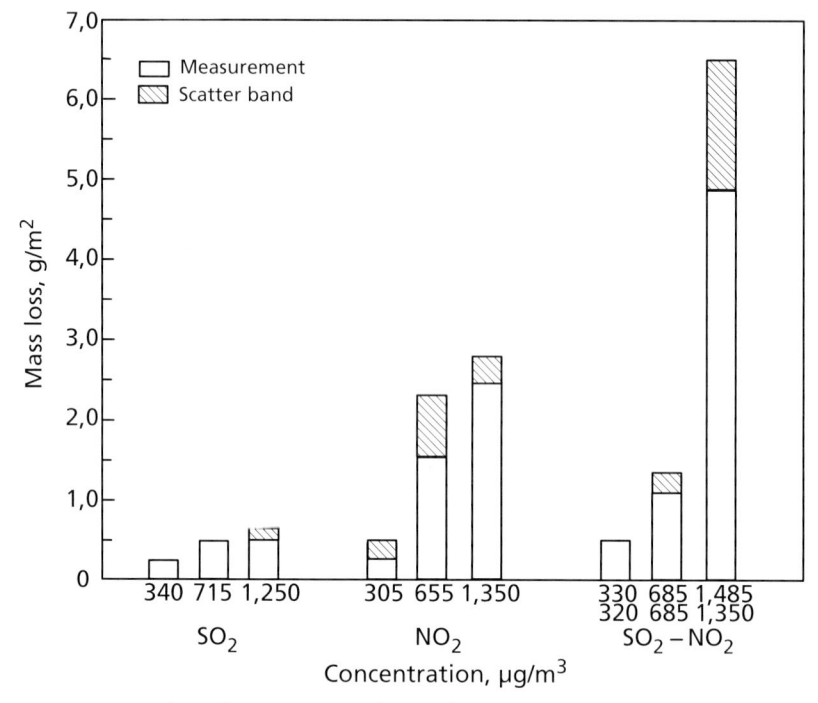

Figure 39: Mass loss of copper in atmosphere polluted with sulfur dioxide and nitrogen dioxide at 295 K (22 °C) with 95 % relative humidity [391]

The combined action of nitrogen dioxide and sulfur dioxide leads to a synergetic effect and increases the corrosion rate to a greater extent than the action of the individual pollutants [393].

The acceleration of the corrosion of the copper is traceable to the significant increase of the immission rate of sulfur dioxide to the doubled rate, and to the tripled rate towards the end of the test as a result of the oxidizing effect of the nitrogen dioxide. With 70 % relative humidity NO_2 and SO_2 produce no significant increase of the corrosion rate compared with pure atmosphere. The mass loss is only about 1/3 of the value obtained with 90 % relative humidity [394].

Complex pollution of the atmosphere with several pollutants significantly increases the corrosion rate compared with the effect of the individual pollutants. Hydrogen sulfide in the presence of atmospheric oxygen leads to the formation of copper sulfide. With 70 % relative humidity at 298 K (25 °C) the corrosion rate increases almost linearly with the concentration of the hydrogen sulfide in the atmosphere. The action of pollutant mixtures reduces the effect of the hydrogen sulfide concentration, but it increases the overall corrosion rate. Ammonia produces a slight inhibition of copper corrosion, whereas ozone progressively increases the corrosion rate with increasing concentration [383]. In the presence of isobutane in the atmosphere the corrosion of copper is inhibited slightly, but with simultaneous presence of sulfur dioxide it is accelerated [395].

Carbon oxysulfide acts similarly corrosive as hydrogen sulfide, but it is present in the atmosphere in higher concentrations. In comparison with hydrogen sulfide and sulfur dioxide it is the chief cause of copper corrosion. It does not react with copper in dry atmosphere. But in moist air at 295 K (22 °C) copper sulfide is produced very rapidly with almost linear corrosion rate [396].

Rainwater flowing down from a bitumen roof onto copper produces copper corrosion with a corrosion rate of > 0.1 mm in seven years.

The cause is that gradual oxidation of the bitumen produces acidic decomposition products. The best surface protection for bitumen roof tracks is a gravel covering with a height of at least 5 cm [397].

Precipitates of volcanic ash accelerate the corrosion attack of copper with increasing deposit thickness. The effect of the ash deposits is shown in Figure 40 [398].

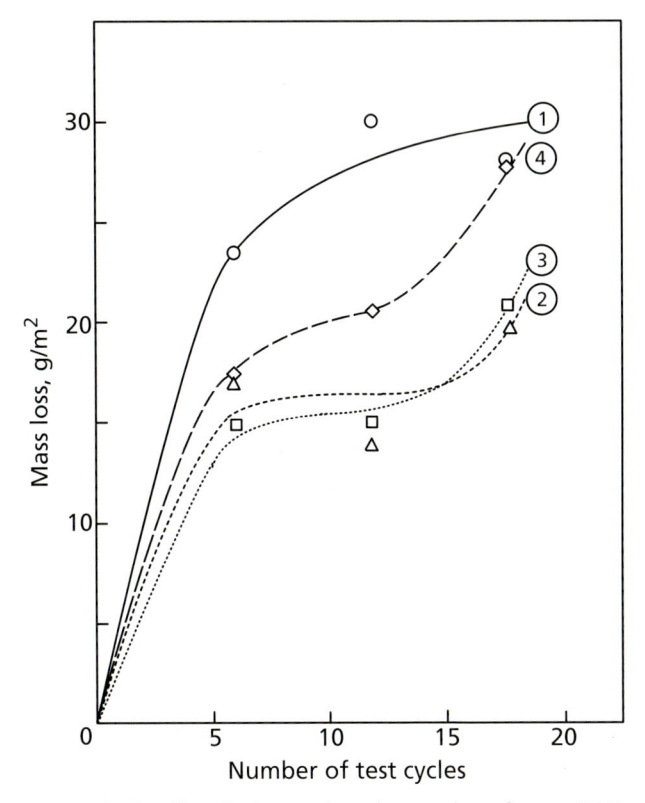

Figure 40: The effect of volcanic ash on the mass loss of copper [398]
① ash free
② ash deposit of 11 g/m^2
③ ash deposit of 22 g/m^2
④ ash deposit of 66 g/m^2

– *Behavior in exterior room climate* –

Small concentrations of hydrogen sulfide and/or sulfur dioxide are corrosion accelerating factors in exterior as well as in interior room climate [356]. High relative humidity also increases the copper corrosion [352].

Experience varies regarding the corrosion rate in exterior room climate without direct rain precipitation and solar radiation. According to the results from 24 test locations along the Spanish west coast in Galicia, the ratio of the corrosion rate of copper in outdoor and exterior room climate is 3.5:1 [362]. According to Russian investigations the corrosion rate is less than in outdoor climate by a factor of 5 to 7 [399].

According to 8 year tests in exterior room climate in Spain, a constant slight decrease of the corrosion rate was found [400]. In the exterior room climate of eight to ten test locations in Madrid the comparison of the thickness losses of copper sheets in exterior room climate and in outdoor climate gave the values reported in Table 63, which prove a significantly smaller corrosion rate in exterior room climate [368].

Climate	Thickness loss, µm		
	After 1 year	After 2 years	After 4 years
Exterior room climate		1.23 ± 0.34	1.07 ± 0.32
Outdoor climate	1.20 ± 0.23	1.72 ± 0.36	2.25 ± 0.43

Table 63: Thickness losses of copper sheet in exterior room climate and outdoor climate at 8 to 10 test locations in Madrid [368]

The corrosion in exterior room climate is affected by the seasonal cycle of the relative humidity and pollutant fraction in the outdoor climate. In the tropical humid climate of Mandapam (India) at a distance of 400 m from the sea coastline the mass loss of copper in exterior room climate was 50 % of that in outdoor climate on the annual average. The separately determined mass losses in the individual quarters of the year were found to increase in the order 1st quarter < 4th quarter < 2nd quarter << 3rd quarter, and their total was 50 % the mass loss of specimens exposed for one year; this clearly indicates the corrosion intensifying effect of a pollutant accumulation [356].

In outdoor climate large insoluble particles with sizes in the range from 1 to 20 µm are washed off the surface. Washing-off is prevented in exterior room climate. The dissolution of the corrosion products takes place in the dry phase of the surface as a result of soot concentration of the electrolyte liquid [361].

Copper materials used in electronics with and without electrolytically or chemically deposited metal coatings of gold, palladium, silver, tin or lead showed an unexpected high corrosion attack in the exterior room climate of Duisburg (Germany, aggressive industrial atmosphere), Ulm (Germany, medium industrial atmosphere) and Cuxhaven (Germany, coastal atmosphere). The corrosion course is affected by the pollutant type and concentration, the duration of moistening and the seasonal time position of the test start. The corrosion attack in coastal atmosphere corresponds to that in industrial atmosphere, but exceeds it because of the high chloride activity for otherwise largely passive materials such as nickel, gold, tin and lead [401].

Single component pollutant gas climates suffice for simulating mild industrial atmosphere. For other exterior room climates addition of several pollutants is more appropriate, in particular hydrogen chloride for coastal atmosphere and maritime climate. Pure hydrochloric acid test climates are too unilateral and aggressive. Multicomponent test climates show synergetic effects [401].

– Behavior in indoor room climate –

The corrosion attack in indoor room climate is significantly less than in outdoor and exterior room climate. Under storage conditions in indoor rooms without air conditioning the mass loss of copper is reported in g/m^2 to be 0.1 to 1.0 after 6 months, 0.2 to 1.5 after 1 year, 0.5 to 3.0 after 2 years and 1 to 7 after 3 years [402]. Traces of hydrogen sulfide already produce tarnishing of the copper [356].
The corrosion rates reported in Table 65 were found in various, partly air conditioned, indoor rooms in North America [383].

The corrosion rate of the copper is affected by the content of sulfur dioxide, hydrogen sulfide, chlorine, hydrochloric acid and ozone in the atmosphere. The corrosion rate of copper is an exponential function of the relative humidity (%) in indoor climate, as tests at 298 K (25 °C) in air with a pollutant content of ($\mu g/m^3$) 810 sulfur dioxide, 940 nitrogen dioxide, 334 ozone, 8.6 chlorine and 21 hydrogen sulfide have shown [383].

The corrosion rate in indoor rooms correlates best with the immission rate of gaseous chlorides on the copper surface, as corrosion tests in indoor rooms with and without air conditioning of the urban atmospheres at Bandung, Singapore, Bangkok and Tokyo have shown. The corrosion attack is manifest as tarnish and pitting [403].

In the Czech Republic the mass loss rates on copper Cu 99.5 reported in Table 64 were found. The specimens were exposed to the exterior room climate, to the outdoor climate and to the indoor room conditions of various factory plants [404].

Climatic condition	Exposure place	Mass loss rate $g/m^2 d$
Exterior room	Aussig (Ustí)	0.036
	Prague-Letnany	0.019
Outdoors	Aussig (Ustí)	0.39
	Prague-Letnany	0.035
Indoor room	Storage hall	0.010
	Sodium chloride production	0.84
	Ammonia synthesis	0.053
	Pickling station	0.004 up to 0.139
	Wire drawing works	0.013
	Rolling mill	0.101
	Dedusting	0.002 up to 0.015

Table 64: Mass loss rates of Cu 99.5 under conditions of exterior room climate, outdoor atmosphere and inside industrial plants in the Czech Republic [404]

Under the aspect of utilization in electrical engineering, some materials were investigated at 295 K (22 °C) with 90 % relative humidity under the influence of pollutant gases in short term corrosion tests. In the presence of 10 ppm sulfur dioxide the addition of 1 ppm hydrogen chloride did not accelerate the corrosion of copper. Additions of ammonia or nitric acid to the model atmosphere containing sulfur dioxide suppressed the corrosion reaction [405].

Indoor room climate leads to tarnish of the surface conditioned by the relative humidity and pollutant gases, changing to corrosion only above 60 to 80 % relative humidity. Parallel short term corrosion tests with 75 % relative humidity at room temperature with addition of 100 times the average atmospheric pollution found in medium industrial atmosphere produced acceleration factors of 40 to 140 for aggressive industrial atmosphere, 80 to 230 for medium industrial atmosphere, 30 to 120 for coastal atmosphere and 100 to 200 for indoor room climate. Single component pollutant gas test climates with 10 ppm sulfur dioxide or 1 ppm hydrogen sulfide are suitable for simulating indoor room climate [401].

Type of atmosphere	Test place	Air conditioning	Corrosion rate µm/a (mpy)
Urban atmosphere	Los Angeles	yes	0.01364 (5.32×10^{-4})
	Chicago	yes	0.01501 (5.85×10^{-4})
	New York	yes	0.01991 (7.77×10^{-4})
Industrial atmosphere	Northern Indiana	yes	0.04179 (1.63×10^{-3})
	South Carolina	yes	0.09231 (3.6×10^{-3})
	New Jersey	no	0.02864 (1.12×10^{-3})
	New Jersey	no	0.06150 (2.4×10^{-3})
Industrial atmosphere, rural	Houston	yes	0.00539 (2.1×10^{-4})

Table 65: Corrosion rates of copper in North American indoor room climates [383]

No corrosion products impairing the insulating effect formed on copper conductors mutually insulated with Al_2O_3 ceramic platelets in the moist room climate of a climate cupboard at 298 ± 1 K (25 ± 1 °C) with 75 ± 5 % relative humidity in the course of a test with a duration of 21 days. However, the addition of 1 ± 0.1 cm³/m³ hydrogen sulfide led to an insulation resistance below the failure limit in 75 % of all cases. No failures have occurred after 29 days test duration when 10 ± 0.3 cm³/m³ sulfur dioxide were added [406].

– *Protection measures against corrosion* –

Copper can be improved with regard to its corrosion resistance or decorative appearance by providing various layers and coatings on it. The natural formation of patina has already been pointed out. It can be accelerated by contact with wine vinegar. Numerous inorganic reagents produce colored layers on copper in the temperature range from 373 to 398 K (100 to 125 °C), whereby the color can be controlled by increasing the pressure of oxygen and reducing the pH-value. Sodium chlorate appears to be the most suitable reagent [351].

The synthetic patina layers produced by adding copper sulfate consist of chiefly Bronchantit ($CuSO_4 \times 3\ Cu(OH)_2$) and provide permanent protection provided that the pH-value is not so low that the basic copper sulfate layers are dissolved, as is the case with acidic rain having a pH-value of less than 3. For colorless organic coatings transparent lacquers on the basis of high molecular weight acrylic resins with the addition of 1.5% benztriazole as corrosion inhibitor are used. Oxidizing solvents such as ketones and aldehydes are unsuitable for this purpose. They lead to peroxide production and to discoloration of the metal surface. The colorless coating substances can be of air or stove drying type and they can also be applied continuously to copper strip before forming [351].

A coating of Incralac® with 25 µm layer thickness protects copper, gold, brass and nickel silver for 3 to 8 years in outdoor climate and much longer in indoor room climate, as 20 years of experience have shown. The roof of the Olympia hall in Mexico City was protected therewith [407].

Stove drying alkyde-amine resin transparent lacquers have also proved to be suitable for coating, whereas the initially well adhering polyurethane lacquers failed after 18 to 24 hours due to embrittlement and cross-linking. Some aliphatic polyurethanes appear to be satisfactory over long periods. The coating material Incracoat® used for strip coating has this basis, with the addition of 4% benztriazole as inhibitor, and was used to protect the "golden" cupola of the Mosque in London. The final coating was made with Tedlar® which is based on polyurethane fluoride. Tedlar® on the basis of fluorocarbon laminates has also been used successfully for protecting copper [351, 407].

Volatile corrosion inhibitors (VCI) have proved to be successful for temporary protection in packagings in the packing industry. These inhibitors provide good corrosion protection only in a largely airtight cover in which an atmosphere saturated with the inhibitor can form. Benztriazole is a suitable volatile corrosion inhibitor. After taking the copper out of the VCI packing, this inhibitor is desorbed from the surface so slowly that the corrosion protection in contact with the atmosphere persists for several days. The protective effect of Dichan (dicyclohexyl ammonium nitrite) and VCI 25® is already strongly reduced a few hours after taking the copper out of the packing material [408].

Packed copper materials can be dependably protected against corrosion, even in the subtropical climate of Batumi (Georgia), by using kraft paper coated with a layer of 4 to 5 g/m^2 benztriazole derivatives. Of seven investigated derivatives, 5-methylbenztriazole and 5-chlorobenztriazole have proved to be most successful in the outdoor exposure test and for storage in the climate cupboard at 313 K (40 °C) with 96 % relative humidity, and in a condensation water test with the addition of 0.1 % of sulfur dioxide at room temperature [409].

Benztriazole and Toluyltriazole are suitable as inhibitors also on oxygen-free copper when they are applied in 0.033 molar alcoholic or aqueous solution. They considerably reduce the corrosion attack with high relative humidity up to temperatures of 343 K (70 °C). They improve solder flow but at the same time reduce the soldering speed. The protective effect is smaller in atmospheres containing sulfur. Specimens treated with benztriazole retain their gloss longer than specimens treated with toluyltriazole [410].

Copper-aluminium alloys

Copper alloys are in general more resistant to atmospheric corrosion than copper. The addition of alloying elements such as aluminum, nickel, zinc, manganese, etc. promotes the formation of more stable top layers on the surface [411].
Copper-aluminium alloys have high strength, excellent corrosion resistance and show good oxidation resistance also at elevated temperatures [412]. CuAlFeNi alloys also show quite good corrosion resistance in contact with the atmosphere [350].

A golden color CuAl alloy with particularly good color stability and high corrosion resistance is obtained by additional alloying of 0.001-1 % Ti, 0.001-1 % Co and 0.001-0.1 % B into the initial alloy (Cu, 5-9 % Al, 0.5-4 % Ni, 0.5-5 % Fe, 0.1-3 % Mn) [413].

Material (Alloy addition, %)	Altoona	New York	Sandy Hook	La Jolla	Key West	State College
	Corrosion rate, m/a (mpy)					
Cu 99.9 (0.03 O_2)	1.52 (0.059)	1.27 (0.05)	0.78 (0.031)	1.27 (0.05)	0.78 (0.031)	0.49 (0.019)
Cu 99.9 (0.02 P)	1.27 (0.05)	1.27 (0.05)	0.78 (0.031)	1.52 (0.059)	0.49 (0.019)	0.49 (0.019)
Cu 95.8 (3.1 Si, 1.1 Mn)	1.76 (0.069)	1.76 (0.069)	1.27 (0.05)	1.52 (0.059)	0.98 (0.038)	0.49 (0.019)
CuAl8 (2.0920)	1.64 (0.064)	1.64 (0.064)	1.19 (0.047)	0.25 (0.0098)	0.12 (0.0047)	0.25 (0.0098)
CuNi20Zn5	2.46 0.096)(	1.72 (0.068)	0.98 (0.038)	0.49 (0.019)	0.25 (0.0098)	0.49 (0.019)
CuNi29Sn1	2.54 (0.1)	1.97 (0.078)	1.23 (0.048)	0.49 (0.019)	0.25 (0.0098)	0.40 (0.016)
CuSn8, CW453K (0.1 P)	2.30 (0.091)	2.54 (0.1)	1.27 (0.05)	2.26 (0.089)	0.78 (0.031)	0.49 (0.019)
CuZn15, CW502L	1.97 (0.078)	1.97 (0.078)	1.23 (0.048)	0.49 (0.019)	0.49 (0.019)	0.49 (0.019)
CuZn30, CW505L	2.91 (0.115))	2.17 (0.085)	0.98 (0.038)	0.25 (0.0098)	0.25 (0.0098)	0.49 (0.019)
CuZn29Sn	1.93 (0.076)	2.42 (0.095)	1.23 (0.048)	0.49 (0.019)	0.49 (0.019)	0.49 (0.019)
CuZn (20-24 Zn, 2.5-5 Mn, 3-7 Al, 2-4 Fe)	10.66 (0.419)	8.20 (0.323)	4.1 (0.161)	1.89 (0.074)	1.89 (0.074)	0.49 (0.019)

Table 66: Corrosion rate of copper materials in 20 year outdoor exposure tests in various types of North American atmospheres [356]

Exposure tests for 20 years showed the corrosion rates reported in Table 66 for copper materials in rural atmosphere (State College, PA), industrial atmosphere (New York), industrial atmosphere (Altoona) and coastal atmosphere (Sandy Hook, La Jolla, CA; Key West, FL) [356].

As is evident from Table 67, the alloy CuAl7Si2 (CW302G) showed no tendency of pitting corrosion after seven years of exposure to the coastal atmosphere of Point Reyes (CA). It also shows the lowest corrosion rate of all investigated alloys [414].

In another test series inter-crystalline corrosion up to a depth of 50 μm was found for the alloy CuAl7Si2 in rural atmosphere (State College, PA), in industrial atmosphere (Newark) and in the coastal atmosphere of Kure Beach after 20 years of exposure, but not in the coastal atmosphere of Point Reyes (CA) [353].

The alloy CuAl5 is very sensitive to stress corrosion cracking in the presence of moist ammonia or under the influence of mercury compounds, and also to the action of water vapor at high stress levels [415].

In coastal atmosphere the alloy CuAl8 (2.0920), in addition to admiralty brass (CW706R, CuZn28Sn1) and the copper-nickel alloy CuNi20Zn5, has the best corrosion resistance [356].

The aluminum-nickel cast bronze CuAl9Ni4.5Fe4Mn1 has a tendency of selective de-alumination in an environment containing chloride. Heat treatment for six hours at 973 K (700 °C) inhibits this effect and also reduces the loss of fatigue strength in contact with the atmosphere by 10 %. But this has no effect on the crack growth [416].

Material (composition)	Corrosion rate μm/a (mpy)	Average pitting depth*, μm	
		Upper side	Underside
CuAl7Si2	0.25 (0.0098)	0	0
CuNi6	0.51 (0.02)	0	0
CuNi9	0.33 (0.0129)	7.6	5.1
CuNi23	0.43 (0.0169)	7.6	5.1
CuNi4Mn13	1.19 (0.0468)	7.6	10.7
CuSn5	0.76 (0.0299)	0	0
CuZn10Sn2	0.51 (0.02)	0	0
CuBe2Co0.25	0.53 (0.0208)	5.1	5.1
CuBe1.75Co0.25	0.53 (0.0208)	7.6	7.6
CuBe0.5CO2.5	0.89 (0.035)	5.1	5.1

* Measured at the 4 deepest pitholes

Table 67: Corrosion rates of copper materials after seven years of exposure to the coastal atmosphere at Point Reyes (CA) and its effect on pitting [414]

Material (composition)	Mass loss, g/m²						Strength loss, %	
	Miraflores			Cristobal			Miraflores	Cristobal
	After years						After 16 years	
	1	4	16	1	4	16		
CuAl4	8.9	20.5	54	17.8	33.8	88	<1	2
CuNi19Zn17	4.5	16.0	63	8.9	27.6	84	0	<1
CuNi30	4.5	17.8	61	7.1	26.7	89	1	1
CuNi67Fe2	<2.7	8.0	32	8.9	16.0	50	1	2
CuSn4.5P0.4	13.4	36.5	81	45.4	89	214	3	3
CuSn3.2	11.6	33.8	98	29.4	65.9	142		
CuZn5.6Sn4.7	11.6	29.4	88	38.3	98	178		
CuZn4Sn6Pb1.7	13.4	29.4	79	38.3	80	151		
CuZn4Sn8	9.8	29.4	98	40.9	98	186		
CuZn10, CW501L	11.6	32.0	72	26.7	49.8	107	<1	3
CuZn20, CW503L	8.9	24.9	66	17.8	35.5	84	<1	3
CuZn30, CW505L	4.5	17.8	56	11.6	24.9	75	1	4
CuZn39Pb0.8	8.9	22.3	72	16.0	32.0	98	4	8
CuZn39Sn0.8	7.1	20.5	63	13.4	29.4	88	1	7
CuZn40Sn0.65	36.5	45.4	116	40.9	67.6	134	4	8
CuSi2.9Zn0.5	36.5	65.9	134	70.3	151	427	2	3

Table 68: Mass and strength loss of copper alloys in the tropical climate of the Panama Canal zone in the course of 16 years of exposure [372]

The mass and strength loss values of copper alloys found in the tropical climate of the Panama Canal zone, in the rural atmosphere of Miraflores and in the coastal atmosphere of Cristobal are shown in Table 68. The maximum pitting depths incurred during 16 years of exposure were less than 125 μm [372].

For quick assessment of the behavior of copper alloys in industrial urban atmosphere the accelerated atmospheric corrosion (AAC) test was developed, in which test body specimens are alternately sprayed under an angle of 30° for 4 minutes with 0.05 % sodium bisulfite solution having a pH-value of 4.75 and then left to dry for 11 minutes exposed to the air. 11 days of subjection to this test in the so-called rain box correspond to three years of exposure to the atmospheric stress in the industrial urban atmosphere of New York. Five copper alloys, thereof three CuAl alloys, were tested by this method. The alloy CuAl8Ni2 showed the best corrosion resistance, compared with CuAl7Fe2, CuZn30, CuSi3 and Cu 99.99 [417].

Copper-nickel alloys

Zinc free copper-nickel alloys have good strength and high corrosion resistance. They are also called copper-nickel. The alloys containing zinc are called nickel silver on account of their bright silver gloss, and they too are resistant to corrosion [412]. The high corrosion resistance of the CuNi alloys is due to the stable top layer which forms [351].

Under the aspect of practical utilization, the CuNi alloys rank among the copper alloys which, with regard to their corrosion behavior, can be utilized under most conditions encountered in rural, industrial and coastal atmospheres [411].

Whereas alloys containing nickel generally resist chloride attack better than attack by industrial atmosphere with corresponding sulfur dioxide content, this is not necessarily so for CuNi alloys. The corrosion rate in industrial atmosphere is rather high compared with other copper alloys, as is evident from Figure 41, above all for the alloys CuNi10Zn25 and CuNi18Zn15 which contain zinc, as well as for the alloy CuNi43Mn2 containing manganese [418].

The corrosion resistance of the CuNi alloys correlates quite well with the nickel content. The corrosion rate becomes smaller with increasing nickel fraction in the alloy, as shown by Figure 42 [418].

The corrosion rates found in outdoor exposure tests for up to 20 years in typical atmospheres, as reported in Table 66, show medium values for CuNi alloys containing tin and zinc, whereby the values found in industrial atmosphere are significantly greater than the values found in the coastal atmosphere. The effect of nickel on the corrosion rate is greatest in the coastal atmosphere. In rural atmosphere the corrosion behavior of CuNi alloys does not differ from that of other copper materials [418].

The mean corrosion rates calculated from exposure for 20 years at State College (PA) for the CuNi alloys CuNi2, CuNi6, CuNi9, CuNi23, CuNi43Mn2 and CuNi4Mn13 lie between 0.53 (0.0208 mpy) (CuNi43Mn2) and 0.83 μm/a (0.0326 mpy) (CuNi2) with an average value of 0.70 μm/a (0.0275 mpy). The corrosion rate increases with the test duration [353].

In urban and medium to heavy industrial atmosphere the mass loss rates have the same order of magnitude of about 1.5 μm/a (0.059 mpy) after 20 years of exposure for all alloys and are thus about twice as high as in other types of atmosphere. The corrosion rate is nearly constant throughout the test duration [353].

In coastal atmosphere the scatter of the corrosion behavior is significantly greater. At Daytona Beach (FL, USA) the mass loss after four years of exposure was less than 5 mg/m^2 for the alloys CuNi, CuNiZn and CuZn, between 5 and 10 g/m^2 for CuZnSn, CuZn10 (Ms90) and CuSnFe, and more than 10 g/m^2 for Cu 99.95 and CuSn8Pb [419].

Of the alloys investigated for up to 20 years in coastal atmosphere, CuAl8, admiralty brass (CW706R, CuZn28Sn1) and the CuNi alloy CuNi20Zn5 have the best corrosion resistance [356]. At Kure Beach (NC) on the American east coast the mean corrosion rate after 20 years of exposure was 0.70 μm/a (0.0276 mpy), and at Point

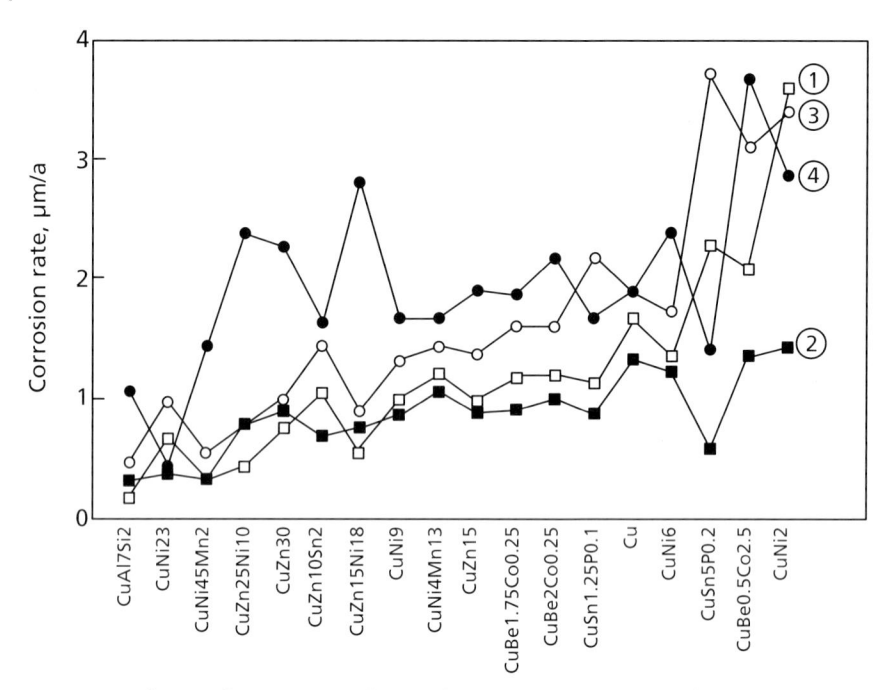

Figure 41: Influence of various types of atmosphere on the corrosion rate of copper and copper alloys [418]
① Rural atmosphere (State College)
② Coastal atmosphere (Point Reyes)
③ Coastal atmosphere, 25 m distant from the coastline (Kure Beach)
④ Urban industrial atmosphere (New York)

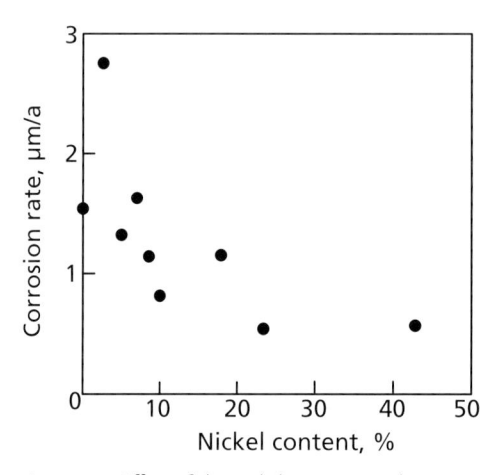

Figure 42: Effect of the nickel content on the corrosion rate of CuNi alloys after exposure for 2 years to the atmospheric conditions specified in Figure 41 [418]

Reyes (CA) on the American west coast it was 0.92 µm/a (0.036 mpy) after 15 years of exposure, but with a considerably greater scatter range from 0.53 (0.0208 mpy) (CuNi43Mn2) to 1.4 µm/a (0.055 mpy) (CuNi9) [353]. After 7 years of exposure at Point Reyes the corrosion rate of the CuNi alloys does not differ appreciably from that of other copper alloys, as is shown in Table 67. Only the alloy CuNi4Mn13 shows a higher corrosion rate [414]. The corrosion rate becomes smaller with increasing test duration [353].

In South Finland no corrosion, but only slight surface discoloration, was found after four years of exposure on copper, brass CuZn20 (CW503L) and CuZn37 (CW508L), tin bronze CuSnPb0.2 and the alloy CuNi25 (CW350H) [366].

The resistance of copper with respect to hydrogen sulfide is reduced by the addition of nickel. CuNi4Sn4 hardly reacts at all with hydrogen sulfide. The alloy CuNi9 merely forms 10 nm thick sulfide layers [381].

The copper coin alloy CuNi25 remains silver-colored in contact with the atmosphere and free from corrosion. More suitable alloys for daily use as coin alloys which do not take-on an unpleasant appearance through tarnishing seem to be the alloys CuNi5Al5Fe1.2Mn0.8 and CuSn6Al2, according to simulation tests [420].

The effect of corrosion attack on the elongation behavior is clearly evident. The greatest reduction of the percentage elongation was found in the industrial atmosphere of New Haven and in the industrial coastal atmosphere of Brooklyn, whereas the mechanical properties changed hardly at all at Daytona Beach 46 m distant from the coastline. The elongation loss is a very sensitive indicator for local corrosion attack [421].

Under the influence of high relative humidity CuNi alloys containing iron show corrosion fatigue. The fatigue strength loss appearing with relative humidity values in the range from 90 to 95 % for the alloy CuNi30Fe0.6 is attributed to the precipitation of iron along the grain boundaries. Iron promotes grain boundary corrosion also for the alloy CuNi10Fe1.5 and leads to a reduction of the fatigue strength in the presence of high relative humidity. Static stress has no effect on the mechanical properties [422, 423].

CuNi alloys containing zinc can show a tendency for plug-type zinc depletion like CuZn alloys. The investigated alloys CuNi18Zn20 and CuNi12Zn24 were found to be resistant with respect to zinc depletion in coastal atmosphere, but showed slight attack in urban and rural atmosphere [424]. The alloy CuNi18Zn27 was damaged in New Haven, Brooklyn and Daytona Beach, like the alloy CuZn30, by plug-type zinc depletion. The zinc depletion plugs showed a silvery color instead of the usual copper color of copper alloys. Pitting corrosion also appeared after four years of exposure at New Haven and Brooklyn, leading to pitholes with a depth of up to 8 µm [421].

CuNi alloys behave differently than copper with regard to the color change of the surface. After four years of exposure in Sweden to rural, urban and coastal atmosphere they showed a stronger discoloration, but which still showed only slight differences with respect to that of copper after seven years of exposure [424].

Copper-tin alloys (bronzes)

CuSn alloys, similar to copper, show good resistance with respect to components of the atmosphere. The resistance increases with increasing tin fraction. Firmly adhering thick top layers are formed [350]. Phosphors bronzes, too, are characterized by high corrosion resistance [412]. Their behavior depends chiefly on the formation of amorphous tin dioxide in the top layer [424]. The significant damage caused by atmospheric pollutants are the result of washout of the tin as soluble stannic acid [425].

The exposure orientation of the specimens affects the course of the corrosion. In comparison with the generally customary orientation of the specimens to the south, specimens of copper bronze CuSn4.5P0.4 were exposed in the tropical coastal atmosphere of Cristobal (Panama) oriented north because of the sea coast located to the north. It was nevertheless found that the corrosion was greater in the land direction (south) [402].

For general technical utilization the copper-tin alloys are assessed to be suitable under most atmospheric conditions [411].

The phosphor bronzes CuSn1, CuSn3, CuSn5 and CuSn9 each containing 0.03 % of phosphorus, and the phosphor bronzes CuSn7 and CuSn9 each containing 0.3 % of phosphorus, confirmed their very good corrosion resistance in exposure tests in Sweden. In urban and coastal atmosphere, but not in rural atmosphere, they discolor rapidly and form a green patina [424].

The patina is somewhat darker in industrial and coastal atmosphere, and more gray in rural atmosphere. It consists predominantly of mixtures of basic copper sulfate and copper nitrate, in coastal atmosphere more of basic sulfate and chloride [426].

The corrosion rates of the investigated alloys are reported in Table 69 together with the corrosion-relevant data for the test locations [424].

Indications of patina formation on CuSn5 were found in comparable tests with various copper alloys in the USA after four years of exposure to the industrial coastal atmosphere of Brooklyn. The patina formation was less pronounced in the coastal atmosphere of Daytona Beach. In contrast, no top layer had yet formed in the industrial atmosphere of New Haven. Consequently the mass loss is also greatest there, and CuSn5 shows the greatest corrosion rates of all examined copper materials. The corrosion rate becomes less with increasing duration of the exposure [421].

These statements are also confirmed by tests in the tropical climate of Mandapam Camp (India) at a distance of 30 m from the coastline on the coast of the Gulf of Bengal. The corrosion rate of the alloy CuSn5P0.4 examined there in wire form is very high. The cumulative corrosion rate decreases exponentially with the test duration. A constant mass loss rate of about 0.23 $g/m^2 d$ is reached after 250 days. The tin and phosphorus fractions in the corrosion product remain constant irrespective of the mass loss rate [427]. After two years the mass loss rate is 54 $g/m^2 a$ [428].

Material composition	Corrosion rate, μm/a (mpy)					
	After two years of exposure			After seven years of exposure		
	1)	2)	3)	1)	2)	3)
Cu 99.0	1.2 (0.047)	1.4 (0.055)	1.3 (0.051)	0.6 (0.024)	1.0 (0.039)	0.8 (0.031)
CuAg0.06	1.2 (0.047)	1.4 (0.055)	1.3 (0.051)	0.6 (0.024)	1.1 (0.043)	0.8 (0.031)
CuP0.4As0.4	0.8 (0.031)	1.2 (0.047)	1.2 (0.047)	0.5 (0.047)	1.0 (0.039)	0.8 (0.031)
CuNi12Zn24Mn	0.4 (0.0157)	1.7 (0.067)	0.7 (0.0275)	0.3 (0.0118)	1.4 (0.055)	0.6 (0.024)
CuNi20Zn20Mn	0.4 (0.0157)	1.7 (0.067)	0.7 (0.0275)	0.3 (0.0118)	1.4 (0.055)	0.6 (0.024)
CuSi3Mn1	0.6 (0.024)	1.4 (0.055)	1.6 (0.063)	0.4 (0.0157)	1.1 (0.043)	0.9 (0.0354)
CuSn1P0.4	0.8 (0.031)	1.4 (0.055)	1.2 (0.047)	0.5 (0.0197)	1.0 (0.039)	0.8 (0.031)
CuSn5P0.03	0.5 (0.0197)	1.2 (0.047)	2.0 (0.0787)	0.4 (0.0157)	1.0 (0.039)	1.1 (0.043)
CuSn6P0.03	0.4 (0.0157)	1.0 (0.039)	1.9 (0.075)	0.3 (0.0118)	1.0 (0.039)	1.0 (0.039)
CuSn7P0.3	0.3 (0.0118)	0.9 (0.0354)	1.6 (0.0629)	0.2 (0.0078)	1.0 (0.039)	1.0 (0.039)
CuSn9P0.3	0.2 (0.0078)	0.7 (0.0275)	1.4 (0.055)	0.2 (0.0078)	0.9 (0.0354)	0.9 (0.0354)
CuSn4Zn4Pb4	0.7 (0.0275)	1.7 (0.067)	1.9 (0.075)	0.4 (0.0157)	1.2 (0.047)	1.0 (0.039)
CuZn8	0.8 (0.031)	1.3 (0.051)	1.1 (0.043)	0.5 (0.0197)	1.1 (0.043)	0.7 (0.0275)
CuZn15	0.6 (0.024)	1.4 (0.055)	1.1 (0.043)	0.4 (0.0157)	1.2 (0.047)	0.8 (0.031)
CuZn28	0.5 (0.0197)	1.5 (0.059)	0.9 (0.0354)	0.4 (0.0157)	1.3 (0.051)	0.7 (0.0275)
CuZn37	0.5 (0.0197)	1.5 (0.059)	0.9 (0.0354)	0.4 (0.0157)	1.7 (0.067)	0.7 (0.0275)

Table 69: Corrosion rates of copper materials determined in Sweden after two year and seven year outdoor exposure in Erken (rural atmosphere (1)), Stockholm (urban atmosphere (2)) and in Bohus Malmön (coastal atmosphere (3)) [424]

Table 69: Continued

Material composition	Corrosion rate, μm/a (mpy)					
	After two years of exposure			After seven years of exposure		
	1)	2)	3)	1)	2)	3)
CuZn40	0.4 (0.0157)	1.9 (0.075)	0.8 (0.031)	0.3 (0.0118)	1.8 (0.071)	0.7 (0.0275)
CuZn50	0.4 (0.0157)	2.2 (0.0866)	0.8 (0.031)	0.3 (0.0118)	2.2 (0.0866)	0.7 (0.0275)
CuZn36Pb	0.4 (0.0157)	1.6 (0.0629)	0.6 (0.024)	0.4 (0.0157)	1.4 (0.055)	0.7 (0.0275)
CuZn28Sn1	0.4 (0.0157)	1.8 (0.0709)	1.0 (0.039)	0.4 (0.0157)	1.4 (0.055)	0.7 (0.0275)
CuZn22Al2	0.3 (0.0118)	1.0 (0.039)	0.7 (0.0275)	0.2 (0.0078)	0.9 (0.0354)	0.6 (0.024)

1) Rural atmosphere: Pollutant content in the air: $\mu g/m^3$ = 8 sulfur, 2 chloride
Pollutant content in the rainfall: mg/m^2 = 50 sulfur, 20 chloride
pH = 4.9

2) Urban atmosphere: Pollutant content in the air: $\mu g/m^3$ = 70 sulfur, 4 chloride
Pollutant content in the rainfall: mg/m^2 = 180 sulfur, 60 chloride
pH = 4.7

3) Coastal atmosphere: Pollutant content in the air: $\mu g/m^3$ = 10 sulfur, 50 chloride
Pollutant content in the rainfall: mg/m^2 = 100 sulfur, 800 chloride
pH = 5.1

Table 69: Corrosion rates of copper materials determined in Sweden after two year and seven year outdoor exposure in Erken (rural atmosphere), Stockholm (urban atmosphere) and in Bohus Malmön (coastal atmosphere) [424]

In the tropical coastal atmosphere of Cristobal (Panama) with simultaneous action of the sea surf breakers, the behavior of the phosphor bronze CuSn4.5P0.4 in the long term corrosion course is poorer than that of the silicon bronze CuSi2.9Zn0.5 (see Section "Other copper alloys") [372].

The alloy CuSn8, too, belongs to the copper materials with above-average high corrosion rates according to the test results for various types of atmosphere in the USA reported in Table 66 [356].

2000 year old Chinese bronze mirrors and weapons from the Chou and Han period (400 BC to 200 AD) still have good corrosion resistance. Naturally formed dark surface layers protect the material CuSn25Pb5. Samples of this ancient material were exposed at 294 K (21 °C) with 95 % relative humidity to the action of 3.6 ppm hydrogen sulfide and 2.5 ppm carbon oxysulfide (COS). The unprotected bronze, like modern bronze, was found to be sensitive with respect to sulfiding. The sulfide layer increases double logarithmically with the hydrogen sulfide content of the atmosphere. The sulfiding behavior of the antique bronze is similar to that of the present day alloy CuNi4Sn.

On coating a 20 to 25 µm thick layer of known permeability onto stove-dried epoxy resin, extremely low diffusion values were obtained with the ancient bronze, so the conclusion is that the modern protection methods are suitable for protecting ancient bronze [429].

In comparison with other copper materials, the mechanical properties of the tin bronze CuSn5 (cf. CW451K) are not affected much, in spite of the high mass losses [421].

For protection of bronze during storage and transportation, packing with kraft paper coated with a layer of 4 to 5 g/m^2 of 5-chlorobenztriazole as inhibitor is suitable. After 3.5 years of outdoor exposure and storage in the exterior room climate of the subtropical coastal atmosphere of Batumi (Georgia), some bronze materials, like copper and brass, remained without corrosion, whereas one bronze showed slight corrosion attack [409].

Copper-tin-zinc alloys (red brass)

The substitution of tin by zinc in tin bronzes improves the casting properties. Thereby the zinc content can hardly exceed 5 % because of the resulting unattractive color and the impairment of the corrosion resistance [412].

The mainly surface area attack by the atmosphere amounts to about 1 mm in 400 years. Compared with the formerly used casting bronzes with high lead fraction, such as CuSn5Pb4, CuSn3Zn7Pb3 and CuSnPb10, the alloys used today G-CuSn5ZnPb, G-CuSn10Zn and G-CuSn10 have better weather resistance [425].

A war monument in Monza (Italy) made of tin bronze with percentage composition = 7.63 Sn, 4.77 Zn, 6.65 Pb, 2.44 Ni, 0.43 S and 0.002 Mn, after 60 years of exposure to a typical industrial atmosphere with chief pollutants NO_x and SO_2, shows hardly appreciable corrosion damage, but only a green top layer of corrosion products [430].

Also in Quebec (Canada) only patina was found on eleven examined bronze monuments, but of three different kinds: one black patina with a tin fraction of less than 1 %, one green patina with a tin content of 8 to 11 % and one green-yellow patina with 17 to 25 % of tin [431].

The results of an extensive international exposure program with copper Cu 99.0 and cast bronze RG 7 corresponding to CuSn7ZnPb with 6 to 8 % Sn, 3 to 5% Zn and 5 to 7 % Pb, after a test duration of four years, are summarized in [352]. According to the results the examined bronze is extremely sensitive to slightly increased sulfur dioxide concentration. The mass loss after exposure outdoors and in exterior rooms for one, two and four years is assigned in Figure 43 to five summarized sulfur dioxide concentrations [380].

In outdoor climate linear regression analysis shows no significant correlation to the environmental data. A relationship between the mass loss and the sulfur dioxide concentration in the atmosphere and the electrolytic conductivity of the rainfalls is discernible. In the exterior room climate of the shelter provided with Venetian blind type walls, the average mass loss was half as great as in outdoor climate. The effect of the relative humidity plays only a subordinate role for the bronze stored with protection. In contrast, the mass loss correlates with the sulfur dioxide concentration in the atmosphere and the electrolytic conductivity of the rainfall.

For example, the high mass loss of specimens exposed to the atmospheres of Venice (Italy), Vlaardingen (Netherlands) and some locations in Great Britain is attributed to the influence of chloride [352]. In contrast the chloride content of the atmosphere at Point Reyes (CA, USA) after seven years of outdoor exposure of the tin bronze CuSn2Zn10 produced a corrosion rate of 0.51 µm/a (0.02 mpy) [414].

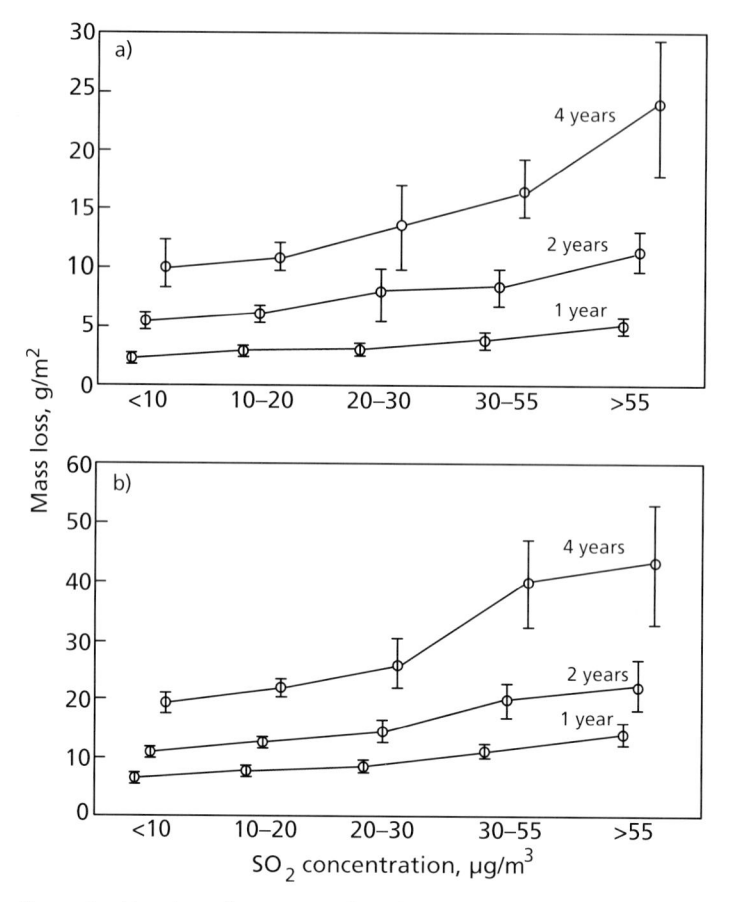

Figure 43: Mass loss of specimens of cast bronze RG 7 (corresponding to CuSn7ZnPb) depending on the sulfur dioxide concentration in the atmosphere [380]
a) in exterior room climate
b) in outdoor exposure

In tropical rural and coastal atmosphere the mass loss of the CuSnZn alloys is comparatively high in comparison with other copper materials and is exceeded only by phosphor bronze and silicon bronze, as shown in Table 68 [372].

The following variants are proposed for the preservation of bronze monuments: Brushing-off the existing patina with brass brushes, subsequent washing with cleaning agent containing detergents, if necessary removal of local chloride enrichments with the help of aqueous silver salt solutions, care or impregnation of the thus cleaned surface with carnauba wax, or sealing of the remaining patina with colorless acrylic resin lacquer such as Incralac®. In the case of inadequate protection with these proposed measures, it is also possible to produce an artificial patina coating with copper sulfate solution, followed by coating with Incralac® [431].

Copper-zinc alloys (brass)

CuZn alloys are the most frequently utilized materials based on copper. They have good corrosion resistance in contact with the atmosphere [412]. Their corrosion behavior is the more similar to that of copper, the greater their copper content is made. α-Brass with a copper content of more than 64% is roughly equivalent to copper [350].

The major effects of the atmosphere on the corrosion of the CuZn alloys are patina formation, uniform surface area attack and zinc depletion. In the course of corrosion readily soluble corrosion products are produced on the brass and are washed out by rain. This effect increases with increasing zinc content of the alloy [424, 426]. Brass types with high zinc content develop no attractive green patina. According to investigations in Sweden, green patina has still not appeared even after 16 years of outdoor exposure in rural atmosphere (Erken). In urban atmosphere (Stockholm) the patina is brown-green, in coastal atmosphere (Bohus Malmön) it is more gray and in rural atmosphere it is black. The corrosion attack of the brass generally increases in the order rural atmosphere < coastal atmosphere < urban atmosphere [427].

The influence of melting ice produces a blueish discoloration within a few days on brass CuZn20 (cf. CW503L), which slowly changes to black. Also in south Finland corrosion tests with CuZn20 and CuZn37 (cf. CW508L) produced only a surface discoloration after four years of exposure, but no corrosion attack [366].

In the strongly industrially influenced coastal atmosphere of Brooklyn (NY, USA) patina starts to appear on CuZn15 (cf. CW502L) already after four years of exposure. After the same test duration the patina formation is less pronounced at Daytona Beach (FL, USA), and at New Haven it is not discernible at all [421].

The average corrosion rate for CuZn alloys is

- in rural atmosphere 0.2 up to 0.6 μm/a
  (0.0078 up to 0.024 mpy) [424, 426]
- in urban atmosphere 0.9 up to 1.3 μm/a
  (0.035 up to 0.051 mpy) [426]

- in urban atmosphere, heavily polluted   up to 2.2 µm/a (0.0866 mpy) [424]
- in industrial atmosphere   1.2 up to 2.3 µm/a
  (0.047 up to 0.090 mpy) [353]
- in coastal atmosphere   0.5 up to 1.1 µm/a
  (0.0197 up to 0.043 mpy) [424, 426].

The maximum corrosion rates found were 1.4 µm/a (0.0551 mpy) at Point Reyes (CA) and up to 1.7 µm/a (0.0669 mpy) at Kure Beach (NC) [353].

The corrosion attack differs for the various brass types. The corrosion rates determined from exposure tests with a duration of many years for certain alloy types and test locations are summarized in Table 70 [356, 421, 424].

The Tables 69 and 70 compare the corrosion behavior of brass types with that of other copper materials. The corrosion rate decreases slightly in the course of the test in rural and urban atmosphere, and to a greater extent in coastal atmosphere [424].

Mass loss determinations alone are insufficient for assessing the corrosion behavior of CuZn alloys. The local corrosion attack is decisive and best recognizable by elongation measurements [419].

The technical utilization possibilities for CuZn alloys are good under most conditions in rural atmosphere and adequate in industrial and coastal atmosphere with slight corrosion performance requirements [411]. Brass types with large copper content show better behavior then alloys with low copper content in industrial and coastal atmosphere. In coastal atmosphere admiralty brass CuZn29Sn1 (cf. CW719R) has the best corrosion resistance [356].

The effect of the zinc content on the corrosion rate in the course of seven years is shown in Figure 44 [424].

With increasing zinc content the corrosion attack increases in urban atmosphere and decreases in rural and coastal atmosphere. In the case of low corrosion rates and high zinc content the influence of the zinc fraction in the alloy is no longer decisive. The β-brass types have the highest corrosion rates in urban atmosphere [424].

In the rural and coastal atmosphere of the tropical climate of Panama the corrosion rate of α-brass decreases with increasing zinc content in the alloy (Table 68) [372]. The corrosion rate decreases almost linearly in the course of the test, as corrosion tests with CuZn30 (cf. CW505L) in the coastal atmosphere of Mandapam Camp (India) have shown. The mass loss in outdoor climate after two years was found to be 18 $g/m^2 a$. In the exterior room climate it is about 70 % of the corrosion rate in outdoor climate. In the tropical climate of India corrosion attack is greater during the months of May through August than during the period December through January because of the greater chloride content and the higher temperature [428].

In atmospheres containing sulfur, copper-zinc alloys are corroded less by a factor of 50 to 100 than pure copper. Partial substitution of zinc by nickel additionally increases the corrosion resistance (see Section "Copper-aluminium alloys") [381].

Atmosphere type, exposure place	Alloys (composition)					
	CuZn15	CuZn30	CuZn12Sn1	CuZn29Sn1	CuZn32 FeAl	CuZn22 Fe3Al5Mn4
	Corrosion rate, μm/a (mpy)					
Rural atmosphere						
State College[1]	0.49 (0.0193)	0.49 (0.0193)		0.49 (0.0193)		0.49 (0.0193)
Erkens[2]	0.4 (0.0157)			0.4 (0.0157)		
Urban atmosphere						
Stockholm[2]	1.2 (0.0472)			1.4 (0.0551)		
Industrial atmosphere						
New Haven[3]	1.05 (0.0413)	1.28 (0.0504)	1.18 (0.0708)		1.05 (0.0413)	
Altoona[1]	1.97 (0.0775)	2.91 (0.114)		1.93 (0.0759)		10.7 (0.421)
New York[1]	1.97 (0.0775)	2.17 (0.0854)		2.41 (0.0948)		8.20 (0.322)
Coastal atmosphere						
Bohus Malmön[2]	0.3 (0.0118)			0.7 (0.0276)		
Key West[1]	0.49 (0.0193)	0.49 (0.0139)		0.49 (0.0193)		1.89 (0.0744)
La Jolla[1]	0.49 (0.0193)	0.25 (0.0098)		0.49 (0.0193)		1.89 (0.0744)
Sandy Hook[1]	1.23 (0.0484)	0.98 (0.0386)		1.23 (0.0484)		4.1 (0.161)
Daytona Beach[3]	0.65 (0.0256)	0.32 (0.0126)	0.99 (0.0389)		0.50 (0.0197)	
Industrial coastal atmosphere		2.12 (0.0835)	2.11 (0.0831)		2.11 (0.0831)	
Brooklyn[3]	1.82 (0.0717)	2.12 (0.0835)	2.11 (0.0831)		2.11 (0.0831)	

1) 20 years test period; 2) 7 years test period; 3) 4 years test period

Table 70: Average corrosion rates of CuZn alloys determined from outdoor exposure test over many years at certain test locations in Sweden and the USA [356, 421, 424]

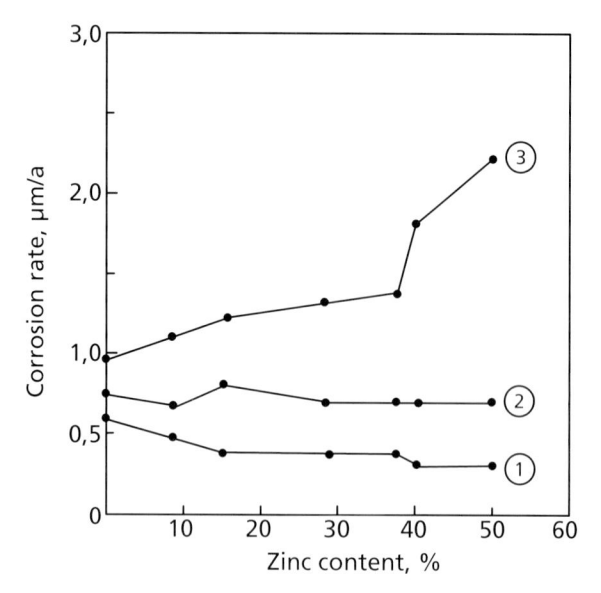

Figure 44: Average corrosion rate of CuZn alloys depending on the Zn content during seven years of exposure [424]
① Rural atmosphere
② Coastal atmosphere
③ Urban atmosphere

In the presence of ammonia the alloy CuZn33.5Pb0.5, which shows good corrosion resistance in all types of atmosphere, is attacked to a significant extent [432]. Brass with zinc contents greater than 15% is damaged chiefly by zinc depletion, and with less than 15% Zn it is damaged chiefly by surface area corrosion attack. Arsenic prevents zinc depletion, particularly in coastal atmosphere. The zinc depletion rate decreases with the duration of the test. The zinc salts produced by the zinc depletion are easily removed by rain [419, 424, 427].

The zinc depletion depth decreases in urban atmosphere with increasing copper content of the alloy, starting with β-brass with a copper content of 50%. Metallographic explorations show that the depth of attack is greatest in binary brass types, but essentially confined to the β-phase [424].

In rural and coastal atmosphere an increase of the copper content up to 92% has no decisive influence on the zinc depletion attack [424]. After 16 years of exposure the zinc depletion depth of binary brass is 90 to 215 μm [426].

Particularly the alloy CuZn30 (Ms70) is attacked by zinc depletion [353]. In coastal atmosphere (Daytona Beach, USA), industrial coastal atmosphere (Brooklyn, USA) and in industrial atmosphere (New Haven, USA) it shows classical plug-type zinc depletion after four years of exposure [421]. However, it is explicitly pointed out that after seven years of exposure at Kure Beach, Point Reyes, New York and State College (USA) no signs of zinc depletion were discernible even in thorough metallurgical examination [418].

Sheets have greater resistance to zinc depletion than pressed profiles. The latter show zinc depletion depths of up to 150 μm, on the average 50 μm in rural atmosphere, and 100 to 150 μm, with a maximum of 200 to 300 μm, in industrial and coastal atmosphere [426].

Zinc depletion strongly affects the mechanical properties of the alloys. Considerable loss of tensile strength and elongation was found for the copper-zinc alloys after four years of exposure [421]. The tensile strength loss after seven years of exposure of the copper alloys is generally less than 5 %, but for β-brass it is 17–32 %, and for some binary brass types, particularly the types alloyed with lead, tin or manganese, the tensile strength loss is up to 12 %. The general elongation loss of copper alloys of less than 10 %, which is independent of the form of the semifinished products, is exceeded by β-brass (CuZn50) with 24 to 41 % and some other brass types (CuZn40, CuZn33Mn, CuZn35SnPbAlMn) with 35 % maximum [424, 426].

Already mild corrosive environment [358], particularly in the presence of traces of ammonia or ammonium salts in the atmosphere [350], can initiate stress corrosion cracking of brass under static stress in the presence of water. Stress corrosion cracking can also be initiated by other compounds which appear in moist atmosphere or electrolyte solution. However, tensional stress, stimulators or stress corrosion cracking and copper ions in solution must be present [433].

Brass types with a zinc content greater than 15 to 20 % are highly sensitive to stress corrosion cracking, and with a zinc content greater than 20 % they are more resistant to stress corrosion cracking [358, 434]. The inherent tensional stresses which appear during cold forming suffice in many case for initiating stress corrosion cracking. Especially cold rolled brass for electrical plug connectors suffers inter-crystalline stress corrosion cracking after several years of outdoor exposure in rural atmosphere [434].

Industrial atmosphere acts more aggressive than coastal atmosphere. Whereas for CuZnPb the stress corrosion cracking shows predominantly inter-crystalline course, CuZnAl and CuZnNi have a tendency for trans-crystalline corrosion course. CuZnSn shows inter-crystalline and trans-crystalline stress corrosion cracking, but only in coastal atmosphere [419].

In exterior room climate stress corrosion cracking takes place in the case of moisture condensation and dust deposits. No damage was found in heated indoor rooms [434].

In short term tests in the rain box (AAC-test, see Section "Copper-aluminium alloys"), of five comparatively investigated copper materials, brass CuZn30 (Ms70) gave relatively poor results on visual inspection and by measuring the polarization resistance [417].

Brass specimens mutually insulated with Al_2O_3 ceramic platelets lost their insulation resistance in 25 % of the examined cases in the climate cupboard with 75 % relative humidity at 398 K (125 °C). The addition of 10 ± 0.3 cm^3/m^3 sulfur dioxide produced the same result [406].

As decorative metal for steel surfaces brass plays an important role, especially by galvanic depositing, in the jewelry and fittings industry.

Other copper alloys

Copper-beryllium alloys with a content of 0.4 to 2% of beryllium can be hardened thermally and show excellent corrosion resistance and at the same time little corrosion fatigue [412].

The corrosion resistance in contact with the atmosphere is better than with other copper materials. Two year outdoor exposure tests in rural atmosphere (State College, PA) showed a corrosion rate of 0.75 µm/a (0.0295 mpy), in the coastal atmospheres at Kure Beach (NC, USA) of 1.5 µm/a (0.059 mpy) and at Point Reyes (CA, USA) of 0.5 µm/a (0.0197 mpy) as well as in industrial atmosphere in Newark (NJ, USA) of 1 µm/a (0.039 mpy) for the alloy CuBe1.9Ni0.6 and 1.25 µm/a (0.0492 mpy) for the alloy CuBe0.6Co2.5 [435]. The corrosion rate decreases with the duration of the test.

In the presence of ammonia and high relative humidity beryllium copper has a tendency for stress corrosion cracking [435].

Silicon bronze CuSi2.9Zn0.5 forms a better protective top layer in coastal atmosphere with simultaneous influence of surf breakers than is the case with phosphor bronze CuSn4.5P0.4. The corrosion course in the tropical climate of Panama has been determined for both materials over a period of 16 years. The almost constant corrosion rate which is reached after eight years is about 13 g/m^2 a for silicon bronze and about 38 g/m^2 a for phosphor bronze [372].

Benzene and Benzene Homologues

Author: K. Hauffe / Editor: R. Bender

Copper

Copper is attacked even by small amounts of sulfur in benzene. The addition of 0.2% 9,10-anthraquinone can inhibit the attack. Copper is completely resistant to pure benzene and toluene [436].

Pure benzene does not affect the course of the decrease in the microhardness of copper with increasing pressure [437].

The electrochemical behavior of the galvanic couple steel (U-4 with 0.42–0.50% C)/copper in benzene is opposite to the behavior in water. According to Table 71, copper is more electronegative than the steel and functions as the anode.

If the benzene contains 1 g/l sulfur, the stationary potential of copper becomes more noble than that of steel, that is, steel is now the anode. An addition of triethanolamine in the absence or presence of sulfur includes copper to function as the anode [438]. In such a case, copper corrodes. At a surface area of 7.5 cm^2 per each metal of the pair, the corrosion current density of the copper at room temperature in the most unfavorable case (benzene + 1 g/l S + 4.48 g/l triethanolamine) is calculated to be 1.79×10^{-14} A/cm^2 which corresponds to a corrosion rate of about 0.2×10^{-7} mm/a (7.87×10^{-7} mpy).

For the reaction in which the hydrogen atoms of benzene were substituted by fluorine between 370 and 520 K (97 and 247 °C), copper and brass reactors were used, which were sufficiently resistant [439].

The observed change in the heat transfer coefficient in the boiling liquids under constant heat flow depends on the increasing ageing of the material. A structure-sensitive characteristic property is the microhardness. Thus, for example, the microhardness of copper, which was 93 before the boiling of toluene, dropped to 62 after boiling for 13 h at a heat flow of 135,000 W/m^2 [440].

To stabilize metallic powder, for example copper powder, in a dispersion and to protect it against corrosion, its surface was coated with oleic acid and napthenic acid by adsorption from a toluene solution. The copper powder thus treated showed detectable oxide amounts on the grain surface only after one year [441].

Conductor plates (copper and semi-conductors) which are exposed to large variations in temperature and to air humidity (for example in aeronautics and astronautics) must not have any impurities between the conductor plates. A solution with a very good cleaning efficiency on conductor plates which were soldered in the presence of activated rosin inter alia, the mixture of (percent by volume) 30 toluene + 5 butyl acetate + 37.5 isopropanol + 27.5 heptane has proved to be suitable. Polystyrene and Makrolon® as substrates must not be cleaned by this solution [442].

Copper-aluminium alloys

The copper-aluminium alloys Cu-(6 to 8)Al-3.5Fe and Cu-(6 to 8) Al-1.0Zn-(1.2 to 2.2)Si are completely resistant to pure benzene and toluene [436].

Copper-nickel alloys

The following copper-nickel alloys are completely resistant to pure benzene and toluene: Cu-30Ni, Cu-10Ni-1.4Fe and Cu-18Ni-17Zn [436].

Copper-tin alloys (bronzes)

The observed change in the heat transfer coefficient of bronze in boiling benzene under 0.1 MPa air at a heat flow of 120,000 W/m^2 could be correlated to the decrease in microhardness of the bronze, which was initially 220 and dropped to 120 after 13 h. This phenomenon is based on changes in the chemical composition of the surface of bronze in the presence of benzene due to diffusion [440].

The copper-tin alloys CuSn5 (CW451K) and CuSn8 (CW453K) are completely resistant to pure benzene and toluene [436].

Copper-zinc alloys (brass)

The microhardness of brass (LS-59-1; Russian type) dropped after 13 hours in boiling benzene under 0.1 MPa air at a heat flow of 250,000 W/m^2 from 220 at the beginning to 120 [440].

With a view to using condensed water from coking plants as feedwater for steam generators, the effect of water impurities, inter alia, by benzene fractions, on the corrosion of brass condenser tubes was investigated. With increasing pressure in the generator, the concentration of copper and zinc in the vapor increased. Thus, for example, at 0.4 MPa the copper and zinc salt concentrations were 220 and 250 µg/kg. To prevent this corrosion, purification of the condensed water is necessary [443].

The following copper-zinc alloys are resistant to pure benzene and toluene: CuZn10 (CW501L), CuZn15 (CW502L), CuZn30 (CW505L), CuZn35.5Pb3, CuZn40Pb3, CuZn20.5Al2, CuZn28Sn1 and CuZn39.25Sn0.75 [436].

Concentration g/l		Current	Internal resistance of the cell	Stationary potential V	
Sulfur	Triethanolamine	10^{-14} A	10^{12} Ω	Cu	Steel
–	–	+0.7	4.2	–0.13	–0.03
1.0	–	–4.0	4.1	+0.23	+0.07
–	4.48	+7.1	1.0	–0.30	–0.23
1.0	4.48	+13.4	0.7	–0.35	–0.26

Table 71: Electrochemical behavior of the galvanic couple steel/copper (7.5 cm^2) in benzene containing additions at room temperature, measured 3 h after the beginning of the test [438]

Other copper alloys

The copper-silicon alloys CuSi1.5 and CuSi3 are resistant to pure benzene and toluene [436].

Bromides

Author: K. John / Editor: R. Bender

Copper

Copper is resistant to sodium, potassium and calcium bromide solutions up to their boiling point, provided air and other oxidizing substances are excluded. In contrast, it is corroded by ammonium bromide at 323 K (50 °C), and by lead bromide at 373 K (100 °C) at the rate of more than 1.3 mm/a (51.1 mpy) [444, 445].

As has been established by galvanostatic investigations with measurements of photoelectric current, photoelectric potential and reflection, in 0.1 mol/l potassium bromide solution 1,2,3-benzotriazole acts as inhibitor of the anodic reaction in the corrosion of copper; the degree of inhibition attains 98 % [446]. The production of a p-type semiconducting surface coating is the cause of this.

In dilute (0.05 mol/l) sulfuric acid corrosion of copper monocrystals (Cu 99.999) by potassium bromide is accelerated or decelerated depending on concentration: with 10^{-8} mol/l potassium bromide, corrosion is decreased by a factor of 2 to 3, whereas with 10^{-5} mol/l potassium bromide the corrosion rate increases somewhat and with potassium bromide concentrations above 5×10^{-3} marked inhibition occurs again. The individual crystal faces exhibit different resistances: the resistance capability decreases in the sequence <110> <100> <111>; see Figure 45 on this point [447].

The corrosion of copper in 0.05 mol/l sulfuric acid at 293 K (20 °C) is accelerated by the addition of 0.1 mol/l sodium bromide, but corrosion is inhibited with higher bromide concentrations, up to 1 mol/l. The same is also true of corrosion in hydrochloric acid. The sequence of the decreasing effect is as follows: $Cl^- > Br^- > F^- > I^-$ [448].

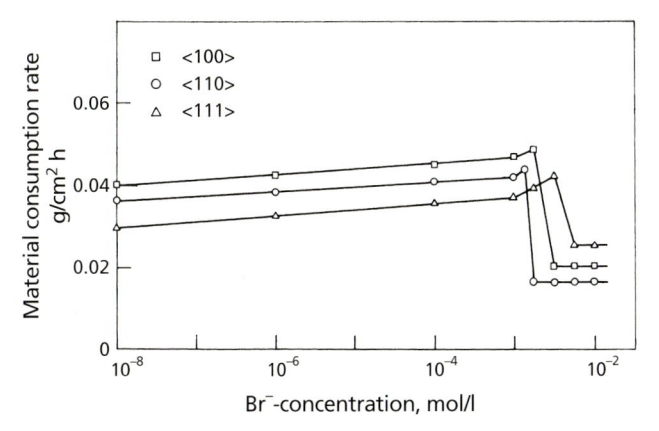

Figure 45: Influence of the concentration of bromide ions on the rate of dissolution of copper monocrystal faces in 0.1 mol/l sulfuric acid at 303 K (30 °C) [447]

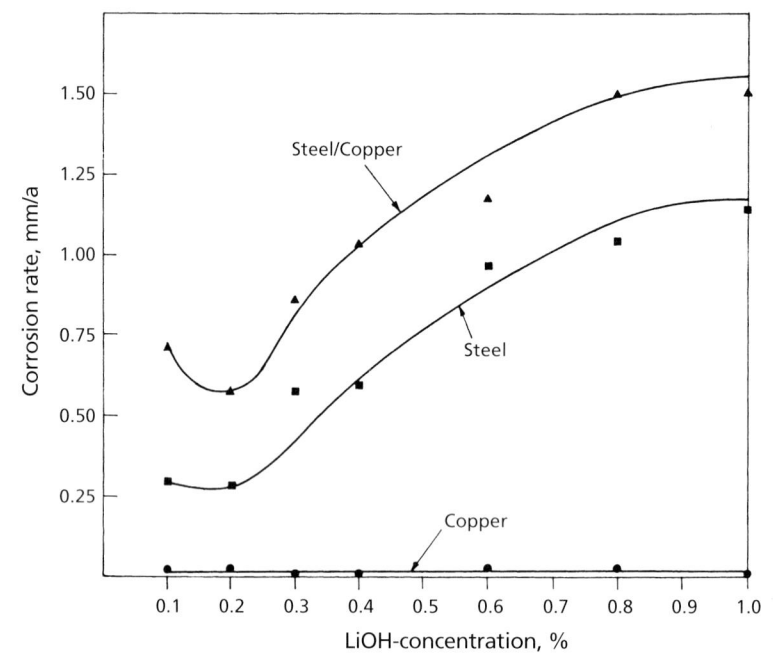

Figure 46: Influence of the lithium hydroxide concentration on the corrosion of metals partially immersed in boiling 65 % lithium bromide solution [449]

Corrosion of copper, steel and the metal couple copper/steel in boiling lithium bromide solution is shown in Figure 46.

The effect of various inhibitors for the metal couples copper/steel is summarized in Table 72 and corresponding results for copper and steel separately in Table 73.

Inhibitor	pH of solution	Inhibitor concentration, %	Corrosion rate mm/a (mpy)
Lithium hydroxide	11.3	0.2	0.75 (29.5)
Lithium hydroxide*	11.3	0.1	0.15 (5.91)
Lithium molybdate	11.3	0.1	0.50 (19.7)
Lithium molybdate*	11.3	0.1	0.15 (5.91)
Lithium molybdate	9.7	0.1	0.63 (24.8)
Lithium nitrite	11.3	0.3	2.78 (109)
Lithium nitrate	9.7	0.3	1.6 (63)
Lithium nitrate	11.3	0.3	1.37 (53.9)

* Tested with CuNi 90 10/steel contact combinations

Table 72: Inhibiting effect of various inorganic salts on the corrosion of the metal couple copper/steel in intermittently boiling 63 % lithium bromide solution [449]

Table 72: Continued

Inhibitor	pH of solution	Inhibitor concentration, %	Corrosion rate mm/a (mpy)
Lithium nitrate	9.7	0.3	1.45 (57.1)
Lithium phosphate	11.3	0.2	0.93 (36.6)
Lithium phosphate	10.0	0.2	0.42 (16.5)
Lithium silicate	10.7	0.3	0.8 (31.5)
Lithium silicate	9.5	0.3	0.45 (17.7)
Lithium borate	11.0	0.3	0.7 (27.6)
Lithium borate	7.3	0.3	1.4 (55.1)
Lithium perchlorate	11.3	0.3	0.68 (26.8)
Lithium perchlorate	9.3	0.3	1.1 (43.3)
Lithium tungstate	11.3	0.1	0.83 (32.7)
Lithium tungstate	8.9	0.1	0.8 (31.5)
Lithium vanadate	11.3	0.1	0.9 (35.4)
Lithium vanadate	9.1	0.1	0.65 (25.6)
Lithium chromate	11.3	0.3	0.05 (1.97)
Lithium chromate	9.6	0.3	0.0025 (0.098)
Lithium chromate*	9.6	0.3	0.0 (0)
Lithium arsenite	11.3	0.2	0.8 (31.5)
Lithium arsenite	9.2	0.2	0.38 (15)
Lithium arsenate	11.3	0.2	0.45 (17.7)
Lithium arsenate	8.0	0.2	0.58 (22.8)
Lithium thiocyanate	11.3	0.3	0.72 (28.3)
Lithium silicofluoride	11.3	0.3	0.58 (22.8))

* Tested with CuNi 90 10/steel contact combinations

Table 72: Inhibiting effect of various inorganic salts on the corrosion of the metal couple copper/steel in intermittently boiling 63 % lithium bromide solution [449]

According to this, of the inhibitors investigated, lithium chromate at pH 10 is the best inhibitor for the metal couple copper/steel [449]. Whereas the corrosion losses increase greatly in potassium chloride solution of 0.5 mol/l after about 4 days, the corresponding curve for potassium bromide solution gradually flattens. This difference is due to the fact that copper ions are produced in the case of potassium chloride solution whereas the sparingly soluble copper bromide is formed in potassium bromide solution [450].

LiBr %	LiOH %	Inhibitor %		pH of solution	Corrosion rate mm/a (mpy)	
					Cu	Steel
54	0.2	–		11.3	0.18 (7.09)	0.005 (0.197)
54	0.2	0.1	Li_2MoO_4	11.3	0.16 (6.3)	0.008 (0.32)
54	0.2	0.26	Li_2CrO_4	11.3	0.023 (0.91)	0.0 (0.0)
54	–	–		10.2	0.068 (2.67)	0.063 (2.48)
54	–	0.1	Li_2MoO_4	10.2	0.14 (5.51)	0.023 (0.091)
54	–	0.26	Li_2CrO_4	10.2	0.0025 (0.098)	0.0 (0.0)
63	0.2	0.1	Li_2MoO_4	11.3	0.0025 (0.098)	0.11 (4.33)
63	0.2	0.1	Li_2MoO_4	11.3	0.0 (0.0)	0.10 (3.94)
63	0.2	0.26	Li_2CrO_4	11.3	0.052 (2.05)	0.005 (0.197)
63	–	–		10.2	0.018 (0.71)	0.10 (3.94)
63	–	0.1	Li_2MoO_4	10.2	0.040 (1.57)	0.077 (3.04)
63	–	0.26	Li_2CrO_4	10.2	0.020 (0.79)	0.015 (0.59)

Table 73: Testing the galvanic corrosion of the metal couple copper/steel in lithium bromide solutions with inorganic inhibitors [449]

In laboratory tests copper was attacked only slightly by powdered sodium bromide at relative humidities of up to 50 %, but was severely attacked with humidities above 60 % [451].

As tests with powdered copper and the vapor of ammonium bromide solutions have shown, copper is also attacked in the gas phase by ammonium bromide [451].

According to photoelectric potential measurements, the surface coating of Cu(I) bromide formed on the copper monocrystals in 0.1 mol/l sodium bromide behaves like a p-type semiconductor: this effect does not occur with polycrystalline copper [452].

The following corrosion rates in Table 74 were measured for copper in potassium bromide melts [453].

Temperature, K (°C)	1,073 (800)	1,123 (850)	1,173 (900)
Corrosion rate, mm/a (mpy)	8.3 (327)	48.5 (1,909)	70.4 (2,772)

Table 74: Corrosion of copper in potassium bromide melts at various temperatures [453]

Copper-aluminium alloys
Copper-nickel alloys
Copper-tin alloys (bronzes)
Copper-zinc alloys (brass)
Other copper alloys

Copper alloys containing tin, aluminium, silicon and nickel are more resistant than copper to corrosion by sodium, potassium and calcium bromide solutions when air is excluded. In contrast to this, some of them are attacked even at room temperature by ammonium bromide, and by lead bromide at 373 K (100 °C) at the rate of more than 1.3 mm/a (51.2 mpy) [444, 445].

Zinc is dissolved less rapidly from brass in 0.5 mol/l sodium bromide solution than in a sodium chloride solution of identical concentration [454].

Stress corrosion cracking of cold-rolled brass (63/37) in an aqueous solution of copper sulfate (0.05 mol/l) and ammonium sulfate (1 mol/l) is inhibited by small quantities of bromide or chloride ions at pH 6.5. These ions, since they are adsorbed on the surface, shift the potential toward more noble values [455].

Figures 47 and 48 show the corrosion rate and the stationary potentials as functions of the composition of copper, CuMg- and MgCu-alloys and magnesium (0 to 99.9 % Cu or Mg).

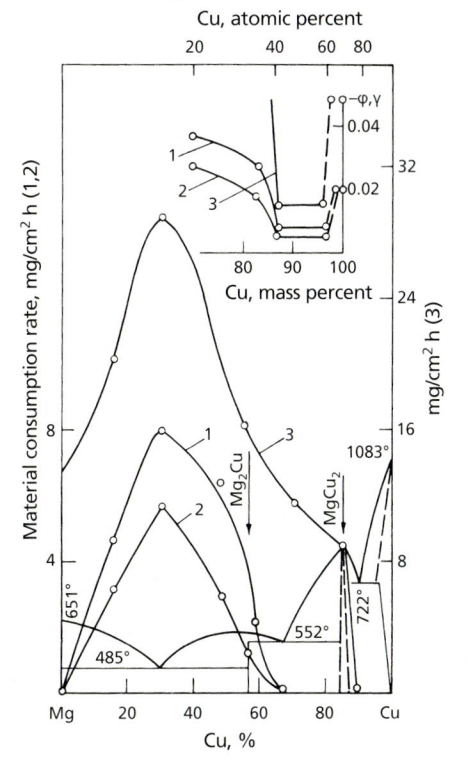

Figure 47: The material consumption rate (mg/cm² h) as a function of the composition of MgCu-alloys in 0.5 mol/l sodium bromide solution [456]
1) 0.5 mol/l NaCl, 2) 0.5 mol/l NaBr, 3) 0.1 mol/l HCl

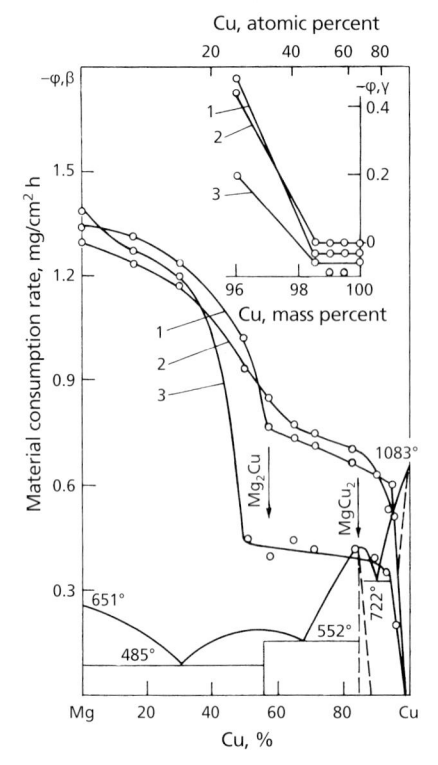

Figure 48: Stationary potentials as a function of the composition of MgCu-alloys in 0.5 mol/l sodium bromide solution [456]

According to these diagrams, corrosion initially increases with increasing copper content and reaches a maximum in the region of the eutectic composition of about 30 % Cu. The corrosion rate then decreases again, and it is possible that this is influenced by intermetallic compounds such as Mg_2Cu and $MgCu_2$ which occur in this region [456].

Bromine

Author: K. John / Editor: R. Bender

Copper

When air is excluded, copper is virtually resistant to dry bromine and is attacked only a little by moist air containing bromine [457]. The corrosion rates found in laboratory tests over 23 days at 294 K (21 °C) are shown in Figure 49 as a function of water content.

Bromination of copper foils in bromine vapor follows a parabolic curve between 373 K and 573 K (100 °C and 300 °C), whereas the plot is linear between 573 K and 773 K (300 °C and 500 °C), since the diffusion process then no longer determines the rate [458].

Corrosion of copper in bromine water and in non-aqueous bromine solutions is considerably reinforced by solar radiation. The effect of the light probably makes it easier for the light-sensitive copper(I) bromide to be formed and also initiates its conversion into copper oxide [459].

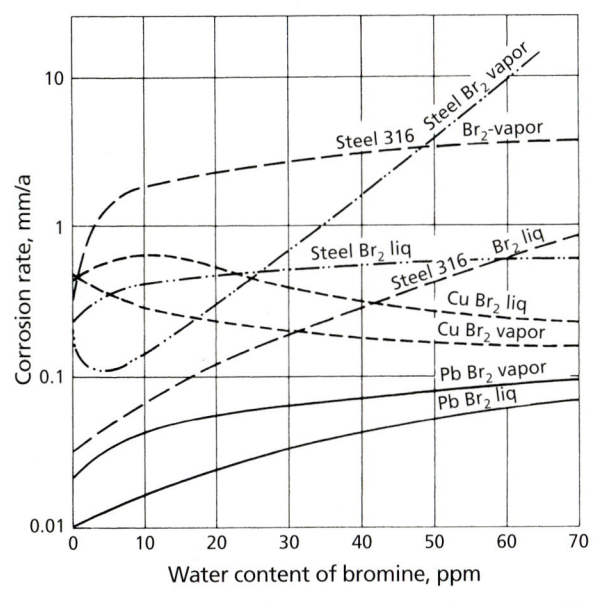

Figure 49: Corrosion rates of copper, chromium-nickel steel and lead in bromine at room temperature as a function of water content [460]

Copper-aluminium alloys
Copper-nickel alloys
Copper-tin-zinc alloys (red brass)
Copper-zinc alloys (brass)
Other copper alloys

Copper alloys containing not less than 85 % copper, but no zinc, are attacked only a little by moist air containing bromine [457]. CuNi-alloys CuNi30 (cf. CM390H) are resistant to bromine at room temperature [461].

Copper-beryllium alloys in dry bromine with air excluded lose less than 0.025 mm/a (0.98 mpy), and copper-silicon alloys, e.g. Everdur® 1000 containing 94.9 % copper, 4 % silicon and 1.1 % manganese also exhibit good resistance [457].

The results given in Table 75 were obtained from aluminium bronze, Muntz metal (60/40 brass) and Ounce metal (red brass) at room temperature in bromine liquid and vapor with varying moisture contents [460].

Medium	Corrosion rate, mm/a (mpy)		
	CuAl	Cu60Zn40	CuZnSn
Bromine, liquid + 7 ppm H_2O	0.0069 (0.27)	0.0089 (0.35)	0.0117 (0.46)
ditto + 57 ppm H_2O	0.0046 (0.18)	0.0203 (0.8)	0.0446 (1.76)
Bromine vapor + 7 ppm H_2O	0.073 (2.87)	0.0202 (0.8)	0.0013 (0.051)
ditto + 57 ppm H_2O	1.695 (66.73)	0.035 (1.38)	0.011 (0.43)

Table 75: Corrosion rates of copper alloys after 23 days at room temperature in bromine liquid and vapor with varying moisture contents [460]

The losses shown in Figure 50 were found for brass 63/37 in bromine water at 308 K (35 °C) after 7 and 15 days.

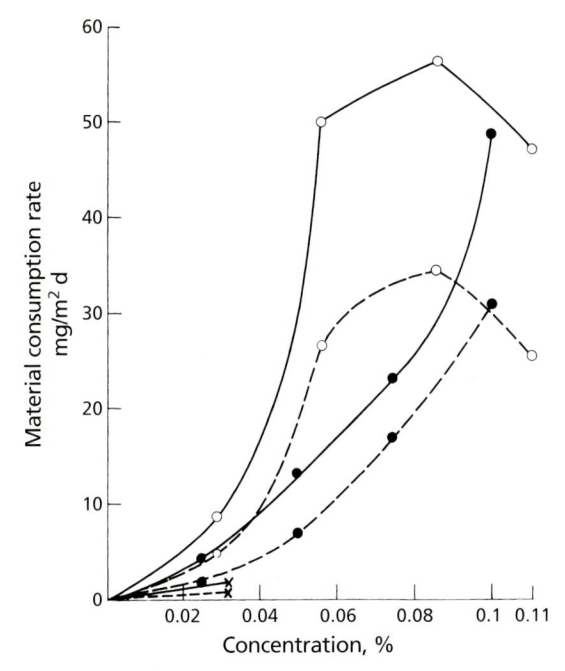

Figure 50: Corrosion loss of brass 63/37 in bromine water at 308 K (35 °C) as a function of the bromine concentration [462]

Carbonic Acid

Author: P. Drodten / Editor: R. Bender

Copper

Copper and copper base alloys are used in many applications due to their good corrosion behavior. Apart from their good corrosion resistance, their further application field in outdoor weather exposure, in fresh water pipelines, fittings, condensers, heat exchangers, in chemical plant construction and many other uses, is based also on their good processability, their good strength properties as well as their high thermal and electrical conductivity.

There is no data available about the behavior of copper towards carbon dioxide gas at higher temperatures, since this material is typically not used under such conditions. At ambient temperatures the use of copper pipes for transporting carbon dioxide, e.g. in medical technology, welding technology and in the beverage industry is permitted [463].

But copper is a material frequently used for pipes and other components exposed to water or aqueous solutions. However, in these cases carbon dioxide dissolved in the fluid may influence the corrosion behavior.

In contact with water the corrosion possibility of copper depends on the type of the cover layers formed by the corrosion products, which consist of Cu_2O or CuO depending on the type of fluid and the corrosion potential [464]. The more this layer prevents an ion and electron exchange between copper and water, the better is its protective effect and the higher is the resistance of copper. The carbon dioxide dissolved in water, the present alkaline earth ions and the carbonate/hydrogen carbonate system play an important role in the formation of the cover layer.

The influence of carbon dioxide in various waters on the corrosion of copper was investigated in connection with damage to drinking water pipes of copper [465]. Specimens from electrolytic copper and hard-drawn copper pipes and the following test solutions were used for the tests:

1. Distilled water with 5 mg/l CO_2 and 10 mg/l SO_4^{2-} added by dissolving the adequate amounts of sodium hydrogen carbonate and calcium sulphate. In several tests 5 mg/l iron in the form of iron(III) chloride were also added to the solution.
2. Drinking water from the Oslo waterworks and water from the supply system purified by means of an ultra-membrane filter. The membrane filter reduced the content of organic carbon compounds to 2.5 mg/l C and the iron content to < 0.02 mg/l Fe.
3. A test solution free from organic substances and manganese, produced by dissolving inorganic salts in distilled water with the same concentrations as in test solution 1.

The tests were performed in a circulation apparatus under thermostatically controlled conditions with a flow rate of 50 ml/h. One, two or three small inflow tubes were provided to pass different, automatically controlled amounts of CO_2 gas into the solution. The contents of free and bound carbon dioxide were calculated using equilibrium equations. Figure 51 depicts the variation of the carbon dioxide content over time and the mean values.

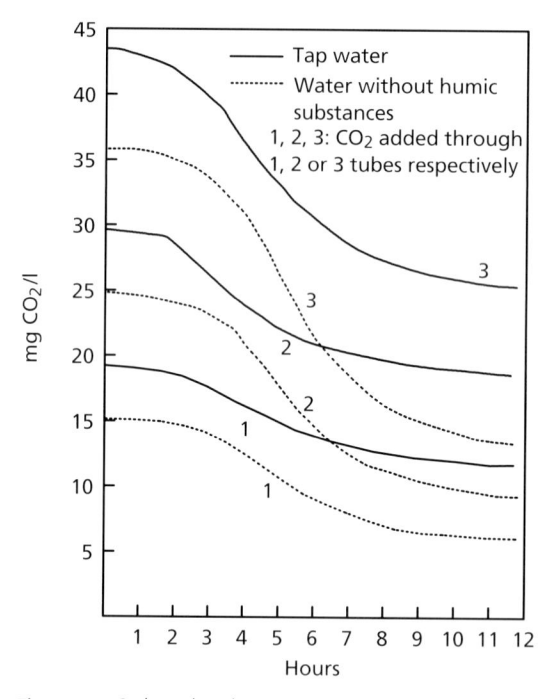

Figure 51: Carbon dioxide contents in test solutions as a function of time and mean contents [465]
Mean values of the CO_2 content (mg/l) for tap water: 1 = 14,6; 2 = 22,8; 3 = 32,9. For water without organic substances: 1 = 10,0; 2 = 16,3; 3 = 23,9

The tests were carried out at temperatures of 303 to 343 K (30 °C to 70 °C). Figure 52 shows the corrosion rates obtained after a test period of three months from tests in the test solution 2 with and without organic substances as a function of the test temperature. In both solutions the addition of carbon dioxide lead to a clear increase in the corrosion rates. Also the presence of organic substances exerted a negative influence on the corrosion of the copper specimens. The lower three curves of the figure might be indicative of a critical temperature in the range of 323 K (50 °C).

Figure 53 depicts the test results in both test solutions after the addition of different amounts of carbon dioxide through one, two or three of the small inflow tubes. A rising carbon dioxide content leads to a clear increase in the corrosion rates of the copper specimens in both solutions.

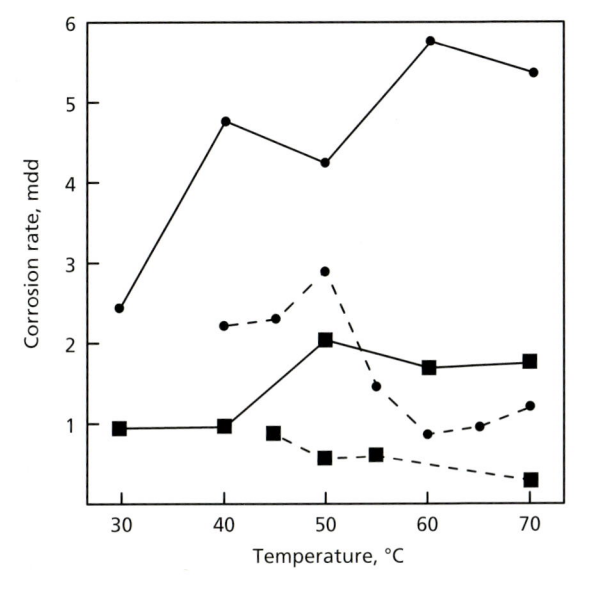

(mdd = mg/dm² d   1 mdd x 0.004 = 1 g/m² h)

Figure 52: Corrosion rates of copper after three months as determined in test solution 2 with and without organic substances as a function of the test temperature [465]
(mdd = mg per decimeter² and day   1 mdd × 0.004 = 1 g/m² h)

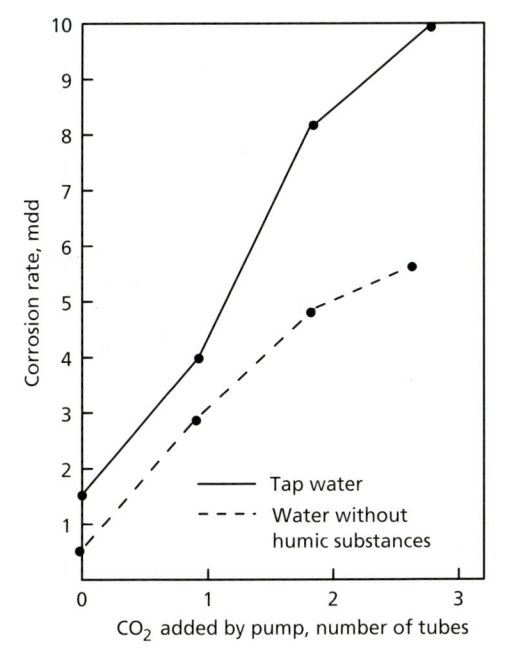

Figure 53: Corrosions rates of copper in drinking water with different carbon dioxide contents [465]

The influence of dissolved carbon dioxide in drinking water on the corrosion of pure copper was also confirmed by investigations in Sweden [466]. These investigations included exposure tests as well as potentiodynamic measurements with rotating disk electrodes in synthetic drinking water with different contents of hydrogen carbonate, calcium and chloride. Double distilled water with $NaHCO_3$ additions of concentrations between 0.5 and 10 mM was chosen as the synthetic drinking water. The pH-value was adjusted either by passing CO_2 into the solution or by adding HCl. Calcium was added as calcium nitrate and chloride as NaCl.

In addition, drinking water samples from copper pipes were randomly taken at various sites and various times and analyzed to determine their content of dissolved copper, calcium and carbon dioxide. The samples were taken each once from water having stagnated over night and once from water after having kept running for 10 minutes (20 l/min). Figure 54 and Figure 55 show the share of water samples with a content of dissolved copper of more than 300 µg/l for the content of free CO_2 and for the ratio of free CO_2 and calcium ions, respectively.

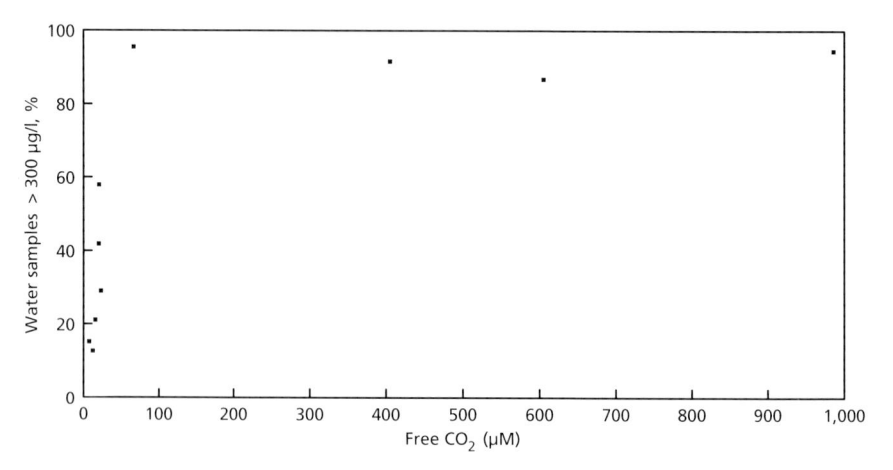

Figure 54: Percentage of water samples with copper contents higher than 300 µg/l as a function of the content of free CO_2 [466]

Comparison of both evaluations reveals that the decisive parameter for copper corrosion in drinking water is the ratio of free CO_2 and calcium ions rather than the content of free carbon dioxide. Figure 56 depicts the results of the electrochemical investigations in synthetic seawater and confirms the effect shown in Figure 55.

Considering the carbonate system in the absence of calcium, the content of free CO_2 is determined by the pH-value and the alkalinity. In Figure 57 the copper content determined in solution following the exposure of the copper specimens to synthetic drinking water as a measure of corrosion is plotted logarithmically against the CO_2 content.

Also the European Standard DIN EN 12502-2 provides information about investigation results and experience regarding the influence of free carbon dioxide and the alkaline earth content in waters obtained from the corrosion of copper materials in

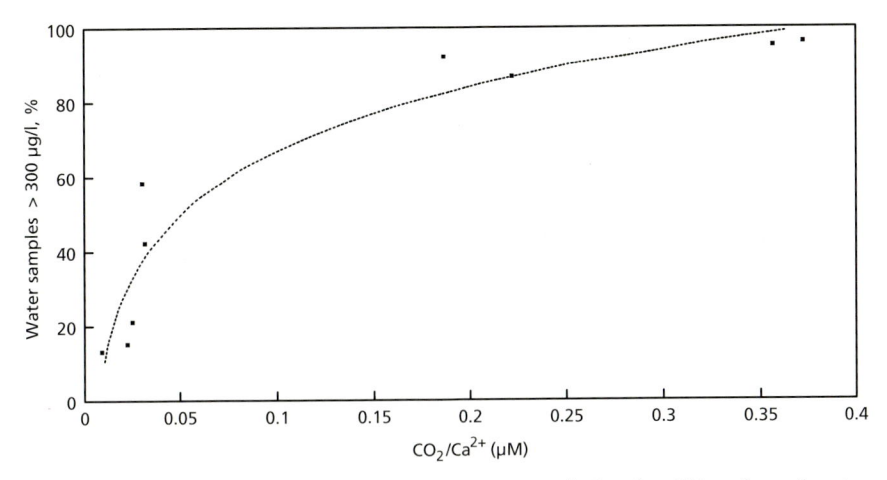

Figure 55: Percentage of water samples with copper contents higher than 300 µg/l as a function of the ratio of free CO_2 and calcium ions [466]

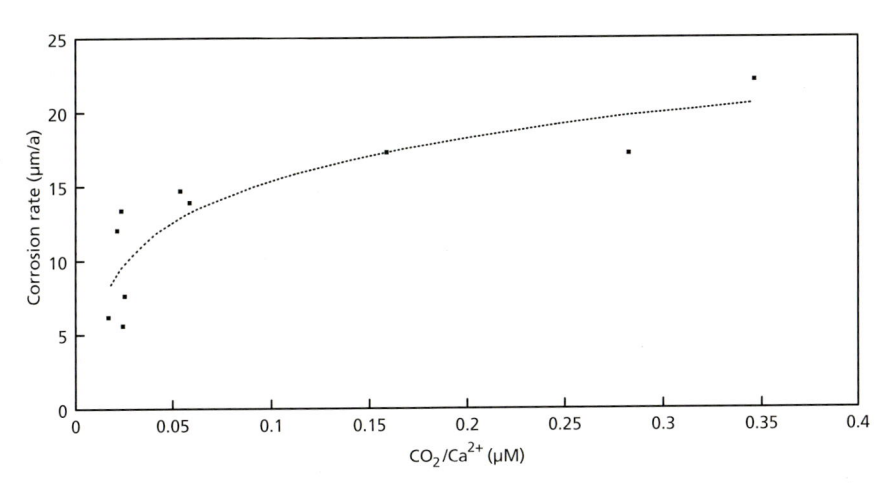

Figure 56: Dependence of the corrosion rate of pure copper in synthetic drinking water on the ratio of free CO_2 and calcium ions [466]

drinking water distribution systems of buildings [467]. By analogy this standard can be applied to other water systems. Under normal conditions drinking water systems of copper and copper alloys are generally resistant to corrosion damage. However, there are certain conditions under which corrosion damage may occur.

The most frequent manifestation is pitting corrosion of copper in water pipes. A distinction is made between two forms of pitting corrosion of copper in water pipe systems. The type of pitting corrosion depends on the water temperature, the composition of the water and the operating conditions.

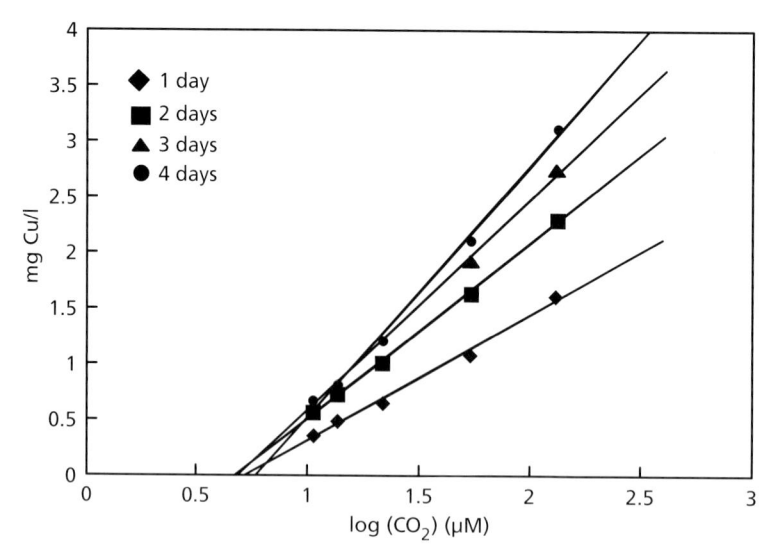

Figure 57: Influence of the CO_2 content of water on the corrosion of pure copper after various exposure periods [466]

Pitting corrosion – type 1 occurs almost exclusively in cold water pipes and is mainly found in moderately hard ground waters and less often in surface waters, which may contain organic substances as potential inhibitors [468].

Typically, the site of attack is characterized by semi-spherical pits with overlying green pustules mainly consisting of basic copper carbonate. Beneath these pustules the pits are always covered by a coherent copper(I) oxide layer. Beneath this layer there are copper oxides and chlorides. Regarding the effect of the water composition, a major role is plaid by the ratio of alkaline earth salts (chlorides, sulfates, nitrates) and alkaline earth hydrogen carbonates. The probability of corrosion decreases as the concentration of hydrogen carbonate ions increases and the concentration of alkaline earth ions decreases. These relations may be described by using a factor Q, whereas:

$$Q = \frac{2 \times c \text{ (alkaline earth ions)}}{c \text{ (hydrogen carbonate ions)}} - 1 \text{ Mol/m}^3$$

There is a low risk of pitting corrosion if $Q < 0{,}15$ and $Q > 2$.
There is a higher risk of pitting corrosion if $Q = 0{,}5$ to 1.

Comprehensive investigations into the influence of the water composition on the occurrence of pitting corrosion in water pipes of copper covering a period of almost 10 years confirm the influence of the ratio of alkaline earth salts and alkaline earth hydrogen carbonates described in said formula [469]. In addition, the corrosion values calculated from this relation and applicable to the evaluation of water depend on the temperature, the content of free carbonic acid and the alkali content.

Pitting corrosion – type 2, which occurs less frequently than type 1, is characterized in that the corroding pipes appear to be intact. Often the inside is covered by pale yellow to yellow-brown amorphous deposits [468]. Beneath there are spatially very limited sites of pitting corrosive attack with irregular internal geometries, leading to holes in the form of pinpricks to the outer side of the tubes if they break up. The pitting corrosion sites are completely filled with copper(1) oxide.

Type 2 pitting corrosion mainly occurs in warm water, in particular in soft and acidic waters (pH 6 to 4.2). Often, those waters encouraging type 2 pitting corrosion contain higher contents of aggressive carbonic acid and, seldom, the carbonate hardness is higher than 1° dH. Also the sulfate content exerts an influence, with the molar ratio of hydrogen carbonate/sulfate playing a role here.

At a molar ratio:

$$S = \frac{c \text{ (hydrogen carbonate ions)}}{c \text{ (sulfate ions)}} \geq 2$$

or at pH values > 7.5 the probability of pitting corrosion is low [467].

As suggested by the investigations described in [470], the contents of aggressive carbonic acid are below 10 mg/l if type 1 pitting corrosion occurs, whereas these contents amount up to 50 mg/l for type 2.

Contrary to the general knowledge that hydrogen carbonates in waters are favorable to counteract the corrosion of copper, [471] suggests that hydrogen carbonate may also have a detrimental effect under certain conditions. Results obtained from measurements of test tubes in the lab and from investigations performed in drinking water supply installations were used to evaluate the connection with water conditions and the release of copper corrosion products. Results obtained from measurements of test tubes in the lab and from investigations performed in drinking water supply installations were used to evaluate the connection with water conditions and the release of copper corrosion products. As turned out, the corrosion rates of copper and the release of corrosion products with higher contents of hydrogen carbonates are higher in fairly new copper pipes. At a constant pH-value there is a linear increase of the concentrations of copper corrosion products with increasing bicarbonate contents. It is assumed that this phenomenon is attributable to the increase of solubility of copper hydroxide caused by the bicarbonate ions. This effect diminishes at higher pH-values. Therefore, desorption of the dissolved carbon dioxide by aeration and hence a higher pH-value are recommended as a countermeasure.

Also in steam systems dissolved carbon dioxide or carbon dioxide carried along in occurring condensates may cause a considerable corrosive attack on copper components and therefore care must be exercised as to carefully treat the water [472].

Copper-aluminium alloys

Older publications in the literature indicate material removal values of 0.006 to 0.02 mm/a (0.24 to 0.79 mpy) for copper-aluminium alloys in carbon dioxide-containing waters depending on the composition of the alloy. In practice these materials are well established also in humid flue gases with carbon dioxide as well as in carbonic mineral waters.

Copper-nickel alloys

As already mentioned in the section "Copper", dissolved carbon dioxide in steam systems or in condensates may cause damage to copper components. [473] describes an incidence of damage to pipes of a copper-nickel material CuNi90/10 (corresponding to CW352H, CuNi10Fe1Mn) in the condenser of a thermal seawater desalination plant. Non-condensable gases, such as CO_2, O_2, N_2, NH_3 from the steam and air entering the system through leaks accumulate in the condensers of such plants. Here, carbon dioxide may stem from both the decomposition of hydrogen carbonates and the transport of CO_2 from the salt solution into the steam. In such gas-loaded condensates the protective effect of the oxide layers formed on the copper materials is not sufficient to prevent a corrosive attack. In the present case the damage was attributed to the dissolved carbon dioxide in the condensate together with low amounts of sulfur-containing compounds. As a corrective measure it is recommended to provide quick and effective ventilation in order to expel the dissolved gases. During periods of downtime the tubes should be flushed and dried to remove the remaining gas-saturated condensate remnants and facilitate the formation of protective copper oxide layers.

Copper-tin alloys (bronzes)

For copper-tin alloys in distilled water saturated with carbon dioxide in a temperature range from 303 to 323 K (30 °C to 50 °C) the indicated material removal values amount from 0.007 to 0.02 mm/a (0.28 to 0.79 mpy).

The information sheets of the German Copper Institute refer to a good corrosion resistance of various copper-tin alloys against carbonic acid. According to these sheets also the good behavior in the atmosphere is not affected by the contents of sulfur dioxide and carbon dioxide [474].

Carboxylic Acid Esters

Author: L. Hasenberg / Editor: R. Bender

Copper-zinc alloys (brass)

Neutral carboxylic acid esters hardly attack copper and its alloys at all [475–512]. However, since carboxylic acid esters do not always occur in pure form, as a result of the production process, impurities must be expected. These are, for example, the carboxylic acids required for preparation of the esters, and the corresponding catalysts, such as sulfuric or hydrochloric acid.

Copper materials are susceptible to corrosion by lower carboxylic acids. However, the corrosive attack by these acids takes place only in the presence of oxygen, so that copper alloys can and are used for their production [513]. Hygienically unacceptable copper compounds are not generally formed. Brasses of high zinc content containing $\geq 25\%$ zinc exhibit the least resistance in the presence of traces of oxygen. Copper materials are very resistant to higher carboxylic acids. The corrosion rates are below 0.01 mm/a (0.39 mpy) [513].

When copper materials are used in the food industry, it should be remembered, for example, that copper is destroyed by traces of ascorbic acid (vitamin C), and an oxidation taste occurs as a secondary effect when milk is boiled [514]. Catalytic decompositions of the ester may also occur due to small copper contents [510].

To gain an overview of the corrosive attack to be expected on copper and its alloys in the presence of carboxylic acids and inorganic acids, Tables 76 to 81 contain a brief list of the corrosion rates in formic, acetic, chloroacetic, oxalic, lactic, hydrochloric and sulfuric acid solutions. It should be remembered that the corrosion rate may be different in mixed media, so that these values cannot be adopted without reservation.

Material	Temperature K (°C)	Concentration %	Aerated		Exclusion of air	
			Agitated	Static	Agitated	Static
			Corrosion rate, mm/a (mpy)			
Copper	293 (20)	0.4	0.01 (0.39))	0.003 (0.12)		
	293 (20)	2.3	0.02 (0.79)	0.007 (0.28)		
	293 (20)	90	0.33 (12.99)			
	Vapour		0.3 (11.8)			
	343 (70)	6	0.2 (7.87)		0.1 (3.94)	
	373 (100)	90	0.3 (11.8)	0.1 (3.94)		

Table 76: Corrosive attack on copper in formic acid [513]

Material	Temperature K (°C)	Concentration %	Aerated		Exclusion of air	
			Agitated	Static	Static	Agitated
			Corrosion rate, mm/a (mpy)			
Copper	293 (20)	0.03	0.5 (19.7)	0.1 (3.94)		0.01 (0.39)
	293 (20)	0.5	1.4 (55.2)	0.5 (19.7)	0.5 (19.7)	0.1 (3.94)
	293 (20)	1.0	2.1 (82.7)	0.6 (23.6)	0.5 (19.7)	0.2 (7.87)
	293 (20)	4.0	36–56 (1,417–2,204)			
	293 (20)	6.4		1.4–4.5 (55.2–177)	0.7 (27.6)	
	293 (20)	20		2.25 (88.6)	0.3 (11.8)	0.25 (9.84)
	293 (20)	35		20.0 (787)		
	293 (20)	conc.				4.1 (161)
	373 (100)	15				1.28 (50.4)
	373 (100)	20				4.8 (189)
	373 (100)	30				0.41–1.2 (16.1–47.2)
	373 (100)	40				approx. 1.2 (approx. 47.2)
CuZn30 (CW505L)	288 (15)	0.004		0.13 (5.12)		0.1 (3.94)
	288 (15)	0.01		0.22 (8.66)		0.1 (3.94)
	288 (15)	0.04	7.6 (299)			0.1 (3.94)
	288 (15)	0.07		0.23 (9.06)	0.13 (5.12)	
	288 (15)	0.36	7.9 (311)			
	291 (18)	1		2.4 (94.5)		
	291 (18)	3.6		2.6 (102)		
	288 (15)	39	10.0 (394)		1	0.1 (3.94)
CuAl8, CuAl9	288 (15)	3.6			0.17 (6.69)	
	288 (15)	5	1.8 (70.9)			0.1 (3.94)
	288 (15)	conc.			1 (39.4)	

* impure

Table 77: Corrosive attack on copper materials in hydrochloric acid [513]

Table 77: Continued

Material	Temperature K (°C)	Concentration %	Aerated		Exclusion of air	
			Agitated	Static	Agitated	
			Corrosion rate, mm/a (mpy)			
CuAl10Ni (G-CuAl10Ni) (G-CuAl10Fe)	293 (20)	10			0.05 (1.97)	0.02 (0.79)
	293 (20)	30			1 (39.4)	0.05 (1.97)
	293 (20)	37			2.6 (102)	0.8 (31.5)
CuSn4 (CW450K)	288 (20)	3.6			approx. 0.5 (approx. 19.7)	
	289 (16)	4.1	1.5 (59.1)	1 (39.4)		
	288 (15)	39			approx. 0.5 (approx. 19.7)	
	373 (100)	15			0.62 (24.4)	
	373 (100)	30			0.62 (24.4)	
CuNi30 (CuNi30Fe)	293 (20)	0.5	0.26 (10.2)	0.2 (7.87)	0.2 (7.87)	0.1 (3.94)
	293 (20)	1	0.5 (19.7)	0.3 (11.8)	0.2 (7.87)	0.1 (3.94)
	298 (25)	13	2–7 (78.7–276)			
CuZn16Si4 CC761S	293 (20)	0.5	0.56 (22)	0.2 (7.87)		0.1 (3.94)
	293 (20)	1	1.2 (47.2)	0.7 (27.6)		0.1 (3.94)*

* impure

Table 77: Corrosive attack on copper materials in hydrochloric acid [513]

Material	Temperature K (°C)	Concentration %	Aerated		Exclusion of air	
			Agitated	Static	Agitated	Static
			Corrosion rate, mm/a (mpy)			
Copper	293 (20)	0.8	0.3 (11.8)	0.11 (4.33)		
	293 (20)	7	0.6 (23.6)	0.3 (11.8)	0.28 (11)	
	298 (25)	60	1.5 (59.1)			
	298 (25)	100	0.6 (23.6)			0.05 (1.97)
	373 (100)	20		1.76 (69.3)		
	373 (100)	conc.	2.5–8 (98.4–315)			
	373 (100)	100			0.03 (1.18)	0.01 (0.39)

Table 78: Corrosion rates on copper materials in acetic acid [513]

Table 78: Continued

Material	Temperature K (°C)	Concentration %	Aerated		Exclusion of air	
			Agitated	Static	Agitated	Static
			Corrosion rate, mm/a (mpy)			
G-CuSn5ZnPb	293 (20)	1.2	0.22 (8.66)		0.14 (5.51)	
	353 (80)	1.2	0.54 (21.3)		0.2 (7.87)	
CuZn30 (CW505L)	293 (20)	0.6	0.2 (7.87)	0.1 (3.94)		
	373 (100)	1.1	0.4 (15.7)	0.1 (3.94)		
CuZn16Si4 CC761S	293 (20)	4	0.15 (5.91)	0.08 (3.15)	0.03 (1.18)	
	373 (100)	4	1.1 (43.3)	0.67 (26.4)	0.23 (9.06)	
CuAl10Fe	295 (22)	10		0.05 (1.97)		0.01 (0.39)

Table 78: Corrosion rates on copper materials in acetic acid [513]

Material	Temperature K (°C)	Concentration %	Aerated		Exclusion of air	
			Agitated	Static	Agitated	Static
			Corrosion rate, mm/a (mpy)			
Copper	288 (15)	20		0.08 (3.15)		
	293 (20)	0.6	0.01 (0.39)	0.01 (0.39)	< 0.01 (< 0.39)	< 0.01 (< 0.39)
	293 (20)	11.4	0.1 (3.94)	0.06 (2.36)		
	341 (68)	5	0.1 (3.94)	0.05 (1.97)	0.03 (1.18)	
CuSn6 (CW452K)	293 (20)	5	0.06 (2.36)	0.03 (1.18)	0.03 (1.18)	0.01 (0.39)
	323 (20)	6.6	0.08 (3.15)	0.04 (1.57)	0.03 (1.18)	0.01 (0.39)
CuSn10	293 (20)	11.2		0.04 (1.57)		
CuZn30 (CW505L)	293 (20)	11.4	0.13 (5.12)	0.07 (2.76)		0.01 (0.39)
	323 (50)	8	0.13 (5.12)		0.04 (1.57)	0.01 (0.39)

Table 79: Corrosive attack on copper materials in oxalic acid [513]

Material	Temperature K (°C)	Concentration %	Aerated		Exclusion of air	
			Agitated	Static	Agitated	Static
			Corrosion rate, mm/a (mpy)			
Copper	293 (20)	1	0.24 (9.45)		0.06 (2.36)	
	293 (20)	1.4	0.25 (9.84)		0.06 (2.36)	
	293 (20)	3.4		0.1 (3.94)		
	338 (65)	1	1 (39.4)		0.3 (11.8)	
CuSn6 (CW452K)	293 (20)	1.3	0.09 (3.54)		0.02 (0.79)	0.01 (0.39)
	293 (20)	1.5		0.04 (1.57)		0.01 (0.39)
	323 (50)	1.3	0.36 (14.2)		0.05 (1.97)	
CuZn30 (CW505L)	293 (20)	1.4	0.34 (13.4)			
	293 (20)	3			0.1 (0.39)	0.05 (1.97)
CuNi30	293 (20)	1.3	0.05 (1.97)			0.01 (0.39)
	293 (20)	0.5–5		0.01 (0.39)	0.01 (0.39)	< 0.01 (< 0.39)

Table 80: Corrosive attack on copper materials in lactic acid [513]

Material	Temperature K (°C)	Concentration %	Aerated		Exclusion of air	
			Agitated	Static	Agitated	Static
			Corrosion rate, mm/a (mpy)			
Copper	293 (20)	6	3–3.9 (118–154)	0.08 (3.15)	0.08 (3.15)	0.04 (1.57)
	293 (20)	10		0.12 (4.72)	0.12 (4.72)	0.1 (3.94)
	293 (20)	20	2.8–3.4 (110–134)	0.27 (10.6)	0.13 (5.12)	0.01 (0.39)
	293 (20)	45		0.3 (11.8)	0.12 (4.72)	0.01 (0.39)
	293 (20)	70	2.6 (102)			
	293 (20)	96.5	0.85–1 (33.4–39.4)	0.3 (11.8)	0.13 (5.12)	0.01 (0.39)
	308 (35)	6	7.7 (303)	0.1 (3.94)	0.17 (6.69)	0.02 (0.79)
	308 (35)	20	4.2 (165)	0.43 (16.9)	0.23 (9.06)	0.03 (1.18)
	308 (35)	96.5	1.0 (39.4)	0.78 (30.7)	0.2 (7.87)	0.02 (0.79)
	313 (40)	10	3.85 (152)			

Table 81: Corrosion of copper materials in sulfuric acid [513]

Table 81: Continued

Material	Temperature K (°C)	Concentration %	Aerated		Exclusion of air	
			Agitated	Static	Agitated	Static
			Corrosion rate, mm/a (mpy)			
	313 (40)	40	1.31 (51.6)			
	313 (40)	96.5	0.8 (31.5)			
	323 (50)	6	8.8 (347)	0.17 (6.69)	0.26 (10.24)	0.08 (3.15)
	323 (50)	20	1.9 (74.8)	0.19 (7.48)	0.2 (7.87)	0.02 (0.79)
	323 (50)	96.5	1.7 (66.9)	0.88 (34.7)	0.8 (31.5)	0.03 (1.18)
	333 (60)	10	3.46 (136)			
	333 (60)	40	1.36 (53.5)			
	333 (60)	96.5	1.2 (47.2)			
	358 (55)	50		0.34 (13.4)		0.016 (0.63)
	395.5 (122.5)	50		2.1 (82.7)		0.1 (3.94)
CuSi3Mn (cf. CW116C)	298 (20)	10	0.2 (7.87)	0.06 (2.36)		
	298 (20)	25		0.035–0.04 (1.38–1.57)		
	298 (20)	70		0.017–0.02 (0.67–0.79)		
	343 (70)	3		0.18 (7.09)		
	343 (70)	25		0.09 (3.54)		
	343 (70)	70		0.01–0.03 (0.39–1.18)		
CuBe2 (CW101C)	333 (60)	10		0.88 (34.6)		
CuNi30Fe (CuNi30)	293 (20)	10	0.1 (3.94)	0.05 (1.97)		
	293 (20)	20	1 (39.4)			
	373–393 (100–120)	10	0.12 (4.72)			0.03 (1.18)
	373–393 (100–120)	50	20.8 (819)			0.52 (20.5)
	373–393 (100–120)	96	131 (5,157)			3.3 (130)

Table 81: Corrosion of copper materials in sulfuric acid [513]

Table 81: Continued

Material	Temperature K (°C)	Concentration %	Aerated		Exclusion of air	
			Agitated	Static	Agitated	Static
			Corrosion rate, mm/a (mpy)			
CuZn15 (CW502L)	293 (20)	3	1 (39.4)	0.2 (7.87)	0.2 (7.87)	
	358 (85)	50		0.3 (11.8)		0.03 (1.18)
	373 (100)	50		0.7 (27.6)		0.2 (7.87)
CuZn30 (CW505L)	293 (20)	4	2.2 (86.6)		0.5 (19.7)	
	293 (20)	10	> 5 (> 197)			
CuSn10Zn	293 (20)	5			0.13 (5.12)	
	293 (20)	95			0.81 (31.9)	
	323 (50)	5			0.36 (14.2)	
	323 (50)	50			0.082 (32.3)	
	323 (50)	95			2.32 (91.3)	
	373 (100)	5			0.34 (13.4)	
	373 (100)	50			0.1 (3.94)	
	373 (100)	95			14.8 (583)	
CuAl10 (CuAl10Ni)	293 (20)	5	0.046 (1.18)		0.03 (1.18)	0.01 (0.39)
	293 (20)	95	0.27 (10.6)		0.01 (0.39)	
	323 (50)	5	0.02 (0.79)			
	323 (50)	50	0.002 (0.079)		0.002 (0.079)	
	323 (50)	95	1.21 (47.6)			
	373 (100)	5	0.25 (9.84)			
	373 (100)	50	0.19 (7.48)			
	373 (100)	95	117.6 (4,629)			

Table 81: Corrosion of copper materials in sulfuric acid [513]

The reaction tank and columns, for example, were made of copper materials for the production of ethyl acetate from acetic acid and ethyl alcohol. Small amounts of sulfuric acid were added as the catalyst. Atmospheric oxygen and other oxidizing agents should be kept away as much as possible in order to keep the corrosion of the material low. Fittings and shut-off valves were made of bronze. CuAl alloys and CuSi alloys have proved to be particularly suitable in acidic, crude ester. The corrosion rate of the materials used was less than 0.1 mm/a (3.94 mpy) according to [478]. Pumps are also made of bronze [490].

In the event of a possible application of copper materials in chloroformic acid esters, it should be ensured that only dry esters are used. Esters containing water form hydrochloric acid by hydrolysis, which is corrosive, depending on the concentration present.

Copper (99.9%) is not recommended as a material in esterification plants for the production of butyl acetate. The corrosion rate is quoted as between 0.13 and 0.24 mm/a (5.12 and 9.45 mpy) [515].

The corrosion rate for copper in the esterification of methanol with acetic acid in the presence of about 1% sulfuric acid as the catalyst is quoted as < 0.3 mm/a (< 11.8 mpy) at 358 K (85 °C) [511].

A corrosion rate of 0.2 mm/a (7.87 mpy) is quoted for copper in an esterification tank for methyl acetate production with a somewhat different medium composition than that just mentioned, but at the same temperature. The esterification tank contained a reaction mixture of 25% alcohol, 10% ester, 3% sulfuric acid and 62% water. The test duration was 168 hours in total. During this period, the reaction mixture was kept at 358 K (85 °C) for periods of 16 hours each and then cooled for 8 hours.

In another experiment to produce methyl acetate, corrosion rates of 0.9 mm/a (35.4 mpy) were determined for copper. The reaction mixture contained 34% acetic acid, 34% ester, 30% alcohol, 0.5% sulfuric acid and 1% water. The test temperature over a period of 8 hours was 373 to 423 K (100 to 150 °C), and dropped to 293 K (20 °C) in the course of 22 hours. The total test duration for determination of the material consumption was 211 hours [511].

These examples show that the corrosive attack on copper in esterification plants is relatively high and varies according to the experimental conditions, so that its use must be given careful consideration with regard to possible pollution of the environment.

Copper materials should not be used if peracetic acid is present, because of the possible risk of explosion [512].

Chloroethanes

Author: *H. G. Spilker* / Editor: *R. Bender*

Copper

Because of its particularly good processability, thermal conduction and corrosion resistance, copper is also used in many branches of industry where solvents are produced or are used.

Copper materials are also used in systems with halogen-containing refrigerants without particular corrosion problems, although it is necessary to keep these systems free from moisture and the ingress of air.

Resistance to organic substances varies widely, and of the chlorinated hydrocarbons in particular, many react with copper, reactions which have started being intensified considerably by moisture, free acid, free chlorine or by decomposition products following the action of sunlight or heat [516].

SF-copper (CW024A) and SW-copper (CW023A) are adequately resistant and can be used in dry and pure chlorinated hydrocarbons, e.g. 1,2-dichloroethane, with moisture contents below 10 ppm at room temperature under normal conditions.

Corrosion rates which are scarcely higher than 0.0025 mm/a (0.09 mpy) are also found in the region of the boiling point [517].

In accordance with the water legislation of 1.7.1981 WHG §§ 19g et seq., approval from the authorities is required for use of single-walled copper pipes in contact with chlorinated hydrocarbons, e.g. 1,1,1-trichloroethane, for laying above ground and underground, because of the risks of water contamination [518].

Copper is a suitable sealing material for use with dry chloroethanes, but is unsuitable with damp chloroethanes.

Corrosion of copper by damp chloroethanes is evidently a matter of the action of hydrogen chloride, which can form by hydrolysis as a result of the water content.

SF-copper is at risk from corrosion in contact with chlorinated hydrocarbons (e.g. 1,1,1-trichloroethane) which contain only 1,000 ppm dichloroacetic acid or dichloroacetyl chloride as degradation products [519].

Mixtures of chloroethanes (e.g. 1,1,1-trichloroethane) with ethers or alcohols are more corrosive than the pure chloroethanes. Copper is severely attacked by mixtures of 1,1,1-trichloroethane and tetrahydrofuran, forming copper chlorides on the metal surface [520].

– *Corrosion in monochloroethane* –

Copper has an adequate resistance to 100% dry monochloroethane at room temperature and also at the boiling point. Copper is resistant to damp monochloroethane at room temperature. As the temperature rises, copper loses resistance, and is not particularly resistant at the boiling point [521].

One producer of metallic corrugated hoses evaluates his products made of copper (SF-Cu, CW024A; old: 2.0090) as being practically resistant to dry monochloroethane and as recommended for applications with 5% aqueous solution at 293 K (20 °C) [522].

Copper is particularly suitable as a sealing material for monochloroethane [523]. Apparatuses made of copper are used for distillation of acid-free monochloroethane.

In spite of having an adequate resistance under normal conditions, long-term or continuous contact of copper with monochloroethane, e.g. during storage, should be avoided as far as possible [524].

– *Corrosion in 1,1-dichloroethane* –

Copper (85 to 99.9%) is resistant in 100% 1,1-dichloroethane up to the boiling point, with corrosion rates of less than 0.05 mm/a (1.97 mpy). At temperatures above the boiling point up to 366 K (93 °C), copper has a moderate resistance with corrosion rates of less than 0.5 mm/a (19.7 mpy) [521].

Copper is also resistant to damp 1,1-dichloroethane at room temperature, and is still moderately resistant at temperatures above room temperature [16]. In this temperature range, copper shows the same resistance in aqueous solutions with low contents of 1,1-dichloroethane [521].

– *Corrosion in 1,2-dichloroethane* –

Copper has an adequate resistance to 100% 1,2-dichloroethane up to the boiling point, is still moderately resistant to damp 1,2-dichloroethane at room temperature, but unstable at higher temperatures.

Corrosion of copper was investigated in pure 1,2-dichloroethane (ethylene chloride) and in dichloroethane covered with a layer of water. The copper specimen was immersed in both phases during the experiment. The corrosion rates [525] and material consumption rates determined at room temperature and at the boiling point are shown in Table 82.

Temperature	100% 1,2-Dichloroethane		1,2-Dichloroethane/water	
	Corrosion rate mm/a (mpy)	Material consumption rate g/m^2 d	Corrosion rate mm/a (mpy)	Material consumption rate g/m^2 d
Room temperature	< 0.005 (< 0.20)	0.02	0.01 (0.39)	0.1
Boiling point	0.07 (2.76)	0.9	0.59 (23.2)	7.3

Table 82: Corrosion of copper in pure 1,2-dichloroethane and in a two-phase mixture of 1,2-dichloroethane/water [525]

The results in Table 82 show that water-containing 1,2-dichloroethane triggers off corrosion of copper.

Copper columns are used for distillation of 1,2-dichloroethane. However, access to atmospheric oxygen can promote an attack on copper by 1,2-dichloroethane.

– Corrosion in 1,1,1-trichloroethane –

Copper is resistant in stabilized 1,1,1-trichloroethane.

Copper wires showed no corrosion after 144 to 168 h in 1,1,1-trichloroethane stabilized with various amounts of methyl 2-propynyl sulfide (1–4 vol%) and additions of nitromethane (< 1 vol%), nitroethane (< 1 vol%) and isopropyl nitrate (2–3 vol%) [526].

The dissolving power of 1,1,1-trichloroethane for copper chlorides is low: for CuCl at 298 K (25 °C) = 20 ppm, at 323 K (50 °C) = 80 ppm, for $CuCl_2$ at 298 K (25 °C) = 30 ppm, at 323 K (50 °C) = 40 ppm. For this reason, the copper chlorides from contaminated surfaces in cleaning baths can hardly be removed by the dissolving power of 1,1,1-trichloroethane, but at best by rinsing effects [527].

In studies on reductive breakdown of environmentally polluting chlorinated hydrocarbons, 1,1,1-trichloroethane was chosen as an example of a saturated, partly chlorinated hydrocarbon. The results of a potential-controlled reduction on copper in 1,1,1-trichloroethane/70 % ethanol solution already showed a breakdown of 71 % at a potential of $U = -1,600$ mV_{SHE} (i = 4.2 mA/cm^2) after 2.5 h. Current/potential curves in 1,1,1-trichloroethane/70 % ethanol solution at room temperature were also plotted. 200 mg/l KOH were added to the solution to improve the conductivity.

A rest potential of –50 mV was found in this solution. The anodic course of the current/potential curve was interpreted as the formation of a passive layer, and from about 450 mV by dissolution of the metal and formation of oxygen [528].

Mixtures of chlorinated hydrocarbons and ethers or alcohols can react more corrosively than the pure substances. The weight losses of SF-copper in various solvent mixtures, in each case containing 67 vol% 1,1,1-trichloroethane, were determined at 323 K (50 °C) after 96 h as a function of the light and water content, see Table 76. Of the mixtures studied, that with tetrahydrofuran shows the severest corrosive attack on SF-Cu, γ-$CuCl_2 \times 3$ $Cu(OH)_2$ and CuCl forming on the surface [520].

The greatest weight loss, measured in tetrahydrofuran in light, of 5 g/m^2 gives, when converted, a material consumption rate of 0.052 $g/m^2 h$, i.e. SF-Cu is resistant under these conditions.

– Corrosion in 1,1,2-trichloroethane –

Contact between copper and 1,1,2-trichloroethane, which is highly corrosive and for which only relatively large amounts of the inhibitors which are suitable for 1,1,1-trichloroethane are effective stabilizers, should be avoided. Owing to its low industrial importance, the available literature contains nothing on the resistance of copper nor reports on findings on uses in 1,1,2-trichloroethane.

1,1,1-Trichloroethane (67 Vol. %) + added solvent + added water		Weight Loss g/m²	
		Dark	Light
Tetrahydrofuran:			
– H₂O content*:	a) 350 mg/l	3.0	5.0
	b) saturated	1.9	4.0
2-Propanol:			
– H₂O content*:	a) 400 mg/l	< 1.0	< 1.0
	b) saturated	< 1.0	< 1.0
Ethyl acetate:			
– H₂O content*:	a) 300 mg/l	0	0
	b) saturated	< 1.0	< 1.0

* initial water content

Table 83: Weight losses of SF-Cu per unit area in solvent mixtures containing 67 vol. % 1,1,1-trichloroethane at 323 K (50 °C) after 96 h as a function of light and water content [520]

– *Corrosion in tetrachloroethane* –

Vigorous reactions can occur when copper comes into contact with tetrachloroethane, and for this reason e.g. copper tanks are not to be used for continuous storage.

Comparable series of experiments on the resistance of copper in boiling tetrachloroethane showed the same results after a test of 10 hours at various moisture contents.

Medium	Material consumption rate g/m² d	Corrosion rate mm/a (mpy)
Tetrachloroethane		
Without moisture	110	4.29 (169)
With moisture	65	2.25 (88.6)

Table 84: Corrosion of copper in boiling tetrachloroethane after 10 h [525]

Moisture contents accordingly have a corrosion-inhibiting effect. However, the corrosion rates in boiling, moist and also dry tetrachloroethane are so high that copper is completely unstable under such conditions [525].

If tetrachloroethane is used for the extraction of fats, copper apparatus must be leaded, since copper is attacked under these conditions.

– Corrosion in pentachloroethane –

Copper is only mildly attacked by pentachloroethane. After use in boiling pentachloroethane for 10 h, material consumption rates of 0.7 to a maximum of 65 g/m² d were found for copper. Copper accordingly is resistant to hardly usable in boiling pentachloroethane.

Copper also has an adequate resistance at room temperature and at the boiling point if the pentachloroethane contains moisture or traces of acid [529].

Stabilized pentachloroethane can be in contact with copper for a long time or stored in copper tanks.

– Corrosion in hexachloroethane –

Copper has an adequate resistance to 100 % hexachloroethane, even at higher temperatures. It is attacked in the presence of moisture.

89 to 99.9 % pure copper is unstable to hexachloroethane with a concentration lower than 100 % [521].

In a plant for the production of hexachloroethane from tetrachloroethylene and chlorine, the centrifuge for centrifuging the crystalline mass is made of copper-lined or Cu-plated steel [530].

Copper-aluminium alloys

The improved resistance of CuAl alloys to corrosion is based on the formation of a protective film of aluminium oxide. The protective film of an aluminium bronze comprises a mixed oxide of copper and aluminium.

Copper-aluminium alloys can be processed as cast or wrought alloys. Cast alloys are less sensitive to dealuminization. Wrought alloys with 4 to 7 % Al are suitable for cold rolling for the production of pipes or sheet metal.

Copper-aluminium alloys are not attacked in dry 100 % chloroethanes under normal conditions. They are corroded in damp chloroethanes.

In resistance tables, the cast aluminium bronzes G-CuAl10Fe (CC331G), G-CuAl9-10(Ni) (cf. CC332G and CC333G) and G-CuAl9Mn are evaluated as having an adequate resistance, with low corrosion rates, to dry chloroethanes at temperatures of up to 333 K (60 °C), and the cast aluminium bronze G-CuAl9-10(Ni) also to damp chloroethanes at 293 to 333 K (20 to 60 °C) [531].

However, during long-term contact between aluminium bronzes and chloroethanes in particular, the start of corrosion must be expected if the penetration of moisture or impurities causes decomposition of the chloroethane.

– Corrosion in monochloroethane –

Aluminium bronzes with 92–95 % Cu and 8–5 % Al have an adequate resistance to 100 % monochloroethane at room temperature and up to the boiling point, with low corrosion rates. If moisture is present, they are still resistant at room temperature but unstable at higher temperatures. The following alloys have a suitable corrosion resistance for applications in contact with 100% monochloroethane at room temperature: Ambraloy® 901 (95.0 Cu, 5.0 Al), Ambraloy® 917 (82.0 Cu, 9.5 Al, 1.0 Mn), alloys 613 (3.0 to 5.0 Fe, 2.0 to 2.2 Al), 637 (1.0 Zn, 6.0 to 8.0 Al, 1.2 to 2.2 Si) and 687 (2.0 Al, 20.5 Zn) and the cast alloys G-CuAl9, G-CuAl10Fe, G-CuAl9-10(Ni) and G-CuAl9Mn [531, 532].

– *Corrosion in dichloroethane* –

According to the tables of the Corrosion Data Survey, aluminium bronze is moderately resistant to 100% 1,1-dichloroethane up to the boiling point, with corrosion rates of up to 1.0 mm/a (39.4 mpy), and resistant to 100% 1,2-dichloroethane up to the boiling point, with corrosion rates of less than 0.1 mm/a (3.94 mpy) [533].

An adequate resistance to dry 1,2-dichloroethane at 293 K (20 °C) results for the cast materials G-CuAl9, G-CuAl10Fe, G-CuAl9-10(Ni) and G-CuAl9Mn [531].

– *Corrosion in trichloroethane and tetrachloroethane* –

No information and also no reports of findings on the properties of copper-aluminium alloys in 1,1,1-trichloroethane, 1,1,2-trichloroethane or tetrachloroethane are to be found in the literature available.

– *Corrosion in pentachloroethane and hexachloroethane* –

Aluminium bronzes are resistant to 100% and slightly less than 100% pentachloroethane up to 423 K (150 °C) [533]. The cast materials G-CuAl9, G-CuAl10Fe, G-Al9-10(Ni) and G-CuAl9Mn are resistant to dry pentachloroethane at 293 K (20 °C) [531].

Aluminium bronzes are resistant to 100% hexachloroethane at 293 K (20 °C) and moderately resistant at high temperatures. They cannot be used at lower hexachloroethane concentrations [533].

Copper-nickel alloys

The usual copper-nickel alloys contain, in addition to copper as the base metal, 5 to 30% nickel and often small amounts of iron, manganese, zinc, tin, silicon or niobium to improve certain properties. Copper-nickel alloys with 10 to 30% Ni are among the most corrosion-resistant copper alloys.

Additions of up to 1% manganese and 1.0 to 2.0% iron improve corrosion resistance further.

Production-related impurities, in particular contents of $FeCl_3$ or $CuCl_2$, in chlorinated hydrocarbons have a corrosive effect on the alloy CuNi30Fe1Mn (CW354H, 2.0882), but without local attack [534].

– *Corrosion in monochloroethane* –

The following copper-nickel alloys are suitable for use in dry monochloroethane:

Supernickel 702	(%, 69.0Cu, 30.0Ni, 0.4Fe)
CuproNickel 706	(%, 88.0Cu, 9.0–11.0Ni, 0.5Zn, 1.3Fe, 1.0Mn, 0.02Pb)
CuproNickel 710	(%, 76.5Cu, 19.0–23.0Ni, 1.0Zn, 1.0Fe, 1.0Mn, 0.05Pb)
CuproNickel 715	(%, 66.5Cu, 29.0–30.0Ni, 1.0Zn, 0.4–0.7Fe, 1.0Mn, 0.05Pb)
CuproNickel 754	(%, 89.25Cu, 10.0Ni, 0.75Fe)
Nickel silver	(%, 65.0Cu, 18.0Ni, 17.0Zn)
CuNi10Fe1Mn	CW352H, 2.0872, CuNi 90/10
G-CuNi30	CC383H, CuNi30Fe1Mn1NbSi-C

Restrictions on the use of copper-nickel alloys are to be found only in the table from one producer of metal hoses and metal bellows, who evaluates his products made from the alloy Cunifer® 30 (old DIN-Material No. 2.0882 with 68.0% Cu, 30.0% Ni, 1.0% Fe, 1.0% Mn) as suitable for dry monochloroethane at 293 K (20 °C) (at expected corrosion rates of 0.11 to 1.1 mm/a (4.33 to 43.3 mpy)) only if no other material with better properties is available. At the boiling point, corrosion rates between 1.1 and 11.0 mm/a (43.3 and 433 mpy) are expected for this material, which therefore is not particularly resistant. Use of these products of Cunifer® 30 is recommended for 5% aqueous solution at 293 K (20 °C), where corrosion rates below 0.11 mm/a (4.33 mpy) are possible [522].

– *Corrosion in dichloroethane* –

Metal hoses of Cunifer® 30 are not recommended for use in dichloroethane even if the product is dry and at room temperature. The corrosion rates are above 1.1 mm/a (43.3 mpy) [522].

Other assessments, however, evaluate the suitability of copper-nickel alloys for dichloroethane as far more favorable. According to [535], these alloys are resistant to 1,2-dichloroethane. The cast material G-CuNi 30 (CW383H) has a good resistance to dry 1,2-dichloroethane at 293 K (20 °C) [531].

Only low corrosion rates were found in experiments with the alloy 55Cu45Ni in 1,2-dichloroethane (ethylene dichloride) under various conditions, see Table 85 [536].

Medium	Temperature K (°C)	Corrosion rate mm/a (mpy)
1,2-Dichloroethane, anhydrous	298–303 (25–30)	4.1×10^{-4} (0.02)
	357 (84) (boiling)	0.0019 (0.75)
1,2-Dichloroethane, with layer of water	298–303 (25–30)	7.1×10^{-4} (0.03)
	357 (84) (boiling)	0.104 (4.09)

Table 85: Corrosion rates of the copper-nickel alloy 55% Cu 45% Ni in 1,2-dichloroethane [536]

– Corrosion in trichloroethane and tetrachloroethane –

Copper-nickel alloys are resistant to adequately stabilized trichloroethane at room temperature. Attack must be expected in nonstabilized or damp trichloroethane.

In addition to the nickel-copper alloy Monel®, the copper-nickel alloys are also preferred materials for applications with trichloroethane, e.g. the construction of reactors and distillation apparatuses.

No information is to be found on the properties of copper-nickel alloys in tetrachloroethane.

– Corrosion in pentachloroethane and hexachloroethane –

Copper-nickel alloys with 66 to 88 % Cu and 11 to 33 % Ni are resistant to 100 % and < 100 % pentachloroethane at room temperature and at the boiling point [533]. The cast material G-CuNi 30 (CC383H, CuNi30Fe1Mn1NbSi-C) is particularly suitable for dry pentachloroethane at room temperature [531].

Copper-nickel alloys with 11 to 33 % Ni are also resistant to hexachloroethane at the melting point, but have only a moderate resistance to damp hexachloroethane.

Copper-tin alloys (bronzes)

The alloys with up to about 14 % tin and sometimes further additions of nickel, zinc, lead or iron are of industrial importance.

As the tin content increases, the corrosion resistance of an alloy increases, and reaches a maximum at about 40 % as an intermetallic compound. However, bronzes of high tin content are scarcely used in industry.

Phosphorus, manganese or silicon in small amounts of between 0.1 and 0.35 % are added to the melt for deoxidation, which prevents formation of tin oxide (SnO_2). The names phosphorus bronze and manganese bronze are sometimes still common for these alloys.

Bronzes with tin contents of up to 8.5 % have a corrosion resistance to organic liquids comparable to that of copper, but additionally a higher mechanical strength.

Chlorinated hydrocarbons in the dry state scarcely attack bronzes. If moisture is present and at higher temperatures, material consumption rates of up to 30 $g/m^2 d$ may arise.

If attack occurs on copper-tin alloys, this usually proceeds as mild uniform general removal of material.

Chlorinated hydrocarbons with contents of breakdown products, which can be formed as chloroacetic acids by the action of air, light and water, have a corrosive effect on bronzes.

Corrosion rates of between 0.14 and 0.32 mm/a (5.51 and 12.6 mpy) were found for the alloy CuSn6 (CW452K, 2.1020), dichloroacetic acid having a more corrosive action than trichloroacetic acid [519].

Only moderate or minor resistance of bronzes must also be expected in the event of contamination with hydrogen chloride.

– *Corrosion in monochloroethane* –

The alloy CuSn6 (CW452K, 2.1020) has a good resistance to dry monochloroethane at room temperature and is moderately resistant at the boiling point. This bronze can also still be used for monochloroethane with limitations if moisture is present [537].

The cast material G-CuSn10–14 is evaluated as having a good resistance to dry monochloroethane at 293 K (20 °C) [531].

– *Corrosion in dichloroethane* –

Bronzes without Zn contents are resistant to dry dichloroethane at room temperature, but probably not particularly resistant to damp dichloroethane.

There are different assessments of the resistance of bronzes to 1,2-dichloroethane. While on the one hand the alloy CuSn6 (CW452K) is evaluated only as being largely resistant to dry 1,2-dichloroethane at 293 K (20 °C), on the other hand bronzes without Zn contents are evaluated as suitable for valves and fittings both at room temperature and at the boiling point [529]. According to [535], copper-tin alloys are resistant up to the boiling point.

The cast material G-CuSn10–14 has a good resistance to dry 1,2-dichloroethane at 293 K (20 °C) [531].

An English hose producer recommends the cast alloy Gunmetal® (red brass) for hose fittings for limited periods of use up to 303 K (30 °C) [538].

– *Corrosion in trichloroethane* –

Copper-tin alloys can be used as a construction material for stabilized 1,1,1-trichloroethane as long as the formation of hydrogen chloride can be excluded under the prevailing operating conditions.

Non-stabilized 1,1,1-trichloroethane also attacks these alloys, and this chlorinated hydrocarbon causes severe corrosion in the presence of moisture or steam.

Contact between copper-tin alloys and 1,1,2-trichloroethane should be avoided as far as possible, since this chlorinated hydrocarbon is highly corrosive and stabilization also is of only limited value.

– *Corrosion in tetra-, penta- and hexachloroethane* –

Gunmetal® (red brass) is suitable for use for tetrachloroethane up to 323 K (50 °C) for a limited period of time [538].

The cast material G-CuSn10–14 is recommended for applications in dry pentachloroethane at room temperature [531].

Copper-tin alloys without Zn contents are resistant to dry hexachloroethane even at elevated temperatures. These alloys are attacked by damp hexachloroethane [529].

Copper-tin-zinc alloys (red brass)

Copper-tin-zinc alloys are standardized according to DIN EN 1982 and contain, as cast alloys, in addition to copper as the main constituent, 4 to 10% Sn, 2 to 7% Zn and usually also 2.5 to 7% Pb. A little nickel is sometimes added to improve strength and toughness. Copper-tin wrought alloys also sometimes contain zinc as a third alloying constituent. Zinc can effect deoxidation and therefore replace phosphorus.

Copper-tin-zinc alloys are adequately resistant to dry and pure chlorinated hydrocarbons at room temperature and also at the boiling point, and corrosion rates scarcely higher than 0.0025 mm/a (0.10 mpy) are reached.

These alloys are attacked by damp chlorinated hydrocarbons, and the corrosion rates increase as the temperatures rise. Nevertheless, the cast alloys G-CuSn10Zn and G-CuSn5–7ZnPb (cf. CC491K, CC492K and CC493K) are also still evaluated as being suitable for damp chlorinated hydrocarbons at 293 to 333 K (20 to 60 °C) [531].

There is the risk of corrosion if copper-tin-zinc alloys come into contact with chlorinated hydrocarbons which contain alcohols, ether, hydrogen chloride, chloroacetic acids, metal chlorides or other impurities, and very high material consumption rates up to complete dissolution of the alloy must sometimes be expected.

Copper-tin-zinc alloys are particularly suitable for use in dry monochloroethane at room temperature, and are also adequately resistant at the boiling point. Only the zinc-rich alloy No. 464, with 60 Cu, 32.25 Zn, 0.75 Sn, is classified as being not particularly resistant to monochloroethane [532].

The cast alloys G-CuSn10Zn and G-CuSn5–7ZnPb are particularly suitable for applications with dry monochloroethane at 293 K (20 °C). They are recommended if no other materials with more favorable properties are available [531].

No detailed information is available on the behavior of copper-tin-zinc alloys to dichloroethane. The cast alloys G-CuSn10Zn and G-CuSn5–7ZnPb have been evaluated as being recommended for uses with dry 1,2-dichloroethane if other materials are not more suitable because of special requirements [531].

These two cast alloys can be used in contact with dry pentachloroethane at 293 K (20 °C) [531].

Copper-zinc alloys (brass)

Of all the copper alloys, the brasses with zinc contents of 5 to 40% are the materials which are used the most and are the most important.

The various qualities are more similar to copper in their corrosion behavior the higher the copper content in the alloy, and at > 85% Cu they can be equated with pure copper, while brasses having high zinc contents are less resistant.

At Zn contents from 37.5 to 46%, a β-structure is present, alongside the α-structure. Brasses with 46 to 50% Zn consist uniformly of the β-phase (β-brass), are extremely brittle and are unsuitable for industrial applications. Selective corrosion is promoted by chloride contents, little electrolyte movement and a simultaneously increased temperature.

The risk of dezincification no longer exists for α-brasses with less than 15% Zn. Dezincification of α-brass is also prevented by addition of As (0.02 to 0.06%), Al or Sn, but these additions have no protective action on the β-structure.

Brass wrought alloys contain up to 37.5% Zn, as tombac up to 30% Zn without further additions, and as lead-containing brass additionally up to 3.5% Pb. Brass cast alloys contain 33 to 42% zinc and up to 3% lead.

Further additions to special brasses are Mn, Ni, Si and Sn. Tin-containing special brasses, e.g. tobin bronze or admiralty brass, have a high corrosion resistance and are used for heat exchanger pipes and steam condensers.

According to a general evaluation of the usual chlorinated hydrocarbons, these attack copper-zinc alloys hardly at all in the dry state, and for this reason brasses are often used under such conditions. On the other hand, attack as a rule occurs when chlorinated hydrocarbons act in the presence of moisture over a relatively long period of time, and especially at higher temperatures [539].

The cast brasses G-Cu65Zn and G-Cu60Zn, the cast special brasses G-Cu55ZnMn, G-Cu55ZnAl1 and G-Cu55ZnAl and the cast silicon brass G-CuZn16Si4 generally prove to be suitable for the usual dry chlorinated hydrocarbons at 293 to 333 K (20 to 60 °C), but hardly usable if moisture is present [531].

Corrosion cracking has been found on brasses in organic chlorides when the chlorides also contain nitrogen-containing compounds [540].

If chlorinated hydrocarbons contain hydrochloric acid as an impurity and breakdown product, brasses with high zinc contents are severely attacked, with dezincification. The corrosion is accelerated by the copper chloride formed. Brasses with Cu contents above 85% are possibly still moderately resistant to hardly usable under these conditions.

Brass pipes are surprisingly evaluated as being suitable for carrying HCl-containing vapors and as heat exchanger pipes for HCl-containing petrol vapors at 573 K (300 °C), the hydrogen chloride originating from decomposition of chlorinated hydrocarbons. However, such a favorable evaluation cannot apply generally to brass materials [541].

Chloroacetic acids or chloroacetyl chlorides can form in chlorinated hydrocarbons, such as e.g. in 1,1,1-trichloroethane, during production, use or storage and transportation due to breakdown reactions under the action of air, light and catalysts. Brass materials are unstable to chlorinated hydrocarbons containing such breakdown products. Red regions, indicating selective corrosion, were sometimes found on the metal surface [519].

– Corrosion in monochloroethane –

Brasses are adequately resistant for applications with 100 % dry monochloroethane at room temperature and also at the boiling point. They are still moderately resistant to not particularly resistant to damp monochloroethane at temperatures up to 293 K (20 °C), but unstable above 293 K (20 °C) [521].

The cast brasses G-Cu65Zn, G-Cu60Zn, G-Cu55Zn(Mn), G-Cu55ZnAl1, G-Cu55ZnAl2 and G-Cu55ZnAl4 are also evaluated as being only moderately resistant to dry monochloroethane at 293 K (20 °C), and only the cast brass G-CuZn16Si4 (CC761S) is classified as having a good resistance [531].

A similar evaluation of the resistance of various brass qualities to monochloroethane is to be found in [542], according to which brass of low Zn content, arsenic-containing admiralty brass and Ambraloy® have a good resistance and are recommended for applications. Brass of high Zn content is evaluated as having only a moderate resistance and being not particularly suitable for use in monochloroethane.

– Corrosion in dichloroethane –

Brasses are only moderately resistant, with corrosion rates of up to 0.5 mm/a (19.7 mpy), to 1,1-dichloroethane at room temperature and also at the boiling point [521].

The cast brasses and cast special brasses G-Cu65Zn, G-Cu60Zn, G-Cu55Zn(Mn), G-Cu55ZnAl1, G-Cu55ZnAl2 and G-CuZn16Si4 are evaluated as having a moderate resistance to dry 1,2-dichloroethane at 293 K (20 °C) [531].

– Corrosion in trichloroethane –

Brasses are sufficiently resistant to well-stabilized trichloroethane as long as the effectiveness of the stabilizer lasts and no decomposition occurs. Brasses are suitable under these conditions.

Brasses are unsuitable for non-stabilized trichloroethane, and they are severely attacked in the presence of moisture or steam.

The times which elapse before corrosion starts were determined for brasses in studies on the effectiveness of stabilization of 1,1,1-trichloroethane. The brass specimens showed no corrosion for up to 144–168 h in 1,1,1-trichloroethane stabilized with various amounts of methyl-2-propynyl sulfide (1–4 vol%) and additions of nitromethane (< 1 vol%), nitroethane (< 1 vol%) and isopropyl nitrate (2–3 vol%) [526].

For brass 67/33 in dry 1,1,1-trichloroethane (100 ppm H_2O), the corrosion rate is < 0.025 mm/a (< 0.98 mpy). This confirms that this material is completely resistant to the industrial quality of this chloroethane [543].

– Corrosion in tetra-, penta- and hexachloroethane –

No information is to be found on the behavior of brasses in tetrachloroethane in the literature available.

Brasses are hardly attacked by 100 % pentachloroethane up to the boiling point. Addition of 0.01 % aniline is recommended for stabilization [529].

The cast brasses and cast special brasses already mentioned on several occasions are recommended for contact with dry pentachloroethane at 293 K (20 °C) [531].

Brass qualities containing at least 85 % Cu are resistant to 100 % hexachloroethane, even at higher temperatures.

According to [521], brasses are resistant to 100 % hexachloroethane up to 366 K (93 °C), but can no longer be used at temperatures above this figure. In the presence of moisture, brasses are attacked by hexachloroethane.

Other copper alloys

The copper alloys dealt with in this section are mainly low-alloy copper materials.

Copper-beryllium alloys with 1.7 to 2.0 % Be have a high tensile strength and heat resistance in the hardened state and are insensitive to stress corrosion cracking. They are generally evaluated as having a good resistance to chlorinated solvents [544].

The cast copper-lead alloy G-CuPb25 and the cast copper-lead-tin alloy G-CuPb5–20Sn are generally evaluated as having an adequate resistance to the usual chlorinated hydrocarbons in the temperature range from 293 to 333 K (20 to 60 °C). These alloys have a good resistance to dry monochloroethane and dry pentachloroethane at 293 K (20 °C) [531].

The cast alloys G-CuAl10Fe, G-CuAl9–10(Ni) and G-CuAl9Mn are suitable in contact with the usual dry chlorinated hydrocarbons at temperatures up to 333 K (60 °C). They are recommended for use in dry monochloroethane and dry pentachloroethane at 293 K (20 °C).

The cast nickel-aluminium bronze G-CuAl9–10(Ni) is a material which can be used in contact with damp chlorinated hydrocarbons up to 333 K (60 °C) [531].

Chloromethanes

Author: H. G. Spilker / Editor: R. Bender

Copper

Copper is utilized for many purposes in industrial fields in which solvents are produced or work is carried out with solvents. Copper materials are also utilized in systems with organic refrigerants containing halogen atoms. The utilization of copper in these fields is generally successful and without special corrosion problems. However, it is necessary to keep the systems free from access of moisture and air.

The resistance of copper to these groups of organic substances is not universal, because many chlorinated hydrocarbons react with copper. Such reactions can be intensified considerably by moisture, free acid, free chlorine or by decomposition products resulting from impacts of sunlight or heat [545].

The corrosion attack of copper materials by chlorinated hydrocarbons takes place inter-crystalline, process during which the corrosion rate differs largely in at the grain boundaries and matrix [546].

Copper can be utilized in contact with dry chlorinated organic liquids with a moisture content below 10 ppm. Also in the vicinity of the boiling point, a hardly significantly greater corrosion rate than 0.0025 mm/a (0.01 mpy) is observed [547].

However, the special reactivity of mixtures of chlorinated hydrocarbons and water must be taken into consideration, because copper is attacked therein with lower or higher corrosion rates depending on the temperature [548].

– Corrosion in methyl chloride –

Copper is generally very resistant to methyl chloride, even in the presence of moisture and at elevated temperatures. An older table specifies a corrosion rate of 0.08 mm/a (3.15 mpy) for copper in liquid methyl chloride at 373 K (100 °C) [549].

According to Corrosion Data Survey, copper with 85 to 99.9 % purity is resistant in 10 to 100 % methyl chloride up to 373 K (100 °C), in 100 % methyl chloride also at 423 K (150 °C), and in the temperature range from 248 to 273 K (–25 to 0 °C), it is resistant with corrosion rates of less than 0.0508 mm/a (2 mpy) [550].

The good resistance of copper (99.9 % Cu and 99.9 % Cu + Ag + 0.02 % P) with respect to dry methyl chloride is confirmed many times in the literature.

For the utilization of copper in contact with moist 100 % methyl chloride at 373 K (100 °C), restrictions are partly made with regard to the durability, but utilization under normal operating conditions is permitted if no other suitable material is available to fulfill further requirements [551].

In refrigeration machines operated with methyl chloride, whose pipes, condensers and evaporators are made of copper, so-called copper cladding sometimes takes place. However, this damage occurs only when an unsuitable lubricating oil is utilized in the compressors. When suitable oils were utilized, no damage occurred even when the methyl chloride was moist.

In tests, resin rich oil was kept in contact with bare copper for 3 weeks at 373 K (100 °C) in a pressurized tube. An organic compound containing copper was produced that led to deposits of copper on copper free plant components after 1 to 3 d.

If this organic compound containing copper is added to the lubricating oil of refrigeration machines the devices fail. It was concluded from this that firstly copper was dissolved of the copper parts of the equipment, forming an organic complex, and then copper was deposited at the sliding locations in exchange for iron.

The iron dissolved in this exchange process appears as iron compound in the oil sludge precipitation and clogs the lubricating system [552].

Copper lined autoclaves are utilized for producing methyl chloride by passing hydrogen chloride into a mixture of methanol and concentrated calcium chloride solution at 423 K (150 °C) [553].

– Corrosion in methylene dichloride –

Dry methylene dichloride does not corrode copper at room temperature. Methylene dichloride containing water corrodes copper, particularly at higher temperatures. The older table already mentioned several times reports corrosion rates of 0.9 mm/a (35.4 mpy) for copper in moist and acidified methylene dichloride [549].

According to Corrosion Data Survey, copper with 85 to 99.9 % purity is resistant to 90–100 % methylene dichloride at 298 K (25 °C) with corrosion rates of less than 0.0508 mm/a (2 mpy), and at 323 K (50 °C) it is moderately resistant with corrosion rates in the range of 0.508 to 1.270 mm/a (20 to 50 mpy).

Copper (CW024A, SF-Cu, 2.0090) is completely resistant to moist 60 % methylene dichloride with corrosion rates of less than 0.11 mm/a (4.33 mpy) [554].

In dry or water saturated methylene dichloride with 10 % methanol or 10 % ethanol, no weight losses of copper were found at room temperature in the vapor phase and in the liquid phase, but within 60 d a tarnish discoloration appeared that was not yet present during the first 8 d [553].

In contact with copper, solutions of acetyl cellulose in methylene dichloride/alcohol mixtures often take on a red color. This coloration is assumed to be due to small amounts of acetic acid still adhering to the acetyl cellulose. However, the weight losses of the copper remained very small.

Under production conditions of triacetate fibers tests were made with several solvents containing methylene dichloride at various temperatures. In a vapor/air mixture of the absorber with 15 to 18 g/m^2 of a solvent mixture (methylene dichloride: ethanol = 9:1) at 313 to 393 K (40 to 120 °C) and in contact with active charcoal, copper parts of the plant corroded with a material consumption rate of 0.04 g/m^2 h [555].

Russian investigations confirm the corrosion of copper in the plant for recovering solvents from the triacetyl cellulose, and the possibility for protection against corrosion by adding benzotriazole (BTA), see Table 86.

The behavior of copper in methylene dichloride solutions at room temperature is summarized in Table 87.

– Corrosion in chloroform –

Copper materials are utilized in manifold applications for producing and recovering chloroform.

According to [556], copper (85 to 99.9 % Cu) is completely resistant to chloroform (100 %) up to 323 K (50 °C) (corrosion rate < 0.0508 mm/a (2 mpy)) and moderately resistant up to 373 K (100 °C) (corrosion rate < 0.508 mm/a (20 mpy)); with respect to chloroform (90 %) it is moderately resistant up to 310 K (37 °C) (corrosion rate < 0.508 mm/a (20 mpy)) and up to 365 K (92 °C) hardly utilizable (corrosion rate above between 0.508 to 1.270 mm/a (20 to 50 mpy)).

Comparative figures are contained in the table of a metal hose manufacturer: Chloroform (dry, 100 %) at 293 K (20 °C) = completely resistant and recommended (corrosion rate less than 0.11 mm/a (4.33 mpy)), up to the boiling temperature = still adequately resistant (corrosion rate 0.1 to 1.0 mm/a (3.94 to 39.4 mpy)); chloroform (moist, 99 %) at 293 K (20 °C) = completely resistant (corrosion rate less than 0.11 mm/a (4.33 mpy)), at the boiling temperature = still adequately resistant (corrosion rate 0.1 to 1.0 mm/a (3.94 to 39.4 mpy)); chloroform (moist, 100 %) containing < 0.03 % H_2O = still adequately resistant [554].

In an older literature the following material consumption rates in g/m^2 d for the attack of copper by chloroform are specified [557]:

- at 293 K (20 °C), dry = 0.05; moist = 0.07
- at 334 K (61 °C), dry = 4.7; moist = 6.5.

Medium	Inhibitor concentration (BTA)	Test temperature	Test period	Material consumption rate	Protective effect of the inhibitor
	g/l		h	g/m^2 h	%
1*		room temperature	144	0.002	
1*	0.25	room temperature	144	0.0005	75
2**		2 h boiling temperature, rest room temperature	19	0.221	
2**	5.0	2 h boiling temperature, rest room temperature	19	0.091	
2**	10.0	2 h boiling temperature, rest room temperature	19	0.0165	93

* Medium 1 = aqueous layer in the heat exchanger, mixture of 6 % ethanol + 0.5 % methylene dichloride + balance water with 14 mg/l NH_3 and 0.3 mg/l Cu salt
** Medium 2 = aqueous layer of the water separator of the drying column, mixture of 20 % ethanol + 1 % methylene dchloride + balance water with 700 mg/l NH_3 and 65 mg/l Cu salt

Table 86: Material consumption rate of copper in the recovery plant for solvents from the production of triacetyl cellulose and the effect of benzotriazole (BTA) as inhibitor [558]

Medium	Corrosion found
Methylene dichloride, technical	dark tarnish patches
Methylene dichloride, dry	individual tarnish patches
Methylene dichloride + 10% ethanol	in liquids silver tarnish
Methylene dichloride + 10% methanol	in liquids violet tarnish
Methylene dichloride, saturated with water + 10% ethanol	yellow tarnish
Methylene dichloride, saturated with water + 10% methanol	violet tarnish

Table 87: Behavior of copper (half immersed) in methylene dichloride solutions at rest at room temperature, test duration 14 d [559]

Hydrogen chloride dissolved in chloroform attacks copper hardly at all, provided that neither air nor water are present. For distilling contaminated acidified chloroform, corrosion can be reduced by adding small amounts of soda or milk of lime. Additions of ammonia are not recommended for this purpose. A mixture of aqueous ammonia with chloroform dissolves copper rapidly already at room temperature [560].

– *Corrosion in carbon tetrachloride* –

Dry carbon tetrachloride is only slightly corrosive with respect to copper, with material consumption rates of around 0.1 g/m^2 d. No changes of the copper were found after exposure for 6 weeks at room temperature [561].

Copper materials containing 85 to 99.9% of copper are also resistant to dry CCl_4 at the boiling temperature, with corrosion rates of up to 0.0508 mm/a (2 mpy). Moisture must be excluded as far as possible.

The presence of water increases the aggressivity of carbon tetrachloride with respect to copper. In 100% CCl_4 the following corrosion rates were found under various conditions [562]:

CCl_4, dry	293 K (20 °C)	no attack
CCl_4, moist	293 K (20 °C)	0.3 mm/a (11.8 mpy)
CCl_4 vapor, moist	350 K (77 °C)	7.1 mm/a (280 mpy)
CCl_4 vapor, dry	350 K (77 °C)	slight attack

In other tests no attack was found in CCl_4-vapor at 350 K (77 °C), but the copper surface became blackened.

When water vapor was introduced into the vapor space above the CCl_4 at room temperature, no or only very slight material consumption resulted, but intense discolorations appeared on the copper surface.

Tests with CCl_4/water mixtures in evacuated melting tubes showed that copper completely immersed in the organic phase was not attacked, but a relatively strong attack took place in the aqueous phase resulting in a blue coloration of the aqueous layer and deposit of copper(I) chloride on the copper.

In hot CCl_4 in the presence of air, copper corrodes rapidly. The corrosion rate can increase by a factor of ten. Therefore, the copper materials containing 85 to 99.9% of copper which are mentioned above as being resistant to dry CCl_4, are assessed to be nonresistant to aerated 90% CCl_4 with corrosion rates > 1.270 mm/a (> 50 mpy) [563].

Copper can be protected against corrosion by adding a small amount of a stabilizer such as ethyl acetate or diphenylamine [564].

In water free CCl_4 copper is adequately resistant, whereas the presence of water causes high corrosion rates. Table 88 shows the material consumption rates of copper in CCl_4 with and without aqueous phase. The specimens were only partly immersed in the liquid during the test.

Temperature	Material consumption rate $g/m^2\,d$	
	without water	with water
Room temperature	0.02	0.4
Boiling point	0.03	87.3

Table 88: Material consumption rates of copper in carbon tetrachloride with and without aqueous phase [85]

Corrosion investigations on SF-Cu (CW024A, 2.0090) were carried out in a) water free CCl_4 (< 0.01% H_2O), b) water saturated CCl_4 (~0,01% H_2O) and c) CCl_4/H_2O mixture (1,500 ml CCl_4 + 150 ml H_2O) at the boiling temperature in glass vessels with reflux cooler. The copper specimens were mounted at various places (vapor space, condensate, organic phase, aqueous phase). In water free and water saturated CCl_4 the corrosion rate was low, i.e. copper is resistant under these conditions. However, in the two phase mixture high corrosion rates were found for copper, particularly in the aqueous phase. Here, Cu was found to be nonresistant; the specimens had dissolved completely after 30 d. The corrosion rates for Cu in the various phases are shown in Figure 58 for the respective H_2O contents. The corrosion of the copper in the two phase mixture of CCl_4/H_2O cannot take place in a direct reaction with HCl as hydrolysis product, because the electrochemical potential sequence shows that direct dissolution of copper with liberation of hydrogen cannot be expected. The mass spectrometric analysis of the gas phase proved that carbon monoxide is formed (> 90% by volume), thus showing that the corrosion process must take place via phosgene [565]:

$CCl_4 + H_2O \rightarrow COCl_2 + 2\,HCl$
$Cu + COCl_2 \rightarrow CuCl_2 + CO$

In a three phase system of CCl_4/water/hydrochloric acid, spontaneous corrosion and pitting corrosion of copper was found at the boiling temperature. Additions of stabilizers and inhibitors can prevent major damage [566].

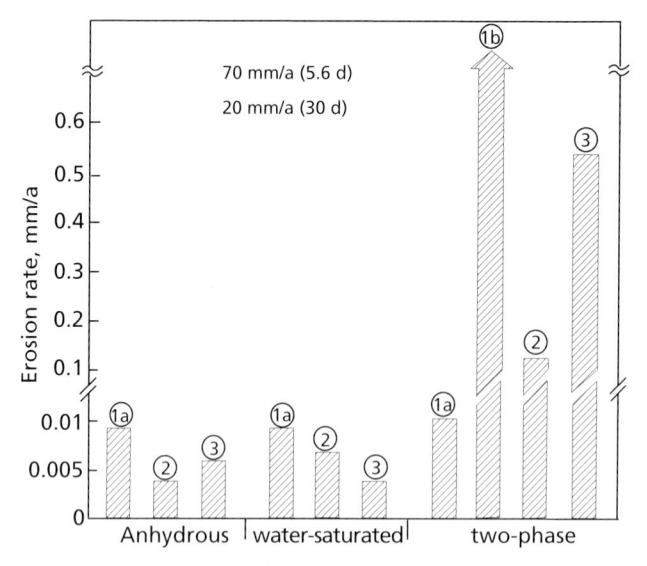

Figure 58: Corrosion rates of copper in a) water free CCl_4 (< 0.01 % H_2O), b) CCl_4 saturated with water (~ 0.01 % H_2O) and c) CCl_4/H_2O mixture (10:1) at the boiling temperature and at various places, test duration 30 d [565]
① organic phase (CCl_4)
① aqueous phase
② vapor phase
③ condensate

SF-Cu (CW024A, 2.0090) is strongly attacked in mixtures of carbon tetrachloride with alcohols, with a mixture with methanol being much more corrosive than mixtures with other alcohols.

In a CCl_4/methanol mixture (500 mg) containing 40 % CCl_4, copper dissolved within 2 weeks at 298 K (25 °C) without air sealing, leaving a yellow-brown solution without solid residue. The reaction of the copper took place slowly and uniformly until complete dissolution. Copper was also attacked in a mixture of 10 mol% CCl_4/ 90 mol% methanol at 298 K (25 °C) [567].

For investigations of the catalytic effect of metals with respect to carbon tetrachloride, the metals were provided as shot, powder, granulate or chips in twice the amount required for quantitative reaction, non-stabilized CCl_4 with more than 99.9 % purity was added, and the mixture was heated at 473 K (200 °C) for 14 d in an oil bath. In this test series copper belonged to the metals with slight reactivity. Of the added amount of CCl_4 (10.4 mmol), 2.270 mmol were converted by the reaction.

The evaluation of the tests showed the presence of the following reaction products for copper in CCl_4:

Hexachloroethane (CCl_3CCl_3)	0.178 mmol
Perchloroethylene ($CCl_2=CCl_2$)	0.255 mmol
Hexachlorobutadiene ($CCl_2=CClCCl=CCl_2$)	0.371 mmol

In the same test series, the reaction of copper with an aqueous mixture of NH₄OH/CCl₄ resulted in chloroform (2.4 %) as the only reaction product and a royal blue coloration of the reaction solution [568].

Copper-aluminium alloys

The contents of aluminium in customary technical CuAl alloys (aluminium bronzes) are restricted to about 12 %. The DIN EN 1982 stipulates 90 to 95 % Cu + 5 to 10 % Al for these alloys. With 4 to 7 % Al the microstructure of the alloys is homogeneous and suitable for cold rolling for producing pipes and sheets.

Apart from binary aluminium bronzes, aluminium bronzes with several alloy components also exist, with an aluminium fraction of about 10 % and additions of iron, nickel, manganese and silicon, microstructure of which is heterogeneous. Cast aluminium bronzes are more resistant to dealuminizing than forged alloys.

Chlorinated hydrocarbons in the dry state do not attack aluminium bronzes. Like all copper alloys, also those containing aluminium are attacked by chlorinated hydrocarbons if they contain moisture.

In resistance tables the cast aluminium bronzes G-CuAl9 (cf. CB330G), G-CuAl10Fe (CC331G), G-CuAl9-10(Ni) (cf. CC332G and CC333G) and G-CuAl9Mn at 293 to 333 K (20 to 60 °C) are assessed as being still resistant to dry chlorinated hydrocarbons with small corrosion rates, and the cast aluminium bronzes GCuAl9-10(Ni) are assessed as also being resistant to moist chlorinated hydrocarbons at 293 to 333 K (20 to 60 °C) [569].

These cast aluminium bronzes are assessed as being resistant to methyl chloride in the temperature range from 293 to 323 K (20 to 50 °C). An aluminium bronze with 8 % aluminium has a good resistance to methyl chloride under most of the operating conditions [570].

The copper alloy 637 (6–8 % Al, 1 % Sn, 1,2–2,2 % Si) is also resistant to dry methyl chloride under most conditions [571].

The already mentioned cast aluminium bronzes are fully resistant to pure methylene dichloride at 273 to 373 K (0 to 100 °C) and are still adequately resistant to moist methylene dichloride [553, 570].

In methylene dichloride containing 10 % methanol or 10 % ethanol, and saturated with water, no material consumption of the aluminium bronzes is found at room temperature, but tarnish appears [553].

Dry chloroform does not attack the cast alloys G-CuAl9, G-CuAl10Fe, G-CuAl9-10(Ni) and G-CuAl9Mn at 293 K (20 °C). G-CuAl10Fe and G-CuAl9-10(Ni) are moderately resistant to moist chloroform at 323 K (50 °C), but the alloy G-CuAl9 can hardly be utilized [569].

The Ampco® alloys, ternary bronzes, in the form of sand casting or rolled, containing 9 to 11 % Al and up to 4.25 % Fe, are resistant to chloroform with corrosion rates of up to 0.15 mm/a (5.91 mpy) at 293 to 334 K (20 to 61 °C) [572].

The aluminium bronze 637 (6–8 % Al, 1 % Sn, 1.2–2.2 % Si, rest Cu) is assessed to be resistant to chloroform under most conditions [571].

Aluminium bronzes are assessed as being resistant to dry and moderately resistant to moist carbon tetrachloride. The same assessment has been made for the cast aluminium bronzes that have already been mentioned several times [569].

Corrosion investigations of G-CuAl10Fe (CC331G, 2.0940) have been made in a) water free CCl_4 (< 0.01 % H_2O), b) CCl_4 saturated with water (0.01 % H_2O) and c) a mixture of CCl_4/H_2O (1,500 ml CCl_4 + 150 ml H_2O) at the boiling temperature in glass vessels with reflux cooler. The specimens were mounted at various places (vapor space, condensate, organic phase, aqueous phase). In water free and water saturated CCl_4 the alloy G-CuAl10Fe was resistant, with corrosion rates of < 0.01 mm/a (0.39 mpy). However, large corrosion rates were found in the two phase mixture. In the aqueous phase the CuAl alloy was found to be nonresistant. There, the specimens had dissolved completely after 30 d. In the two phase mixture selective corrosion of the aluminium was observed, indicated by a copper-red metal surface. Specimens in the condensate showed a greater corrosion rate (0.50 mm/a (19.7 mpy)) and a more intense red coloration than the specimens that were exposed to the organic phase (corrosion rate 0.25 mm/a (9.84 mpy)). The corrosion rates for the alloy G-CuAl10Fe in the various phases are plotted in Figure 59 for the various H_2O contents.

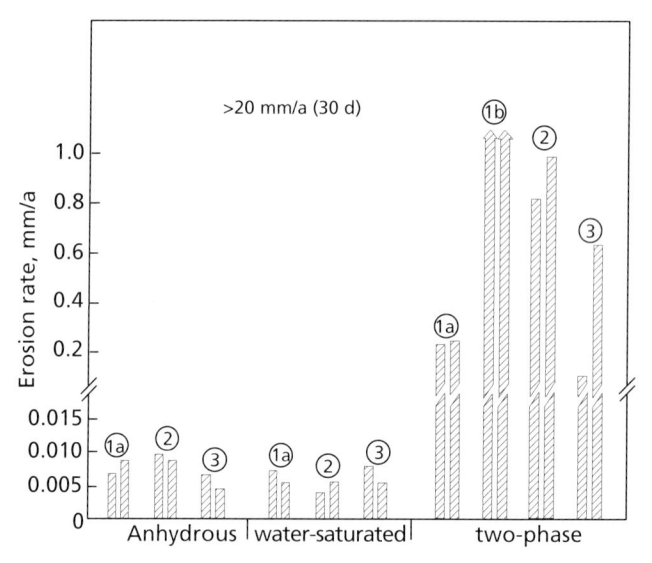

Figure 59: Corrosion rates of G-CuAl 10 Fe in a) water free CCl_4 (< 0.01 % H_2O), b) water saturated CCl_4 (~ 0.01 % H_2O) and c) CCl_4/H_2O mixture (10:1) at the boiling temperature at various places, test duration 30 d [565]
① organic phase (CCl_4)
① aqueous phase
② vapor phase
③ condensate

The mass spectroscopic gas analysis above the boiling CCl_4/H_2O mixture verified the presence of CO that is produced by a corrosion process via phosgene ($COCl_2$) [565].

For Ampco® alloys corrosion rates in carbon tetrachloride of less than 0.15 mm/a (5.91 mpy) were found [572]. The aluminium bronze 637 was assessed as being utilizable with dry CCl_4 under most operating conditions and as still adequately resistant to moist CCl_4 [571].

Copper-nickel alloys

Copper and nickel form homogeneous mixed crystals in any ratio. Copper-nickel alloys crystallize in the cubic face centered lattice. In these alloys copper is the base metal containing a nickel fraction of 5 to 30 % and often additional small fractions of iron, manganese, zinc, silicon, niobium.

Copper-nickel alloys with additions of iron or manganese belong to those materials which have the best corrosion resistance, and they are applied as alloys with 30 % nickel for pipes of evaporators and heat exchangers that are exposed to attack by corrosive chemicals such as chloromethanes.

In the absence of moisture methyl chloride attacks the copper-nickel alloys only very slightly, with low corrosion rates of less than 0.0254 mm/a (1 mpy). However, moisture and higher temperatures can lead to corrosion rates of up to 0.508 mm/a (20 mpy) [573].

The following copper-nickel alloys are resistant to dry methyl chloride:

Supernickel® 702 (%, 69Cu, 30Ni, 0.40Fe)
Cupro Nickel® 706 (%, 88.5Cu, 9–11Ni, 1Zn, 1.3Fe, 1Mn, 0.05Pb)
Cupro Nickel® 710 (%, 76.5Cu, 19–23Ni, 1Zn, 1Fe, 1Mn, 0.05Pb)
Cupro Nickel® 715 (%, 66.5Cu, 29–30Ni, 1Zn, 0.4–0.7Fe, 1Mn, 0.05Pb)
Cupro Nickel® 754 (%, 89.25Cu, 10Ni, 0.75Fe)
Nickel-silver alloy 752 (%, 65Cu, 18Ni, 17Zn)
CuNi 10 Fe(CW352H)
G-CuNi 30 (CW383H).

At temperatures of 273 to 373 K (0 to 100 °C), the cast alloy G-CuNi 30 is assessed to be resistant to pure and moist methylene dichloride [86].

A manufacturer of metal hoses and metal folding bellows assesses his products, which are made of the alloy Cunifer® 30 (CW354H, CuNi30Mn1Fe, 2.0882), as being resistant with corrosion rates of less than 0.11 mm/a (4.33 mpy) and as being suitable for the utilization in contact with dry methylene dichloride at 293 K (20 °C) and in 60 % methylene dichloride (moist) at 293 K (20 °C) and at the boiling temperature [554].

Copper-nickel alloys are generally resistant to dry chloroform. These alloys are assessed as resistant to 100 % chloroform up to the boiling temperature with corrosion rates of less than 0.0508 mm/a (2 mpy) [556].

For the materials CuNi 10 Fe and Cunifer® 30 corrosion rates of less than 0.11 mm/a (4.33 mpy) in 100 % chloroform (< 0.03 H_2O) are reported [551, 554].

The Cupro Nickel® and the nickel silver grades, too, are resistant to dry chloroform. For the cast alloy G-CuNi 30 the following durability information is available [569]:

Chloroform, dry   completely resistant at 293 K (20 °C)
Chloroform, moist   completely resistant at 293 K (20 °C)
Chloroform, moist   moderately resistant at 323 K (50 °C)
Chloroform, gas   fully resistant.

The corrosion rates of the CuNi alloy 55 45 in chloroform are shown in Table 89.

Medium	Temperature K (°C)	Corrosion rate mm/a (mpy)
Chloroform, water free	298–303 (25–30)	0.00028 (0.01)
Chloroform + water layer (partially immersed)	298–303 (25–30)	0.0007 (0.03)
Chloroform, water free	334 (boiling) (61)	0.0094 (0.37)
Chloroform + water layer	334 (boiling) (61)	0.0584 (2.30)

Table 89: Corrosion rates of the CuNi alloy 55 45 in chloroform at various temperatures [90]

Customary copper-nickel alloys are practically not attacked by water free carbon tetrachloride. The maximum observed corrosion rates are reported to be 0.025 mm/a (0.98 mpy) [562, 569].

For the copper-nickel alloy CuNi20Fe (cf. UNS C71000), the corrosion rates in carbon tetrachloride at 350 K (77 °C) were found to be 0.012 mm/a (0.47 mpy) without water and 0.05 mm/a (1.97 mpy) with water [547].

The presence of moisture and a temperature rise increase the attack of CuNi alloys by CCl_4 [574]:

CCl_4, water free   293–303 K (20–30 °C)   0.0002 mm/a (0.008 mpy)
with aqueous phase   293–303 K (20–30 °C)   0.0015 mm/a (0.06 mpy)
CCl_4, water free   350 K (77 °C)   0.012 mm/a (0.47 mpy)
with aqueous phase   350 K (77 °C)   0.05 mm/a (1.97 mpy).

For the alloy Ambrac® (nickel silver with 75 % Cu, 20 % Ni, 5 % Zn), no corrosion was found in dry CCl_4 at room temperature and at the boiling temperature. In moist CCl_4 a material consumption rate of 2.29 g/m² d at room temperature, and of 1,740 g/m² d at 340 K (67 °C) was found, as well as a green skin on the material [575].

In water/carbon tetrachloride mixtures Ambrac® suffered the small consumption of 0.000003 mm within 24 h, and in a water/carbon tetrachloride vapor mixture the high consumption of 0.00205 mm [575].

Copper-tin alloys (bronzes)

Of the tin bronzes, the alloys with α-phase up to about 14% Sn are of technical interest; bronzes with large amounts of Sn have hardly any technical applications.

As wrought or rolled alloys, these bronzes contain up to 8.5% of Sn and have a homogeneous microstructure on which good alternating load strength, formability and general corrosion resistance are based.

Tin bronzes and phosphor bronzes with 91 to 99% of copper and 1 to 9% of tin have a corrosion resistance to organic media that is roughly comparable to that of copper, but they have a higher mechanical strength [576].

In the literature the tin bronzes and phosphor bronzes are unanimously reported as being resistant to dry methyl chloride at 293 K (20°C). The cast tin bronze G-CuSn 10–14 is reported to be resistant to methyl chloride up to 323 K (50°C) [569].

The bronzes without zinc fraction are even reported to be resistant to dry and moist methyl chloride from the boiling temperature to 423 K (150°C), and they are used as material for valves, pumps and screens [553].

Fittings made of copper-tin alloys are also used for pressurized gas cylinders and in refrigeration equipment.

The behavior of tin bronzes in contact with methylene dichloride is the same as that of copper. This means that dry methylene dichloride at room temperature does not corrode tin bronzes, but in the presence of moisture, particularly at higher temperatures, it attacks these materials.

Cast tin bronze G-CuSn 10–14 is reported to be resistant to pure 100% methylene dichloride and to be moderately resistant to moist methylene dichloride up to 373 K (100°C) [569].

If there is a contact between solutions of acetyl cellulose in methylene dichloride/ alcohol mixtures and bronzes, often a red coloration of the solution is found. This effect is due to the fact that the acetyl cellulose is not completely free from adhering acetic acid. The weight consumptions of the bronzes remained slight, in spite of the discoloration [558].

The copper-tin alloys are resistant to dry chloroform (< 0.03% H_2O) up to the boiling point, they are still resistant to moist chloroform at room temperature, and they are only moderately resistant or hardly utilizable at the boiling temperature.

The bronzes are assessed to be resistant to 100% chloroform up to the boiling point (corrosion rate < 0.0508 mm/a (2 mpy)), to be moderately resistant to 90% chloroform up to 311 K (38°C) (corrosion rate < 0.508 mm/a (20 mpy)), and to be hardly utilizable at temperatures of up to 366 K (93°C) (corrosion rate 0.508 to 1.270 mm/a (20 to 50 mpy)) [24].

Cast tin bronzes have a similar corrosion behavior [569]:

Chloroform, dry	293 K (20°C)	resistant
Chloroform, moist	293 K (20°C)	resistant
Chloroform, moist	323 K (50°C)	hardly utilizable
Chloroform, gas		resistant.

As material for housings of shut off valves, cast bronzes are suitable for utilization in contact with chloroform [577].

Dry carbon tetrachloride, liquid at 288 K (15 °C) or as vapor at 349 to 350 K (76 to 77 °C) does not attack copper-tin alloys. In CCl_4/water mixtures corrosion rates of 0.146 mm/a (5.75 mpy) were found at room temperature and of 6.935 mm/a (273 mpy) at the boiling temperature [575].

Tin bronzes are assessed to be resistant in 100% dry CCl_4 at up to 353 K (80 °C) (corrosion rate < 0.0508 mm/a (< 2 mpy)), to be moderately resistant at 378 K (105 °C) (corrosion rate < 0.508 mm/a (20 mpy)) and not be resistant in 90% CCl_4 at the boiling temperature (corrosion rate > 1.270 mm/a (> 50 mpy)) [556].

For bronzes without zinc fraction, material consumption rates of around 2 g/m² d were found with air sealing. The consumption rates may increase by a factor of up to 10 if air is admitted. The addition of small amounts of ethyl acetate may prevent the corrosion [578].

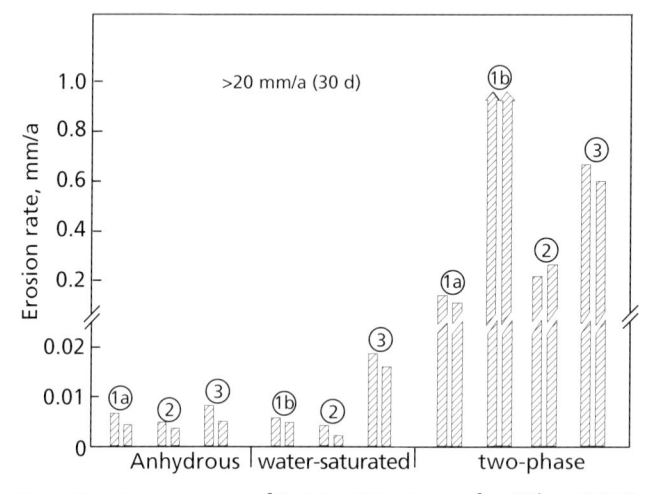

Figure 60: Corrosion rates of G-CuSn 10 in a) water free CCl_4 (< 0.01 % H_2O), b) water saturated CCl_4 (~ 0.01 % H_2O) and c) CCl_4/H_2O mixture (10:1) at the boiling temperature at various places, test duration 30 d [565]
① organic phase (CCl_4)
① aqueous phase
② vapor phase
③ condensate

The influence of the temperature and moisture on the corrosion of phosphor bronzes, manganese bronzes, Tobin bronze (60% Cu, 39.25% Zn, 0.75% Sn) and Ambrac® (75% Cu, 20% Ni, 5% Zn) in carbon tetrachloride is shown in Table 90.

Corrosion investigations with G-CuSn10 (CC480K, 2.1050) were carried out in a) water free CCl_4 (< 0.01 % H_2O), b) water saturated CCl_4 (about 0.01 % H_2O) and c) a CCl_4/H_2O mixture (1,500 ml CCl_4 + 150 ml H_2O) at the boiling temperature in glass vessels with reflux cooler. The specimens were mounted at various places (vapor space, condensate, organic phase, aqueous phase). In water free and water saturated

CCl_4 the G-CuSn10 alloy was resistant with corrosion rates of < 0.01 mm/a (0.39 mpy). Higher corrosion rates of (0.6 mm/a (23.6 mpy)) were found in the two phase mixture CCl_4/H_2O. In the aqueous phase, the CuSn-alloy was found to be nonresistant. The specimens in this phase dissolved completely in less than 30 d. The corrosion rates for the alloy G-CuSn10 in the various phases are plotted in Figure 60 for the various H_2O contents [565].

As material for housings of shut off valves, cast bronzes are suitable for utilization in contact with carbon tetrachloride [577].

In Russian investigations the friction coefficients and the abrasion of several metallic friction pairs were determined in carbon tetrachloride, dimethyl sulfoxide (DMSO) and their mixtures. For the friction pair bronze/bronze with a friction area of 0.85 cm² and a contact area metal/media of 68.5 cm², the following values were obtained:

Medium	Friction coefficient	Abrasion
CCl_4	0.31	0.0486 g
DMSO	0.18	0.2637 g
CCl_4/DMSO	0.06	0.9077 g

CuZn alloy	CCl_4	Temperature K (°C)	Material consumption rate g/m² d
Phosphor bronze	dry	room temperature	–
	dry	350 (77)	–
	humid	room temperature	34.8
	humid	340 (67)	1,610
Manganese bronze	dry	room temperature	–
	dry	350 (77)	–
	humid	room temperature	11.87
	humid	340 (67)	787
Tobin bronze	dry	room temperature	–
	dry	350 (77)	–
	humid	room temperature	3.4
	humid	340 (67)	704
Ambrac®	dry	room temperature	–
	dry	350 (77)	–

Table 90: Material consumption rates of various copper/tin alloys in carbon tetrachloride [575]

In the 1:1 mixture of CCl_4/DMSO 0.0143 g were dissolved from the friction pair bronze/bronze. In this mixture high abrasion was found, in spite of a low friction coefficient, that probably is attributable to corrosion. Compared to the friction pairs steel/steel and copper/steel that were also investigated, the friction coefficient and the abrasion of bronze/bronze was much greater [579].

Copper-tin-zinc alloys (red brass)

The copper-tin-zinc alloys, like the copper-tin alloys, are standardized according to DIN EN 1982. Being cast alloys (red casting), they contain, in addition to the main constituent copper, 4 to 10 % Sn, 2 to 7 % Zn and often also 2.5 to 7 % Pb. Some nickel is partly added to improve the strength and toughness.

The lead added to most alloys allows the production of pressure tight alloys. The significance of the zinc content is based on the shift of the phase boundary of the α-region towards lower tin contents and the deoxidation of the melts, which means that additions of phosphorus are not necessary for this purpose [580].

On the whole, the red casting grades G-CuSn10Zn (old 2.1086.01), G-CuSn5ZnPb (CC491K, old 2.1096.01) and G-CuSn7ZnPb (CC493K old 2.1090.01)) are assessed as being conditionally to moderately resistant to chlorinated hydrocarbons in the temperature range of 293 to 333 K (20 to 60 °C) [569].

With respect to dry methyl chloride, copper-tin-zinc alloys are resistant at 293 K (20 °C) and up to 323 K (50 °C). CuSnZn alloys are comparable to brass, i.e. they are resistant to dry and moist methyl chloride in the range between the boiling temperature and 373 K (100 °C) [553].

Fittings made of red casting are commonly utilized in applications involving contact with methyl chloride for pressurized gas cylinders and refrigeration units.

At 273 K (0 °C) and up to 373 K (100 °C), the cast alloys G-CuSn10Zn, G-CuSn5ZnPb and G-CuSn7ZnPb are resistant to pure methylene dichloride, and moderately resistant to moist methylene dichloride [569].

According to [553], CuSnZn alloys are assessed to have better behavior in contact with methylene dichloride than pure copper.

The copper-tin-zinc alloys are resistant to dry and moist chloroform at 293 K (20 °C). The following corrosion behavior is to be expected for the cast alloys G-CuSn10Zn, G-CuSn5ZnPb and G-CuSn7ZnPb [569]:

Chloroform, dry	293 K (20 °C)	resistant
Chloroform, moist	293 K (20 °C)	resistant
Chloroform, moist	323 K (50 °C)	hardly utilizable
Chloroform, gas		resistant.

Red casting is suitable as material for fittings in contact with chloroform [581].

The corrosivity of carbon tetrachloride for the copper-tin-zinc alloys, as well as for copper and the copper-tin alloys is higher than that of the other chloromethanes. The alloys are resistant to dry CCl_4 at room temperature and up to the boiling point.

The cast alloys G-CuSn10Zn, G-CuSn5ZnPb and G-CuSn7ZnPb and Arsenical Admirality Brass (70 % Cu, 28.96 % Zn, 1 % Sn, 0.04 % As) are assessed to be moderately resistant to moist CCl_4, and to be utilizable under some conditions.

For Tobin bronze, an alloy with small Sn content of 0.75 %, material consumption rates were found in moist carbon tetrachloride that showed a better resistance of this material compared to phosphor bronze, manganese bronze and Ambrac® (see Table 90) [575].

Copper-zinc alloys (brass)

Among the copper alloys, brass is the most widely used material. The corrosion behavior of brass depends largely on the zinc content. Brass with high Zn content (e.g. 30 %) is less resistant. Alloys with max. 37.5 % Zn have a uniform microstructure based on a cubic face centered lattice (α-brass). With zinc contents between 37.5 and 46 % α-β-microstructure is present in addition to the α-microstructure (α-β-brass).

Under corrosive conditions, the zinc-enriched β-phase in the heterogeneous microstructure is attacked preferentially. Alloys with 46 to 50 % Zn consist uniformly of β-brass. γ-Brass types with more than 50 % Zn are extremely brittle and unsuitable for technical applications.

As cast alloy, brass contains up to 60 % Cu, and as wrought alloy, it contains 58 to 90 % Cu. These wrought alloys can consist of brass without any further alloying components, or of brass with lead, or also of special brass [582].

Halogenated hydrocarbons do not attack brass in the dry state, but an attack can take place when brass is exposed to them for a long time in the presence of moisture, particularly at higher temperatures [570].

As an example of preferentially intercrystalline corrosion, the zinc depletion of some brass types by halogenated hydrocarbons is pointed out during which the attacked phases have collected on grain boundaries [546].

For α-brass with 65 % Cu and small fractions of Fe, Sn, Pb and Ni no stress corrosion cracking was found in tests with halogenated organic media, but rapid zinc depletion of the brass took place [583].

According to an information with regard to anti corrosion methods and materials, corrosion cracking takes place in organic materials containing chlorine when compounds containing nitrogen are participating [584].

– Corrosion in methyl chloride –

The brass cast alloys G-Cu55Zn(Mn), G-Cu55ZnAl1-2, G-Cu55ZnAl4, G-CuZn16Si4 are completely resistant to methyl chloride up to 323 K (50 °C).

Brass with 80 % Cu is resistant to dry and moist methyl chloride at the boiling temperature and up to 373 K (100 °C) [553].

Fittings for pressurized gas cylinders and refrigeration units, screens and filters of refrigeration machines are partly made of Tombak (CuZn20, CW503L, 2.0250) [552].

– Corrosion in methylene dichloride –

Brass containing > 80 % Cu is resistant to dry and water saturated methylene dichloride at room temperature and up to the boiling point.

The following statements are made with regard to the resistance of cast alloys with respect to methylene dichloride in the temperature range from 273 K (0 °C) to 373 K (100 °C) [570]:

- G-Cu55Zn(Mn)

  – Methylene dichloride, pure = fully resistant
  – Methylene dichloride, moist = hardly utilizable

- G-Cu55ZnAl1-2

  – Methylene dichloride, pure = fully resistant
  – Methylene dichloride, moist = hardly utilizable

- G-Cu55ZnAl4

  – Methylene dichloride, pure = fully resistant
  – Methylene dichloride, moist = hardly utilizable

- G-CuZn16Si 4

  – Methylene dichloride, pure = fully resistant
  – Methylene dichloride, moist = suitable under normal operating conditions.

In corrosion tests with volatile solvents for selecting suitable materials for regenerating plants, for brass a mixture of 25 % methylene dichloride, 3 % ethanol and 72 % steam was found that produced a corrosion rate of 0.037 mm/a (1.46 mpy) at 398 K (125 °C). A spread discoloration appeared on the metal surface [585].

– Corrosion in chloroform –

The brass alloys are all assessed to be fully resistant to dry chloroform (< 0.03 % H_2O) at room temperature. α-brass containing > 85 % Cu and < 80 % Cu, α-β-brass and α-β-special brass are assessed to be resistant to dry chloroform [570].

Brass with 70 to 80 % Cu (+ Zn, Sn or Pb) is resistant to 100 % chloroform at room temperature with corrosion rates of < 0.0508 mm/a (2 mpy), and it is moderately resistant up to the boiling point with corrosion rates of < 0.508 mm/a (20 mpy). Brass with 59 to 93 % Cu (+ Al, Zn or As) is resistant to 100 % chloroform up to the boiling point with corrosion rates of < 0.0508 mm/a (2 mpy), moderately resistant to 90 % chloroform at 323 K (50 °C) with corrosion rates of < 0.508 mm/a (20 mpy), but nonresistant at the boiling temperature with a corrosion rate of 1.270 mm/a (50 mpy) [563].

– Corrosion in carbon tetrachloride –

The behavior of brass in carbon tetrachloride is similar to that of copper, but there are considerable differences in the corrosion rates with moist CCl_4. Dry carbon tetrachloride is not corrosive for brass. Table 91 shows the corrosion rates of Cu and CuZn alloys in moist CCl_4, measured after 24 h at various temperatures. The material consumption rates of CuZn alloys in CCl_4 are shown in Table 92.

Brass with a zinc content greater than 20% suffers selective corrosion in moist carbon tetrachloride, during which zinc depletion takes place at high rates until dissolution.

While CuZn10 (MS90, CW501L) and CuZn15 (CW502L) can still be utilized in contact with moist CCl_4 under normal operating conditions, CuZn40 (Ms60, CW509L), CuZn35.5Pb3 and CuZn40Pb3 are nonresistant [571].

The cast alloys G-Cu55Zn(Mn), G-Cu55ZnAl1-2, G-Cu55ZnAl4 and G-CuZn16Si4 are completely resistant to dry CCl_4, but hardly utilizable anymore with moist CCl_4 [569].

In corrosion tests of copper materials in carbon tetrachloride the copper-zinc alloys CuZn15 and CuZn36 (CW507L) were placed in boiling dry and water saturated CCl_4, in a boiling CCl_4/water mixture in the organic phase, in the aqueous phase, in the vapor space and in the condensate. The material CuZn15 is resistant in water free and water saturated boiling carbon tetrachloride with corrosion rates of less than 0.01 mm/a (0.39 mpy). In the organic phase of the two phase mixture CuZn15 can still be considered as resistant with a corrosion rate of approx. 0.02 mm/a (0.79 mpy), but in the vapor space the corrosion rate increases to 2.0 mm/a (78.7 mpy). This brass is nonresistant in the aqueous layer, the specimens dissolved in less than 30 d.

Material	Material consumption rate, mm/a (mpy)	
	293 K (20 °C)	Boiling temperature
Cu	3.28×10^{-3} (0.13)	7.55×10^{-2} (2.97)
CuZn10 (Ms90)	2.55×10^{-3} (0.10)	1.43×10^{-1} (5.63)
CuZn20 (Ms80)	7.25×10^{-1} (28.5)	2.93 (115)
CuZn30 (Ms70)	4.38×10^{-3} (0.17)	6.68 (263)
CuZn40 (Ms60)	1.46×10^{-3} (0.06)	18.25 (719)

Table 91: Corrosion rates of copper and copper-zinc alloys in moist carbon tetrachloride, test duration 24 h [575]

Brass CuZn36 (Ms62, CW507L) suffered greater corrosion rates than CuZn15 in water free and water saturated CCl_4, but is still assessed to be resistant. In the two phase mixture high material consumption rates were found for this alloy; CuZn36 was nonresistant in the organic phase, in the vapor phase and in the condensate. Very large material consumption rates were found in the aqueous layer; the specimens had dissolved after a few days. The corrosion of this brass type was combined with zinc depletion.

Material	Temperature K (°C)	Material consumption rate g/m² d	Comment
CuZn40 (Ms60)	room temperature	3.42	brown deposit
	340 (67)	4.25×10^4	red deposit
CuZn30 (Ms70)	room temperature	10.70	brown deposit and green patches
	340 (67)	1.96×10^4	red deposit
CuZn20 (Ms80)	room temperature	16.90	brown deposit
	340 (67)	6.82×10^3	purple film and green precipitate
CuZn10 (Ms90)	room temperature	6.18	iridescent purple deposit and green patches
	340 (67)	3.33×10^2	green precipitate

Table 92: Material consumption rates of copper-zinc alloys in moist carbon tetrachloride [575]

The detection of carbon monoxide (> 90 % by volume) in the gas above the boiling CCl_4/water mixture proves that the corrosion in the aqueous phase takes place via phosgene. The corrosion rates of CuZn15 (CW502L, 2.0240) and CuZn36 (CW507L, 2.0335) in carbon tetrachloride with various water contents at the boiling temperature are shown in the Figures 61 and 62 [565].

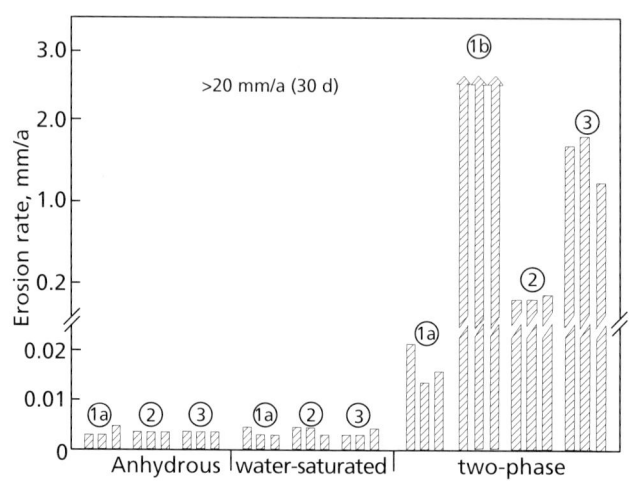

Figure 61: Corrosion rates of CuZn15 in a) water free CCl_4 (< 0.01 % H_2O), b) water saturated CCl_4 (~ 0.01 % H_2O) and c) CCl_4/H_2O mixture (10:1) at the boiling temperature at various places, test duration 30 d [565]
① organic phase (CCl_4)
① aqueous phase
② vapor phase
③ condensate

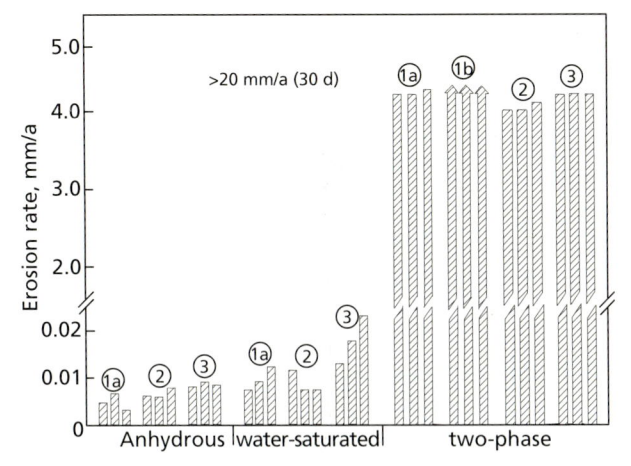

Figure 62: Corrosion rates of CuZn36 in a) water free CCl_4 (< 0.01 % H_2O), b) water saturated CCl_4 (~ 0.01 % H_2O) and c) CCl_4/H_2O mixture (10:1) at the boiling temperature in various places, test duration 30 d [565]
① organic phase (CCl_4); ① aqueous phase; ② vapor phase; ③ condensate

A comparison of the corrosion of various copper materials in the boiling two phase mixture CCl_4/H_2O (10:1) is shown in Figure 63. CuZn36 (Ms62, CW507L) suffered large corrosion rates under test conditions. The surface of these specimens took on a copper-red color and therewith indicated zinc depletion. This zinc depletion led to a spongy brittle state; it was possible to crumble the specimens. By qualitative X-ray investigations the corrosion products were identified as alkaline zinc chloride ($ZnCl_2 \times 4\ Zn(OH)_2$), Paratacamite ($Cu_2Cl(OH)_3$) and Nantokite (CuCl) [565].

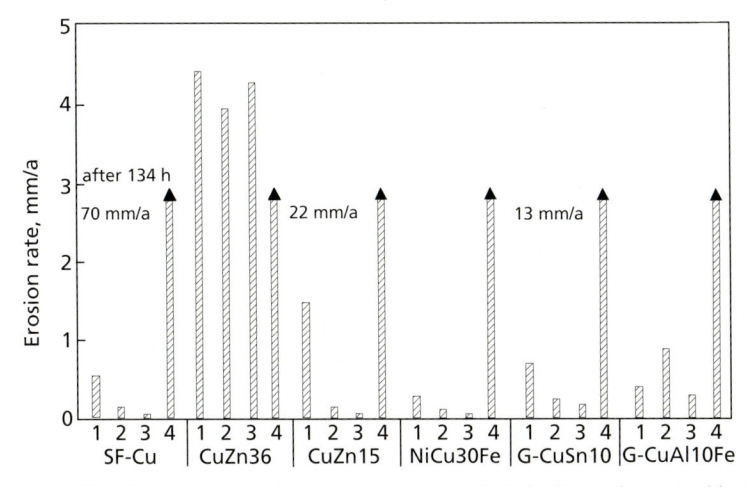

Figure 63: Corrosion rates of various copper materials in boiling carbon tetrachloride/water mixtures in various places, test duration 30 d [565]
1 = condensate; 2 = vapor; 3 = sump (organic); 4 = sump (aqueous); ↑ = specimen dissolved in 30 d or corrosion rates in mm/a after 134 h

The addition of 0.001 to 0.1 % of aniline is recommended as an effective corrosion inhibitor for brass in moist carbon tetrachloride [586].

Other copper alloys

Copper-beryllium alloys containing 1.7 to 2.0 % Be in the hardened state have high tensile strength and temperature resistance, and they are generally not sensitive to stress corrosion cracking. Their resistance to chlorinated solvents is assessed to be good [587].

Copper-silicon alloys containing up to 5 % Si and partly additional components like manganese and sometimes zinc, are copper-rich materials with higher strength, but which can be formed and forged. These alloys – silicon bronze (98.5 % Cu, 1.5 % Si), high silicon content bronze (95.0 % Cu, 3.0 % Si, 1.0 % Mn), Everdur® (91.0–98.25 % Cu, 0.4–3.1 % Si, Mn, Sn, Al) – are resistant to dry methyl chloride. These alloys are also resistant to dry methylene dichloride. At the boiling temperature silicon bronzes are slightly attacked by methylene dichloride/water mixtures.

The silicon bronzes 651, high silicon content bronze, Everdur® 1010 (95.8 % Cu, 3.1 % Si, 1.1 % Mn) and Everdur® 1015 (98.5 % Cu, 1.5 % Si, 0.2 % Mn) are resistant to 100 % chloroform at 293 K (20 °C). The latter alloys are also completely resistant at 293 K (20 °C) to dry, but only moderately resistant to moist carbon tetrachloride. A less favorable assessment is found in Corrosion Data Survey, where silicon bronze is reported to be nonresistant to 90 % CCl_4 at 293 K (20 °C) [588].

Copper-lead alloys (bimetal) and copper-lead-tin alloys (trimetal) have very special technical significance as materials for crankshaft bearing shells of motor vehicle engines. The content of lead, that is present in the solid state in almost insoluble and finely distributed particles in the microstructure, extends up to 26 %. The copper-lead and copper-lead-tin cast alloys G-CuPb25 and G-CuPb5-20Sn (cf. CC495K, CC496K, CC497K) can be assessed as being generally resistant to moderately resistant up to 333 K (60 °C) to chlorinated hydrocarbons. The following assessments are made for the chloromethanes [569]:

Methyl chloride	293–323 K (20–50 °C)	completely resistant
Methylene dichloride, pure	273–373 K (0–100 °C)	completely resistant
Methylene dichloride, moist	273–373 K (0–100 °C)	moderately resistant
Chloroform, dry	293 K (20 °C)	completely resistant
Chloroform, moist	293 K (20 °C)	completely resistant
Chloroform, moist	323 K (50 °C)	hardly utilizable
Chloroform, gas		completely resistant
Carbon tetrachloride, dry		completely resistant
Carbon tetrachloride, moist		moderately resistant.

Chlorine and Chlorinated Water

Author: K. Hauffe / Editor: R. Bender

Copper

The chlorine contained, in addition to other impurities, in industrial atmospheres also attacks copper which is used as a protective coating on electronic components [589]. Even chlorine contents below 2 ppm in moist air (about 75 % relative humidity (RH)) attack the surface rapidly enough for the wettability (and therefore the solderability) of the copper to be drastically impaired after 21 days and the contact resistance to have increased by 3 to 6 powers of ten at the same time.

Although no systematic studies on the chlorination of copper to form cuprous or cupric chloride have been carried out between 423 and 723 K (150 and 450 °C), it may be concluded – in analogy to the formation of silver chloride on silver and on the basis of electrochemical studies (current density/potential measurements in the system: Cu/CuCl/graphite) – that such processes occur. The rate of corrosion of copper in chlorine or that of the formation of cuprous chloride surface layers should be reduced by alloying the copper with metals which form bivalent ions (for example, cadmium, nickel and indium) [590]. No experiments have, however, been carried out so far in this direction.

Some first quantitative studies on the attack of chlorine on copper in moist air were published already in 1923; at that time, a kinetic law was proposed. In that case, copper had been exposed, with rapid heating (from 288 to 503 K (15 to 230 °C) in 10 minutes), to the action of air which had been passed through chlorinated water (2×10^{-4} mol/l chlorine). The reported average rate of chlorination was 2.3×10^{-4} cm^2/h [591]. A paper which appeared in the same year deals with the kinetics of the chlorination of copper at 293 K (20 °C) and with the morphology of the resulting corrosion product layer [592]. Figure 64 illustrates the time dependence of the chlorination of copper (and, for comparison, also of silver: curves 4 and 5) in moist and dry chlorine-containing nitrogen (nitrogen passed through 1:4 chlorinated water) at 293 K (20 °C). As can be seen in Figure 64, water vapor in the atmosphere increases the rate of corrosion considerably. Whereas a transparent cuprous chloride layer, invisible to the eye, is formed during the first reaction period, grayish-white layers are formed during the subsequent stages and turn dark-gray during storage in daylight (photochemical effect), especially in moist air or moist nitrogen. The grayish-white color shade reappears on storage in the dark.

Experiments with single crystals of pure copper in chlorine (1.33×10^{-3} bar, 717 K (444 °C)) have shown, moreover that the crystal faces (111), (011), (012) and (001) are corroded with varying intensity, but the first two most severely [593]. Copper chloride sublimates at this high temperature and leaves a clean corroded copper surface behind. Kinetic studies of the reaction of copper with chlorine within the low pressure range between 743 and 1,023 K (470 and 750 °C) showed a high impact yield of the chlorine molecules impinging on copper, with little influence of temperature. Thus, for example, at 743 K (470 °C), every third to fourth impact leads to a reaction [594]. Increasing the temperature to 1,040 K (767 °C) increases the impact yield, and therefore the reaction rate, only by one and a half times. Some experimental results are compiled in Table 93.

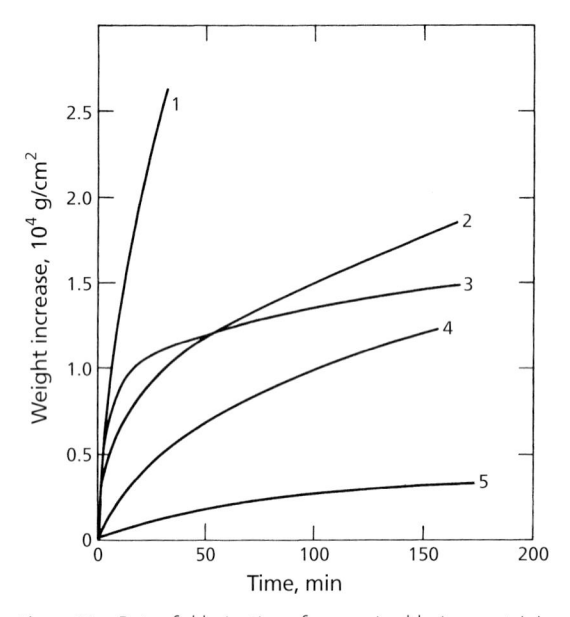

Figure 64: Rate of chlorination of copper in chlorine-containing nitrogen at 293 K (20 °C) [592]
1: with water vapor copper; 2: gas dried over $CaCl_2$ copper; 3: gas dried over H_2SO_4 copper; 4: with water vapor silver; 5: dried over $CaCl_2$ silver

Temperature K (°C)	p_{Cl_2} 10^{-6} bar	n/n_0	Reaction rate mol/cm² h
743 (470)	1.3	0.23	3.2×10^{-4}
743 (470)	3.7	0.23	9.1×10^{-4}
973 (700)	8.6	0.13	1.2×10^{-3}
1,023 (750)	1.1	0.36	4.1×10^{-4}
1,023 (750)	39.9	0.42	1.8×10^{-2}

Table 93: Impact yield (n/n_0 = reacted copper atoms per chlorine molecule impact) and rate of chlorination as a function of chlorine pressure and temperature [594]

The activation energy of the chlorination, calculated from the high impact yields and the low temperature coefficient is remarkably low (10.9 kJ).

As a consequence of the pronounced heat evolution during the reaction of copper with chlorine at elevated pressures, the reaction rate increases so much above 50 K that the metal starts glowing; the attack proceeds most intensely at edges and curved areas [595, 596].

Some experimental data on the corrosion of copper in chlorine (approx. 1 bar, 423 to 573 K (150 to 300 °C)) are compiled in Table 94. According to these data, the flow rate has a marked influence, at constant gas composition, on the rate of corrosion. While, at 15 liters per hour, the copper was completely oxidized under the experimental conditions [597, 598], the rate of chlorination at 2.5 liters per hours was reduced by about two powers of ten.

Temperature K (°C)	Flow rate ml/h	Weight loss g Cu/cm^2 h	Estimated loss of Cu mm/a (mpy)
423 (150)	1.5×10^4	1.02×10^{-4}	0.88 (34.6)
443 (170)	1.5×10^4	5.7×10^{-4}	5.0 (197)
493 (220)	1.5×10^4	1.62×10^{-3}	14 (551)
523 (250)	1.5×10^4	9.60×10^{-3}	84 (3,307)
558 (285)	1.5×10^4	2.10×10^{-2}	180 (7,087)
573 (300)	1.5×10^4	2.64	23,000* (905,512)
573 (300)	2.4×10^3	2.64×10^{-2}	230 (9,055)
573 (300)	2.1×10^2	6.60×10^{-3}	57 (2,244)

* only for 1 hour

Table 94: Mean corrosion rates of copper in a chlorine stream (about 1 bar) after 6 hours [597, 598]

Chlorine attack is accelerated by oxygen and hydrogen between 473 and 523 K (200 and 250 °C), but lowered at higher temperatures [599, 600]. The ignition temperature is shifted at the same time from 573 to about 625 K (300 to about 352 °C).

In order to be able to decide whether cupric chloride appears in addition to cuprous chloride, cuprous chloride was chlorinated between 348 and 403 K (75 and 130 °C) [601]:

$2 CuCl + Cl_2(gas) \rightarrow 2 CuCl_2$

Following an initial indistinct reaction phase, the growth of the surface layer of cupric chloride, building up on the cuprous chloride, may be described by a parabolic law (energy of activation: 47.3 kJ/mol). It is to be expected, according to these studies, that, at high chlorine partial pressures, the corrosion layer being formed on copper will consist of an outer layer of cupric chloride and an inner layer of cuprous chloride. Whereas predominantly copper ions diffuse in the cuprous chloride layer, the mechanism of the diffusion and growth of the cupric chloride layer is still largely unknown at present.

It is also worth mentioning that, as experiments with copper in 0.1 mol chlorinated water at 302 K (29 °C) in the presence of air have shown, copper corrodes in chlorinated water more quickly under the influence of light than in the dark. During experiments lasting 7 hours, a rate of corrosion of 1.9×10^{-4} g/cm^{2}h was determined in the dark and one of 2.6×10^{-4} g/cm^{2}h on exposure to light [602]. Presumably, chlorine atoms are created by the action of light:

$$Cl_2(gas) \xrightarrow{h\nu} 2\,Cl(gas)$$

as a result of which the formation of chloride is accelerated:

$$Cu + Cl(gas) \rightarrow CuCl$$

the extent of formation of cuprous chloride is, however, diminished by its simultaneous photochemical decomposition.

Copper is also mentioned among the materials that are suitable for spray devices injecting chlorine into reactors; other materials mentioned in this connection are tantalum, steel, platinum, CrNi-steel 18 8, NiCu-alloy (Monel®), NiMo-alloys (Hastelloy B and C), NiCr-alloys (Inconel® 600) and nickel. The rate of corrosion depends, in this case, on the flow rate and the size of the liquid droplets formed [603].

Copper alloys

Because of the small number of publications on the action of gaseous chlorine and chlorinated water on copper alloys, so this alloys are dealt with jointly in this section.

According to studies in a sea water desalination unit, copper alloys (Table 95) are attacked to varying extents by chlorinated sea water (chlorine prevents sludge formation) [604]. The chlorine contents are 3 ppm in these cases, during short periods (45 minutes), 18 to 20 ppm. Figure 65 shows a typical corrosion curve for brass (Admiralty). This shows that, within one hour after discontinuing the additional supply of chlorine, the corrosion rate has dropped to the rate corresponding to the normal chlorine content of 3 ppm. Addition of chromate or phosphate reduces corrosion by about 80 % under these conditions (from 38×10^{-3} to 7×10^{-3} mm/a (1.50 to 0.28 mpy)). It should be noted in this connection that, as expected, chlorine attack is intensified with decreasing pH. Figure 66 shows the mean corrosion rate in the heat exchanger and in the vertical evaporators which operate at different temperatures, before and after addition of some ppm of chromate.

Alloy	Cu	Ni	Al	Fe	Mn	Sn	Pb	Zn
CDA 122	99.9	0.01	0.01	0.01	–	–	–	–
CDA 706	88.0	10.0	0.01	1.25	0.5	0.01	0.01	–
CDA 715	68.0	30.4	0.01	0.60	0.06	0.01	–	–
Al-Brass	77.5	0.01	2.20	0.01	–	0.03	0.01	20
Brass (Admiralty)	72	0.01	–	0.06	–	1.0	–	26.5

Table 95: Composition of copper alloys tested in a desalination unit (percent by weight) [604]

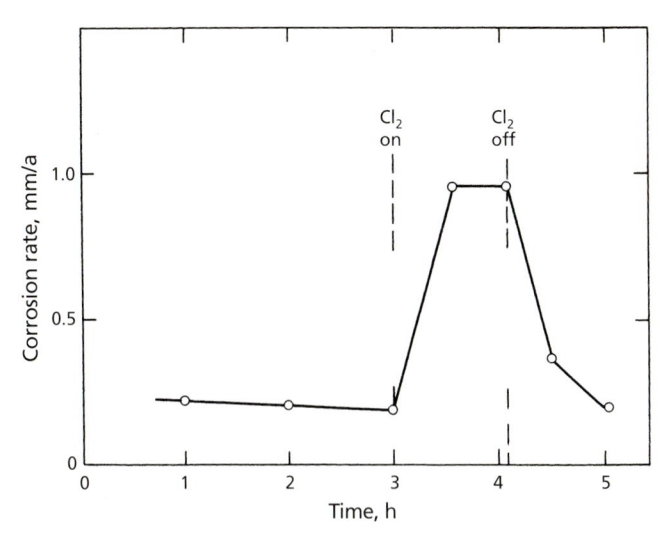

Figure 65: Influence of 18 to 20 ppm chlorine on the rate of corrosion of brass (Admiralty) in sea water in the first heat exchanger (320 K (47 °C)) of a desalination unit [604]

The feasibility of decreasing the corrosion of brass by addition of inhibitors to chlorinated solutions was studied also in simulated cooling water (mg/l: 42.1 magnesium sulfate, 70.2 calcium sulfate, 68.5 sodium bicarbonate, 13.4 sodium chloride, 26.4 calcium chloride) containing 10 mg/l of chlorine (maintained by supplying chlorine gas or by adding Chloramine® T or sodium hypochlorite) (Table 96). The results showed that benzotriazole and tolyltriazole yield high inhibition efficiency, whereas mercaptobenzothiazole is unsuitable: Excess mercaptobenzothiazole inactivates chlorine as a biocide, while excess chlorine inactivates the inhibitor (Figures 67 and 68).

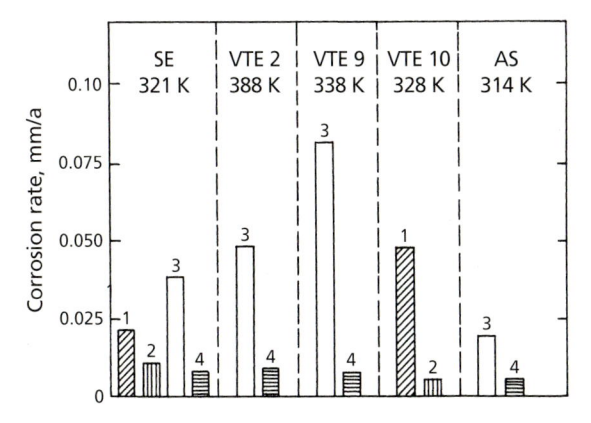

Figure 66: Mean rate of corrosion of copper and brass (Admiralty) in the sea water heater (SE), the evaporators (VTE) and the original brine (AS) before and after addition of chromate [604]
1: brass before addition of the inhibitor; 2: brass after addition of the inhibitor; 3: copper before addition of the inhibitor; 4: copper after addition of the inhibitor

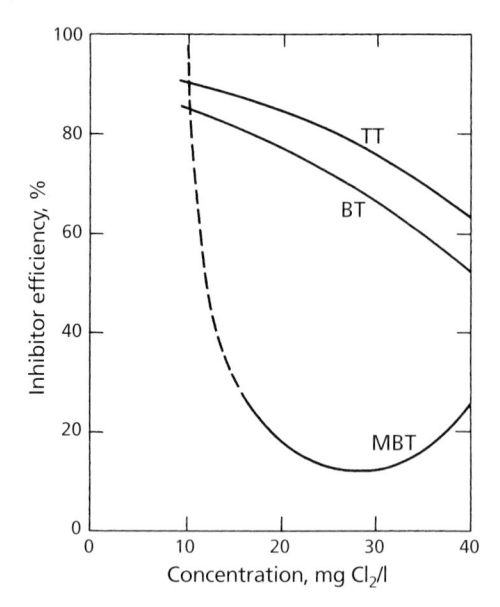

Figure 67: Relationship between chlorine concentration and inhibitor efficiency (IE, %) in the corrosion of brass in simulated cooling water containing 10 mg/l inhibitor at 323 K (50 °C) [605]
BT = benzotriazole
MBT = mercaptobenzothiazole
TT = tolyltriazole

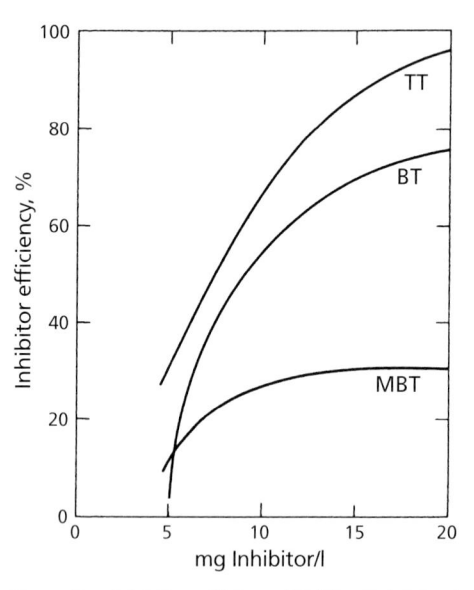

Figure 68: Inhibiting efficiency (IE, %) of tolyltriazole (TT), benzotriazole (BT) and mercaptobenzothiazole (MBT) on the corrosion of brass in sea water containing 40 mg/l chlorine at 323 K (50 °C) [605]

Chlorine source	Inhibitor*	Chlorine mg/l		Inhibitor efficiency %
		Start	End	
Chloramine® T	BT	10.0	5.9	96
	TT	10.2	5.8	95
	MBT	9.7	2.5	25
NaOCl	BT	9.9	2.7	74
	TT	9.4	3.2	90
	MBT	8.8	0.5	27
Chlorine (Gas)	BT	10.4	4.6	84
	MBT	10.6	0.4	15

* BT = benzotriazole
TT = tolyltriazole
MBT = mercaptobenzothiazole

Table 96: Effect of inhibitors on the corrosion of brass (Admiralty) at 323 K (50 °C) in simulated cooling water containing chlorine originating from different sources. Chlorine concentration: 10 mg/l; inhibitor concentration: 5 mg/l [605]

The corrosion of domestic water fittings made of brass (Ms 63 37) is intensified by the chlorination of drinking water, too; however, chlorine has a considerably weaker action in this case than other oxidizing agents (potassium persulfate, ammonium perchlorate and bromine) [606].

The corrosion behavior of aluminium brass is distinctly different; because the latter (for example Cu20Zn2.2Al) might be an inexpensive alternative to the preferentially used CuNi-alloys, studies were carried out to see whether it might be a suitable material for heaters and heat exchangers in desalination units. The studies carried out in 45 desalination units (total capacity: 750,000 m^3 per day) in the Middle East showed that

- aluminium brass is, in fact, resistant under these conditions and has cost advantages relative to the CuNi-alloys
- aluminium brass should not be used when the sea water is strongly chlorinated and
- aluminium brass is usually more sensitive than CuNi-alloys to impingement attack.

This sensitivity to impingement attack can, however, be markedly reduced by adding ferrous ions to the sea water (Figure 69) [607, 608].

Damage to aluminium brass pipes in condensers cooled with sea water in Japanese power stations is attributed to impingement attack caused by sea water containing chlorine and also manganous ions: manganese dioxide, formed in the presence of chlorine, is deposited on the internal wall of the pipe and, being an active cathode, intensifies the corrosion as soon as the surface film is locally destroyed by erosion or other mechanical action. The problem could be partially solved by using AP bronze pipes and stopping chlorine addition to a large extent [609].

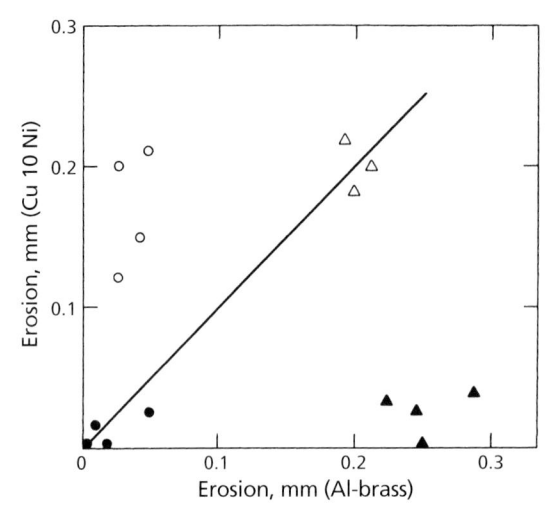

Figure 69: Impingement attack in sea water containing no chlorine and containing 0.2 ppm of chlorine with and without sponge ball cleaning (SC) (10 balls/week) of aluminium brass and Cu10Ni-alloy after 362 days [607]
● : no chlorine and no SC
○ : no chlorine, with SC
▲ : containing chlorine, no SC
△ : containing chlorine, with SC

Even more resistant than aluminium brass are CuNi-alloys whose resistance can be improved still further by addition of iron as an alloying element [610, 611]. As the data given in Table 97 show, however, the iron content in the alloy should not exceed 0.7 % since the corrosion resistance deteriorates again at higher contents.

It is interesting that under certain conditions CuNi-alloys are more sensitive to impingement attack than aluminium brass; this is the case when the piping is cleaned with the aid of sponge balls and regardless of whether the sea water is chlorinated or not.

During short-term contact with gaseous chlorine (4.5×10^{-4} bar; 673 K (400 °C); 5 min), a surface layer consisting predominantly of cuprous chloride is formed on CuAl-alloys containing 8.3 % Al and CuSi-alloys containing 3.1 % Si [612]. In this case, the layer formation is greatly influenced by the dislocation structure of the surface, which gives rise to internal stresses in the chloride layer, so that corrosion may be accelerated owing to the inevitable crack formation in the surface layer.

All bronze slide valves are offered by a valve manufacturer as "suitable for dry chlorine" [613].

Alloy	Corrosion rate, mm/a (mpy) Chlorine content			
	None	0.25 ppm	0.5–1.0 ppm*	0.5 ppm Cl_2**
Cu10Ni-0.7Fe	0.0058 (0.23)	0.0091 (0.36)	0.0079 (0.31)	0.0061 (0.24)
Cu10Ni-1.5Fe	0.0069 (0.27)	0.0140 (0.55)	0.0119 (0.47)	0.0071 (0.28)
Admiralty Brass (Cu28Zn1Sn)	0.0074 (0.29)	0.0132 (0.52)	0.0150 (0.59)	0.0099 (0.39)
Al-Brass (Cu21.3Zn,2.2Al)	0.0030 (0.12)	0.0066 (0.26)	0.0018 (0.07)	0.0036 (0.14)
Cu30Ni-0.47Fe-0.45Mn	0.0028 (0.11)	0.0025 (0.10)	0.0020 (0.08)	0.0020 (0.08)

* 1 hour's action of chlorine after every 6 hours
** 1/2 hour's action of chlorine after every 8 hours

Table 97: Influence of chlorine content in flowing sea water on the corrosion of Cu10Ni- and other copper alloys with varying iron contents; results from tests (62 days, May to July) at Harbor Island, North Carolina (USA) [611]

Copper alloys (CuZn37 (CW508L), CuAu50Cd7, CuSn24Zn18, CuNi18Zn21 and CuBe2 (CW101C)) are sometimes employed as coatings on electronic components and are said to be very resistant to chlorine-containing atmospheres; however, they are not yet fully equivalent substitutes for noble metals as contact materials.

In this connection, it is of interest that only cuprous chloride is formed during the action of chlorine on CuAg-alloys, because the reaction

$$Cu + AgCl \rightarrow CuCl + Ag$$

proceeds (as has been confirmed by experiment) only in that direction owing to its ΔG value [614, 615].

Chlorine Dioxide

Author: L. Hasenberg / Editor: R. Bender

Copper

Copper is attacked in aqueous solutions in the presence of oxidizing agents [616].

Since chlorine dioxide is a potent oxidizing agent, only a low resistance in media containing ClO_2 is to be expected of copper and its alloys. However, it should be remembered that because of its tendency to explode and its high activity, for example as a disinfectant, ClO_2 is used industrially only in low concentrations. Copper, like its alloys, is not resistant to pure chlorine dioxide. The corrosion rates at room temperature are stated as being more than 1 mm/a (39.4 mpy).

Copper and its alloys are also not very resistant in chlorites, hypochlorites and bleaching agents generally [617], especially in acid solutions. On the other hand, it can be used in alkaline solutions at room temperature.

The corrosion rates in bleaching agents such as calcium hypochlorite and sodium hypochlorite depend on the concentration, but as a rule are too high for industrial use, especially at a higher concentration [618, 619]. The decisive factor, however, is probably the fact that chlorites are destroyed by copper materials, and the corrosion products lead, for example, to discoloration of the bleached material.

The material consumption rate of copper at 353 K (80 °C) in a 5 g/l $NaClO_2$ bleach (pH 4) is low, at 1.9 $g/m^2 d$ (0.08 mm/a). The test duration was 7 to 10 days.

Precision etching of printed circuit boards of copper can be undertaken by means of ammoniacal sodium chlorite solution. $NaClO_2$ acts as an oxidizing agent here. The pH of the solution is between 8 and 10 and the temperature is 299 K (26 °C) [620].

Alkaline solutions for decopperization also work on the basis of chlorites as oxidizing agents [621].

– Use of copper in water supply –

Copper materials have found a wide field of use in water containing the most diverse additives and at the most diverse temperatures, for example as heat exchangers and in domestic installations. They have proved particularly suitable under these conditions, although corrosion damage cannot be excluded. An important prerequisite for the resistance of copper in water is the formation of a protective Cu(I) oxide coating. Disturbances in the external build-up of the coating and external influences of a local nature may lead to pitting corrosion. In addition to other causes, there are chlorides above all which trigger off this type of corrosion. Their influence is poorly or scarcely quantifiable. Such damage can be caused, for example, by chlorine dioxide as a disinfectant. A precondition for this is a local concentration of chlorides. Low pH-values have a promoting effect. For the suitability of copper materials in drinking water, with the corrosion rates to be expected at a known composition of the water, see the following sections and Table 98 [622].

Alloy	CDA[1]	UNS No.	DIN-Mat. No.	Chemical composition, %								Immersion time years	Corrosion rate[2] mm/a (mpy)	Max. pitting depth mm
				Cu	Zn	Sn	Al	Ni	Fe	Mn	Pb	Si			
Aluminium bronze	614	C61400	CW303G 2.0932	balance	–	0.3	7	–	2	–	–	–	11.5	0.001 (0.039)	0.79
	952	C95200	CB331G 2.0941	88	–	–	9	–	3	–	–	–	12.2	0.002 (0.079)	0.15
	953	C95300	–	88	–	–	10	–	1.2	–	–	–	12.2	–[3]	0.05
	954	C95400	–	84.5	–	–	10	–	5	0.5	–	–	13.3	0.007 (0.28)	2.54[4]
	955	C95500	CB334G 2.0981	80.5	–	–	9.5	5	4	0.5	–	–	13.3	0.001 (0.039)	0.69
	958	C95800	CB333G 2.0976	79.5	–	–	9.5	5	3	3	–	–	13.3	0.002 (0.079)	0.41

1) Copper Development Association
2) average of at least 2 samples
3) not determined
4) including dealuminization below the pit
5) dezincification in the form of pitting
6) sample fracture due to dezincification

Table 98: Results from corrosion tests according to ASTM G4-68 on copper alloys in New York drinking water (see Table 99) over a period of up to 14 years [622]

Table 98: Continued

Alloy	CDA[1]	UNS No.	DIN-Mat. No.	Chemical composition, %									Immersion time years	Corrosion rate[2] mm/a (mpy)	Max. pitting depth mm
				Cu	Zn	Sn	Al	Ni	Fe	Mn	Pb	Si			
Manganese bronze	675	C67500	–	57	41.0	0.5	0.2	–	0.9	0.3	0.1	–	13.3	0.008 (0.31)	1.52[5]
	862	C86200	–	67.0	22.0	–	5.2	–	2.1	3.6	–	–	9.2	0.006 (0.24)	> 4.01[5,6]
	865	C86500	CB765S 2.0602	57.6	40	0.6	0.5	–	0.4	0.7	0.2	–	11.5	0.007 (0.28)	0.38[5]
	868	C86800	CB764S 2.0606	balance	36.7	0.1	0.5	3.8	1.2	3.9	–	–	13.3	0.006 (0.24)	0
Nickel silver	752	C75200	CW409J 2.0740	65	16.5	–	–	18.5	–	–	–	–	8	0.006 (0.24)	0
Muntz metal	280	C28000	CW509L 2.0360	61.0	38.9	–	–	–	–	–	–	–	13.3	0.010 (0.39)	0.48[5]

1) Copper Development Association
2) average of at least 2 samples
3) not determined
4) including dealuminization below the pit
5) dezincification in the form of pitting
6) sample fracture due to dezincification

Table 98: Results from corrosion tests according to ASTM G4-68 on copper alloys in New York drinking water (see Table 99) over a period of up to 14 years [622]

Table 98: Continued

Alloy	CDA[1]	UNS No.	DIN-Mat. No.	Chemical composition, %								Immersion time years	Corrosion rate[2] mm/a (mpy)	Max. pitting depth mm
				Cu	Zn	Sn	Al	Ni	Fe	Mn	Pb	Si			
Tin brass	420	C42000	–	88	10	2	–	–	–	–	–	–	13.3	0.004 (0.16)	0
Lead bronze	836	C83600	CB491K 2.1097	85	5	5	–	–	–	–	5	–	13.3	0.004 (0.16)	0
Lead-tin bronze	922	C92200	–	88.4	3.5	6.4	–	–	–	–	1.5	–	8	0.005 (0.20)	0
	932	C93200	CB493K 2.1091	83.7	2.8	6.6	–	–	–	–	6.8	–	10.5	0.004 (0.16)	0
	937	C93700	CB495K 2.1177	81.3	–	9	–	–	–	–	9.7	–	10.5	0.003 (0.12)	0
	945	C94500	–	73	–	8	–	–	–	–	18.9	–	10.5	0.002 (0.079)	0

1) Copper Development Association
2) average of at least 2 samples
3) not determined
4) including dealuminization below the pit
5) dezincification in the form of pitting
6) sample fracture due to dezincification

Table 98: Results from corrosion tests according to ASTM G4-68 on copper alloys in New York drinking water (see Table 99) over a period of up to 14 years [622]

Table 98: Continued

Alloy	CDA[1]	UNS No.	DIN-Mat. No.	Chemical composition, %									Immersion time years	Corrosion rate[2] mm/a (mpy)	Max. pitting depth mm
				Cu	Zn	Sn	Al	Ni	Fe	Mn	Pb	Si			
Nickel-tin bronze	947	C94700	–	88	2	5	–	5	–	–	–	–	13.3	0.006 (0.24)	0
	948	C94800	–	87	2	5	–	5	–	–	1	–	13.3	0.006 (0.24)	0
	949	C94900	–	80	5	5	–	5	–	–	5	–	13.3	0.006 (0.24)	0
Copper-nickel	706	C70600	CW352H 2.0872	88.1	0.1	–	–	10	1.4	–	–	–	8	0.009 (0.35)	0
	715	C71500	CW354H 2.0882	68.9	–	–	–	30.1	0.6	–	–	–	8	0.007 (0.28)	0
	962	C96200	CC380H 2.0815.01	86.2	–	–	–	10	1.4	1.3	–	–	8	0.010 (0.39)	0

1) Copper Development Association
2) average of at least 2 samples
3) not determined
4) including dealuminization below the pit
5) dezincification in the form of pitting
6) sample fracture due to dezincification

Table 98: Results from corrosion tests according to ASTM G4-68 on copper alloys in New York drinking water (see Table 99) over a period of up to 14 years [622]

Electrochemical studies in hot water also confirm the promoting influence of chloride ions on pitting corrosion under these conditions, this influence is increasing as the concentration increases [623, 624].

At this point, it should be remembered that electrochemical detection of pitting corrosion as a function of the Cl⁻-concentration presents relatively few problems in the laboratory, but monitoring in practice in the corresponding plants is difficult, even though installations and equipment are available for this.

Corrosive attack on copper in hypochlorite solutions occurs locally as pitting corrosion with corrosion rates of 0.51 mm/a (20.08 mpy).

Copper-aluminium alloys

Approximately the same as for copper applies to the resistance of aluminium bronzes. They cannot be used in bleaching and hypochlorite solutions. The corrosion rates depend on the concentrations [618], and are more than 1 mm/a (39.4 mpy) [619]. Furthermore, discoloration of the products may occur under these conditions [617].

The corrosion rates of Al bronze in bleaching solutions at room temperature are stated as being up to 13 mm/a (512 mpy). In the same work, Al bronze blades are used in hollander machines, although in 2% NaOCl solution at room temperature.

The use of recycled paper in bleaching and the use of closed water circulations, for environmental reasons, and also increasing temperatures in the processes has led to greater stress on the suction roll materials in the paper industry. The aluminium bronze GC-CuAl9.5Ni is insensitive to chlorides as a suction jacket material. The pH should be above 4.5, the S^{2-}-content should not exceed 2 ppm, the $S_2O_4^{2-}$-content should not exceed 10 ppm and the solution should not contain mercury or ammonia [625].

For corrosion damage to be largely excluded, the material should be checked frequently. GC-CuAl9.5Ni has proved to be a suitable suction roll material in screen water at ambient temperature, consisting of 20 mg/l Na_2SO_3, 60 mg/l $Na_2S_2O_3$, 270 mg/l Na_2SO_4 and 20 mg/l NaCl and having a pH, adjusted with sulfuric acid, of 3.5. The corrosion current density in such a solution was determined as 13 $\mu A/cm^2$ by electrostatic corrosion studies. This corresponds to a corrosion rate of 0.17 mm/a (6.69 mpy). This alloy thus proved to be the most resistant copper material studied [626].

According to [627], suction rolls are a weak point in the paper industry, which is why they are given particular attention. Intercrystalline stress corrosion cracking was the reason for the failure of the material CuAl10Ni3Fe2-C (CC332G, old: 2.0970.01). According to detailed studies, the cause of this damage was a low mercury content in the mixing water (no composition data).

As a rule, felting screen jackets (felting screens on card-board machines) are made of the aluminium bronze GC-CuAl9.5Ni. CrNi steels are used only in those cases where ammonium ions are to be expected in the water, since these attack the aluminium bronze [628].

The use of aluminium bronzes as components in drinking water treated with ClO_2 has been considered.

Studies in New York drinking water (Rondout Reservoir) over a period of up to 14 years have shown that the corrosion rate of the copper materials studied is very low, sometimes less than 0.01 mm/a (0.39 mpy) (Tables 98 and 99). However, pittings reaching depths of up to 2.6 mm were found in this period in particular on the aluminium bronzes studied.

Water conditions	
Alkalinity as $CaCO_3$	5–14 mg/l
Calcium	3–6.3 mg/l
Carbon dioxide	1.8–4.0 mg/l
Chlorides	4–7 mg/l
Copper	0.01–0.10 mg/l
Dissolved oxygen	13.7–14.3 mg/l
Hardness as $CaCO_3$	19–23 mg/l
Sulfates	6–14 mg/l
Silicon as SiO_2	0.5–2.5 mg/l
Solids content	40–54 mg/l
pH	6.8–7.5
Conductivity	60–73 µS/cm
Temperature (winter/summer)	275/289 K (2/16 °C)
Flow rate	0.03 m/s

Table 99: Compilation of data obtained from corrosion tests on New York drinking water (Rondout Reservoir, Grahamsville) (Table 98) [622]

The pit depth data include the additional dealuminization in the base of the pit. In comparison with seawater, in which Ni improves the corrosion resistance of the bronze, it has the opposite effect in this drinking water. According to Table 98, use of aluminium bronzes for a prolonged period of time should therefore be excluded [628].

Copper-nickel alloys

CuNi alloys are attacked hardly at all by dry chlorine [618], but severe attack is to be expected by ClO_2. Increased corrosive attack is to be expected from damp gas [618]. The corrosion rate in chlorine dioxide at room temperature is in excess of 1.27 mm/a (50 mpy). Copper-nickel alloys therefore cannot be used here.

Corrosion rates of more than 1 mm/a (39.4 mpy) are stated for CuNi30-Fe1Mn1NbSi-C (CC383H, old: 2.0835.01) at room temperature in aqueous sodium hypochlorite solution [619]. CuNi is resistant in 2% NaOCl solution at room temperature, with maximum corrosion rates of 0.008 mm/a (0.31 mpy).

If used, in-house corrosion studies are advisable, depending on the corrosion conditions, and it can be assumed that copper-nickel alloys have a good resistance when dealing with commercially available chlorine dioxide (low concentrations), especially at room temperature.

In chloride-containing drinking water, as may occur after treatment with ClO_2, CuNi alloys exhibit practical resistance with maximum corrosion rates of 0.01 mm/a (0.39 mpy), without pitting corrosion occurring, as can be seen from Table 98 [622].

According to drinking water legislation, the enrichment of Cu in water should be remembered in respect of the usability of such alloys.

Copper-tin alloys (bronzes)

The adverse effect on bleached material in many cases renders the use of copper materials impossible.

An adequate resistance of CuSn alloys can be expected in bleaching solutions and hypochlorite solutions (no precise data) [618].

According to [619], corrosion rates of more than 1 mm/a (39.4 mpy) are to be expected. This material also cannot be used in pure ClO_2 gas.

CuSn alloys can be used in alkaline or very dilute neutral bleaching solutions, as in 2% sodium hypochlorite solutions, at room temperature with corrosion rates of less than or about 0.01 mm/a (0.39 mpy).

In an $NaClO_2$ solution (5 g/l) at pH 4 and 353 K (80 °C), material consumption rates of about 0.4 g/m²d (test duration 1 day) are reported. This approximately corresponds to a corrosion rate of 0.02 mm/a (0.79 mpy). Catalytic decomposition of the bleaching solution, or degradation of sodium chlorite, is associated with this corrosion. The corrosive attack is characterized by pitting corrosion.

Material trade name / composition	Corrosion rate mm/a (mpy)	Observations
88 10 2 bronze (88Cu, 10Sn, 2Zn)	46.0 (1,811)	samples destroyed
Duriron® (14.5Si, 0.66Mn, 0.8C, balance Fe)	0.18 (7.11)	etched
Durichlor® (≥ 14Si, ≥ 3Mo, balance Fe)	0.08 (3.15)	the beginnings of pitting corrosion

Table 100: Results from corrosion tests on metals under operating conditions during the bleaching of cellulose at the top of the chlorine dioxide reaction tower (medium consists of conc. H_2SO_4, 32% $NaClO_3$, CH_3OH, ClO_2-gas, test duration 351 h at 330 K (57 °C)) – no other details given – [629]

Table 100: Continued

Material trade name / composition	Corrosion rate mm/a (mpy)	Observations
AISI 430 (1.4016) (X6Cr17)	11.70 (461)	samples destroyed
AISI 304 (1.4301) (X5CrNi18-10)	14.0 (551)	samples destroyed
AISI 316 (1.4401) (X5CrNiMo17-12-2)	3.81 (150)	samples perforated (0.79 mm)
Durimet® 20, CN-7M (GX7NiCrMoCuNb25-20)	0.69 (27.2)	pitting (max. depth 1.5 mm)
Nickel	12.2 (480)	samples destroyed
Inconel® 600 (2.4816) (NiCr15Fe)	9.4 (370)	samples destroyed
LaBour R55® (52Ni, 23Cr, 8Fe, 6Cu, 4Mo, 4Si, 2W, 0.2–0.3C)	0.051 (2)	no pitting
Hastelloy® C (cast) (NiMo 16 Cr)	0.33 (13)	no pitting
Chlorimet® 3 (UNS N30107) (NiMo18Cr)	0.51 (20.1)	no pitting
Monel® 400 (2.4360) (NiCu 30 Fe)	12.20 (480)	samples destroyed
Lead (99.93Pb, 0.07Cu, 0.005Ag)	0.16 (6.3)	etched
Tantalum	0.003 (0.12)	no attack

Table 100: Results from corrosion tests on metals under operating conditions during the bleaching of cellulose at the top of the chlorine dioxide reaction tower (medium consists of conc. H_2SO_4, 32 % $NaClO_3$, CH_3OH, ClO_2-gas, test duration 351 h at 330 K (57 °C)) – no other details given – [629]

The bronze CuSn10-C (CC480K, old: 2.1050.01) is found to be insensitive to chlorides in washing solutions and is used as a suction roll material if the pH of the solution is above 4, and the solution contains no NH_3 or Hg and contains less than 2 ppm and 10 ppm of S^{2-} and $S_2O_4^{2-}$ respectively [625].

Bronze can tend to suffer vigorous corrosion in waters in the paper industry when an increasing sulfide content coincides with simultaneous contact and erosion corrosion [628].

In chloride-containing model screen water of pH 3.5 (established with H_2SO_4) and containing 20 mg/l Na_2SO_3, 60 mg/l $Na_2S_2O_3$, 270 mg/l Na_2SO_4 and 20 mg/l NaCl, a corrosion current density of 0.026 mA/cm^2 (corresponding to 0.30 mm/a) is established in CuSn10-Cu and of 0.042 mA/cm^2 (corresponding to 0.50 mm/a) in the alloy CuSn12-C (CC483K, old: 2.1052.01) during potentiostatic testing. The resistance of these alloys under such conditions increases as the Sn content decreases [626].

In drinking water of pH 7.8 (no data on its composition), a corrosion rate on bronze of 0.003 mm/a (0.12 mpy) is to be expected. Chlorine contents increase the corrosion rate to 0.013 mm/a (0.51 mpy). In contaminated drinking water, such as is present, for example, during paper manufacturing, it is greatly increased at 0.54 mm/a (21.3 mpy), in particular by 40 ppm alum (potassium aluminium sulfate) at a pH of 3.3. In such waters, the addition of ClO_2 and the pH should be monitored [630].

Copper-tin-zinc alloys (red brass)

The same applies to these alloys as to copper and other alloys dealt with above. Alloys with about 5% Zn are no longer used in circulation water in the paper industry having a high chloride content, since embrittlement of these alloys has been found as a consequence of dezincification [628].

Their resistance depends on the particular corrosion conditions [618, 619].

Red brass should be used only in very dilute neutral bleaching solutions at room temperature. The corrosion rate of an alloy with 27% Zn, 5.5% Sn and Cu as the balance is about 0.01 mm/a (0.39 mpy). In this case too, it is less the linear than the local corrosion rate and the effect on the bleaching agent and the materials to be bleached which are of importance.

In chloride-containing synthetic screen water of pH 3.5 and containing 20 mg/l Na_2SO_3, 60 mg/l $Na_2S_2O_3$, 270 mg/l Na_2SO_4 and 20 mg/l NaCl, a corrosion current density of 0.026 mA/cm^2 was found on the alloy G-CuSn10Zn under potentiostatic conditions. This corresponds to a corrosion rate of 0.32 mm/a (12.6 mpy). In this case too the resistance of the copper materials rises as the pH of the solution increases [626].

An overview of the corrosion behavior of a CuSnZn alloy with 10% Sn and 2% Zn under operating conditions for pulp bleaching can be seen in Tables 100 to 103 (No. 1). According to these studies, the alloy at best can be used in the spent ClO_2 in the ventilation lines of the head of the bleaching tower (Table 103). However, the corrosion rates of 0.14 mm/a (5.51 mpy) with local attack are too high or too unreliable for use over a period of several years [629]. In HCl-operated plants for ClO_2 production (more advantageous than H_2SO_4), even more vigorous corrosion is to be expected, since HCl is more aggressive than H_2SO_4.

Table 104 contains a summary of the various stages of the bleaching process. Amongst these, ClO_2, which is replacing chlorine more and more, is the most aggressive constituent. In view of environmental protection, the safest material should always be used.

Material trade name / composition	Corrosion rate mm/a (mpy)	Observations
88 10 2 bronze (88Cu, 10Sn, 2Zn)	17.27 (680)	severely etched
Duriron® (14.5Si, 0.66Mn, 0.8C, balance Fe)	15.75 (620)	severe attack
Durichlor® ($\geq$ 14Si, $\geq$ 3Mo, balance Fe)	0.08 (3.15)	no pitting
AISI 430 (1.4016) (X6Cr17)	10.16 (400)	perforated samples
AISI 304 (1.4301) (X5CrNi18-10)	8.38 (330)	perforated samples
AISI 316 (1.4401) (X5CrNiMo17-12-2)	7.37 (290)	perforated samples
Durimet® 20, CN-7M (GX7NiCrMoCuNb25-20)	4.83 (190)	perforated samples
Nickel	8.89 (350)	etched samples
Inconel® 600 (2.4816) (NiCr15Fe)	7.62 (300)	perforated samples
LaBour R55® (52Ni, 23Cr, 8Fe, 6Cu, 4Mo, 4Si, 2W, 0.2–0.3C)	11.68 (460)	pitting (max. depth 2.41 mm)
Hastelloy® C (cast) (NiMo16Cr)	1.17 (46.1)	no pitting
Chlorimet® 3 (UNS N30107) (NiMo18Cr)	0.38 (15)	no pitting
Monel® 400 (2.4360) (NiCu30Fe)	9.14 (360)	etched samples
Lead (99.93Pb, 0.07Cu, 0.005Ag)	0.77 (30.3)	no pitting
Tantalum	0.003 (0.12)	no attack

Table 101: Results from corrosion tests on metals under operating conditions during the bleaching of cellulose (see also Table 100) in chlorine dioxide gas lines leading to the absorber (8–10% ClO_2 gas and ClO_2 solutions in the form of condensates, flow rate 425 m³/h, 339 K (66 °C), test duration 350 h) [629]

Material trade name / composition	Corrosion rate mm/a (mpy)	Observations
88 10 2 bronze (88Cu, 10Sn, 2Zn)	8.64 (340)	severely etched
Duriron® (14.5Si, 0.66Mn, 0.8C, balance Fe)	0.003 (0.12)	no pitting
Durichlor® (≥ 14Si, ≥ 3Mo, balance Fe)	0.003 (0.12)	no pitting
AISI 430 (1.4016) (X6Cr17)	0.64 (25.2)	perforated samples
AISI 304 (1.4301) (X5CrNi18-10)	0.13 (5.12)	incipient pitting
AISI 316 (1.4401) (X5CrNiMo17-12-2)	0.003 (0.12)	no pitting
Durimet® 20, CN-7M (GX7NiCrMoCuNb25-20)	0.018 (0.71)	no pitting
Nickel	8.13 (320)	etched intercrystalline corrosion
Inconel® 600 (2.4816) (NiCr15Fe)	2.79 (110)	perforated samples
LaBour R55® (52Ni, 23Cr, 8Fe, 6Cu, 4Mo, 4Si, 2W, 0.2–0.3C)	0.09 (3.54)	no pitting
Hastelloy® C (cast) (NiMo16Cr)	0.13 (5.12)	no pitting
Chlorimet® 3 (UNS N30107) (NiMo18Cr)	0.36 (14.2)	macro-etching
Monel® 400 (2.4360) (NiCu30Fe)	12.19 (480)	samples destroyed
Lead (99.93Pb, 0.07Cu, 0.005Ag)	5.33 (210)	no pitting
Tantalum	0.003 (0.12)	no attack

Table 102: Results from corrosion tests on metals under operating conditions during the bleaching of cellulose (see also Table 100), in the chlorine dioxide receiver (4–5 g/l, pH 2–3.5, test duration 351 h at 275 K (2 °C)) [629]

CuZnSn alloys show a very good resistance in drinking water (Table 99). The corrosion rates are less than 0.01 mm/a (0.39 mpy), and no pitting corrosion was detectable after a test duration of 13.3 years [622].

Copper-zinc alloys (brass)

The corrosion resistance of these alloys in particular depends on the content of the base metal zinc. Wrought alloys show a better resistance than cast alloys. Brass containing 60 to 90% copper can be used in water with a chlorine content of 1 mg/l with corrosion rates of up to 0.01 mm/a (0.39 mpy). These alloys are not resistant to damp chlorine, and the same is also to be expected for ClO_2 solutions.

In hypochlorite-solutions, they can likewise be used only at room temperature and at a concentration below 2% NaOCl. Their use is no longer possible in solutions of higher concentration [618, 619].

After a test duration of 24 h in a solution of 5 g/l $NaClO_2$ at 353 K (80 °C) and pH 4, the following material consumption rates (corrosion rates) resulted:

CuZn 40 12 g/m² d (0.52 mm/a)
CuZn 33 10 g/m² d (0.43 mm/a)
CuZn 28   5 g/m² d (0.22 mm/a).

Material trade name / composition	Corrosion rate mm/a (mpy)	Observations
88 10 2 bronze (88Cu, 10Sn, 2Zn)	0.132 (5.2)	localized etching
Duriron® (14.5Si, 0.66Mn, 0.8C, balance Fe)	0.010 (3.94)	incipient pitting
Durichlor® (≥ 14Si, ≥ 3Mo, balance Fe)	0.003 (0.12)	incipient pitting
AISI 430 (1.4016) (X6Cr17)	0.023 (0.91)	pitting
AISI 304 (1.4301) (X5CrNi18-10)	0.015 (0.59)	pitting
AISI 316 (1.4401) (X5CrNiMo17-12-2)	0.013 (0.51)	pitting
Durimet® 20, CN-7M (GX7NiCrMoCuNb25-20)	0.005 (0.19)	no attack
Nickel	0.018 (0.71)	slightly etched
Inconel® 600 (2.4816) (NiCr15Fe)	0.018 (0.71)	1 perforation
LaBour R55® (52Ni, 23Cr, 8Fe, 6Cu, 4Mo, 4Si, 2W, 0.2–0.3C)	–	not tested

Table 103: Results from corrosion tests on metals under operating conditions during the bleaching of cellulose, in the vent line from the top of the bleach tower (spent ClO_2 gas) (see also Table 100) (341 K (68 °C), 338 h) [629]

Table 103: Continued

Material trade name / composition	Corrosion rate mm/a (mpy)	Observations
Hastelloy® C (cast) (NiMo16Cr)	0.005 (0.19)	no attack
Chlorimet® 3 (UNS N30107) (NiMo18Cr)	0.008 (0.31)	no attack
Monel® 400 (2.4360) (NiCu30Fe)	0.015 (0.59)	incipient pitting
Lead (99.93Pb, 0.07Cu, 0.005Ag)	0.36 (14.2)	pitting (max. 0.89 mm)
Tantalum	0.003 (0.12)	no attack

Table 103: Results from corrosion tests on metals under operating conditions during the bleaching of cellulose, in the vent line from the top of the bleach tower (spent ClO_2 gas) (see also Table 100) (341 K (68 °C), 338 h) [629]

As can be seen from this summary, according to expectations the corrosion rate increases as the Zn-content increases. The corrosive attack manifested itself exclusively as pitting corrosion with a pitting density of 100 pittings per 100 cm^2.

Use of zinc-containing alloys over a prolonged period of time should be avoided in chlorine-containing drinking water (Table 99). The corrosion rates are relatively low, but pitting corrosion occurs, with deep pitting depths (Table 98) [622].

Other copper alloys

These can be expected to have a similar corrosion behavior to that of other copper materials in bleaching solutions and hence comparably in ClO_2 solutions. As the pH increases, their resistance will increase, but in solutions containing Cl^- pitting corrosion is to be expected. Multicomponent alloys as a rule show less favorable corrosion chemistry properties.

In synthetic screen water (see Sections copper-aluminium alloys, copper-tin alloys (bronzes), copper-tin-zinc alloys (red brass)), corrosion current densities of about 0.014 mA/cm^2 were established on the alloy CuSn5Zn5Pb5-B (CB491K) in potentiostatic corrosion studies. These current densities correspond to corrosion rates of 0.18 mm/a (7.09 mpy) [626]. Use under such conditions is acceptable only with reservations.

In sodium chlorite solution of 5 g/l, pH 4 and 353 K (80 °C), the nickelsilver alloys Maillechort (CuNi45Zn15 and CuNi25Zn12) show material consumption rates of 0.4 $g/m^2 d$, but the alloy with 45 % Ni and 15 % Zn and Cu as the balance shows a rate of 9.3 $g/m^2 d$. For the first two alloys, this approximately corresponds to a corrosion rate of 0.02 mm/a (0.79 mpy), hardly any pitting corrosion being observed on the first alloy and exclusively pitting corrosion being observed on the second alloy. The alloy with 15 % Zn corrodes at a rate of about 0.38 mm/a (15 mpy), but by pitting corrosion to only a minor degree. No data are available from this reference on the development of the pitting corrosion or on the pitting depths. The alloy containing more Ni and less Zn shows less catalytic decomposition than the alloys with 25 % Ni and 12 % Zn. The alloy with 45 % Ni and 15 % Zn proves to be even more unfavorable here.

The corrosion behavior of the alloys dealt with in this section in drinking water varies (Tables 98 and 99).

Although the Ni-free manganese bronzes exhibit low corrosion rates after a test duration of 9.2 years, they show a marked tendency to suffer from pitting corrosion with pitting depths of up to almost 6 mm (CDA 670) [622].

The CDA alloys 752, 836, 922, 947–949 and 658 are resistant and exhibit no pitting corrosion after 8 to 13 years [622].

Bleaching sequences	
Common bleaching sequences	
C-E-H	C-E-H-D-H
C-E-H-H	C-E-H-D-E-D
C-E-D-H	C-D-E-H-H
C-E-H-D	C-D-E-D-E-D
C-E-D-E-D	D-E-D-E-D
Using peroxide	
(C)-P/E-H-D-H	(C)-P/E-H-D
(C)-P/E-H-D	(C)-P/E-H
Using oxygen	
O-D-E-D	O-C-D-E-D-E-D
O-D-P-D	O-O/D-E-D-E-D
O-D-O-D	O-C/D-O-D
O-D-E-D-E-D	O-C-E-D-E-D
O-C/D-O-D	O-C-E-D
O-C-D-E-H-D	

C chlorine
E extraction with alkali
P/E peroxide and alkali
D chlorine dioxide
H hypochlorite
O oxygen

Table 104: Overview of the general bleaching sequences (2–7 stages) resulting from the available raw material and the desired degree of whiteness [631]

Drinking Water

Author: G. Heim, K. Reeh / Editor: R. Bender

Copper

Water pipes made of copper are widely used in domestic installations. Copper can be regarded as resistant to drinking water in almost all applications. This statement is applicable without restriction to uniform surface corrosion of copper pipes. In these cases, protective layers of copper(I) oxide form on the inner surfaces.

There are, however, cases of corrosion damage in which these layers are locally penetrated, i.e. pitting occurred. Typical pitting phenomena are known, which have hitherto been described in the literature and in [632] as Type I and Type II. Type I occurs almost exclusively in cold water, while Type II is predominantly restricted to hot water. Type I is by far the most frequent form of damage [633]. Damage according to Type II has become much less frequent in recent years, which can be attributed, inter alia, to reduced water temperatures due to energy-saving measures and the decrease in local supplies of acidic water.

The materials listed in Table 105 can be used for drinking water lines [633].

Experience in practice has, however, shown that copper pipes produced from oxygen-containing copper (E-Cu58 and E-Cu57) are more sensitive to pitting than the oxygen-free copper grades. In Germany and other West European countries, oxygen-free copper SF-Cu (CW024A), deoxidized with phosphorus, is used or prescribed [633]. Copper pipes are supplied in rod form (extended length) and in rings according to DIN EN 1057 [634]. According to [632], they are marked by embossing in the longitudinal direction. The marking contains the following data: pipe diameter, wall thickness, DIN EN 1057, manufacturer's name, if appropriate DVGW mark with registration number and the quality mark of the Copper Pipe Quality Association [635].

The state of the inner surfaces of the copper pipes has a marked influence on the corrosion resistance. Campbell found as early as 1950 [636] that a high proportion of copper pipes which contained carbon deposits on the inner surfaces were destroyed by pitting. The same phenomenon was also observed in the Netherlands [637] and in Germany [638]. Carbon deposits are formed by annealing soft pipes.

The lubricants used for drawing are converted into carbon deposits in this process. Hard copper pipes are drawn but not annealed, which means that no carbon deposits are formed on the copper pipes nowadays supplied in Germany. In the region of brazed joints, however, locally restricted carbon deposits can form in the event of high temperatures and prolonged treatment [638]. In cold waters, carbon deposits cause Type I pitting only if the following conditions apply:

- certain types of water [639]
- the carbon deposit must have been formed as a continuous film on the inner surface of the copper pipe.

Material number	Old designation		Composition %	Density kg/dm³
CW004A CW005A	Oxygen-containing copper E-Cu58	2.0065	Cu ≥ 99.90 oxygen 0.005 to 0.040[1]	8.9
–	E-Cu57	2.0060	Cu ≥ 99.90 oxygen 0.005 to 0.040[1]	8.9
CW008A	Oxygen-free copper, not deoxidized OF-Cu	2.0040	Cu ≥ 99.95	8.9
CW020A CW021A	Oxygen-free copper, deoxidized with phosphorus SE-Cu[2]	2.0070[2]	Cu ≥ 99.90 P ≈ 0.003[3]	8.9
CW023A	SW-Cu	2.0076	Cu ≥ 99.90 P 0.005 to 0.014	8.9
CW024A	SF-Cu	2.0090	Cu ≥ 99.90 P 0.015 to 0.040	8.9

1) Local exceeding of the upper limit is permissible; restrictions within the given limits should be mentioned in the order.
2) SE-Cu is generally delivered in the soft state with an electrical conductivity of ≥ 57.0 m/Ω mm²; by arrangement it can also be delivered with an electrical conductivity of ≥ 58.0 m/Ω mm² and a low phosphorus content.
3) Phosphorus can be replaced totally or partially by other oxidizing agents.

Table 105: Copper materials suitable for use in drinking water according to [640] (old designations according to DIN 1787)

A definitive method of assessing the water type mentioned is not yet available. There is, however, agreement that it depends on the water composition. In fact, it can be proved that pitting attacks have not occurred in copper pipes with heavy carbon deposits in many drinking waters.

The chemical composition of these inhibitors is at present not precisely known. Campbell [641] presumes that they are amino acids which are present in surface waters as a consequence of biological processes. This view is in line with experience, according to which pitting corrosion is considerably less likely in surface waters which are more heavily polluted with organic substances. This behavior can be explained by the properties of the copper(I) oxide layers formed primarily on the inner surfaces [642]. These layers are semiconductors, which means that they can act as the cathode of an electrochemical cell, if current conduction in this layer is possible. This is possible in the case of copper(I) oxide layers which grow without disturbance, that is to say in the absence of the inhibitors mentioned. If the growth of these layers is disturbed, for example by inhibitors, as is the case in surface waters, they have a very high electrical resistance. Cathodic reactions do not then proceed, and no anodic dissolution takes place at the pitting sites. In groundwaters, there are no inhibitors, which means that the necessary conditions for the formation of an electrochemical cell and hence for pitting apply.

The influencing parameters which lead to corrosive attack on copper pipes in cold waters have been investigated in field tests lasting five years [643]. The composition of the copper materials investigated is shown in Table 106. The result found was that some influencing parameters must interact for the occurrence of Type I pitting corrosion, while each of these influencing parameters by itself does not necessarily cause pitting corrosion. It was found in these studies, amongst other things, that the critical pitting potential $U_H \geq 410$ mV and the potential transition region $U_H = 340$ to 410 mV were not reached. The non-occurrence of pitting corrosion is attributed to this, amongst other things. The critical pitting potential was given as 410 mV in [644]. The importance of the pitting potential was confirmed by tests based on this [645, 646]. In the potential transition region, pitting may or may not occur according to [647].

Element	Pipes in strength class	
	F 22	F 37
Copper	99.93%	99.92%
Phosphorus	0.029%	0.034%
Oxygen	46 ppm	34 ppm
Lead	10 ppm	8 ppm
Tin	29 ppm	12 pmm
Nickel	94 ppm	115 ppm
Silver	95 ppm	85 ppm
Arsenic	119 ppm	153 ppm
Antimony	33 ppm	44 ppm
Iron	6 ppm	4 ppm
Selenium	30 ppm	42 ppm
Sulfur	13 ppm	15 ppm

The elements Bi, Te, Mn, Si, Cr, Co, Be, Al, Mg, Cd, Zn, Zr were below the detectable limit.

Table 106: Composition of copper materials used in field tests according to [643]

A further important result from the field tests gives an indication of the effect of the water velocity. As already described in [648], pitting preferentially occurs at low flow velocities. The non-occurrence of Type I pitting in the field tests is also ascribed to the high flow velocities (0.5 m/s).

In the field tests, it could already be decided after an operating time of one year whether the pipes develop pitting or show deposits which could develop into pitting areas.

According to the current state of knowledge, the probability of the occurrence of Type II pitting in warm water is largely determined by the bicarbonate/sulfate molar ratio in the drinking water. If this is less than 2, there is a high probability of damage, especially at low pH values [633, 649].

Flow rates which are too high locally have a clearly detectable influence on the corrosion resistance of copper pipes, since these can lead to erosion corrosion. The flow rates should, therefore, not exceed 2 m/s [633].

Erosion corrosion may occur above all in the circulation pipes of water heaters. Circulation pumps must be adjusted so that the flow rates are between 0.3 and 0.5 m/s, in order to avoid erosion corrosion [650].

As with all materials in which the corrosion resistance is based on the formation of protective surface layers, the initial conditions are also critical in the case of copper pipes. Conditions for later pitting can be created with the first filling with water. These include the introduction of foreign particles and only partial filling of pipeline conduits for a prolonged period of time. In the last case in particular, the formation of the surface layer on the copper/water/air phase boundary can be disturbed to the extent that pit nucleating sites develop here. The later conditions such as the composition of the water and the operating conditions, decide whether the pitting progresses at the pit nucleating sites to the extent of breaking through the pipe wall, or remains static. The latter is known as repassivatable pitting, e.g. in the case of stainless steels.

Apart from the forms of corrosion so far mentioned, pitting corrosion due to excessive use of soft solder auxiliaries (fluxes), the so-called soldering pastes, was often reported in the sixties [651]. It is pointed out that corrosion damage to copper pipes is in a large number of cases due to incorrect handling of fluxes and the lack or incorrect preparation of solder areas. The pitting corrosion caused by fluxes has been greatly reduced by the development of suitable fluxes in accordance with GW 2 [652].

From the hygiene point of view, copper is not harmful to health in drinking water if pipes are installed properly, and usually does not impair the taste [653].

Copper-tin-zinc alloys (red brass)
Copper-zinc alloys (brass)
Other copper alloys

Copper alloys are preferably used for valves and fittings in drinking water installations. According to [640], the materials listed in Table 107 are used for this purpose. The alloying constituents of the copper alloys are shown in percent by weight in this table.

Corrosion phenomena may occur on CuZn alloys as selective attacks known as dezincification. The latter is characterized by a depletion of zinc at the surface. The yellow brass color changes visibly to a copper color. The phenomenon may occur generally (layer-type dezincification) or locally in the form of plug-type dezincification. Dezincification preferentially occurs on lead-containing, two-phase copper-zinc alloys of high zinc content (α-, β-brass). Generally, the probability of dezincification decreases as the copper content increases. Examples of failure are described in [654–656].

According to [657], a changeover to the corrosion-resistant material red brass Rg 5 (Table 107, first line) is recommended in areas where dezincification occurs. According to current knowledge, dezincification depends on the water parameters of chloride content and acid capacity. The pH evidently has little influence. The graph in Figure 70 shows the influence of the parameters mentioned [658, 659]. The applicability of this graph has been confirmed in studies on failure [660].

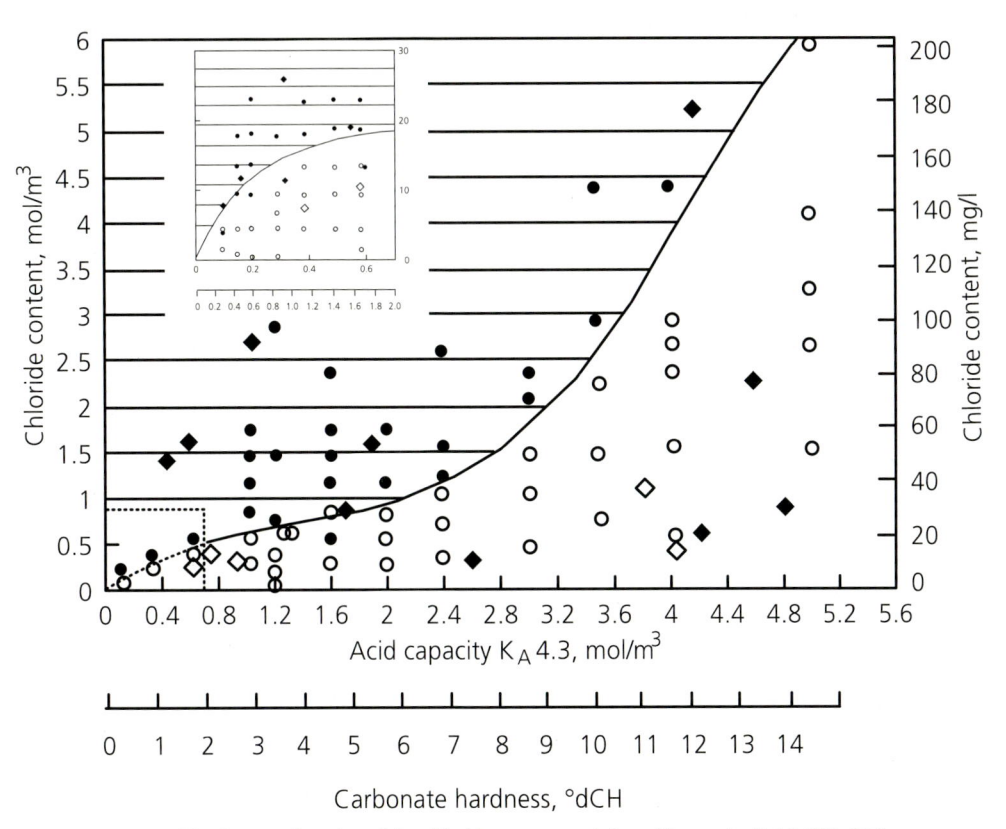

Figure 70: Dezincification as a function of the chloride content and the acid capacity $K_A4.3$ [658, 659]

Material designation	No.	old designation	Cu	Sn	Zn	Pb	Ni	Al	Use	V	F
CuSn5Zn5Pb5-C	CC491K	G-CuSn5ZnPb 2.1096.01	84–86	4–6	4–6	4–6	2.0	–	red brass Rg5	X	X
CuZn39Pb1Al-C	CC754S	GK-CuZn37Pb 2.0340.02	58–64	(1.0)	balance	(2.0)	–	1.0	GK-Ms60	X	X
		GD-CuZn37Pb 2.0340.05	58–64	(1.0)	balance	(2.0)	–	1.0	GD-Ms60	X	–
CuZn33Pb2-C	CC750S	G-CuZn33Pb 2.0290.01	63–67	(1.0)	balance	1.0–3.3	–	(0.1)	G-Ms65	X	–
CuZn39Pb3	CW614N	2.0401	57.5–59.0	–	balance	2.5–3.3	(0.4)	–	Ms58	X	–
CuZn40Pb2	CW617N	2.0402	57.5–59.0	–	balance	1.5–2.5	(0.4)	–	Ms58	X	X

V: valves
F: fitting

Table 107: Copper materials used for fittings in drinking water systems [640]

Ferrous Chlorides (FeCl$_2$, FeCl$_3$)

Author: A. Werner / Editor: R. Bender

Copper

– Iron(III) chloride –

Copper undergoes moderate to severe attack in iron(III) chloride [661].

Corrosion rates of copper in iron(III) chloride solutions are given in Table 108 [662] and in Figures 71 and 72 [663].

Medium	Mass loss rate, g/m^2 h	
	without sunlight	with sunlight
0.5 % FeCl$_3$ × 6H$_2$O + 10 ml/l conc. HCl	4.58	5.44
1.0 % FeCl$_3$ × 6H$_2$O + 20 ml/l conc. HCl	4.72	
2.0 % FeCl$_3$ × 6H$_2$O + 40 ml/l conc. HCl	5.14	7.40
2.5 % FeCl$_3$ × 6H$_2$O + 50 ml/l conc. HCl	22.00	29.20
5.0 % FeCl$_3$ × 6H$_2$O + 100 ml/l conc. HCl	90.80	99.30

Table 108: Mass loss rate of copper in static solutions of FeCl$_3$ × 6H$_2$O with HCl, with or without exposure to sunlight, at 305 K ± 1 (32 °C ± 1), exposure time 2 h [662]

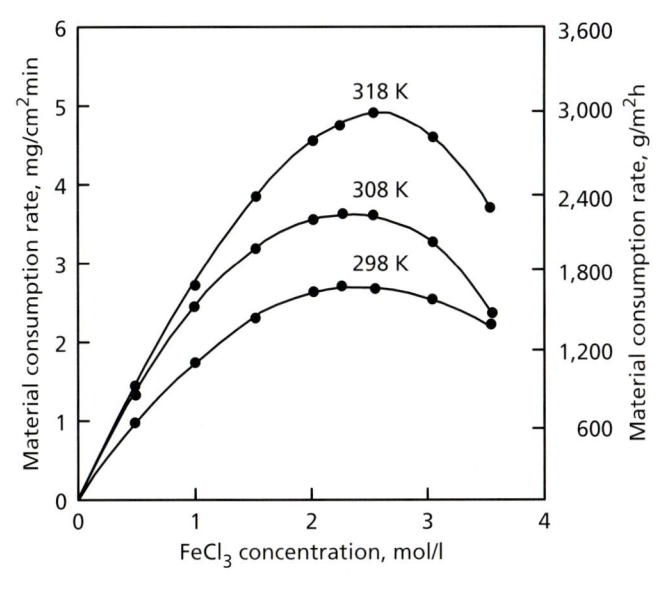

Figure 71: Dependence of the mass loss rate of copper on the iron(III) chloride concentration and the temperature. Flow velocity for the given geometry: 1,450 ml/min, exposure time 60 min [663]

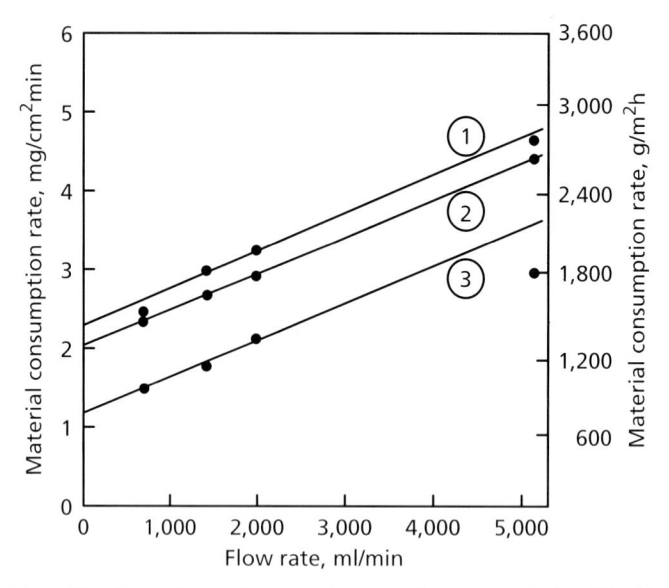

Figure 72: Dependence of the mass loss rate of copper on the iron(III) chloride concentration and flow velocity at 298 K (25 °C), exposure time 60 min [663]
① 2.15 mol/l $FeCl_3$
② 3.00 mol/l $FeCl_3$
③ 1.00 mol/l $FeCl_3$

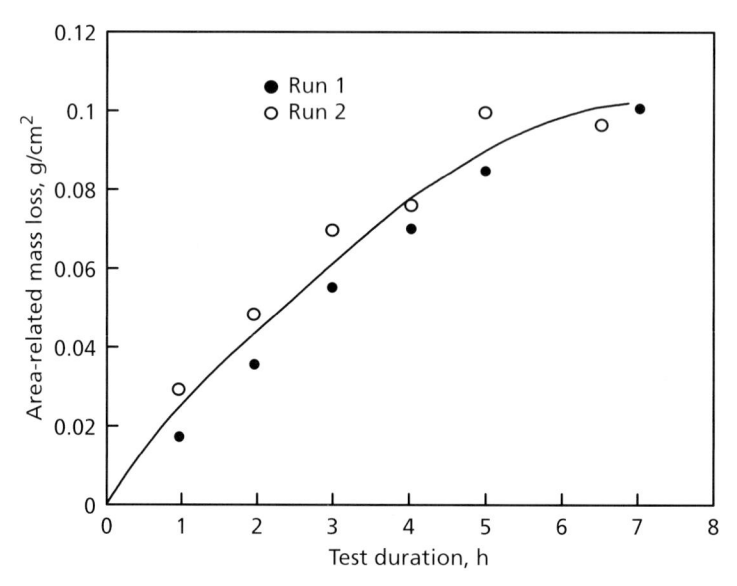

Figure 73: Mass loss of copper (99.9 % Cu) in a solution of 10 % $FeCl_3$ (pH 1) at room temperature, 2 trials (○, ●) [664]

As the iron(III) chloride concentration increases, the mass loss rate of copper initially increases and then decreases again at iron(III) chloride concentrations of 2 to 3 mol/l (see Figures 71 and 72). Increasing the flow velocity increases the mass loss rate (see Figure 72) [663].

Exposure to sunlight can increase the corrosion of copper in iron(III) chloride solutions (see Table 108).

Commercially available pure copper (99.9 % Cu) exposed to a solution of 10 % $FeCl_3$ at pH 1 and room temperature exhibited a mass loss rate of approximately 2×10^3 to 7×10^3 g/m^2 d, which decreased somewhat over time, see Figure 73 [664].

Owing to the high rates of material consumption, iron(III) chloride solutions are used to etch copper, e.g. chemical milling [665]. It is also used for pickling copper [666].

Intergranular corrosion was found in pure copper (99.999 % Cu) after exposure to a solution of 1 g/l $FeCl_3 \times 6H_2O$ and 12 % HCl at 263 K (–10 °C) [667].

– Inhibitors –

To reduce the corrosion rate of copper alloys in solutions containing iron(III) chloride and copper(II) chloride solutions, inhibitors I-1-A, I-1-E and I-1-V (Russian) and reducing agents were added to the medium [668–670].

Copper-aluminium alloys

– Iron(III) chloride –

Copper-aluminium alloys undergo moderate to severe attack in iron(III) chloride solutions [661, 671].

Copper-nickel alloys

– Iron(II) chloride –

Copper-nickel alloys are reported to be resistant to iron(II) chloride [672].

– Iron(III) chloride –

No copper-nickel alloys have been reported that can be regarded as being resistant to iron(III) chloride solutions. The mass loss rates as a function of the $FeCl_3$ concentration are shown in Figure 74 for a CuNi 90 10 alloy ((%) 88 Cu, 9.5 Ni, 1.1 Fe, 0.6 Mn, 0.04 Pb, 0.7 Zn) [673].

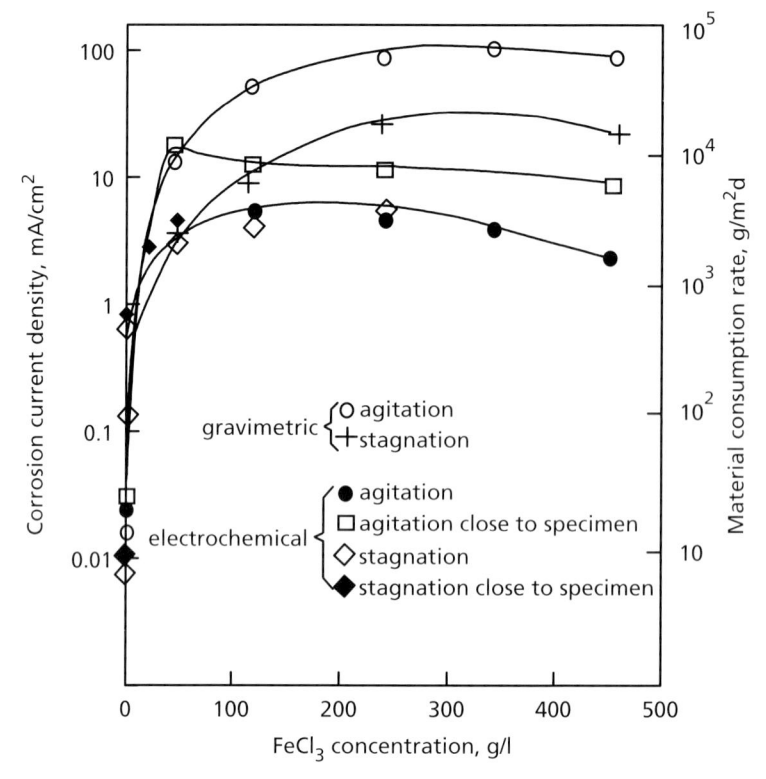

Figure 74: Mass loss rates and corrosion current densities of CuNi 90 10 in stirred and unstirred iron(III) chloride solutions as a function of the concentration at 298 K (25 °C). The solution was stirred with a magnetic stirrer at a low speed [673]

Further details are reported in the literature as follows:
Copper-nickel alloys are attacked by iron(III) chloride [672].

Monel® 411 ((%) 64.00 Ni, 0.20 C, 0.80 Mn, 1.00 Fe, 1.50 Si, 31.50 Cu) and Monel® 505 ((%) 63.00 Ni, 0.10 C, 0.80 Mn, 2.00 Fe, 4.00 Si, 29.50 Cu) are reported to be nonresistant to iron(III) chloride [674, p. 208].

The mass loss rate of CuNi 90 10 in iron(III) chloride solutions at 298 K (25 °C) increases with the $FeCl_3$ concentration strongly at first and then drops slightly at high concentrations. In this case, the mass loss rates in a stirred medium (stirred with a magnetic stirrer at a low speed) were much higher than in an unstirred medium. The corrosion current densities behaved similarly, see Figure 74. The corrosion potentials of this CuNi alloy are shifted in a more positive direction as the $FeCl_3$ concentration increases. Stirring increases this shift, see Figure 75 [673, Fig. 2, Fig. 5].

CuNi 90 10 and Monel® ((%) 30 Cu, 67 Ni, 1.1 Mn) exposed to a 10 % solution of $FeCl_3 \times 6H_2O$ (pH 1.6) at room temperature exhibited severe uniform corrosion in crevices. The crevice was produced by pressing the base of a Teflon cylinder onto the surface of the specimen [675].

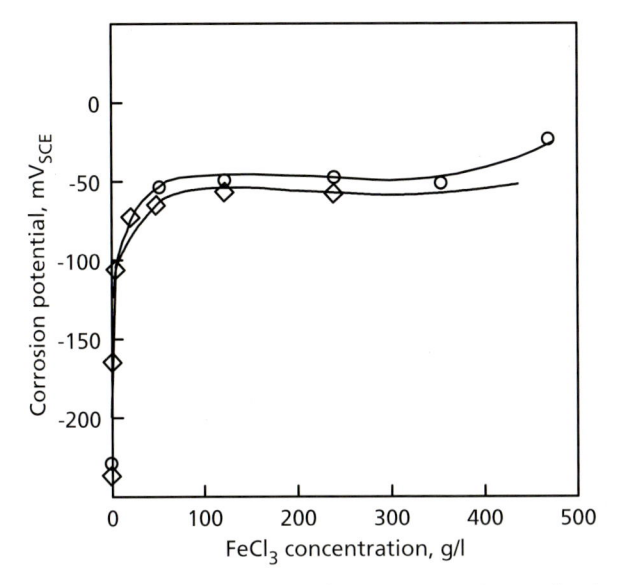

Figure 75: Corrosion potentials of CuNi 90 10 in (○) stirred and (◇) unstirred iron(III) chloride solutions of different concentrations at 298 K (25 °C) [673]

The attack can also take place preferentially at grain boundaries. This is the case for e.g. a CuNi alloy with (%) 89.53 Cu, 9.34 Ni, 0.927 Fe, and 0.17 Zn in a 1 mol/l solution of iron(III) chloride. There are indications here that dissolution of nickel is taking place [676].

Monel® 400 (2.4360) is not resistant to gaseous iron(III) chloride at 423 K (150 °C) or at 693 K (420 °C) [677].

Alloys NiCu30Fe (2.4360) and NiCu30Al (2.4374) are reported to be nonresistant to oxidizing acidic salts such as iron(III) chloride [678].

Copper-tin alloys (bronzes)

The concentration-dependent corrosion behavior of a bronze ((%) 84.6 Cu, 7.1 Sn, 6.1 Pb, 1.6 Zn, 0.08 Fe, 0.38 Sb) in iron(III) chloride solutions is shown in Figure 76. According to this, the mass loss rates of this bronze in iron(III) chloride solutions at 298 K (25 °C) is so high that this material cannot be used in this medium [673].

Copper-tin-zinc alloys (red brass)

These alloys undergo moderate to severe attack in iron(III) chloride solutions [661, 671].

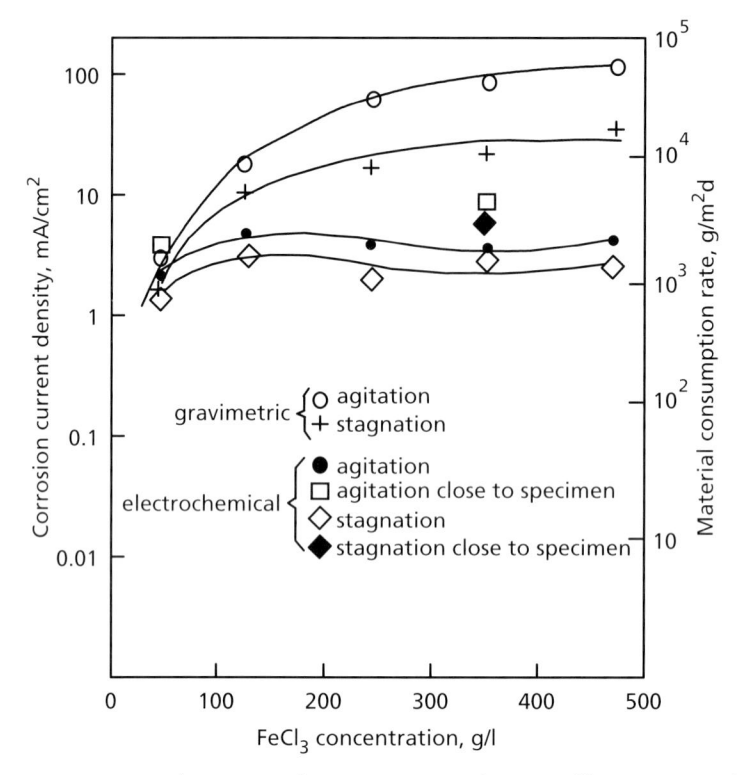

Figure 76: Mass loss rates and corrosion current densities of bronze in stirred and unstirred iron(III) chloride solutions of differing concentrations at 298 K (25 °C) [673]

Copper-zinc alloys (brass)

– Iron(III) chloride –

In an iron(III) chloride-containing 10 % NaCl solution, brass is only moderately resistant (mass loss rates of e.g. 0.24 and 0.37 g/m² h), and often unsuitable (> 3 g/m² h, see Table 109) [679].

Material	Mass loss rate, g/m² h
So Ms60 K	0.24
So Ms60	0.37
Ms60	1.12

Table 109: Mass loss rate of brass in a stirred solution of 1 % FeCl₃ and 10 % NaCl at room temperature, exposure time 35 d [679]

Similar to copper, iron(III) chloride-containing solutions are also used to etch brass, e.g. chemical milling [680].

Exposure to sunlight can intensify the corrosion of brass in iron(III) chloride with HCl (see Table 110) [662].

Medium	Mass loss rate, $g/m^2\,h$	
	without sunlight	with sunlight
0.5 % $FeCl_3 \times 6H_2O$ + 10 ml/l conc. HCl	3.52	4.36
1.0 % $FeCl_3 \times 6H_2O$ + 20 ml/l conc. HCl	7.64	9.72
2.0 % $FeCl_3 \times 6H_2O$ + 40 ml/l conc. HCl	16.66	28.10
2.5 % $FeCl_3 \times 6H_2O$ + 50 ml/l conc. HCl	24.76	34.40
5.0 % $FeCl_3 \times 6H_2O$ + 100 ml/l conc. HCl	66.30	83.70

Table 110: Mass loss rate of brass ((%) 63.2 Cu, 36.6 Zn, 0.001 Pb) in static solutions of $FeCl_3 \times 6H_2O$ with HCl and with or without exposure to sunlight at 305 K ± 1 (32 °C ± 1), exposure time 2 h [662]

Iron(III) chloride solutions with and without hydrochloric acid can dissolve zinc selectively from brass [681, 682]. This also applies to iron(III) chloride + copper(II) sulfate. Such solutions are thus used to determine the resistance to zinc dissolution, see Table 111 [683–685].

Admiralty brass ((%) 71 Cu, 28 Zn, 1 Sn) and cartridge brass ((%) 70 Cu, 30 Zn) exhibited severe uniform corrosion in crevices on exposure to a 10 % solution of $FeCl_3 \times 6H_2O$ (pH 1.6) at room temperature. The crevice was produced by pressing the base of a Teflon cylinder onto the surface of the specimen [675].

Brass (CuZn 70 30) was found to be susceptible to stress corrosion cracking in iron(III) chloride solutions with a concentration of > 0.01 and < 0.1 mol/l. The time-to-rupture greatly decreased as the iron(III) chloride concentration increased, see Figure 77 [686]. Stress corrosion cracking of brass with Zn contents of approx. ≥ 30 % can be expected in a 2 % iron(III) chloride solution. Uniform corrosion generally predominates at lower Zn contents [687].

Brass is susceptible to intergranular corrosion in a 2 % iron(III) chloride solution [687].

Material	Chemical composition, %												Zinc removal depth
	Cu	Zn	Sn	Pb	Ni	Fe	Sb	S	P	Mn	As	Al	Si	mm
CDA 642[1)]	90.21	–	0.13	0.01	0.46	0.65		0.009	0.010	0.19	0.003	6.45	1.83	0
CDA 836	84.71	4.55	4.43	5.22	0.64	0.15	0.20	0.040	0.002	0.01	0.012	0.002	0.002	0
CDA 320	83.47	14.55	–	–	–	–	–	–	–	–	–	–	–	0
CDA 844	82.15	7.10	2.92	6.71	0.72	0.15	0.15	0.035	0.002	0.01	0.012	0.002	0.002	0
CDA 848	78.05	12.2	2.52	6.09	0.71	0.18	0.16	0.030	0.010	0.01	0.010	0.002	0.002	0
CDA 879	65.29	33.52	0.12	0.01	0.02	0.037	–	0.008	0.013	0.02	0.003	0.002	0.94	0.089
CDA 879[1)]	65.29	33.52	0.12	0.01	0.02	0.037	–	0.008	0.013	0.02	0.003	0.002	0.94	0.089
Proprietary alloy A	64.79	30.13	0.10	3.15	0.015	0.07	–	0.010	0.014	0.29	0.074	0.64	0.69	0
Proprietary alloy B1	64.79	30.33	0.18	2.25	0.51	0.23	0.052	0.008	0.003	0.50	0.005	0.26	0.83	0
Proprietary alloy B1[2)]	64.79	30.33	0.18	2.25	0.51	0.23	0.052	0.008	0.003	0.50	0.005	0.26	0.83	0
Proprietary alloy B2	64.16	31.32	0.30	2.07	0.46	0.35	0.05	0.016	0.005	0.36	0.006	0.18	0.68	0.089
Proprietary alloy B3	63.31	31.93	0.11	2.24	0.51	0.24	0.052	0.008	0.002	0.47	0.007	0.24	0.82	0.038
Proprietary alloy B3[2)]	63.31	31.93	0.11	2.24	0.51	0.24	0.052	0.008	0.002	0.47	0.007	0.24	0.82	0
Proprietary alloy C[3)]	62.33	35.80	–	1.71	0.01	0.03	–	0.009	0.002	0.01	0.035	0.001	0.002	0
CDA 360	61.45	35.64	–	2.61	0.07	0.12	–	0.010	0.002	0.01	0.005	0.001	0.002	0.30
CDA 377	59.84	38.12	–	1.75	0.08	0.12	–	0.006	0.002	0.01	0.002	0.001	0.002	0.64
CDA 377[4)]	59.84	38.12	–	1.75	0.08	0.12	–	0.006	0.002	0.01	0.002	0.001	0.002	0.64
CDA 367[2)]	59.71	39.54	0.01	0.63	–	0.01	0.05	0.009	0.004	–	–	0.002	0.001	0.46

1) Heat treatment: 1 h/866 K (593 °C)/air (< 93 K/h)
2) 5 h/827 K (554 °C)/cooling rate 107 K/h
3) 4 h/811 K (538 °C)/air (< 93 K/h)
4) 1 h/827 K (554 °C)/cooling rate 107 K/h

Table 111: Zinc removal depths in CuZn alloys in a solution of 13.5 g/l $FeCl_3 \times 6H_2O$ + 18.7 g/l copper(II) sulfate hydrate + 1.5 ml/l HCl at room temperature, exposure time 17 d [685]

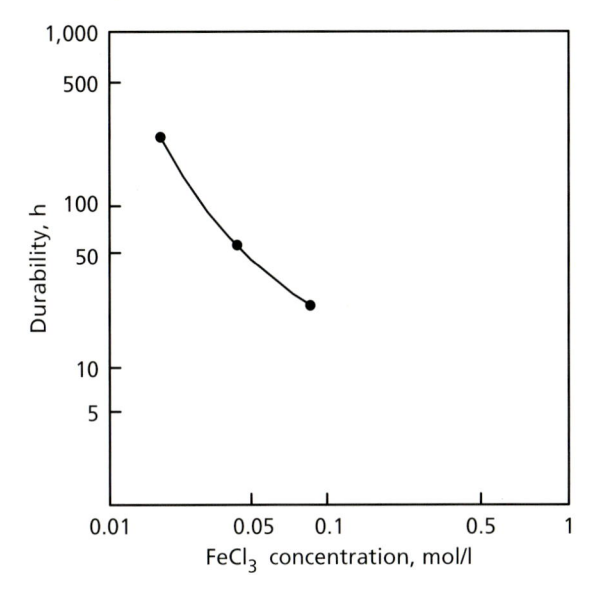

Figure 77: Influence of the FeCl₃ concentration on the susceptibility of brass (CuZn 70 30) to stress corrosion cracking, measured as the time until SCC damage occurred [686]

– Inhibitors –

Hydrochloric acid is often used as a cleaning agent to remove oxides from the surface of condenser tubes. This generally leads to a large increase in the content of iron(III) and copper ions in the cleaning solution, which results in a corresponding increase in corrosive attack of brass and other copper alloys [669]. Inhibitors, e.g. ascorbic acid, are thus recommended for such cleaning solutions. The addition of inhibitors (e.g. I-1-V, I-1-A and I-1-E) and reducing agents can considerably reduce the corrosion rate of L-68 (brass with 67 to 70 % Cu) in hydrochloric acid solutions containing iron(III) chloride and copper(II) chloride [670].

Uniform corrosion of commercially available brass ((%) 70 Cu, 30 Zn) is only insufficiently inhibited in a 0.03 mol/l iron(III) chloride solution by the addition of triazole, benzotriazole, or naphthotriazole at a concentration of 0.001 mol/l at room temperature [668].

Other copper alloys

– Iron(III) chloride –

Unlike Brass Ms70 without Au, Ms70 containing 2 at.% Au is susceptible to stress corrosion cracking in a 2 % iron(III) chloride solution [688].

CuNiPd alloys with 2 to 40 at.% Pd exhibit considerable mass loss rates in a 2 % FeCl₃ solution at room temperature, see Figure 78 and Table 112. In this case, the corrosion losses of the alloy series with 2 at.% Pd were the highest: about 1,000 g/m²d for high-copper alloys, which dropped to about 400 g/m²d as the nickel content increased. In contrast, alloys with Pd contents of 10 and 40 at.% exhibited mass loss rates of 200 to 300 g/m²d, see Figure 78.

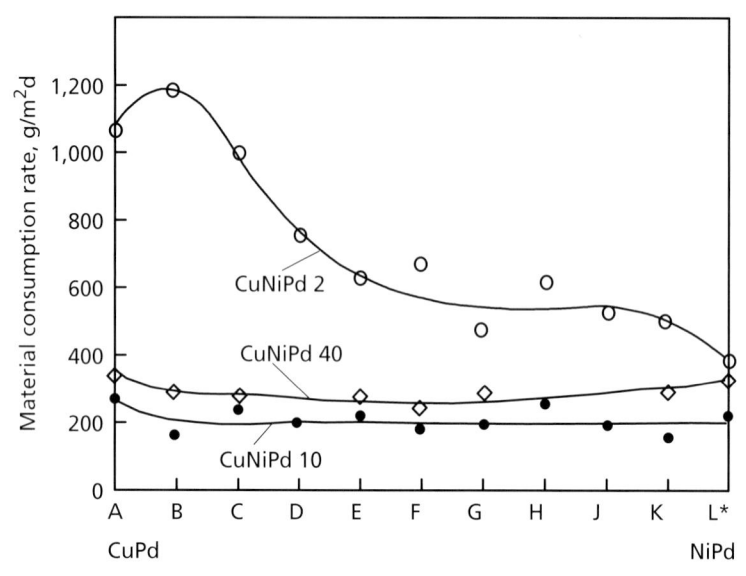

Figure 78: Mass loss rates of copper-nickel-palladium alloys in 2% FeCl$_3$ at room temperature with constant palladium contents of 2, 10, and 40 at.% depending on the nickel content [689] Exposure times: 24 h for 10 and 40 at.% Pd; 7 h for 2 at.% Pd (* A to L see Table 112)

The corrosion potentials in a 2% FeCl$_3$ solution were relatively independent of the Ni:Cu ratio for Pd contents of 10 to 40 at.% Pd. On the other hand, at a Pd content of 2 at.%, the corrosion potential dropped considerably as the proportion of Ni increased. Tensile specimens of these alloys subjected to a tensile stress of 80% of the tensile strength exhibited a strong susceptibility to stress corrosion cracking on exposure to a 2% FeCl$_3$ solution at room temperature. The lifetimes were greatly dependent on the Pd content and the Cu:Ni ratio, see Figure 79 [689].

Copper alloys with (%) 3–12 Al, 3–6 Fe, 0.3–5 Si, and 4–7 Ni exposed to an air-bubbled solution of 18% HCl and 2 g/l Fe(III) ions at 308 K (35 °C) exhibited a mass loss rate that was approximately 30–40% lower than that of a copper alloy with (%) 10 Al, 5 Ni, and 5 Fe [690].

Designation	CuNiPd 2			CuNiPd 10			CuNiPd 40		
	Composition, at.%								
	Cu	Ni	Pd	Cu	Ni	Pd	Cu	Ni	Pd
A	98	–	2	90	–	10	60	–	40
B	89	9	2	81	9	10	54	6	40
C	79	19	2	72	18	10	48	12	40
D	69	29	2	63	27	10	36	24	40
E	59	39	2	54	36	10			
F	49	49	2	45	45	10	30	30	40
G	39	59	2	36	54	10	24	36	40
H	29	69	2	27	63	10			
I	19	79	2	18	72	10			
K	9	89	2	9	81	10	6	54	40
L	–	98	2	–	90	10	–	60	40

Table 112: Composition of CuNiPd alloys in Figure 78 [689]

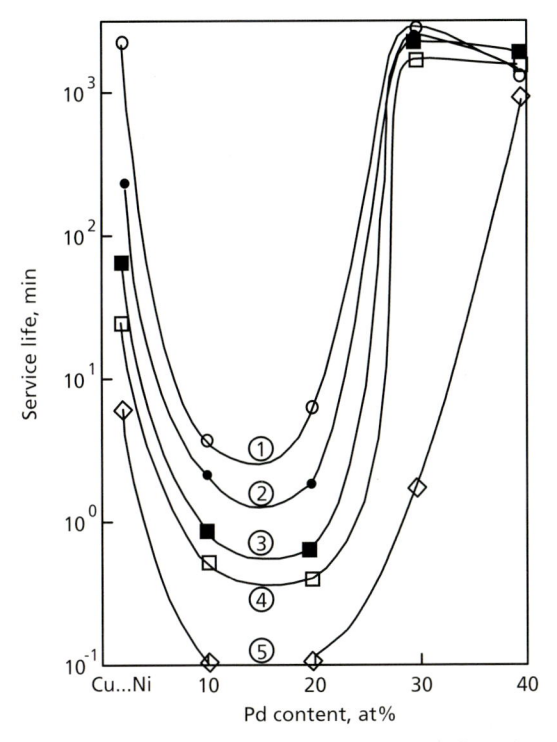

Figure 79: Stress corrosion cracking resistance of tensile specimens of copper-nickel-palladium alloys in a 2% FeCl₃ solution (room temperature), each with a constant Cu/Ni ratio, as a function of the palladium content; mechanical load: 80% of the tensile strength [689]
① Ni-Pd   ② Cu_2Ni_8-Pd   ③ CuNi-Pd   ④ Cu_7Ni_3-Pd   ⑤ Cu-Pd

Fluorides

Author: K. Hauffe / Editor: R. Bender

Copper

Copper is attacked in anhydrous hydrofluoric acid, containing 0.01 mol/l $NaHF_2$, at 5.0 V(Hg/HgF_2) at the rate of 1.3 $g/m^2 h$ (about 1.3 mm/a) [691]. Copper would seem to be satisfactorily resistant to fluoride-containing solutions; the influence of alternating current density on the corrosion of copper in sodium fluoride solutions is also slight [692]. Copper is rapidly dissolved in an eutectic melt consisting of 0.8 mol/mol lithium hydroxide and 0.2 mol/mol lithium fluoride at 1,073 K (800 °C) [693].

Copper-nickel alloys

Alloy CuNi30Be1 (Berylco® 717) loses about 2.2 $g/m^2 h$ (about 2.2 mm/a) in a melt of 0.8 mol/mol lithium hydroxide and 0.2 mol/mol lithium fluoride at 1,073 K (800 °C).

Copper-zinc alloys (brass)

Corrosion of brass (CuZn30, CW505L) in 0.1 mol/l potassium fluoride is inhibited by 0.001 mol/l triazole, benzotriazole and naphthotriazole; of these, benzotriazole is the most effective, whereas triazole may even accelerate corrosion. According to Table 113, however, the corrosion rate of brass, even in the absence of an inhibitor, is so low that it may be regarded as resistant to potassium fluoride, even at elevated concentrations [694].

Inhibitor	pH	Inhibition efficiency %	Corrosion loss	
			$g/m^2 h$	mm/a (mpy)
–	6.2–7.3	–	8.5×10^{-4}	8.8×10^{-4} (0.032)
Triazole	6.2–7.2	–26	0.0011	0.0011 (0.043)
Benzotriazole	6.2–6.4	85	1.3×10^{-4}	1.3×10^{-4} (5.12×10^{-3})
Naphthotriazole	6.2–6.4	70	2.5×10^{-4}	2.6×10^{-4} (0.01)

Table 113: Inhibition of the corrosion of brass in 0.1 mol/l KF solution by some triazoles (7 days; room temperature; 0.001 mol/l inhibitor) [694]

Fluorine, Hydrogen Fluoride, Hydrofluoric Acid

Author: K. Hauffe / Editor: R. Bender

Copper

In pure fluorine with 0.02 % oxygen, 0.02 % nitrogen, a CuF_2 film is formed on copper (99.999) over a wide pressure range at various temperatures. Through this film the copper ions and electrons migrate in an ambipolar current to the phase boundary CuF_2/F_2, where more CuF_2 is formed [695]. This fluoride starts to evaporate above 670 K (397 °C). Above 920 K (647 °C), the evaporation becomes more rapid than the reformation of the fluoride so that in this case a decrease in weight occurs (Figure 80) [696].

At 973 K (700 °C), the vapor pressure of CuF_2 is 1.66×10^{-6} MPa [696]. Copper(I) fluoride is not stable in the direct fluorination of copper [697].

According to the experimental results which are available in the literature on the reaction between copper and fluorine, some of which are listed in Table 114, the kinetics of the fluorination of copper below 620 K (347 °C) seems to follow a logarithmic rate law of the form

$$\Delta\xi = k_e \ln(t_0 + a\,t)$$

and between 670 and 770 K (397 and 497 °C) an approximately parabolic rate law of the form

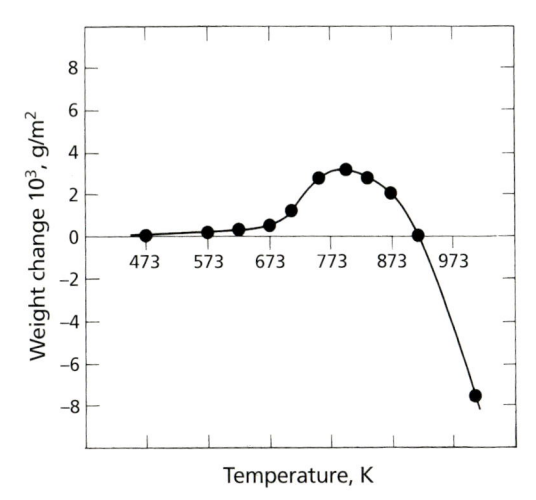

$$\Delta\xi^n = k_n t \quad (\text{where } 1.5 \leq n \leq 3)$$

Figure 80: Kinetics of the fluorination of copper in fluorine of 0.1 MPa as a function of temperature, plotted as weight change after 2 hours [695]

In these formulae, $\Delta\xi$ is the thickness of the fluoride film and t is the time; k_e and k_n are the logarithmic and parabolic rate constants respectively, which are given in Table 114 as penetration depth in mm/a; a and t_0 are constants to be determined experimentally [698].

Between 300 and 600 K (27 and 327 °C), the growth of the CuF_2-film as a function of time can be described approximately by a logarithmic rate law of the form

$$\Delta m = k \ln(t_0 + a \cdot t)$$

The rate constant k and the constants t_0 and a can be seen from Table 114. Δm is the weight increase per unit area (for example cm^2) and t is the test duration (for example h). Instead of Δm, the increase in the film thickness can also be used.

Since, in contrast to the parabolic rate law, the logarithmic rate law cannot be described by a uniform mechanism, the constants t_0 and a can only be determined empirically by means of measured results. Despite this shortcoming, the formula which contains constants t_0 and a has a practical value because with its help the extent of the expected corrosion can be extrapolated over a longer period of time.

While a compact and almost pore-free CuF_2 film is formed below 670 K (397 °C), the film becomes porous and brittle above this temperature. At even higher temperatures, evaporation of CuF_2 begins, this becomes more and more extensive, and between 920 and 1,020 K (647 and 747 °C) the fluorination even proceeds combustion-like [699]. This combustion of copper can be prevented by the addition of oxygen to the fluorine in an amount which is sufficient for the formation of CuO.

However, severe complications take place in the reaction mechanism if the fluorine contains small amounts of oxygen, which frequently cannot be avoided. The reason is that a thin Cu_2O film is formed on the copper and only then on top of it a CuF_2 film, as a result of which the corrosion rate on the whole becomes higher, because the Cu_2O is formed very rapidly [695]. The values listed in Table 114 also contain those of n, k_n, a and t_0 so that the above relationship can be used.

Different values are given for temperature and thickness of CuF_2 film over the range of the logarithmic rate law. Thus, for the film thickness of about 670 Å 530 K (257 °C) is given in one case [700], "below 570 K (297 °C)" in another [701], in which the mechanism according to Evans [702] is discussed, which is based on the extensive growth of pores and cracks in the resulting film. Above 670 K (397 °C), the diffusion of copper ions through the developing CuF_2 film is, as proposed earlier, probably decisive for the continuation of the fluorination [695, 703]. For the pressure dependence of the fluorination of copper at 723 K (450 °C), the logarithmic rate constant k is found to be $7.68 \times 10^{-7} \, p_{F_2}^{0.78}$ [704]. The distinct decrease in the corrosion rate in static fluorine atmosphere at higher temperatures is noteworthy, especially since in a stream of fluorine noticeable evaporation of CuF_2 occurs. The CuF_2 evaporation is determined by the mass transport through the boundary layer, that is to say, by the diffusion of the CuF_2 molecules. The thickness of the boundary layer reaches a maximum value in a static atmosphere, the decrease in the corrosion rate in a static atmosphere is therefore understandable; the favorable formation of the CuF_2 film slows down further attack of fluorine and causes chemical passivation.

In high-purity fluorine at 77 K (−196 °C) copper is absolutely resistant (1.5 × 10^{-4} mm/a or 1.5 × 10^{-4} g/m^2 h), as could be confirmed by one-year storage [705, 706].

Impurities in fluorine and the resulting local cell formation can increase the corrosion rate considerably. Thus, for example, water vapor in fluorine considerably accelerates the corrosion at 300 to 800 K (27 to 527 °C), since CuF_2 is hygroscopic and compounds such as, for example, $Cu(OH)F \times CuF_2$, $CuF_2 \times 2\,H_2O$ and $Cu(OH)_2$ are formed then, which destroy the protective CuF_2-film. Short-term corrosion tests can therefore result in large errors because passive films which decrease the corrosion rate are formed only gradually [707–709].

At 1,023 K (750 °C) and above, the fluorination proceeds in combustion-like manner due to the extensive evaporation of the fluoride [710].

In fluorine which has relatively high oxygen contents, thick films are formed between 670 and 1,020 K (397 and 747 °C). There a CuF_2 film is on the outside and a CuO film under it, which gives the surface a black appearance [711]; the red appearance of the film at oxygen contents between 0.1 and 0.5 % is an indication for the formation of Cu_2O under the CuF_2 film [712].

The fluorination of Cu(I) oxide takes place in two steps [713]

$$Cu_2O + F_2(\text{gas}) \rightarrow CuF_2 + CuO$$

and

$$CuO + F_2(\text{gas}) \rightarrow CuF_2 + 1/2\,O_2(\text{gas}).$$

Temperature K (°C)	F_2-pressure kPa	Test duration h	a* 1/min	t_0	Depth of attack mm/a (mpy)	Literature
354 (81)	liquid	3	1.8 × 10^4	0	0.0033 (0.1)	[705]^a
295 (22)	0.8 to 8	5	1.6	32	0.029 (1.1)	[714]^a
300 (27)	100	3.3	67	0	0.022 (0.9)	[715]^a
303 (30)	100	3.5	3.3	0	0.023 (0.9)	[705]^a
373 (100)	0.8 to 8	5	0.12	2.4	0.16 (6.3)	[714]^a
473 (200)	0.8 to 8	5	0.15	3.0	1.07 (42.1)	[714]^a
573 (300)	100	4	1.0	0	26.2 (1,031.5)	[701]^a
673 (400)	100	4	–	–	2.23 (87.8)	[701]^b

a* = constant in equation $\Delta\xi = K_e \ln(t_0 + a\,t)$ and $\Delta m = k \ln(t_0 + a\,t)$
a = logarithmic rate law
b = parabolic rate law
c = linear rate law
d = exponential rate law $\Delta\xi = [1 - \exp(t_0 + a \cdot t)]$
e = in a statically closed system
f = extrapolated values are too low due to evaporation of CuF_2

Table 114: Compilation of some kinetic data for the fluorination of copper

Table 114: Continued

Temperature K (°C)	F_2-pressure kPa	Test duration h	a* 1/min	t_0	Depth of attack mm/a (mpy)	Literature
700 (427)	26.3	3.3	–	–	3.07 (120.9)	[716][b,c]
700 (427)	26.3	5.3	-4.4×10^{-3}	0	475.3 (18,712.6)	[716][a,e]
723 (450)	100	4	–	–	2.62 (103.1)	[701][b]
723 (450)	1.3	1	4.1	0	8.73 (343.7)	[704][c,e]
723 (450)	1.3	4	–	–	3.30 (129.9)	[704][b,c]
743 (470)	100	4	–	–	3.60 (141.7)	[701][b]
755 (482)	26.3	3.3	–	–	42.1 (1,657.5)	[716][b,c]
783 (510)	100	4	–	–	16.0 (629.9)	[701][b,f]
811 (538)	26.3	5	-3.5×10^{-2}	0	611 (24,055.1)	[716][d,f]
866 (593)	26.3	2	−0.064	0	630 (24,803.1)	[716][d,e]
922 (649)	26.3	3	−0.052	0	727 (28,622)	[716][d,e]
1,013 (740)	100	2	–	–	−40.7 (1,602.4)	[695][c,f]

a* = constant in equation $\Delta\xi = K_e \ln(t_0 + a\,t)$ and $\Delta m = k \ln(t_0 + a\,t)$
a = logarithmic rate law
b = parabolic rate law
c = linear rate law
d = exponential rate law $\Delta\xi = [1 - \exp(t_0 + a \cdot t)]$
e = in a statically closed system
f = extrapolated values are too low due to evaporation of CuF_2

Table 114: Compilation of some kinetic data for the fluorination of copper

Since the experiments were carried out using copper oxide powder, the reaction rates are not directly comparable to those obtained with a copper specimen coated with Cu_2O or CuO. Nevertheless, the conversion with time, plotted as $\Delta m/\Delta m_{100}$ versus time (Figure 81), where Δm is weight change measured and Δm_{100} is the corresponding value for complete conversion to CuF_2, gives a certain impression of the extent and rate of the CuF_2 formation.

As a result of the relatively high vapor pressures of CuF_2 at 873 and 973 K (600 and 700 °C) (5×10^{-8} and 1.7×10^{-6} MPa), the rate of evaporation of CuF_2 at 1.33×10^{-2} MPa of fluorine is at about 5×10^{-6} at 873 K (600 °C) and 5×10^{-4} g/cm² h at 973 K (700 °C) also relatively high.

An interesting case of the application of copper and brass is as a reactor for the direct fluorination of benzene, in which the two reactants are injected through two concentric tubes: the benzene is added in the tube jacket; the fluorine diluted with nitrogen is added in the inner tube designed as a nozzle. Thus, a rapid reaction is possible. Attack on the reactor wall is largely prevented. The sealing material used is a fluorocarbon elastomer (Viton®) or polytetrafluoroethylene (Teflon®); this material also serves together with stainless steel as carrier material for the valves [717].

Copper is also suitable as material for the construction of containers for fire extinguishers which contain tetrafluorodibromomethane, by which it is virtually not attacked [718].

Contrary to its behavior in fluorine, copper is very resistant to hydrogen fluoride. The resistance is comparable with that of nickel between 770 and 870 K (497 and 597 °C) [711]; hydrofluoric acid hardly causes pitting corrosion [719].

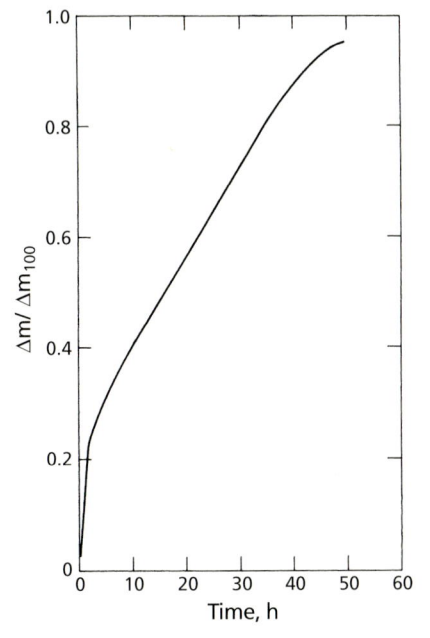

Figure 81: Progress of fluorination of CuO_2 to CuF_2 with time at 445 K (172 °C) in fluorine of 2.66×10^{-2} MPa. The specific surface area of the Cu_2O powder was initially 29.3 m^2/g [713]

The potential of copper increases with increasing concentration of hydrofluoric acid from about 0.25 V at 2 % HF to about 0.58 V at 98 % HF, thus reaching the range of less electropositive potential [720].

The use of hydrofluoric acid as additive to other acids is recommended in pickling agents for copper and copper alloys [721].

Copper reinforced with tantalum fibers, which is used for contact in space technology, is selectively attacked in solutions containing 55 % nitric acid and 45 % hydrofluoric acid, in the course of which tantalum is dissolved by hydrofluoric acid, while copper remains virtually unattacked. Because of the relatively close proximity of the potentials of copper and tantalum, local cell formation is not expected, so that this selective corrosion cannot yet be explained. However, this is dangerous due to the loss in strength caused by the dissolution of the fiber, particularly as it cannot be detected by visual inspection [722].

Copper-aluminium alloys

Only few investigations on the corrosion behavior of CuAl-alloys in fluorine and hydrofluoric acid are available, probably because aluminium improves the corrosion resistance of copper only slightly and often even makes it worse.

Some results obtained with the alloy Cu-Al9Mn2 in hydrofluoric acid (6 to 60%, 293 to 353 K (20 to 80 °C)) are shown in Table 115. As can be seen, the corrosion rate of the alloy is somewhat lower than that of copper. With both materials, the corrosion susceptibility is greatest in 40% hydrofluoric acid; fluorosilicic acid acts as corrosion promoter [723].

Material	Temperature K (°C)	Hydrofluoric acid, %			
		6	20	40	60
		Corrosion rate, mm/a (mpy)			
Copper	293 (20)	0.20 (7.87)	0.26 (10.2)	0.43 (16.9)	0.19 (7.48)
	323 (50)	0.45 (17.7)	0.62 (24.4)	0.87 (34.3)	0.37 (14.6)
	353 (80)	0.69 (27.2)	0.95 (37.4)	1.37 (53.9)	0.85 (33.4)
Cu-Al9Mn2	293 (20)	0.13 (5.12)	0.10 (3.94)	0.20 (7.87)	0.12 (4.72)
	323 (50)	0.13 (5.12)	0.56 (22.0)	0.73 (28.7)	0.32 (12.6)
	353 (80)	0.28 (11.0)	0.87 (34.3)	0.81 (31.9)	0.62 (24.4)

Table 115: Corrosion rate of copper and Cu-Al9Mn2 in hydrofluoric acid between 293 and 353 K (20 and 80 °C) [723]

Copper-nickel alloys

CuNi-alloys which contain less than 30% nickel are corroded in fluorine at room temperature two to three times more quickly than pure copper [724].

Apart from the NiCu-alloy Monel® copper containing 30% nickel can also be used for containers and equipment which come into contact with hydrofluoric acid, if the air content of the hydrofluoric acid is kept to a minimum. Aeration is corrosion-promoting [725, 726].

Copper-zinc alloys (brass)

The resistance of copper to fluorine, hydrogen fluoride and hydrofluoric acid is impaired by tin and zinc. At room temperature, brass with 15% zinc and bronzes behave in fluorine similar to copper; brass with 35 to 40% zinc, however, is attacked about five times more quickly than copper [699, 709]. In this attack, the zinc content is important: while brass with 15% zinc is attacked by fluorine at 473 K (200 °C) only at the same rate as copper, the corrosion rate of brass with zinc is about twice as high [709].

Moist fluorine severely attacks all copper alloys without exception so that they cannot be used [727].

In hydrofluoric acid up to 70%, not only brass with 15% zinc (CA-230) but also admirality metal (CuZn28Sn1) are resistant up to the boiling point of the acid [728].

Pickling solutions for copper alloys consist of nitric and sulfuric acid containing hydrofluoric acid and/or dichromate. Hydrofluoric acid acts as an inhibitor and guarantees uniform removal for Cu-5.3Sn3.9Ti0.32P [729].

As can be seen from Figure 82, the corrosion rate of CuZn-alloys (CuZn21-Al2As0.05) at 353 K (80 °C) in 12 mol/l nitric acid containing 2 mol/l hydrofluoric acid strongly decreases during the first few hours and then reaches a stationary value [730].

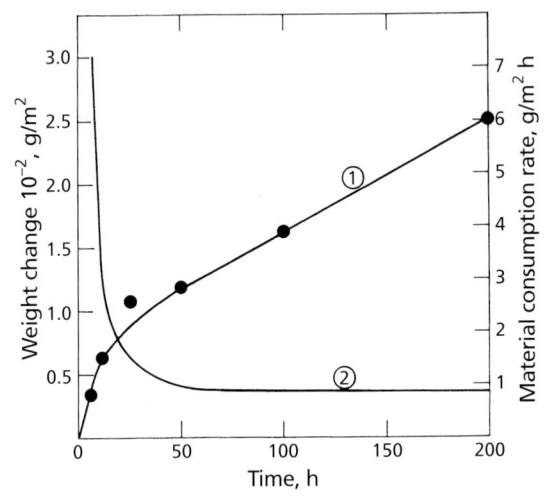

Figure 82: Weight increase ① and material consumption rate ② with time of CuZn21Al2As0.05 in a solution of 12 mol/l HNO_3 and 2 mol/l HF at 353 K (80 °C) [730]

Other copper alloys

CuSi-alloys are attacked in fluorine about twice as fast as pure copper.

Formic Acid

Author: H. Leyerzapf / Editor: R. Bender

Copper

Copper is severely attacked by formic acid in the presence of oxygen, the corrosion rate depending on the oxygen partial pressure and on the rate at which oxygen is transferred to the surface of the metal [731].

It was already recognized at an early stage [732, 733] that oxygen is the cause of the increased corrosion of copper and copper alloys in media containing formic acid. The primary remedy recommended is deaeration [734].

The following findings of laboratory and operational tests illustrate the influence of oxygen on corrosion of copper in formic acid.

– Operational test in 90% formic acid –

Corrosion rates of 0.97 mm/a (38.2 mpy) in the liquid and 0.15 mm/a (5.91 mpy) in the vapor space were determined for copper in 90% formic acid at 373 K (100 °C) in a distillation column. The corrosion rates in the vapor space of a storage vessel of copper at an ambient temperature of 298 K (25 °C) were of the same order of magnitude, whereas they were lower, at 0.33 mm/a (13 mpy), in the liquid [735].

– Laboratory tests with oxygen excluded –

These tests were carried out on 12 different materials, including SF copper (see Table 116) with exposure to 50% boiling formic acid with air excluded by continuous gassing with ultra-purity nitrogen (99.996 % N_2) [736].

HCOOH-concentration	Temperature	Corrosion rate, mm/a (mpy)
50 %	boiling	0.0022 (0.087)

Table 116: Corrosion rate of SF-Cu (CW024A, old DIN Material No. 2.0090) in boiling 50% formic acid (test duration 14 days) [736]

The results for the remaining 11 materials in this series of tests, modified CrNiMo-steels, nickel-based alloys and titanium, are quoted under the relevant groups of materials.

– Laboratory test with and without aeration –

Table 117 lists results of laboratory tests on copper in formic acid with and without aeration [735].

The results of a laboratory test listed in Table 118 provide information on the effect on copper foil of formic acid vapors saturated with water [737].

HCOOH	Corrosion rates, mm/a (mpy)			
	Not aerated		Aerated	
%	liquid	vapor space	liquid	vapor space
10	0.15 (5.91)	0.15 (5.91)	no data	no data
50	0.18 (7.09)	0.36 (14.2)	no data	no data
90	0.1–0.6 (4–24)	0.08–0.28 (3–11)	5.59–22.86 (220–900)	0.38–0.89 (15–35)

Table 117: Corrosion rates of copper in boiling formic acid at various concentrations with and without aeration (test duration 48 hours) [735]

HCOOH concentration above the acid solution		Corrosion rate mm/a (mpy)
Percent by volume	ppm	
0.01	0.6	0.03 (1.18)
0.10	6	0.04 (1.57)
1.0	60	0.14 (5.51)

Table 118: Corrosion rates of copper foil in the vapor space above aqueous formic acid solutions at 303 K (30 °C) and a relative humidity of 100 % (test duration 3 weeks) [737]

The corrosive attack of decomposition products containing formic acid, from plastic-based insulating coatings, on copper coils and copper contacts in the electrical equipment; this corrosion which is similar to the damage described for noble metal coatings, merits attention [738, 739].

According to a press bulletin, a Japanese telephone company uses cables which incorporate a flame-inhibiting insulation material containing magnesium hydroxide to protect against the decomposition products of polymers [740].

An incident involving damage to electrical railway signalling switches was described as follows [741]:

The damage occurred to an alarming extent to carefully inspected signal switches on a certain section of a London suburban railway. Investigation of the damage showed corrosion of the varnished copper coils at points of contact with the bakelite housing. Neither the materials used nor the ambient atmosphere appeared in any way abnormal. The only deviation in this section in comparison with others was in the operation of a safety switch in the normal testing operations of the signals, when a heavy current was switched on. The cause of corrosion was therefore suspected to be the formation of formic acid from the ozone formed during the switching operation (due to sparks) and its effect on a plastic component of the relay. It was difficult to demonstrate this, since the only material available for investigation was the broken end of a copper wire with a diameter of 0.1 mm. Using the electron microscope, however a crystal of corrosion product was discovered on the end of the wire which was identified as copper formate by electron diffraction. The ozone and formic acid attacked a plastic constituent of the varnish used in conjunction with the bakelite housing. To prevent further damage, the manufacturers of the relay were instructed to discontinue using this varnish [741].

Copper-aluminium alloys
Copper-nickel alloys
Copper-tin alloys (bronzes)
Copper-tin-zinc alloys (red brass)
Copper-zinc alloys (brass)

The behavior in contact with formic acid of copper alloys such as CuAl, CuNi, CuSn (tin bronze), CuSnZn (red-brass) and silicon bronze largely corresponds to that of copper. Grades of brass with high zinc content are susceptible to dezincification in organic acid [735]. The use of copper alloys in contact with formic acid presupposes that oxygen is excluded at the point of use.

Tables 119 and 120 list the results of laboratory and plant corrosion tests with boiling 90 % formic acid at 373 K (100 °C) with various copper alloys [735, 742].

According to [743], isolation valves made of tin bronze are classified as completely unsuitable for use in formic acid (50 % at 313 K (40 °C) 80 % at 333 K (60 °C) and 90 % at 293 K (20 °C)).

Material	Corrosion rate mm/a (mpy)
Copper, deoxidized	0.30 (11.8)
Aluminium bronze	0.06 (2.36)
Copper-nickel alloy	0.12 (4.72)
Phosphorus bronze (tin bronze)	0.04 (1.57)
Red brass (CuSnZnPb 85-5-5-5)	0.15 (5.91)
Everdur® 1010 (Cu-3.1Si-1-1Mn)	0.30 (11.8)

Table 119: Corrosion rates of various copper alloys in boiling 90 % formic acid (laboratory test duration 240 hours) [735]

Material	Corrosion rate mm/a (mpy)
Copper	0.97 (38.1)
Phosphorus bronze (tin-bronze)	1.17 (46.1)
Everdur® (3–4 % Si, about % Mn, bal. Cu)	0.89 (35)

Table 120: Corrosion rates of various copper alloys in boiling 90 % formic acid (test duration 7.6 days) [742]

Hot Oxidizing Gases

Author: K. Hauffe / Editor: R. Bender

Copper

Like its alloys, copper is used not only in the cable industry but also in the construction of apparatusses and equipment and for many special technical applications. In this context the materials are often exposed to oxidizing atmospheres at elevated temperatures, which means that adequate resistance to oxidizing gases such as air and oxygen is required.

First reports on detailed quantitative investigations into the oxidation rate of copper, both sheet and wire, in oxygen between 673 and 1,273 K (400 and 1,000 °C) and in air between 1,073 and 1,273 K (800 and 1,000 °C) were published as long ago as 1923 [744], and the chronological pattern of oxidation described by a parabolic growth law which was then later propounded by Wagner [745]. In the low-temperature region below 473 K (200 °C) however, considerable deviations from this growth law occur, and this then approaches a logarithmic law [746].

In an investigation [747] the growth and the thickness of the oxide coating on copper between 323 and 423 K (50 and 150 °C) was investigated in pure air and in typical laboratory air by means of Auger spectroscopy. In general, the oxidation rate of copper in laboratory air was greater by a factor of 3 to 8 than that in pure air. Between 398 and 423 K (125 and 150 °C) oxidation followed a parabolic and between 323 and 348 K (50 and 75 °C) approximately a cubic growth law. As shown from tests lasting 1,000 hours, the oxidation rate below 348 to 370 K (75 to 97 °C) is independent of temperature [747].

On the basis of investigations into the structure and studies of the mechanical properties of the oxide coatings produced on copper under the various test conditions it was shown [748] that adequate ductility of the oxide film cannot be attained below 873 K (600 °C), with the consequence that it ruptures. If copper is oxidized at elevated temperatures in the oxygen partial pressure region in which the CuO phase exists, the oxide coating then mainly consists of Cu_2O with only a very small proportion of CuO provided. This builds up densely and without pores on the Cu_2O, which frequently does not appear to be the case. Moreover, the oxidation rate under these conditions is independent of the oxygen pressure [749]. These findings have a direct connection with the disorder in the Cu_2O and/or CuO lattice [745]: whereas Cu_2O is a small p-type oxide with relatively high concentration of copper ion vacant sites, which facilitates transport of copper ions and electrons from the metal to the phase boundary between the Cu_2O and O_2, because of its substantially lesser disorder, the CuO builds up considerably more slowly [750]. Since the oxidation rate of copper even at high oxygen partial pressures is often determined only by the formation of the Cu_2O, independently of the external pressure, at an oxygen partial pressure which corresponds to the equilibrium pressure when both Cu_2O and CuO are present, the independence of the oxidation rate from the oxygen partial pressure even at relatively high oxygen pressures, can be understood. A number of experimental results are summarized in Table 121 for the purpose of illustration.

With the aid of diffusion tests with radioactive ^{64}Cu-ions it was quantitatively demonstrated that the oxidation rate of copper is determined by the rate of diffusion through the Cu_2O lattice by way of copper ion vacant sites [751, 752].

Temperature K (°C)	Oxygen pressure MPa	Parabolic rate constant $g^2/cm^4 h$
873 (600)	0.1	1.2×10^{-6}
	2.04	1.2×10^{-6}
1,073 (800)	0.1	2.7×10^{-5}
	0.60	2.8×10^{-5}
	1.36	2.8×10^{-5}
	2.72	2.8×10^{-5}
1,173 (900)	0.1	1.1×10^{-4}
	2.04	1.0×10^{-4}

Table 121: Oxidation rate constants for copper at various oxygen partial pressures [753]

A number of authors have studied the mechanism and extent of the simultaneous formation of CuO and Cu_2O during the oxidation of copper [750, 754–756]. In contrast to what would be expected from the growth rate constants of the Cu_2O and CuO layers it was found that at 1,223 K (950 °C) and 0.1 MPa oxygen the oxide coating contains 10 % CuO, at 1,143 K (870 °C) 17 % CuO, at 923 K (650 °C) 38 % CuO and at 573 K (300 °C) as much as 90 % CuO [754, 756]. These findings are understandable only if a porous CuO-layers is built up in which the oxygen can diffuse along the pores toward the growing Cu_2O layer. As has been demonstrated [750, 757], the oxidation of Cu_2O to form CuO does not follow a parabolic but a cubic growth law; the cause of this is assumed to be an aging process in the growing CuO-layer during oxidation [757]. Figure 83 shows a number of experimental results of the oxidation of a copper foil oxidized to Cu_2O.

When copper is oxidized between 1,073 and 1,273 K (800 and 1,000 °C) in carbon dioxide and/or in a mixture of carbon dioxide and carbon monoxide, the formation of Cu_2O follows a linear growth law with an initially larger rate constant [758]. Table 122 lists the linear rate constants for this reaction.

The dissociative chemisorption of carbon dioxide in accordance with the following equation is regarded as the rate-determining step:

$$CO_2 \text{ (gas)} \rightarrow CO \text{ (gas)} + O^- \text{ (ads)} + |e|^-$$

where O^- (ads) designates a negatively charged chemisorbed oxygen atom and $|e|^-$ a defect electron in the Cu_2O.

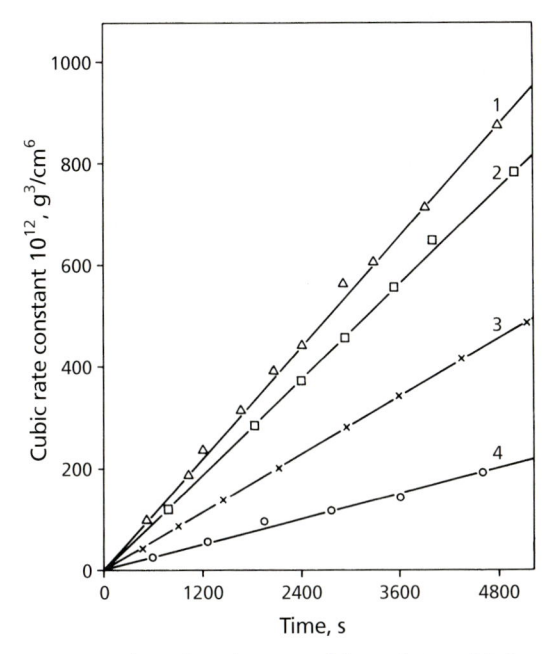

Figure 83: Chronological pattern of the oxidation of CuO_2 at 1,083 and 1,121 K (810 and 848 °C) at various oxygen partial pressures [750]
1: 1,121 K (848 °C) and 0.1 MPa oxygen ($k_c = 1.8 \times 10^{-13}$)
2: 883 K (610 °C) and 0.1 MPa oxygen ($k_c = 1.6 \times 10^{-13}$)
3: 883 K (610 °C) and 5.98×10^{-3} MPa oxygen ($k_c = 9.3 \times 10^{-14}$)
4: 883 K (610 °C) and 6.5×10^{-4} MPa oxygen ($k_c = 4.1 \times 10^{-14}$)

Temperature K (°C)	Linear rate constant, $g/cm^2 h$	
	(1)	(2)
1,073 (800)	3.4×10^{-6}	–
1,173 (900)	1.5×10^{-5}	1.2×10^{-5}
1,273 (1,000)	2.0×10^{-5}	1.5×10^{-5}

Table 122: Mean linear oxidation rate constants for pure copper at various temperatures in carbon dioxide at 0.1 MPa prior to (1) and after 50 hours (2) [758]

There have been earlier reports on similar investigations with the same result [759, 760].

It has long been known that additions of sulfur dioxide and/or sulfur dioxide and hydrogen sulfide to air substantially increase the oxidation rate of copper [761, 762]. On the other hand, the oxidation of copper in pure, dry sulfur dioxide between 973 and 1,173 K (700 and 900 °C) takes place much more slowly than in dry air [763]. On the basis of thermodynamics, sulfur dioxide should not attack copper, which means that the observed oxidation by sulfur dioxide is probably due to traces of sulfur trioxide and oxygen in the gas. The influence of sulfur dioxide on the oxidation rate of copper in oxygen is shown in Table 123 [764].

SO_2 pressure MPa	Parabolic rate constant, $g^2/m^4 h$	
	0.01 MPa O_2	0.05 MPa O_2
0	2.1×10^{-5}	2.1×10^{-5}
0.5×10^{-2}	2.1×10^{-5}	3.8×10^{-5}
0.01	2.5×10^{-5}	1.6×10^{-4}
0.014	4.4×10^{-5}	3.4×10^{-3}
0.02	6.0×10^{-4}	8.6×10^{-3}
0.03	1.9×10^{-3}	1.5×10^{-2}

Table 123: The influence of sulfur dioxide on the oxidation rate of copper in oxygen at 1,073 K (800 °C) [764]

As investigations of the structure have shown, the scale layer consists of an inner Cu_2O coating on which there is a CuO coating and above that an irregular coating of the basic sulfate $CuO \cdot CuSO_4$. This basic sulfate, which forms a low-melting eutectic at about 988 K (715 °C) with CuO, is the cause of the marked increase in the rate of oxidation. It was further observed that considerable quantities of sulfur are dissolved in both Cu_2O and CuO. These results were confirmed by further investigations [765]. Whereas at 773 K (500 °C) after oxidation in an atmosphere consisting of 40 vol% of Ar + 30 vol% of O_2 + 30 vol% of SO_2 the outermost zone of the layer consists of $CuSO_4$ and the inner of Cu_2O, a scale layer obtained between 873 and 1,073 K (600 and 800 °C) has the following structure:

$Cu/Cu_2O\text{-}Cu_2S/Cu_2O/CuO/CuO \cdot CuSO_4/CuSO_4$

In this case too [765], the devastating part played by the eutectic melt consisting of CuO and $CuO \cdot CuSO_4$ is also pointed out.

The conditions leading to the formation of the CuO phase were clarified by measuring the adhesive strength of the oxide coatings formed on copper [766]. The decrease in the coefficient of friction of copper produced by superficial oxidation is noteworthy [767].

It is essential to research further into the structure and disorder of the oxide coating as it builds up, in view of their importance in determining the steps to take for protection against oxidation. For example, the method of nucleation and the growth of an oxide coating building up on a copper monocrystal under an oxygen pressure of 2.7×10^{-2} MPa between 573 and 673 K (300 and 400 °C) were investigated using the electron microscope and electron diffraction [768]. On the basis of the results obtained, the oxide crystals are some distance apart at the beginning of the oxidation and are oriented parallel to the <110>-direction of the copper substrate. The crystals continue to build up in the oxygen-free environment, probably as a result of the oxygen dissolved in the metal. Moreover, as shown electron diffraction and scanning electron microscope investigations, Cu_2O monocrystal platelets, with epitaxial structure, form on the <100> and <111> monocrystal surfaces of copper between 673 and 973 K (400 and 700 °C) in oxygen at 10^{-9} MPa, for which the dissolved oxygen in the copper is also used. The surface of the metal between the individual oxide platelets is free from oxide [769].

A number of researchers have studied the first stage of oxidation of copper [770–773]. It was found, for instance, that copper partially covered with a Cu_2O film gradually loses the oxide coating when subsequently heat treated at 723 K (450 °C) in high vacuum (~ 1.3×10^{-11} MPa). Since the rate of degradation of the oxide coating increases with increasing oxide-free surface area, it seems reasonable to conclude that the degradation of the oxide must take place at the phase boundary between Cu and Cu_2O. In this context the Cu_2O dissociates and diffuses as copper and atomic oxygen into the metallic substrate, from which the dissolved oxygen is then desorbed into the atmosphere in molecular form [774]. The desorption process its thought to be the rate-determining step in oxide degradation during the period in high vacuum. This solubility of gas in copper, which also includes hydrogen, may lead to embrittlement of the metal.

Such embrittlement has even been observed in copper materials which are normally resistant to hydrogen embrittlement, when these were annealed at high temperatures in air and then in hydrogen. This type of embrittlement is initiated by penetration of oxygen into the copper during annealing in air. The depth of penetration of the oxygen depends on the temperature and duration of annealing and is not influenced by subsequent annealing in hydrogen. It becomes more marked with decreasing phosphorus content of the copper and with increasing duration and temperature of annealing in air [775].

Technically interesting tests were carried out between 923 and 1,023 K (650 and 750 °C) in oxygen with specimens made of compressed powdered copper [776]. In the first phase of oxidation, characterized by a linear growth law, oxide formation takes place remarkably quickly because of the large number of pores present in the compressed specimen. During the subsequent second phase, which may be described by a parabolic growth law, oxide formation takes place considerably more slowly, since the pores are then apparently largely packed with oxide. As would be expected, oxide formation increases with increasing oxygen partial pressure. In continuation of these investigations the kinetics of the oxidation of compressed powdered copper specimens was investigated with additions of MgO, Al_2O_3, Cr_2O_3 and ThO_2 [777]. In this case, too, after lengthy periods of oxidation a parabolic growth law for oxidation was found; the additions of the oxides bring about a decrease in the rate of oxidation.

As already mentioned, at a temperature of 673 K (400 °C) and below a logarithmic growth law applies to the growth of the oxide coating. In this temperature range marked mechanical stresses occur between the copper and the oxide, and this was directly illustrated by torsion measurements during oxidation between 473 and 673 K (200 and 400 °C) [778]. In a more recent investigation the conditions determining the structure of the oxide coating were investigated more precisely than previously [779] by using a magnetic balance which allows simultaneous determination of mass and susceptibility, in oxidation tests between 473 and 813 K (200 and 540 °C). When the concentrations of copper ion vacancies in the Cu_2O were not too high, $Cu_{1.5}O$ (= Cu_3O_2) was detected in addition to CuO, the molar fraction at 813 K (540 °C) being 0.2 although it may increase to 1 with falling temperature [779]. Periodic changes in the composition of the layer which were followed precisely are explained by rearrangements taking place within the oxide coating.

Copper-aluminium alloys

In contrast to pure copper, the aluminium bronzes are distinguished by relatively good resistance to oxidation at elevated temperatures. Because of the only slight susceptibility to oxidation even at high temperatures, aluminium bronzes, such as cast and wrought alloys to DIN EN 1982, are used, for example, for valve seats in internal-combustion engines. The relatively high oxidation resistance of the aluminium bronzes is attributable to the formation of a protective Al_2O_3 coating. This thin, dense oxide coating, which is also present just below the melting point of copper, consists almost exclusively of Al_2O_3 and contains only relatively small quantities of copper oxides. This barrier effect of Al_2O_3 was recognized earlier [780–782] and later confirmed by more detailed experiments [783, 784].

It may therefore be assumed that the observed decrease in the rate of oxidation of CuMn- and CuNi-alloys (7 % Mn or Ni) containing 13 % aluminium must also be due to the barrier effect of the Al_2O_3 formed during oxidation [785]. It is intended to discuss the oxidation-inhibiting effect of additions of aluminium to CuNi-alloys and Constantan® (CuNi45Mn1) [786] in section CuNi-alloys. Summary papers are available on the kinetics of the oxidation of CuAl-alloys [787–789].

In addition to resistance to oxidation in air, aluminium bronzes exhibit good resistance to various furnace gases containing sulfur dioxide and hydrogen sulfide; the Al_2O_3 coating also formed in this case offers good protection, so that aluminium bronzes are often superior in corrosion behavior to other copper alloys. On the basis of the experimental results available, all aluminium bronzes may be regarded as sufficiently resistant in dry and moist air at 673 K (400 °C) [783]. Whereas a copper alloy containing 2 % aluminium is still significantly attacked in air containing 0.1 % sulfur dioxide and even substantially attacked in air containing 5 % sulfur dioxide, the aluminium bronzes containing 10 % aluminium and 2 % iron or 12 % aluminium are still resistant to corrosion by atmospheres containing sulfur dioxide at 673 K (400 °C). Al_2O_3 coatings formed on CuAl5 by selective oxidation, however, result in no additional improvement in resistance to oxidation [790]. Aluminium bronzes are not attacked by hydrogen and other reducing gases, which is why the familiar "hydrogen sickness", as it is sometimes called, does not occur in these alloys. Exposure to an atmosphere of hydrogen and nitrogen (3:1), however, at a pressure of 100 MPa at 773 K (500 °C) for 1,535 hours leads to a deterioration in the mechanical properties, when aluminium is simultaneously removed from the surface layer, this being attributed to the reaction of the nitrogen with aluminium. The use of aluminium bronzes is therefore not to be recommended under such conditions [791].

Work is still being carried out on the clarification of the complex oxidation mechanism of aluminium bronzes, so that steps to improve oxidation resistance can be improved. Comparative measurements of the oxidation behavior of copper, CuAl7.5Si2 and CuAl2.5Si2.5 have shown [792], for instance, that even after 8 months at room temperature in laboratory air no complete oxide film but only single oxide crystals have been formed which were identified as Cu_2O crystallites. In contrast to this, the alloy containing 2.5 % Al and 2.5 % Si exhibits behavior similar to that of pure copper with formation of an oxide coating about 4 nm thick which then continues to build up only very slowly [792, 793]. The oxide coating present on the CuAl7.5Si2 alloy after oxidation for 12 hours in air at 1,073 K (800 °C) consists principally of γ-Al_2O_3 with α-Al_2O_3 inclusions [794]. As a consequence of the slight permeability of this coating to oxygen Cu_2O is formed on the substrate alloy. Significant diffusion of copper through the Al_2O_3 coating is not observed. From comparative oxidation tests with a Cu5Al and a Cu8Al alloy it can be suspected that the addition of 2 % silicon improves the oxidation resistance of the CuAl-alloys [794], whereas the addition of silicon to copper alone produces no significant protection against oxidation, since an amorphous SiO_2 coating permeable to gas is then formed.

In order further to research the mechanism of high-temperature oxidation of CuAl-alloys, alloys containing up to 4.5 % aluminium were investigated at 1,223 K (950 °C) in oxygen in the pressure range between 1.3×10^{-5} and 0.1 MPa. For orientation purposes, a number of experimental results are shown in Figure 84; it may be seen from these that even 4.5 % aluminium decreases the oxidation rate of the alloy at 1,223 K (950 °C) and 0.1 MPa oxygen partial pressure from 2.0×10^{-4}, for pure copper, to 2.9×10^{-9} $g^2/cm^4 h$. Microprobe measurements further showed that aluminium is enriched in the oxide coating, which indicates the presence of Al_2O_3 [795]. No significant improvement in resistance to oxidation is obtained by prior selective oxidation of the CuAl-alloy by heat treatment at 1,273 K (1,000 °C) and $pO_2 < 1.3 \times 10^{-4}$ MPa, when only the following reactions can take place [795]:

$3 Cu_2O + 2 Al$ (alloy) $\rightarrow Al_2O_3 + 6 Cu$ (alloy)

and

$3 O$ (alloy) $+ 2 Al$ (alloy) $\rightarrow Al_2O_3$.

The internal oxidation of low-alloy copper alloys which is observed in particular in the case of aluminium bronzes with low aluminium content is noteworthy. This phenomenon, which is of general interest for all copper alloys, was earlier intensively studied [796–799]. Since internal oxidation may modify the mechanical properties of the copper alloys in both desirable and undesirable ways, the conditions under which the alloy is used must be precisely known; only then the correct copper alloy be selected, for example when a ductile alloy is required not harden during use. Due to comprehensive work on the mechanism of internal oxidation and on the transition from external to internal oxidation, this objective is at least partially attainable [800], since the rate and extent of internal oxidation of copper alloys can be theoretically estimated with the aid of clear relationships. The following prerequisites for the occurrence of internal oxidation of copper alloys must be defined as qualitative guidelines [801]:

- The diffusion rate of oxygen in the alloy must be greater than that in the alloying metal (e.g. aluminium).
- Internal oxidation of a copper alloy is always to be expected when the concentration of the less-noble alloying element, e.g. aluminium, is low.
- When the oxygen partial pressure of the foreign oxide is lower than the equilibrium pressure of oxygen between the Cu_2O and Cu phases, internal oxidation must be expected.
- If the external oxygen partial pressure is higher than the equilibrium pressure in contact with Cu_2O phase, an external oxide coating will first be preferentially formed which may consist of Cu_2O and the foreign oxide, e.g. Al_2O_3, which means that internal oxidation can be retarded.

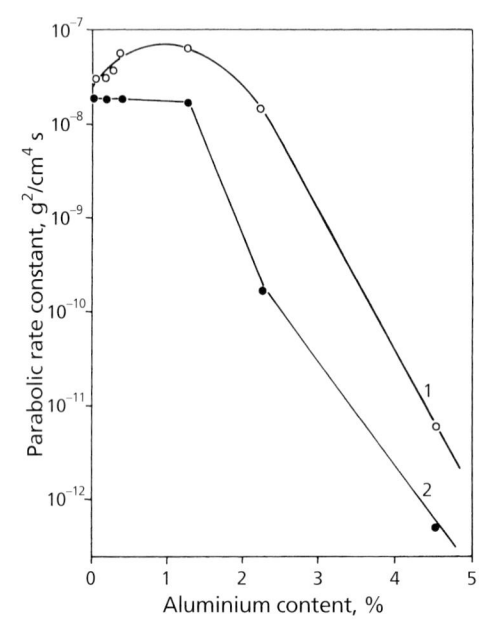

Figure 84: The oxidation rate constants of CuAl-alloy as a function of the aluminium content as 1,223 K (950 °C) [795]
1: 9.9×10^{-2} MPa oxygen; 2: 10.4×10^{-4} MPa oxygen

The objective of further investigations therefore must be the development of materials based on copper alloys on which a dense and pore-free oxide coating is formed in use. This prevents the penetration of oxygen into the alloy.

Copper-nickel alloys

When selecting CuNi-alloys for use in oxidizing atmospheres at elevated temperatures the possibility of internal oxidation must also be borne in mind, and this can be prevented or drastically decreased by the addition of aluminium or beryllium.

CuNi-alloys with relatively high nickel contents are used, for example, as materials for heat exchangers in power plants. Reference to standards which give the technical data for the CuNi-alloys on the market enables a preliminary selection to be made of alloys exhibiting the required mechanical and physical properties. With regard to the oxidation behavior, the aspects described below should then be noted, as these are based on the most recent comprehensive experimental data.

Although copper and nickel form a homogeneous range of mixed crystals, their oxides are not mutually soluble [802].

It was already recognized at an early date [803, 804] that when the nickel contents, were too low, 7% Ni for example, the oxide coating formed on the alloy on oxidation between 573 and 973 K (300 and 700 °C), which is similar to that formed on pure copper, consists of Cu_2O with a thin surface coating of CuO. Beneath that is a coating of NiO, which should be of a critical thickness, since it severely retards the formation of Cu_2O. In further investigations, in addition to the kinetic data the structure of the scale layer was studied [805, 806]; this was found to exhibit the following composition when formed on CuNi-alloys with nickel contents below 10%:

CuNi-alloy/NiO/NiO + Cu_2O + pores/Cu_2O/CuO.

With nickel contents exceeding 15% NiO is preferentially formed, and this severely inhibits further attack of the alloy by oxygen. From theoretical estimations it was possible to show [807] for which composition of the alloy, taking into account temperature and oxygen partial pressure, the oxide coating consists of pure NiO. For instance, a nickel content of not less than 75% is necessary at 1,223 K (950 °C). In further investigations on the oxidation behavior of alloys with rather high nickel contents (CuNi30 (CM390H) and CuNi50 (CM239E)) [808] the structure of the scale layer was studied by modern spectroscopic methods such as ESCA and Auger spectroscopy [809]. According to the results obtained, in the region of elevated temperatures a scale layer of the type mentioned above is actually obtained; its composition is determined by the following substitution reaction:

Cu_2O + Ni (alloy) → NiO + 2Cu (alloy).

At room temperature, and probably also at slightly higher temperatures up to 353 K (80 °C), the oxide film consists entirely of NiO. At higher temperatures internal oxidation also occurs, and this leads to mechanical changes in the material [809]. Paper [810] reports on the mechanism of oxidation of CuNi3, CuNi4.6 and CuNi50 (CM239E) alloys between 723 and 1,103 K (450 and 830 °C) at very low oxygen partial pressures of about 5×10^{-7} MPa. Using a high-resolution electron microscope it was established that Cu_2O and NiO continue to build up independently of each other on all CuNi-alloys after they start to form. The growth rate and the crystal form of Cu_2O in this context are similar to those when growth occurs on pure copper. In analogous fashion, the NiO growing on the CuNi50 alloy during oxidation corresponds in structure and growth rate to the oxide film growing on pure nickel. In order to permit estimation of the extent of oxide formation on CuNi-alloys, Figure 85 shows the chronological pattern of oxidation of a number of CuNi-alloys and, for comparison, the patterns for pure copper and nickel [811]. As may be seen, substantial resistance to oxidation can be attained with nickel contents from about 30%. On the basis of a model evaluation, the rate-determining step is the surface diffusion of the copper atoms via the porous internal coating of NiO. The structure of the oxide coatings forming on CuNi-alloys containing 20, 40, 60 and 80% Ni at 623, 773 and 923 K (350, 500 and 650 °C), containing oxygen with a partial pressure of 1×10^{-8} MPa, was investigated by means of microprobe Auger spectroscopy [812].

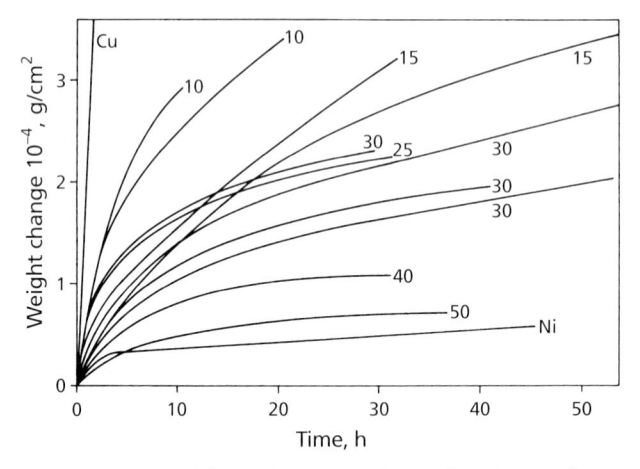

Figure 85: Curves of the oxidation rate of a number of CuNi-alloys at 773 K (500 °C) [811]. The numbers on the curves indicate the nickel content in percent

Copper-tin-zinc alloys (red brass)

The original copper bronze is a CuSn-alloy with tin contents of 1 to 10%. Only for special purposes, e.g. for casting bells, is up to 20% tin used because of the requirement that the alloy should have good sound characteristics. The oxidation resistance of such alloys in the temperature range from 263 to 323 K (–10 to 50 °C) is so good, provided there is no exposure to industrial atmospheres containing sulfur dioxide, that even after hundreds of years no oxide coatings occur that adversely affect the value of the bell. Small quantities of copper phosphide are added to remove the oxygen from the bronze, especially during casting and heat treatment.

Lead-tin bronzes containing 8% tin and small quantities of lead are used as material for highly-stressed bearings. Because of the development of elevated temperatures during operation, defined oxide coatings are formed which serve as the actual running surface. Normal tin bronzes consist of 1 to 2% tin, 0 to 0.05% lead, 0 to 0.1% phosphorus and 0 to 3% zinc. There are also, however, tin bronzes with tin contents between 3 and 8%. Among the CuSn-alloys there are also the SnPb-casting bronzes (to DIN EN 1982) and the CuSn-wrought alloys (to DIN EN 12168) to be considered.

Kinetic data on oxidation are of interest because of the various potential applications for the tin bronzes in air and oxygenic gases at room temperature and in the average temperature range, e.g. as material for pipes used for transporting dry and saturated steam in steam-raising plants. Data on this, however, are quite meagre and are also obtainable only from old literature references. The oxidation rate of CuSn-alloys in air at 473 to 1,073 K (200 to 800 °C) reaches about half of that found for pure copper when the tin content is 5 %, and with a tin content of 20 % it reaches about 1/5 to 1/6 of that for pure copper [780, 783, 813].

Internal oxidation is also seen in CuSn-alloys [801]. For example, the rate constant for internal oxidation, k_i, is 2.7×10^{-7} cm^2/h in a Cu0.3Sn alloy at 873 K (600 °C) in air [798]. However, since CuSn-alloys exhibit only low high-temperature strength, their use at temperatures higher than 473 K (200 °C) is no longer to be recommended [814]. Additions of magnesium, aluminium and beryllium are said to decrease the oxidation rate to a similar degree to that of pure copper. The fact that electron diffraction measurements in the outer zone of the oxide coating detected no SnO_2 and only small quantities of this oxide immediately in the vicinity of the phase boundary between the CuSn-alloy and the oxide coating indicates that tin diffuses only slowly [803, 815].

CuSn-alloys are attacked by gases containing sulfur dioxide with a degree of severity similar to that experienced by unalloyed copper [816, 817]. Since too little definitive experimental material is available on this subject if it should appear that these materials can be recommended by virtue of their other properties, the corrosion behavior of tin bronzes must first be tested, including testing in gases containing sulfur dioxide. At all events mentioned here, the addition of magnesium or aluminium could have a favorable effect, too.

Copper-zinc alloys (brass)

Copper-zinc alloys are also of great technical interest in the middle temperature range because of their outstanding cold working properties and good plasticity at elevated temperatures as well as their good thermal conductivity. Brass according to DIN EN 1982 is recommended as construction material for condensers in steam-generating plants or for applications in which good thermal conductivity is required. For example, the thermal conductivity of the CuZn22Al4 alloy at 873 K (600 °C) is 22 % of the value for pure copper; this improves the corresponding value for alloy steels by a factor of 10 [818]. These properties make it possible to use brass at elevated temperatures, provided sufficiently high oxidation resistance can be guaranteed.

According to earlier investigations [819], zinc in quantities of up to 10 and 15 at.% – i.e. in the α-range of brass – does not significantly influence the oxidation rate of copper. The oxide coating mainly consists of Cu_2O with minor ZnO inclusions. Only when the zinc content exceeds 15 at.% does a significant decrease in the oxidation rate begin, since ZnO is then formed. At the same time, not more Cu_2O is found, but mainly CuO. Due to the protective ZnO coating, diffusion of the copper ions through the oxide coating to the reaction site is inhibited so greatly that the Cu_2O is oxidized to form CuO, this also contributing to an improvement in oxidation resistance (this phenomenon does not occur in CuNi-alloys because of the insufficient mobility of the nickel ions). However, the oxidation resistance of brass containing 10 at.% zinc can be significantly improved if, for example, after several hours of oxidation at 973 K (700 °C) giving rise to an oxidation rate of about 2.3×10^{-6} $g^2/cm^4 h$, heat treatment in an argon atmosphere is carried out for 2 hours; this decreases the oxidation rate to about 1.9×10^{-7} $g^2/cm^4 h$ [791], as additional zinc oxide is formed according to the following exchange reaction:

$$Cu_2O + Zn \text{ (brass)} \rightarrow ZnO + 2 Cu \text{ (brass)}.$$

Above 20 at.%, the oxidation rate of brass is again independent of the zinc content; the rate in this case is about one power of 10 lower than for brass containing up to about 10 at.% zinc. It has been found that the addition of even 4 at.% aluminium to brass (CuZn20, CW503L) further improves the oxidation resistance [819] and also has a favorable effect on the mechanical properties of the brass.

The protective effect of aluminium against the attack of oxygen on various CuZn-alloys has been investigated in more recent work [813, 820]. Some experimental results on the chronological pattern of oxidation at 1,023 K (750 °C) and 0.1 MPa oxygen are shown in Figure 86, the more slowly oxidizing ternary alloys in particular

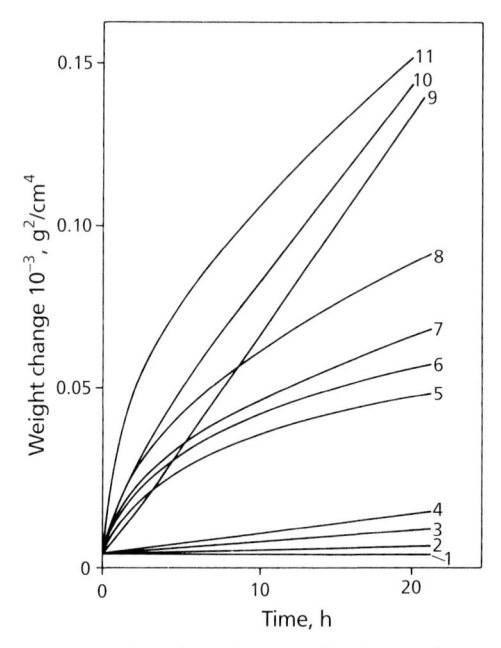

Figure 86: Chronological pattern of oxidation of a number of CuZnAl-alloys (contents in atomic percent) at 1,023 K (750 °C) in oxygen at 0.1 MPa [820]
1 – 22Zn8Al; 2 – 22Zn6Al; 3 – 17Zn9Al; 4 – 17Zn12Al; 5 – 18Zn5Al; 6 – 9Zn11Al; 7 – 15Zn6Al; 8 – 10Zn10Al; 9 – 23Zn12Al; 10 – 18Zn3Al; 11 – 8Zn7Al

having been selected. Starting with the CuZn18Al3 alloy it can be seen that even a slight increase in the aluminium content by 2 at.% considerably decreases the oxidation rate. A particularly low oxidation rate is exhibited by the alloys containing 22 at.% zinc and 4, 6 and 8 at.% aluminium. The high oxidation rate of the CuZn22Al12, however, cannot be explained on these grounds. The oxide coatings formed on the ternary alloys sometimes also contain, in addition to CuO and ZnO, substantial quantities of spinels such as $CuAlO_2$ and $CuAl_2O_4$ [820] which may be regarded as a further cause of the good oxidation resistance. According to investigations with the scanning electron microscope, X-ray and electron diffraction studies, the thin oxide coating on the oxidation-resistant alloy CuZn22Al8 principally consists of flaky Al_2O_3. Due to the high zinc and aluminium alloying contents, the occurrence of the internal oxidation generally typical of copper alloys is prevented [807]. It may be generally stated that only 3 to 4 at.% aluminium is sufficient for the formation of a dense Al_2O_3 coating on alloys of the CuZn22 type.

The adhesion of the oxide coatings exposed to varying temperatures is of great interest in the use of oxidation-resistant alloys. Figure 87 shows the chronological oxidation patterns of a number of alloys exposed to heating/cooling cycles (25 minutes heating periods alternating with 5 minutes cooling intervals). As can be seen, only the ternary alloys with aluminium contents of not less than 8 at.% form thermally stable oxide coatings with good adhesion. In the case of the CuZn22Al4 and CuZn22Al6 alloys rupture of the initially protective Al_2O_3 coating, followed by increasing exfoliation on cooling, leads to substantial formation of scale with high weight loss. Experience shows, however, that adhesion can be improved by the addition of very small quantities of cerium, thorium and yttrium [821, 822], and this has been confirmed by experimental investigations with CuZnAl-alloys [823]. Although it is true that in a test lasting 50 hours, the addition of cerium and yttrium increases the oxidation rate of the CuZnAl-alloys at 1,023 K (750 °C) in air at 0.1 MPa in comparison with the alloys without these additions. The adhesion of the oxide coatings is then better, this having a favorable effect on the oxidation resistance with oxidation times in excess of 50 hours, there being no exfoliation.

As can be seen from Figure 88, 0.2 % cerium or yttrium has varying effects on the oxidation rate of the ternary alloys. The curves shown in Figure 88 were obtained with intermittent oxidation of 60 minutes at 1,023 K (750 °C) and 10 minutes at 298 K (25 °C) in air at 0.1 MPa. When about 1 % sulfur dioxide is added to the air or oxygen, the oxidation rate is increased by several orders of magnitude. Probably the formation of a dense Al_2O_3 coating is thereby prevented, Cu_2S- and Al_2S_3-coatings being formed instead.

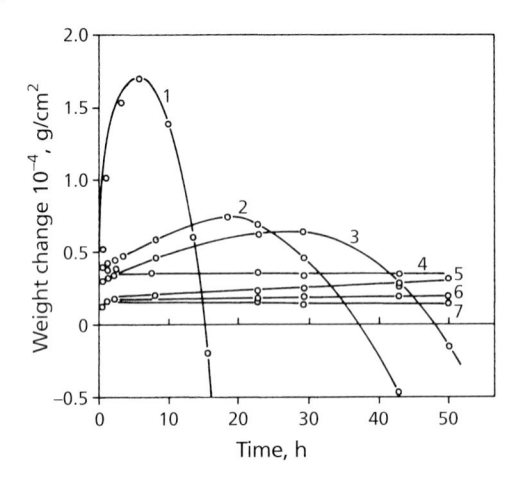

Figure 87: Chronological pattern of the change in weight of a number of CuZnAl-alloys (composition in at.%) during oxidation in oxygen at 0.1 MPa with alternate heating and cooling periods [820]
1 – CuZn18Al5 at 1,123 K (850 °C)
2 – CuZn22Al4 at 1,123 K (850 °C)
3 – CuZn22Al6 at 1,123 K (850 °C)
4 – CuZn22Al6 at 1,023 K (750 °C)
5 – CuZn17Al9 at 1,123 K (850 °C)
6 – CuZn22Al8 at 1,123 K (850 °C)
7 – CuZn22Al8 at 1,123 K (850 °C) with continuous oxidation at constant temperature

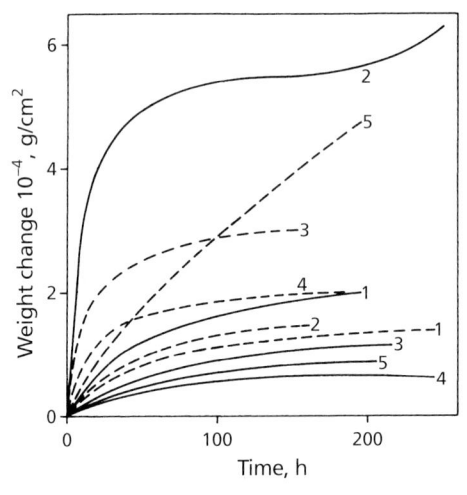

Figure 88: Chronological pattern of the change in weight with intermittent oxidation of a number of CuZnAl-alloys containing 0.2 at.% cerium (– – –) or yttrium (–) in air at 0.1 MPa at 1,023 K (750 °C) [823]
1 – CuZn17Al8
2 – CuZn17Al12
3 – CuZn22Al4
4 – CuZnAl6
5 – CuZn22Al8

Other copper alloys

In a similar manner to that described in the preceding sections, other copper alloys with interesting mechanical properties have also been investigated in respect of their oxidation behavior at elevated temperatures. Information about the type and degree of internal oxidation always arouses great interest.

For example, in the case of the CuTi-alloys containing 0.25 to 0.91 % titanium the adhesion of the outer oxide coating is improved by internal oxidation [824]. CuZr-alloys containing 0.3 to 1% zirconium can be effectively dispersion hardened by internal oxidation at 1,173 K (900 °C) in a gas mixture containing 2 parts helium and 1 part carbon dioxide by volume [825]. The best behavior on the basis of the best combination of hardness and electrical conductivity is exhibited by a CuZr-alloy containing 0.6 % zirconium. In this context, the additional electrical resistance caused by internal oxidation is substantially less than that, for example, of a comparable CuBe-alloy also subjected to oxidizing heat treatment.

The internal and external oxidation of CuSi-alloys, with silicon contents between 0.04 and 4.75 %, in the elevated temperature range have been studied in particular detail [798, 826–828]. According to these reports, with silicon contents in excess of 2 % internal oxidation with precipitation of SiO_2 particles no longer occurs and only of SiO_2 coating is formed on the alloy [827]. With silicon contents in excess of 1 % the kinetics of oxidation become irregular and not very reproducible [828]. Because of the relatively high mobility of the silicon, a protective SiO_2 film is initially formed on the alloy containing 4.75 % silicon, but because of the decrease in the silicon concentration at the phase boundary between the CuSi and the oxide coating, this is no longer stable at a later stage, so that the film ruptures and the oxidation rate again increases. In the temperature range between 1,073 and 1,273 K (800 and 1,000 °C) the silicon dioxide formed may be amorphous [826] or it may occur as tridymite [829] and also as α-cristobalite [830].

In the case of the CuSi-alloy containing 0.04 % silicon, after oxidation in oxygen at 10^{-3} MPa at 1,273 K (1,000 °C), no silicon dioxide, but only Cu_2O, was found on the surface of the alloy. The rate of oxidation in this case was comparable with that of pure copper [831].

The results of investigations into the extent of internal and external oxidation are also available for CuMn-alloys containing up to 20 % manganese [832]. According to these findings, the distribution of the oxides precipitated is dependent on time; as oxidation proceeds, penetration of the reaction site decelerates, since diffusion of the oxygen is hindered by the presence of precipitated particles. In the later phases of oxidation for periods exceeding 24 hours, the lateral growth of the particles predominates.

As may be seen from Table 124, the oxidation rate of copper-manganese alloys in oxygen at 0.1 MPa between 823 and 1,123 K (550 and 850 °C) is not significantly decreased until the rather high manganese contents of 23.85 and 34.85 % are reached [833]. As a consequence of the MnO coating immediately formed on the alloy, outward diffusion of copper is slower than the diffusion through the Cu_2O

coating which is formed on alloys with lower Mn content. Due to the slow transport of copper to the surface of the oxide, CuO is exclusively formed; internal oxidation, however, occurs in all alloy specimens.

CuBe-alloys may also form BeO coatings, which provide protection against oxidation, in a similar manner to CuAl-alloys, provided the beryllium content of the alloy is sufficiently high. Whereas the oxidation rate of a CuBe-alloy containing 6.6 at.% beryllium is the same as that of pure copper, the oxidation rate is significantly decreased with a beryllium content of 12.6 at.%, this being attributable to the formation of a thin, dense BeO coating [834]. Whereas below the critical beryllium concentration the oxidation rate at 1,123 K (850 °C) in air at 0.1 MPa is about 3.5×10^{-5} $g^2/cm^4 h$, whereas above this concentration it is only about 7×10^{-9} $g^2/cm^4 h$.

The scale layer formed consists of a dense BeO coating and a mixture of 3 phases: Cu_2O + BeO + pores. Wagner [835] has studied the problem of the formation of such oxide coatings and those with regular structure and composition and elucidated it by taking CuAu-alloys as examples [836].

Manganese content %	Temperature K (°C)			
	823 K (550 °C)	923 K (650 °C)	1,023 K (750 °C)	1,123 K (850 °C)
	Parabolic rate constant $g^2/cm^4 h$			
–	1.45×10^{-7}	1.24×10^{-6}	8.9×10^{-6}	3.5×10^{-5}
2.27	–	1.94×10^{-6}	8.8×10^{-6}	5.1×10^{-5}
4.92	–	2.81×10^{-6}	1.26×10^{-5}	6.2×10^{-5}
10.05	5.2×10^{-8}	–	–	3.6×10^{-5}
23.85	1.13×10^{-7}	4.60×10^{-7}	1.44×10^{-6}	5.9×10^{-6}
34.85	–	1.94×10^{-7}	1.75×10^{-6}	1.14×10^{-5}
100	2.7×10^{-7}	1.83×10^{-6}	8.53×10^{-6}	3.02×10^{-5}

Table 124: Parabolic oxidation rate constants of copper and CuMn-alloys in oxygen at 0.1 MPa [833]

The oxidation of gold-rich copper alloys, in both oxygen and air, follows the same pattern as pure copper and obeys a parabolic growth law [837], whereas the oxidation of CuAu-alloys in carbon dioxide follows a linear growth law and is dependent on the carbon dioxide partial pressure and the velocity of the gas flow [838]. In Table 125 a number of results has been assembled to assist in orientation.

CO_2 pressure	Linear rate constant, $g/cm^2 s$		
MPa	Cu	Cu + 5.3 %* Au	Cu + 10 %* Au
0.1	6.9–7.1	5.0–5.3	2.3–2.5
0.08	4.8–4.9	2.6–2.8	1.3
0.04	2.0	0.9	0.4

* at.%

Table 125: Linear rate constants for the oxidation of copper and CuAu-alloys at 1,273 K (1,000 °C) in carbon dioxide [838]

Comparable investigations into oxidation have also been carried out for CuPt- and CuPd-alloys [839]. With increasing Pd- or Pt-content, the oxidation rate of the copper alloy between 1,123 and 1,273 K (850 and 1,000 °C) in oxygen at 0.1 MPa decreases by up to four powers of ten in comparison with that of pure copper. These results are in agreement with the predictions based on the theory of oxidation of metal alloys [835, 840, 841].

Hydrochloric Acid

Author: A. Bäumel, P. Drodten / Editor: R. Bender

Copper

Copper has a slightly more noble potential [$E_o(Cu/Cu^{2+})$ = +0.33 V] than a standard hydrogen electrode and should therefore be resistant in non-oxidizing acids. The attack by hydrochloric acid is thus not only dependent on the concentration and the temperature but, in particular, on the presence of oxidizing agents. Copper is strongly attacked by aerated hydrochloric acid.

The behavior of copper in mixtures of oxidizing and reducing acids is important with regard to pickling and electropolishing. In reference [842], the dissolution behavior of copper in acid mixtures HCl-HNO_3, HCl-H_2CrO_4 and H_3PO_4-HNO_3 was investigated with thermometrics, by measuring the rate of mass loss and the free mass loss at 298 K (25 °C). The total concentration of each acidic solution was 5 mol/l.

Figure 89 and Figure 90 gives the results for the three acid mixtures. In HNO_3-H_3PO_4 mixtures, the removal rate constantly increases from zero in pure phosphoric acid to the corrosion rate in pure nitric acid (Figure 91).

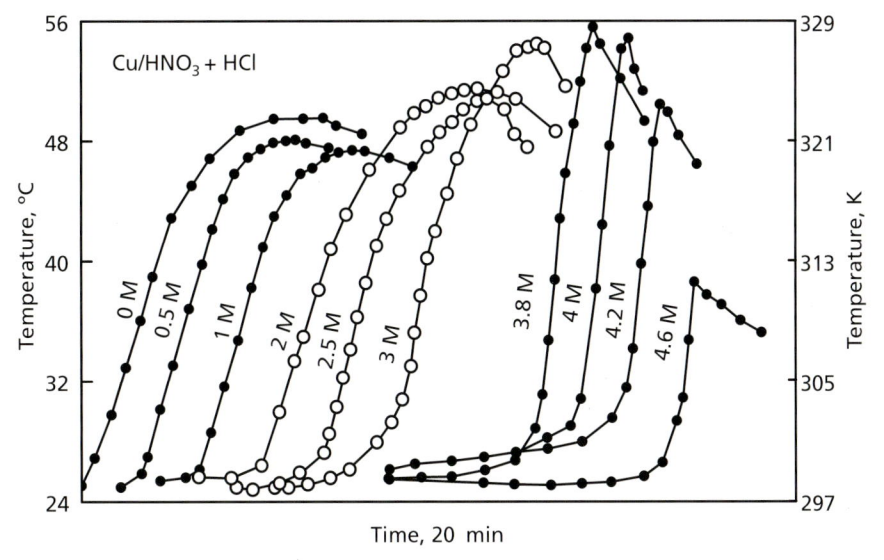

Figure 89: Temperature change in the copper/aqueous HNO_3-HCl system. The numbers on the curves denote the molarity of HCl [842]

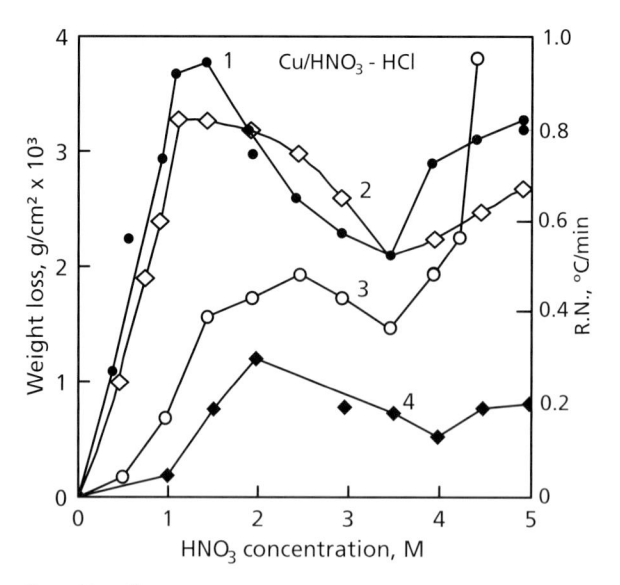

Figure 90: Changes:
1) of the reaction number (RN),
2) of the thermometrically determined mass loss, and
3) of the free mass loss in a stirred solution,
4) in an unstirred solution as a function of the HNO_3 concentration [842]

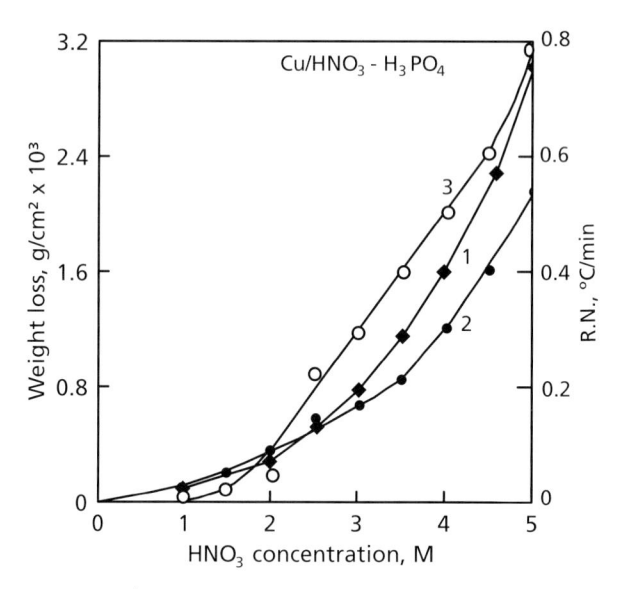

Figure 91: Changes:
1) of the reaction number (RN),
2) of the thermometrically determined mass loss, and
3) of the free mass loss as a function of the HNO_3 concentration (remainder H_3PO_4; total concentration 5 mol/l) [842]

In HCl-H$_2$CrO$_4$ mixtures, even for a low addition of chromic acid, the corrosion rate is greatly increased, and already reaches a maximum below 1 mol/l. It then drops slowly at first and then rapidly drops towards zero above 3 mol/l H$_2$CrO$_4$. The strongly oxidising chromic acid passivates the copper in spite of the Cl$^-$ ions still present (Figure 92).

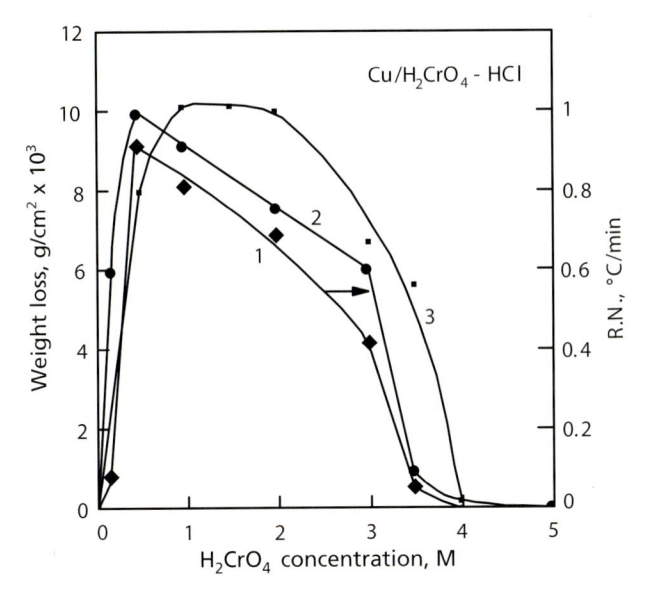

Figure 92: Changes:
1) of the reaction number (RN),
2) of the thermometrically determined mass loss, and
3) of the free mass loss as a function of the H$_2$CrO$_4$ concentration (remainder HCl; total concentration 5 mol/l) [842]

In HCl-HNO$_3$ mixtures, there is a similar tendency initially, namely a maximum at a relatively low addition of HNO$_3$ followed by a decrease. However, the values increase again between 3 to 4 mol/l HNO$_3$ up to the value in pure nitric acid. Nitric acid in the concentration used here is, however, not able to passivate copper.

If the results from the current density/potential curves that were also recorded are included, it can be seen that the optimum acid mixture to pickle copper consists of 3 to 3.5 mol/l HCl and 1.5 to 2 mol/l H$_2$CrO$_4$. A mixture of the acids H$_3$PO$_4$ and HNO$_3$ is especially suitable for electropolishing of copper.

Copper-aluminium alloys

The addition of iron and manganese to aluminium bronze with approx. 7% aluminium improves the mechanical properties; however, it slightly decreases the corrosion resistance. The addition of tin can largely compensate this. In the study [843], comparative corrosion tests were carried out in HCl, H$_2$SO$_4$, NaCl and seawater on an alloy with the percentage composition 89.5 Cu, 7.0 Al, 2.0 Sn, 1.0 Fe, 1.0 Mn and on a corresponding alloy without tin. The corrosion rate was determined by the rates of mass loss and by means of the current density/potential curves. In 5 and 10 % solutions of the investigated media at 289 and 353 K (16 °C and 80 °C), the corrosion rate was approximately halved by the addition of 2 % Sn. The heat-treated state of the material, which affects the microstructure, also had a slight influence (Table 126).

	5 % HCl	10 % HCl	2 % NaCl	5 % NaCl	10 % NaCl	5 % H_2SO_4	10 % H_2SO_4	Seawater
Initial state	0.254 (10)	0.203 (8.0)	0.086 (3.4)	0.172 (6.8)	0.183 (7.2)	0.058 (2.3)	0.107 (4.2)	0.076 (3.0)
Annealed*	0.205 (8.1)	0.178 (7.0)	0.078 (3.1)	0.147 (5.8)	0.178 (7.0)	0.056 (2.2)	0.119 (4.7)	0.089 (3.5)
Alloy without added tin**	0.226 (8.9)	0.064 (2.5)	0.203 (8.0)	0.305 (12)	0.366 (14.4)	0.147 (5.8)	0.203 (8.0)	0.127 (5.0)

* slowly cooled from 1,223 K (950 °C); ** initial state

Table 126: Corrosion rate of aluminium bronze in mm/a (mpy) at 289 K (16 °C) [843]

Aluminium bronzes

The study [844] investigated how the corrosion behavior of an aluminium bronze (type B-150) with the base composition 91.2 Cu, 6.9 Al, 1.8 Fe changes when small amounts of Ta, La and Nd are added. The results of the mass losses in 1 mol/l HCl at 303 K (30 °C) are summarized with respect to the test duration in Table 127. A low addition of 0.1 % Ta and La and of 0.05 % Nd improved the corrosion resistance by approx. 60 %. A higher addition of Ta and La decreased the favorable effect.

In H_2SO_4 and HNO_3, the addition of this element only gave a much smaller improvement in the corrosion resistance.

The same authors as in [843] investigated in [845] the effect of the addition of the rare earth metals La, Ce and Nd to an aluminium bronze of the base composition 90.1 Cu, 9.5 Al and 0.2 Mn in air-saturated 1 mol/l HCl at 303 K (30 °C). The result is summarised in Table 128. As in [843], the lowest concentrations of additives gave the greatest improvement in the corrosion resistance. Compared to the results in [843], the small improvement of the corrosion resistance by the addition of 0.05 % Nd stands out. The addition of 0.1 % Nd even decreased the corrosion resistance.

Amount added Mass%	Testing time, h						
	24	48	72	96	120	144	168
	Mass loss rate, g/m² d						
Basic material (B-150)	5.90	6.35	6.07	5.90	5.62	4.92	4.27
Ta 0.1	2.30	2.75	2.43	2.05	1.74	1.90	1.88
	(61)	(57)	(60)	(65)	(69)	(61)	(56)
0.2	4.80	4.40	3.64	3.83	3.09	2.90	2.60
	(19)	(31)	(40)	(35)	(45)	(41)	(39)
La 0.1	3.90	3.25	2.30	2.25	1.90	2.02	2.08
	(34)	(49)	(62)	(62)	(66)	(59)	(51)
0.25	4.40	3.50	2.80	2.87	2.58	2.42	2.33
	(25)	(45)	(54)	(51)	(54)	(51)	(46)
Nd 0.05	5.90	3.20	2.27	1.87	1.54	1.63	1.64
		(49)	(62)	(68)	(73)	(67)	(61)

Numbers in brackets denote an improvement of the corrosion resistance, stated in percent

Table 127: Influence of the elements Ta, La and Nd on the mass loss rates of aluminium bronze of type B-150 in 1 mol/l HCl at 303 K (30 °C) [844]

Amount added Mass%	Testing time, h				
	24	48	72	96	120
	Mass loss rate, g/m² d				
Base alloy	3.30 ± 0.12	1.85 ± 0.07	1.30 ± 0.04	1.20 ± 0.02	1.66 ± 0.04
La 0.05	2.50 ± 0.10	1.05 ± 0.05	0.70 ± 0.03	0.85 ± 0.02	0.98 ± 0.02
	(24)	(43)	(46)	(29)	(41)
La 0.15	2.10 ± 0.08	1.45 ± 0.04	0.96 ± 0.04	1.12 ± 0.01	1.22 ± 0.02
	(36)	(22)	(26)	(6)	(26)
La 0.2	2.70 ± 0.10	1.50 ± 0.04	0.83 ± 0.02	0.75 ± 0.02	1.20 ± 0.04
	(18)	(19)	(36)	(37)	(28)
Ce 0.2	3.0 ± 0.07	1.50 ± 0.07	1.0 ± 0.05	0.75 ± 0.02	0.94 ± 0.04
	(9)	(19)	(23)	(37)	(43)
Ce 0.3	3.0 ± 0.06	1.75 ± 0.02	1.33 ± 0.04	1.10 ± 0.02	1.26 ± 0.02
	(9)	(5)	(–)	(8)	(24)
Nd 0.05	3.40 ± 0.10	1.70 ± 0.03	1.17 ± 0.02	0.97 ± 0.03	1.36 ± 0.03
	(–)	(8)	(10)	(19)	(18)
Nd 0.1	3.50 ± 0.11	2.15 ± 0.10	1.60 ± 0.04	1.57 ± 0.02	2.08 ± 0.05
	(–)	(–)	(–)	(–)	(–)

Numbers in brackets denote the improvement of the corrosion resistance, stated in percent

Table 128: Results of the investigations of the mass loss rates in air-saturated 1 mol/l HCl, 303 K (30 °C) [845]

The change of the mass losses of propeller bronze of the base composition 82 Cu, 9 Al, 4 Fe, 4 Ni, 1 Mn by the addition of Ta and Nd in 4% HCl at 303 K (30 °C) is given in Table 129. As the concentrations of the additives increased, the corrosion rate also increased compared to that of the base alloy [846]. However, in a later investigation by the same authors on the effect of addition of Ta and Nd to the base alloy 91.3 Cu, 6.9 Al, 1.8 Fe (type B-150) in 1 mol/l HCl at 303 K (30 °C), up to 60% lower corrosion rates were obtained compared to the base alloy [844].

Amount added Mass%	Testing time, h						
	24	48	72	96	120	144	168
	Mass loss, mg/cm^2						
Base material	0.36	0.68	1.05	1.57	2.40	3.43	4.61
Tantalum							
0.10	0.30	0.65	0.78	1.50	2.08	5.38	8.23
0.20	0.37	0.50	0.74	1.14	5.46	8.26	12.64
0.30	0.32	0.69	1.48	2.23	4.20	6.15	9.24
Neodymium							
0.05	0.19	0.70	0.90	1.40	2.34	3.84	6.05
0.10	0.30	0.80	1.22	2.06	3.42	9.32	15.72
0.15	0.47	0.95	1.55	2.70	5.12	10.8	16.51

Table 129: Influence of added tantalum and neodymium on the mass loss of propeller bronze in 4% HCl at 303 K (30 °C) [846]

Copper-nickel alloys

The attack of CuNi alloys with 10 to 30% Ni in hydrochloric acid is moderate to strong. The presence of oxidizing substances noticeably enhances the attack.

In multistage evaporator systems to produce drinking water from seawater, one of the main problems is the formation of deposits in the condenser pipes, which are frequently made of CuNi 70/30. These deposits must be periodically removed to avoid a reduction of the heat transfer and flow rate. The best method is to subject the system to acid washing. For this, seawater acidified with HCl with a controlled pH is circulated through the system. The dissolution of $CaCO_3$ and $Mg(OH)_2$, the main components of the deposits, consumes HCl. This is replaced by adding more HCl. During acid washing of a seawater flash distillation plant in Abu-Dhabi in the United Arab Emirates, the metallic components of the plant were protected from corrosion by the acidic wash solution by the addition of dibutylthiourea as an inhibitor [847]. A number of different metallic materials are used in such a plant, such as titanium, stainless steels, CuNi alloys, bronze, aluminium bronze as well as large quantities of steel. Because no exact knowledge of the effectivity of the inhibitor was available, investigations were carried out with the main materials used in the plant, namely CuNi 70/30 and carbon steel. The following materials were available for the investigation:

CuNi 70/30: 66.47 % Cu, 29.54 % Ni, 1.92 % Fe

Steel: 0.21 % C, 0.10 % Si, 0.35 % Mn, 0.025 % P, 0.014 % S, 0.02 % Cr, 0.27 % Ni.

The mass losses were determined in pure HCl solutions at pH 1.8 to 2.0 and in a correspondingly acidic HCl solution made from seawater from the Gulf of Arabia that had a total salt content of 55,000 ppm. The inhibitor was added in concentrations of 20 to 300 ppm.

Figure 93a gives the chronological progression of the mass loss depending on the temperature in pure hydrochloric acid. After a certain incubation time, which decreases with increasing temperature, the linear removal of material starts, the rate of which becomes increasingly faster as the temperature rises. Figure 93b gives the chronological progression of the mass loss rates at 333 K (60 °C), which is the usual temperature for acid washing, in dependence on the inhibitor concentration. For the curves marked with an "A", the same solution was used for the whole test period. However, for the curves marked with a "B", a new solution was used each time the mass was determined. The different shapes of the curves for the two methods show that the inhibitor used was consumed in a relatively short time and thus became ineffective (A curves). However, if the concentration of additive was maintained at a level of at least 200 ppm, then the inhibition efficiency was sufficient over a long period.

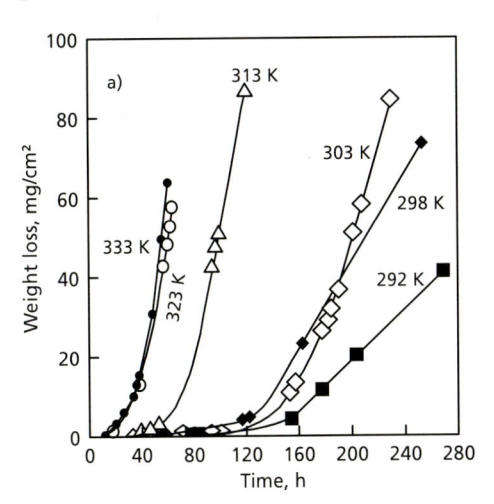

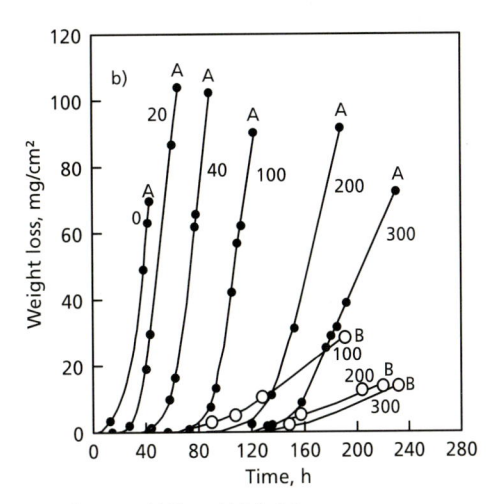

Figure 93: Time-dependent mass loss for CuNi 70/30 immersed in pure HCl at pH 1.8–2.0:
a) at temperatures between 292 and 333 K (19 and 60 °C);
b) at 333 K (60 °C) with a specified concentration of dibutylthiourea (numbers on the curves [ppm]) after mass determination in A: the same solution and B: immersed in fresh solution [847]

Figure 94 gives the corresponding inhibition curves in HCl solution with seawater from the Gulf. Under these conditions with an additional amount of 30,000 ppm Cl⁻ ions, that originate from the seawater, the inhibition efficiency is low and is effective at higher concentrations only after longer periods. Because acid washing is carried out within a period of 6 to 8 h, the suitability of dibutylthiourea as the inhibitor for CuNi 70/30 is questionable.

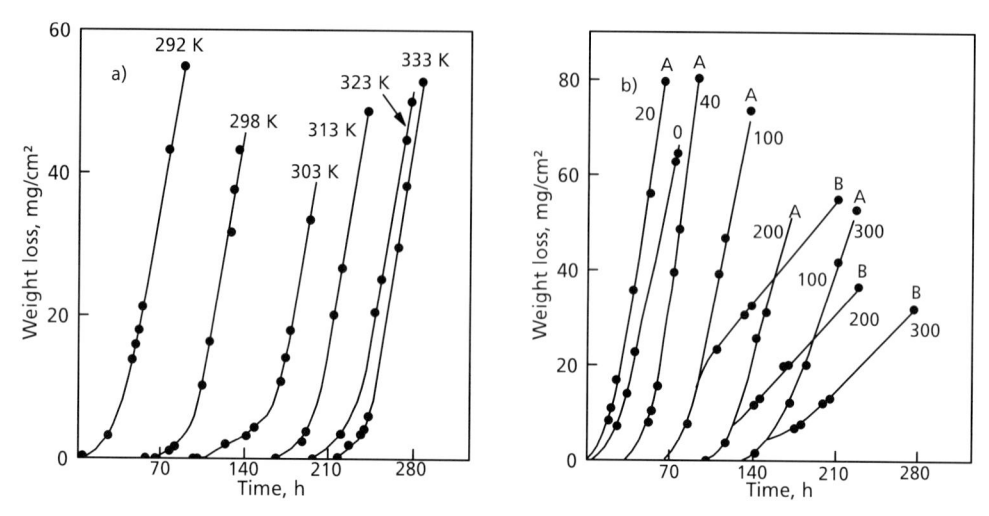

Figure 94: Time-dependent mass losses for CuNi 70/30 immersed in seawater acidified with HCl to a pH of 1.8–2.0
a) at temperatures between 292 and 333 K (19 and 60 °C);
b) at 333 K (60 °C) with a specified concentration of dibutylthiourea (numbers on the curves [ppm]) after mass determination in A: the same solution and B: fresh solution [847]

The time-dependence of the mass losses for steel are shown in Figure 95 and Figure 96. From the start of immersion of the samples in the solution there is a linear time-dependence of the mass losses. The corrosion rates in the pure HCl solution are greater than those in the acid of the same strength but made up with seawater from the Gulf. The inhibition efficiency in the pure acid determined using method A at concentrations from 100 to 300 ppm is 85 %, on average, and in the acidic Gulf water solution it is 70 %. Method B shows almost complete inhibition in both solutions. For steel, dibutylthiourea is thus a serviceable inhibitor for acid washing of the seawater desalination plant.

Copper-tin alloys (bronze)
Copper-tin-zinc alloys (red bronze)

The copper-tin alloys and copper-tin-zinc alloys have only low resistance in hydrochloric acid and are not suitable for continuous exposure.

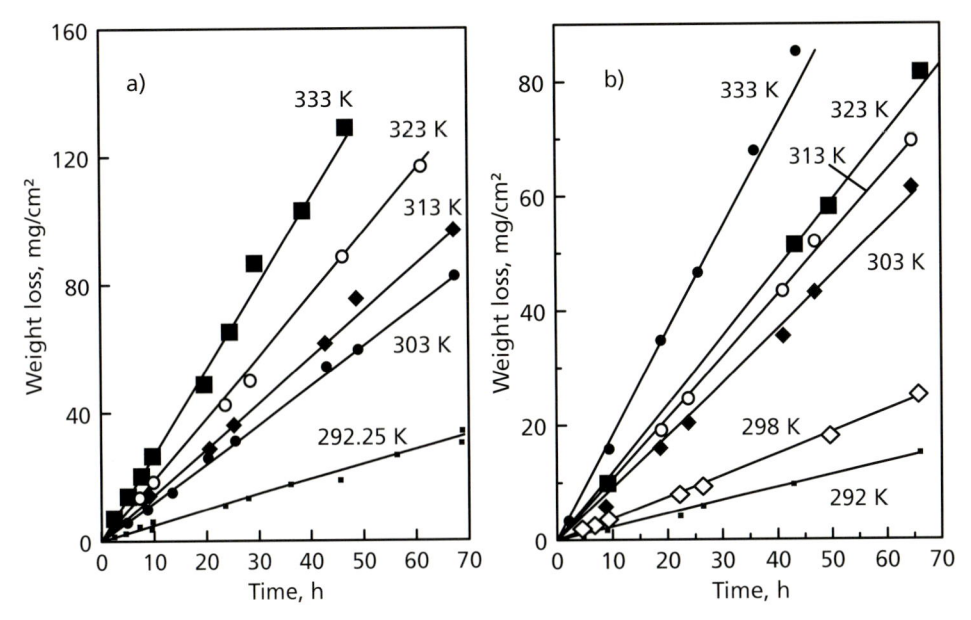

Figure 95: Time-dependent mass losses for steel in solutions with pH values from 1.8 to 2.0
a) pure HCl;
b) in seawater acidified with HCl
(numbers on the curves correspond to the individual temperature) [847]

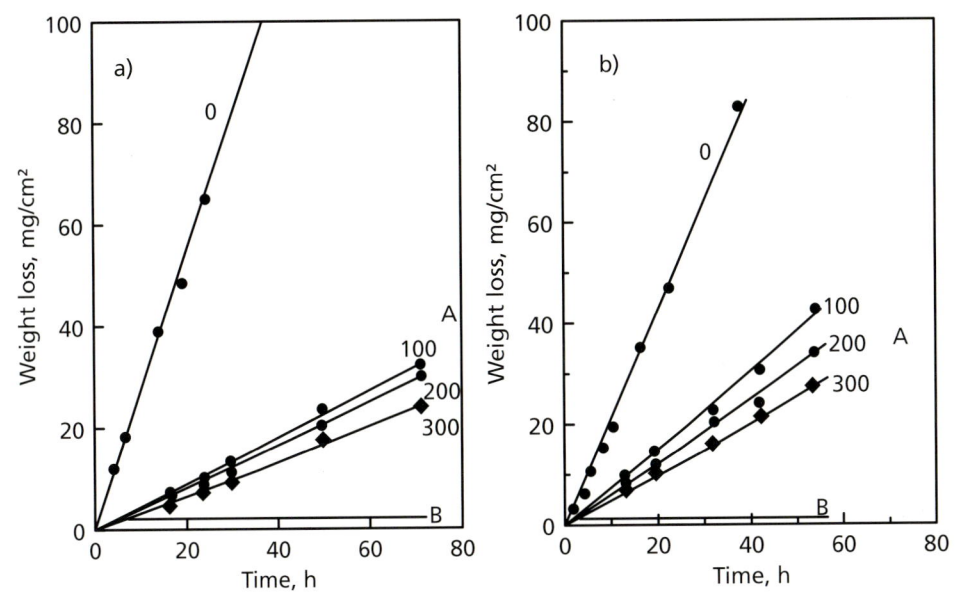

Figure 96: Time-dependent mass losses for steel immersed in solutions at 333 K (60 °C) with pre-specified concentrations of dibutylthiourea (numbers on the curves [ppm]) and pH values of 1.8 to 2.0 according to the mass determination in A: the same solution and B: in fresh solution [847]
a) pure HCl;
b) with seawater acidified with HCl

Copper-zinc alloys (brass)

Hydrochloric acid (HCl) and sulphamic acid (SA) are used to remove salt deposits in heat exchangers made of CuZn alloys (brass). Sulphamic acid is easy to handle and dissolves carbonate and phosphate deposits well. However, it has been shown that in the presence of oxidising Cu^{2+} ions, which are enriched during the pickling process, hydrogen and oxygen are absorbed by brass and have a detrimental effect on the behavior of the material.

The study in [848] investigated the corrosion of α-brass (CuZn37, DIN-Mat. No. 2.0321) in 5 % HCl and sulphamic acid with additions of $CuCl_2$ and $CuSO_4$ at ambient temperature. Before and after the corrosion tests, the oxygen and hydrogen contents in the material were determined by hot extraction. The results are given in Table 130. The mass loss rates, which are also given, are approx. 10 times greater in hydrochloric acid than in sulphamic acid (information on the duration of the tests is not given). However, with regard to the uptake of the gases, the two acids behave conversely. Thus the oxygen content is 15 to 35 times greater in hydrochloric acid and 25 to 55 times greater in sulphamic acid compared to the initial content of the brass samples. The uptake of the gas destroys the intergranular bonding and thus impairs the mechanical behavior. In the initial state, the samples can be bent backwards and forwards by 180° for 14 to 15 times until they fracture. After the corrosions tests, the samples immersed in hydrochloric acid fracture after 4 to 5 bendings and in sulphamic acid after only 1 to 2 bendings.

	Mass loss rate $g/m^2 h$	[O] mass%	[H] mass%
Brass sample (initial value)	–	0.003–0.009	0.0002–0.0003
Brass sample after testing in HCl + 13 g/l $CuCl_2 \times 2\ H_2O$	34.4	0.082–0.117	0.0008–0.0012
Brass sample after testing in SA + 20 g/l $CuSO_4 \times 5\ H_2O$	3.2	0.140–0.236	0.0020–0.0052

Table 130: Oxygen and hydrogen content in the brass samples before and after the corrosion tests in 5 % HCl- and SA-solutions in the presence of divalent copper ions [848]

The reason for the impairment of the mechanical properties is the formation of compounds that contain oxygen, hydrogen and sulphur. The formation of Cu_2O is known; it forms a galvanic element with copper, in which the Cu_2O acts as a cathode. This leads to destruction of the material. Cu_2O also reacts with hydrogen, which is formed by acid corrosion and penetrates into the material, according to the equation

Equation 16 $\quad Cu_2O + 2\ H \rightarrow 2\ Cu + H_2O.$

This process also leads to destruction of the material. Since these processes are favored in SA, HCl should be preferred over SA for acid washing of brass components in spite of the higher mass loss rates. Unfortunately, the report does not provide information on the duration of the tests.

Brass grades with high zinc contents are selectively corroded by HCl with dezincing.

Other copper alloys

Silicon bronze

Silicon bronzes with 1 % Si (e.g. UNS C64900) or 3 % Si (e.g. UNS C65500 formerly referred to as CuSi3Mn, DIN-Mat. No. 2.1525) have a superior resistance in HCl than the other copper-based alloys because they form a silicon-containing protective layer. At room temperature, these bronzes are regarded as sufficiently resistant to HCl in concentrations of up to approximately 20 %.

Hydrogen Chloride

Author: H. Barkholt / Editor: R. Bender

Copper

Copper is hardly attacked by dry hydrogen chloride (HCl) and is to be regarded as practically resistant up to 505 K (232 °C); however, at 589 K (316 °C) marked corrosion takes place [849].

Moist hydrogen chloride gas attacks copper considerably; the temperature used must therefore always be regulated to 5 to 9 K above the dew point [850, 851]. Copper loses 180 to 200 g/m² in ethanol with 0.36 g/l of HCl in the course of 40 days at 294 K (21 °C). HCl dissolved in acetone or acetic acid dissolves about 100 g/m² copper under the same conditions [852].

The influence of HCl on the corrosion behavior of pure copper (99.99 %) in polluted air was investigated. Laboratory tests in synthesized environments containing SO_2, NO_2, H_2S, O_3, Cl_2 and HCl were performed at 298 K (25 °C).

Copper corrosion rates increased with an increase in the HCl or Cl content (partial pressure) in single polluted environments. In complex environments the influence was small.

It was found out that there is a direct dependence of the relative humidity on the corrosion rates [853].

Copper-aluminium alloys

Practically the same applies to these alloys as for copper.

Copper-nickel alloys

The copper-nickel alloy CuNi30 is resistant to dry HCl at room temperature but is severely attacked when traces of moisture are present [854].

Copper-tin alloys (bronzes)

Copper-tin alloys correspond to copper in their behavior towards dry HCl. They are unsuitable, for example, as material in valves for hydrogen chloride gas [855] in the presence of moisture.

The corrosion rates of phosphor bronze in SO_2 or H_2S (10 ppm) atmosphere with 90 % relative humidity is suppressed by the addition of 1 ppm HCl [856].

The suitability as material of construction for HCl is discussed in [857].

Copper-tin-zinc alloys (red brass)

The behavior of this material towards HCl also resembles that of copper. Less than 0.05 mm/a (1.97 mpy) is corroded by a mixture of HCl and water vapor at temperatures below 373 K (100 °C) [858]; red brass is unsuitable for HCl at higher temperatures.

Copper-zinc alloys (brass)

Brass with approximately 70 % Cu and 30 % Zn behaves like copper towards HCl. Brass is practically resistant up to 505 K (232 °C), but at 530 K (257 °C) marked corrosion occurs [849]. A mixture of hydrogen chloride and water vapor leads to a corrosion rate of 0.05 mm/a (1.97 mpy) at 373 K (100 °C) [858, 859].

Vapors containing HCl, for example from the chlorination of benzene, can be discharged in brass pipes [860]. Similarly brass is suitable as a heat exchanger tube for gasoline vapors containing HCl at 573 K (300 °C). The HCl originates from the decomposition of chlorinated hydrocarbons [861].

Other copper alloys

Cast silicon bronze is practically resistant in HCl up to 477 K (204 °C) but is severely attacked at 589 K (316 °C) [849]. In other respects its behavior corresponds to that of copper.

CuAg-films are less suitable for the protection of alumino-borosilicate and soda lime glass in wet HCl vapor at 303 and 323 K (30 and 50 °C) than pure Ag-films [862].

The influence of HCl (1 ppm) on the corrosion behavior of beryllium copper is investigated in SO_2- and H_2S-atmosphere (10 ppm) under 90 % relative humidity. In both cases the corrosion rates were suppressed by the addition of HCl. Methods for testing the behavior of materials for electronic elements were developed under atmospheric conditions [856].

Hypochlorites

Author: L. Hasenberg / Editor: R. Bender

Copper

With regard to the use of copper materials, it must be pointed out that Cu^{2+} ions catalyze the decomposition of hypochlorite solutions, which only have slight-to-low stability in neutral and acidic pH values anyway. Cu^{2+} ions are undesirable in textile bleaches owing to product discoloration and to faster decomposition of the bleach on aging [863–866].

Copper and its alloys are attacked by hypochlorite solutions [867].
Copper is resistant at room temperature in alkaline solutions with up to 2 % sodium hypochlorite and in alkaline calcium hypochlorite solutions. The corrosion rates do not exceed 0.1 mm/a (3.94 mpy). However, it is not resistant in neutral and acidic solutions. The corrosion rates are greater than 3 mm/a (118 mpy) [866]. According to [868], copper cannot be used at room temperature because of corrosion rates of up to 0.5 mm/a (19.7 mpy) in hypochlorite solutions and < 1.3 mm/a (< 51.2 mpy) in sodium hypochlorite solutions.

Copper generally undergoes local attack by hypochlorite solutions with corrosion rates greater than 0.5 mm/a (19.7 mpy), which corresponds to a pit growth rate of 0.5 mm/a (19.7 mpy) [869].

A summary of the corrosion rates of cast copper-based materials is given in Table 131 [870].

G-Cu (cast copper materials) refers to all materials that contain at least 98 % copper. It must be mentioned at this point that wrought copper alloys exhibit almost the same corrosion behavior. As Table 131 shows, sodium hypochlorite as a solution is the most aggressive salt. The conditions are not mentioned in this reference. The cast materials exhibit corrosion rates of < 0.5 to > 1.0 mm/a (< 19.7 to > 39.4 mpy) in hypochlorite solutions, which excludes their use in nearly all cases [870].

Samples of copper M 3 exhibited uniform corrosion with mass loss rates of 0.0042 g/m^{2}d (2×10^{-4} mm/a (7.8×10^{-3} mpy)) in an alkaline sodium hypochlorite solution with 11 % free chlorine at 303 K (30 °C) after an exposure time of 125 h [871].

Corrosion tests are recommended before the materials described in this section are used. This can be carried out by electrochemical methods as well as electrochemical monitoring methods to determine the ranges at which pitting corrosion occurs so that they can be subsequently avoided.

Electrochemical corrosion tests in calcium hypochlorite solutions at 303 K (30 °C) at pH 11.4 to 11.6 showed that corrosion rates of up to 0.3 mm/a (11.8 mpy) are possible for pure copper (completely immersed in the solution) in flowing media [872].

In water with pH values < 6, corrosion of SF copper (2.0090) can be expected if the chloride content exceeds 1,000 mg/l or the free chlorine exceeds 5 mg/l. Of course, other substances in the water can promote corrosion, such as sulfates, sulfides, nitrates, and particularly ammonium and aggressive carbonic acid.

Hypochlorites

DIN-EN Mat. No.	DIN-EN designation	DIN-Mat. No.	DIN designation	Temperature K (°C)	Corrosion resistance* in Ca(OCl)$_2$	KOCl	LiOCl	NaOCl
–	–	2.0075	G-Cu	293–353 (20–80)	2			
–	–	2.0082	G-Cu	293 (20)		4	3	4
CC040A	Cu-C	2.0085	G-Cu	293–353 (20–80)	2			
–	–	2.0109	G-Cu	293 (20)		4	3	4
CC140C	CuCr1-C	2.1292	G-Cu	293 (20)		4	3	4
CC332G	CuAl10Ni3Fe2-C	2.0970	G-CuAl9	293–353 (20–80)	2			
				293 (20)		4	3	4
CC331G	CuAl10Fe2-C	2.0940	G-CuAl10Fe	293–353 (20–80)	2			
				293 (20)		4	3	4
CC332G	CuAl10Ni3Fe2-C	2.0970	G-CuAl9-10(Ni)	293–353 (20–80)	1			
				293 (20)		4	3	4

* 1: < 0.05 mm/a (< 1.97 mpy); 2: > 0.5 mm/a (> 19.7 mpy); 3: 0.5–1.0 mm/a (19.7–39.4 mpy); 4: > 1.0 mm/a (> 39.4 mpy)

Table 131: Corrosion behavior and comparison of the resistance of cast copper materials in hypochlorite solutions [870]

Table 131: Continued

DIN-EN Mat. No.	DIN-EN designation	DIN-Mat. No.	DIN designation	Temperature K (°C)	Ca(OCl)$_2$	KOCl	LiOCl	NaOCl
CC333G	CuAl10Fe5Ni5-C	2.0975	G-CuAl9-10(Ni)	293–353 (20–80)	1			
				293 (20)		4	3	4
	—	2.0962	G-CuAl8Mn	293–353 (20–80)	2			
				293 (20)		4	3	4
CC383H	CuNi30Fe1Mn1NbSi-C	2.0835	G-CuNi30	293–353 (20–80)	2			
				293 (20)		4	3	4
CC480K	CuSn10-Cu	2.1050	G-CuSn10	293–353 (20–80)	2			
				293 (20)		4	3	4
CC483K	CuSn12-Cu	2.1052	G-CuSn12	293–353 (20–80)	2			
				293 (20)		4	3	4

* 1: < 0.05 mm/a (< 1.97 mpy); 2: > 0.5 mm/a (> 19.7 mpy); 3: 0.5–1.0 mm/a (19.7–39.4 mpy); 4: > 1.0 mm/a (> 39.4 mpy)

Table 131: Corrosion behavior and comparison of the resistance of cast copper materials in hypochlorite solutions [870]

Table 131: Continued

DIN-EN Mat. No.	DIN-EN designation	DIN-Mat. No.	DIN designation	Temperature K (°C)	Corrosion resistance* in			
					Ca(OCl)$_2$	KOCl	LiOCl	NaOCl
–	–	2.1086	G-CuSn10Zn	293–353 (20–80)	2			
				293 (20)		4	3	4
CC491K	CuSn5Zn5Pb5-C	2.1096	G-CuSn5ZnPb	293–353 (20–80)	2			
				293 (20)		4	3	4
CC493K	CuSn7Zn4Pb7-C	2.1090	G-CuSn7ZnPb	293–353 (20–80)	2			
				293 (20)		4	3	4
CC750S	CuZn33Pb2-C	2.0290	G-CuZn33Pb	293–353 (20–80)	2			
				293 (20)		4	4	4
CC754S	CuZn39Pb1Al-C	2.0340	G-CuZn37Pb	293–353 (20–80)	2			
				293 (20)		4	4	4

* 1: < 0.05 mm/a (< 1.97 mpy); 2: > 0.5 mm/a (> 19.7 mpy); 3: 0.5–1.0 mm/a (19.7–39.4 mpy); 4: > 1.0 mm/a (> 39.4 mpy)

Table 131: Corrosion behavior and comparison of the resistance of cast copper materials in hypochlorite solutions [870]

Table 131: Continued

DIN-EN Mat. No.	DIN-EN designation	DIN-Mat. No.	DIN designation	Temperature K (°C)	Corrosion resistance* in			
					Ca(OCl)$_2$	KOCl	LiOCl	NaOCl
CC765S	CuZn35Mn2Al1Fe1-C	2.0592	G-CuZn35Al1	293–353 (20–80)	2			
				293 (20)		4	4	4
CC764S	CuZn34Mn3Al2Fe1-C	2.0596	G-CuZn34Al2	293–353 (20–80)	2			
				293 (20)		4	4	4
CC762S	CuZn25Al5Mn4Fe3-C	2.0598	G-CuZn25Al5	293–353 (20–80)	2			
				293 (20)		4	4	4
CC761S	CuZn16Si4-C	2.0492	G-CuZn15Si4	293–353 (20–80)	2			
				293 (20)		4	4	4
–	–	2.1170	G-CuPb5Sn	293–353 (20–80)	2			
				293 (20)		4	3	4

* 1: < 0.05 mm/a (< 1.97 mpy); 2: > 0.5 mm/a (> 19.7 mpy); 3: 0.5–1.0 mm/a (19.7–39.4 mpy); 4: > 1.0 mm/a (> 39.4 mpy)

Table 131: Corrosion behavior and comparison of the resistance of cast copper materials in hypochlorite solutions [870]

Table 131: Continued

DIN-EN Mat. No.	DIN-EN designation	DIN-Mat. No.	DIN designation	Temperature K (°C)	Corrosion resistance* in			
					Ca(OCl)$_2$	KOCl	LiOCl	NaOCl
CC495K	CuSn10Pb10-C	2.1176	G-CuPb10Sn	293–353 (20–80)	2			
				293 (20)		4	3	4
CC496K	CuSn7Pb15-C	2.1182	G-CnPb15Sn	293–353 (20–80)	2			
				293 (20)		4	3	4
CC497K	CuSn5Pb20-C	2.1188	G-CuPb20Sn	293–353 (20–80)	2			
				293 (20)		4	3	4
–	–	2.1166	G-CuPb22Sn	293–353 (20–80)	2			
				293 (20)		4	3	4

* 1: < 0.05 mm/a (< 1.97 mpy); 2: > 0.5 mm/a (> 19.7 mpy); 3: 0.5–1.0 mm/a (19.7–39.4 mpy); 4: > 1.0 mm/a (> 39.4 mpy)

Table 131: Corrosion behavior and comparison of the resistance of cast copper materials in hypochlorite solutions [870]

Because water and the substances it may contain are of great interest with regard to the resistance of copper materials, any information on the resistance of copper materials must be taken into account. For this, see [873] and also DIN EN 12502, Parts 1 and 5. These reports can only serve as an overview of possible corrosion processes in water, and should be used only for orientation purposes.

In swimming pool halls whose water has been treated with hypochlorites, galvanized steel parts can corrode much faster if they are in contact with copper materials (potential difference) [874]. Copper is the more noble material. However, this process only occurs in electrolytes, i.e. in the presence of water. The same applies to steel and other materials that are less noble than.

Copper-aluminium alloys

Copper alloys can contain up to 14 % aluminium. In particularly pure α-alloys (4–9 % Al), aluminium improves the corrosion resistance of copper with increasing content because, as a result of exposure to the medium, it forms firmly adhering protective layers on the surface of the material. They are also relatively resistant to local corrosion (pitting corrosion).

Copper-aluminium alloys are also attacked by hypochlorites [867].

Copper-aluminium alloys (aluminium bronzes) are regarded as being resistant in alkaline solutions of hypochlorite at low concentrations and at room temperature. In neutral and acidic solutions, corrosive attack increases with increasing temperature. For CuAl alloys as well, it must be pointed out that Cu^{2+} ions in the solution have a catalytic effect, and if they are present as residues in bleaching solutions, they increasingly destroy the material being bleached over time. The corrosion rates at room temperature are summarized in Table 132. Here too, sodium hypochlorite in solution is the most aggressive substance. It decomposes more quickly.

Medium	Corrosion rate mm/a (mpy)	Comment	Reference
Bleaching solution ($CaCl_2$ + $Ca(OCl)_2$)	≤ 13 (≤ 510)	room temperature, neutral solution	[875]
Calcium hypochlorite solution	≤ 0.5 (≤ 19.7) ≤ 1.3 (≤ 51.2)	room temperature no data	[876] [877, 878]
Potassium hypochlorite solution	≤ 0.5 (≤ 19.7)	room temperature	[876]
Sodium hypochlorite solution	> 1.3 (> 51.2) 0.11 (4.3)	room temperature room temperature, 2% solution	[876] [875, 879]

Table 132: Corrosion behavior of copper-aluminium alloys in hypochlorite solutions (see also Table 131)

The corrosion rates of cast materials in hypochlorite solutions are summarized in Table 131. According to [870], G-CuAl9 is certainly the alloy with the material number 2.0970, and G-CuAl9-10 (Ni) represents the alloys 2.0970 and 2.0975, whereas G-CuAl8Mn is the alloy 2.0962.

Copper-nickel alloys

Nickel as an alloying element extends the range of the α-phase in copper. Single-phase alloys are less susceptible to selective corrosion on account of their homogeneous microstructure. The corrosion resistance of the CuNi alloys (Ni content 5–45 %) increases as the Ni content increases [880]. These materials are also known as German silver. They are characterized by a high resistance in neutral to slightly alkaline salt solutions.

An overview of the corrosion behavior of G-CuNi 30 (CW383H) in water containing other substances (impurities) is given in [873] and DIN EN 12502, Parts 1 and 5.

Nickel ions, like copper ions, have a decomposing effect on hypochlorite solutions.

The corrosion behavior of CuNi alloys can be compared with that of other copper materials on the basis of Table 131 [870]. The alloy G-CuNi 30 (2.0835) exhibits a corrosion rate of 0.05 mm/a (1.97 mpy) in calcium hypochlorite solution. On the other hand, in solutions of potassium and sodium hypochlorite, the corrosion rates already exceed 1 mm/a (39.4 mpy) at room temperature [870].

If the corrosion behavior in hypochlorite solutions is compared to that in damp chlorine, CuNi alloys (2.0852, 2.0872, 2.0878 and 2.0882) cannot be used at all or only for a short time in these media as materials for screws and nuts. The corrosion rates lie between 0.12 and > 1.22 mm/a (4.72 and > 48 mpy) [881]. In neutral calcium hypochlorite solutions, CuNi alloys may exhibit corrosion rates of up to 1.3 mm/a (51.2 mpy). Noticeably less corrosion can be expected in alkaline solutions.

In potassium hypochlorite solutions at room temperature, corrosion rates of up to 0.5 mm/a (19.7 mpy) are reported for CuNi alloys [868].

CuNi alloys can be expected to have a good resistance in sodium hypochlorite solutions of up to 2 %. These low concentrations, however, do not generally occur in practice and also require a very strict monitoring of the solutions.

Corrosion tests under operating conditions are also recommended for the use of CuNi alloys in hypochlorite solutions, especially for high concentrations and temperatures.

Copper-tin alloys (bronzes)

Wrought copper-tin alloys may contain up to 9 % Sn. Single-phase alloys exhibit the best corrosion resistance. Copper-tin alloys are essentially resistant to aqueous solutions of salts and carbonic acid. They exhibit mostly uniform corrosion, and are essentially insensitive to stress corrosion cracking. The cast alloys exhibit a similar corrosion behavior to that of the wrought alloys.

These materials are used in seawater on account of their good resistance [880].

Bronzes, like all other copper alloys, decompose hypochlorites owing to the presence of Cu^{2+} ions or other heavy metals in solution.

These materials can be assumed to be resistant in dilute hypochlorite solutions with < 2 % NaOCl at room temperature.

In solutions containing approximately 3 % active chlorine, mass loss rates of approximately 10 g/m²d (0.4 mm/a (15.7 mpy)) can be expected [863, 866].

Although these materials were regarded as being resistant at that time on the basis of these corrosion rates, nowadays, they are regarded as being moderately resistant.

Corrosion rates of < 1 mm/a (< 39.4 mpy) at room temperature are reported for G-CuSn 10–14 in sodium hypochlorite solutions [870]. According to [868], the corrosion rates in sodium hypochlorite solutions exceed 1.27 mm/a (50 mpy), and are less than 0.5 mm/a (19.7 mpy) in potassium hypochlorite solutions at room temperature.

These high corrosion rates exclude the use of these materials in hypochlorite solutions on the basis of today's requirements, and even more so because they also decompose the solutions.

Copper-tin-zinc alloys (red brass)

Approximately the same applies to CuSnZn alloys as for pure copper-tin alloys (see section Copper-tin alloys). Alloys with a heterogeneous microstructure are more susceptible to selective corrosion than homogeneous alloys.

According to Table 131, copper-tin alloys exhibit only a moderate resistance in hypochlorite solutions.

The corrosion rates of alloys G-CuSn 10 Zn (2.1086) and G-CuSn 5–7 ZnPb (2.1096 and 2.1090) are less than 0.5 mm/a (19.7 mpy) only in calcium hypochlorite solutions, otherwise they are greater [870].

These copper materials also decompose hypochlorite solutions.

Copper-zinc alloys (brass)

The corrosion behavior of copper-zinc alloys is largely determined by the proportion of zinc and thus by the microstructure. Alloys containing up to 37.5 % zinc are known as α-brasses. They have a homogeneous, cubic face-centered crystal lattice. Their corrosion behavior is similar to that of copper (see section Copper). Alloys containing between 37.5 to approximately 46 % zinc have a heterogeneous (α + β) microstructure. This mixed structure (heterogeneous microstructure) is synonymous with a lower corrosion resistance. The β-phase, which is richer in zinc, represents the anode (dissolution anode) in this heterogeneous microstructure, and is preferentially attacked when exposed to the medium. This selective type of corrosion is known as "dezincing". In this type of corrosion, the β-mixed crystal is preferentially attacked or dissolved. A high concentration of copper ions arises at the phase boundary, and this can lead to the deposition of copper on the surface of the metal under certain conditions. This process is accelerated autocatalytically. For this reason, corroded copper-zinc alloys with a pure β-microstructure exhibit predominantly zinc compounds in their corrosion products, whereas copper (dissolved out of the β-mixed crystal) is mostly deposited as a metal (sponge-like) on the "surface of the material".

These α-β-brasses must be protected against corrosion or should not be used under these conditions.

The corrosion resistance of α-brass in the atmosphere and in water can be improved by alloying with aluminium, manganese, nickel, and zinc. For example, the alloy CuZn20Al (2.0460) exhibits good resistance to seawater. The alloy CuZn28Sn (2.0470) is frequently used in river water [880].

The two alloys have very good resistance in water containing free chlorine and chlorides [873]. According to this report, in some cases, these materials have a superior resistance to SF-Cu (2.0090) and CuNi 10 Fe (2.0872).

Tolyltriazole is an effective inhibitor in cooling water for the alloy CuZn28Sn (2.0470). If a biocide is added, here sodium hypochlorite (NaOCl), this inhibitor retains its effectiveness and is hardly or insignificantly decomposed.

The results of corrosion tests in simulated cooling water with a pH of 6.5 and a total hardness of 110 ppm $CaCO_3$ at 323 K (50 °C) are summarized in Table 133. Samples of CuZn28Sn were immersed into the aerated cooling water (glass vessel with a capacity of 3.5 l) and rotated at a speed of 52 rpm. The pH value was adjusted by adding sulfuric acid. Measurements were taken every 48 h. The results, given in Table 133, show that 3.0 ppm tolyltriazole decreased the corrosion rate from 0.01 to 0.002 mm/a (0.39 to 0.079 mpy). Additions of NaOCl increased the corrosion in every case, whereas the presence of tolyltriazole inhibited it. Increasing the concentration of NaOCl in the cooling water led to decomposition of the tolyltriazole inhibitor; however, it also led to hardly any increase in the corrosion of samples of the alloy Cu28ZnSn [882].

Stress corrosion cracking (in hypochlorite solution)

Stress corrosion cracking (in hypochlorite solution) cannot be excluded in brasses. This can occur on work-hardened parts, in particular, if the residual stresses have not been essentially eliminated by heat treatment (stress-relief annealing).

The resistance of copper-zinc alloys is moderate in hypochlorite solutions. According to Table 131, the corrosion rates of the alloys G-CuZn15Si4 (2.0492), G-CuZn25Al5 (2.0598), G-CuZn33Pb (2.0290), G-CuZn34Al2 (2.0596), G-CuZn35Al1 (2.0592) and G-CuZn37Pb (2.0340) are less than 0.5 mm/a (19.7 mpy) only in calcium hypochlorite solutions. In all other hypochlorite solutions, the rates lie above this value, and in sodium hypochlorite solution they even exceed 1 mm/a (39.4 mpy) [870].

NaOCl as Cl$_2$	Tolyltriazole, ppm		Corrosion rate	Comment
	Initial	Final	mm/a (mpy)	
0	0	0	0.010 (0.39)	
0	3	3	0.002 (0.079)	
0.3	0	0	0.020 (0.79)	
0.3	3	3	0.004 (0.16)	
1.0	0	0	0.026 (1.02)	
1.0	3	2.8	0.005 (0.2)	
0.3	3	3	–	24 h
64–35.2	3	1.5	–	66 h
75–20	3	2	–	69 h
150	150	150	–	4 h

Table 133: Results of corrosion tests with the alloy CuZn28Sn in simulated cooling water with and without an inhibitor or biocide. Temperature 323 K (50 °C), pH 6.5, total hardness of the water 110 ppm CaCO$_3$ [882]

According to [868], brass has corrosion rates > 1.27 mm/a (> 50 mpy) in sodium hypochlorite solution and cannot be used.

Other copper alloys

This section deals with copper alloys that have alloy compositions that differ from those described so far. Because of their composition, they have a fairly inhomogeneous microstructure and thus mostly exhibit a tendency to selective corrosion.

The alloys listed under "Other copper alloys" in Table 131 cannot be used or are only suitable for intermittent use in neutral and acidic hypochlorite solutions. In slightly alkaline solutions and in solutions containing a maximum of 2 % hypochlorite, they should be suitable, otherwise, corrosion rates < 1 mm/a (< 39.4 mpy) are to be expected. Because the exact medium used to investigate the corrosion rates given in Table 131 is not known, it must always be assumed to be sodium hypochlorite, so that the maximum corrosion rate is 1 mm/a (39.4 mpy) for the alloys G-CuPb5–20Sn (CC495K, CC496K, CC497K and old 2.1170) and G-CuPb22Sn (old 2.1166) [870].

Lithium Hydroxide

Author: K. John / Editor: R. Bender

Copper

A certain insight into the corrosion behavior of copper is provided by papers on the oxidation kinetics and the growth of the copper hydroxide film in alkali metal hydroxide solutions [883, 884]. Stationary and rotating copper disks were used as electrodes at 294 K (21 °C) in 1 mol/l lithium hydroxide solution and the growth of oxide and hydroxide films determined under potentiostatic and galvanostatic conditions.

In the potential range from −0.17 to −0.22 V with 0.1 C/cm^2 a material consumption rate of about 0.4 g/m^2 h ($\approx$ 0.44 mm/a) was obtained after 8 hours. In this context it is known that bivalent copper ions are built into the surface film of $Cu(OH)_2$ and that CuO is produced and also that there is dissolution of copper [884]. From information available to date, it cannot yet be unequivocally decided whether Cu_2O and CuO occur simultaneously.

As shown in Figure 97, the corrosion potential of copper in 1 mol/l lithium hydroxide solution initially falls from −0.250 V to −0.325 V, remains constant there for about 24 hours and approaches a stationary value of about −0.560 V. Cu_2O is then formed exclusively from $Cu(OH)_2$ [885]. With increasing concentration of lithium hydroxide the second change in potential (to −0.560 V) takes place after shorter times with simultaneous marked reduction in the rate of corrosion, due to the formation of a dense and pore-free Cu_2O layer. Formation of Cu_2O slows down with high flow velocities because of the rapid removal of the copper ions, whereas corrosion is increased [885]. Under the experimental conditions used here, from the Coulomb vs. time curve a material consumption rate for copper of about 1.5 g/m^2 h ($\approx$ 1.66 mm/a) is obtained.

The dissolution of copper is already inhibited on the occurrence of the second monomolecular layer of Cu_2O formed in 0.1 mol/l lithium hydroxide solution at about −300 mV [886].

The relationships derived from photoelectric polarization measurements in 0.1 mol/l sodium hydroxide solution regarding oxidation and passivation of copper can probably be applied to the corrosion of copper in lithium hydroxide solutions: according to this, it would increase with the intensity of the light radiation received [887].

Figure 98 shows the dependence on temperature of the corrosion rate calculated from 4 hour tests on copper in molten lithium hydroxide between 773 and 873 K (500 and 600 °C). According to this, corrosion of the copper increases with temperature and is substantial at 873 K (600 °C) 80 g/m^2 h, equivalent to about 88 mm/a (3,465 mpy), which is why copper is unsuitable as material for molten lithium hydroxide [888, 889]. It is noteworthy that the corrosivity of the molten alkalis increases with increasing radius of the alkali metal ion (Li < Na < K). For example, the material consumption rate of copper at 773 K (500 °C) in molten lithium hydroxide is

7.2 g/m² h (≈ 8 mm/a) and in molten sodium hydroxide and potassium hydroxide about 120 and 1150 g/m² h respectively (≈ 132 and 1,270 mm/a), so that "rapid dissolution" may be used [889].

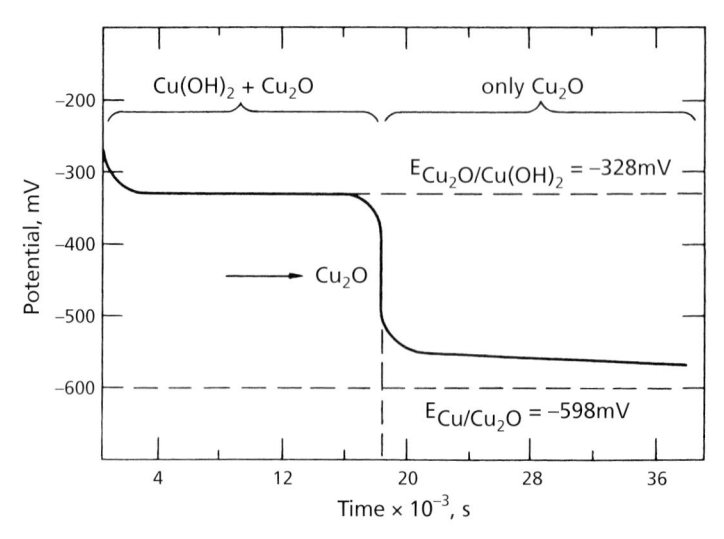

Figure 97: Potential of copper after corrosion in 1 mol/l lithium hydroxide solution at 293 K (20 °C) in static electrolytes. After 5 hours the corrosion product layer consists only of Cu_2O [885]

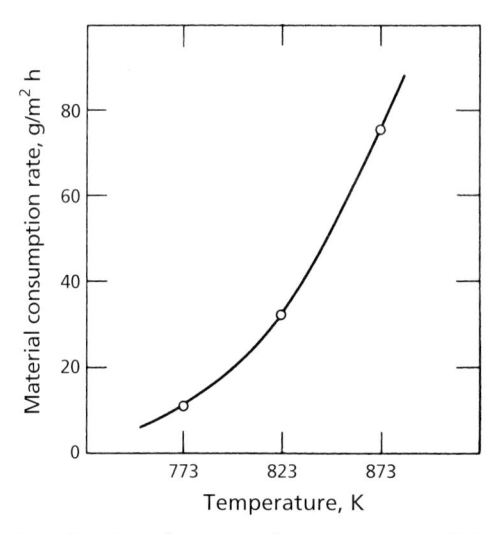

Figure 98: Rate of corrosion of copper in molten LiOH as a function of temperature [889]

Similar considerations apply to a melt consisting of 80 mol percent lithium hydroxide and 20 mol percent lithium fluoride at 1,073 K (800 °C) (140 g/m² h, equivalent to about 155 mm/a), according to which lithium fluoride has no noticeable influence on the corrosion of copper [890].

Copper may therefore be used as material only for lithium hydroxide solutions which do not exceed 1 mol/l in concentration up to 360 K (87 °C), and is not serviceable in molten lithium hydroxide.

Copper-nickel alloys

Because of the low resistance of copper to corrosion, corrosion tests with copper alloys in lithium hydroxide solutions and molten lithium hydroxide have evidently not even been carried out.

One exception to this is the alloy Berylco® 717 (CuNi30Be1), an alloy which is attacked remarkably little for a copper alloy in a melt consisting of 80 mol percent lithium hydroxide and 20 mol percent lithium fluoride at 1,073 K (800 °C) (2 g/m² h, equivalent to about 2.2 mm/a). This value, however, is based only on a test duration of 24 hours and would have to be confirmed by more lengthy tests [890].

Methanol

Author: H. G. Spilker / Editor: R. Bender

Copper

Copper is of great importance for production, further processing and direct use in methanol in many fields of application and under various conditions.

Alloying with other metals for controlled modification of properties is dealt with in sections of copper alloys.

The Cu contents of primary copper are 99.0 % to 99.9 %.

Oxygen-containing copper grades tend to become brittle during welding or hard soldering (hydrogen disease). Only the oxygen-free grades are resistant to hydrogen [891].

Oxygen-free, phosphorus-deoxidized grades, abbreviated names according to DIN SF-Cu (2.0090, CW024A) and SW-Cu (2.0076, CW023A), which have phosphorus contents of 0.005 to 0.040 % and are readily weldable, are used for apparatus construction and comparable applications [892].

The reasons for the good corrosion behavior of copper lie, in particular, in its natural nobleness. In contrast, the protective action of a copper oxide layer is of less importance [893]. The low reaction enthalpy is also decisive for the resistance. Corrosion which proceeds with evolution of hydrogen is impossible under non-oxidizing conditions. General corrosion under neutral conditions usually does not cause functional impairment, but corrosion damage may occur due to pitting [894].

Copper is moderately resistant to 10 to 100 % methanol up to the boiling point, with a corrosion rate of < 0.64 mm/a (< 25.2 mpy) [895].

In its "Corrosion Resistance Charts", Du Pont evaluates copper as being satisfactorily resistant to methyl alcohol [896].

In the "Data Charts", Part II, copper is evaluated as having a limited resistance (< 0.508 mm/a (< 20 mpy)) to 25 and 100 % methanol at 298 K (25 °C) [897].

Manufacturers of flexible metal tubes evaluate copper as being completely resistant in contact with 98 % methanol at 293 K (20 °C) and at the boiling point [898, 899].

A previous corrosion atlas gives a corrosion rate of 0.08 mm/a (3.15 mpy) for copper in methanol generally and consequently evaluates it as resistant [900].

Copper is used in methanol synthesis as a catalyst and catalyst additive and also as a material for plant components which are exposed to high temperatures during the process.

The copper-containing catalysts give a high selectivity and, because side reactions such as the formation of methane, dimethyl ether, higher alcohols or higher hydrocarbons are avoided, a product of the highest possible purity.

The high activity of these catalysts has promoted the development of the low-pressure processes, which are more economical than those employing high pressure, since 1970 (ICI, Lurgi) [901].

In addition to the pure catalysts zinc oxide and, to a limited extent, also copper oxide, a mixture of copper or copper oxide with zinc oxide and chromium oxide have proved to be particularly suitable [902].

Mixed contacts containing 8 to 28 % Cu gave the highest yields of methanol.

However, copper-containing catalysts can be used only with practically sulfur-free synthesis gas, since hydrogen sulfide acts on copper to form CuS [903].

Copper, as a good conductor of heat, promotes the distribution of heat for the reaction in these contacts [904].

Pure copper used as a catalyst gave methane instead of methanol in a process at 536 to 878 K (263 to 605 °C) under a pressure of 11.7 to 41.4 MPa [903].

Cu catalysts rapidly lose their activity in the high-pressure processes in methanol synthesis [901].

As well as the hydrogen resistance of the materials employed for the synthesis of methanol from carbon monoxide and hydrogen, their resistance to carbon monoxide is also of importance, in order to avoid the formation of carbonyls, which would lead further to the formation of methane.

For this reason, apparatuses, pipes and other components which come into contact with the hot synthesis gas are either made of or lined with copper, or made of materials and alloys which do not form carbonyls [905].

To increase its strength and improve forgeability, copper is alloyed with a small amount of 1.2 to 1.8 % Mn and should be oxygen-free [906].

In the previously commonly employed high-pressure processes, steels resistant to high temperature and high-pressure hydrogen and lined with copper had to be used [907].

Damage to the copper sheet by cracking occurred as a result of the differing thermal expansions of copper and steel, and the lining collapsed as a result of expansion of gases which had penetrated between the jacket and lining [904].

Copper-plated sheet steels are, therefore, used in apparatus construction at high temperature and exposure to pressure [908].

The cast-iron stills used for dry distillation of wood have copper transition pieces. A crude acid condensate containing about 3 % wood spirit (methanol and acetone), in addition to wood tar, acetic acid, phenols and water, is obtained in the cooled receivers during this distillation [909].

Copper is also used for the washers employed in processing this crude acid.

As the ring of a pump plunger, copper showed no detectable corrosion in a mixture of methanol and acetone at room temperature [910].

Copper is recommended for sealing rings in a rectification column for the distillation of methanol/acetone mixtures with 4 % to 6 % water and a small content of hydrochloric acid, since the alloy CuZn42 had an unsatisfactory service life.

Formaldehyde was already obtained industrially in 1889 by passing methanol over copper gauze [911].

Nowadays, the major proportion of formaldehyde is obtained by catalytic oxidation of synthetic methanol with catalysts of copper or silver.

If methanol vapor is passed over heated copper together with air, an aqueous formaldehyde solution is obtained [912].

Copper is resistant at 381 K (108 °C) in the heater for methanol vapor [913].

In the oxidative dehydrogenation process of methanol using Cu catalysts, mainly in the USA and GB, the processes of endothermal dehydrogenation and exothermic combustion of the hydrogen formed proceed synchronously over copper catalysts, and yields of up to 92 % are achieved with conversions of 30 to 50 % [911].

Contact apparatuses are made of copper pipes. If a product which is free from metal ions is required, pipes of tin-plated copper are used [914].

If silver catalysts are used, these are deposited on fine-mesh copper wire gauzes.

In the ICI plants, the reactor is protected by a copper flash-back safeguard [914].

Reactors of 5 to 6 mm thick copper sheet are used for the catalytic conversion of a methanol/air mixture to formaldehyde.

Copper is resistant to mixtures of methanol, formaldehyde and water. Even if formic acid is present, the corrosion rates are not more than 0.05 mm/a (1.97 mpy) [915].

Studies on the resistance of copper in a mixing tank containing (%) 40 formaldehyde, 10 methanol, 0.1 formic acid, balance water, with moderate aeration and turbulence, resulted in corrosion rates of 0.0508 mm/a (2 mpy) after a test duration of 28 days at room temperature.

Copper specimens in a storage tank containing (%) 37 formaldehyde, up to 1 methanol, remainder water, oxygen content not stated, were corroded at a corrosion rate of 0.0965 mm/a (3.80 mpy) after a test duration of 87 days at about 309 K (36 °C), the mixture being stirred once daily for 15 to 30 min for temperature compensation.

A corrosion rate of 0.0076 mm/a (0.30 mpy) was determined on specimens in the still of a fractionation plant in a solution with the composition (%) 2 formaldehyde, 0.2 formic acid, small amounts of methanol, ketones and other aldehydes, after a test duration of 139 days at about 393 K (120 °C) with no aeration and severe turbulence [915].

Copper is used for autoclaves in the production of methyl chloride from methanol + HCl.

Russian studies show that the corrosion of copper in methanolic hydrochloric acid increases with the degree of acidity and is reduced by the addition of water [916].

According to a previous table, copper is unstable to 10 % HCl in alcoholic solution at 293 K (20 °C) [917].

Copper-lined autoclaves are used at temperatures of 573 K (300 °C) under 25 MPa for the catalytic reaction of methanol with CO to give acetic acid (CH_3COOH) [913].

Table 134 shows some peculiarities in the behavior of copper in methanol from studies on the corrosion behavior of various metals in contaminated anhydrous alcohols.

In anhydrous methanol without impurities, the material consumption rate both at 293 K (20 °C) and 323 K (50 °C) was <0.01 g/m² h.

The addition of 50 ppm Cl^-, sodium acetate or sodium formate also caused no noticeable attack.

Methanol

However, copper is attacked by formic acid (HCOOH) to a very much greater degree than by CH_3COOH.

This manifests itself in particular in the O_2-saturated solution [918].

The time/weight loss curves of Figure 99 also show the influence of acidity. The curves show that copper is a resistant material in N_2-saturated solutions [918].

Concentration mol/l	Room temperature			323 K (50 °C)	
	N_2	O_2	Air	N_2	O_2
	Material consumption rate*, $g/m^2 h$				
No impurity	0	0	0	0	0
50 ppm Cl^-	0.01	0.02			
CH_3COONa					
0.1	0.01			0.01	0
CH_3COOH					
0.1	0.03	0.3	0.22	0.05	4.10
HCOONa					
0.1	0.01	0.01		0.01	0
HCOOH					
0.1	0.04	1.66	0.34	0.03	0.16
0.05		1.64			
0.01		0.83			
0.005		0.51			
0.001		0.11			

* 0 corresponds to a material consumption rate of < 0.01 $g/m^2 h$

Table 134: Corrosion behavior of copper in methanol as a function of the gases present and various impurities [918]

The effect of the concentration on the rate of corrosion in methanolic solutions containing formic acid and acetic acid is non-linear [919].

As can also be seen in Table 134, copper is largely resistant to chlorides in methanolic solutions. This is said to apply in particular to concentrations above 100 mg/l, because the formation of a protective layer than takes place at the same time as the wearing away of copper [920].

In methanol solutions of sulfuric acid, an increasing acid concentration has the effect of an increasing tendency towards passivation. Copper has a range of active dissolution up to 5 mol/l H_2SO_4 at 298 K (25 °C). Increasing the concentration above 5 mol/l leads to passivity.

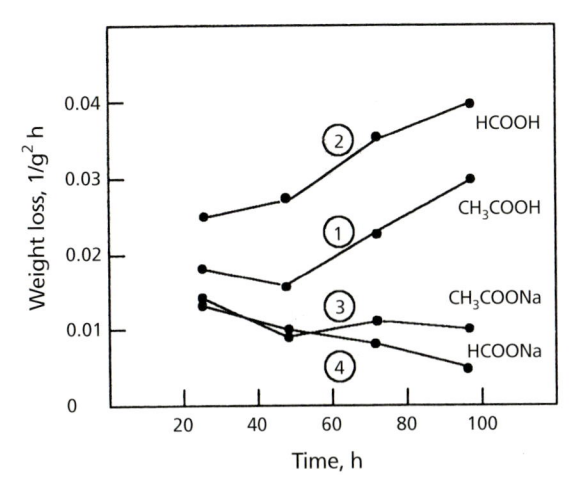

Figure 99: Time-dependence of the weight loss of copper in methanol containing 0.1 mol/l acetic acid 0.1 mol/l formic acid 0.1 mol/l sodium acetate 0.1 mol/l sodium formate (The solutions were saturated with N_2) [918]

A reduction in the temperature stabilizes the passivity. For the same H_2SO_4 concentrations, a higher passivity is achieved with lower temperatures, and lower values result for the critical passivation current density [921].

No active/passive transition was achieved for copper up to a potential of +2.0 V with a concentration of 0.05 mol/l H_2SO_4 in methanol [922].

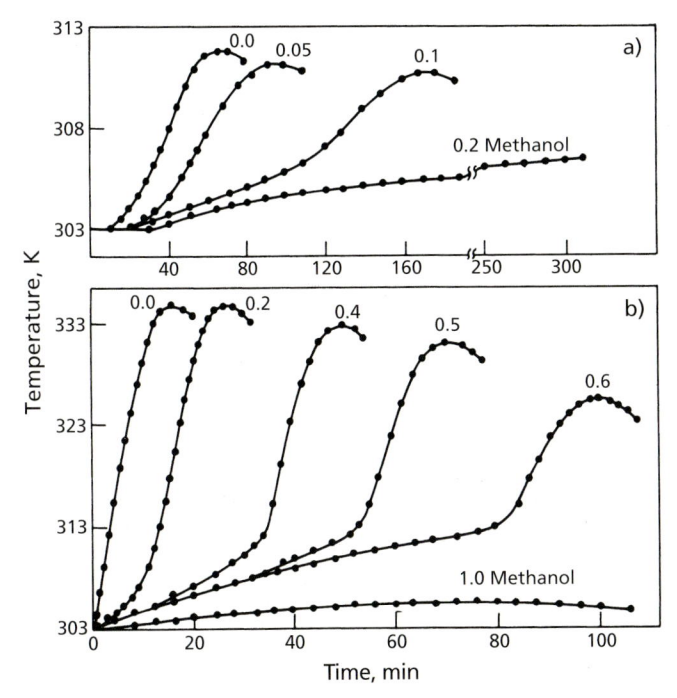

Figure 100: Effect of the addition of methanol on the dissolution of copper in a) 3 mol/l and b) 6 mol/l HNO_3 [923]

Figure 100 shows the effect of various additions of methanol (ml) to nitric acid.

The addition of even tiny amounts of methanol results in a decreasing dissolution of copper. This is associated with an increasing incubation time and decreasing maximum temperature.

Compared with Figure 100 a, Figure 100 b shows comparable curves at a higher acid concentration, but also shows that greater additions of methanol are necessary to achieve the same effect. The two diagrams show that in the acid concentrations given here, additions of methanol inhibit copper dissolution [923].

Copper was dissolved in a methanol/40 % CCl_4 mixture at 298 K (25 °C) without exclusion of air within 2 weeks, and a yellow-brown solution without solid residues remained [924].

The corrosion of SF-Cu in methanol/perchloroethylene mixtures depends on the concentration.

Figure 101 shows the maximum material consumption in 30 vol.% methanol after a test duration of 50 days at 323 K (50 °C) [925].

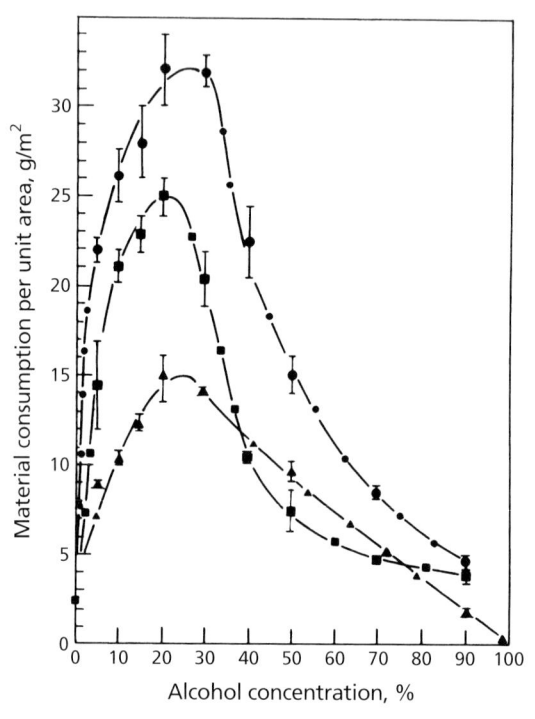

Figure 101: Material consumption per unit area of SF-Cu in perchloroethylene/alcohol mixtures as a function of the alcohol concentration at 323 K (50 °C) (test duration 30 d) [925]
● with methanol
■ with ethanol
▲ with isopropanol

Studies on the corrosion behavior of Cu in rhodamine grades dissolved in alcohol showed an average material consumption rate of 1.2 mg/dm^2 d (0.12 g/m^2 d) for rhodamine 6G dissolved in methanol, the copper becoming dark at the same time [926].

While copper is completely resistant to fuel or gasoline, it is attacked by gasoline/methanol mixtures.

Local corrosion was found on motor vehicle tank level indicators, which operate under direct current, when the vehicles were operated with methanol [927].

In a fuel mixture containing 15% methanol, both corrosion with a change in color of the surface of copper components and a yellowish discoloration of the fuel were found.

Corrosion was accelerated when sodium chloride was added to gasoline/methanol fuel mixtures as a simulated impurity in deicing salt, and the fuel turned blue in color (see Table 135) [928].

The blue coloration of the fuel was due to the presence of copper ions, which could be detected from the dark spots which developed on insertion of a zinc plate.

These copper ions in the fuel can lead to problems due to precipitation on less noble components of the fuel system [928].

Corrosion was also triggered off on copper components of fuel injection systems by the fuel M 100 (94.5% methanol + 5.5% isopentane) [918].

Results of studies by the University of Miami on the suitability of automobile materials to methanol fuel mixtures are listed for copper (99.9% Cu) in Table 136 [929]. No pitting corrosion occurred.

The corrosion rates of copper in gasoline are increased in gasoline/methanol mixtures, especially if water is present.

The condition of the metal surface is also of importance for the corrosion resistance of copper. Improvements can be achieved by various treatment methods, both mechanical and chemical.

Production-related films of carbon from drawing agents, especially on pipes, are thus the reason for the occurrence of pitting corrosion, which can be prevented by brushing, grinding or sand-blasting [930].

Acid chromate/chloride solutions and chromate/chromic acid solutions also cause the formation of protective layers.

Passivation in a chromate/chromic acid solution has so far proved to be the best corrosion protection.

The best gloss effects on the copper surface have been achieved in electrolytes of nitric acid + methanol. No information on the mixing ratio is given [931].

A protective layer also forms slowly on copper in static water.

These layers on copper are susceptible to damage, however.

Oxygen-containing and oxygen-free copper is usually resistant to stress corrosion cracking. However, oxygen-containing copper tends to undergo hydrogen embrittlement, which is caused by welding or hard soldering and can lead to stress corrosion cracking [932].

Copper-aluminium alloys

Aluminium is alloyed to copper to improve the wear resistance, tensile strength, hardness and corrosion resistance. At Al contents of 4 to 9%, the structure of the alloy is homogeneous, but at higher contents it is heterogeneous and further additions such as Fe, Mn and Ni are made [933].

	Methanol fuel[1] vol.%	Temperature K (°C)	Water vol.%	NaCl mg/l	Test duration d	Material consumption rate mg/m² d
a)	0	298 (25)	–	0	15	0
	15	298 (25)	0	0	40	14
	15	308 (35)	0	0	15	33
b)	15	308 (35)	0	0	40	38
	15	308 (35)	0.1	0	40	25
	15	308 (35)	0.1	10	40	116

1) 90% methanol + 2% ethanol + 3% propanol + 5% isobutanol

Table 135: Corrosion behavior of copper in gasoline/methanol fuel mixtures a) air-free and b) air-saturated [928]

Medium	Gasoline	Methanol	G/M 90:10	G/M 90:10 + 1% water
	Material consumption rate, g/m² d			
297 K (24°C), static solution	0.0240	0.0015	0.0279	0.0431
321 K (48°C)	0.0010			0.0035
agitated solution	0.0172			0.0090

Table 136: Corrosion behavior of copper 99.9 in gasoline, methanol and mixtures thereof with or without a water content at 297 K (24°C), test duration 1 year [929]

The alloying additions, which are always less noble than the base material copper, cannot shift the potential of the alloy to nobler values. Nevertheless, the addition of aluminium is advantageous for the corrosion behavior.

The excellent corrosion resistance of the CuAl alloys (aluminium bronzes) is based on a coupling of the resistance of copper as the base material with the ability of aluminium to replace the less resistant and sensitive copper oxide layer with its own resistant and impermeable aluminium oxide layer, which regenerates itself rapidly in the event of damage. This is the case at an alloying content of 5% or more [893].

CuAl alloys can be processed as both cast and wrought alloys.

They can be welded, hard- and soft-soldered, but the welding process must be capable of eliminating the aluminium oxide formed during welding. Special fluxes are needed during soldering, in order to dissolve the chemically resistant aluminium oxide layer [934].

The old name steel bronze and also their previous use as gunmetal indicates the particular hardness and strength of the CuAl alloys. These alloys are of particular importance as cladding materials for steel [935].

They are moderately resistant to stress corrosion cracking and should, therefore, be used in the non-stressed state. They are more sensitive to crevice corrosion than copper. Dealuminization may occur at high temperatures under adverse conditions [936].

Aluminium bronze containing 8% Al is described as being resistant to methanol [937].

US grades are described as being suitable for use in methanol under most conditions [938, 939]. Aluminium bronze is described as having a resistance to methanol which is equivalent to that of copper [897].

CuAl alloys are quoted as being resistant in methanol at room temperature and up to the boiling point, and also when mixed with 10% acetic acid and 1% sulfuric acid [913].

However, aluminium bronze promotes the formation of dimethyl ether in methanol synthesis. The material should, therefore, not be used for plant components in the various processes of this synthesis [940].

On the other hand, mixed catalysts of Cu/Al_2O_3 gave the best yields in methanol synthesis, while contacts of pure copper or pure Al_2O_3 gave methane instead of methanol [903].

For a summary of CuAl alloys in methanol, see also Table 140.

Copper-nickel alloys

CuNi alloys contain copper as the base material, nickel with contents of 4 to 50%, as a rule 5 to 30%, as the main alloying constituent, and often Zn, Mn, Fe, Sn, Si or Nb to improve certain properties.

As the nickel content increases up to 30%, the corrosion resistance is improved by passivation peculiar to nickel.

A simultaneous increase in the tensile strength, extensibility and heat resistance is accompanied by a decreasing conductivity for heat and electricity [941].

CuNi alloys are homogeneous materials. The range of the α-phase is extended by alloying with nickel, and the color changes to white at 15% Ni [933].

A classical copper alloy is the CuNiZn alloy (nickel silver), the corrosion resistance of which is considerably better than that of the CuZn alloys and which tends to undergo neither dezincification nor stress corrosion cracking [942].

It is also true for these alloys that the corrosion resistance gets better as the nickel content increases.

In a table on the "Relative Corrosion Resistance of Standard Copper and Copper Alloys", the alloys Cu70Ni30 and Cu65Ni18Zn17 are described as suitable for methanol under most conditions [943].

CuNi alloys are also generally claimed to be resistant to methanol [937]. A resistance to methanol equivalent to that of copper can be assumed.

CuNi alloys which also contain iron and manganese additions have a better erosion resistance, as well as an improved corrosion resistance. They are used as cladding materials and for condenser pipes.

Other materials resistant to methanol are: CuNi 10 Fe at 293 K (20 °C) in 98 % methanol [898], the particularly corrosion-resistant alloy CuNi20Fe used as a cladding material [933], CuNi10Fe1Mn, CuNi30Mn1Fe and CuNi30Fe2Mn, and the Cupro®-nickel grades 706 ((%) 88.5 Cu, 9–11 Ni, 1 Zn, 1.3 Fe, 0.05 Pb) [944], 702 ((%) Cu-30 Ni-0.40 Fe) and 754 ((%) Cu-10 Ni-0.75 Fe) [938, 945].

The CuNi alloy with varying contents of Cu, Ni, Pb and Zn, with 18 % Ni in this case, known as nickel silver is described as having a very good resistance [946].

The only available result on the resistance of CuNi alloys to methanol/formaldehyde mixtures is that from an experiment with a low methanol content.

Specimens which were kept in a still in a fractionation plant solution containing (%) 2 formaldehyde, 0.2 formic acid, small amounts of methanol, ketones and other aldehydes, at about 393 K (120 °C) without aeration and under severe turbulence for 138 days, showed corrosion rates of 0.0025 mm/a (0.10 mpy) [947]. A comparison of the corrosion rates of this series of experiments shows that the CuNi alloy with a rate of 0.0025 mm/a (0.10 mpy) was more resistant than copper with a rate of 0.0076 mm/a (0.30 mpy) [948].

CuNi alloys can be regarded as resistant to methanol/formaldehyde mixtures under the usual operating conditions.

In methanol/perchloroethylene mixtures, impurities from production, or from degreasing of the metal, have a corrosion-accelerating effect.

The presence of chlorides, in particular, leads to an increased attack even on the highly corrosion-resistant materials CuNi30Fe and NiCu30Fe [925].

Other alloyed elements have the task of improving certain properties, but have no adverse effects on the resistance of the alloys to corrosion by methanol.

Tin (2 %), in particular, increases the fatigue resistance and is, therefore, of special importance for spring materials.

Manganese has the effect of deoxidizing the melt, binding sulfur and improving the casting properties.

Silicon improves castability, has a deoxidizing effect and increases the strength.

Lead improves the hot deformability of wrought alloys, but impairs their weldability.

Niobium improves the weldability and increases the tensile and yield strength, chromium increases the strength and the resistance to erosion, titanium improves the weldability and aluminium increases the strength and scale resistance [941].

For all CuNi alloys it can be said generally that, along with the electrolytic copper grade, they have the highest resistance to stress corrosion cracking, pitting corrosion is hardly ever found and the risk of selective corrosion is very low because of the homogeneous structure [933].

Copper-tin alloys (bronzes)

CuSn alloys, also called tin bronzes or composition bronzes, contain up to 20 % tin and in some cases additions of zinc, lead and phosphorus.

As wrought alloys, tin bronzes contain up to 9 % tin, the content of which influences the properties of the alloys.

The wrought alloys have a homogeneous structure, which explains the good fatigue strength, deformability and corrosion resistance.

Additions of small amounts of phosphorus, hence the previous name phosphorus bronze, reduce the metal oxides which act as impurities in the melt, and improve the strength and corrosion resistance [933].

Cast alloys (cast tin bronzes) contain up to 13 % tin, or 20 % tin as bell bronze and 9 % tin as gunmetal.

At tin contents above 8 %, the structure is heterogeneous, and the cast alloys are electrochemically baser than the homogeneous wrought alloys. The higher tin contents cause the formation of an even denser oxidic surface film and in this way have the effect of increasing corrosion resistance. An increase in strength is also achieved by higher tin contents, and can be improved even more by further additions [893].

Lead contents in the alloys (cast tin-lead bronzes) improve the compactness of the casting, but impair weldability.

Some CuSn alloys have a small content of alloyed nickel, which increases the toughness and corrosion resistance [949].

The tin bronzes are comparable to the pure copper materials in chemical resistance.

Cast tin bronzes are amongst the most resistant copper materials.

The bronze grade 2.1020 (CuSn6, CW452K) is classified as being completely resistant to 98 % methanol at 293 K (20 °C) and up to the boiling point [898, 899].

Other literature completes the evaluation of the corrosion resistance of tin bronzes in methanol.

Described as resistant are:

- bronze (without further data) [896],
- bronze (containing 10 to 20 % Sn) in 100 % methanol at 333 to 343 K (60 to 70 °C) as a material for shut-off valves [950],
- cast bronze (G-Bz) [937],
- bronze (without zinc) [913],
- tin bronze in 25 % and 100 % methanol [897],
- phosphorus bronze [946],
- phosphorus bronze 5 % (95Cu-5Sn-0.03–0.35 P) [943] and
- phosphorus bronze 8 % (92Cu-8Sn-0.03–0.35 P) [943].

Weight increases of 0.4 g/m^2 on phosphorus bronze containing 4% Sn and 0.57 g/m^2 on phosphorus bronze containing 8% Sn were found after exposure for 1 month to boiling methanol.

After exposure to methanol at room temperature for 6 months, the weight increases were 0.41 g/m^2 for phosphorus bronze (4% Sn) and 0.35 g/m^2 for phosphorus bronze (8% Sn) [951].

Phosphorus bronzes (A) 351, 4% (grade A) 903, leaded phosphorus bronze 5% (grade B) 976, special free-cutting phosphorus bronze 610 and phosphorus bronze 314 are resistant to methanol. Also mentioned as being resistant are phosphorus bronzes (D) 354, 8% (grade C) 353, 316 and 310 [938], gunmetal at 293 K (20 °C) and for lower concentrations at 333 K (60 °C) [952], and CuSn and CuPbSn alloys [953].

If tin bronzes suffer mild attack by methanol, this, as a rule, takes place as uniform general corrosion.

The alloys are largely insensitive to pitting and stress corrosion cracking [944].

The Power Gas Corp. Ltd. reports on the use of a copper-bronze alloy in formaldehyde synthesis as a flashback safe-guard against flashback of the methanol/air/vapor mixture from the reactor to the evaporator [954].

To protect antique bronze statues from decay by corrosion, known as "bronze cancer", a method has been developed in which the statues are treated first in a bath of 10% ammonia in methanol, and then in a solution of 10 parts 24% hydrogen peroxide and 90 parts methanol. In this procedure, the ammonia dissolves the copper chlorides of the pitting areas, and hydrogen peroxide oxidizes the copper hydroxide which forms a layer in the pits right down to the bronze [955].

Copper-tin-zinc alloys (red brass)

CuSnZn-alloys are classified in accordance with DIN EN 1982 and as cast alloys (red brass) contain, in addition to copper, 4 to 10% Sn, 2 to 7% Zn and usually up to 7% Pb. A little nickel is sometimes added to increase the strength and toughness [953].

The importance of the zinc content lies in its shifting of the phase boundary of the α-region to lower tin contents and its deoxidation of the melts, so that phosphorus is not needed [956].

Lead contents facilitate production of a pressure-tight casting. Centrifugal casting achieves a particularly dense and fine structure [957].

These cast materials are rarely soft-soldered or welded, since lead-containing alloys are not fusion-weldable. If red brass is used for pipe flanges, it is preferably hard-soldered with silver solders [934].

The corrosion resistance of the CuSnZn cast alloys corresponds to that of copper and CuSn cast alloys.

Data on the resistance of these alloys to methanol are rarely found in the literature.

A table on the resistance of copper alloys evaluates CuSnZn alloys as being resistant to alcohols [953].

Red brass (an alloy containing Sn and Zn) is quoted as being resistant to methanol and having a behavior like that of copper [913].

Copper-zinc alloys (brass)

Of all the copper alloys, the brasses are used the most.

The corrosion behavior depends largely on the zinc content and the structure [958].

At a zinc content of 20 % or more, the protective action due to the formation of a zinc oxide surface layer is more important than the nobility of the metal [893].

Alloys containing up to 37.5 % Zn have a face-centered cubic, uniform structure (α-brass) and are ductile and tough. Their corrosion resistance corresponds to that of copper. At zinc contents of 37.5 % up to 46 %, β-structure is present alongside the α-structure (α-β-brass).

These alloys are particularly sensitive to dezincification and are unsuitable for corrosive media [958].

Alloys containing 46 to 50 % Zn consist uniformly of the β-phase (β-brass). γ-Brasses containing more than 50 % Zn are unsuitable for industrial use because of their extreme brittleness [959].

Brass wrought alloys contain up to 37.5 % Zn, as tombac up to 30 % Zn without further additions, as lead-containing brass up to 3.5 % Pb and as high-strength brass further alloying additions.

Normal brass and lead-containing brass can be readily soft-soldered, and hard-soldered where mechanical stresses occur.

The suitability for fusion welding is poor because of the low boiling point of zinc. It vaporizes and forms pores.

The MIG and WIG processes are of only limited suitability.

Gas welding can be used because a zinc oxide layer develops due to the excess oxygen, preventing further vaporization of the zinc [960].

Lead as a heterogeneous constituent has a corrosion-inhibiting action, improves machineability, but impairs weldability. High-strength brasses or multicomponent alloys based on CuZn contain Al, Fe, Mn, Si or Sn as further additions.

The high-strength brasses are superior to the normal brasses in strength, sliding and wear properties and corrosion resistance, and are used in chemical engineering [959].

Both brasses containing 33 to 42 % Zn and < 3 % Pb, and also high-strength brasses are cast alloys [957].

Aluminium-containing, high-strength brasses are, in some cases, superior to pure copper in corrosion resistance because of the formation of an aluminium oxide layer [961].

Iron improves the strength and grain fineness of the alloy, and the usual contents are between 0.5 and 1.5 %.

High-strength manganese brasses can contain up to 5 % Mn, which improves the strength and corrosion resistance, and, if Al and Si are added at the same time, offers a high wear resistance [959].

Nickel contents of up to 3 % are usual in high-strength brasses. Higher Ni contents are dealt with under CuNi alloys (section Copper-nickel alloys).

Nickel improves the strength, also at a higher temperature, and the deformation capacity.

Silicon reduces the susceptibility to stress cracking in α-brasses and improves the strength, tarnish resistance and castability of α-β-brasses.

Tin-containing, high-strength brasses, tobin bronze and admiralty brass (SoMs71, 71Cu-28Zn-1Sn) also contain traces of other metals in addition to Sn. They have a high corrosion resistance and are used for heat exchanger pipes and steam condensers [962]. Brass is more resistant to pitting corrosion than other copper alloys. However, there is the danger of dezincification, a particular form of brass corrosion which occurs both uniformly over the surface and in the form of plugs, and can lead to ruptures [936].

Most at risk are the α-β-brasses, in which the β-structure is preferentially attacked and dissolved out, especially at elevated temperature with low electrolyte agitation and aeration. A soft matrix of porous copper remains.

As dezincification progresses, the shape of a component does not change, but increasing embrittlement takes place and the color changes from brass-yellow to the darker copper color [936].

Pure α-brasses have a good resistance to dezincification, which can be prevented entirely by addition of a small amount of arsenic (As) (0.02 to 0.06 %). However, As has no dezincification-preventing effect on the β-structure [935, 939].

Stress corrosion cracking is a problem for brass amongst the copper alloys. As the Cu content increases, the sensitivity to this form of corrosion decreases.

Brass containing more than 80 % Cu is regarded as not particularly susceptible, high-strength α-brasses as moderately susceptible, and high-strength α-β-brasses and brasses containing less than 80 % Cu as susceptible to stress corrosion cracking [932].

Other literature assesses brasses and high-strength brasses containing more than 15 % Zn as sensitive to stress corrosion cracking [893].

The process of stress corrosion cracking proceeds rapidly on brass. Even external stresses cause the surface layer to tear at individual points, from which corrosion starts [963].

The alloys α-brass (Cu ≥ 85 %), α-brass (Cu ≥ 80 %), α-β-brass, high-strength α-brass S 20, high-strength α-brass S 28 and high-strength α-β-brass are evaluated as being resistant to methyl alcohol [937].

Tombac Ms85, CuZn15 (2.0240, CW502L) and CuZn20 (2.0250, CW503L) are regarded as being resistant to 98 % methanol at 293 K (20 °C) and up to the boiling point [898, 899, 963].

The alloys CuZn10 (2.0230, CW501L), CuZn15 (2.0240, CW502L), CuZn30 (2.0265, CW505L), CuZn40 (2.0360, CW509L), CuZn35.5Pb3, CuZn40Pb3, CuZn28Sn1 and CuZn39.25Sn0.75 are said to be usable under most conditions [943].

An information sheet on screws and nuts made of CuZn alloys refers to the CuZn37 (2.0321, CW508L), CuZn39Pb3 (2.0401, CW614N) and CuZn39Pb2 (2.0380, CW612N) grades as being completely resistant (< 0.1 mm/a (< 3.94 mpy)) to methanol [964].

A manufacturer of flexible metal tubes evaluates brass (brass 80 20, grade A) as resistant (0.01 mm/month) [965].

The brass grades 70–80 Cu + Zn, Sn or Pb and 59–93 Cu + Al, Zn or As are resistant to all concentrations of methanol up to 373 K (100 °C) [895].

The brass grades listed below are usable in methanol under most conditions [938].

Brass of low Zn content:
commercial bronze 14 (90 Cu-10 Zn),
leaded commercial bronze 201 (89.6 Cu-10 Zn-0.4 Pb),
leaded commercial bronze 202 (88.5 Cu-10 Zn-1.5 Pb),
high-strength commercial bronze 286 (90.25 Cu-6.9 Zn-1.75 Pb-1 Ni-0.1 P),
Orcide® 420 (87.25 Cu-11.5 Zn-1.25 Sn),
red brass 24 (85 Cu-15 Zn),
silicon red brass 1027 (82 Cu-17 Zn-1 Si) and
trumpet brass 435 (81 Cu-18 Zn-1 Sn).

Brass of relatively high Zn content:
cartridge brass 42 (70 Cu-30 Zn),
low brass 32 (80 Cu-20 Zn),
yellow brass 59 (65 Cu-35 Zn),
leaded brass 211 (69 Cu-29.5 Zn-1.5 Pb),
yellow brass 218 (66.5 Cu-33 Zn-0.5 Pb),
free-cutting tube brass 282 (66.5 Cu-31.9 Zn-1.6 Pb),
special threading brass 223 (65 Cu-34.75 Zn-0.25 Pb),
butt brass 226 (64.25 Cu-35.25 Zn-0.5 Pb),
butt brass 229 (64 Cu-35 Zn-1 Pb),
extruded architectural bronze (56 Cu-41.5 Zn-2.5 Pb),
muntz metal 66 (60 Cu-40 Zn),
yellow brass 61 (63 Cu-37 Zn),
rule brass 238 (62.5 Cu-35 Zn-2.5 Pb),
free-cutting yellow brass 271 (62 Cu-35 Zn-3 Pb),
clock brass 243 (61.5 Cu-37 Zn-1.5 Pb),
leaded muntz metal 274 (60 Cu-39.5 Zn-0.5 Pb),
free-cutting muntz metal 293 (60 Cu-39 Zn-1 Pb) and
forging brass 250 (60 Cu-38 Zn-2 Pb).

Special brass:
tobin bronze (60 Cu-39.25 Zn-0.75 Sn),
naval brass 452 (60 Cu-39.25 Zn-0.75 Sn),
leaded naval brass 605 (60 Cu-38.55 Zn-0.75 Sn-0.7 Pb),
leaded naval brass 612 (60 Cu-37.25 Zn-0.75 Sn-2 Pb),
manganese bronze 937 (59 Cu-39 Zn-0.7 Sn-0.5 Mn-0.8 Fe),
economy bronze,
anaconda 997 – low-fuming bronze,

manganese bronze 984,
tobin bronze 481,
arsenical admiralty 439 (70 Cu-28.96 Zn-1 Sn-0.04 As)
and
Ambraloy® 927 (76 Cu-21.96 Zn-2 Al-0.04 As).

For admiralty bronze (70 Cu-29 Zn-1 Sn), weight increases of 0.48 g/m² were found after storage for 1 month in boiling methanol, and a weight decrease of 1.08 g/m² was found after 7 months at room temperature [951].

In methanol synthesis, the jackets of furnaces and heat exchangers are protected by pipes of brass or copper pressed tightly against the jackets [966].

In this synthesis, the brass lining of the tubular furnaces (60 to 90 % Cu) is resistant at 698 to 973 K (425 to 700 °C), the prevention of methane formation due to carbonyls also being of importance [905, 940].

The corrosion behavior of the brass grades mentioned here, containing 60 to 90 % Cu, in methanol is equated with that of copper, which is described as resistant [913].

In rectification columns for the distillation of methanol/acetone mixtures with a water content of 4 to 6 % and small contents of hydrochloric acid, sealing rings of brass wire (Ø 520 × 5.0 mm) led to leakages in the columns and malfunctions after discontinuous operating times of 1 to 2.5 years.

Some of the wires had assumed the reddish color of copper, and in some cases a residual core of yellow-colored brass remained.

These rings broke or disintegrated into small pieces under the slightest flexural stress.

This phenomenon indicates the occurrence of selective corrosion (dezincification) and can be attributed to the chloride content, low degree of agitation and simultaneously increased temperature.

The heterogeneous structure of the brass used, CuZn42, also promoted corrosion [967].

Dezincification takes place in particular due to chloride in the medium and also due to acid liquids [932].

In the course of a study on the behavior of materials in seawater, a brass disk (76 Cu, 22 Zn, 2 Al, 0.03 As) was pickled in an agitated solution of 33 % nitric acid in methanol at room temperature and at a potential of 6 to 7 V using a platinum cathode. The entire region of the disk exposed dissolved [968].

The curves in Figure 102 show the behavior of CuZn alloys in 3 mol/l HNO_3 as the methanol concentration increases.

The curve 0.0 shows the dissolution rate of the brass containing 67.5 % Cu and 32.5 % Zn in 3 mol/l nitric acid (HNO_3), which corresponds to that of copper [923].

Addition of methanol causes decreasing corrosion rates, shown by a decrease in the maximum temperatures and an increase in the incubation times.

As Figure 102 shows, methanol has an effect on the corrosion of CuZn alloys similar to that on copper. However, the same increase in the methanol content under otherwise identical conditions has the effect of a smaller decrease in the corrosion rates for the alloy (see also Figure 100).

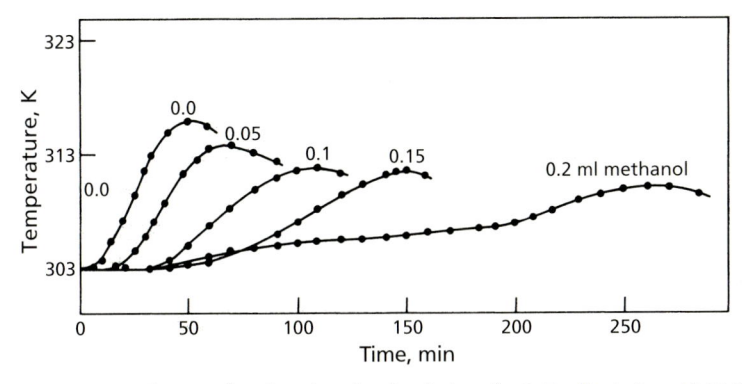

Figure 102: Influence of methanol on the dissolution of a CuZn alloy in 3 mol/l HNO$_3$ [923]

As is also the case for copper, methanol inhibits the corrosion of CuZn alloys in nitric acid [923].

As with copper, brass suffered only mild attack in methanol containing rhodamine 6 G, as can be seen from a material consumption rate of 0.36 g/m^2 d [926].

As Table 137 shows, the presence of air in the fuel has the greatest influence on the material consumption rate.

Addition of water seems to lower the corrosion rate, which perhaps indicates passivation.

The resistance of brass is of great importance for the operation of motor vehicles with methanol-containing fuels, since brass components are customary in the fuel supply system. Gauze filters for gasoline pumps and carburettors are thus always made of brass wire (Ø 0.12 mm) with a mesh width of 0.125 to 0.15 mm [969].

While gasoline by itself causes an insignificant change in the weight of brass, changes in the color of the fuel occur during operation with gasoline/methanol fuel mixtures (see Table 137). It turns yellow or, if NaCl impurities are present, green [970].

Sodium chloride in the fuel, which can originate from contamination by de-icing salt, increases the material consumption rate further [928].

Studies by the University of Miami on the suitability of various materials for motor vehicle fuels showed comparable results for yellow brass (CuZn30, cf. CW505L) (see Table 138) [929]. Nevertheless, pitting corrosion was found in the gasoline/methanol mixtures.

The values for mixtures containing water are striking here. While an addition of 0.1 % water seems to reduce the corrosion rates in Table 137, an addition of 1 % increases the values.

The results of a corrosion test on free-cutting brass (SAE CA-360) from studies on the suitability of materials for motor vehicle construction with respect to previously unusual fuels are shown in Table 139 [971].

	Methanol fuel[1] vol.%	Temperature K (°C)	Water vol.%	NaCl mg/l	Test duration d	Material consumption rate mg/m² d
a)	0	298 (25)	0	0	15	6.8
	15	298 (25)	0	0	40	26
	15	308 (35)	0	0	15	35
b)	15	308 (35)	0	0	40	112
	15	308 (35)	0.1	0	40	79
	15	308 (35)	0.1	10	40	93

1) 90 % methanol + 2 % ethanol + 3 % propanol + 5 % isobutanol

Table 137: Corrosion behavior of brass in gasoline/methanol fuel mixtures a) air-free and b) air-saturated [970]

Other copper alloys

The copper alloys dealt with here are mainly low-alloy copper materials in which one or more properties are improved by small additions of other elements, but other properties are retained [972].

CuMn alloys are of particular importance in methanol synthesis in which hot components are lined with CuMn sheet or made of such a material.

Reaction furnaces made of heat-resistant steel and contact pipes are lined with CuMn sheet.

Regenerators consist of a bank of tubes made of CuMn alloy.

Catalysts are located on grids of MnCu alloys [973].

A copper alloy containing 1.2 to 1.8 % Mn has proved suitable for the pipe material of heat exchangers, the connecting lines and the furnace inserts [974].

The use of CuMn alloys is a safe way of protecting all the hot components in methanol synthesis from attack by hydrogen and carbonyl formation [966].

CuMn alloys usually contain up to 5 % Mn. Up to about 20 % Mn, the alloys have a homogeneous structure.

As a result of the deoxidizing action of Mn, the alloys are oxygen-free and thus insensitive to hydrogen.

Mn improves the tensile strength and corrosion resistance of copper.

The alloys have a high resistance to heat and are insensitive to stress corrosion cracking [893].

For manganese bronze, a weight increase of 0.52 g/m² was found after storage in boiling methanol for 1 month, and a weight loss of 0.59 g/m² after storage at room temperature for 6 months [951].

Alloying with silicon also improves the strength and corrosion resistance of copper [893].

Copper-silicon containing 95.8 % Cu is mentioned in the literature as having a good resistance to alcohols [946].

The behavior of CuSi alloys in methanol is equal to that of copper, and the resistance described as very good, with a corrosion rate of 0.008 mm/a (0.31 mpy) [975].

The copper alloys 651 (CuSi1.5) and 652 (CuSi3.0) and the high-silicon bronze (CuSi3Mn1, CW116C) are usable in respect of their corrosion resistance under most conditions [943].

CuSiMn alloys containing up to about 3.6 % Si have a homogeneous structure, improved strength and corrosion resistance and good weldability.

CuNiSi is an age-hardenable alloy which has a high strength, wear resistance, good sliding properties and corrosion resistance [976].

Medium	Gasoline	Methanol	G/M 90:10	G/M 90:10 +1 % water
		Material consumption rate, g/m² d		
297 K (24 °C), static solution	0.0218	0.0003	0.0532*	0.0843*
321 K (48 °C)	0.0019			0.0014
agitated solution	0.0290			0.0162

*pitting corrosion

Table 138: Corrosion behavior of yellow brass in gasoline, methanol and gasoline/methanol mixtures with and without water at 297 K (24 °C), test duration 1 year [929]

Fuel designation	Gasoline	Methanol	Higher alcohols	Water[2]	Corrosion rate mm/a (mpy)	
		Composition, %			297 K (24 °C)	321 K (48 °C)
P	100				< 0.005 (< 0.20)	< 0.005 (< 0.20)
P-1	100			1	< 0.005 (< 0.20)	< 0.005 (< 0.20)
MH 18	82	13.5	4.5		< 0.005 (< 0.20)	< 0.005 (< 0.20)
MH 18-1	82	13.5	4.5	0.25	< 0.005 (< 0.20)	< 0.005 (< 0.20)
MH 18-2	82	13.5	4.5	1.34	< 0.005 (< 0.20)	< 0.005 (< 0.20)
M 93	7	93			< 0.005 (< 0.20)	< 0.005[1] (< 0.20)
M 93-1	7	93		4.8	< 0.005 (< 0.20)	< 0.005 (< 0.20)

1) insignificant pitting
2) the water was added to the anhydrous fuel

Table 139: Corrosion behavior of free-cutting brass (SAE CA-360) in various fuels, test duration 24.5 weeks [971]

CuPb alloys have very good wear and emergency running properties, as a result of the hard matrix with embedded soft constituents, as the lead content increases.

As CuPbSn alloys, they are used as sliding materials in engine construction [953]. They are resistant to alcohols.

In experimental motor vehicles, crankshaft bearings of CuPb bimetal showed selective corrosion of the lead phase when operated with M 50 fuels. However, impurities in the engine oil after regulation of the air/fuel mixture may also be the cause of this corrosion.

Motor vehicles used in the same test series with CuPbSn (trimetal) bearings showed satisfactory results when operated with M 15 and M 85 fuels [977].

CuFe alloys containing 2.1 to 2.6 % Fe and a small addition of phosphorus have a high thermal and electrical conductivity coupled with tensile strength.

Iron increases the temperature limits for application and improves the corrosion resistance of copper by formation of a protective film of iron hydroxide, which is rapidly repaired when damaged.

Alloying with chromium increases the breaking strength and corrosion resistance of copper.

Chromium-containing copper 999 (99.05 Cu-0.10 Si-0.85 Cr) is usable in methanol under most conditions [938].

In recent developments, CuCr alloys are improved in strength, creep strength, temperature resistance and processability by addition of zirconium.

CuZr alloys contain 0.1 to 0.3 % Zr, are oxygen-free and therefore insensitive to hydrogen [960].

They have a good strength and creep strength and a high electrical conductivity [935].

CuBe alloys (CuBe1.7 (CW100C, old 2.1245), CuBe2 (CW101C, old 2.1247)) have a high tensile strength and temperature resistance in the age-hardened state.

The corrosion resistance corresponds to that of oxidized copper. The alloys are insensitive to stress corrosion cracking. The resistance to alcohols is given as good [946].

A lead-containing CuBe alloy CuBe2Pb (CW102C, old 2.1248) with improved cutting properties is a new inclusion in DIN.

A pickling process for removal of very tough oxide layers of CuCoBe (Cu-0.4–0.7Be-2.3–2.7Co) and CuNiBe (Cu-0.4–0.7Be-1.2–2.5Ni) in an alkali chloride/methanol solution is described in a patent specification [978].

Medium and concentration	Temperature, K (°C)	Material	Evaluation	Literature
Methanol, 10 to 100 %	≤ boiling point	Cu	moderately resistant, < 0.64 mm/a (25.2 mpy)	[895]
Methanol, 25 to 100 %	298 (25)	Cu	resistant, < 0.51 mm/a (< 20.1 mpy)	[897]
Methanol, 25 to 100 %	–	tin bronze	resistant	[897]
Methanol, low concentrations	333 (60)	gunmetal	resistant	[952]
Methanol, all concentrations	373 (100)	brass	resistant	[895]
Methanol, 98 %	293 (20)	(70–80 Cu + Sn or Pb)	resistant	[898, 899]
		Cu, CuNi10Fe	resistant	[933]
		CuNi20Fe	resistant	[963]
		bronze CW452K, CuZn20 (CW503L)	resistant	[963]
		tombac Ms85	resistant	[963]
Methanol, 98 %	boiling	Cu, bronze CW452K	resistant	[898, 899]
		Cu 20, tombac Ms85	resistant	[963]
Methanol, 100 %	297 (24)	Cu 99.9	resistant, 0.0015 g/m^2 d	[929]
	333–343 (60–70)	bronze (20 % Sn)	resistant	[950]
Methanol, in general		Cu	resistant, 0.08 mm/a (3.15 mpy)	[900]
		CuAl8, CuNi30	resistant	[937]
		CuNi18Zn17	resistant	[942]
		CuNi10Fe1Mn (CW352H)	resistant	[945]
		CuNi30Mn1Fe (CW354H)	resistant	[938]
		CuNi30Fe2Mn2 (CW353H)	resistant	[938]

1) The corrosion rates of copper and copper alloys in methanol/gasoline mixtures allow an evaluation of resistant under normal operating conditions. However, they are not suitable for use in fuel-conducting components in cars.
* see text
** vol.%

Table 140: Evaluation of the corrosion resistance of copper and copper alloys to methanol and methanol-containing mixtures (for more detailed conditions, see text)

Table 140: Continued

Medium and concentration	Temperature, K (°C)	Material	Evaluation	Literature
Methanol, in general		cupro-nickel 706, 702, 754	resistant	[938]
		phosphor bronzes	resistant	[946]
		gunmetal, bronze (without Zn)	resistant	[913]
		CuSn, CuPbSn alloy	resistant	[959]
		CuSnZn alloy	resistant	[953]
		α-brass (Cu 80, Cu 85)	resistant	[953]
		α-brass 520	resistant	[953]
		α-β-brass		
		α-β-high-strength brass	resistant	[937]
		CuZn10 (CW501L), CuZn15 (CW502L)	resistant	[937]
		CuZn30 (CW505L), CuZn40 (CW509L)	resistant	[937]
		CuZn35.5Pb3	resistant	[942]
		CuZn40Pb3	resistant	[942]
		CuZn28Sn1	resistant	[942]
		CuZn39.25Sn0.75	resistant	[942]
		CuZn37 (CW508L), CuZn39Pb3 (CW612N)	resistant, < 0.1 mm/a (< 3.94 mpy)	[964]
		CuZn39Pb2, CW612N	resistant	[964]
		brass 80–20, brass	resistant, 0.01 mm/month*	[896]
		brass with slight Zn content	resistant	[896]
		brass with high Zn content	resistant	[896]
		high-strength brass	resistant	[938]

1) The corrosion rates of copper and copper alloys in methanol/gasoline mixtures allow an evaluation of resistant under normal operating conditions. However, they are not suitable for use in fuel-conducting components in cars.
* see text
** vol.%

Table 140: Evaluation of the corrosion resistance of copper and copper alloys to methanol and methanol-containing mixtures (for more detailed conditions, see text)

Table 140: Continued

Medium and concentration	Temperature, K (°C)	Material	Evaluation	Literature
Methanol, in general		brass 60 to 90 % Cu	resistant	[940]
		CuZn30, CW505L	resistant, 0.0219 g/m²d	[929]
		CuMn, Cu95.8Si	resistant, 0.008 mm/a (0.31 mpy)	[975]
		Alloy 651 (CuSi 1.5)	resistant	[975]
		Alloy 652 (CuSi 3)	resistant	[975]
		high-silicon bronze (CuSi3Mn1, CW116C)	resistant	[975]
		CuPbSn, CuCr alloys	resistant	[953]
		chromium copper 999	resistant	[938]
		CuBe alloys	resistant	[946]
Methanol and impurities (10 % acetic acid, 1 % sulfuric acid)		Cu	resistant	[913]
Methanol + 0.1 mol/l CH_3COOH,				
N_2-saturated	293 (20)	Cu	resistant, 0.03 g/m²h	[918]
O_2-saturated	293 (20)	Cu	moderately resistant, 0.3 g/m²h	[918]
air	293 (20)	Cu	moderately resistant, 0.22 g/m²h	[918]
N_2-saturated	323 (50)	Cu	resistant, 0.05 g/m²h	[918]
O_2-saturated	323 (50)	Cu	not resistant, 4.10 g/m²h	[918]
Methanol + 0.1 mol/l HCOOH,				
N_2-saturated	293 (20)	Cu	resistant, 0.003 g/m²h	[918]

1) The corrosion rates of copper and copper alloys in methanol/gasoline mixtures allow an evaluation of resistant under normal operating conditions. However, they are not suitable for use in fuel-conducting components in cars.
* see text
** vol.%

Table 140: Evaluation of the corrosion resistance of copper and copper alloys to methanol and methanol-containing mixtures (for more detailed conditions, see text)

Table 140: Continued

Medium and concentration	Temperature, K (°C)	Material	Evaluation	Literature
O_2-saturated	293 (20)	Cu	hardly usable, 1.66 g/m²h	[918]
air	293 (20)	Cu	moderately resistant, 0.34 g/m²h	[918]
N_2-saturated	323 (50)	Cu	resistant, 0.03 g/m²h	[918]
O_2-saturated	323 (50)	Cu	moderately resistant, 0.16 g/m²h	[918]
Methanol + 0.1 mol/l CH_3COONa,				
N_2-saturated	293 (20)	Cu	resistant, 0.01 g/m²h	[918]
N_2-saturated	323 (50)	Cu	resistant, 0.01 g/m²h	[918]
O_2-saturated	323 (50)	Cu	resistant, < 0.01 g/m²h	[918]
Methanol + 0.1 mol/l $HCOONa$,				
N_2-saturated	293 (20)	Cu	resistant, 0.01 g/m²h	[918]
O_2-saturated	293 (20)	Cu	resistant, 0.01 g/m²h	[918]
N_2-saturated	323 (50)	Cu	resistant, 0.01 g/m²h	[918]
O_2-saturated	323 (50)	Cu	resistant, < 0.01 g/m²h	[918]
Methanol, formaldehyde, water and small amounts of formic acid		Cu	resistant, 0.05 mm/a (1.97 mpy)	[948]
10 % methanol, 40 % formaldehyde, 0.1 % formic acid		Cu	resistant, 0.0508 mm/a (2.00 mpy)	[948]
Methanol/acetone mixture + 4 to 6 % water + small amounts of HCl		Cu	hardly usable, dezincification	[967]

1) The corrosion rates of copper and copper alloys in methanol/gasoline mixtures allow an evaluation of resistant under normal operating conditions. However, they are not suitable for use in fuel-conducting components in cars.
* see text
** vol.%

Table 140: Evaluation of the corrosion resistance of copper and copper alloys to methanol and methanol-containing mixtures (for more detailed conditions, see text)

Table 140: Continued

Medium and concentration	Temperature, K (°C)	Material	Evaluation	Literature
Methanol + 40 % CCl_4, without air exclusion	298 (25)	Cu	unsuitable	[924]
CCl_4 + 30 %** methanol		SF-Cu (CW027A)	unsuitable	[925]
Methanol + 50 ppm Cl^-	293 (20)	Cu	resistant, 0.01 g/m² h	[913]
10 % HCl in methanol	293 (20)	Cu	unsuitable	[917]
3 mol/l HNO_3 + < 0.1 ml methanol		Cu	unsuitable	[923]
3 mol/l HNO_3 + > 0.2 ml methanol		Cu	moderately resistant	[923]
6 mol/l HNO_3 + < 0.6 ml methanol		Cu	unsuitable	[923]
6 mol/l HNO_3 + > 1.0 ml methanol		Cu	moderately resistant	[923]
33 % HNO_3 + methanol	293 (20)	Cu 76 Zn 22 Al 2 As 0.3	unsuitable	[968]
3 mol/l HNO_3 + < 0.15 ml methanol	303 (30)	Cu 67.5 Zn 32.5	unsuitable	[923]
3 mol/l HNO_3 + > 0.2 ml methanol	303 (30)	Cu 67.5 Zn 32.5	moderately resistant	[923]
< 5 mol/l H_2SO_4 in methanol	298 (25)	Cu	hardly usable	[921]
> 5 mol/l H_2SO_4 in methanol	293 (20)	Cu	moderately resistant	[921]
Rhodamine 6 G in methanol		Cu	resistant, 0.12 g/m² d, discoloration	[926]

1) The corrosion rates of copper and copper alloys in methanol/gasoline mixtures allow an evaluation of resistant under normal operating conditions. However, they are not suitable for use in fuel-conducting components in cars.
* see text
** vol.%

Table 140: Evaluation of the corrosion resistance of copper and copper alloys to methanol and methanol-containing mixtures (for more detailed conditions, see text)

Table 140: Continued

Medium and concentration	Temperature, K (°C)	Material	Evaluation	Literature
Rhodamine 6 G in methanol		brass	resistant, 0.36 g/m² d	[926]
M 15 fuel, air exclusion	298 (25)	Cu	14 mg/m² d, discoloration	[928]
	298 (25)	brass	6.8 mg/m² d [1)]	[928]
	308 (35)	Cu	33 mg/m² d [1)]	[928]
	308 (35)	brass	35 mg/m² d [1)]	[928]
M 15 fuel, air-saturated	308 (35)	Cu	38 mg/m² d [1)]	[928]
	308 (35)	brass	112 mg/m² d [1)]	[928]
M 15 fuel + 0.1 %** water, air-saturated	308 (35)	Cu	25 mg/m² d [1)]	[928]
	308 (35)	brass	79 mg/m² d [1)]	[928]
M 15 fuel + 0.1 %** water, 10 mg/l NaCl, air-saturated	308 (35)	Cu	116 mg/m² d [1)]	[928]
	308 (35)	brass	93 mg/m² d [1)]	[928]
Gasoline/methanol mixture 90/10	297 (24)	Cu 99.9	0.0279 g/m² d [1)]	[929]
Gasoline/methanol mixture 90/10 + 1 % water*	297 (24)	Cu 99.9	0.0431 g/m² d [1)]	[929]
Gasoline/methanol mixture 90/10	297 (24)	Cu 70 Zn 30	0.0532 g/m² d [1)]	[929]
Gasoline/methanol mixture 90/10 + 1 % water*	297 (24)	Cu 70 Zn 30	0.0843 g/m² d [1)]	[929]
13.5 % methanol + 82 % gasoline + 4.5 % higher alcohols	297 (24)	free-cutting brass	resistant	[971]
	321 (48)	free-cutting brass	resistant	[971]
13.5 % methanol + 82 % gasoline + 4.5 % higher alcohols + 0.25 % water*	297 (24)	free-cutting brass	resistant	[971]
	321 (48)	free-cutting brass	resistant	[971]

1) The corrosion rates of copper and copper alloys in methanol/gasoline mixtures allow an evaluation of resistant under normal operating conditions. However, they are not suitable for use in fuel-conducting components in cars.
* see text
** vol.%

Table 140: Evaluation of the corrosion resistance of copper and copper alloys to methanol and methanol-containing mixtures (for more detailed conditions, see text)

Table 140: Continued

Medium and concentration	Temperature, K (°C)	Material	Evaluation	Literature
93 % methanol +7 % gasoline	297 (24)	free-cutting brass	resistant	[971]
	321 (48)	free-cutting brass	resistant, insignificant pitting	[971]
93 % methanol +7 % gasoline +4.8 % water*	297 (24)	free-cutting brass	resistant	[971]
	321 (48)	free-cutting brass	resistant	[971]
M 100 fuel (94.5 % methanol +5.5 % isopentane)		Cu	resistant to moderately resistant	[918]

1) The corrosion rates of copper and copper alloys in methanol/gasoline mixtures allow an evaluation of resistant under normal operating conditions. However, they are not suitable for use in fuel-conducting components in cars.
* see text
** vol.%

Table 140: Evaluation of the corrosion resistance of copper and copper alloys to methanol and methanol-containing mixtures (for more detailed conditions, see text)

Mixed Acids

Author: M. B. Rockel / Editor: R. Bender

Copper

Copper and its alloys are strongly attacked by hot concentrated mixed acids, and dilute mixed acids attack and all copper alloys at room temperature, which is why they are used for preliminary pickling [979]. Fittings used in the handling of 1 to 2% scrubbing acids at 373 K (100 °C), for example, those used to wash nitrocellulose, are made of phosphor bronze. In more highly concentrated mixed acids, such as 94% H_2SO_4 + 1% HNO_3 + 5% H_2O, the corrosion rate at room temperature is already more than 2 mm/a (79 mpy). Aluminium bronze with 10% Al is the most resistant and is used for stuffing boxes and bolts in acid pumps. The corrosion rate at room temperature in an anhydrous mixed acid consisting of 55% H_2SO_4 + 40% HNO_3 + 5% SO_3 was determined to be < 0.1 mm/a (< 3.94 mpy) and in 94% H_2SO_4 + 1.3% HNO_3 + 4.7% H_2O it was approximately 0.2 mm/a (7.87 mpy); however, it increases by more than 15-fold in the latter mixture at 323 K (50 °C).

Pure copper has very low resistance in all inorganic acids and thus also in all mixtures, even at high dilution and at RT. According to the resistance tables of the Deutsches Kupferinstitut (DKI, German Copper Institute) [980], the corrosion behavior in pure acids is given generally as "moderate" to "not resistant". This publication also gives the evaluations of the resistance made by two US institutes: the Copper Development Association (CDA) and the American Brass Company.

Very dilute mixed acids can be formed in condensates that originate during the combustion of natural gas. Investigations using the amount of metal ions that have entered the solution as a function of the acidity of natural gas condensates prove that sulfuric and nitric acids primarily determine the corrosion of aluminium and stainless steel, but not the corrosion of copper. The relatively strong corrosive attack of copper is attributed to the oxygen content of the condensates and thus to classical oxygen corrosion. The acidity of the condensates only has an indirect effect here: the decrease in the pH increases the solubility of the oxidic reaction products of copper and thus prevents the formation of the corrosion protection layer. Also in the case of increased formation of condensates when the temperature is continually below the dew point, the corrosion of copper is considerable, especially in combination with a lower temperature because this increases the solubility of oxygen. The use of unprotected copper is therefore not recommended.

The influence of inhibitors based on urea derivatives or glycolic acid on the corrosion resistance of pure copper in 0.5 M H_2SO_4 and 1 M HNO_3 has been investigated at 303 K (30 °C). While the corrosion rate was 135 mg/dm^2 d (mdd) in dilute sulfuric acid, it decreased to 43 mdd on addition of only 10 ppm diphenylthiourea. The inhibition is even more pronounced in the case of 1 M HNO_3, if the extremely high corrosion rate of 5470 mdd in non-inhibited acid is taken into account. However, the inhibitor dithioglycolic acid is much more effective in nitric acid: the corrosion rate decreased to 50 mdd with only 10 ppm, i.e. 99 % inhibition.

The corrosion behavior of pure copper was determined in binary mixtures of $HNO_3 + HCl$, $HNO_3 + H_3PO_4$ as well as $H_2CrO_4 + HCl$ at room temperature and for total concentrations in the range 0.5–5 M [981]. A thermometric measuring method was used, supplemented by potentiodynamic polarisation curves as well as determinations of the mass loss. The results are not discussed in detail at this point because they are more interesting for electrochemists and are of less use for practical chemists, and because exact corrosion rates cannot be derived from them since no exact times are given so that the area-related mass loss cannot be determined. The corrosion rates can only be estimated if the time is assumed to be a few hours. Even if a time of max. 10 h is taken, the values given in the diagrams indicate very high corrosion rates, which can be estimated for the individual mixed acids as follows:

		$g/cm^2 \times 10^3$	Corrosion rate, mm/a (mpy)
$HCl + HNO_3$	above 0.5 M	0.2	> 10 (> 390)
	above 2 M	2	> 100 (> 3,900)
$HNO_3 + H_3PO_4$	up to 1.5 M	0.1	> 5 (> 197)
	above 3 M	1	> 50 (> 1,970)
$H_2CrO_4 + HCl$	above 0.5 M	10	> 500 (> 19,700)
	above 4 M	< 0.1	> 5 (> 197)

On the basis of these extremely high corrosion rates, the term "resistant" cannot be used. The data was confirmed by the passive current densities, which can be taken from the potentiodynamically determined current density-voltage curves. They all lie in the range of several mA/cm^2 and even up to 20 and 100 mA/cm^2. The electrochemist knows that a current density of approximately 0.1 mA/cm^2 corresponds to a corrosion rate of 1 mm/a (39.4 mpy).

The corrosion rates determined here are only estimates, although exact data are not useful because of the extremely high corrosion rates. Pure copper is not resistant in the above-mentioned mixed acids.

Copper-aluminium alloys

Copper-aluminium alloys are the most corrosion resistant of the copper alloys. This is essentially due to the additional element aluminium, which forms a protective oxide layer in oxidising media. In CuAlNi alloys, nickel provides a further positive contribution. Thus the alloy CuAl10Ni5Fe4 with 9–11 % Al, 5–6 % Ni, 1.5–6 % Fe is very resistant in sulfuric acid of all concentrations up to 90% up to a temperature of 343 K (70 °C) with corrosion rates below 0.1 mm/a (3.94 mpy) (see also [980]).

Copper-nickel alloys

Copper-nickel alloys, like the copper-aluminium alloys, belong to the most corrosion-resistant copper materials. Typical representatives, such as CuNi10Fe1Mn, CuNi30Mn1Fe, CuNi25 and CuNi44Mn1, are, however, used in chloride-containing cooling water and in seawater because they are essentially resistant in corrosive atmospheres. They are resistant towards stress corrosion cracking and pitting corrosion. They are relatively resistant in very dilute non-oxidising acids. Thus, according to the DKI tables [980], they have very good resistance in boric and oxalic acids; good resistance in hydrofluoric, hydrofluorosilicic acid and phosphoric acids, as well as in dry hydrochloric and sulfuric acids up to 80%. Copper-nickel alloys are not resistant in nitric and chromic acids. Cast alloys behave similarly. It is possible to transfer their behavior to mixed acids, but it must be tested for each individual case.

Copper-nickel-zinc alloys (German silver, Alpaka) have a much better corrosion resistance in neutral chloride-containing media than the binary copper-zinc alloys. This is particularly valid for alpha alloys with higher nickel contents that do not have a tendency to dezinc and to stress corrosion cracking. In contrast, copper-nickel-zinc alloys are attacked by all inorganic acids and so are their mixtures, even if they are very dilute and at room temperature. In dilute sulfuric and hydrochloric acids, their behavior clearly depends on the amount of oxygen or on the aeration (for individual evaluations, see [980]).

Copper-tin alloys (bronzes)

Copper-tin alloys belong to the most corrosion-resistant copper alloys, for example, the friction bearing material CuSnP (DIN-Mat. No. 2.1830) with 90–92.5 Cu, 7.5–9.0 Sn and 0.1–0.4 P. However, this refers to the great number of media used in technical applications, rather than the acids. They are definitely not suitable for use in inorganic acids. Therefore, according to the tables published by the DKI [980], their resistance is evaluated as follows: "suitable" in the media boric, oxalic and phosphoric acids, "limited suitability" in hydrofluoric and hydrofluosilicic acids as well as in sulfuric acid up to 80%, "not suitable" in chromic, nitric and hydrochloric acids as well as sulfuric acid above 80%.

The corrosion behavior of tin-containing copper cast materials is similar. Higher contents of up to 40% tin as well as up to 20% lead improve the resistance. Thus, Cu-Pb-Sn-bronzes are used in acidic sulfide solutions in the paper industry and in sugar manufacturing.

Copper-tin-zinc alloys (red brass)

The same (as bronzes) applies to copper-tin-zinc alloys, which are generally only CuSnZn casting materials.

Copper-zinc alloys (brass)

The corrosion resistance of copper-zinc alloys is primarily determined by the zinc content. Thus, the zinc-rich beta-phase of brass is less resistant than the copper-rich alpha phase. The addition of further alloying elements, such as aluminium, tin, manganese, lead and arsenic, improves the resistance in water and seawater. An evaluation of the resistance in the most important inorganic acids is given in Table 141, the data was taken from the DKI tables [980].

Corrosive medium	A	B	C	D	E
Boric acid	1	1	2	2	1
Hydrobromic acid	3	3	4	4	3
Chromic acid	3	3	4	4	3
Hydrofluoric acid	3	3	4	4	4
Oxalic acid	2	2	3	3	2
Phosphoric acid	3	3	4	4	3
Nitric acid	4	4	4	4	4
Hydrochloric acid	3	3	4	4	4
Sulfuric acid, 40–80 %	2	2	4	4	3
Sulfuric acid, 80–95 %	3	3	4	4	3

1 = complete resistance (< 0.1 mm/a; < 3.94 mpy); 2 = sufficient resistance (up to 1 mm/a; 39.4 mpy); 3 = moderate resistance (1–3 mm/a; 39.4–118 mpy); 4 = not resistant (> 3 mm/a; > 118 mpy)

Table 141: Qualitative behavior of copper and copper-zinc alloys in various media [980]
Key to the alloys : A) copper; B) brass Ms90 (CuZn10), Ms85 (CuZn15); C) brass Ms72 (CuZn28), Ms60 (CuZn40), Ms58 (CuZn40Pb2); D) special brass SoMs58, manganese-containing (CuZn40Mn); E) special brass SoMs71 (CuZn28Sn), SoMs76 (CuZn20Al).

The relatively unfavorable behavior in pure acids certainly also applies to all variations of mixed acids. In general, brass is also susceptible to dezincing as well as to stress corrosion cracking.

Brass is not resistant in mixtures of nitric and sulfuric acids, even at room temperature, even in very dilute solutions. For example, laboratory tests in 0.1 N mixed acid at 303 K (30 °C) gave a mass loss of approximately 100 mdd, which corresponds to a corrosion rate of 0.4 mm/a (15.7 mpy) [982]. Although the corrosion rates can be decreased by the addition of various inhibitors based on urea and its derivates, in the present case to approximately 40–60 mdd, i.e. to 0.15 mm/a (5.91 mpy). At higher concentrations, such as 1 N, the inhibitors are still effective; however, the corrosion rates are already > 1 mm/a (39.4 mpy), so that brass cannot be used on a permanent basis in inorganic mixed acids.

Nitric Acid

Author: K. Hauffe / Editor: R. Bender

Copper

Copper is not resistant in nitric acid. It is attacked by 15 mg/l HNO_3 at 288 K (15 °C) with a material consumption rate of 20 g/m²h after 1 h and 160 g/m²h after 5 h. This rate could be reduced to about 27 g/m²h after 4 h by 2 mg/l maize extract (pH inhibitor, Russian grade, consisting of (40–52) % protein + (22–27) % soluble hydrocarbons + (1–3) % fat + < 0.5 % starch + (0.7–1.1) % lactic acid), which corresponds to an inhibition efficiency of 74 % [983]. At a concentration of 10 mg/l nitric acid, the inhibition efficiency at 2 g/m²h was 82 %.

The severe corrosion on copper in nitric acid can also be seen from current density/potential measurements. At a polarization voltage of +0.23 V_{SHE}, the current density at room temperature could be reduced to < 10^{-3} mA/cm² [984], which corresponds to a corrosion rate of < 0.01 mm/a (0.39 mpy), if the current density was stationary.

To clarify the influence of the crystal structure of copper on its corrosion in nitric acid, dissolution tests were carried out on copper monocrystal beads in nitric acid (0.037 to 0.21 mol/mol) between 298 K (25 °C) and the boiling point [985]. The corrosion rate on the (100), (111) and (311) planes was between 1 and 10 µm/min.

The corrosion rate of copper in nitric acid was reduced in a magnetic field by 10 % at pH 2 and by 40 to 50 % at a pH 5 [986].

The inhibition efficiency of dimethylol thiourea on the corrosion of copper in 0.182 mol/l nitric acid between 293 and 333 K (20 and 60 °C) at a concentration of 1 g/l was 90 and 78 %. The corrosion rate is thereby reduced from 0.093 to 0.011 mm/a (3.66 to 0.43 mpy) at 293 K (20 °C) and correspondingly from 0.30 to 0.065 mm/a (11.81 to 2.56 mpy) at 333 K (60 °C) [987].

The inhibition of copper corrosion in 1 to 5 mol/l nitric acid by amine-containing inhibitors after 0.5 and 2 h is summarized in Table 142 [988]. The inhibition efficiency is not sufficient for industrial use.

The inhibition efficiency of monosaccharides (glucose, fructose, mannose and galactose) regarding the corrosion of electrolytic copper in 0.01 to 1.0 mol/l nitric acid between 288 and 308 K (15 and 35 °C) was completely inadequate even at 1 mol/l of the saccharide [989, 990].

The inhibition efficiency of 0.12, 0.5 and 1.0 g/l benzotriazole in 0.1 mol/l nitric acid at room temperature after 5 days was 61, 84 and 97 % [991]. Benzotriazole is unsuitable in corrosive solutions with a complexing or oxidizing action.

The inhibition efficiency of 0.001 mol/l tolyltriazole at 293 K (20 °C) in 0.1 mol/l nitric acid of 66 %, i.e. 0.0336 g/m²h, is greater than that of 0.001 mol/l benzotriazole (0.0494 g/m²h), based on the value without an inhibitor of 0.0991 g/m²h. Further information on the action and structure of the inhibitors can be found in a review [992].

HNO$_3$ concentration mol/l	Inhibitor	Inhibitor concentration %	Material consumption rate g/m^2 h	Inhibition efficiency %
1	–	–	2.41	–
	hexamine	2.0	0.45	81.0
2	–	–	65.9	–
	hexamine	2.0	8.2	87.5
	α-naphthylamine	0.2	1.87	97.2
	benzidine	0.1	0.42	99.4
4	–	–	1,604	–
	hexamine	2.0	150	90.6
	α-naphthylamine	0.5	13.6	99.2
	benzidine	0.2	843	46.7
5	–	–	2,335	–
	α-naphthylamine	1.0	6.6	99.7

Table 142: Corrosion of copper in HNO$_3$ at 308 K (35 °C) in the presence of amine-containing inhibitors [988]

Aromatic amines, such as aniline, o-toluidine, o-chloroaniline and o-phenetidine, show an inhibition efficiency of up to 99 % in 2 to 5 mol/l nitric acid, but do not reduce the material consumption rate of copper to a value which is of industrial interest. The lowest material consumption rate in 2 mol/l nitric acid of 2.21 g/m^2 h was obtained with 17.4 ml/l o-phenetidine [993].

Thiourea is unable to reduce the corrosion of copper in 3.5 mol/l nitric acid to values of industrial interest. Stimulation even occurs at higher concentrations, probably due to hydrolysis to diaminocarbonyl disulphide, which is reduced to thiocarbamic acid in the presence of the dissolving metal at cathodic sites [994].

According to Table 143, the inhibition efficiency of gallic acid regarding the corrosion of copper (Cu-0.0015P-0.003Cd-0.0005Zn-0.001Pb-0.0015S) in 1 mol/l nitric acid at room temperature was superior to that of tannic acid [989]. In contrast, in an acid mixture of 1 mol/l (5 ml HCl + 2 ml HNO$_3$ + 95 ml H$_2$O) the inhibition efficiency of 0.5 % tannic acid of 66 % was somewhat better than that of gallic acid of 62 %.

The material consumption rates of copper in 0.5 and 2 mol/l nitric acid at 298 K (25 °C) of 0.468 and 46.1 g/m^2 h respectively can be reduced to 0.025 and 0.169 g/m^2 h by 2 g/l inhibitor KMA, which is based on waste products of coking plants [995].

To protect copper wires from corrosion in an HNO$_3$-containing atmosphere (0.01 %) at room temperature, they were provided with a gold layer 5 μm thick [996]. To bypass additions of HNO$_3$ as an oxidizing agent to sulfuric acid for pickling copper pipes, hydrogen peroxide is used, which reduces the loss of metal [997].

Inhibitor	Concentration %	Material consumption rate g/m² h	Inhibition efficiency %
Gallic acid	–	2.1	–
	0.1	0.45	79
	0.2	0.38	81
	0.5	0.35	82
Tannic acid	0.25	1.0	47
	0.5	0.83	60
	1.0	0.55	73

Table 143: Inhibition of the corrosion of copper in 1 mol/l HNO_3 by gallic and tannic acid at room temperature after 400 h [989]

Oxide films generated by an alternating current of 50 Hz on copper in 15 g/l sodium nitrate and in 40 g/l sodium carbonate at about 330 K (57 °C) and a current density of 15, 6 or 0.5 A/dm² over a period of 10 min improve the corrosion behavior in solutions of 5 % acetic acid + nitric acid [998].

Before copper is coated with aluminium in an ester- and hydride-containing bath, the surface must be etched. The following mixture is recommended: (vol.%) 58 to 60 % conc. HNO_3 + 2 to 3 % conc. HCl + 37 to 40 % 12 mol/l H_2SO_4 [999].

The following solution is recommended for cleaning and polishing of copper at room temperature: (parts by volume) 1 HNO_3 (63 %) + 2 H_2SO_4 (98 %) + 2 water + 1 ml/l HCl (36 %) [1000].

To improve the uniform removal of copper during electro-polishing in a viscous electrolyte consisting of an aqueous HNO_3-containing glycerol solution at 293 K (20 °C), the following solution is recommended for increasing the solubility of the $Cu(NO_3)_2 \times 3\ H_2O$ formed for the purpose of optimum polishing conditions: (percent by volume) 75 % glycerol (d = 1.225 g/ml) + 25 HNO_3 (d = 1.342 g/ml) [1001].

The dissolution rate of copper (99.99 %) in 0.1 and 1 mol/l nitric acid was drastically increased by additions of mercury nitrate in short-term tests of 15 min duration as shown in Table 144. This solution is used as a quick test for the susceptibility of copper alloys to stress corrosion cracking. No cracking occurs on unalloyed copper [1002].

The wear resistance and corrosion resistance of copper and bronze in nitric acid were improved by diffusion layers of aluminium at 1,220 K (947 °C) and chromium at 1,173–1,203 K (900–930 °C) [1003].

The action of resorcinol, o- and p-aminophenol, pyrocatechol, o-cresol and salicylaldehyde as inhibitors of copper corrosion is described in [1004].

Addition of polyvinyl alcohol, which improves chemical polishing of copper in dilute nitric acid and dilute mixtures of nitric and phosphoric acids, reduces the solubility of copper and increases the uniformity of material removal as a result of an increase in the solution viscosity. Some material consumption rates are summarized in Table 145 [1005].

Hg(NO$_3$)$_2$ mol/l	0	0.001	0.01	0.1	0.4
Dissolution rate, mm/h	< 1 × 10^{-5}	4.9 × 10^{-4}	5.9 × 10^{-4}	0.018	0.048

Table 144: Dissolution rate of copper in 1 mol/l HNO$_3$ as a function of the Hg(NO$_3$)$_2$ content at room temperature [1002]

Electrolyte, %*	Polyvinyl alcohol, g/l				
	0	0.1	1.0	10	50
	Material consumption rate, g/m^2 h				
5 HNO$_3$ + 95 H$_2$O	2	0.1	0.1	0.1	0
10 HNO$_3$ + 90 H$_2$O	4	1.0	0.8	0.3	0.1
20 HNO$_3$ + 80 H$_2$O	16	32	40	52	12
40 HNO$_3$ + 60 H$_2$O	1,072	1,570	1,683	1,651	643
5 HNO$_3$ + 5 H$_2$O + 90 H$_3$PO$_4$	94	134	224	182	0.1
5 HNO$_3$ + 10 H$_2$O + 85 H$_3$PO$_4$	128	170	269	147	0.1
5 HNO$_3$ + 20 H$_2$O + 75 H$_3$PO$_4$	263	365	402	279	0.37
10 HNO$_3$ + 5 H$_2$O + 85 H$_3$PO$_4$	132	222	383	303	2.4
10 HNO$_3$ + 20 H$_2$O + 70 H$_3$PO$_4$	414	497	625	402	1.1
15 HNO$_3$ + 5 H$_2$O + 80 H$_3$PO$_4$	353	312	480	393	7.4

* vol.%

Table 145: Corrosion of copper in HNO$_3$ and HNO$_3$/H$_3$PO$_4$ mixtures at room temperature with and without polyvinyl alcohol [1005]

Copper-aluminium alloys

Some corrosion rates of copper-aluminium alloys with additions of manganese, iron, tin and nickel in 5 % nitric acid in the presence of air are summarized in Table 146. These alloys are not resistant to dilute nitric acid [1006].

Alloy	Temperature K (°C)	Corrosion rate mm/a (mpy)
Cu-6.0Al-2.0Ni-1.95Sn-7.98Mn-2.5Fe	293 (20)	5.4 (213)
	343 (70)	44.3 (1,744)
Cu-8.0Al-1.95Sn-8.0Mn-2.5Fe	293 (20)	18.6 (732)
	343 (70)	30.0 (1,181)
Cu-10.6Al-3.1Fe-0.8Mn	293 (20)	79.9 (3,146)

Table 146: Corrosion rates of some CuAl alloys in 5 % HNO$_3$ containing air [1006]

Copper-nickel alloys

Additions of mercury nitrate increase the corrosion rate of the copper-nickel alloy (Cu-29.57Ni-0.88Mn-0.60Fe-<0.10Zn-0.005Pb-0.029P-0.036C) in 0.1 and 1 mol/l nitric acid. Some test results from short-term tests of 15 min duration are summarized in Table 147 [1002].

$Hg(NO_3)_2$ concentration mol/l	0.1 mol/l HNO_3	1.0 mol/l HNO_3
	\multicolumn{2}{c}{Corrosion rate, mm/h}	
0	3.7×10^{-4}	3.7×10^{-4}
0.001	4.4×10^{-4}	5.0×10^{-4}
0.01	0.0025	0.0012
0.1	0.0044	0.0044
0.25	0.0051	–
0.4	–	0.013

Table 147: Dissolution rate of the CuNi alloy in HNO_3 as a function of the $Hg(NO_3)_2$ content at room temperature [1002]

The corrosion resistance of copper-nickel-chromium layers at room temperature in a solution of 5 ml/l HNO_3 + 10 g/l $NaNO_3$ + 1.3 g/l NaCl was demonstrated by coulometric measurements [1007].

Copper-tin alloys (bronzes)

According to Table 148, the high corrosion rate of the copper-tin alloys BrCh08 and SP19 (no composition data) in concentrated nitric acid at room temperature is drastically reduced by aluchromizing of the alloys at 1,253 K (980 °C) for 6 and 9 h respectively [1008].

Alloy	Duration of aluchromizing h	Test duration h	Material consumption rate $g/m^2 h$
BrCh08	–	0.67	3,130
	6	3	67
	9	50	0.36
SP19	–	1	2,330
	6	4	7.9
	9	100	0.12

Table 148: Material consumption rate of the CuSn alloys BrCh08 and SP19 in concentrated HNO_3 with and without aluchromizing to a thickness of about 4–7 µm [1008]

Copper-zinc alloys (brass)

The Cu-37Zn alloy (Cu-36.6Zn-0.07Sn-0.01Pb) corrodes in 0.5 mol/l nitric acid at 308 K (35 °C) with a material consumption rate of 0.22 g/m² h. The inhibitors 2-mercaptobenzimidazole and 2-mercaptobenzothiazole, suitable for reducing the corrosion of brass in 0.5 mol/l hydrochloric acid, are unusable for corrosion in nitric acid (stimulation) [1009, 1010]. Brass casting alloys are not resistant to nitric acid even at room temperature [1011].

The corrosion behavior of different CuZn alloys in varying concentrations of HNO_3 at various temperatures can be seen from Table 149. The inhibition efficiency of most substances is not sufficient to reduce the corrosion rates, in particular in more highly concentrated nitric acid, to values of industrial interest.

HNO_3 mol/l	Material	Temperature K (°C)	Inhibitor	Inhibitor Concentration mol/l	Efficiency %	Material consumption rate g/m² h	Literature
0.1[1]	Cu-30Zn	RT	none	–	–	0.0864	[1012]
			triazole	0.001	–30[2]	0.112	
			benzotriazole	0.001	60	0.0351	
			naphthotriazole	0.001	30	0.0609	
0.1	Cu-40Zn	303 (30)	none	–	–	114	[1013]
			chloroacetic acid	0.001	97	3.42	
			dichloroacetic acid	0.001	99.5	0.57	
			trichloroacetic acid	0.001	> 99.9	0.114	
			bromoacetic acid	1.0×10^{-5}	89	12.54	
0.1	Cu-40Zn	303 (30)	none	–	–	105	[1014]
			diphenyl thiourea	1.0×10^{-4}	> 99.9	0.105	
			allyl thiourea	1.0×10^{-4}	< 95.2	< 1	
			phenyl thiourea	1.0×10^{-5}	94.5	5.6	
			dimethyl thiourea	1.0×10^{-5}	84	16.7	
0.5	Cu-37Zn	303 (30)	none	–	–	0.22	[1015]
			p-thiocresol[3]	0.016	73	0.059	

1) pH 1.4 to 1.7
2) stimulation
3) stimulating action in 1 mol/l HNO_3
4) inhibitor based on waste products from coking plants
5) higher inhibitor additions produce no improvement (≤ 200 ppm)
6) 30 min tests
7) low-copper brass
RT = room temperature

Table 149: Corrosion and inhibition of corrosion using different CuZn alloys in nitric acid

Table 149: Continued

HNO$_3$ mol/l	Material	Temperature K (°C)	Inhibitor	Inhibitor Concentration mol/l	Efficiency %	Material consumption rate g/m^2 h	Literature
0.5	Cu-40Zn	303 (30)	none	–	–	311	[1013]
			chloroacetic acid	0.001	76	74.64	
			dichloroacetic acid	0.001	85	46.65	
			trichloroacetic acid	0.001	91	27.99	
			bromoacetic acid	1.0×10^{-5}	76	74.64	
0.5	Cu-36Zn	298 (25)	none	–	–	4	[1016]
			KMA[4]	2 g/l	> 99.2	0.029	
1.0	Cu-40Zn	RT	none	–	–	18	[1017]
			dithioglycolic acid	20 ppm[5]	99	0.18	
1.0	Cu-40Zn	303 (30)	none	–	–	437	[1013]
			chloroacetic acid	0.001	62	166.06	
			dichloroacetic acid	0.001	76	104.88	
			trichloroacetic acid	0.001	80	87.4	
			bromoacetic acid	1.0×10^{-5}	67	144.21	
1.5	Cu-40Zn	303 (30)	none	–	–	602	[1013]
			chloroacetic acid	0.001	51	294.98	
			dichloroacetic acid	0.001	65	210.7	
			trichloroacetic acid	0.001	69	186.62	
			bromoacetic acid	0.00001	59	246.82	
2.0	Cu-30Zn	RT	none	–	–	72	[1018]
			o-phenylenediamine	0.05 %	95	3.4	
				0.2 %	96	2.8	
			m-phenylenediamine	0.2 %	95	3.4	
			p-nitroaniline	0.2 %	96	2.6	
			m-toluidine	4.35 ml/l	94	4.6	
				43.5 ml/l	95	3.6	
			p-toluidine	1.0 %	96	2.6	

1) pH 1.4 to 1.7
2) stimulation
3) stimulating action in 1 mol/l HNO$_3$
4) inhibitor based on waste products from coking plants
5) higher inhibitor additions produce no improvement (≤ 200 ppm)
6) 30 min tests
7) low-copper brass
RT = room temperature

Table 149: Corrosion and inhibition of corrosion using different CuZn alloys in nitric acid

Table 149: Continued

HNO$_3$ mol/l	Material	Temperature K (°C)	Inhibitor	Inhibitor Concentration mol/l	Efficiency %	Material consumption rate g/m^2 h	Literature
			m-aminobenzoic acid	0.5 %	99	0.4	
			p-aminobenzoic acid	0.05 %	99	0.4	
			o-chloroaniline	8.7 ml/l	96	2.6	
			p-chloroaniline	0.05 %	99	0.64	
				0.1 %	99.5	0.34	
2.0[6]	Cu-30Zn	RT	none	–	–	71.6	[1019]
			p-nitroaniline	0.2 %	96	2.56	
	Cu-37Zn	RT	none	–	–	63.4	[1019]
			phenylhydrazine	4.35 ml/l	96.5	2.22	
			thiourea	0.2 %	98.7	0.82	
	Cu-40Zn		none	–	–	52.0	[1019]
			p-anidisine	2.0 %	92	4.2	
2.0	brass[7]	RT	none	–	–	30.95	[1020]
			hydrazine sulfate	0.05 %	95.8	1.3	
2.0	Cu-37Zn	RT	none	–	–	24.4	[1021]
			o-toluidine	0.024	94	1.56	
			acridine	0.02	99	0.22	
2.0	Cu-37Zn	300 (27)	none	–	–	78.9	[1022]
			p-chloroaniline	0.01 %	98	1.51	
				0.1	98.4	1.25	
			p-aminobenzoic acid	0.025	99.4	0.47	
				0.1	99.6	0.36	
2.0	Cu-36Zn		none	–	–	101	[1016]
			KMA[4]	2 g/l	99.8	0.169	

1) pH 1.4 to 1.7
2) stimulation
3) stimulating action in 1 mol/l HNO$_3$
4) inhibitor based on waste products from coking plants
5) higher inhibitor additions produce no improvement (≤ 200 ppm)
6) 30 min tests
7) low-copper brass
RT = room temperature

Table 149: Corrosion and inhibition of corrosion using different CuZn alloys in nitric acid

Table 149: Continued

HNO₃ mol/l	Material	Temperature K (°C)	Inhibitor	Inhibitor Concentration mol/l	Efficiency %	Material consumption rate g/m² h	Literature
2.0	Cu-40Zn	303 (30)	none	–	–	710	[1013]
			chloroacetic acid	0.001	45	390.5	
			dichloroacetic acid	0.001	50	355	
			trichloroacetic acid	0.001	60	284	
			bromoacetic acid	0.00001	50	355	
3.0	Cu-37Zn	300 (27)	none	–	–	1,067	[1022]
			p-chloroaniline	0.01 %	86.5	143	
				0.1	97.6	25.2	
			p-aminobenzoic acid	0.025	99.8	0.4	
				0.1	99.9	0.24	
4.0[6]	Cu-30Zn	RT	none	–	–	1,090	[1019]
			p-nitroaniline	0.2 %	99.9	1.28	
	Cu-37Zn	RT	none	–	–	1,359	[1019]
			phenylhydrazine	4.35 ml/l	> 99.9	0.38	
			thiourea	0.2 %	> 99.9	0.44	
	Cu-40Zn		none	–	–	1,972	[1019]
			p-anidisine	2.0 %	99.9	2.78	
4.0	Cu-37Zn	308 (35)	none	–	–	2,740	[1015]
			N-methylaniline	17.4 ml/l	98.6	38.2	
				43.5 ml/l	99.4	15.4	

1) pH 1.4 to 1.7
2) stimulation
3) stimulating action in 1 mol/l HNO₃
4) inhibitor based on waste products from coking plants
5) higher inhibitor additions produce no improvement (≤ 200 ppm)
6) 30 min tests
7) low-copper brass
RT = room temperature

Table 149: Corrosion and inhibition of corrosion using different CuZn alloys in nitric acid

Meta-substituted aromatic amines are highly effective inhibitors of corrosion of Cu-30Zn in 2 mol/l nitric acid on the basis of current density/potential measurements (inhibition efficiency 98 to 99 %). m-Chloroaniline is also worth mentioning [1023]. Of the heterocyclic amines 2-aminothiazole, 2-aminopyridine and 4-aminophenazone, only the first-mentioned is suitable (but only from 1 %) as an efficient inhibitor with Cu-30Zn. The inhibition efficiency at a material consumption rate of 0.1 g/m² h is > 99 % [1024].

10^{-4} mol/l trichloroacetic acid can be recommended as an effective inhibitor of the corrosion of Cu-40Zn at 300 K (27 °C) in dilute nitric acid (< 0.1 mol/l) [1013].

o-Phenetidine, α-naphthylamine, o-toluidine and o-anisidine are unusable as inhibitors of the corrosion of brass in nitric acid [1025].

Phenylthiourea is also not effective enough as an inhibitor in HNO_3 [1026].

High-alloy types of brass, such as, for example, cast brass Cu-28.3Zn-3.8Pb-0.9Sn-0.2Fe, suffer from high corrosion rates in nitric acid, e.g. about 119 mm/a (4,685 mpy) in 5 % HNO_3 at 293 K (20 °C) [1006].

Pipes made of the brasses Cu-28Zn-1Sn and Cu-20Zn-2Al are resistant to solutions of 0.5 mol/l NaCl + 0.02 HNO_3 at room temperature [1027].

According to 1,300 h tests at 408 K (135 °C), the copper alloy No. 445 (phosphorus-containing admiralty brass (Cu-28Zn-1Sn-P)) can be used in a mixture of organic acids with nitric acid (80 % propionic acid + 15.4 % higher organic acids + 2.5 % butyric acid + 0.1 % acetic acid + 2 % nitric acid). Its corrosion rate was 0.19 mm/a (7.48 mpy). Arsenic-containing Admiralty brass is destroyed [1028].

The rate of dissolution of the brass types 220 (Cu-10.05Zn-0.004Pb-0.006Fe) and 260 (Cu-30.01Zn-0.02Pb-0.01Fe) in 0.1 and 1 mol/l nitric acid is greatly increased by additions of mercury nitrate of 0.1 mol/l or greater (Table 150) [1002].

HNO_3 concentration mol/l	$Hg(NO_3)_2$ content, mol/l		
	None	0.01	0.1
	Corrosion rate, mm/h		
0.1	0.0011	0.0023	0.054
	(3.9×10^{-4})*	(5.9×10^{-4})*	(0.127)*
1.0	–	8.9×10^{-4}	0.021
	(6.0×10^{-4})*	(9.7×10^{-4})*	(0.031)*

*alloy 260

Table 150: Dissolution rate of the CuZn alloys 220 and 260 in HNO_3 at room temperature as a function of the $Hg(NO_3)_2$ content [1002]

Figure 103 shows the embrittlement coefficient as a measure of the stress corrosion cracking. The latter is defined as the ratio of the fracture resistances of specimens after pre-exposure in the corrosive environment to that without pre-exposure. According to this figure, the decrease in the embrittlement coefficient with the duration of exposure on α-brass specimens along the direction of rolling is greater in comparison with specimens perpendicular to the direction of rolling [1029].

The dependence of the dissolution rate of 0.30 mm specimens of Cu-15Zn in nitric acid at room temperature on the degree of rolling is also remarkable. While a maximum material consumption rate of 43 g/m²h is observed at a degree of rolling of about 30 %, a minimum of about 24 g/m²h occurs between 50 and 53 %, followed by another maximum of about 40 g/m²h at 65 % [1030].

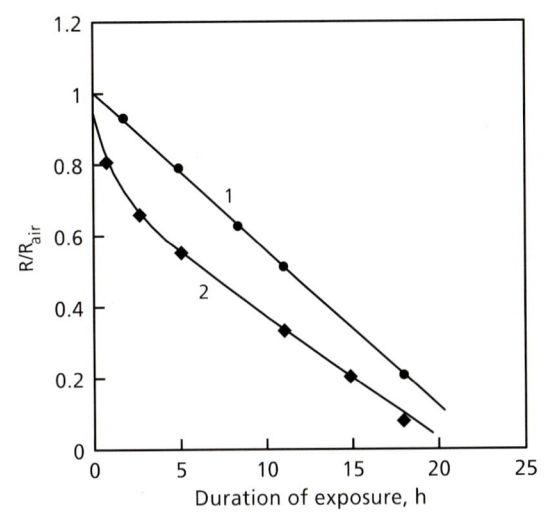

Figure 103: Dependence of the embrittlement coefficient R/R_{air} of α-brass specimens as a function of the exposure time in 4.5 mol/l HNO_3 at room temperature [1029]
1) R_{air} (perpendicular to the rolling direction) 20 kJ/m^2
2) R_{air} (along the direction of rolling) 30 kJ/m^2

These results are understandable bearing in mind previous studies on the time to fracture carried out on brass (containing 15 at.% Zn) in HNO_3-containing and copper tetramine solutions under defined rolling conditions [1031]. In the soft state, cracking is transcrystalline and the sensitivity to stress corrosion cracking is low.

According to Figure 104, an optimum resistance to stress corrosion cracking of loop samples of the brasses Ms-72 and Ms-80 in 0.1 % nitric acid without alloying measures is achieved solely by choosing specific annealing temperatures or degrees of cold working. Cold-worked sheet metal usually shows increasing resistance to stress corrosion cracking with increasing deformation. Tempering up to about 570 K (297 °C) drastically increases the life; but only a perfectly recrystallized specimen (sheet 1 mm thick) shows the maximum values [1032]. Higher copper contents in principle increase the life of the brass as regards stress corrosion cracking in 0.1 mol/l nitric acid. However, regardless of the copper content of the brass, the resistance to stress corrosion cracking can be increased by a factor of 100 or more (i.e. from a life of 50 min to one of > 2,600 min) by suitable heat treatment and production measures (degree of cold rolling), that is to say without complex alloying measures [1033].

The start and course of stress corrosion cracking of a cold-drawn pipe made of α-brass cleaned by dipping in 40 % nitric acid, washing and drying could be monitored by acoustic emission analysis as a function of time by means of a piezoelectric sound source [1034].

Formed parts of copper and copper alloys, e.g. of brass, are polished chemically by mixed solutions of, inter alia, nitric and sulfuric acid. The treatment time can be between a few seconds and a few minutes, depending on the nature of the surface and the rate of polishing in the solution [1035].

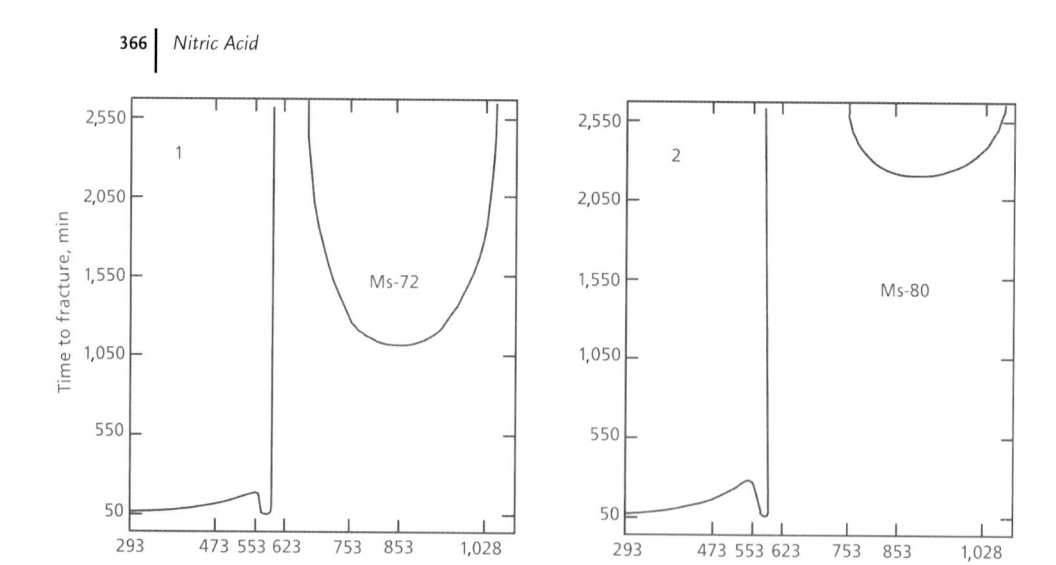

Figure 104: Time to fracture of loop samples in 0.1 mol/l HNO_3 at room temperature as a function of the annealing temperature of the brass sheets [1032]

For surface treatment of aluminium brass (Cu-22Zn-2Al-0.03As), electropolishing is carried out in agitated 33% nitric acid in methanol at room temperature and at 6 to 7 V using platinum as the cathode [1036].

Other copper alloys

According to Table 151, of the phenylenediamines only o-phenylenediamine is of any interest as an inhibitor for copper-cadmium alloys in 0.1, 0.2 and 0.3 mol/l nitric acid at 303 K (30 °C) [1037]. However, this inhibitor is of only limited use for industrial applications.

Alloy	Inhibitor concentration, %	Inhibition efficiency, %		
		0.1 mol/l HNO_3	0.2 mol/l HNO_3	0.3 mol/l HNO_3
Cu-6Cd	0.005	70	29	82
	0.010	85	55	89
Cu-20Cd	0.005	95	91	73
	0.010	92	94	87

Table 151: Inhibition of corrosion on CuCd alloys in HNO_3 at 303 K (30 °C) by o-phenylenediamine, test duration 18 h [1037]

As has already been mentioned for the brass types, urea and thiourea cannot be used as inhibitors for reducing the corrosion rates of copper-cadmium alloys in 0.1 and 0.2 mol/l nitric acid at 303 K (30 °C) because their action is too slight [1038].

The corrosion rate of amorphous and crystalline samples of copper containing 50 at.% Ti or Zr in 1 mol/l nitric acid at 303 K (30 °C) can be seen from Table 152. According to this table, the amorphous CuTi alloy has a higher resistance than the crystalline alloy [1039].

Alloy	Structure	Corrosion rate, mm/a (mpy)
Cu-50Ti	amorphous	0.0022 (0.09)
	crystalline	0.017 (0.67)
Cu-50Zr	amorphous	4.0×10^{-4} (0.02)
	crystalline	4.0×10^{-4} (0.02)

Table 152: Corrosion of Cu containing 50 at.% percent Ti or Zr in 1 mol/l HNO_3 at 303 K (30 °C) [1039]

According to Figure 105, the sensitivity of mono- and polycrystalline copper-gold alloys to stress corrosion cracking in 1:1 dilute aqua regia at room temperature reaches a maximum with 27 at.% gold. The time to fracture was determined on mono- and polycrystalline circular specimens (Ø 3 mm) [1031].

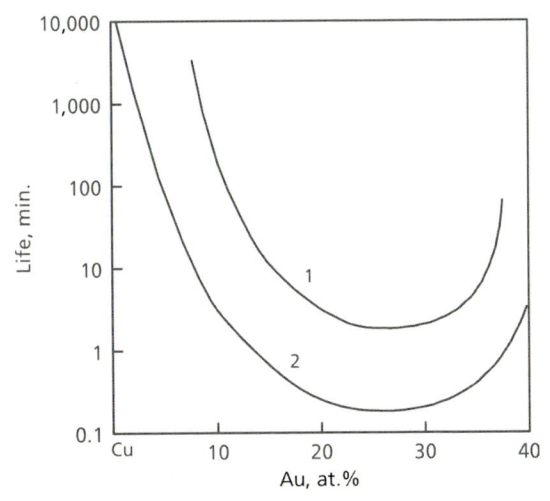

Figure 105: Time to fracture of 1) mono- and 2) polycrystalline circular specimens of CuAu alloys in 1:1 dilute aqua regia under load (50 % of the tensile strength) at room temperature [1031]

According to potentiodynamic current density/potential curves, anodic dissolution of the alloys Cu-3Pd, Cu-3Y, Cu-1Pd-1Y and Cu-2Pd-2Y (additions in at.%) in 10 % nitric acid above +1 V_{SHE} with 0.9 A/cm^2 is the same as that of unalloyed copper [1040].

Phosphoric Acid

Author: L. Hasenberg / Editor: R. Bender

Copper

The corrosion behavior of copper and its alloys under exposure to phosphoric acid, like that of hardly any other metal, depends quite substantially on the particular conditions. The limits of use of copper, for example, are outlined in Figure 107.

Oxygen above all, and in general the presence of oxidizing agents, has a corrosion-promoting effect. Fluorides furthermore promote corrosion, as does agitation, concentration of the acid and heat. Deposits can be the reason for pitting corrosion taking place underneath, especially in multicomponent alloys.

The corrosion rates listed in the following sections are always to be considered from the viewpoint that the in-house conditions will always be somewhat different.

After a preliminary choice of material has been made, use in practice in phosphoric acid should therefore always be preceded by in-house corrosion studies.

Low corrosion rates reported in the literature should never be adopted without consideration.

The corrosion behavior of copper in inorganic and organic acids depends largely on the presence of oxidizing agents. Copper corrodes hardly at all in dilute non-oxidizing acids. On the other hand, small amounts of oxygen cause a noticeable increase in corrosion, even in weak acids. Since phosphoric acid is one of the weak non-oxidizing acids, a varying corrosion behavior is to be expected in phosphoric acid, also as a function of its contamination [1041–1048].

From these comments, it must be deduced that corrosion rates for copper and copper materials cannot be adopted without reservation, but are merely a guideline and must be proved under concrete conditions by in-house studies.

Figure 106 shows the corrosion rates to be expected for copper in phosphoric acid. The test conditions for this graph were such that specimens were immersed in solutions of pure phosphoric acid, which were neither aerated nor treated with other gas streams. Air had access to the surface of the test solution. The test duration is not known.

As can be seen from Figure 106, corrosion rates above 0.5 mm/a (19.7 mpy) are already to be expected in pure phosphoric acid solutions of less than 15 % strength. As the temperature increases, increasing corrosion rates are to be expected quite generally. Partly acceptable corrosion rates of 0.2–0.5 mm/a (7.90–19.7 mpy) exist in pure acid in the concentration range of between about 20 and 60 % (14.5–43.5 % P_2O_5) at temperatures of not more than 373 K (100 °C) [1049]. At higher concentrations and relatively low temperatures, copper is to be regarded as resistant in phosphoric acid (max. 0.1 mm/a (39.4 mpy)) [1049].

Figure 107 shows the possible range of use of copper and a number of its alloys in the concentration field 1, which marks the corrosion limit of about 0.5 mm/a (19.7 mpy) in thermal phosphoric acid for the materials mentioned there [1050].

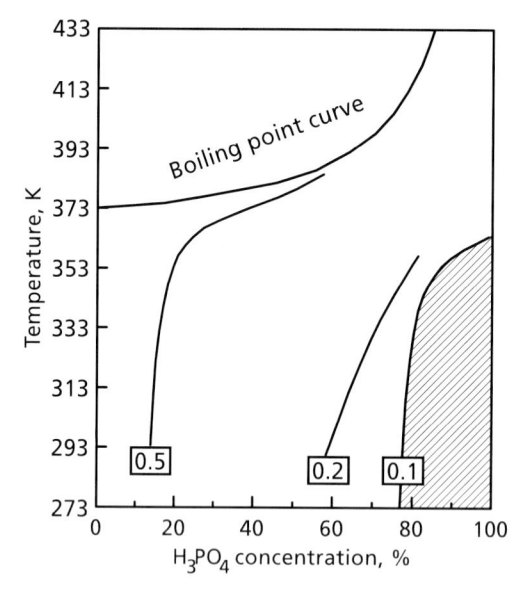

Figure 106: Isocorrosion curves (mm/a) for copper in pure phosphoric acid solutions (non-aerated) [1049]
0.1 mm/a (3.9 mpy)   0.2 mm/a (7.9 mpy)   0.5 mm/a (19.7 mpy)

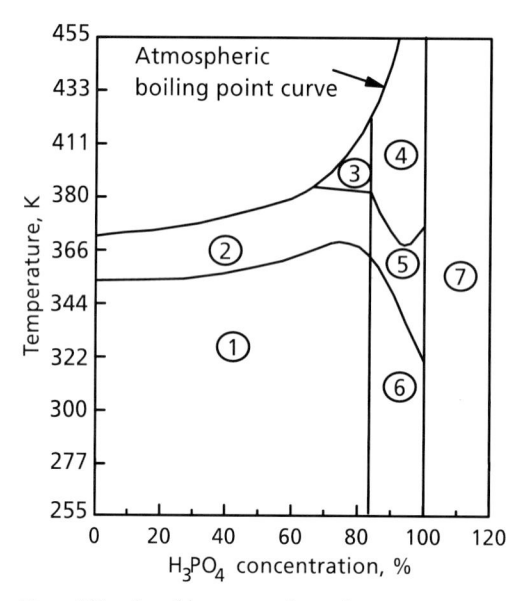

Figure 107: Possible ranges of use (isocorrosion curves about 0.5 mm/a (20 mpy)) of various materials (Table 153) in thermal phosphoric acid (Ta, Mo, W 0–85 % H_3PO_4 up to 463 K (190 °C) and Zr 0–50 % H_3PO_4 up to 463 K (190 °C) [1050]

A number of corrosion rates are contained in [1041, 1048, 1051, 1052]. Extracts of these are contained in Table 154.

Material	DIN-Mat. No.	Designation	Use range (see figure 15)
Ceramic	–		①, ②, ③, ④, ⑤, ⑥, ⑦
Hastelloy® B[1)]	cf. 2.4882	NiMo30†	①, ②, ③, ⑤, ⑥, ⑦ (⑦ up to 477 K (204 °C))
Hastelloy® C	cf. 2.4883	NiMo16Cr†	①, ②, ③ ⑤, ⑥
Durimet® 20	cf. 1.4500	CN-7M (ACI), cf. GX7NiCrMoCuNb25-20	①, ②, ③, ⑤, ⑥, ⑦
AISI 316[1)]	cf. 1.4401	X5CrNiMo17-12-2	①, ②, ⑤, ⑥, ⑦
Lead (up to 369 K (96 °C))	PB990R	Pb 99.99	①, ②
	PB985R	Pb 99.985	
		Pb 99.9 Cu	
Copper[2)]	CW024A[3)] 2.0090†	Cu 99.90	①, ⑥
Monel® 400[2)]	2.4360	NiCu30Fe	①, ⑥, ⑦
Monel® K-500[2)]	2.4375	NiCu30Al	
Duriron®[1)]	–	G-X 70 Si 15*	①, ②, ③, ④, ⑤, ⑥
Impervious graphite	–	carbon + binder	①, ②, ③
Haveg® 41	–	phenol resin + asbestos	①, ②
Rubber (up to 338 K (65 °C))	–	–	①, ⑥ (⑥ up to 302 K (29 °C))
PVC (up to 338 K (65 °C))	–	PVC	①, ⑥ (⑥ up to 302 K (29 °C))
Polyesters	–	UP	①, ②, ⑤ (①, ② up to 65 %)
Epoxy	–	EP	①, ②
Polyethylene (up to 333 K (60 °C))	–	PE	①
Glass	–	glass	①, ②, ③, ④, ⑤, ⑥, ⑦
Cl-polyether	–	chlorinated polyether	①, ②
Saran® (up to 296 K (23 °C))	–	PVDC	①
Ni-O-Nel® Alloy 825	2.4858	NiCr21Mo	①, ②, ③, ⑤, ⑥

† old designation; 1) attacked in the presence of fluorine compounds; 2) air-free solutions; 3) according to DIN EN 12165 (04/1998)

Table 153: Overview of materials which can be used in thermal phosphoric acid in the ranges identified in Figure 107 – the isocorrosion curves approximately correspond to a corrosion rate of 0.5 mm/a (19.7 mpy) [1050]

Temperature K (°C)	H₃PO₄ %	Corrosion rate, mm/a (mpy)				Comments
		Air access		Air exclusion		
		Agitated	Static	Agitated	Static	
288–294 (15–21)	20				0.1 (3.9)	chemically pure
	25	8.8 (346)			0.4 (15.8)	industrially pure
	40				0.14 (5.5)	chemically pure
	42		0.4 (15.8)	0.4 (15.8)	0.4 (15.8)	industrially pure
	42		1.2 (47.2)			commercial
	42	40 (1,575)				impure
	60				0.03 (1.2)	pure
	75				0.1–1 (4–40)	industrially pure
	76		0.2 (7.9)	0.3 (11.8)	0.18 (7.1)	industrially pure [1051, 1053]
	85				0.12 (4.7)	industrially pure
323 (50)	20				0.09 (3.5)	industrially pure
	40				0.11 (4.3)	industrially pure
	42.3		0.5 (19.7)		0.2 (7.9)	industrially pure
	60				0.035 (1.9)	industrially pure
348 (75)	10				2.0 (78.7)	industrially pure [1041]
	20				0.4 (15.8)	industrially pure
	25				1.7 (66.9)	industrially pure [1041]
	40				0.1 (3.9)	industrially pure
	60				0.08 (3.2)	industrially pure
	75				0.1–1 (3.94–39.4)	industrially pure
368 (95)	25		8.3 (327)		0.42 (16.5)	commercial
	42		1–3 (40–118)		<1 (<39.4)	commercial
	76		ca. 1 (40)		0.1 (3.94)	commercial
	85				0.1–1 (3.94–39.4)	commercial
523 (250)	85				1.5 (59)	commercial

Table 154: Comparison of the corrosion behavior of copper (99.9%, CR024A) in phosphoric acid of various concentrations, temperatures and purities [1041, 1048, 1051, 1053]

In summary, it can be said that copper shows an adequate to good corrosion resistance in pure, non-aerated phosphoric acid, especially at temperatures of about room temperature and below. The sometimes very low corrosion rates in Table 154 should be checked without fail for use in a practical case.

Even with sometimes very low copper corrosion rates, copper materials are to be excluded from cases where biochemical reactions are to proceed.

Corrosion studies in 85 % thermal phosphoric acid (non-aerated) showed corrosion rates of between 3.13 and 4 mm/a (123 and 158 mpy) on various grades of copper at 431 K (158 °C). The test duration here was 500 h. The corrosion rates in this acid at 353 K (80 °C) were between 0.6 and 2.3 mm/a (23.6 and 90.6 mpy). In aerated acid at 431 K (158 °C) the corrosion rate rose as expected to 24–27 mm/a (945–1,063 mpy). Under these conditions, the test duration was only 24 h [1052].

Although of little interest for industrial use, Figure 108 shows the very high weight loss per unit area of copper in 100 % phosphoric acid above 373 K (100 °C). The corrosion rate is accordingly about 10 mm/a (400 mpy) at 373 K (100 °C) 90 mm/a (3,550 mpy) at 473 K (200 °C) and 600 mm/a (23,600 mpy) at 573 K (300 °C) [1054]. The test durations were only 1–4 h. No linear corrosion course was found within this period [1054].

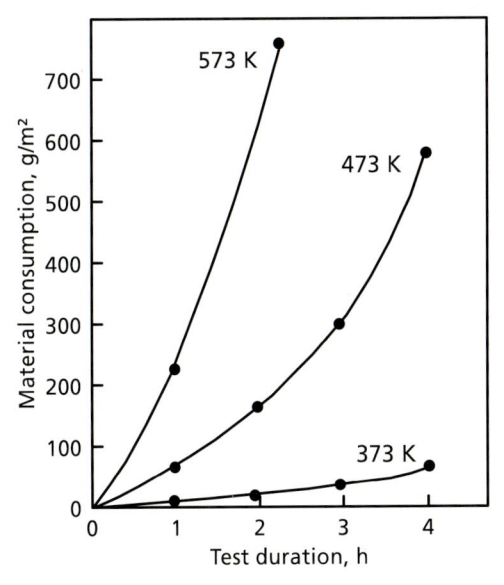

Figure 108: Corrosion behavior of copper in 100 % strength phosphoric acid at various temperatures [1054]

Fluorine compounds in particular in phosphoric acid have a corrosion-intensifying effect on copper materials (see Figure 107). Table 155 shows, without comparison with pure acid, the corrosion behavior of copper materials, copper, Monel® 400 and unalloyed steel in fluorine-containing acid (no precise data). Further details are also unknown, and it can only be expected that the test conditions were the same in

Material		3 to 14% H_3PO_4[1]	70 to 80% H_3PO_4[2]	70 to 80% H_3PO_4[3]	75 to 80% H_3PO_4[4]	85 to 95% H_3PO_4[5]	85 to 95% H_3PO_4[6]	Phosphorus[7]	H_3PO_4 vapor[8]
		Corrosion rate, mm/a (mpy)							
Beryllium bronze CuBe2 (2.02% Be, 0.21% Ni)	CW101C* C17200 (2.1247†)	0.05 (2.0)	0.21 (8.27)	14.2 (559)	0.16 (6.30)	1.50 (59.1)	0.15 (5.91)	0.03 (1.18)	1.10 (43.3)
Phosphorus bronze CuSn8P	CW459K* C52100 (2.1030†)	0.05 (1.97)	0.29 (11.4)	20.8 (819)	0.17 (6.69)	2.91 (114.6)	0.22 (8.66)	0.03 (1.18)	2.26 (89.0)
Copper (99.9% Cu)	CW024A** (2.0090†)	0.02 (0.79)	0.34 (13.4)	16.6 (654)	0.25 (9.84)	2.11 (83.1)	0.56 (22.0)	0.03 (1.18)	1.64 (64.6)
Aluminium bronze CuAl5As	CW300G (2.0918†)	–	–	–	6.90 (272)	–	–	–	8.10 (319)
Brass CW507L	CuZn36, UNS C27000 (2.0335†)	0.11 (4.33)	0.60 (23.6)	2.45 (96.5)	0.45 (17.7)	3.40 (134)	0.50 (19.7)	0.02 (0.79)	1.35 (53.1)
Cupronickel (20% Ni, 5% Zn)		0.07 (2.76)	0.20 (7.87)	18.17 (715)	0.61 (24.0)	1.13 (44.5)	0.18 (7.09)	0.02 (0.79)	1.13 (44.5)
Silicon bronze (3.1% Si, 1.1% Mn)		0.12 (4.72)	0.17 (6.69)	1.75 (68.9)	0.13 (5.12)	1.10 (43.3)	0.14 (5.51)	0.03 (1.18)	0.80 (31.5)
Monel® 400	2.4360	0.07 (2.76)	1.51 (59.4)	–	0.61 (24.0)	1.90 (74.8)	0.35 (13.8)	0.02 (0.79)	1.12 (44.1)
Unalloyed steel (0.24% Cu)	cf. 1.8962 cf. A 588	0.80 (31.5)	completely unsuitable	–	0.09 (3.54)	–	0.21 (8.27)	0.10 (3.94)	0.92 (36.2)

* according to DIN EN 12163 (04/1998); ** according to DIN EN 12165 (04/1998); † old designation
1) 339 K (66 °C) with small amounts of fluorine compounds; 2) 358–373 K (85–100 °C) with small amounts of fluorine compounds; 3) 368–383 K (95–110 °C) dripping acid, small amounts of fluorine compounds in an H_3PO_4 mist; 4) 348 K (75 °C); 5) 373–388 K (100–115 °C) dripping acid, small amounts of fluorine compounds in an H_3PO_4 mist; 6) 348–358 K (75–85 °C) with small amounts of fluorine compounds; 7) elemental phosphorus, stored at 338–343 K (65–70 °C); 8) 358–373 K (85–100 °C) steam with small amounts of phosphoric acid mist and traces of fluorine compounds

Table 155: Comparison of the corrosion behavior (corrosion rates in mm/a (mpy), converted from g/m²d) of copper materials, unalloyed steel and Monel® 400 in phosphoric acid, phosphorus and phosphoric acid vapor contaminated with fluorine [1055]

all cases, and pure phosphoric acid was probably used. The corrosion rates listed are consequently to be used only as comparison values for preselecting a material, and precise values should be determined under the particular operating conditions.

Copper-aluminium alloys

The corrosion behavior of aluminium bronzes is similar to that of copper in phosphoric acid. Increasing corrosive attack is caused by impurities in the acid and increasing temperature. Correct heat treatment (largely homogeneous structure) is a basic prerequisite for a good to adequate resistance of these materials. Pitting corrosion (in H_3PO_4) can never be excluded on inhomogeneous materials. The possible uses, if any, lie in regions 1 and 6 in Figure 107 [1041, 1043, 1045, 1048, 1050].

The best resistance of these materials, as with copper itself, is achieved in non-aerated, generally oxidant-free phosphoric acid. Figure 109 gives an overview of the corrosion rates to be expected [1049].

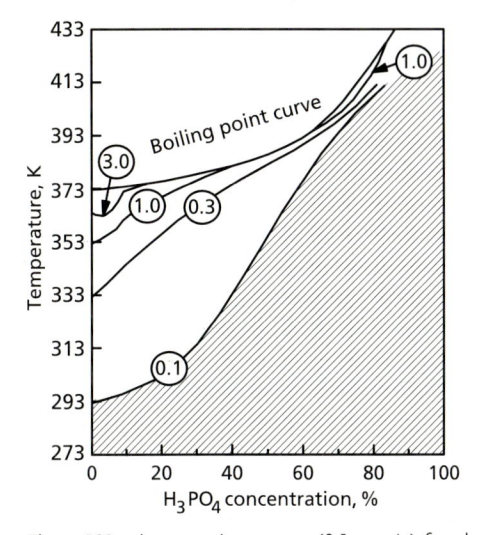

Figure 109: Isocorrosion curves (0.1 mm/a) for aluminium bronze (90% Cu, 7% Al, 3% Fe) in static, pure phosphoric acid solution, non-aerated, with air access on the test solution surface [26]
0.1 mm/a (3.94 mpy) 1.0 mm/a (39.4 mpy)
0.3 mm/a (11.8 mpy) 3.0 mm/a (118 mpy)

The corrosion rates were determined in pure acid solution (completely immersed) with access of air to the surface of the test solution. The test medium or the specimens were not agitated. As can be seen from this graph, the corrosion resistance of the aluminium bronze with 90% Cu, 7% Al and 3% Fe is considerably better than that of pure copper under these conditions. The use range is shifted to higher temperatures, in particular at high concentrations, and the 0.1 mm/a (3.94 mpy) isocorrosion curve is shifted to lower concentrations (compare with copper in Figure 106). At temperatures above 353 K (80 °C) only limited use is possible.

A number of corrosion rates under various conditions are summarized in [1041, 1048]. Table 156 contains an evaluating and summarizing compilation of corrosion rates for aluminium bronzes. These corrosion rates will very probably be achievable in practice. In-house studies are always necessary, however, in working acids.

The aluminium bronze with 5–6.5 % Al (UNS C60800), like other representatives of this alloy group, shows more severe attack by non-oxidizing acids, as in phosphoric acid here, as the content of oxidizing impurities increases, and in particular at temperatures above 328 K (55 °C) [1056].

The resistance of aluminium bronzes, and also toward phosphoric acid, can be modified and improved by added alloying elements. Addition of 0.5 % Cr increases the resistance to 80 % phosphoric acid at 408–413 K (135–140 °C) [1057]. No further information is given here.

The influence which alloying elements in these alloys have on corrosion resistance is shown in Figure 110. In particular, the alloying elements Si and Mn increase the resistance to sulfuric and phosphoric acid [1058]. The modification or improvement in corrosion resistance is detectable on the one hand by decreasing corrosion rates (Table 157), and electrochemically by the increase in potential to more positive values [1058].

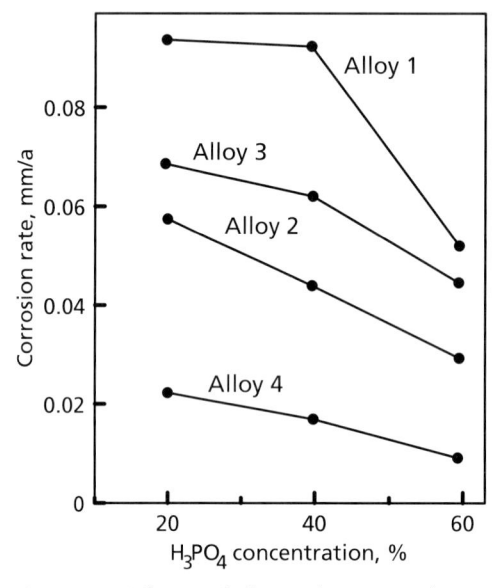

Figure 110: Influence of alloying elements on the corrosion behavior of aluminium bronzes (see Table 157) at room temperature in phosphoric acid solutions [1058]

It can be seen from these two illustrations that alloy 4 with 2 % Si and 0.2 % Mn has the best resistance, if resistance can even be referred to at the low corrosion rates. The influence of Si is greater (alloy 2) than that of Mn (alloy 3), although here the amount also plays a role.

Alloy	DIN-Mat. No.[1]	H_3PO_4 %	Temperature K (°C)	Corrosion rate mm/a (mpy)	Literature
CuAl5As	CW300G 2.0918†	7	293 (20)	0.08 (3.15)	[1048]
CuAl8	2.0920†	10	353 (80)	2.9 (114)	[1055]
		25	368 (95)	22.0 (866) (aerated)	[1041]
		25	369 (96)	8.2 (323)	[1055, 1059]
		60–75	423 (150)	0.12 (4.72)	[1055]
		80	413 (140)	0.12 (4.72)	[1055]
		conc.	293–303 (20–30)	0.18 (7.09)	[1055]
		conc.	353 (80)	0.12 (4.72)	[1055]
CuAl10Fe3Mn2	CW306G 2.0936†	1, 5, 10, 20	295 (22)	0.05 (1.97)	[1055]
		1, 5, 10, 20	323 (50)	0.15 (5.91) (max.)	[1055]
		20	288 (15)	0.06 (2.36)	[1041]
		20	323 (50)	0.10 (3.94)	[1041]
		20	348 (75)	0.25 (9.84)	[1041]
		60	288 (15)	0.01 (0.39)	[1041]
		60	323 (50)	0.01 (0.39)	[1041]
		60	348 (75)	0.00 (0)	[1041]
		60	boiling	0.25 (9.84)	[1041, 1059]
		80	363 (90)	0.05 (1.97)	[1041]

† old designation
1) Since the composition sometimes deviates, the corresponding DIN-Mat. No. has been chosen

Table 156: Summary of the corrosion rates of aluminium bronzes in phosphoric acid of varying concentration and temperature under various conditions

H_3PO_4 concentration %	Alloy 1[1]	Alloy 2[1]	Alloy 3[1]	Alloy 4[1]
	Corrosion rate, mm/a (mpy)			
20	0.09 (3.54)	0.06 (2.36)	0.07 (2.76)	0.022 (0.87)
40	0.08 (3.15)	0.04 (1.57)	0.06 (2.36)	0.017 (0.67)
60	0.05 (1.97)	0.03 (1.18)	0.04 (1.57)	0.007 (0.27)

1) chemical composition, weight percent

	Cu	Al	Si	Mn
Alloy 1	94.0	6.0	–	–
Alloy 2	92.0	6.0	2.0	–
Alloy 3	93.8	6.0	–	0.2
Alloy 4	91.8	6.0	2.0	0.2

Table 157: Influence of alloying elements on the corrosion rates of aluminium bronzes in phosphoric acid (chemically pure, aerated and stirred, test duration 100 h) at room temperature [1058]

These comments underline the need for corrosion studies with the corresponding material in the particular acid.

Phosphoric acid is used as a pickling agent by itself or in combination with sulfuric acid. Table 158 contains the corrosion rates to be expected on various metallic materials in such pickling solutions. In the solutions mentioned, the aluminium bronze is corroded at a rate of 2.9 and, respectively, 8 mm/a (114 and 315 mpy) [1060, 1061].

Designation		DIN-Mat. No.	Test 1[1]	Test 2[1]	Test 3[1]
			Corrosion rate, mm/a (mpy)		
AISI 304	X5CrNi18-10	1.4301	< 0.003[2] (0.12)	0.003 (0.12)	< 0.003 (0.12)
AISI 316	X5CrNiMo17-12-2	1.4401	0.003[2] (0.12)	0.003 (0.12)	0.003 (0.12)
AISI 317	X3CrNiMo18-12-3	1.4449	0.005 (0.20)	0.003 (0.12)	0.003 (0.12)
Hastelloy® C alloy	NiMo 16 Cr†	2.4883	0.033 (1.30)	0.064 (2.52)	0.025 (0.98)
High-purity lead		PB990R, PB985R	0.360 (14.17)	> 13.0[3] (512)	
UNS N06600	NiCr15Fe	2.4816	1.30 (51.18)	> 11.5[3] (453)	
UNS N04400	NiCu30Fe	2.4360	1.75 (68.9)	5.20 (205)	
Hastelloy® B alloy	NiMo30†	2.4810	> 2.35[3] (92.52)	7.80 (307)	
CW306G	CuAl10Fe3	2.0936†	2.90 (114)	8.00 (315)	

† old designation
1) Test conditions:

	Test 1	Test 2	Test 3
Medium:	40 % H_3PO_4, 5 % H_2SO_4, 0.25 % wetting agent, 0.05 % inhibitors	26 % H_3PO_4, wetting agent	26 % H_3PO_4, wetting agent
Test site:	specimens immersed in the pickling tank	specimens immersed in the pickling tank	specimens immersed in the pickling tank
Duration:	62 days	12.2 days	61 days
Temperature:	361–369 K (88–96 °C) (average: 366 K (93 °C))	355–366 K (82–93 °C) (average: 364 K (91 °C))	355–366 K (82–93 °C) (average: 364 K (91 °C))
Agitation:	slight	slight	slight
Aeration:	none	none	none

2) severe intercrystalline corrosion at the weld seams (the particular stabilized grades or those of low carbon should therefore be preferred)
3) specimens completely destroyed

Table 158: Corrosion rate of metallic materials in pickling solutions containing phosphoric acid under operating conditions [1060, 1061]

Copper-nickel alloys

The corrosion behavior of these alloys in phosphoric acid is generally comparable to that of copper.

A good to very good resistance is to be expected in air-free or oxidant-free acid, above all at about room temperature [1041, 1043, 1045, 1046, 1048, 1060–1063].

The copper-nickel alloys CuNi5Fe, CW352H (CuNi10Fe1Mn, 2.0872†), CuNi20Fe and CW354H (CuNi30Mn1Fe, 2.0882†) are reported as being of limited suitability or unsuitable as combining elements in 25% H_3PO_4 at 303 K (30 °C) and in 30–80% acid at 363 K (90 °C) [1063].

The corrosion rates are in each case stated as being up to 1.2 or more than 1.2 mm/a (47.2 mpy) [1063]. No further details are available in this context here.

As on copper, heavily contaminated acid and its solutions have an increasingly corrosive action, especially at high temperatures. Table 159 contains a summary of corrosion rates for copper-nickel alloys in phosphoric acid. According to [1043], copper-nickel alloys are attacked to the extent of more than 1.27 mm/a (50 mpy) in the entire concentration range under aeration and at 283 K (10 °C). In non-aerated acid, the corrosion rates are lower. The corrosion rate decreases as the concentration increases. According to this reference, corrosion rates of more than 1.27 mm/a (50 mpy) in 20% acid at 366 K (93 °C) 0.51–1.27 mm/a (20.1–50 mpy) at 283 K (10 °C) in 10% acid and at 339 K (66 °C) in 20 and 80% acid and 0.05–0.51 mm/a (1.97–19.7 mpy) at 339 K (66 °C) in 10 and 90% acid and between 283 and 339 K (10 and 66 °C) in 40% acid are reported. CuNi alloys corrode in 60% acid at 283–339 K (10 and 66 °C) and at 311 K (38 °C) in 80% phosphoric acid to the extent of less than 0.05 mm/a (1.97 mpy) [1043]. (Compare also Figures 106, 107, 109 and Table 155).

In non-agitated phosphoric acid, pitting corrosion (in H_3PO_4) may occur on these alloys under deposits caused by corrosion [1041].

Of the copper-nickel alloys, only the high-copper alloy CW354H (CuNi30Mn1Fe, 2.0882†) will be used in plants in contact with phosphoric acid [1060, 1061].

Copper-tin alloys (bronzes)

The corrosion resistance of copper-tin alloys toward phosphoric acid is comparable to that of copper [1041–1046, 1048, 1059].

The corrosion rates of these copper alloys also vary to a greater or lesser degree with the purity of the acid and the temperature. Deposits of corrosion products increase the corrosion rates and may lead to pitting corrosion (in H_3PO_4) [1041].

Copper-tin alloys are just as vigorously attacked as copper in phosphoric acid (see under section Copper) [1043]. Bronzes cannot be used as shut-off elements in 30% H_3PO_4 at 313 K (40 °C) [1064]. Table 160 contains a summary of corrosion rates of copper-tin alloys in phosphoric acid.

Phosphoric Acid

Table 159: Corrosion rates of copper-nickel alloys in phosphoric acid

Alloy	H$_3$PO$_4$ %	Temperature, K (°C)											Literature
		288 (15)	RT	323 (50)	328 (55)	333 (60)	339 (66)	348 (75)	368 (95)	373 (100)	393 (120)	453 (180)	
		Corrosion rate, mm/a (mpy)											
CuNi5Sn5	25								33.0 (1,299)				[1055]
CuNi14.5Al2.5	20	0.61 (24.0)		0.17 (6.69)				0.58 (22.8)					[1041, 1048]
	40	0.03 (1.18)						0.11 (4.33)					[1041, 1048]
	60	0.01 (0.4)						0.03 (1.18)					[1041, 1048]
CuNi16Al5.5Fe3.7	25								44.0 (1.732)				[1041, 1055]
CuNi20Zn5	1–14						0.07 (2.76)						[1055]
	70–80									0.20 (7.87)			[1055]
	116						0.004 (0.16)						[1055]
CW354H (CuNi30Mn1Fe, 2.0882†)	116					0.003 (0.12)					0.052 (2.05)	0.67 (26.4)	[1060, 1061]
						0.025$^{1)}$ (0.98)					0.29$^{1)}$ (11.4)	2.02$^{1)}$ (79.5)	
Cu alloys with 10, 20 and 25 % Ni	8.4		0.6 (23.6)										[1041]
Cu alloys with 33 and 45 % Ni	8.4		< 0.5 (< 20)										[1041]

† old designation; 1) aerated   RT = room temperature

Since it cannot always be clearly seen from the literature what conditions existed, it is rather to be assumed that pure phosphoric acid was used, and in particular non-aerated, if the corrosion rates are unexpectedly low. According to [1043], the corrosion rates are above 1.27 mm/a (50 mpy) at 283 K (10 °C) in aerated H_3PO_4. See also Figure 106, 107 and 109 for the possible applications.

Alloy	Temperature K (°C)	H_3PO_4 %	Air access		Air exclusion		Literature
			Agitated	Static	Agitated	Static	
			Corrosion rate, mm/a (mpy)				
CuSn5Ni5	368 (95)	25[1]		32 (1,260)			[1041]
CW452K (CuSn6, 2.1020[†])	293 (20)	5[1]	0.9 (35.4)	0.3 (11.8)	0.3 (11.8)	0.1 (3.94)	[1048]
	293 (20)	20[1]	3.4 (134)	0.3 (11.8)	0.5 (19.7)	0.2 (7.87)	[1048]
G-CuSn6ZnPb (cf. 2.1096.01[†])	RT	5[2]	0.63–1.25 (24.8–49.2)				[1059]
CW459K (CuSn8, 2.1030[†])	458 (185)	1[1]			0.02[3] (0.79)		[1055]
	339 (66)	3–14[4]			0.05[2] (1.97)		[1055]
	373 (100)	70–80[4]			0.23[2] (9.06)		[1055]
	333 (60)	80[1]			0.01 (0.39)	< 0.01 (< 0.39)	[1041, 1048]
	383 (110)	80[1]			0.21 (8.27)		[1041, 1048]
	393 (120)	80[1]			0.6 (23.6)		[1041, 1048]
	400 (127)	90[1]	unsuitable[2]				[1055]
CuSn10[†]	288 (15)	6.7[1]	0.8 (31.5)	0.3 (11.8)		0.1 (0.4)	[1048]
G-CuSn10Zn	RT	5[2]	0.63–1.25 (24.8–49.2)				[1059]

[†] old designation
RT = room temperature
1) pure or industrially pure
2) no further details
3) under 1.13 MPa
4) contains fluorine

Table 160: Corrosion rates of copper-tin alloys in phosphoric acid solutions

Copper-tin-zinc alloys (red brass)

Copper-tin-zinc alloys correspond to copper in their corrosion behavior, but are more similar to the copper-tin alloys. The probability of precipitation in these multi-component alloys is also greater than in two-component alloys for example, which in turn can lead to local corrosion.

As is also the case with other copper materials, the lowest corrosive attack is to be expected in oxidant-free solutions (for example air-free), and here in turn in pure acid.

Without further information on the acid, corrosion rates of 1.2 mm/a (47.2 mpy) are reported for completely immersed specimens of the alloys CuSn7Zn4, CuSn10Zn4 and CuZn9Sn6 in 5% H_3PO_4 at room temperature. One exception is the alloy CuSn10Zn7Pb1, which has a corrosion rate under these conditions of only 0.63 mm/a (24.8 mpy) [1055, 1059]. These corrosion rates are hardly acceptable for prolonged use. Corrosion rates of 1.3 mm/a (51.2 mpy) at 328 K (55 °C) 3.4 mm/a (133.9 mpy) at 348 K (75 °C) and 8.2 mm/a (322.8 mpy) at 373 K (100 °C) are reported in 85% phosphoric acid produced by the thermal process [1055].

Because of these high corrosion rates, corrosion studies under the actual conditions are advisable in all cases if these alloys are used, and air ingress and a content of oxidizing agents in the solution should be excluded.

Copper-zinc alloys (brass)

Copper-zinc alloys are also similar to copper in their corrosion behavior in phosphoric acid (see section Copper) [1041, 1043, 1045, 1048].

The best resistance is to be expected in pure acid with exclusion of air and at low temperatures. The resistance, of course, depends on the particular alloy composition. Table 161 contains a summary of corrosion rates for brass. These data show that corrosion studies on the particular alloy under concrete conditions are necessary if the alloy is used.

According to [1043], brasses corrode at a rate of more than 1.27 mm/a (50 mpy) in phosphoric acid solutions aerated at 283 K (10 °C).

According to [1059], copper alloys are also of only little use in phosphoric acid solutions. The sometimes very vigorous corrosion in thermal phosphoric acid at high temperatures, and above all the dependence of this on the aeration, is shown in Table 162. This table shows that aeration excludes the use of these materials, as is general for copper materials.

The influence of the production process and aeration on the corrosion of CW502L (CuZn15, UNS C23000) in 85% phosphoric acid is shown in Table 163.

Alloy	Temperature K (°C)	H$_3$PO$_4$ %	Air access		Air exclusion	
			Agitated	Static	Agitated	Static
			Corrosion rate, mm/a (mpy)			
CW505L (CuZn30, UNS C26000) 2.0265†	288 (15)	20			0.27 (10.6)	0.1 (3.94)
	293 (20)	4	0.4 (15.8)	0.21 (8.27)		0.1 (3.94)
	323 (50)	20			0.14 (5.51)	0.1 (3.94)
	348 (75)	20			0.27 (10.6)	0.1 (3.94)
	288 (15)	40			0.11 (4.33)	< 0.1 (< 3.94)
	323 (50)	40			0.1 (3.94)	< 0.1 (< 3.94)
	348 (75)	40			0.09 (3.54)	< 0.1 (< 3.94)
	298 (25)	53	0.3 (11.8)	0.2 (7.87)		
	288 (15)	60			0.02 (0.79)	
	323 (50)	60			0.03 (1.18)	
	348 (75)	60			0.04 (1.57)	
CuZn10.5	369 (96)	25[2)]		0.6 (23.6)		
CuZn32.5	369 (96)	25[2)]		0.8 (31.5)		
CuZn23Mn3Fe2	369 (96)	25[2)]		0.8 (31.5)		
CuZn39Sn1	369 (96)	25[2)]		1.9 (74.8)		
Brass[1)]	328 (55)	85		0.9 (35.4)		
	348 (75)	85		3.0 (118)		
	373 (100)	85		12.5 (492)		
CuZn16Si4	293 (20)	5	1.2 (47.2)	0.5 (19.7)	0.6 (23.6)	0.2 (7.87)
	293 (20)	20	3.8 (150)	0.5 (19.7)	0.8 (31.5)	0.2 (7.87)

† old designation
1) no further information, possibly CW507L (CuZn36, UNS C27000)
2) produced by the wet process, aerated

Table 161: Summary of corrosion rates for copper-zinc alloys in industrially pure phosphoric acid [1041, 1048]

Phosphoric Acid

Alloy	Temperature K (°C)	H$_3$PO$_4$[1] %	Corrosion rate mm/a (mpy)		Comments
			Aerated	Non-aerated	
CW502L (CuZn15, UNS C23000)	366 (93)	1	6.3 (248)	1.7 (66.9)	
	458 (185)	1	0.04 (1.57)		1.13 MPa
	458 (185)	1		0.03 (1.18)	1.13 MPa + 5 % sugar
	366 (93)	3	7.8 (307)	3.3 (130)	
	366 (93)	5	20.5 (807)	7.0 (276)	
	366 (93)	10	16.0 (630)	3.6 (142)	
	366 (93)	20	35.0 (1,378)	1.5 (59.1)	
	366 (93)	40	38.8 (1,528)	1.3 (51.2)	
	366 (93)	60	23.3 (917)	0.83 (32.7)	
	377 (104)	85	15.8 (622)	0.53 (20.9)	
	386 (113)	85	47.5 (1,870)	0.88 (34.7)	
	397 (124)	85		2.5 (98.4)	
	422 (149)	85	138.0 (5,433)	2.1 (82.7)	
CW507L (CuZn36, UNS C27000)	327 (54)	70		0.06 (2.36)	
	366 (93)	78		0.28 (11.0)	
	386 (113)	85		0.15 (5.91)	
	397 (124)	85		0.03 (1.18)	
	400 (127)	90	unsuitable		
CW706R (CuZn28Sn1, UNS C44300)	400 (127)	90	unsuitable		

1) phosphoric acid from the thermal plants of the Tennessee Valley Authority
For phosphoric acid of 85 % the following analysis is valid:

P$_2$O$_5$	62.9 %	Al	0.04 %	As	0.003 %	SO$_3$	0.003 %
K	0.28 %	Pb	0.015 %	Mn	0.002 %	Cu	0.002 %
Ca	0.20 %	F	0.024 %	Si	0.01 %		
Na	0.08 %	Fe	0.01 %	C	0.003 %		

Table 162: Corrosion behavior of copper-zinc alloys in thermal phosphoric acid[1] at high temperatures and various concentrations [1055]

Aeration c.f.m.[1]	Thermal phosphoric acid		Wet phosphoric acid	
	371 K (98 °C)	411 K (138 °C)	371 K (98 °C)	411 K (138 °C)
	Corrosion rate, mm/a (mpy)			
0.00	0.53 (20.9)	0.88 (34.7)	0.25 (9.84)	2.7 (106)
0.14	6.0 (236)	17.0 (669)	8.3 (327)	35.0 (1,378)
0.28	9.5 (374)	45.0 (1,772)	8.5 (335)	53.0 (2,086)
0.57	15.0 (591)	60.0 (2,362)	12.0 (472)	60.0 (2,362)
0.85	17.0 (669)	73.0 (2,874)	18.0 (709)	78.0 (3,071)
1.13	19.0 (748)	83.0 (3,268)	19.0 (748)	165.0 (6,496)

1) c.f.m. = cubic feet per minute; 1 c.f. = 28.3 l

Table 163: Comparison of the corrosion behavior of CW502L (CuZn15, UNS C23000) in 85 % H_3PO_4 as a function of the production process and the aeration [1055]

According to this table, as the aeration increases, the corrosion rate increases in both acids. While the corrosion rates at 371 and 411 K (98 and 138 °C) virtually correspond to one another, they differ considerably at 1.13 l/min and 411 K (138 °C). The corrosion rate in wet acid is accordingly about twice as high [1055].

Additions of fluorides as K_2SiF_6, Na_2SiF_6, KF, NaF, HF or H_2SiF_6 to 85 % H_3PO_4 at 366 K (93 °C) have a corrosion-promoting effect, in particular in amounts of 0.085–0.1 %. Corrosion rates of at least 1.4 mm/a (55.1 mpy) are reported for CW502L (CuZn15, UNS C23000) without aeration, and up to 7 mm/a (276 mpy) with aeration [1055].

Corrosion on brass can be reduced by a number of inhibitors [1064, 1065]. An inhibition efficiency of 99 % is achieved by addition of 0.001 mol/l (0.169 g/l) 1,2-naphthotriazole in 0.03 mol/l H_3PO_4 solution at room temperature [1065].

Table 164 contains a summary of the inhibition efficiency of quinoline (C_9H_7N) and $SnCl_2$ in phosphoric acid solutions.
As can be seen from this summary, corrosion can be prevented completely under certain conditions [1055, 1066].

Brass Ms60 (CuZn40, CW509L, 2.0360†) is attacked less in a conductive combination with cast iron in phosphoric acid solutions than brass by itself, as can be seen from Table 165 [1055, 1066]. It behaves cathodically in this combination.

In addition to this finding, the fact that although brass is the cathode in this corrosion system, it nevertheless also corrodes itself, is of importance. This fact should never be ignored in practice. In steam containing small amounts of phosphoric acid mist and fluorine compounds, corrosion rates of more than 1 mm/a (39.37 mpy) are to be expected at 358–373 K (85–100 °C) on copper materials, as is the case here with brass. In steam containing 13 % P_2O_5 and 250 ppm F_2 at a steam flow rate of 9 m/s at 413–423 K (140–150 °C) the corrosion rate for CW502L (CuZn15, UNS C23000) is 0.88 mm/a (34.7 mpy) [1055].

H_3PO_4 mol/l	Additive	Inhibition efficiency, %			
		After 1 d	3 d	7 d	14 d
0.32	2 % quinoline	15	22	33	43
0.32	5 % $SnCl_2$	100	100	100	100
0.32	2 % quinoline + 5 % $SnCl_2$	100	100	100	100
0.63	2 % quinoline	7	20	45	49
0.63	5 % $SnCl_2$	50	66	85	87
0.63	2 % quinoline + 5 % $SnCl_2$	100	100	100	100
0.93	2 % quinoline	31	17	38	44
0.93	5 % $SnCl_2$	64	68	67	74
0.93	2 % quinoline + 5 % $SnCl_2$	70	68	82	93

Table 164: Efficiency of inhibitors on corrosion of Ms60 (CuZn40, CW509L, 2.0360†) in various concentrated phosphoric acid solutions at room temperature [1055, 1066]

H_3PO_4 %	Brass in contact with cast iron	Brass alone
	Corrosion rate, mm/a (mpy)	
0.17	0.004 (0.16)	0.115 (4.53)
0.33	0.006 (0.24)	0.121 (4.76)
0.67	0.007 (0.28)	0.183 (7.2)

Table 165: Corrosion rates for brass Ms60 (CuZn40, CW509L, 2.0360†) in contact with cast iron at room temperature, test duration 6 days [1055, 1066]

Brass CW507L (CuZn36, UNS C27000) is corroded in phosphoric acid containing 84 % P_2O_5 at 333 K (60 °C) at a rate of 0.008 mm/a (0.31 mpy) (non-aerated) and 0.024 mm/a (0.94 mpy) (aerated). At 393 K (120 °C) the corrosion rate is 0.36 mm/a (14.17 mpy) (aerated), and at 453 K (180 °C) 0.48 mm/a (18.9 mpy) (non-aerated) [1055]. Other copper materials have corrosion rates of the same order of magnitude, apart from the alloy CuZn23Al6Mn4 (2.0500) at 453 K (180 °C) with a rate of 4.0 mm/a (157 mpy).

Other copper alloys

These alloys also exhibit a comparable corrosion behavior to copper and the other alloys described above under exposure to phosphoric acid. Here also, aeration and oxidizing agents such as fluorides above all have a deteriorating effect on the resistance. In this context, see, for example, Tables 162 and 163.

In ethanol synthesis plants, the alloy CuMn2 (2.1363†) exhibited an attractive corrosion resistance when exposed to hot phosphoric acid [1067]. Corrosion data in this context are not available in this reference.

The hardenable spring alloys CuBe2 (CW101C, UNS C17200, 2.1247†) and CuCoBe (CW104C, 2.1285†) are corroded at a rate of up to 0.1 mm/a (3.94 mpy) in 10 and 89% H_3PO_4 at 293 K (20 °C). At 373 K (100 °C) the corrosion rates for CW101C in 10% acid and for 2.1285 in 10 and 89% acid are between 1 and 10 mm/a (39.4 and 394 mpy). The corrosion rate for CW101C in 89% acid at 373 K (100 °C) is reported as being between 0.1 and 1 mm/a (3.94–39.4 mpy) [1068].

No details on the corrosion conditions or the purity of the acid and the aeration, for example, are available in this reference.

Table 166 contains a summary of the corrosion rates for silicon bronzes.

Alloy	Temperature K (°C)	H_3PO_4 %	Air access		Air exclusion		Literature
			Agitated	Static	Agitated	Static	
			Corrosion rate, mm/a (mpy)				
CuSi3Mn	339 (66)	3–14[1]		0.13 (5.12)			[1055]
	288 (15)	20		0.14 (5.51)		< 0.1 (< 3.94)	[1048]
	293 (20)	24	0.23 (9.06)		0.10 (3.94)	< 0.1 (< 3.94)	[1048]
	323 (50)	20		0.05 (1.97)			[1048]
	348 (75)	20		0.36 (14.2)	0.2 (7.87)		[1048]
	368 (95)	25	35.0 (1,378)				[1041]
	288 (15)	40		0.035 (1.38)			[1048]
	288 (15)	40		0.40 (15.8)			[1041]
	323 (50)	40		0.05 (1.97)			[1041, 1048]
	348 (75)	40		0.17 (6.69)			[1041, 1048]
	288 (15)	60		0.003 (0.12)		< 0.002 (< 0.08)	[1041, 1048]
	323 (50)	60		0.01 (0.39)			[1041, 1048]
	348 (75)	60		0.07 (2.76)			[1041, 1048]
	373 (100)	70–80[1]		0.18 (7.09)			[1055]

1) plant studies, no further information

Table 166: Corrosion rates for CuSi alloys in phosphoric acid solutions under various conditions

Table 166: Continued

Alloy	Temperature K (°C)	H$_3$PO$_4$ %	Air access		Air exclusion		Literature
			Agitated	Static	Agitated	Static	
			Corrosion rate, mm/a (mpy)				
	333 (60)	84		0.03 (1.18)		0.005 (0.20)	[1048]
	393 (120)	84		0.27 (10.6)		0.05 (1.97)	[1048]
	453 (180)	84		1.9 (74.8)		0.21 (8.27)	[1048]
	397 (124)	85				0.58 (22.8)	[1055]
	400 (127)	90	unsuitable				[1055]
	333 (60)	116[1]		0.03 (1.18)		0.004 (0.16)	[1055]
	393 (120)	116[1]		0.28 (11.0)		0.04 (1.57)	[1055]
	453 (180)	116[1]		1.76 (69.2)		0.25 (9.84)	[1055]
CuSi2Mn	339 (66)	3–14[1]			0.06 (2.36)		[1055]
	373 (100)	70–80[1]			0.24 (9.45)		[1055]

1) plant studies, no further information

Table 166: Corrosion rates for CuSi alloys in phosphoric acid solutions under various conditions

For all the copper materials within this section, corrosion studies should also be carried out under the particular conditions before they are used, and above all if oxidizing agents, fluorides and contaminated acid are present. The data in the tables should serve only as a guideline, and more so than in other cases, since the resistance of these materials in phosphoric acid depends very greatly on the conditions.

Corrosion rates of greater than or equal to 1 mm/a (39.4 mpy) are to be expected in steam containing small amounts of phosphoric acid mist and fluorine compounds at temperatures of between 358 and 373 K (85 and 100 °C). The corrosion rates are of the same order of magnitude in steam at 413–423 K (140–150 °C) containing 250 ppm F$_2$ and 13 % P$_2$O$_5$, at a steam flow rate of about 9 m/s [1055].

Polyols

Author: G. Elsner / Editor: R. Bender

Copper and copper alloys

Copper and copper alloys exhibit good resistance to ethylene glycol, its homologues, polyglycols and their aqueous solutions up to their boiling points [1069, 1070].

In coolants of ethylene glycol and/or mixtures of ethylene glycol and water, copper is not attacked substantially differently at room temperature than in pure water; the decisive factor is the oxygen content, while aeration exacerbates corrosion [1071].

In the presence of air, autoxidation in glycols produces acids which slightly attack copper and copper alloys, mainly at the phase boundary with air, which is why solar collectors may be increasingly corroded [1072, 1073].

The soldered joints of soldered brass radiators are particularly at hazard, although this can be largely suppressed by using inhibitors [1069].

As is to be expected, corrosion of copper and brass in ethylene glycol solutions containing hydrochloric acid increases with the acid concentration, but is inhibited by the addition of 10 % water. In contrast to brass, copper is subject to intergranular attack. With increasing temperature corrosion also decreases because of the decrease in the oxygen content, and this occurs with increasing hydrochloric acid concentration [1074].

Additions of ethylene glycol, glycerin, xylitol and sorbitol to 5 mol/l sodium hydroxide shift the potential of copper toward more negative values and therefore make passivation more difficult. On transition of the copper into the passive state, the alcohols are oxidized to form the corresponding aldehydes which react with Cu^{2+} ions, thus leading to decomposition of the cuprite/polyol solution and the precipitation of copper oxide [1075].

The really quite slight corrosion of copper and brass by glycols is completely suppressed by the addition of 3 % triethanolamine phosphate + 0.23 % mercaptobenzothiazole [1076, 1077]. It must be borne in mind that triethanolamine phosphate used alone increases the severity of corrosion of copper, which is why it is used together with sodium mercaptobenzothiazole; distilled water should not be used for this preparation, since the hardness-forming agents promote inhibition [1071, 1077].

Copper is precipitated on aluminium or iron surfaces from such copper-containing ethylene glycol solutions produced by corrosion of copper by triethanolamine phosphate but not in the case of anhydrous ethylene glycol, and then promotes corrosion of these metals [1071]. The effect of the copper complex compound can be eliminated by adding 0.2 to 0.3 % of the sodium salt of 2-mercaptobenzothiazole.

Mixtures of ethylene glycol, propylene glycol and water, in the proportion 63:33:5, inhibited with the alkaline arsenates K_2HAsO_4, Na_2HAsO_4, are said not to attack copper and brass [1078]. Sodium tellurite and sodium selenite are also said to inhibit corrosion of copper and copper alloys; a typical formulation contains sodium tellurite or selenite, borax, aliphatic carboxylic acids (C_9–C_{10}) or benzoic acid or catechin and EDTA, which forms complexes with calcium ions [1079].

Ethylene glycol antifreeze agents inhibited with tannin mixtures are also said to prevent corrosion of copper [1080].

Brass containing lead (%) Cu-3.92Pb-0.31Sn-0.31Fe-37.66Zn is said to be attacked neither by ethylene glycol solutions inhibited with triethanolamine phosphate + sodium mercaptobenzothiazole nor by ethylene glycol solutions inhibited with sodium benzoate + sodium nitrite [1081].

Negligible corrosion of copper and copper/brass soldered joints was found (10^{-8} mm/a) in a coolant consisting of a mixture of 91 % ethylene glycol, 5 % sodium benzoate, 0.05 % sodium nitrite, 1.53 % Na_2HPO_4, 1.45 % NaH_2PO_4 and 0.6 % water diluted with four times the quantity of water when tested in practical driving operation with a test duration of 222 days [1082].

Combinations of sodium cinnamate + benzotriazole provide excellent corrosion protection for copper and brass in 33 to 50 % ethylene glycol/water solutions [1083].

As was found from polarization measurements in investigations into the effect of inhibitors in antifreeze coolants containing ethylene glycol for automobile engines in apparatus simulating actual operating conditions, additions of borax (1; 1.5 %), triethanolamine (1.5 %), sodium chromate (0.1; 0.25 %), sodium silicate (0.8 %) and sodium benzoate (1.5 %) in 30 % ethylene glycol solution at 298 K (25 °C) primarily act as anodic inhibitors of corrosion of copper. The effect of 1 % borax, 0.25 % sodium chromate, 0.8 % sodium silicate, 1.5 % triethanolamine or 0.1 % mercaptobenzothiazole (this + 1 % borax) lasted 2,000 hours under laboratory conditions. With 1 % sodium nitrite in 30 % ethylene glycol solution, copper was severely corroded after 1,081 hours, and 1.5 % sodium benzoate in 30 % ethylene glycol solution was also not very good; 0.5 % hydroquinone in 30 % ethylene glycol even accelerates corrosion [1084]. The above-mentioned duration of protection of 2,000 hours would be equivalent to driving a distance of more than 100,000 km and therefore suffice for the entire service life of a vehicle in many cases.

0.15 to 0.25 % sodium tolyltriazole (COBARTEC® TT 100) in engine coolants containing glycol effectively protects copper and copper alloys, even against galvanic corrosion [1085].

Corrosion of copper in an antifreeze mixture consisting of 94.8 % ethylene glycol with 2.5 % triethanolamine phosphate and 2.5 % water can be decreased by more than 99 % by the addition of 0.2 % benzotriazole monoethanolamine [1086].

The corrosion rate of copper (Cu 122) is more than about 100 µm/a (3.94 mpy) and about 80 µm/a (3.15 mpy) in solar collectors operated with non-inhibited mixtures of equal volumes of ethylene glycol and water or propylene and water. Contact corrosion, but not crevice corrosion, is to be expected on contact with chromium steel (AISI 444, chromium steel 182Mo). Since glycol solutions become acidic by oxidation under operating conditions (373 K (100 °C)) in the presence of air, such systems for contact with copper must always contain neutralizing inhibitors [1087].

Simulated laboratory tests with solar heat-transfer liquids, including non-inhibited mixtures of 50 % ethylene glycol and 50 % water as well as 50 % propylene glycol and 50 % water (as per ASTM D 1384), showed, for example, that copper

(Cu 110) is completely dissolved with rapid fall in the pH-value when the two glycol solutions are aerated, whereas it merely tarnishes slightly in the solutions through which nitrogen is passed. When air is excluded, the corrosion rates of copper in both solutions (no temperatures stated, probably boiling temperature) are negligible 10^{-5} mm/a (3.94×10^{-4} mpy) [1088].

The risk of galvanic corrosion in aqueous heat-transfer media containing glycol when copper and aluminium are used in the system is largely eliminated by the addition of Preventol® Cl-2 [1089].

According to an American publication, phosphorus-deoxidized copper (UNS alloy C12200) in solar energy systems suffers generalized attack by heat-transfer media – a total of 26 American commercial products were examined – and in some cases suffers local attack after 6 months. Copper corrodes at the rate of 25 μm/a (0.98 mpy) in ethylene glycol solutions without inhibitors, and the pit depth may reach 100 μm. Since the inhibitors are consumed with time, they must be regularly replenished. Table 167 shows some of the results obtained in this investigation [1090].

Modern brake fluids usually consist of mixtures of glycols + glycol ethers + 0.1 to 5 % inhibitors. The most suitable materials for brake lines are phosphorus-deoxidized copper containing 0.024 % phosphorus, so-called copper iron CuFe containing 2.3 % Fe, and stainless steel SIS 2343 ((%) Fe-(16.6–19.0)Cr-(10.5–14.0)Ni-(2.5–3.0)Mo < 0.005C, similar to DIN-Material No. 1.4436). Although steel pipes brazed with copper are frequently used, they do not exhibit sufficient resistance in the salt-spray test [1091]. VOLVO, the Swedish automobile manufacturer, therefore uses brake lines of CuNi10Fe, since these are resistant to both brake fluids and de-icing salt [1092].

The corrosion rates of the Ampco® alloy ((%)Cu-(8–14)Al-(0–3)Fe) in ethylene glycol are below 0.15 mm/a (5.91 mpy) [1093].

Bronze (BROS-10-10) is recommended as material for systems with non-combustible hydraulic fluids, such as aqueous solution of 30 % polyethylene glycol and glycerin inhibited with 1.4 % disodium monohydrogen phosphate + 0.4 % benzotriazole [1094].

Corrosion of brass and other copper alloys by hydraulic fluids can be prevented by, for example, 1,2,3-benzotriazole or tolyltriazole [1095].

Copper, bronze and brass are to be regarded as resistant to glycerin at 294 K (21 °C) [1070]; brass is attacked, however, by boiling glycerin [1096].

Copper 99.9 lost in pure glycerin, for example, when immersed at the surface of a storage tank, 7.5×10^{-4} mm/a (0.03 mpy) at room temperature in 250 days; cast bronze ((%) Cu-5.0Pb-5.0Sn-5.0Zn) lost 0.001 mm/a (0.04 mpy); both these figures are negligible, provided product purity is not the decisive factor [1097].

Copper is also resistant to boiling glycerin and loses less than 8.0×10^{-4} mm/a (0.03 mpy) in glycerin containing 5 % water at 353 K (80 °C). It is therefore used for boilers, distillation apparatus, piping and heating coils; heating coils in evaporators, however, are corroded at the rate of 0.38 mm/a (15.0 mpy) by glycerin at 433 K (160 °C) (vigorous motion); in vacuum evaporators (slight motion) the corrosion rate at 423 K (150 °C) is about 0.08 mm/a (3.15 mpy) [1098].

Heat transfer medium	Material consumption mg/cm²			
	1 Month	2 Months	3 Months	4 Months
I	0.35	0.58	0.88	1.74
II	0.68	1.20	1.32	4.00
III	0.94	1.67	2.01	2.16
IV	0.57	(0.73)	0.70	0.76
V	0.86	2.10	5.36	12.84
VI	0.17	0.23	0.21	0.85
VII	0.18	0.25	0.44	0.84
VIII	0.31	0.34	0.30	0.44
IX	0.12	0.14	0.14	0.51
X	0.32	0.24	0.64	0.73
XI	0.80	1.25		2.86
I:	Dowtherm® SR-1 (Dow Chem. Co.) inhibited ethylene glycol for protection of multi-metal systems; test concentration 1:1 in NHPW (New Haven Potable Water: pH 7.0; 17.5 ppm Cl; 52 ppm $CaCO_3$; hardness 66 ppm $CaCO_3$; 93 ppm total dissolved solids as NaCl)			
II:	Drewsol® TM (Drew. Chem. Corp.) for protection of multi-metal systems; inhibited heat-transfer media; test concentration 100 %			
III:	Foodfreeze® 35 (Union Carbide Corp.) for protection of multi-metal systems; inhibited propylene glycol; test concentration 1:1 in NHPW			
IV:	Nutek® 835 (Nuclear Technology Corp.) for protection of aluminium; inhibited propylene glycol; test concentration 100 %			
V:	Propylene glycol USP (J. T. Baker Chem. Co.) non-inhibited propylene glycol; test concentration 1:1 NHPW			
VI:	PIPG phosphate-inhibited propylene glycol; test concentration 1:1 NHPW			
VII:	PIPG buffered with borax; test concentration 1:1 NHPW + 625 ppm sodium tolyltriazole			

Table 167: Material consumption of phosphorus-deoxidized copper (C12200) in commercial heat-transfer media ("Simulated Solar Service Test") [1090]

Table 167: Continued

Heat transfer medium	Material consumption mg/cm²			
	1 Month	2 Months	3 Months	4 Months
VIII:	PIPG buffered with borax; test concentration 1:1 NHPW + 2000 ppm sodium molybdate			
IX:	PIPG buffered with borax; test concentration 1:1 NHPW + 2000 ppm sodium molybdate + 625 sodium tolyltriazole			
X:	Prestone® II (Union Carbide Corp.) for protection of multi-metal systems; inhibited ethylene glycol; test concentration 1:1 NHPW			
XI:	Sun Sol® 60 (Sun Works Co.) for protection of copper; inhibited propylene glycol; test concentration 1:1 NHPW			

Table 167: Material consumption of phosphorus-deoxidized copper (C12200) in commercial heat-transfer media ("Simulated Solar Service Test") [1090]

Copper and copper alloys, with the exception of brass, are also resistant to mixtures of glycerin, sodium chloride and water at 389 K (116 °C) [1097].

Bronze and aluminium bronze, such as Corrix Metal® (%) Cu-8.7Al-3.1Fe, are resistant to boiling glycerin and behave in a similar way to copper [1096, 1098].

Corrosion of copper in glycerin solutions acidified with hydrochloric acid increases with increasing acid concentration [1099], and this should be taken into consideration when solutions containing glycerin are used in the foodstuffs industry [1100].

CuNi-alloys are reasonably resistant to glycerin up to its boiling point; for example, Cu30Ni in a vacuum evaporator containing aqueous glycerin at 423 K (150 °C) lost less than 0.06 mm/a (2.36 mpy) in the liquid phase and less than 0.12 mm/a (4.72 mpy) in the vapor phase [1098].

Bronze, containing no zinc, behaves like copper in glycerin, and is used for pumps and fittings and is suitable for the bodies of isolation valves for 100 % glycerin at 353 K (80 °C). Cast bronze loses, for example, about 0.001 mm/a (0.04 mpy) in glycerin containing 5 % water with a pH of 6 to 7 at 327 K (54 °C) [1098, 1101].

Red brass (Cu-alloy containing Sn and Zn) behaves like copper in glycerin and is used for pumps and fittings [1098].

Brass containing less than 30 % zinc behaves like copper in glycerin. The corrosion rate is less than 0.10 mm/a (3.94 mpy) at 423 K (150 °C) in vacuum evaporators for aqueous glycerin. Brass is also slightly attacked by cooling brines containing glycerin [1098].

CuSi-alloys behave like copper in glycerin [1098].

Copper in the liquid phase lost 0.23 mm/a (9.06 mpy), and in the vapor phase 0.40 mm/a (15.7 mpy), in an aqueous solution of initially 5 % glycerin + 10 %

sodium chloride and finally 80 % glycerin + 8 % sodium chloride, balance water, in a vacuum evaporator [1098].

Copper is used, inter alia, for vessels for the manufacture of glyptal and alkyd resins from glycerin and phthalic acid anhydride or copal acid at 573 to 673 K (300 to 400 °C) [1098].

A 40 % glycerin solution inhibited with 0.2 % sodium nitrite + 0.8 % sodium monohydrogen phosphate does not attack copper. This inhibitor mixture is said to be far more effective than the combination of triethanolamine + sodium mercapto-benzothiazole [1102]. An electrolyte consisting of (vol.%) 75 glycerin and 25 nitric acid at 293 K (20 °C) is said to be very suitable for electropolishing copper [1103], as is also an electrolyte consisting of (%) 35 to 70 phosphoric acid, 8–64 glycerin, 1–12 lactic acid, 5–10 glutamic acid and 5–30 water for copper alloys [1104].

Copper is also resistant to boiling sorbitol solutions in all concentrations; however, it is not often used for handling sorbitol because even traces of copper may have a disturbing effect in the cosmetics industry [1105].

Bronze containing no zinc is resistant to sorbitol solutions in all concentrations at room temperature [1105].

CuSi-alloys are resistant to sorbitol solutions in all concentrations and behave like copper [1105].

The reaction vessels about 18 m high and 0.6 to 0.8 m in diameter used for the manufacture of 1,4-butanediol and 1,3-butanediol in which 2-butyne-1,4-diol is hydrogenated to form 1,3-butanediol, are constructed either of stainless steel, unalloyed steel or low-alloy CrMo-steel with lining of electrolytic copper [1106–1109].

Potassium Chloride

Author: L. Hasenberg / Editor: R. Bender

Copper

Copper and its alloys show good resistance to aqueous solutions of potassium chloride. The corrosion rates are about 0.005 mm/a (0.20 mpy) in non-agitated and non-aerated solutions. The corrosive attack increases with aeration or agitation of the solutions.

The corrosion rate in potassium chloride solutions up to 20 % at temperatures up to 373 K (100 °C) is less than 0.05 mm/a (1.97 mpy). Stress corrosion cracking must be expected [1110].

Corrosion rates of 0.012 mm/a (0.47 mpy) were obtained at room temperature on test durations of 96 hours in saturated potassium chloride solution. They increased to 0.13 mm/a (5.12 mpy) after the same test duration when air is introduced and to 0.06 mm/a (2.36 mpy) without the introduction of air [1111]. Aqueous carnallite solutions when air-free cause a maximum corrosion rate of copper of 0.008 mm/a (0.31 mpy) [1112].

Table 168 provides the corrosion rates of copper and copper alloys.

On storage of one year copper sheets sprinkled with potassium chloride exhibited a dull surface with small punctiform, blue-green stains under the influence of atmospheric humidity [1113]. The same result was obtained by sprinkling with carnallite. The coating with this blue layer of corrosion products was irregular. Where the layer was dissolved, a dull red color was visible, which seemed to suggest a uniform corrosive attack. Under these conditions of storing potassium chloride and salts containing potassium chloride, pure copper is more suitable than the copper alloys [1113], with bronze having similar resistance.

In the potash industry, the dissolving liquors (concentrated solutions of potassium and sodium chlorides) are passed through pipelines made of copper and copper alloys for heating in the liquor preheaters in order to process the crude salts. The external heating is done by steam. Marked localized wear often leads to the breakdown of the whole pipe system or individual parts.

Results of corrosion tests on copper as a function of the concentration of the potassium chloride solutions at 368 K (95 °C) are shown in Figure 111.

The weight loss in Figure 111 shows a distinct dependence on the concentration of the solution. With increasing concentration of potassium chloride, the weight loss of copper increases and reaches a maximum at 15 % potassium chloride, then decreases again. The test duration for these experiments was 75 minutes. The corrosive medium was dripped on to the copper shavings [1114].

The corrosion rate as a function of the concentration in an agitated medium is shown in Figure 112. Copper samples attached to an agitator blade were agitated at 368–373 K (95–100 °C) in 1,000 cm^3 of liquid of varying concentrations. The surface of the samples was 27.5 cm^2 [1114]. The erratic values obtained with high concentrations are attributed to changes in temperature and above all to the influence of air, which increases the corrosion rate. A corrosion maximum in an agitated medium can also be deduced from Figure 112. It is located at 15–20 % potassium chloride solutions.

Medium	Temperature, K (°C)	G-CuSn10-14 (cf. CC480K, CC483K)	G-CuSn10Zn	G-CuSn5Znpb (cf. CC491K) G-CuSn7Znpb (cf. CC493K)	G-CuPb 25	G-CuPb 5-20 Sn (cf. CC495K, CC496K, CC497K)	G-CuAl9	G-CuAl10Fe (cf. CC331G)	G-CuAl9Ni (cf. CC332G) G-CuAl10Ni (cf. CC333G)	G-CuAl9Mn	G-Cu 65 Zn	G-Cu 60 Zn	G-Cu 55 Zn (Mn)	G-Cu 55 ZnAl 1-2	G-Cu 55 ZnAl 4	G-Cu Zn 16 Si 4	G-CuNi30 (cf. CC383H)
Potassium chloride	293 (20)	2	2	2	2	2	2	2	2	2	3	3	3	3	3	3	2
Potassium chloride	373 (100)	2	2	2	2	2	2	2	2	2	3	3	3	3	3	3	2
Potassium chloride	solutions	2	2	2	2	2	2	2	2	2	4	4	4	4	4	4	2

2: less than 0.5 mm/a (19.7 mpy) 3: 0.5–1.0 mm/a (19.7 to 39.4 mpy) 4: over 1.0 mm/a (39.4 mpy)

Table 168: Corrosion behavior of copper materials in potassium chloride [1114]

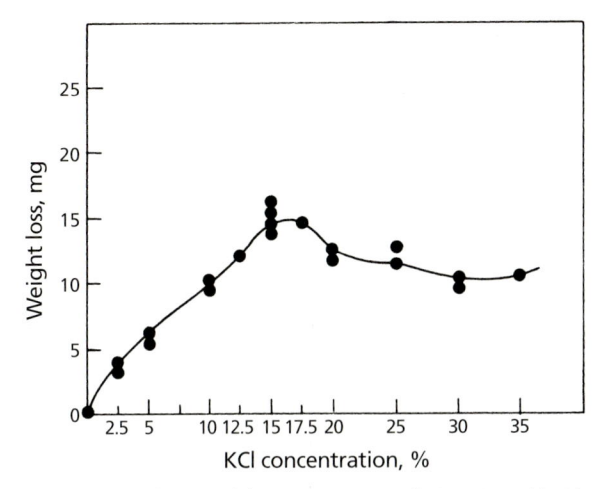

Figure 111: Influence of the concentration of potassium chloride on the weight loss of copper at 368 K (95 °C) (dropping test with 500 cm^3 of liquid in each case, test duration 75 min) [1115]

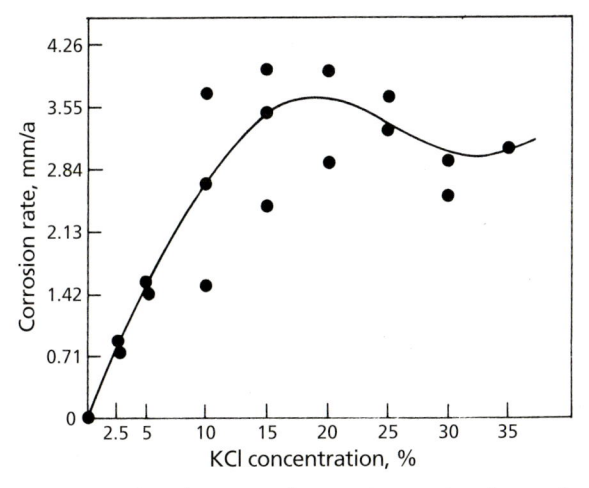

Figure 112: Corrosion rates of copper in an agitated potassium chloride solution (600 rpm, 368–373 K (95–100 °C), 1,000 cm^3 of liquid, test duration 150 min) [1115]

Corrosion tests in a flowing medium correspond the most to the conditions in practice. The corrosion behavior of copper in potassium chloride solutions of various concentrations is shown in Figure 113. The curve also passes through a maximum. As in the case of the corrosion conditions discussed above, it is at about 15–20 % potassium chloride. The curve representing the corrosion rate of copper in 10 % potassium chloride solution shows a constantly rising parabolic course of the corrosion as the temperature increases (Figure 114). According to this curve the corrosion rate reaches about 0.45 mm/a (17.7 mpy) for copper at a flow velocity of 23.1 cm/s at 353 K (80 °C) [1115] and falls to 0.13 mm/a (5.12 mpy) at 293 K (20 °C) in 10 %

solution. The test duration was 6 hours. Copper tubes with an internal diameter of 35 mm and a surface dimension of 348 cm^2 were the sample material dealt with in Figures 113 and 114.

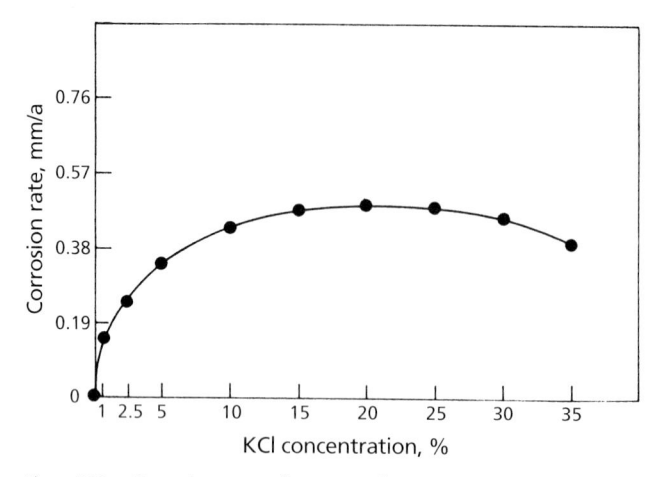

Figure 113: Corrosion rates of copper in flowing potassium chloride solution (flow rate 32.1 cm/s, 340 K (67 °C), test duration 3 h) [1115]

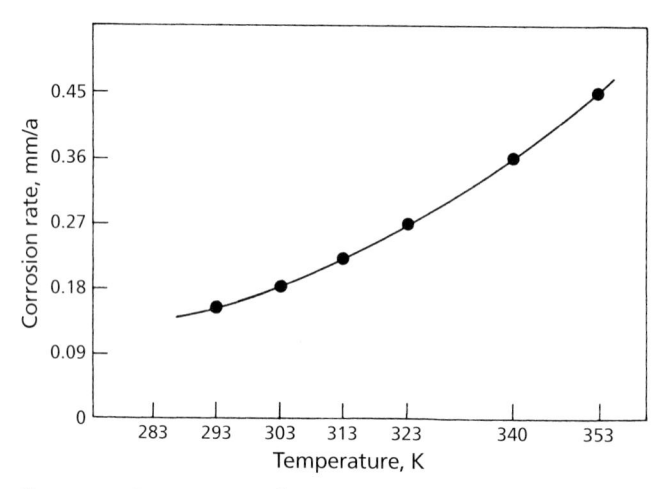

Figure 114: Corrosion rates of copper in 10 % potassium chloride solution in relation to the temperature (flow rate 23.1 cm/s, test duration 6 h) [1115]

The corrosion rates of copper materials in potassium chloride solutions rise with increasing flow velocity, as can be seen from Figure 115. According to these tests they seem to approach an upper limit. The material used in this case is a copper-tin alloy with 1.5 % tin [1116].

The corrosive attack is intensified by oxygen. It increases with the oxygen content and the salt concentration of the solution. The solubility of oxygen at 282 K (9 °C) in potassium chloride solution is given in Figure 116 and the temperature dependence can be referred from Figure 117.

Figure 116 and 117 show that the oxygen content decreases as the concentration of the salt solution or the temperature increases.

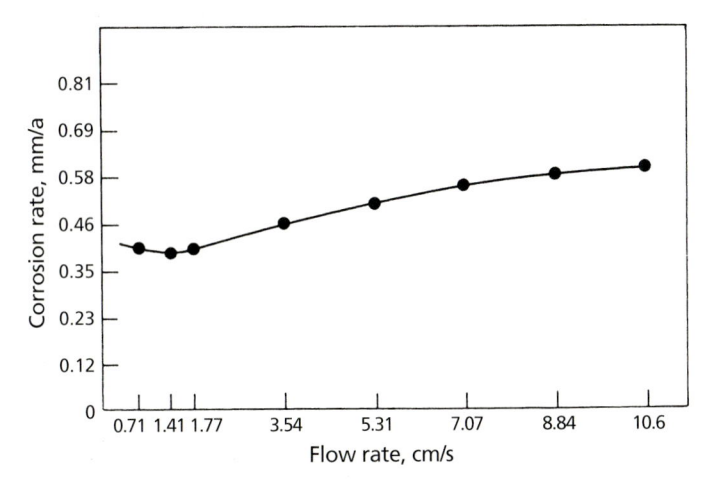

Figure 115: Corrosion behavior of copper-tin alloys (1.5 % Sn) in 10 % potassium chloride solution at different flow rates (343 K (70 °C), pipe internal diameter 10 mm, test duration 2 h) [1116]

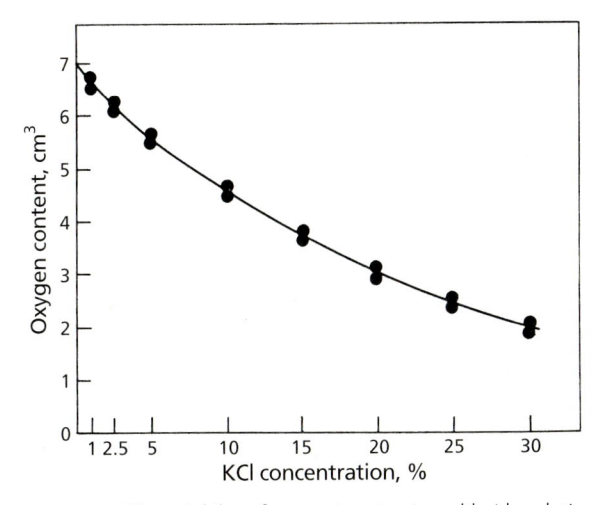

Figure 116: The solubility of oxygen in potassium chloride solutions at 292 K (19 °C) [1116]

The corrosion rates of copper in the presence of air at 358 K (85 °C) were 0.97 mm/a (38.2 mpy) in 25 % KCl solution, 3.28 mm/a (129 mpy) in 10 % solution and 3.17 mm/a (125 mpy) in 20 % solution (see also Figure 112). The corrosive attack in oxygen-free solutions is very slight. Copper and copper alloys are virtually resistant to such solutions [1116].

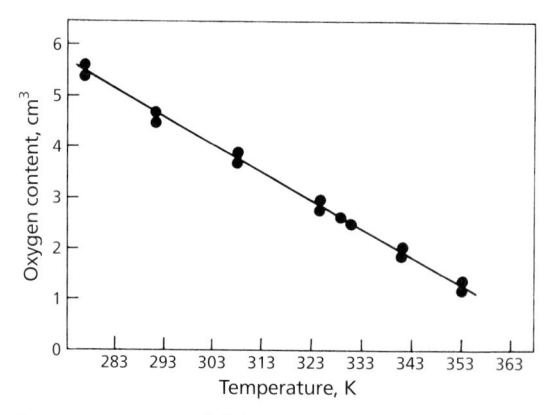

Figure 117: Oxygen solubility in 10% potassium chloride solution at different temperatures [1116]

The Figures 118 to 120 show the corrosion behavior of copper and copper alloys. It can be seen from these curves that copper is less resistant to potassium chloride solutions than its alloys, with the exception of tin bronze (Figure 120).

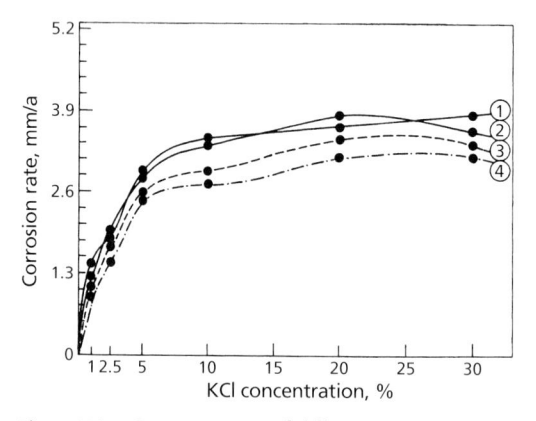

Figure 118: Corrosion rates of different copper materials in potassium chloride solutions (test duration 2 h, agitated medium, air, 353 K (80 °C)) [1117]
① electrolytic copper 1; ② electrolytic copper 2; ③ tin bronze (1.5% of Sn); ④ aluminium bronze (4% of Al)

The erosion of copper by solid salts in the potash industry is insignificant and can therefore be ignored compared with the corrosion caused by aqueous solution.

The corrosion attack by 10–20% potassium chloride solutions is generally higher than the one in dissolving liquors of the potash industry [1117]. Various copper materials were tested in a dissolving liquor containing (g/l) 80 KCl, 100 NaCl, 50 $MgSO_4$ and 120 $MgCl_2$, in a large-scale experiment close to practical conditions over a period of 5.5 days. Pipes of 20/26 mm in diameter were used as the samples. Heating and pumps were shut down during the night, so that the samples were tested for a total of 60 hours at 333 K (60 °C) in a flowing medium; the rest of the time they remained in the cooled or cold, still medium.

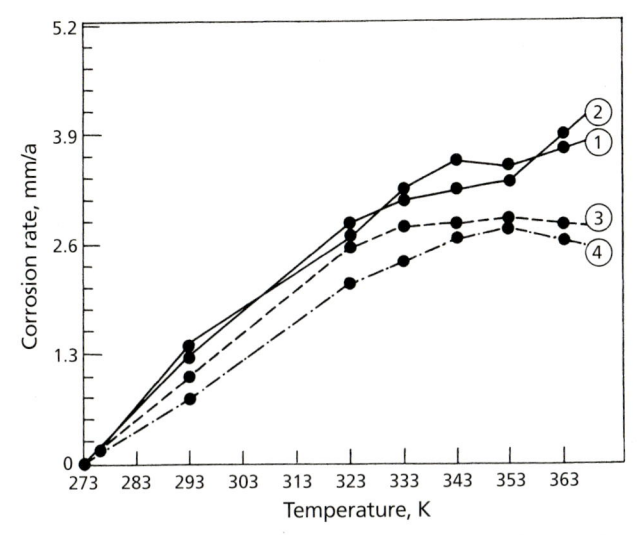

Figure 119: Corrosion rates of different copper materials in 10% potassium chloride solution (test duration 2h, agitated medium, air) [1117]
① electrolytic copper 1
② electrolytic copper 2
③ tin bronze (1.5% of Sn)
④ aluminium bronze (4% of Al)

Table 169 summarizes the corrosion behavior [1118].

The corrosion behavior of electrolytic copper is far worse than that of aluminium bronze, but slightly better than that of tin bronze. The corrosion rate of copper is 0.45 or 0.59 mm/a (17.7 or 23.2 mpy) (Table 169) at a flow rate of 0.9 m/s.

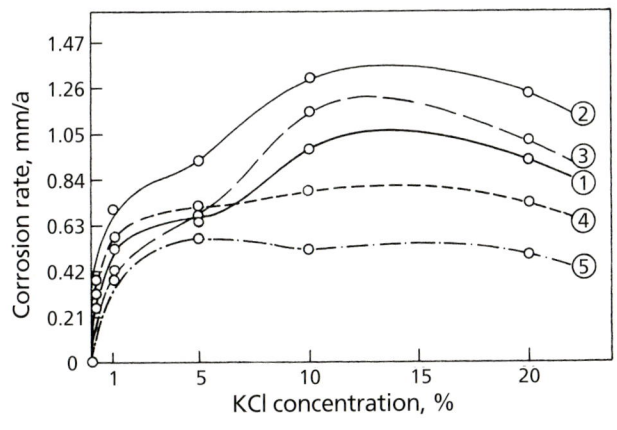

Figure 120: Corrosion rates of different copper materials in potassium chloride solutions of varying concentrations (test duration 2 h, flow rate 3.53 cm/s, 358 K (85 °C), sealed system) [1117]
① electrolytic copper
② commercial copper
③ Sn-bronze (1.5% of Sn)
④ Al-bronze (4% of Al)
⑤ brass (30% Zn)

Material	Test Duration d	Flow Rate m/s	Corrosion Rate mm/a (mpy)
E-Cu	5.5	0.9	0.45 (17.7)
E-Cu	5.5	0.9	0.59 (23.2)
CuSn(1.5Sn)	5.5	0.9	0.61 (24.0)
CuSn(1.5Sn)	5.5	0.9	0.63 (24.8)
CuAl(4.0Al)	5.5	0.9	0.07 (2.76)
CuAl(4.0Al)	5.5	0.9	0.10 (3.94)
CuAl(4.0Al)	5.5	0.9	0.08 (3.15)
CuAl(4.0Al)	5.5	1.17	0.16 (6.30)

Table 169: Corrosion rates of copper materials in a potash dissolving liquor (80 g KCl, 100 g NaCl, 50 g MgSO$_4$ and 120 g MgCl$_2$ in 1,000 cm^3) at 333 K (60 °C) for 60 hours of the total test duration [1118]

When magnesium chloride-containing potash liquors are processed, it should be noted that at temperatures above 373 K (100 °C) hydrochloric acid is formed, which increases the corrosivity of the solution considerably.

The solubility of copper in potassium chloride solutions is affected by light. Copper is dissolved more slowly in 0.1 mol/l KCl solution in the dark than in the light. The difference in the dissolution rate is stated to be about 40 %. The potential fluctuations are about 10 mV. The corrosion rate under illumination with 25 watt from a distance of 15 cm was about 0.021 mm/a (0.83 mpy) after a testing period of 15 days. In the dark it is half as high with 0.011 mm/a (0.43 mpy) [1119]. If about 1 ml 0.001 mol/l FeSO$_4$ solution is added to this KCl solution, the corrosion of copper can be inhibited in both, light and dark. The effect of sunlight on the corrosion behavior of copper in a parent solution (50 g/l FeCl$_3$ × 6 H$_2$O + 25 ml/l concentrated HCl + 100 g/l KCl) and its dilution is shown in Table 174. In the pure solution copper is destroyed in the dark and on exposure to light. The corrosion rate is 69.90 mm/a (2,752 mpy) in the dark and 80.87 mm/a (3,184 mpy) in sunlight at a temperature of 305 ± 1 K (32 ± 1 °C). Copper is not resistant even in a 10 % solution of the mentioned parent solution. The corrosion rates at the same temperature are 3.79 and 5.25 mm/a (149 and 207 mpy) [1120].

Copper is corroded by molten potassium chloride at 1,133 K (860 °C) at a corrosion rate of 3.1 mm/a (122 mpy) (test duration 8 h) (see Table 170).

Metal	Corrosion Rate, mm/a (mpy)
Copper	3.1 (122)
Carbon steel	1.4 (55.1)
18 8 chromium-nickel steel	0.3 (11.8)
15 6 chromium-aluminium steel	0.6 (23.6)
Nickel	1.3 (51.2)

Table 170: Corrosion rates of metallic materials in molten potassium chloride (test duration 8 hours at 1,133 K (860 °C)) [1121]

Copper-aluminium alloys

The corrosion resistance of copper is improved by alloying with aluminium. The maximum permissible aluminium content in α-bronzes is about 7.5–8 % in practice.

Nickel as another alloying component of the aluminium bronzes has a favorable effect on their corrosion behavior. Aluminium bronzes are resistant to various types of neutral salts, namely from 293 to 368 K (20 to 95 °C). Their use with potassium chloride is recommended. The corrosion rate is less than 0.15 mm/a (5.91 mpy) [1122].

Aluminium bronzes show better resistance to potassium chloride solutions and dissolving liquors containing potassium chloride in the potash industry than copper. The aluminium content in the alloys used in the potash industry is 5 %. The relevant parts should be stress-free when used. The corrosion rate for CuAl-alloys in potassium chloride solutions up to 373 K (100 °C) is less than 0.5 mm/a (19.7 mpy) (see Table 168) [1114].

Copper alloys with 4 % Al are corroded at a rate of 0.16 mm/a (6.30 mpy) in a caustic potash solution containing (g/l) 80 KCl, 100 NaCl, 50 $MgSO_4$ and 120 $MgCl_2$ at 333 K (60 °C) and at a flow rate of 1.17 m/s. The corrosion rate is reduced to 0.07 mm/a (2.76 mpy) as the flow rate decreases to 0.9 m/s (see Table 169) [1118].

The highest corrosion attack is to be expected from 10–20 % potassium chloride solutions, as can be seen from Figures 118–120. The corrosion rates of 3 mm/a (118 mpy) in 20 % solution at 353 K (80 °C) (see Figure 118) would exclude use under these conditions. However, the values are calculated from a test duration of only 2 hours, which include the generally high initial corrosion. In Figure 120 the values are about 0.8 mm/a (31.5 mpy) for the same test duration and temperature and a flow rate of 3.53 cm/s.

In 10 % potassium chloride solution the corrosion rates at 273 K (0 °C) are close to zero and at 293 K (20 °C) reach values of about 0.5 mm/a (19.7 mpy) (see Figure 119).

The presence of oxygen increases the corrosion of copper materials in salt solutions. At 353 K (80 °C) for example a 10 % potassium chloride solution can still absorb about 1.5 cm^3 oxygen (see Figure 117). In sylvinite solution containing oxygen the corrosion rate on copper is 0.0198 mm/a (0.78 mpy) and decreases in oxygen-free solution to 0.0016 mm/a (0.06 mpy). It can also be referred from these values that the corrosion attack by pure potassium chloride solutions is greater than in potash salt solutions (compare Table 169) [1117]. CuAl-alloys are more resistant to corrosion in potassium chloride solutions and in potash salt solutions than pure copper [1118]. Aluminium bronze can be used for tubes and tube sheets for high velocity preheaters for dissolving liquors. This material is also suitable for the manufacture of radiators in large capacity, preheaters and for heating tube bundles for evaporation of liquors, for the manufacture of screens, as filter fabric and as pump material.

The copper-aluminium alloy CuAl5 is a material which is used internationally chiefly for the manufacture of tubes in preheaters and surface condensers in the potash industry. Table 171 contains the chemical composition of comparable alloys in different countries.

The addition of 0.02–0.04 % arsenic is also permissible for the use of CuAl5 for condenser tubes in the USA. Arsenical aluminium bronzes exhibit a finer grain under the same heat treatment as alloys without arsenic. This can be explained by the inhibiting effect on the grain growth as a result of the presence of the finely dispersed arsenic phase in the aluminium bronze [1123].

Damages as a result of fatigue fractures of the arsenic-free material for the above-mentioned applications in the potash industry were the reason for investigations as to what extent grain size and aluminium content influence the fatigue strength of this material and whether the corrosive medium can cause corrosion fatigue [1124].

The investigations were carried out on alloys, with compositions according to Table 172. They were all arsenic-free, because only these materials can be employed by these users. Table 173 provides the average grain diameters. A potash salt solution with the following composition was used as the corrosive medium: (g/l) 70 KCl, 230 $MgCl_2$, 55 NaCl, 20 $MgSO_4$ and 0.5 $CuSO_4$. The corrosion tests were carried out at room temperature on flat specimens. A lower bending fatigue strength is to be expected with these than with round specimens.

With increasing Al-content of the alloy the bending fatigue strength $\sigma_{bw}(10^7)$ increases almost linearly over the entire concentration range if specimens with roughly the same structure – homogeneous α-mixed crystal and approximately the same grain size – are compared (see Figure 121).

The bending fatigue strength $\sigma_{bw}(10^7)$ is reduced by about 20 %, when the recrystallization grain size increases from 30 to 70 μm. Fine grained material (less than 10 μm average grain diameter), which is not yet recrystallized, has a substantially higher bending fatigue strength than corresponds to the grain size dependence in the recrystallized range (see Figure 122).

The action of the salt solution reduced the bending fatigue strength to a greater extent as the aluminium content increases (see Figure 123).

The decrease in the fatigue strength with round specimens as a result of additional corrosion exposure starts later with the rotating bending fatigue specimens than with the flat fatigue strength specimens, namely only after about 5×10^5 fatigue cycles. As can be seen from Figure 124, it reaches a difference of 15 % with 10^7 fatigue cycles.

The influence of further exposure to the potash salt solution on the fatigue limit of CuAl-alloys with varying aluminium contents is shown in Figure 125.

It can be seen that the fatigue strength of the alloys decreases as the aluminium content increases.

As a result of additional corrosive attack by the mentioned caustic potash solution the fatigue strength of the alloy CuAl5 with a maximum of 6 % Al (arsenic-free) can be estimated to be at most 20 % below the value in air. The fatigue strength of the flat samples tested in air decreases from 200 MPa to 170 MPa as a result of corrosive attack.

Standard	Country	Alloy designation	Alloying components, %				
			Al	Mn	Ni	Fe	Cu
StRGW 377-76	RGW*	BrA5	4.0...6.0	–	–	–	balance
TGL 35488	GDR	CuAl5	4.0...6.0	–	–	–	balance
PN-G9/H67050	Poland	BA5	4.0...6.0	–	–	–	balance
GOST 18175-78	USSR	BrA5	4.0...6.0	–	–	–	balance
BDS 788-71	Bulgaria	CuAl5	4.0...6.0	–	–	–	balance
MSZ 711/1-77	Hungary	CuAl5	4.0...6.0	–	–	–	balance
STAS 203-75	Rumania	CuAl5	4.0...6.5	0...0.5	0...0.8	–	balance
DIN 17665	FRG	CuAl5	4.0...6.0	–	–	–	balance
DIN 1785	FRG	CuAl5 As	4.0...6.0	–	–	–	92.5...96.0
CSN 423042	CSSR	CuAl5	4.0...6.0	–	–	–	balance
BS 378:1963	GB	CA 102	6.0...7.5	–	1.0...2.5	–	balance
ASTM-B 111-68	USA	alloy 608	5.0...6.5	–	–	–	93

* CMEA (Council for Mutual Economic Assistance)

Table 171: Comparison of standards for the chemical composition of CuAl5 [1124]

Table 171: Continued

Standard	Maximum permissible impurities, %								
	Mn	Ni	Fe	Zn	Pb	Si	P	Sn	As	Total
StRGW 377-76	0.5	–	0.5	0.5	0.03	0.1	–	0.1	0.4	1.1
TGL 35488	0.5	0.5	0.5	0.5	0.03	0.1	–	0.1	2)	1.1
PN-G9/H67050	0.5	0.5	0.5	0.5	0.03	–	0.01	0.1	0.1	1.1
GOST 18175-78	0.5	–	0.5	0.5	0.03	0.1	0.01	0.1	–	1.1
BDS 788-71	0.5	0.5	0.5	0.5	0.03	–	0.01	–	–	1.1
MSZ 711/1-77	0.5	–	0.5	0.5	0.03	0.1	0.01	0.1	–	1.1
STAS 203-75	–	–	0.5	0.5	0.1	–	–	–	–	0.8
DIN 17665	0.3	0.8	0.4	0.5	–	0.2	0.02	0.3	–	–
DIN 1785	0.2	0.2	0.2	0.3	–	–	0.02	0.1	0.4	–
CSN 423042	–	–	0.4	–	0.03	–	–	0.1	–	0.5
BS 378:1963	–	–	–	–	–	–	–	–	–	0.5
ASTM-B 111-68[1]	–	–	0.1	–	–	–	0.1	–	0.35	–

1) Mn, Ni and Fe: 1.0 to 1.5 %, others in Σ 0.5 %
2) An arsenic content of 0.02 to 0.04 % is permissible for the use of CuAl5 for condenser tubes.

Table 171: Comparison of standards for the chemical composition of CuAl5 [1124]

Alloy	Al %	Ni %	Fe %	Mn %	As ppm
CuAl3	2.92	0.005	0.017	< 0.01	< 1
CuAl4	3.92	0.004	0.016	< 0.01	< 1
CuAl5.5	5.44	0.004	0.019	< 0.01	< 1
CuAl7	7.10	0.004	0.016	< 0.01	< 1

Table 172: Chemical composition of aluminium alloys to determine the influence of potash salt solutions on the fatigue strength of CuAl-alloys [1124]

Alloy	Average grain diameter, μm		
	0.5 h, 673 K (400 °C) 0.5 h, 773 K (500 °C)	0.5 h, 853 K (580 °C)	0.5 h, 923 K (650 °C)
CuAl3	< 10	25	70
CuAl4	< 10	30 (15.30)*	70
CuAl5.5	< 10	30 (14.30)*	70
CuAl7	< 10	35	90

* check tests

Table 173: Average grain diameter of the test alloys (see Table 172) after respective heat treatment [1124]

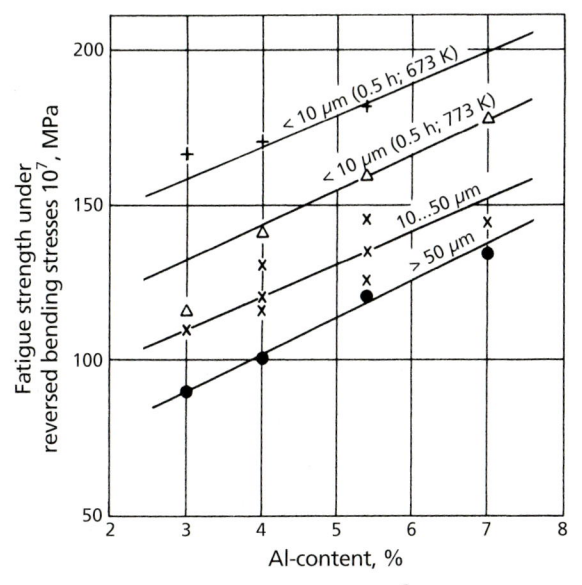

Figure 121: Bending fatigue strength $\sigma_{bw}(10^7)$ of sheet specimens of CuAl-alloys (see Table 172) with varying recrystallization grain size as a function of the aluminium content [1124]

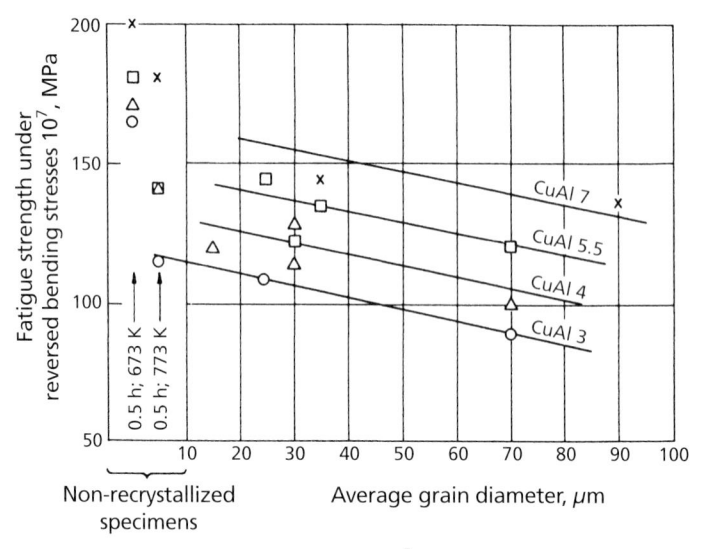

Figure 122: Bending fatigue strength $\sigma_{bw}(10^7)$ of sheet specimens of CuAl-alloys (see Table 172) as a function of the recrystallization grain size [1124]

The investigations showed that there is no influence of the temperature on the fatigue limit at 343 K (70 °C) as a result of corrosion exposure comparable with that encountered in the potash industry.

An average grain size of less than 70 µm is required for optimum corrosion resistance with respective fatigue strength.

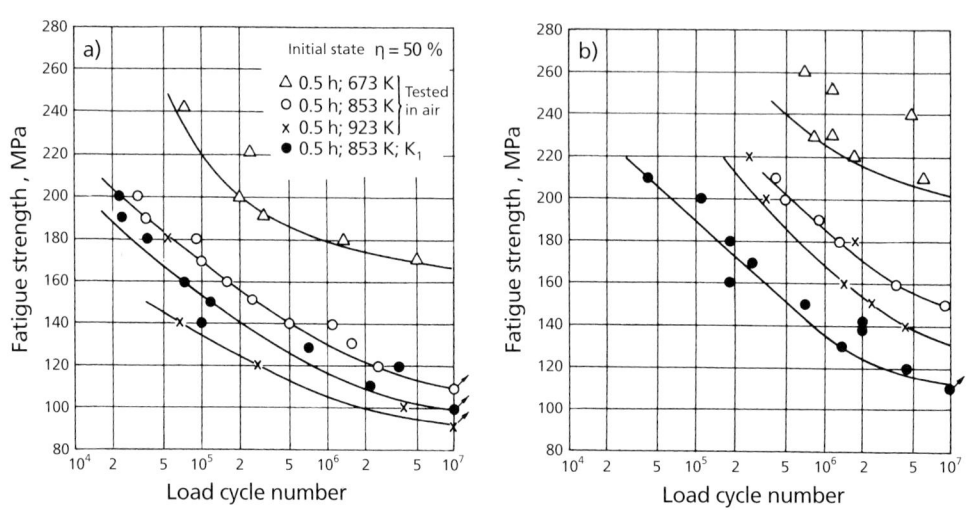

Figure 123: Wöhler curves of CuAl3 a) and CuAl7 b) sheet specimens for the bending fatigue test [1124]
η = 50 % (degree of cold deformation)
K_1 = corrosion test with potash salt solution ((g/l) 230 $MgCl_2$, 70 KCl, 55 NaCl, 20 $MgSO_4$ and 0.5 $CuSO_4$)

De-aluminization under the conditions of the potash industry is not to be expected with this alloy [1124].

Copper-nickel alloys

As in the case of the other copper materials, corrosion in neutral potassium chloride solutions is largely dependent on the presence of oxidizing substances, that is especially of air. In acidic or alkaline solutions the corrosion resistance is analogous to that found with the corresponding acid or caustic solution.

Copper nickel alloys are not attacked by dry potassium chloride. They are more resistant to moist salt than iron-containing copper or nickel alloys. Corrosion investigations on a copper nickel alloy with 33 % Ni showed on exposure for 1 year relatively small sites of attack on the surface of specimens sprinkled with potassium chloride. The relatively vigorous corrosion compared to that with dry salt can be attributed to the moisture absorption of the salt. The corrosive attack under these conditions is judged to be stronger than that in aqueous solution at the same temperature [1125]. The corrosive attack by potassium chloride moist from atmospheric humidity is not uniform but characterized by a relatively large number of punctiform sites of attack (pitting). The same corrosion attack is found with carnallite.

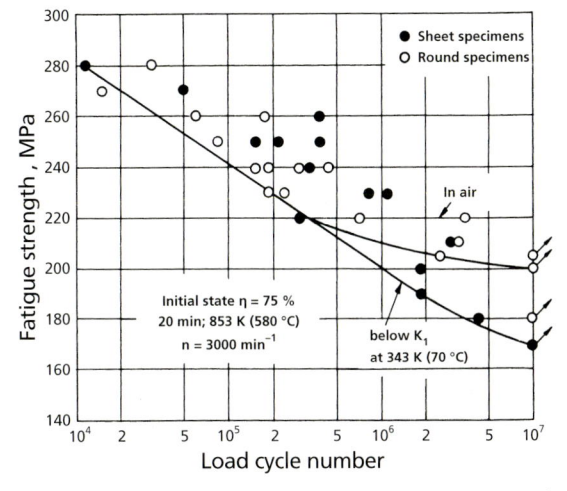

Figure 124: Wöhler curves of CuAl5 round specimens according to TGL 35488 for rotating bending fatigue tests at room temperature in air and at 343 K (70 °C), with sprinkling with caustic potash solution [1124]
$\eta = 75\%$ (degree of cold deformation)
K_1 = corrosive medium

The copper-nickel alloys are not suitable for this type of exposure. The corrosive attack is caused by the copper content and increases with the Cu-content of the alloy. The number and the area of the localized corrosion attacks increase [1113]. The corrosion behavior of copper-nickel alloys in caustic potash solutions is worse than that of the nickel-copper alloys.

After test durations of 4 months a corrosion rate of 3.3×10^{-3} mm/a (0.13 mpy) was calculated from the weight loss in gas-free sylvinite liquor and only 1.5×10^{-3} mm/a (0.06 mpy) in carnallite liquor at room temperature. It can be deduced from the corrosion rates that the attack by pure potassium chloride solution is highest, since sylvinite is the richer in KCl than carnallite.

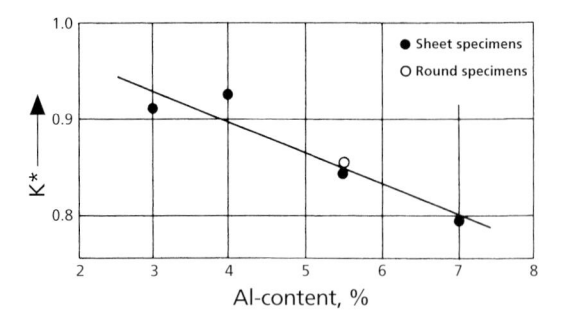

Figure 125: Influence of the corrosion attack by exposure to potash salt solution on CuAl-alloys (see Table 172) with varying aluminium contents [1124]
$K^* = \sigma_{bw}$ (caustic potash solution)$/\sigma_{bw}$ (air)

Virtually no corrosion loss is found at room temperature on samples completely immersed in sylvinite solutions (2.0×10^{-4} mm/a (0.01 mpy)). The corrosion rate increases as the temperature rises; it is 0.11 mm/a (4.33 mpy) in sylvinite liquor at 363 K (90 °C) and 0.03 mm/a (1.18 mpy) in carnallite liquor. The test duration in this case was only 8 days. Since uniform corrosion is expected, these values can be used for dimensioning.

Due to this corrosion loss in sylvinite solution, the resistance of this alloy is in the range of the low-alloy steels.

A uniform, very thin coating is formed in oxygen-free solution at room temperature. The use of the alloys can be recommended under these conditions.

A corrosion rate of less than 0.5 mm/a (19.7 mpy) is given in Table 168 for the cast alloy G-CuNi 30 in potassium chloride solutions. According to [1126] the alloy is suitable for use under the majority of conditions.

Copper-tin alloys (bronzes)

Corrosion in neutral potassium chloride solutions is dependent on the oxidizing agents contained therein, also in the case of copper-tin alloys.

Accelerated corrosive attack generally occurs in acid potassium chloride solutions with copper and its alloys [1127].

CuSn-alloys are not attacked by dry, anhydrous potassium chloride. Corrosion attack by potassium chloride, which was scattered on samples and started corrosion as a result of moisture absorption from the air, revealed small, punctiform, blue-green stains on the surface of the samples after 1 year. The alloy contained 3.7 % Sn. Under these conditions it showed the best corrosion behavior of any of the copper materials examined, similar to electrolytic copper.

The corrosive attack by potassium chloride is roughly comparable with that by sodium chloride, but is less than that by magnesium chloride. Pitting must always be expected with copper materials when handling moist salts [1113]. The corrosion rates in aqueous potassium chloride solutions are stated to be less than 0.5 mm/a (19.7 mpy). The corrosion rate of tin bronze with 3.7 % Sn in aqueous solutions containing potassium chloride, as used in the potash industry, is less than 5.0×10^{-3} mm/a (0.20 mpy) at room temperature in solutions free of or low in oxygen. In solutions enriched with oxygen the corrosion rate is 0.022 mm/a (0.87 mpy). This corrosion rate applies to sylvinite liquor; it is slightly more in carnallite liquor with 2.6×10^{-3} mm/a (0.10 mpy). As the temperature rises, the corrosion rate increases to 0.064 mm/a (2.52 mpy) in sylvinite mother liquor and to 0.04 mm/a (1.57 mpy) in carnallite mother liquor at 363 K (90 °C). The corrosion products formed are adherent and greenish-white. The corrosion solutions become faintly greenish.

As a result of the action of the sylvinite solution, which causes the strongest attack amongst the test solutions, fine reddish-brown coatings, an irregular coating with light copper etching and an irregular layer with pitting are formed in oxygen-free solution at room temperature. In a solution rich in oxygen a thicker layer of whitish corrosion products is formed. The surfaces of the samples were severely etched by the corrosive attack at 363 K (90 °C) after a test duration of 8 days. The corrosion rate in a potash dissolving liquor consisting of (g/l) 80 KCl, 100 NaCl, 50 $MgSO_4$ and 120 $MgCl_2$ at 333 K (60 °C) is 0.6 mm/a (23.6 mpy) (Table 169). The corrosion rates in pure potassium chloride solutions are given in Figures 118 to 120. The maximum corrosion of copper materials is in 10–20 % solutions. Figure 117 shows the possible oxygen content of a 10 % potassium choride solution as a function of the temperature. The corrosion rate of an alloy with 1.5 % Sn is about 3.6 mm/a (142 mpy) in an agitated, aerated medium at a maximum temperature of 333 K (60 °C) (test conditions, see section Copper).

This material cannot be used under these conditions. The corrosion rate is still about 1.1 mm/a (43.3 mpy) in a sealed system at 358 K (85 °C) and with a flow rate of 3.53 cm/s.

In an aerated potassium chloride solution the corrosion rate increases roughly linearly with the temperature rising (Figure 119).

This 10 % potassium chloride solution is more corrosive than the potash dissolving liquors. These contain more salt and are therefore less capable of absorbing oxygen [1117, 1118].

The copper-tin alloys are less suitable materials for potassium chloride solutions than copper-aluminium alloys.

Copper-tin-zinc alloys (red brass)

The corrosion behavior of copper-tin-zinc alloys under the influence of potassium chloride is more or less comparable with that of the copper-tin alloys (see section CuSn-alloys (bronzes)).

The corrosion resistance of these materials is largely dependent on the homogeneity of the structure, little or no grain-boundary deposits and freedom from stress of the components.

According to Table 168, the corrosion rate is less than 0.5 mm/a (19.7 mpy). Copper-aluminium alloys should be preferred in the case of exposure to attack by potassium chloride-containing media and mixed salts of the potash industry (see section CuAl-alloys).

The material loss of CuSnZn-alloys, like that of the other copper materials, is dependent on the oxygen content of the solutions. With rising oxygen content the corrosion rates increase. These materials are very suitable for use when oxygen and other oxidizing agents can be completely excluded in potassium chloride solutions. This applies above all to temperatures of about 293 K (20 °C). At room temperature a copper-tin alloy with 3.7 % tin is only corroded at a rate of 0.0015 mm/a (0.06 mpy) in sylvinite mother liquor and therefore exhibits good corrosion behavior similar to copper-zinc alloys.

CuSnZn-alloys are supposed to be resistant under the reported conditions, too. In the presence of oxygen the corrosion rate increases to about 0.02 mm/a (0.79 mpy) at room temperature. With increasing temperature, an even higher corrosion rate is to be expected. It increases to about 0.2 mm/a (7.87 mpy) in a sealed system at 363 K (90 °C). The corrosion loss in aqueous potassium chloride solutions is to be expected to be even higher, as can be referred from Figures 118–120. The corrosion rate for a copper alloy with 1.5 % Sn increases to about 3.12 mm/a (123 mpy) at 353 K (80 °C) in 5–25 % potassium chloride solutions in the presence of air (Figure 118, 119) and is slightly over 1 mm/a (39.4 mpy) in a sealed system without any further air supply (Figure 120). About the same corrosion rates are to be expected for copper-tin-zinc alloys.

These are not attacked by pure, dry potassium chloride. Moist potassium chloride causes punctiform attack, as in the case of the above-mentioned binary alloys, but which is not to be regarded as pitting corrosion.

Copper-zinc alloys (brass)

Copper-zinc alloys are said to have good resistance when exposed to potassium chloride if their Zn-content is ≤ 15 %. It is not recommended using them when having higher zinc contents [1126].

As in the case of other copper materials the corrosion rates are increased by the oxygen content, by an increase in the temperature and by agitating the corrosive medium.

Pure, dry potassium chloride does not attack copper-zinc alloys.

The alloys are locally attacked by moist potassium chloride, which cannot yet be described as pitting but which is stronger for these materials than for pure copper. They are slightly less attacked by potash salts than by pure potassium chloride or sodium chloride.

Pure brass is less attacked than brass with α- and β-mixed crystals. The corrosion attack on pure brass is punctual, contrary to the more uniform attack on the latter alloy [1113].

The corrosion rate of copper-zinc alloys in sylvinite liquor (air-free) at room temperature is slightly depending on the Zn-content and has a maximum of 0.002 mm/a (0.08 mpy). The copper content decreases from 72–59% in the sequence of the references numbers. In the presence of oxygen the corrosion rate increases by a factor of ten to 0.02 mm/a (0.78 mpy) in the otherwise static medium. Brass, however, is attacked locally under these conditions (pitting).

It is to be expected that the corrosive attack increases with increasing temperature and zinc content of the materials. The rise in the corrosion rate to about 0.1 mm/a (3.94 mpy) on average confirms this tendency.

The highest corrosion rates are to be expected with brass and other copper-containing materials in 10–20% potassium chloride solutions. The corrosion rate at 353 K (80 °C) can be as high as 3 mm/a (118 mpy) in agitated potassium chloride solution and is 1 mm/a (39.4 mpy) in flowing solution in a sealed system.

Pitting, above all at relatively high temperatures and oxygen contents, cannot be excluded when using brass with potassium chloride solutions.

The corrosion rates for brass are said to be 0.19 mm/a (7.48 mpy) after a test period of 7 days in air-free sylvinite mother liquor in sealed vessels at room temperature and 0.11 mm/a (4.33 mpy) under the same conditions in carnallite mother liquor. In open vessels the corrosion rate increases to 0.23 mm/a (9.06 mpy) in sylvinite solution and to 0.17 mm/a (6.69 mpy) in carnallite solution. When oxygen is introduced, the corrosion rates in open vessels are 0.81 and 0.30 mm/a (31.8 and 11.8 mpy) respectively.

The corrosion rate in both solutions is 0.64 mm/a (25.2 mpy) at 368–373 K (95–100 °C). The test duration for this temperature was 7 days, the evaporated water being constantly replaced. In this relatively short test period the brass surface was uniformly corroded [1128].

The corrosion of brass (63% Cu, 37% Zn) in chloride-containing solutions is increased under the influence of light [1129].

The corrosion rates in an aqueous parent solution (g/l) 50 $FeCl_3 \times 6H_2O$ + 100 KCl + 25 ml/l HCl (concentrated) at 305 ± 1 K (32 ± 1 °C) were found to be 69.9 mm/a (2,752 mpy) in the solution without exposure to sunlight and 80.87 mm/a (3,184 mpy) under the influence of sunlight on copper samples after a test period of 2 hours. The corrosion rates on brass 63/37 were significantly lower under the same conditions; 43.93 mm/a (1,730 mpy) without exposure to light and 49.13 mm/a (1,934 mpy) on exposure to light. Brass cannot be used in this solution or its dilutions. The corrosion rates are given in Table 174 [1129].

The increase in the corrosion rate as a result of the influence of light is smaller in the case of brass than with copper. The corrosion rate caused by the pure parent solution without exposure to light is about 37%, and with exposure to light about 39% below the corrosion rates for copper (99.9%).

Concentration of the parent solution %	Corrosion rate mm/a (mpy)	
	Without sunlight	With sunlight
Copper		
10	3.79 (149)	5.25 (207)
20	1.61 (63.4)	6.19 (244)
40	20.39 (803)	19.90 (783)
50	28.83 (1,135)	34.17 (1,345)
100	69.90 (2,752)	80.87 (3,184)
Brass 63/67		
10	4.21 (166)	5.19 (204)
20	4.41 (174)	5.66 (223)
40	15.41 (607)	17.49 (689)
50	19.97 (786)	23.84 (939)
100	43.93 (1,730)	49.13 (1,934)

Table 174: Corrosion rates of copper and brass (63.2% Cu, 36.6% Zn, 0.01% Pb) in a parent solution (50 g/l $FeCl_3 \times 6 H_2O$ + 25 ml/l concentrated HCl + 100 g/l KCl) and its dilutions, with and without exposure to sunlight at 305 ± 1 K (32 ± 1 °C) in mm/a (test duration 2h) [1129]

The presence of other salts has a significant effect on the corrosion behavior of brass in potassium chloride solutions. Their influence in 0.1 mol/l potassium chloride solution is described in Table 175. Whereas with other salts the corrosion rate remains significantly below 0.4 mm/a (15.7 mpy). When $K_2S_2O_8$ is present, the corrosion rates are about 32 mm/a (1,260 mpy), and the brass is destroyed [1130].

Other copper alloys

The corrosion behavior of alloys, which, like for example beryllium bronze, contain only small amounts of alloying constituents, is almost the same as that of copper with respect to corrosion in potassium chloride solutions and mixed salt solutions (see section Copper).

They are resistant to dry potassium chloride and hardly attacked by dilute solutions at room temperature with exclusion of air or in the absence of strong oxidizing agents. As in the case of copper, the maximum corrosion will occur in 10–20% solutions.

Salt	Brass 80/20		Brass 59/41	
	Corrosion rate mm/a (mpy)	Ratio Cu:Zn	Corrosion rate mm/a (mpy)	Ratio Cu:Zn
$K_2S_2O_8$	32.10 (1,264)	1.40	32.81 (1,292)	0.55
KI	0.34 (13.4)	4.80	0.29 (11.4)	2.0
KCl	0.15 (5.91)	0.41	0.12 (4.72)	0.38
KBr	0.03 (1.18)	–	0.04 (1.57)	–
K_2SO_4	0.14 (5.51)	0.89	0.15 (5.91)	0.38
KNO_3	0.13 (5.12)	1.5	0.13 (5.12)	0.55
$KHCO_3$	0.30 (11.8)	2.7	0.35 (13.8)	1.4
K_2CO_3	0.02 (0.79)	–	0.04 (1.57)	–
KNO_3	0.05 (1.97)	–	0.10 (3.94)	–
$KMnO_4$	0.06 (2.36)	–	0.07 (2.76)	–
K_3PO_4	0.02 (0.79)	–	0.04 (1.57)	–
K_2CrO_4	0.01 (0.39)	–	0.03 (1.18)	–
$K_2Cr_2O_7$	0.01 (0.39)	–	0.03 (1.18)	–

Table 175: Corrosion behavior of brass in 0.1 mol/l KCl solution at 308 ± 2 K (35 ± 2 °C) on the addition of small amounts of salts, and given Cu:Zn ratio in the corrosion products [1130]

Potassium Hydroxide

Author: P. Drodten / Editor: R. Bender

Copper

The simplified Pourbaix diagram in Figure 126 shows that copper is resistant to alkaline solutions only at negative potentials. At positive potential a passivated range exists only in the pH range from about 7.5 to 11.5, whereas copper is active and attacked in acidic as well as in strongly alkaline solutions.

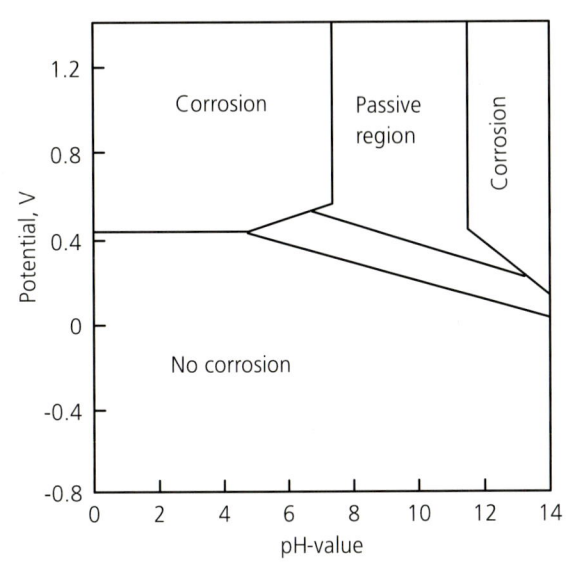

Figure 126: Pourbaix diagram for copper [1131]

Thus with copper as material for handling alkali hydroxide solutions of higher concentration, considerable corrosive attack must be contended with already at room temperature.

Also in alkali hydroxide melts copper is considerably more susceptible to corrosion than iron and nickel (Figure 127). In sodium hydroxide melts at 723 K (450 °C), the corrosion rate of iron is 4.5 g/m^{2}h) corresponding to 4.9 mm/a (193 mpy) and that of nickel is 1.8 g/m^{2}h corresponding to 1.7 mm/a (66.9 mpy), whereas copper is attacked with a corrosion rate of approximately 126 mm/a (4,961 mpy). In potassium hydroxide solution the corrosion rate even reaches a value of almost 600 mm/a (23,622 mpy), (Figure 128). The temperature dependence of the corrosive attack in KOH melt is shown for three materials in Figure 129 [1132].

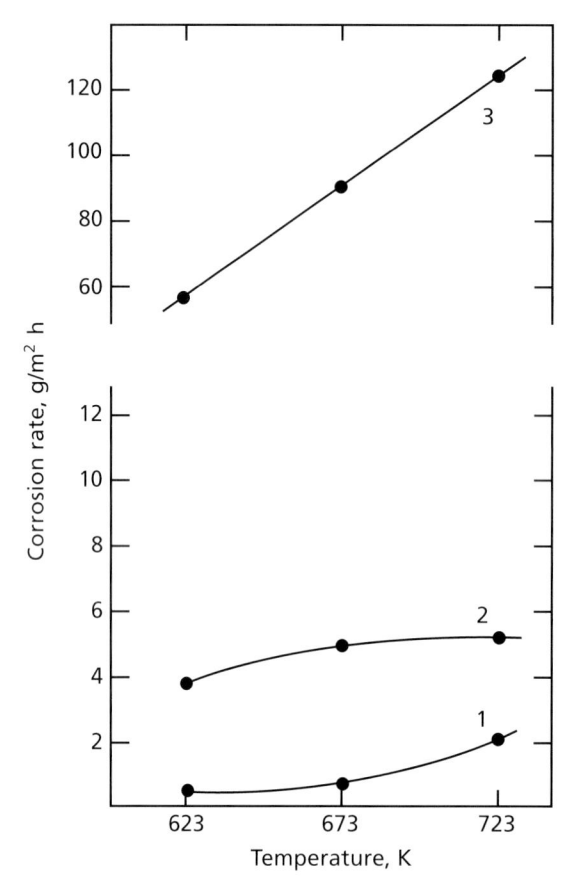

Figure 127: Temperature dependence of the corrosion rate of nickel (1), iron (2) and copper (3) in sodium hydroxide melts [1132]

Thus copper is unsuitable as material for handling potassium hydroxide melts.

However, copper is cleaned with alkaline solutions, e.g. with sodium gluconate as inhibitor [1133].

Copper can be colored in a defined manner by anodic oxidation in 0.05 mol/l alkaline solution at 333 K (60 °C) under the operating conditions specified below, that produce Cu_2O layers of various thicknesses [1134].

- Alternating current with frequency of 50 Hz
  Current density: 0.02 A/cm^2
  Potential: $U_H = -0.3$ to -0.5 V
  Duration: 10 to 60 minutes

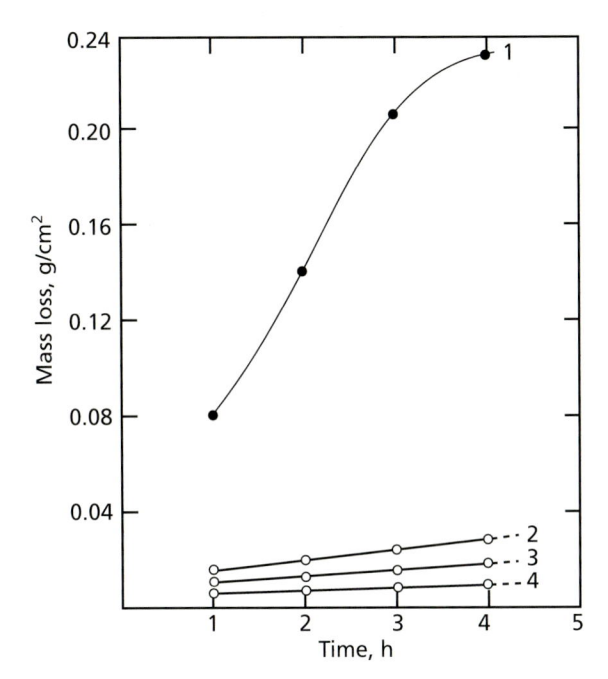

Figure 128: Time dependence of the corrosion of copper in KOH melts at 773 K (500 °C) (1), 723 K (450 °C) (2), 673 K (400 °C) (3) and in NaOH melts at 723 K (450 °C) (4) [1132]

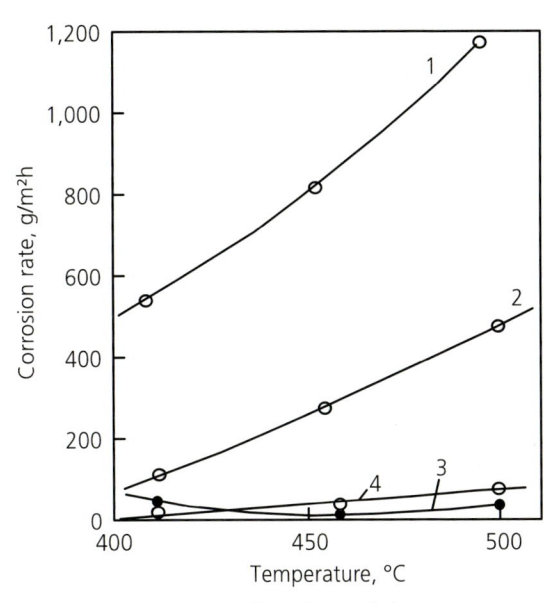

Figure 129: Temperature dependence of the corrosion rate of copper (1), iron (2), Ch18N9T (corresponding to Material No. 1.4541) (3) and nickel (4) in potassium hydroxide melts [1132]

Copper-aluminium alloys

Although some Cu-Al alloys show adequate resistance in NaOH solution [1135], these materials like copper are not adequately resistant in potassium hydroxide solution.

Copper-nickel alloys

The good corrosion resistance of the Cu-Ni alloys, particularly those with nickel contents greater than 20%, such as CuNi30 and CuNi40, that can additionally contain 0.5 to 5.75% Fe, up to 1.7% Mn, up to 0.05% Pb and up to 1% Zn, with respect to alkaline solutions is well known [1136, 1137]. CuNi30 suffers only little attack, for example between 328 and 339 K (55 and 66 °C), even with high potassium hydroxide concentrations (up to 95%) (< 0.025 mm/a (< 0.98 mpy)) and can therefore be utilized without reservations in waters containing potassium hydroxide [1131].

The corrosion resistance of the Cu-Ni alloys increases with increasing Ni content, and the copper-nickel alloys containing more than 20% of nickel can be utilized, as shown by good experience in NaOH solution, probably also as material coming into contact with solutions containing potassium hydroxide, also at higher temperatures.

Copper-tin alloys (bronze)

There is little data in the literature concerning the corrosion behavior of copper-tin alloys in alkaline solutions. The corrosion rate of copper-tin alloys with 8% Sn in 1 to 2 mol/l NaOH solutions is reported as being below 0.25 mm/a (9.84 mpy) at room temperature and approximately 0.5 mm/a (19.7 mpy) at higher temperatures [1137]. However, it must be assumed that the values in potassium hydroxide solution are still higher, so that utilization of these materials is probably ruled out.

Copper-tin-zinc alloys (red bronze)
Copper-zinc alloys (brass)

The corrosion resistance of brass in alkaline solutions is considerably better than that of pure copper, but not adequate for practical applications [1138]. By adding inhibitors it was found to be possible to reduce the corrosion rates of the two brass types CuZn20 (CW503L) and CuZn41 in 1 mol/l sodium hydroxide solution significantly [1139], and brass types with higher zinc content (greater than 35%) were found to be adequately resistant in 1 mol/l sodium hydroxide solution even without inhibitor, but it is questionable whether these results also apply for the behavior in potassium hydroxide solution.

The fact must be taken into consideration that under certain conditions, stress corrosion cracking has been observed in α-brass in NaOH solution.

Other copper alloys

Copper-beryllium alloys are utilized as annealable alloys for springs. With the commercially available alloys Beryvac-200® (Cu-2Be) and as beryllium conductor bronze Beryvac-60® (Cu-CoBe), a corrosion rate at 293 K (20 °C) of about 0.01 mm/a (0.39 mpy) in 50% sodium hydroxide solution is reported and can probably serve as rough indication of the behavior in potassium hydroxide solution [1140].

Seawater

Author: *P. Drodten* / Editor: *R. Bender*

Copper

A number of copper materials have proved of value in seawater applications. The resistance of these copper materials to seawater is determined by the formation and maintenance of a highly adhesive protective layer. This protective layer comprises mainly copper(I)- and copper(II)-oxides. Depending on the composition of the water, other compounds may also be deposited in the covering layer forming over it.

Figure 130 shows a sketch of the structure of a protective layer on copper alloys as it develops in seawater applications.

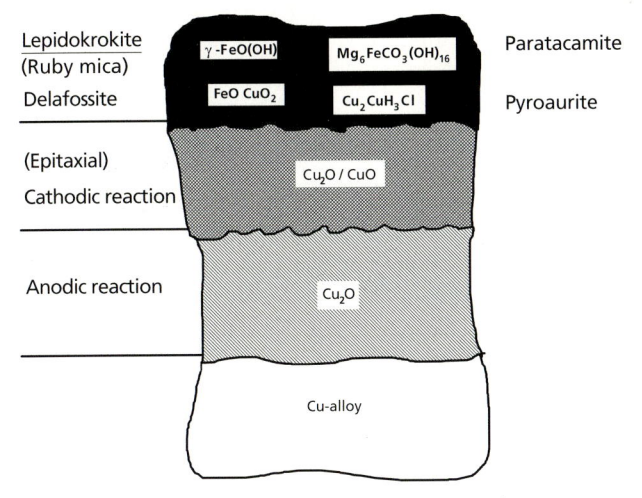

| Lepidokrokite (Ruby mica) | γ-FeO(OH) | $Mg_6FeCO_3(OH)_{16}$ | Paratacamite |
| Delafossite | FeO CuO_2 | Cu_2CuH_3Cl | Pyroaurite |

(Epitaxial) Cathodic reaction — Cu_2O / CuO

Anodic reaction — Cu_2O

Cu-alloy

Figure 130: Protective layer structure in copper materials in seawater

The primary oxidic protective layer forms quickly and uniformly only on clean surfaces in pure, well-aerated seawater. For this reason, the water conditions and surface state are decisive for the corrosion behavior of a structural element made of copper materials over its entire service life. Data on the structure of the covering layers, which form depending on medium and potential on the various copper materials and their effect on corrosion behavior are contained in [1141, 1142]. Table 176 lists the most important copper types suitable for use in seawater.

The pure copper types show only moderate resistance in seawater. They are not used for structural elements directly exposed to seawater. For such structural elements, copper alloys of varying alloy basis are used [1143]. Of the brass types CuZn (see section Copper-zinc alloys), DIN-Mat. No. 2.0460 (CW702R), corresponding to CuZn20Al2 (naval brass) is used as a standard material for seawater pipelines on ships and in coastal power plants. Also frequently used are alloys based on CuNiSi (see section Copper-nickel alloys); CuSn (tin bronzes) (see section Copper-tin alloys),

CuSnZn (red brass, see section Copper-tin-zinc alloys) and CuAl (aluminium bronzes, see section Copper-aluminium alloys).

DIN-Mat. No.	Abbreviated designation	Standard
2.0060	E-Cu57	DIN 1787 [1144]
CW004A (2.0065)	Cu-ETP (E-Cu58)	DIN 1787 [1143]
CW020A (2.0070)	Cu-PHC (SE-Cu)	DIN 1787 [1143]
CW024A (2.0090)	Cu-DHP (SF-Cu)	DIN 1787 [1143]

Table 176: Copper types for use in seawater

In all of the copper materials named above, the corrosion rate is very low due to uniform surface corrosion and is negligible in practical applications of these materials. These materials may, however, show local corrosion in quiet or slow-moving seawater. This is caused mainly by concentration elements, especially oxygen concentration differences. Therefore, the main areas at risk are segments beneath deposits, under corrosion products and under fouling, as well as crevices in which oxygen access is hindered.

Stress corrosion cracking does not occur in copper materials in pure seawater. However, in the presence of contaminations with ammonia or ammonium salts, which may come from discharge of sewage or decomposition of organic compounds, copper-zinc alloys (see section Copper-zinc alloy) are susceptible to stress corrosion cracking. The testing of copper alloys for susceptibility to stress corrosion cracking in the presence of ammonia is described in DIN 50916-1 [1145] and DIN 50916-2 [1146]. CuZn alloys with more than 80 % Cu show stress corrosion cracking in exceptional cases only. In some copper materials under certain conditions, selective corrosion in the form of dezincification or dealumination may occur. This tends to occur mainly in sections with incomplete or damaged protective layers. Deposits and contact with higher-potential materials encourage the development of selective corrosion. Suitable alloying measures can be applied to effectively prevent both types of selective corrosion.

When certain, material-dependent, seawater flow rates are exceeded, the covering layers may be eroded and the protective layer destroyed. This results in exacerbated, shallow pit-like corrosion. Structural elements such as pipelines, pump casings, pump impellers and drive propellers, in which copper materials are preferred, may be at risk. The relevant regulations (DIN 81249-2) provide information on the critical threshold rates as dependent on material and structural element type. General information on corrosion tests in flowing liquids is provided by DIN 50920-1.

Table 177 lists the chemical compositions, critical flow rates and the shear stress determined for a number of copper materials that result in the destruction of the covering layer and thus to incipient corrosion [1147]. The addition of alloying elements, especially nickel, effectively increases the resistance in flowing seawater. This is also seen in Table 178, in which the critical flow rates in seawater for the most important copper pipe materials, which rates also depend on the pipe diameter.

Material	% Cu	% Ni	% Fe	% Mn	% Cr	% Al	% Zn	% Other	Temperature K (°C)	Critical rate m/s	Critical shear stress N/m²
C12200 cf. SF-Cu, 2.0090	99.9							0.024 P	290 (17)	1.3	9.6
C68700 cf. CuZn20Al2, CW702R, 2.0460	77.7		<0.01			2.1	20.1	0.04 As	285 (12)	2.2	19.2
C70600 cf. CuNi10Fe1Mn, CW352H, 2.0872	87.4	11.0	1.58	0.42					300 (27)	4.5	43.1
C71500 cf. CuNi30Mn1Fe, CW354H, 2.0882	68.9	30.6	0.53	0.65					285 (12)	4.1	47.9
C72200	balance	16.3	0.80		0.52			0.018 Ti 0.12 Zr	300 (27)	12.0	196.9

Table 177: Chemical composition of copper materials and the critical values determined for these materials for flow rate and shear stress strain in synthetic seawater [1147]
Test conditions: pH 8, T = 291 K (18 °C), c_{Cl^-} = **19 g/l**, c_{O_2} = **6.4 mg/l**
Test period: 36 days

Material designation	EN	Old DIN-Mat. No.	Calculated flow rate, m/s	
			DN < 50 mm	DN ≥ 50 mm
CuZn20Al2	CW702R	2.0460	max. 2.8	max. 3.0
CuNi10Fe1.6Mn	–	2.1972	max. 2.5	max. 3.5
CuNi10Fe1Mn	CW352H	2.0872	max. 2.5	max. 3.5
CuNi30Mn1Fe	CW354H	2.0882	max. 3.1	max. 4.5
CuNi30Fe2Mn2	CW353H	2.0883	max. 4.5	max. 6.0

Table 178: Maximum flow rates for pipes made of copper materials at different pipe diameters DN [1148]

Due to their good heat conductivity, copper materials are candidates for use in heat exchangers, evaporators, etc.

Figure 131 shows the behavior of some copper materials under the conditions applying in a multistage evaporator in salt solutions at given content levels of NaCl, $CaCl_2$ and $MgCl_2$ [1149]. A temperature rise of 313 to 380 K (40 °C to 107 °C) increases the corrosion rate of unalloyed copper from 0.28 mm/a to 0.46 mm/a (11.02 to 18.11 mpy). In CuNi alloys, the temperature increase is less noticeable, as it is with Monel® as well (cf. DIN-Mat. No. 2.4360).

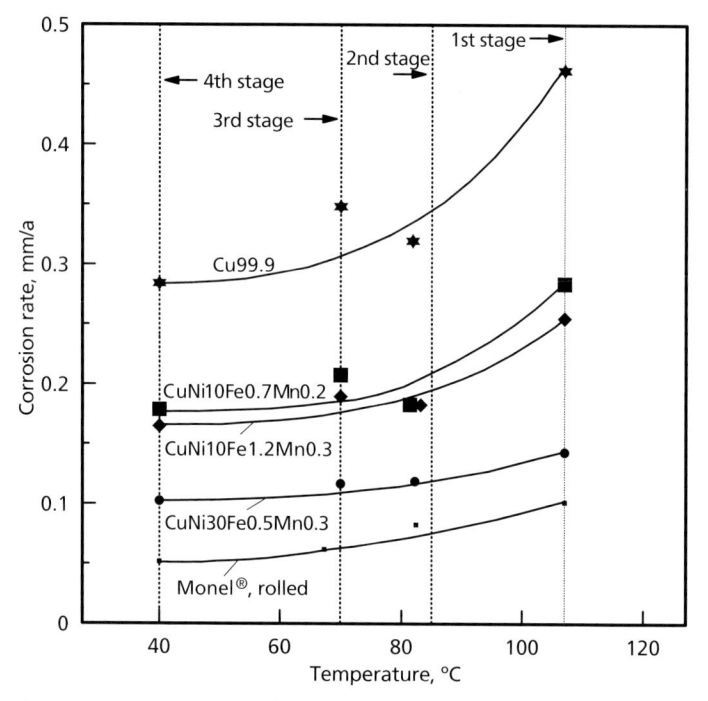

Figure 131: Corrosion rates of copper, copper-nickel and nickel-copper alloys in salt solutions in a four-stage evaporator at different temperatures after an operational period of 185 days [1149]

In connection with applications in seawater desalinization plants, the corrosion rates in Table 180 were measured in tests of pipes made of the three copper materials, the chemical compositions of which are listed in Table 179 in flowing, warm seawater with an oxygen content of less than 6 ppb [1150].

No.	Material	Cu	Fe	Sn	As	P	Pb	Zn
1	C12200 cf. SF-Cu, CW024A	99.9				0.015–0.040		
2	C14200 cf. CuAsP, old 2.1491	99.4			0.15–0.50	0.015–0.040		
3	C19400 cf. CuFe2P, CW107C	97.0	2.1–2.6	0.03		0.01–0.04	0.03	0.05–0.20

Table 179: Chemical composition of copper materials tested in flowing seawater (mass%) [1150]

No.	Test duration, d								
	90	170	697	90	170	697	90	170	697
	Flow rate, m/s								
	1.2			2.4			4.8		
	Corrosion rate, mm/a (mpy)								
1	0.024 (0.94)	0.029 (1.14)	0.025 (0.98)	0.029 (1.14)	0.038 (1.50)	0.054 (2.13)	0.057 (2.24)	0.059 (2.32)	0.11 (4.33)
2	0.029 (1.14)	0.039 (1.54)	0.033 (1.30)	0.033 (1.30)	0.042 (1.65)	0.057 (2.24)	0.066 (2.60)	0.062 (2.44)	0.12 (4.72)
3	0.021 (0.83)	0.025 (0.98)	0.024 (0.94)	0.025 (0.98)	0.033 (1.30)	0.058 (2.28)	0.56 (22.05)	0.055 (2.17)	0.11 (4.33)

Table 180: Mean erosion levels for the copper materials in Table 179 in flowing seawater at 394 K (121 °C) [1150]

In pipes made of different copper materials (Table 179), only a very low level of wall thickness reduction of 0.007 mm was determined in a salt solution containing 1.2 times as much salt as seawater at a temperature of 377 K (104 °C) and a flow rate of 1.5 m/s after 5 months of operation [1150].

In tests carried out in the laboratory on copper (99.8 %) in biologically inert seawater, pH 7.8 at solute oxygen content levels of 7 ppm at different pressures and temperatures of 278, 283 and 293 K (5 °C, 10 °C and 20 °C), it was seen that the general corrosion of copper increases, both with increasing temperature and with increasing pressure. At temperatures of 283 K (10 °C) and lower, the influence of the pressure is negligible. The pressure, however, has a pronounced influence at temperatures of 293 K (20 °C) and higher [1151].

Copper and copper-based materials are characterized in seawater not only by good corrosion resistance, but also by their resistance to fouling. They are therefore used successfully in equipment, structural elements and pipelines exposed to seawater to prevent cross-sectional reductions due to adherent fouling [1144, 1152].

In cases in which greater wall thicknesses or higher strength levels are required, plated structural elements offer technical economic advantages. CuNi10Fe1Mn (CW352H, DIN-Mat. No. 2.0872) has proved effective as a plating material [1153]. This material has also been used outside plating of hulls to take advantage of both the corrosion resistance and antifouling properties [1154, 1155]. The fouling resistance of copper materials is based on the precipitation of small amounts of Cu ions in the reaction with seawater, which ions are toxic for the fouling organisms. Since precipitation of the Cu ions only occurs in the presence of free corrosion, any degree of cathodic polarization results in the loss of fouling resistance. For this reason, all contact with metals bearing a more negative potential must be avoided.

Copper-aluminium alloys

The seawater-exposed copper-aluminium, wrought and cast alloys with further additive elements (formerly: aluminium bronzes) are listed in Table 181. The free corrosion potentials of these materials in seawater are within the range $U_H = -0.04$ V to -0.20 V and their mean corrosion rate due to uniform surface corrosion is 0.02 mm/a–0.04 mm/a (0.79 mpy–1.57 mpy). The values increase with increasing temperatures and in brackish water. The multiple substance aluminium bronzes, especially those containing more than 4% Ni, are superior to the binary alloys.

The corrosion resistance of these alloys (binary and multiple substance bronzes) increases with their aluminium content. The homogeneous alloys with α-phase, which contain up to about 7% aluminium, and the two phase α/β- or α/γ-alloys with higher aluminium content levels, show little practical difference in terms of corrosion behavior. They show little tendency to develop pitting corrosion and are also effectively resistant to boiling seawater. They are susceptible to fouling due to the low level of Cu dissolution that accompanies the corrosion. The emission of toxic copper compounds is not sufficient to prevent fouling. Up to content levels of about 5% aluminium they can be readily drawn to make pipes. At higher aluminium contents and higher strength levels accordingly they are used only as forged or cast parts.

The corrosion resistance is reduced when the β-phase disintegrates into α- and γ_2-phase during gradual cooling from 923 K (650 °C), e.g. with greater wall thicknesses. The γ_2-phase shows pronounced anodic qualities against the matrix and results in selective corrosion in the form of dealumination, particularly around welding seams with an integral network of γ_2-phase. The disintegration of the β-phase can be prevented by stabilizing additions of iron, nickel (> 4%), silicon and manganese, which at the same time will improve the mechanical property values by means of precipitation hardening [1156].

Designation	Old DIN-Mat. No.	Abbreviated designation	Standard
CW306G	2.0936	CuAl10Fe3Mn2	DIN 17665 [1157]
–	2.0957.01/ 2.0957.91	G-CuAl8Mn8	BWB WL 2.0957 [1158]
–	2.0958	CuAl8Mn	BWB WL 2.0958 [1159]
–	2.0960	CuAl9Mn2	DIN 17665 [1156]
–	2.0962.01	G-CuAl8Mn	DIN EN 1982 [1160]
CW307G	2.0966	CuAl10Ni5Fe4	DIN 17665 [1156]
–	2.0967	CuAl9Ni7	BWB WL 2.0967 [1161]
–	2.0968.01	G-CuAl9Ni7	BWB WL 2.0968 [1162]
CC333G	2.0975.01	G-CuAl10Ni	DIN EN 1982 [1159]
CW308G	2.0978	CuAl11Ni6Fe5	DIN 17665 [1156]
CC334G	2.0980.01	G-CuAl11Ni	DIN EN 1982 [1159]
–	2.1357	G-CuMn9Al6Zn4Ni2Fe2	BWB WL 2.1357 [1163]
–	2.1358	G-CuMn10Zn8Al6Ni2Fe2	BWB WL 2.1358 [1164]

Table 181: Copper-aluminium alloys used in seawater (formerly aluminium bronzes)

Specially developed alloys with 2% to 3% silicon, (e.g. 6.5% Al + 2.2% Si; 5.8% Al + 2.6% Si; 5% Al + 2.8% Si; 4.2% Al + 3.2% Si), in the forged, unquenched state show, with a corrosion depth of 0.10 mm to 0.14 mm after 600 hours, the same resistance to impingement corrosion as a water-quenched binary aluminium bronze with 8.9% Al. The optimum composition is considered to be an alloy with 6% to 6.4% Al; 2.0% to 2.4% Si and 0.8% to 1.0% Fe (cf. CW301G, CuAl2Si2Fe) [1165–1170].

In a CuAl6 alloy subjected to testing in aerated seawater (100 h at 293 K (20 °C)), additions of 2% Si and 0.2% Mn reduced the corrosion rate of 0.092 mm/a to 0.029 mm/a (3.62 to 1.14 mpy) [1171].

The corrosion rates of a CuAl7Fe1Mn1 cast alloy in seawater can be reduced by addition of 2% Sn at 289 K (16 °C) from 0.127 mm/a to 0.076 mm/a (5.00 to 2.99 mpy) and at 353 K (80 °C) from 0.185 mm/a to 0.134 mm/a (7.28 to 5.28 mpy). The alloying elements iron and manganese result in excellent mechanical properties and allow for use of the material in heat exchanger systems and other maritime applications [1172].

Table 182 shows results of short-term tests in moving seawater of two copper-aluminium bronzes in the cast state at 289 and 353 K (16 °C and 80 °C). Addition of silicon has a negative influence on the corrosion behavior of the alloy. Different heat treatment states also showed no improvements over the cast state [1173].

Tests on the influence of contaminations and bacterial microfouling on the behavior of aluminium bronzes and CuNi10 (DIN-Mat. No. 2.0811), which materials are used on a large scale in seawater-exposed heat exchangers, showed that bacteria and

Alloying content (mass%)					Temperature K (°C)	Corrosion rate g/m² d
Al	Fe	Mn	Cr	Si		
10.6	3.2	0.8	0.3		289 (16)	1.5
10.6	3.2	0.8	0.3		353 (80)	2.1
10.0	3.0	0.8	0.5	1.0	289 (16)	2.1
10.0	3.0	0.8	0.5	1.0	353 (80)	2.4

Table 182: Corrosion behavior of cast CuAl alloys in seawater, tests on rotating discs [1173]

sulfides result in lasting damage to the surface passivation layer and thus to increased corrosion. The combination of bacterial growth on the surface and anaerobic conditions produces an extremely aggressive corrosion system to which none of the copper materials tested showed a satisfactory level of resistance [1174].

Table 201 also lists corrosion rates of aluminium-multiple substance bronzes in rapidly flowing seawater. The aluminium bronzes are, accordingly, clearly superior to brasses under these conditions [1175].

Tests of a cast alloy CuAl10Ni3Fe2-C (CC332G, G-CuAl9Ni) in a 3% NaCl solution under conditions of erosion corrosion in comparison with other materials commonly used in seawater pumps resulted in different incubation times and corrosion rates, which are listed in detail in [1176].

Applications

Aluminium bronzes with 5–10% aluminium have proved useful in condensers and heat exchangers in fixed installations on land using tidal water, contaminated seawater and brackish water for cooling purposes. Under the influence of slow water movement 0.8 m/s they are not insensitive to pitting corrosion. The higher cavitation and erosion resistance levels compared to other copper materials are turned to advantage in pumps, impellers, valves, spindles and ship propellers [1177–1181].

In a comparison of condenser pipes in flowing seawater, the aluminium bronze CuAl7Ni1Fe0.3Cr0.05 showed better behavior than CuAl6.1, tin bronze CuSn5.1 and Cu99.99. In condenser pipes in a power plant in Japan that were exposed to soiled harbor water, aluminium bronze showed pitting corrosion, as did aluminium and admiralty brass as well as copper-nickel alloys, whereas a tin bronze CuSn6Al1-Si0.2 showed comparatively good behavior [1182, 1183].

The cast bronze CuAl9Ni5Fe4Mn1.5 proved superior to the cast bronze CuAl8Mn8.2-12.8Fe2.7-3.8Ni2.7 and the 13% chromium steels in terms of resistance to the eroding stress load on the turbine vanes and fastening elements in the French tidal power plant in the mouth of the Rance [1184]. The cavitation-resistant multiple substance bronzes with iron and nickel are used for larger or more stressed ship propellers [1185].

Copper-aluminium and copper-zinc materials with the chemical compositions listed in Table 183 were subjected to long-term testing under conditions as they apply in seawater desalinization plants in a test loop heat exchanger. The mean corrosion rates determined are summarized in Table 184. In tests in seawater 1.2 times as much salt concentration at 377 K (104 °C) and a flow rate of 1.5 m/s, these copper materials also showed very low corrosion levels after 5 months [1186].

No.	Material	Cu	Al	Zn	Fe	Sn	As	Pb
1	C60800 cf. CuAl5As, 2.0918	balance	5–6.5		0.10		0.02–0.35	0.10
2	C61300	balance	6.8		3.5	0.2–0.5		
3	C68700 cf. CuZn20Al2	76–79	1.8–2.5	balance	0.06		0.02–0.1	0.07
4	C44300 cf. CuZn28Sn1, 2.0470	70–73		balance	0.06	0.8–1.2	0.02–0.1	0.07

Table 183: Chemical compositions of copper materials tested under simulated conditions of a seawater desalinization plant (mass%) [1186]

	Test duration, d								
	90	170	697	90	170	697	90	170	697
	Flow rate, m/s								
	1.2			2.4			4.8		
No.	Corrosion rate, mm/a (mpy)								
1	0.026 (1.02)	0.030 (1.18)	0.019 (0.75)	0.030 (1.18)	0.036 (1.42)	0.044 (1.73)	0.057 (2.24)	0.055 (2.17)	0.054 (2.13)
2	0.023 (0.91)	0.030 (1.18)	0.018 (0.71)	0.026 (1.02)	0.033 (1.30)	0.022 (0.87)	0.046 (1.81)	0.049 (1.93)	0.097 (3.82)
3	0.012 (0.47)	0.022 (0.87)	0.015 (0.59)	0.019 (0.75)	0.028 (1.10)	0.038 (1.50)	0.044 (1.73)	0.046 (1.81)	0.070 (2.76)
4	0.015 (0.59)	0.028 (1.10)	0.016 (0.63)	0.002 (0.08)	0.028 (1.10)	0.047 (1.85)	0.050 (1.97)	0.048 (1.89)	0.093 (3.66)

Table 184: Mean corrosion rates for the copper materials from Table 183 in flowing seawater at 394 K (121 °C) [1186]

Practical experience with the alloy C68700 (cf. CuZn20Al2, DIN-Mat. No. 2.0460) revealed satisfactory levels of resistance to seawater with oxygen contents up to 200 ppb at a temperature of 350 K (77 °C). At 377 K (104 °C), resistance to seawater with higher oxygen content levels was sharply reduced. Since the presence of oxygen generally reduces the service life of the pipes, an efficient method of oxygen elimination is important for successful use of copper alloys in seawater desalinization plants. CuAl bronze showed satisfactory behavior under all the environmental conditions of a desalinization plant [1186]. CuAl bronze is also recommended as a material for bolts, screws, nuts and washers in seawater and in saltwater cooling towers [1187, 1188].

In large-scale pump engineering, the combination of cast parts made of CuAl alloys with welded semifinished components is often an optimum solution. Examples of applications of CuAl alloys in underwater engineering, pipelines and flanges in offshore structures are found in [1189].

Comparative corrosion tests of materials used frequently in pumps, G-CuAl10Ni (CC333G, 2.0975.01) and G-CuSn10 (CC480K, 2.1050.01), to determine the influence of flow rate and ammonium or sulfide content in the seawater revealed a clear increase in corrosion rates in the moving medium and with increasing content levels of ammonium and sulfide for both cast materials. The material G-CuAl10Ni showed superior behavior to G-CuSn10 by a considerable margin [1190].

Copper-nickel alloys

The seawater-exposed copper-nickel wrought and cast alloys with additional alloying elements are listed in Table 185. The free corrosion potentials of these materials in seawater are within the range $U_H = -0.03$ V to $+0.03$ V and their mean corrosion rates due to uniform surface corrosion are between 0.01 mm/a and 0.05 mm/a (0.39 and 1.97 mpy).

Old DIN-Mat. No.	Abbreviated designation	Standard
2.0872	CuNi10Fe1Mn	DIN 17664 [1191]
2.0880	CuNi17Mn5Al2Fe	BWB WL 2.0880
2.0882	CuNi30Mn1Fe	DIN 17664 [1191]
2.0883	CuNi30Fe2Mn2	DIN 17664 [1191]
2.1504	CuNi14Al3	BWB WL 2.1504
2.1972	CuNi10Fe1.6Mn	BWB WL 2.1972

Table 185: Copper-nickel alloys used in seawater [1148]

In the system copper-nickel, increasing nickel content levels improve resistance to seawater, brackish water and dilute or concentrated chloride solutions. As Figure 132 shows, the corrosion rate of the unalloyed coppers by seawater, approximately 1.5 g/m²d, is reduced in CuNi alloys with 45 %–67 % nickel to 0.12 g/m²d [1192].

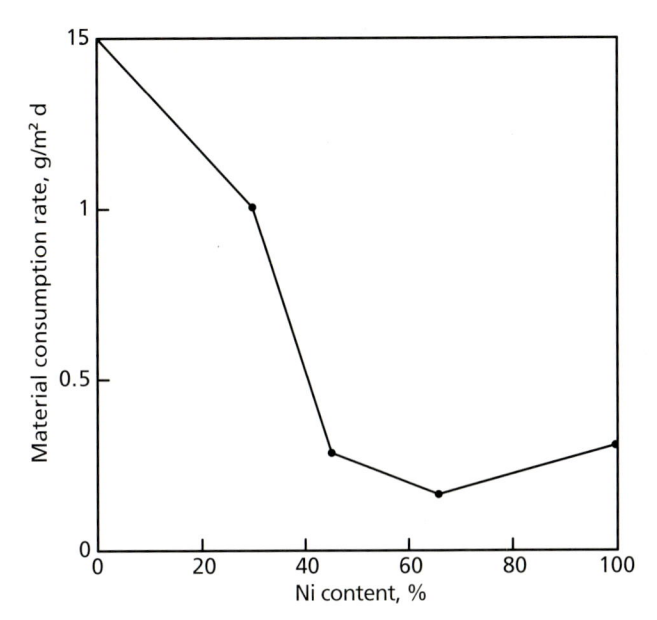

Figure 132: Influence of the nickel content in copper-nickel alloys on corrosion in seawater [1192]

Tests of copper-nickel alloys with the Cu/Ni ratios 90/10 (CuNi10), 70/30 (CuNi30), 50/50 (CuNi50 or NiCu50) and 30/70 (NiCu30) in the laboratory, including electrochemical methods, in 3.5 % NaCl solution with pH 8.1, showed lower levels of mass loss than in similar tests in seawater. With increasing copper content, the difference between the corrosion rates in the two mediums also increases. The only exception is the 30/70 alloy (NiCu30), which also showed the lowest mass losses. The cause of these different results is apparently the formation of covering layers consisting in seawater mainly of nickel and copper oxides, whereas in the NaCl solution higher proportions of copper chlorides are integrated in the layer, to which a better protection is ascribed. In the corrosion reaction, the copper-nickel alloys with nickel contents of 50 % and more show a tendency to dissolution of the copper and accumulation of nickel at the surface. At alloying contents of less than 50 % Ni, as in the materials CuNi10 and CuNi30, the two components enter into solution equally [1193–1195].

The system of copper-nickel alloys includes important material groups with numerous maritime applications due to their resistance to corrosion by seawater and brackish water. These materials include CuNi5, CuNi10, CuNi30 and CuNi40.

The modified types iron and manganese additions, CuNi10Fe1Mn (CW352H, 2.0872) and CuNi30Mn1Fe (CW354H, 2.0882), are the most important. The alloy CuNi5Fe (formerly DIN-Mat. No. 2.0862) is used in seawater pipelines, albeit with some limitations. A previously used alloy CuNi20Fe (C71000, formerly DIN-Mat. No. 2.0878) has now been completely replaced by CuNi10Fe1Mn and CuNi30Mn1Fe for other reasons. The alloy CuNi40Fe, with its higher levels of strength and oxidation resistance, is used in special cases, but not in applications involving direct seawater exposure.

The engineering binary alloys show improved corrosion resistance due to alloying additions that enhance their tendency to protective layer formation and improve and stabilize the properties of the protective layers. In this respect, iron and manganese have proved particularly effective.

Experiential data on comparisons of materials such as copper, nickel, nickel-copper and other materials commonly used with seawater in a four-stage vacuum evaporator in a rock salt refinery confirm the technical and economic advantages of the two alloys CuNi10Fe1Mn (C70600, CW352H, formerly DIN-Mat. No. 2.0872) and CuNi30Mn1Fe (C71500, CW352H, formerly DIN-Mat. No. 2.0882) [1149]. This experience has been confirmed by systematic tests with copper, CuNi10Fe1Mn, CuNi30Mn1Fe and NiCu30Fe (Monel® type, e.g. Monel® 400, DIN-Mat. No. 2.4360) in a multistage seawater evaporator plant. In the highest temperature stage at 380 K (107 °C), the corrosion of CuNi10Fe1Mn is half, that of CuNi30Mn1Fe a third and that of NiCu30Fe a fourth of the level of 0.46 mm/a (18.1 mpy) for copper (Figure 131) [1196].

The potential-pH diagrams for the alloys CuNi10 with 1.74% Fe (C70600, cf. DIN-Mat. No. 2.0872) and CuNi30Mn1Fe with 0.61% (C71500, cf. DIN-Mat. No. 2.0882) Fe in aerated and moving seawater at 298 K (25 °C) show that at potentials of $U_H \geq 0.2$ V the material CuNi10 at pH < 8.5 and the material CuNi30 at pH < 7.8 is resistant in seawater [1197].

Copper-nickel alloys are recommended as material for heat exchangers under seawater conditions. To increase their corrosion resistance, the types alloyed with iron are preferred, such as CuNi10Fe1Mn (DIN-Mat. No. 2.0872) with 10% Ni and 1.5% Fe as well as CuNi30Mn1Fe (DIN-Mat. No. 2.0882) with 30% Ni and 0.7% Fe or CuNi30Fe2Mn2 (DIN-Mat. No. 2.0883) with 2% Fe. These alloys are unsusceptible to stress corrosion cracking and are also effective in contaminated water. CuNi30Mn1Fe is suitable for higher water flow rates up to 3.5–4.0 m/s and also shows the highest resistance level in the presence of ammonium ions. CuNi10Fe1Mn is readily suitable at lower water flow rates and is also fouling-resistant. The best resistance results are obtained in all alloys with blasted surfaces [1198].

Influence of iron and manganese

The favorable influence of iron on the corrosion resistance of alloys of the type CuNi30 was described in [1199] and confirmed in [1200]. 0.5% Fe is recommended as a useful content level. The detailed tests in [1201] revealed efficient iron content levels between 0.3% and 1%. In [1202] the optimum efficiency was measured at 1.0% to 1.5% iron with manganese content levels of between 0.25% and 2%. Manganese supports the effect of the iron, especially at low iron content levels. Its influence on the corrosion resistance is, however, minor compared to that of the iron. On the other hand, iron content levels above 2% encourage crevice corrosion. At content levels exceeding 4% Fe the adhesion of the protective layers is compromised and both uniform surface corrosion and erosion increase.

The beneficial influence of iron also applies to the copper-nickel alloys with 10% and 20% Ni. The material CuNi10Fe1Mn (C70600, DIN-Mat. No. 2.0872) has attained to by far the highest status. Material loss due to erosion, as well as the risk of pitting corrosion, are about 1/3 less than with CuNi30.

Figure 133 shows that the favorable influence of low levels of iron and manganese also applies in the presence of increased water movement, thus increasing the range of permissible flow rates [1203, 1204].

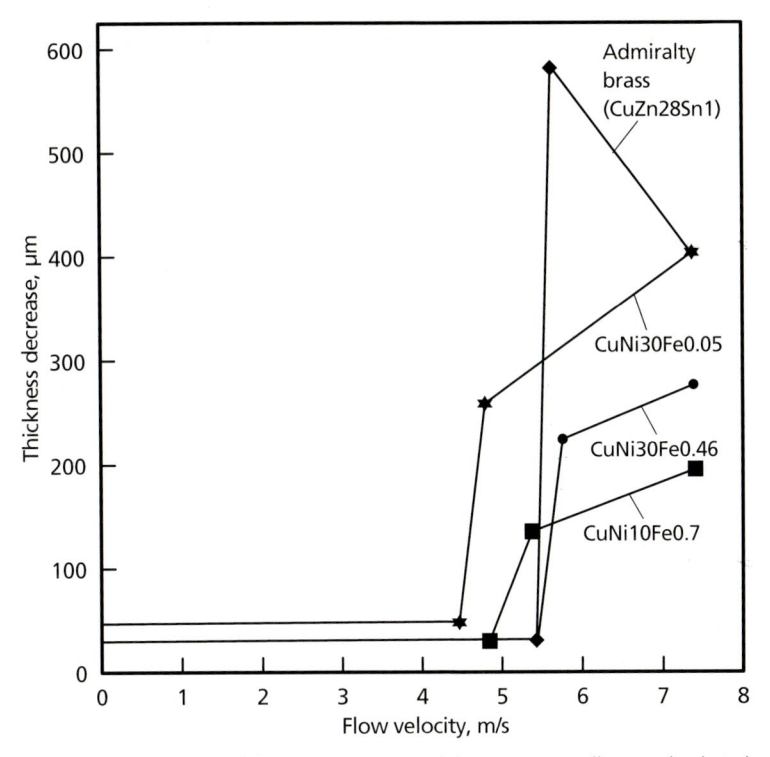

Figure 133: Influence of flow rate on material loss in CuNi alloys and admiralty brass (cf. CuZn28Sn1, CW706R, DIN-Mat. No. 2.0470) caused by seawater after 60 days at 299 K (26 °C) [1203, 1204]

Tests to determine the nature of the covering layers that develop on CuNi10Fe1Mn alloys, the conditions necessary for their formation and dissolution are described and discussed using the example of the material CuNi10Fe1Mn (CW352H, DIN-Mat. No. 2.0872) in [1205].

The connection between corrosion behavior and iron content is illustrated in Figure 134. The tests were carried out with rotating samples in fresh seawater. Even low-level iron additions of between 0.8% and 1.2% to CuNi10Fe1 reduce the corrosion in flowing seawater in a pronounced manner.

Figure 134 also contains two comparative values for the ferrous alloy CuNi30 with 0.46% Fe [1203].

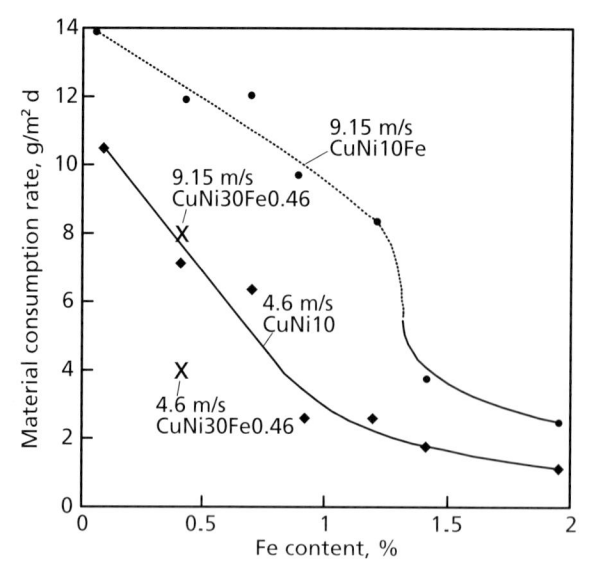

Figure 134: Influence of iron content on corrosion of CuNi10Fe1Mn in flowing seawater at 296 K (23 °C) [1203]

Exposure tests in the immersion zone at Helgoland, a German North Sea Island, confirm that iron content levels greater than about 2% cause the corrosion behavior of the materials CuNi10Fe1Mn (CW352H, DIN-Mat. No. 2.0872) to deteriorate again, even in seawater that is practically stagnant (Figure 135) [1206].

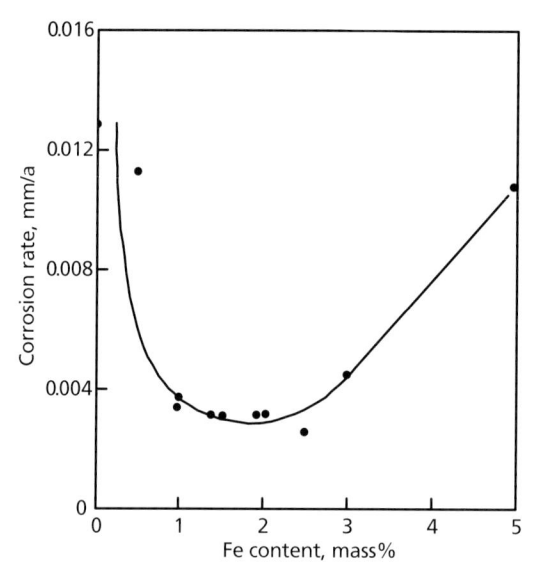

Figure 135: Influence of iron content on corrosion behavior in CuNi10Fe1Mn after exposure for 2 years in the Helgoland immersion zone [1206]

When heated to 773–823 K (500–550 °C), the engineering CuNi alloys with iron contents of 0.8–1.2 % can precipitate an iron-rich phase that occurs in area of welding seams or after other heat treatments when the solubility threshold is exceeded.

The precipitations are then redissolved by a heat treatment at higher temperatures. Condenser pipes are therefore homogenized at 1,173 K (900 °C) and then quenched. The effect of the precipitated iron phases is noticed, but the corrosion is not exacerbated to the point that the zones affected by welding require thermal aftertreatment.

Detailed tests on the influence of iron content and microstructure on corrosion behavior of CuNi 10 Fe alloy were carried out on the alloys listed in Table 186 [1207]. The tests were done in natural seawater at 303 K (30 °C), both at a flow rate of 1.5 m/s in test pipes and in rotating cylindrical samples at 1,500 rpm. Different microstructures were established by means of different heat treatments. In all of the alloys the highest corrosion rates were observed in the presence of discontinuous precipitations. In the presence of homogeneous or continuous precipitations, the corrosion rates did not differ significantly. The best behavior was observed in the homogeneous, single-phase alloys, on which highly adhesive, thin protective layers formed. In comparison to the CuNi 10 Fe alloys, the CuNi5Fe alloys with higher aluminium and manganese contents showed what were clearly the highest corrosion resistance levels.

Material group	Designation	Fe	Al	Mn	Ni	Cu
CuNi10Fe1Mn 2.0872	CuNi10Fe1.2	1.20		0.70	10.01	balance
	CuNi10Fe1.36	1.36		0.72	9.9	balance
	CuNi10Fe1.51	1.51		0.68	10.1	balance
	CuNi10Fe1.78	1.78		0.74	10.7	balance
CuNi 10 Fe (CW352H) (pure)	CuNi10Fe1.5	1.5		0.65	9.8	balance
	CuNi10Fe2.0	1.9		0.74	9.5	balance
	CuNi10Fe2.5	2.4		0.76	9.6	balance
CuNi5Fe (pure)	37	3.3	5.1	2.1	5.0	balance
	43	3.3	5.2	5.4	5.0	balance
	47	3.3	3.2	<0.1	5.0	balance
	48	3.2	7.1	<0.1	5.1	balance
	50	4.2	7.2	<0.1	5.0	balance
CuNi 10 Fe (pure)	CuNi10Fe1.4	1.44	<0.4	0.57	10.1	balance

Table 186: Chemical composition of the CuNi 10 Fe alloys tested in natural seawater (mass%) [1207]

Figure 136 also reveals the favorable influence of a homogenizing heat treatment on corrosion behavior of the alloys CuNi10Fe1Mn (CW352H, former DIN-Mat. No. 2.0872) and CuNi5Fe. A solution heat treatment at 1,123 K (850 °C) followed by quenching in water increases corrosion resistance compared to the heterogeneous state. The minimum rate of corrosion was at iron content levels of 1 %–2 % [1201].

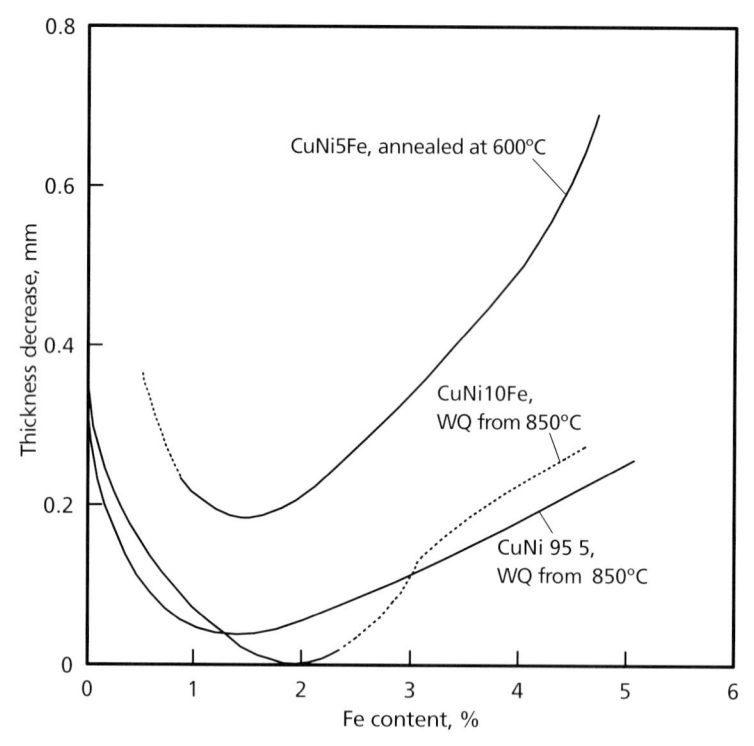

Figure 136: Influence of iron content and heat treatment on corrosion of CuNi10Fe1Mn and CuNi5Fe in seawater [1201]

Influence of chromium

Resistance to rapidly flowing seawater is important in operation of condensers with higher flow rates. Alloying additions of chromium can give copper-nickel alloys higher strength levels and better resistance to flow-induced corrosion due to precipitation hardening. Such a copper-nickel alloy with 30 % nickel and chromium content between 2.6 % and 3.2 %, developed under designation IN-732 (CA-719), is particularly suitable as a material for condenser pipes. Its advantage is expressed in higher water flow rates above 6 m/s up to approximately 90 m/s [1208, 1209].

The chemical composition of this alloy and its mechanical properties in comparison to material CuNi30Mn1Fe (CW354H) are seen in Table 187 and Table 188.

	Ni	Cr	Fe	Mn	Zr	Ti	C	Cu
IN-732	29–32	2.6–3.2	≤ 0.25	0.5–1.0	0.08–0.2	0.02–0.08	≤ 0.2	balance
CuNi30Mn1Fe	30		0.5	0.6				balance

Table 187: Chemical composition of alloys CuNi30Mn1Fe (CW354H, DIN-Mat. No. 2.0882) and IN-732 (mass%)

	Hardness	Tensile strength	0.2% yield point	Elongation	Reduction of area	E-modulus	Shear modulus
	HRB	N/mm^2	N/mm^2	%	%	N/mm^2	N/mm^2
IN-732	87	615	385	33	65	15.4×10^7	15.4×10^7
CuNi30Mn1Fe		430	350	140	45		6.0×10^7	5.8×10^7

Table 188: Mechanical properties of alloys CuNi30Mn1Fe (CW354H, DIN-Mat. No. 2.0882) and IN-732

Even smaller proportions of 0.3 %–0.6 % chromium suffice to increase the corrosion resistance of copper-nickel alloys in flowing seawater. The test alloys IN-838 and IN-848 with nominal 16 % or 30 % nickel contain chromium additions at this level. Table 189 shows the composition of these alloys. The best effect of a chromium content level of 0.3 %–1.0 % is achieved in copper-nickel alloys with nickel content levels of 15 %–30 % nickel as shown in Figure 137 [1210].

		Ni	Cr	Fe	Mn	Si	Ti	C	Cu
IN-838	nominal	16	0.4	0.8	0.5				balance
	range	15–18	0.3–0.6	0.7–1.0	0.4–0.9	≤ 0.03	≤ 0.03	≤ 0.03	balance
IN-848	nominal	30	0.4	0.3	0.7				balance
	range	28–32	0.3–0.6	≤ 0.5	0.5–1.0	≤ 0.05	≤ 0.05	≤ 0.05	balance

Table 189: Chemical composition of chromium alloys IN-838 and IN-848 (mass%) [1210]

At low water flow rates, the corrosion resistance of alloys containing chromium is the same as, or little better than, chromium free alloys. A comparison of resistance to seawater moving at higher rates in relative values is provided by Figure 138. The corrosion resistance of copper is reduced sharply at comparatively low flow rates. In the chromium free CuNiFe alloys, corrosion resistance drops at rates above approximately 3–4 m/s. At higher flow rates, the superiority of the copper-nickel alloys containing chromium becomes evident [1210].

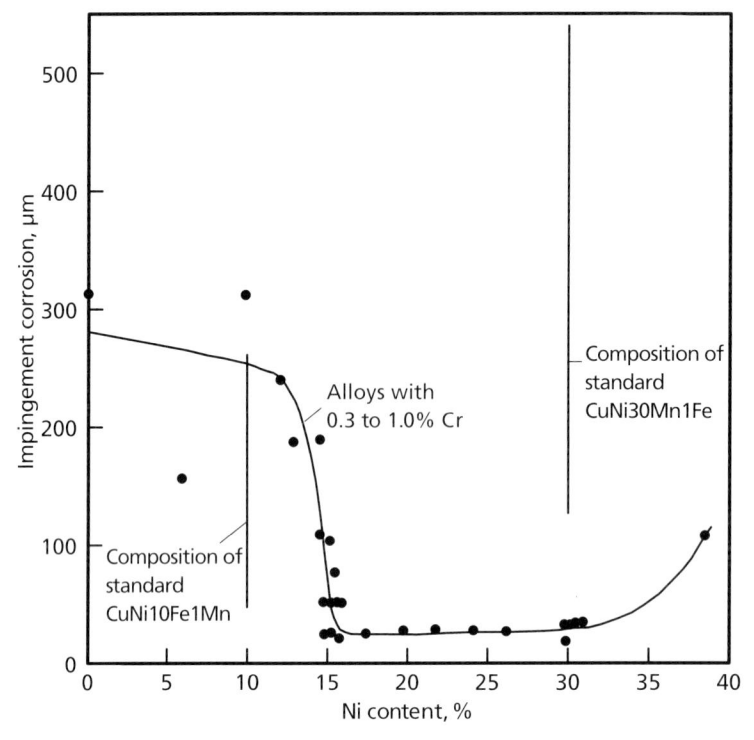

Figure 137: Influence of addition of 0.3 %–1.0 % Cr in CuNi alloys on corrosion in seawater at a flow rate of 7.6 m/s in the temperature range 280 to 300 K (7 °C to 27 °C) after 56 days [1210]

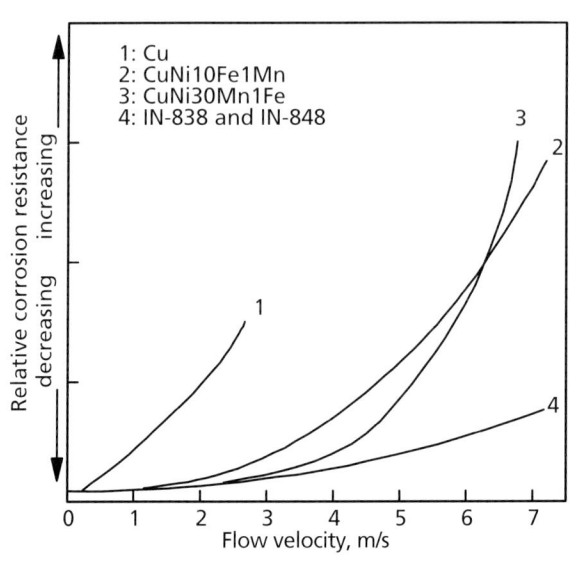

Figure 138: Relative corrosion resistance of copper and copper-nickel alloys as dependent on flow rate [1210]

The alloy CuNi5Fe, with 0.9–1.5 % Fe, 0.3–0.8 % Mn, is not used for condenser pipes, but for seawater pipelines and as a replacement for copper eroded at higher water flow rates or in turbulent currents. In the nineteen forties in England and the US, pipelines made of CuNi5Fe (C70400, formerly DIN-Mat. No. 2.0862) for seawater were installed on ships, but then had to be replaced after suffering damage at flow rates above 1.7 m/s by CuNi10Fe1Mn (C70600, DIN-Mat. No. 2.0872), which latter material is still installed in new ships today. In the seawater supply lines on an aircraft carrier, after 5,000 hours of operation downstream from turbulence-producing elements (T elements, lids, sharp bends, salient flanges, etc.), severe damage was observed on CuNi5Fe, whereas CuNi10Fe1Mn and CuNi30Mn1Fe (C71500, DIN-Mat. No. 2.0872) showed no damage. CuNi5Fe tends to show precipitations when exposed to heat up exceeding 873 K (600 °C) followed by gradual cooling as in laying projects. The corrosion resistance is reduced in the affected areas, usually in turbulence zones. Stagnating, contaminated seawater and brackish water also increase the likelihood of corrosion.

In the alloy with 7 % Ni, a restricted iron content range of 0.6–0.9 % is intended to prevent the susceptibility to pitting caused by iron-rich precipitations. The practical recommendation is, at rates above 1.7 m/s and with possible turbulence, to replace the alloy CuNi5Fe (C70400, formerly DIN-Mat. No. 2.0862) with CuNi10Fe1Mn (C71500, DIN-Mat. No. 2.0872) [1165, 1211–1215].

Despite the good resistance of these CuNi materials, the supply of copper ions is sufficiently high to prevent fouling.

In the CuNi10 and CuNi30 alloys, only the types with additions of iron and manganese have become technically important that have demonstrated excellent levels of general corrosion resistance and good resistance to pitting corrosion and erosion. The alloy CuNi30Fe2Mn2 (UNS C71640, CW353H, 2.0883) shows a higher resistance to erosion and erosion corrosion than the alloy CuNi30Mn1Fe (UNS C71500, CW354H, 2.0882). This alloy is specifically designed for seawater pipelines that carry solids as well [1216]. The material CuNi10Fe1Mn (CW352H, DIN-Mat. No. 2.0872) provides, in addition to excellent resistance in seawater, very good fouling resistance as well. This material is therefore also frequently used as plating material for seawater pipelines and has been used and tested for the outer plating of boats [1155].

CuNi10Fe1Mn and CuNi30Mn1Fe show suitable properties even at the higher temperatures generated by the heating elements in seawater desalinization plants after the distillation process. Table 190 lists a number of figures on the corrosion behavior of CuNi alloys in hot flowing seawater [1217].

The CuNi alloys listed in Table 191 with their chemical compositions were tested under the simulated conditions present in the saline heater of a multistage desalinization plant. The mean corrosion rates thus obtained are summarized in Table 192 [1150, 1186].

Materials	Materials	Corrosion rate, mm/a (mpy)		
at flow rate, m/s		1.2	2.5	5.0
Copper	C12200, cf. SF-Cu, 2.0090 (99.9 Cu, 0.015-0.04 P)	0.030 (1.18)	0.037 (1.46)	0.058 (2.28)
	C14200, cf. former 2.1491 (99.4 Cu, 0.15-0.05 As, 0.015-0.04 P)	0.038 (1.5)	0.041 (1.61)	0.062 (2.44)
	C19400, cf. CuFe2P, 2.1310 (97.0 Cu, 2.1-2.6 Fe, 0.01-0.04 P)	0.025 (0.98)	0.032 (1.26)	0.055 (2.17)
Admiralty brass[1]	C44300, cf. CuZn28Sn1, 2.0470 (70-73 Cu, 0.8-1.2 Sn, 0.02-0.1 As balance Zn)	0.028 (1.10)	0.028 (1.10)	0.048 (1.89)
Aluminium bronze	C60800, cf. CuAl5As, 2.0918 (5-6.5 Al, 0.1 Fe, 0.35 As, balance Cu)	0.029 (1.14)	0.036 (1.42)	0.055 (2.17)
	C61300 (6-8 Al, 3.5 Fe, 0.2-0.5 Sn, balance Cu)	0.029 (1.14)	0.032 (1.26)	0.048 (1.89)
Aluminium brass	C68700, cf. CuZn20Al2, 2.0460 (76-79 Cu, 1.8-2.5 Al, 0.02-0.1 As)	0.022 (0.87)	0.028 (1.10)	0.045 (1.77)
CuNi10Fe1Mn	C70600, cf. CuNi10Fe1Mn, 2.0872 (9-11 Ni, 1.0-1.8 Fe, 1.0 Mn)	0.017 (0.67)	0.026 (1.02)	0.046 (1.81)
	C70600[2], cf. CuNi10Fe1Mn, 2.0872 (9-11 Ni, 1.0-1.8 Fe, 1.0 Mn)	0.017 (0.67)	0.026 (1.02)	0.047 (1.85)
CuNi30Mn1Fe	C71500, cf. CuNi30Mn1Fe, 2.0882 (29-33 Ni, 0.4-0.7 Fe, 1.0 Mn, balance Cu)	0.015 (0.59)	0.023 (0.91)	0.040 (1.57)
	C71640[2], cf. CuNi30Fe2Mn2, 2.0883 (29-33 Ni, 1.8-2.2 Fe, 1.8-2.2 Mn, balance Cu)	0.015 (0.59)	0.022 (0.87)	0.039 (1.54)
	C71640, cf. CuNi30Fe2Mn2, 2.0883 (29-33 Ni, 1.8-2.2 Fe, 1.8-2.2 Mn, balance Cu),	0.014 (0.55)	0.023 (0.91)	0.042 (1.65)
	29-33 Ni, 4.8-5.8 Fe, 1.0 Mn, balance Cu	0.015 (0.59)	0.023 (0.91)	0.042 (1.65)
CuNi40Fe	CuNi40Fe (40-43 Ni, 1.5-2.5 Fe, 0.05-1.7 Mn, balance Cu)	0.015 (0.59)	0.020 (0.79)	0.038 (1.50)

1) butt-welded pipe
2) welded pipe

Table 190: Corrosion rate in pipe samples made of copper materials in flowing seawater at pH 7.4, 394 K (121 °C) and oxygen content between 0.007 ppm and 0.003 ppm after a test period of 170 days [1217]

Materials	Fe	Ni	Pb	Mn	Zn	Cu
C70600, cf. CuNi10Fe1Mn, 2.0872	1.0–1.8	9–11	0.05	1.0	1.0	balance
C71500, cf. CuNi30Mn1Fe, 2.0882	0.4–0.7	29–33	0.05	1.0	1.0	balance
C71640, cf. CuNi30Fe2Mn2, 2.0883	1.8–2.2	29–33	0.05	1.8–2.2	1.0	balance
Alloy No. 716	4.8–5.8	29–33	0.05	1.0	1.0	balance
Alloy No. 720	1.5–2.5	40–43	0.05	0.5–1.7	0.30	balance

Table 191: Chemical composition of the tested CuNi alloys (mass%) [1150]

Materials	Test period, d								
	90	170	697	90	170	697	90	170	697
	Flow rate, m/s								
	1.2			2.4			4.8		
	Corrosion rate, mm/a (mpy)								
C70600, cf. CuNi10Fe1Mn, 2.0872	0.013 (0.51)	0.017 (0.67)	0.018 (0.71)	0.020 (0.79)	0.026 (1.02)	0.032 (1.26)	0.047 (1.85)	0.046 (1.81)	0.087 (3.43)
C71500, cf. CuNi30Mn1Fe, 2.0882	0.009 (0.35)	0.015 (0.59)	0.012 (0.47)	0.018 (0.71)	0.032 (1.26)	0.037 (1.46)	0.037 (1.46)	0.040 (1.57)	0.078 (3.07)
C71640, cf. CuNi30Fe2Mn2, 2.0883	0.011 (0.43)	0.014 (0.55)	0.010 (0.39)	0.020 (0.79)	0.024 (0.94)	0.034 (1.34)	0.043 (1.69)	0.042 (1.65)	0.026 (1.02)
Alloy No. 716	0.009 (0.35)	0.015 (0.59)	0.009 (0.35)	0.018 (0.71)	0.023 (0.91)	0.032 (1.26)	0.038 (1.50)	0.042 (1.65)	0.063 (2.48)
Alloy No. 720	0.008 (0.31)	0.015 (0.59)	0.008 (0.31)	0.016 (0.63)	0.020 (0.79)	0.029 (1.14)	0.038 (1.50)	0.038 (1.50)	0.072 (2.83)

Table 192: Mean corrosion rates for the CuNi alloys listed in Table 191 at 394 K (121 °C) and at different flow rates [1150]

Pitting corrosion

Compared to other materials frequently used in a seawater environment, for example brass, nickel-copper or high-alloyed chromium-nickel steels, CuNi10Fe1Mn (CW352H, DIN-Mat. No. 2.0872) and CuNi30Mn1Fe (CW354H, DIN-Mat. No. 2.0882) are not readily susceptible to pitting corrosion. Some pitting corrosion with corrosion depths of 0.025–0.130 mm/a (0.98–5.12 mpy) is observed only in stagnating and contaminated waters. The alloys of the type CuNi10Fe1Mn with iron contents up to 1 % are better resistant to pitting corrosion than the alloys with higher iron content levels. On the other hand, resistance to erosion corrosion increases as the iron content rises to 1.5 % [1218]. The iron content of the engineering alloys is then a compromise between resistance to pitting and erosion corrosion.

Because of their good antifouling and corrosion properties, pipes made of CuNi alloys are frequently used in condensers, heat exchangers and pipeline systems with seawater exposure. The influence of higher temperatures on the corrosion behavior was investigated on standard commercial pipes made of CuNi30Mn1Fe under free exposure and with electrochemical methods in natural seawater [1219]. The composition of the pipes, 68.7 % Cu, 0.85 % Mn, 0.70 % Fe, 0.30 % Zn, balance Ni, corresponded to the normal values for an alloy CW354H (2.0882). The seawater used in the tests was taken from the Mediterranean Sea at a depth of 2 m. The salt content 3.5 g/kg and the pH level 8.2 were kept constant during the tests. The tests were carried out in stagnant solutions at different temperatures. The solute oxygen content levels, dependent on temperature, were:

T = 293 K (20 °C)   7.0 ppm O_2
T = 313 K (40 °C)   6.0 ppm O_2
T = 333 K (60 °C)   4.5 ppm O_2
T = 353 K (80 °C)   3.0 ppm O_2.

The tests demonstrated that with increasing temperature both the general corrosion and the tendency to pitting corrosion are reduced because the composition of the passivation layer changes as the temperature increases. Whereas at 293 K (20 °C) a significant portion of basic salts (oxichlorides, oxicarbonates) is observed, at temperatures above 333 K (60 °C) the passivation layer comprises mainly oxides and few chlorides. Crevice corrosion sensitivity increases with increasing temperature, because covering layers form in aerated and less aerated areas.

Tests of CuNi10Fe1Mn (CW352H, formerly DIN-Mat. No. 2.0872) in flowing natural seawater at a flow rate of 1.5 m/s also showed a better protection effect of the covering layers formed at temperatures above 313 K (40 °C) [1220].

Erosion corrosion

The corrosion rates of condenser pipes made of CuNi10Fe1Mn (CW352H, DIN-Mat. No. 2.0872) and CuNi30Mn1Fe (CW354H, DIN-Mat. No. 2.0882) cooled with seawater increase with the flow rate. Higher content levels of chloride ions or sulfide ions raise corrosion rates. In flowing seawater containing sulfide ions at a flow rate of 1.1 m/s, protective covering layers did not form readily and local corrosion was observed under these layers on both materials [1221].

The sensitivity to flowing seawater of the CuNi alloys in comparison to copper and other copper alloys is illustrated by Figure 139 [1222].

Figure 140 shows that with increasing iron content the resistance of the alloy CuNi10Fe1Mn (CW352H) to flowing seawater also increases. In the solution heat-treated state, the effect of the iron content at low flow rates begins at about 1 % Fe. At higher flow rates, at least about 1.4 % Fe are required to reduce the corrosion rates noticeably.

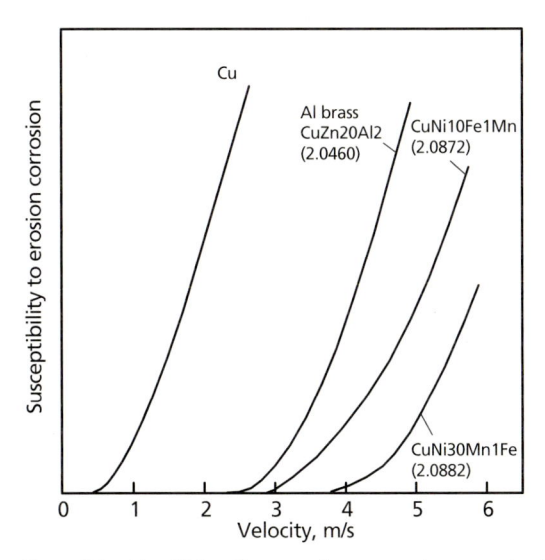

Figure 139: Sensitivity of copper alloys to erosion corrosion in seawater as dependent on the flow rate [1222]

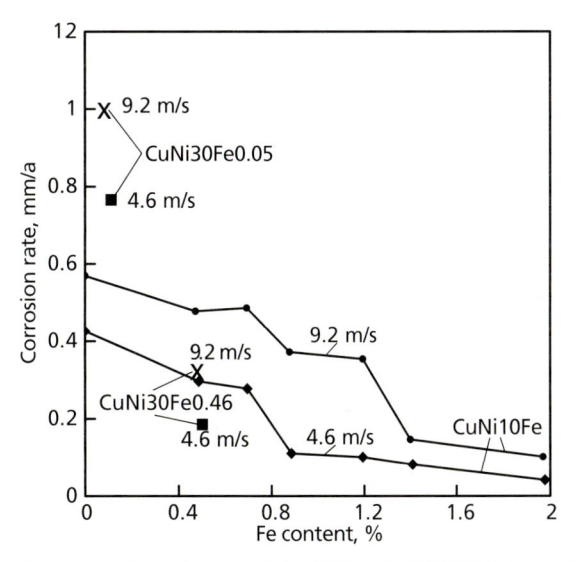

Figure 140: Corrosion rate of CuNi10Fe1Mn (CW352H) in solution heat-treated state (quenched from 1,173 K (900 °C)) as dependent on iron content, results with rotating samples at circumferential rates of 4.6 m/s and 9.2 m/s in seawater [1203]

For the tempered state with precipitated iron-rich phases, the influence of the iron content is no longer as pronounced and is much lower, as Figure 141 makes clear. The values for the material CuNi30Fe in Figure 140 and Figure 141 show that for this alloy iron content levels of only 0.5 % result in a clear improvement in its behavior in flowing seawater [1203].

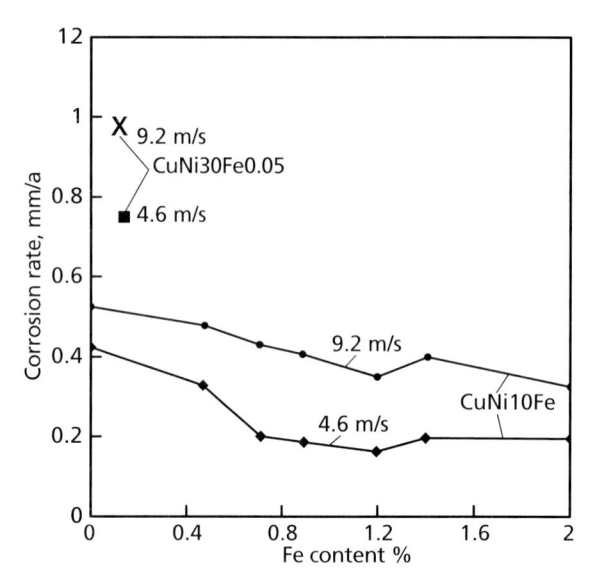

Figure 141: Corrosion rate of CuNi10Fe1Mn (CW352H) after 8 hours of tempering at 923 K (650 °C) as dependent on iron content, results with rotating samples at circumferential rates of 4.6 m/s and 9.2 m/s in seawater [1203]

Table 193 summarizes results obtained with rotating discs made of CuNi10Fe (CW352H) in laboratory tests in seawater. The positive effect of higher iron contents is lost again for the most parts in precipitation hardening. At high iron content levels, the solution heat treatment raises the threshold flow rate at which erosion corrosion begins considerably and reduces the drop rate of mass loss [1203]. The results of tests in Table 194 obtained for CuNi alloys and CuZn bronzes in flowing seawater also demonstrate the favorable influence of iron content in CuNi alloys on corrosion rates. They also show that CuNi10Fe1Mn with higher iron contents is superior to CuNi30Fe [1203].

Alloy	Composition mass%				Threshold rate m/s	Mass loss g/m² d
	Cu	Ni	Fe	Zn		
CuNi10Fe0.7	86.1	9.8	0.71	0.20	5.0	14.9
CuNi10Fe1.5[1]	88.4	9.6	1.51		4.95	14.7
CuNi10Fe3.52[1]	85.7	10.2	3.52		4.95	19.4
CuNi10Fe3.52[2]	85.7	10.2	3.52		7.2	5.7

1) precipitation hardened, 8 hours at 973 K (700 °C)
2) solution heat-treated (hardened) at 1,200 K (927 °C), quenched in water

Table 193: Chemical composition, threshold rates for occurrence of erosion corrosion and corrosion rates of CuNi10Fe1Mn (CW352H) alloys in flowing seawater at 299 K (26 °C), test period 60 days [1203]

Alloy	Chemical composition mass%							Flow rate m/s	
								3.7	4.6
	Cu	Ni	Fe	Mn	Al	Zn	Sn	Max. corrosion depth mm/a (mpy)	
CuNi10Fe0.7Mn0.2Zn0.2	89.06	9.84	0.71	0.19		0.20		0.9 (35.4)	0.3 (11.8)
CuNi10Fe2Mn0.5	87.18	10.25	2.07	0.46		0.04		0.3 (11.8)	0.6 (23.6)
CuNi30Mn0.5Zn0.2Fe0.06	69.05	30.26	0.06	0.48		0.15		1.8 (70.9)	1.2 (47.2)
CuNi30Fe0.5Mn0.5	68.92	30.16	0.47	0.45				0.6 (23.6)	0.9 (35.4)
CuZn28Sn1, CW706R, 2.0470	71.11					27.87	0.98	3.3 (129.9)	7.5 (295.3)
CuZn20Al2, CW706R, 2.0460	76.55				2.18	21.25		2.1 (82.7)	1.2 (47.2)

Table 194: Chemical composition and corrosion rates of CuNi10Fe1Mn (CW352H) alloys as well as of two copper bronzes in flowing seawater [1203]

The behavior of CuNi alloys under stress load from erosion corrosion depends on the flow rate as well as on a number of other parameters, among them the following seawater properties:

- pH level
- oxygen content
- temperature
- solute gases and solids
- chloride ion content

and material properties:

- alloying content
- surface condition.

[1223] provides an overview of the influential parameters, their effects and the test methods.

Applications of CuNi10Fe1Mn (UNS C70600, CW352H, 2.0872)

Examples of applications of CuNi10Fe1Mn:

- In shipbuilding for pipelines in the *cooling water circuit*, in the sanitary and fire extinguishing systems, for deck cleaning and for scoop inlets and outlets, also possibly for linings
- For compact water chambers or as linings on cast metal or steel
- For condenser pipes
- For anchoring and drag cables
- For buoys, floating islands, tackle, dragnets, bridge spans [1224]

- In seawater desalinization plants for pipeline bottoms and condenser pipes (but not in first stages), pipelines with seawater and enriched saline solutions, water containers, and to line steel containers in the high-temperature zone, for water chambers and evaporator corpuses as lining material when low iron content levels are required and for heat exchangers in heat recovery systems
- For industrial plants using seawater or brackish water for cooling – as in shipbuilding.

Examples of applications of copper-nickel alloys in construction of ship and boat hulls are to be found in [1225]. For the hull of the 20 m shrimp boat "Copper Mariner" the material CuNi10Fe1Mn was selected to prevent fouling and the resulting increased surface roughness. Compared to the other boats in a fishing fleet in Nicaragua, the boat's hull made of CuNi10Fe1Mn was free of fouling after 4.5 months and showed a maximum corrosion of 0.05 mm, whereas its three sister boats with steel hulls showed copious fouling and greater fuel consumption accordingly [1226].

Further areas of application for CuNi10Fe1Mn can be found in [1227]. An exhaustive overview of applications of CuNi10Fe1Mn in seawater taking particular account of use in offshore installations can be read in [1228].

Applications of CuNi30Mn1Fe (UNS C71500, CW354H, 2.0882)

The standardized alloys with iron and manganese content have become important engineering materials, see DIN EN 12451 [1229].

Applications of CuNi30Mn1Fe should be limited to temperatures up to 393 K (120 °C) and flow rates of 3.6 m/s. The maximum permissible limits are 423 K (150 °C) and 4.5 m/s. The corrosion rates are around 0.043 mm/a (1.7 mpy) in the temperature range from 393 K (120 °C) to 405 K (132 °C) and a flow rate of 3 m/s [1177].

Figure 142 shows the pattern of corrosion in the immersion and tidal zones in tropic seawater off the Panama Canal over a period of 16 years [1230].

The values obtained in these tests in the immersion zone are shown in Figure 143 in comparison to other materials; the pitting corrosion depths as determined are listed in Table 195 [1230].

Material	NiCu30Fe1.8 Monel® B, cf. 2.4360	CuNi30Mn1Fe C71500 2.0882	CuAl5As CW300G 2.0918	EN AW-6061 AA 6061 (AlMg1SiCu)	Lead	Zinc
			Pitting depth, mm			
Mean value for 20 deepest corrosion sites	1.000	0.300	0.125	0.575	0.700	1.500
Maximum pitting depth	1.400	0.925	0.525	1.975	1.200	2.675

Table 195: Pitting depths in different materials after 16 years of exposure in the immersion zone off the Panama Canal [1230]

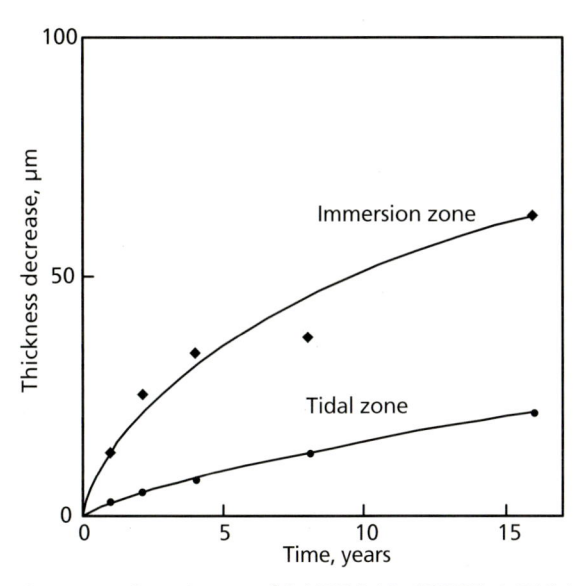

Figure 142: Corrosion rate of CuNi30Mn1Fe (C71500, 2.0882) in seawater off the Panama Canal [1230]

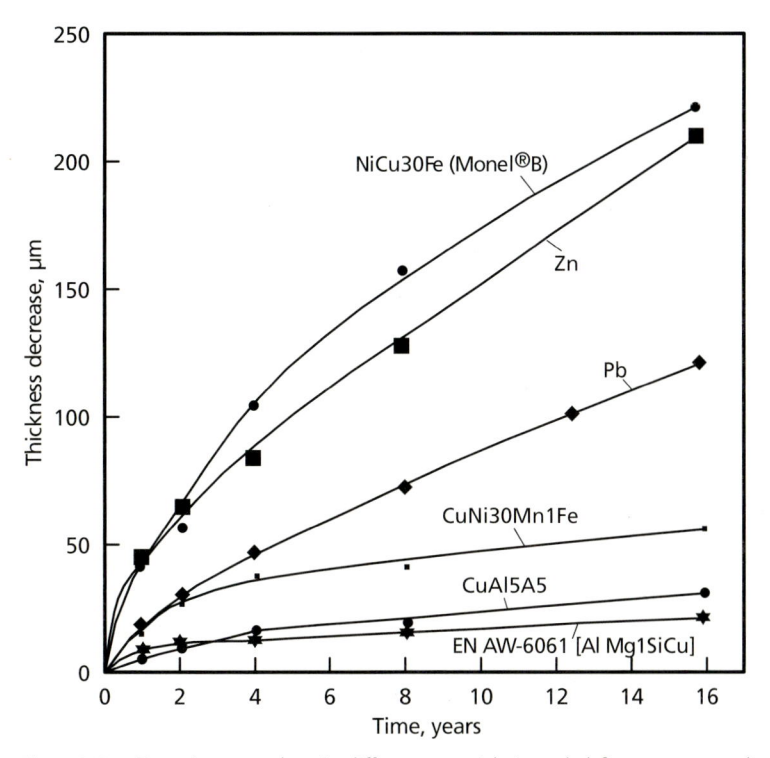

Figure 143: Corrosion mass loss in different materials intended for seawater applications after 16 years of exposure in the immersion zone off the Panama Canal [1230]

CuNi30Mn1Fe (DIN-Mat. No. 2.0882) is used

- for cooling water and condenser pipes as well as for water chambers
- in seawater desalinization plants for pipes in the first evaporator stages, pipes and pipeline bottoms in heat recovery applications.

Heavy-duty parts of condensers are frequently made of CuNi30Mn1Fe (C71500), whereas CuNi10Fe1Mn (C70600) is used for the other pipes. CuNi30Fe2Mn2 (CW353H, 2.0883) is preferred in waters with higher saline content and at higher water temperatures [1231].

[1232] reports on damage to cooling water pipes made of CuNi30Mn1Fe resulting from pronounced reduction of wall thickness with deposits.

[1233] provides an overview of the different fields of application of CuNi10Fe1Mn (C70600, 2.0872) and CuNi30Mn1Fe (C71500, 2.0882) in seawater, including numerous examples. The report also contains a list of ships or boats in which CuNi10-Fe1Mn alloys were used for the hull plating (Table 196).

Boat	Length	Year built	Country	Hull thickness	Operating territory
Asperida II	16	1968	Netherlands	4	USA
Ilona	16	1968	Netherlands	4	Curacao
Copper Mariner	22	1971	Mexico	6	Nicaragua
Pink Lotus	17	1975	Mexico	4	Sri Lanka
Pink Jasmine	17	1975	Mexico	4	Sri Lanka
Pink Rose	17	1975	Mexico	4	Sri Lanka
Pink Orchid	17	1975	Mexico	4	Sri Lanka
Copper Mariner II	25	1977	Mexico	6+2*	Nicaragua
Sieglinde Marie	21	1978	UK	6	Caribbean
Pretty Penny	10	1979	UK	3	UK
Sabatino Bocchetto	21.5	1984	Italy	6+2*	Naples
Romano Rosati	21.5	1984	Italy	6+2*	Genoa
Aldo Filippini	21.5	1985	Italy	6+2*	Ancona
VF 54	21.5	1985	Italy	6+2*	Bari
Pilot Boat	14.4	1988	Finland	7+2.5*	Baltic
Akisushima	10	1991	Japan	4	Japan
Cupro	6.5	1991	Japan	4.5+1.5*	Asano

* = steel plated with CuNi

Table 196: Boats with CuNi hull plating

Contaminated or treated seawater

The influence of ammonium salts in seawater on the corrosion behavior of pipes made of CuNi10Fe1Mn (CW352H, DIN-Mat. No. 2.0872) was thoroughly tested in circulation apparatus with water from the Persian Gulf [1234]. The salt contents of the different waters used are listed in Table 197.

Ion type	Na^+	Ca^{2+}	Mg^{2+}	SO_4^{2-}	Cl^-	CO_3^{2-}	HCO_3^-	Total
mg/l	19,186	704	2,400	4,894	34,080	18	189	61,471

Table 197: Salt contents of waters from the Persian Gulf

Figure 144 shows the time dependency of corrosion in this aerated water. The maximum corrosion rate, at 0.045 mm/a (1.77 mpy), is reached after about 168 h. When the covering layer has formed, the corrosion rate drops after 600 h to 0.011 mm/a (0.43 mpy). The final value of 0.004 mm/a (0.16 mpy) is reached after about 2,000 h.

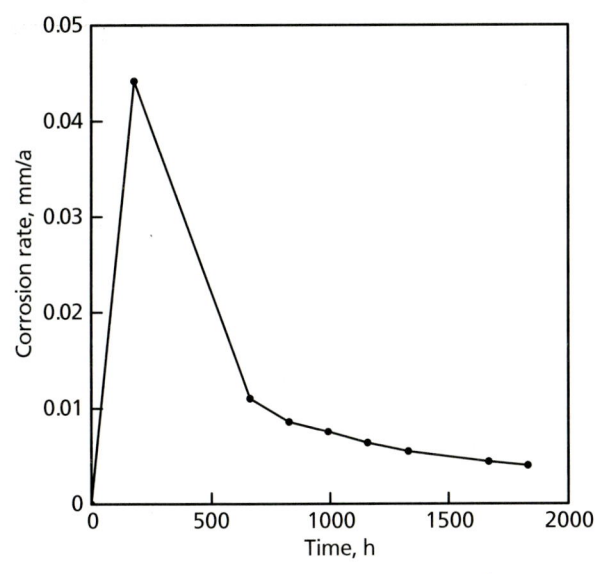

Figure 144: Time curve of the corrosion rate of CuNi10Fe1Mn (CW352H, DIN-Mat. No. 2.0872) in aerated, moderately moving seawater in the Persian Gulf [1234]

Table 198 shows the corrosion rates of CuNi10Fe1Mn (CW352H, DIN-Mat. No. 2.0872) in the seawater of the Persian Gulf at different flow rates and different ammonium nitrate content levels after a test duration of 200 hours. With an increasing flow rate, the corrosion rate in ammonium ion-free water increases only minimally up to a flow rate of about 6 m/s. Increasing ammonium ion content raises the corrosion rate markedly, even at comparatively low flow rates. At ammonium content levels beginning at 800 mg/l, the effect of increasing flow rates becomes stronger. Therefore, satisfactory performance of the material CuNi10Fe1Mn in the water of the Persian Gulf can only be expected if the content level of ammonium ion does not exceed the value of 800 mg/l and the flow rate is kept down to about 1.6 m/s (Reynolds' number < 60,000).

Flow rate m/s	Reynolds' number x 10^4	NH_4NO_3 mg/l	NH_4NO_3 mg/l	NH_4NO_3 mg/l	NH_4NO_3 mg/l
		0	80	800	8,000
		Corrosion rate, mm/a (mpy)			
1.62	6.1	0.016 (0.63)	0.065 (2.56)	0.164 (6.46)	0.840 (33.1)
2.37	7.3	0.021 (0.83)	0.052 (2.05)	0.300 (11.8)	1.872 (73.7)
3.45	8.9	0.022 (0.87)	0.088 (3.46)	0.362 (14.3)	2.897 (114.1)
4.97	10.6	0.051 (2.01)	0.082 (3.23)	0.670 (26.4)	6.433 (253.3)

Table 198: Corrosion rates in CuNi10Fe1Mn (CW352H, DIN-Mat. No. 2.0872) in the seawater of the Persian Gulf at different flow rates and different ammonium nitrate content levels, test duration: 200 hours

Treatment of seawater with chlorine (1–4 mg/l) has no significant influence on the general corrosion behavior in the alloys CuNi10Fe0.5 and CuNi30Fe1.4 [1235]. In flowing seawater, however, more corrosion is observed at lower flow rates than in pure seawater, which fact is attributed to a change in the covering layer composition. Chlorine doses of 0.5 mg/l are considered non-critical for CuNi10Fe1Mn [1236].

The influence of ammonia and chlorine in seawater on corrosion of CuNi alloys was tested on the following materials:

- CuNi10Fe1Mn (CW352H, DIN-Mat. No. 2.0872)
- CuNi30Mn1Fe (CW354H, DIN-Mat. No. 2.0882)
- CuNi30Fe2Mn2 (CW353H, DIN-Mat. No. 2.0883)
- C72200 with 83.22 % Cu, 15.56 % Ni, 0.54 % Fe, 0.5 % Mn.

Content levels up to 2 ppm ammonia had no influence whatever on the corrosion behavior. At content levels of 2 ppm ammonia with 0.5 ppm chlorine only a slight increase in corrosion in the two first alloys was observed, whereas the two last showed significantly more corrosion under these conditions. Pitting corrosion in crevices occurred in all alloys in the presence of 2 ppm ammonia, but not after the addition of 0.5 ppm iron(II) ions [1237].

In flowing seawater, sulfides cause the corrosion levels rise in the same alloys. The amount of increase rises with the sulfide contents. CuNi10Fe1Mn shows the least increase among the alloys tested [1237].

In contrast to this, [1238] determined that the material corrosion in CuNi10Fe1Mn and CuNi30Mn1Fe in flowing seawater with sulfide contents of 0.2 g/m^3 was lower than in aerated, non-contaminated seawater (Figure 145).

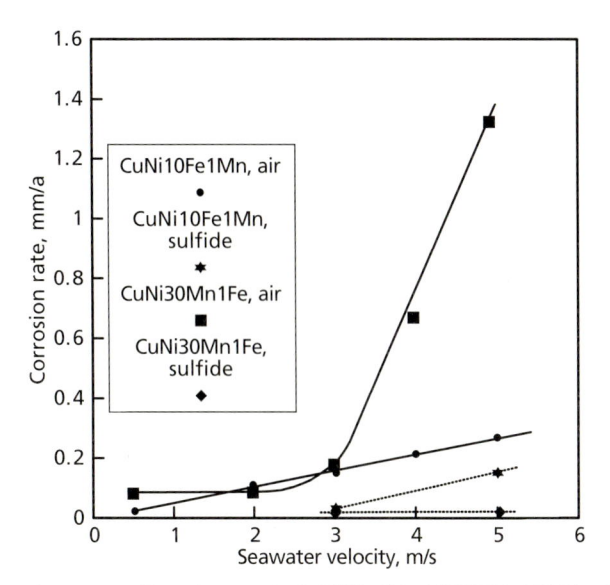

Figure 145: Corrosion rates in CuNi10Fe1Mn (CW352H, DIN-Mat. No. 2.0872) and CuNi30Mn1Fe (CW354H, DIN-Mat. No. 2.0882) in aerated and non-aerated seawater containing sulfide as dependent on the flow rate, test period 230 h [1238]

In aerated seawater, the corrosion rate in CuNi30Mn1Fe (CW354H, DIN-Mat. No. 2.0882) rises above a flow rate of 3 m/s appreciably. In this range, pitting corrosion also increases markedly. The difference between the two alloys is explained as follows: In CuNi10Fe1Mn (CW352H, DIN-Mat. No. 2.0872), the oxygen increases the resistance of the Cu$_2$O covering layer, whereas in the Ni-richer alloy the formation of Cu$_2$(OH)$_3$Cl deposits reduces the resistance.

In addition to the results of these tests, the literature regularly reports that corrosion rates in both CuNi alloys increase with the concentration of solute sulfides in the seawater. [1239] contains a collection of data from the literature that agrees closely with the test carried out by this author. Figure 146 and Figure 147 illustrate the connection between the sulfide ion concentration in synthetic seawater and the corrosion of CuNi10Fe1Mn (C70600) and CuNi30Mn1Fe (C71500) in both stagnating and moving seawater.

The corrosion rates of both alloys generally increase with increasing sulfide concentration. The differences in the behavior of CuNi10Fe1Mn and CuNi30Mn1Fe are ascribed to the more stable covering layers that form on CuNi30Mn1Fe.

In seawater containing sulfides, CuNi alloys can be protected against corrosion by means of the following methods [1240]:

- Precipitation of the sulfides by continuous addition of iron(II) sulfate

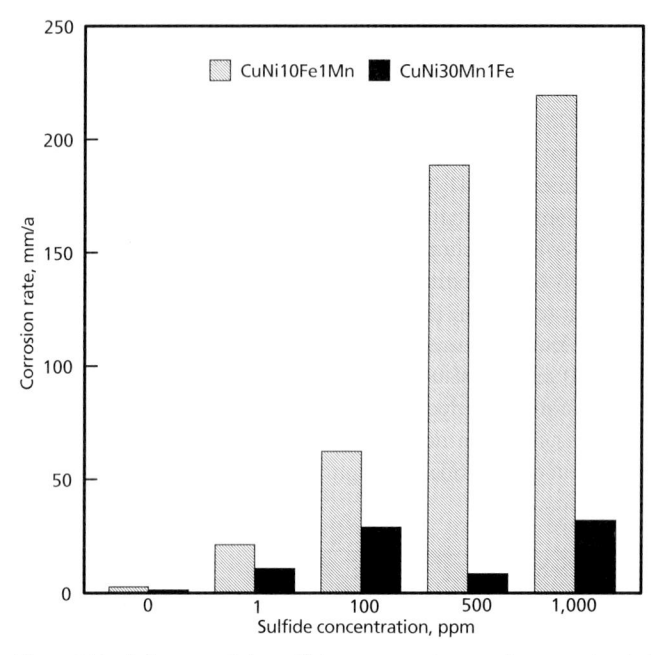

Figure 146: Influence of the sulfide concentration on the corrosion behavior in CuNi10Fe1Mn (C70600) and CuNi30Mn1Fe (C71500) in aerated and moving seawater [1239]

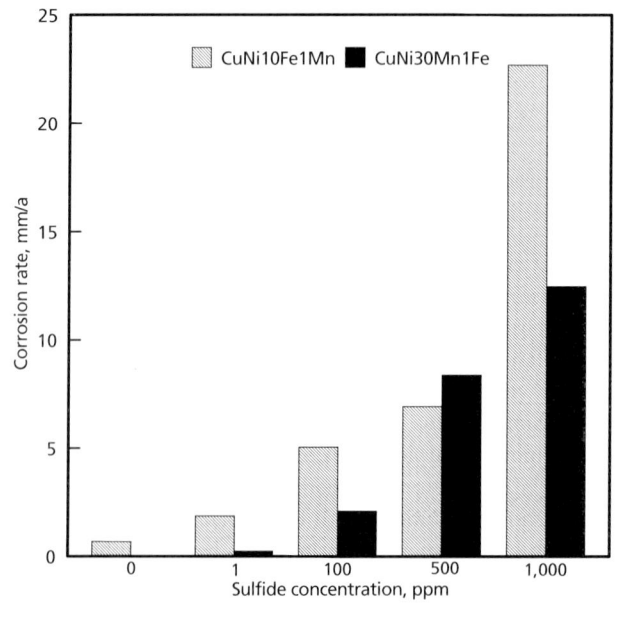

Figure 147: Influence of the sulfide concentration on the corrosion behavior in CuNi10Fe1Mn (C70600) and CuNi30Mn1Fe (C71500) in aerated and stagnating seawater [1239]

- Electrochemical oxidation of the sulfides
- Cathodic corrosion protection with galvanic anodes, in particular iron anodes or external current at potentials below that of the $Cu/Cu_2S/HS$-balance, approximately $U_H = -0.4$ V.

In the presence of sulfides in seawater, the corrosion resistance of all copper materials is disturbed [1241]. Under these conditions, the CuNi alloys show better behavior than aluminous brasses. In sulfide-free seawater, on the other hand, the behavior of aluminium brasses is more favorable up to flow rates of 3 m/s, even in the presence of ammonia [1241].

Under the particularly critical conditions of gradual elongation, both CuNi10Fe1Mn (DIN-Mat. No. 2.0872) and CuNi30Mn1Fe (DIN-Mat. No. 2.0882) can become susceptible to stress corrosion cracking in seawater in the presence of sulfides. Tests in seawater with sulfide contents of 200 ppm–3,120 ppm in the temperature range from 298 to 343 K (25 °C to 70 °C) showed that at an elongation rate of 1.4×10^{-6} 1/s for both materials, maximum sensitivity is at 298 K (25 °C), whereas the level of sulfide content is secondary [1242].

Corrosion inhibition

Under the conditions of seawater desalinization, chromate-phosphate mixtures prove to be effective inhibitors for CuNi10Fe1Mn (CW352H, DIN-Mat. No. 2.0872) and CuNi30Mn1Fe (CW354H, DIN-Mat. No. 2.0882). Table 199 show the measured corrosion rates, with and without addition of an inhibitor (1 ppm CrO_4^{2-} + 9 ppm PO_4^{3-}) are listed [1243]. The corrosion rates in the inhibited system were very low, even with the longer test period of 101 days.

With an inhibitor based on a polycarboxylate polymer, steam-heated CuNi10-Fe1Mn pipes in a multistage evaporator can be effectively protected, even at a temperature of 393 K (120 °C) [1244].

Alloy	No inhibitor		Inhibited	
	Recycle flow 383 K (110 °C) < 25 ppb O_2	Heater exit 394 K (121 °C) 200 ppb O_2	Recycle flow 383 K (110 °C) < 25 ppb O_2	Heater exit 394 K (121 °C) 200 ppb O_2
	Corrosion rate, mm/a (mpy)			
CuNi10Fe1Mn	0.16 (6.3)	0.45 (17.7)	0.007 (0.28)	0.01 (0.39)
CuNi30Mn1Fe	0.10 (3.94)	0.06 (2.36)	0.002 (0.08)	0.005 (0.20)

Table 199: Corrosion rates in samples in a seawater desalinization plant with and without addition of a chromate-phosphate inhibitor, test period 30 days [1243]

The inhibition of CuNi10Fe1Mn (CW352H, DIN-Mat. No. 2.0872) and CuNi30Mn1Fe (CW354H, DIN-Mat. No. 2.0882) in sulfide-contaminated seawater by means of addition of iron(II) sulfate is described in [1245]. Tests of CuNi10Fe1Mn confirm that at content levels of 10 ppm sulfide ions in the seawater the corrosion rate of the CuNi material can be reduced to the values in sulfide-free seawater by adding 10 ppm iron(II) sulfate. When cathodic protection is used, the potential range from $U_H = -0.050$ V ($U_{SCE} = -0.30$ V) to $U_H = -0.75$ V ($U_{SCE} = -1.00$ V) should be avoided in sulfide-contaminated seawater [1246].

In an evaluation of a number of different treatment methods aimed at reducing corrosion of CuNi10Fe1Mn at the start of exposure to seawater at 323 K (50 °C), sodium dichromate proved particularly effective in reducing corrosion rates, not only at the start but also for longer operating periods [1247].

Copper-tin alloys (bronze)

The seawater-exposed copper-tin wrought and cast alloys with tin contents of about 5–14 % are listed in Table 200. The free corrosion potentials of these materials in seawater are within the range from $U_H = -0.10$ to -0.04 V and their mean corrosion rate due to uniform surface corrosion is 0.01–0.03 mm/a (0.39–1.18 mpy).

Abbreviated designation	DIN-Mat. No.	Standard
CuSn6	CW452K	2.1020	DIN CEN/TS 13388 [1248]
CuSn8	CW453K	2.1030	DIN CEN/TS 13388 [1248]
G-CuSn10	CC480K	2.1050.01	DIN CEN/TS 13388 [1248]
G-CuSn12	CC483K	2.1052.01	DIN CEN/TS 13388 [1248]
G-CuSn12Ni	CC484K	2.1060.01	DIN CEN/TS 13388 [1248]

Table 200: Seawater-exposed copper-tin alloys (formerly tin bronzes) [1148]

Corrosion over a period of up to 16 years in the tropical waters of the Caribbean Sea off the Panama Canal in the immersion zone and in the tidal zone is presented in Figure 148 [1249]. The corrosion decreases as time passes.

In the range from 273 to 313 K (0 to 40 °C), the linear corrosion rate doubles in response to a temperature increase of 20 K (20 °C). Higher rates also lead to higher corrosion rates.

Figure 149 shows the temperature-dependence of corrosion of CuSn5, CuAl8, CuSn5Fe, CuAl10, CuZn16Si4 and CuAl10Ni5Fe4 in seawater compared to other copper alloys [1250].

The materials are moderately susceptible to pitting and crevice corrosion. They are therefore resistant to both seawater and brackish water. The cast alloys contain, for reasons relating to casting, small amounts of zinc and lead and are frequently used in fixtures and pumps in shipbuilding.

In Japan, under the name AP bronze (anti-pollution bronze), a CuSn8Al1Si0.2 alloy was developed and used in power plant condenser pipes cooled with highly contaminated seawater or brackish water. Table 201 lists test results on the behavior of this alloy in comparison to other copper materials [1251].

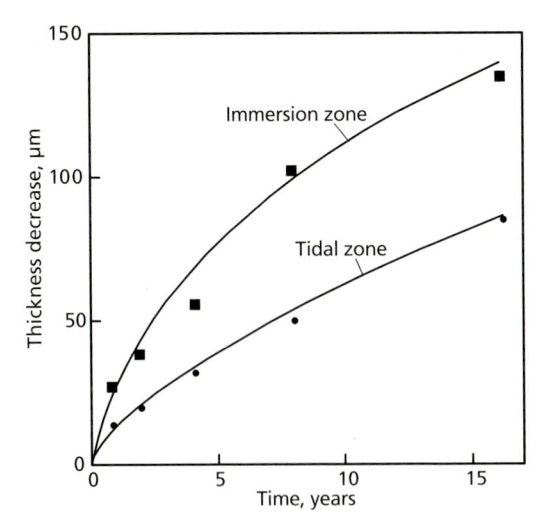

Figure 148: Corrosion of a tin bronze (cf. CuSn4, CW450K, 2.1016) with 95.5 % Cu, 0.25 % P, 4.21 % Sn and 0.30 % Zn in tropical seawater in the immersion and tidal zones of the Caribbean off the Panama Canal [1249]

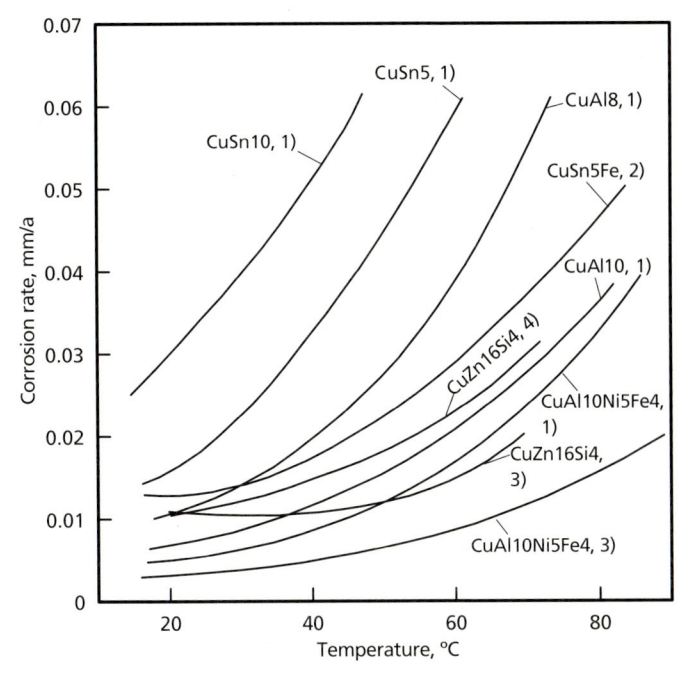

Figure 149: Corrosion rate in mm/a of copper materials in seawater as dependent on temperature [1250]
1) = air access, moving about 1 m/s
2) = air access, stagnating
3) = no air access, moving
4) = no air access, stagnating

Flow rate	Test duration h	Corrosion rate mm/a (mpy)		
		CuZn20Al2 CW702R, 2.0460	CuNi30Mn1Fe CW354H, 2.0882	AP bronze
Rotating disc 2 m/s	3,000	0.24 (9.45)	0.25 (9.84)	0.20 (7.87)
Rotating disc 7 m/s	720 870 1,000	0.45 (17.7) 0.33 (13.0) 0.21 (8.27)	0.46 (18.1) 0.29 (11.4) 0.31 (12.2)	0.45 (17.7) 0.30 (11.8) 0.19 (7.48)
Blasting test with air 5 m/s	200	0.33 (13.0)	0.21 (8.27)	0.32 (12.6)

Table 201: Corrosion rates for AP bronze, naval brass and CuNi30Mn1Fe in moving seawater [1251]

Copper-tin-zinc alloys (red brass)

The copper-tin-zinc-(lead) cast alloys behave in seawater similarly well to zinc-free copper-tin alloys and have the advantage of being more readily castable. The corrosion rate in terms of uniform surface corrosion of these materials is in the range 0.01–0.03 mm/a (0.39–1.18 mpy). The free corrosion potentials are U_H = –0.10 V to –0.04 V.

The most important alloys in this material group are listed in Table 202.

Abbreviated designation		DIN-Mat. No.	Standard
G-CuSn10Zn	–	2.1086.01	DIN CEN/TS 13388 [1248]
G-CuSn7ZnPb	CC493K	2.1090.01	DIN CEN/TS 13388 [1248]
G-CuSn5ZnPb	CC491K	2.1096.01	DIN CEN/TS 13388 [1248]

Table 202: Seawater-exposed copper-tin-zinc alloys (formerly red brass) [1148]

The higher lead contents do affect the mechanical properties and corrosion resistance, but both of these negative influences can be compensated by means of low-level nickel additions.

The red brass alloys are highly resistant to erosion corrosion and ensure long and reliable operation in pump casings, impellers, valves and inserts as well as in water chambers, also in contact with Monel® and aluminium bronze.

G-CuSn5ZnPb (CC491K, 2.1096.01) has been used successfully in an Italian seawater desalinization plant at Brindisi for pump bodies, valves and water chambers [1252].

Old cannon barrels retrieved from the sea floor were covered by a layer of basic copper chloride and generally show only slight uniform corrosion. Old French cannons made of a copper alloy with 4.5 % Sn, 1.5 % Zn, 0.9 % Pb and 0.002 % P, from a ship that was sunk in 1707 in the Spanish War of Succession near the Scilly Islands, and an old Swedish cannon barrel made of an alloy with 87.8 % Cu, 7.6 % Sn, 0.4 % Zn and 2.6 % Pb, which had been in the sea for over 300 years, showed corrosion of less than 5 µm/a (0.20 mpy) [1253, 1254].

Copper-zinc alloys (brass)

Table 203 lists the main groups of copper-zinc alloys.

Group	Abbreviated designation	Old DIN-Mat. No.		Standard
CuZn	CuZn37	CW508L	2.0321	DIN CEN/TS 13388 [1248]
CuZnAl	CuZn20Al2	CW702R	2.0460	DIN CEN/TS 13388 [1248]
	CuZn37Al1	CW716R	2.0510	DIN CEN/TS 13388 [1248]
CuZnSi	CuZn31Si1	CW708R	2.0490	DIN CEN/TS 13388 [1248]
	G-CuZn15Si4	CC761S	2.0492.01	DIN EN 1982 [1255]
CuZnSn	CuZn28Sn1	CW706R	2.0470	DIN CEN/TS 13388 [1248]
CuZnNi	CuZn35Ni2	CW710R	2.0540	DIN CEN/TS 13388 [1248]
CuZnPb	G-CuZn33Pb	CC750S	2.0290.01	DIN EN 1982 [1255]
	GK-CuZn37Pb	CC754S	2.0340.02	DIN EN 1982 [1255]
	CuZn39Pb3	CW614N	2.0401	DIN CEN/TS 13388 [1248]
	CuZn40Pb2	CW617N	2.0402	DIN CEN/TS 13388 [1248]

Table 203: Seawater-exposed copper-zinc cast and wrought alloys

Of these materials only CW702R, CuZn20Al2 (formerly naval brass) is suitable for direct use in seawater, i.e. for structural elements that carry or are washed by seawater. The other alloys should only be used to build structural elements without direct seawater exposure.

The copper-zinc alloys with high copper contents are approximately comparable to copper in their corrosion behavior in seawater. With increasing zinc content the materials become more resistant. The best corrosion behavior is observed in alloys with 20–40 % zinc. The behavior can be further improved by additions of other alloying elements. For instance, additions of tin, antimony or arsenic increase the resistance of the alloys to dezincification. Additions of aluminium improve the resistance to abrasion.

A copper alloy with 21.7 % Zn, 1.04 % Al, 1.02 % Ti and 0.03 % As (TIBRAL) shows, after solution heat treatment, a corrosion resistance in flowing and in stagnating seawater at room temperature comparable to the traditional aluminium bronzes [1256].

In addition to uniform surface corrosion, selective corrosion by means of **dezincification** may also occur in highly zinciferous copper-zinc alloys, particularly when exposed in contaminated seawater, brackish water, harbor water or bilge water. In highly cupriferous brass with pure α-phase, for example CuZn20Al2 (DIN-Mat. No. 2.0460), the sensitivity to dezincification is compensated by small amounts of phosphorus and/or arsenic.

Brasses with more than 85 % copper do not tend to show dezincification in contrast to the α-brasses with lower copper content. The tendency to dezincification can also be eliminated in the single-phase α-alloys by means of small alloying additions, e.g. 1 % tin, 0.25 % antimony or 0.02–0.05 % arsenic. These alloys are called "inhibited brasses," which is not quite accurate since an inhibitor, by definition, is understood to be an additive to the aggressive medium that hinders or delays the corrosion process. These above-named additions are only effective in the α-phase alloys, not in the two-phase α+β-alloys and the β-phase [1196, 1257].

Unexpected damage due to dezincification on arsenic-stabilized condenser pipes made of brass types CuZn28Sn1 (CW706R, DIN-Mat. No. 2.0470) and CuZn20Al2 (CW702R, DIN-Mat. No. 2.0460) was ascribed to magnesium, which had been used in the manufacturing process as a deoxidant. Magnesium forms an intermetallic phase Mg_3As_2 with arsenic, which then decomposes in the cooling water, whereby arsenic trihydride is formed. This process removes arsenic from the metal and the "inhibiting" effect is lost. It is recommended that the magnesium content be limited to 0.005 %, which recommendation has been included in the new edition of the relevant standard [1258].

The more cupriferous CuZn types do not show a tendency to dezincification, but they cannot be recommended for applications involving direct seawater contact due to their comparatively low levels of corrosion and erosion resistance.

The α-brasses see the most widespread practical use. The alloy CuZn30 (CW505L, DIN-Mat. No. 2.0265), formerly used for condenser pipes, is not longer considered suitable if brackish water and seawater are used as the coolants. It is also not advisable to use CuZn37 (CW508L, DIN-Mat. No. 2.0321) for condenser pipes on ships [1218].

The alloy CuZn28Sn1 (CW706R, DIN-Mat. No. 2.0470, (formerly naval brass 71, admiralty brass)), which was formerly used frequently for pipes in cooling water systems, is rendered resistant to dezincification by addition of arsenic, but shows only limited corrosion and erosion resistance. The cooling water flow rate should not exceed 2 m/s. Pipes made of admiralty brass have lasted for 10–20 years in ship condensers, from which figure an corrosion rate of about 0.085–0.170 mm/a (3.35–6.69 mpy) can be derived. Under exacerbated conditions, however, corrosion reaches 1.7 mm/a (66.9 mpy) and reduces the service life to about 1 year [1259].

The aluminiferous naval brass types, e.g. CuZn20Al2, are resistant to corrosion, dezincification and erosion. This alloy shows good corrosion resistance in clean seawater at flow rates of 1–3 m/s. They are used for heat exchanger pipes and pipeline systems on ships and in power plants [1260].

A number of reports have been published of good results with cooling pipes made of aluminiferous CuZn materials, also under lower stress loads in seawater desalinization plants [1177, 1196, 1217, 1261].

Experience with heat exchangers has shown that the material CuZn22Al2 shows the best corrosion behavior in seawater of all brass types. The flow rate should be around 0.8–2 m/s and abrasive blasting of the surface is recommended [1198].

The corrosion behavior of pipes made of CuZn20Al2 (CW702R, DIN-Mat. No. 2.0460) in synthetic seawater was investigated in a test loop under conditions of constant flow with periods of additional stress load in the form of solid particles. Even brief stress loads due to solid materials cause local corrosion by removing the surface passivation layer. After the particle stress, the corrosion rate also rises significantly, since there is no repassivation [1262].

Tests of a brass material (CuZn20Al2) in seawater and in 3% NaCl solution at temperatures of 298 to 333 K (25 °C to 60 °C) showed that at the onset of the stress load the corrosion rates increase with the temperature, and that under longer exposure, after dense covering layers have formed, they decrease as the temperature increases [1263]. The structure and composition of the covering layers, which form on CuZnAl materials in seawater dependent on the exposure time, was investigated in [1264].

At content levels of 1–4 mg/l, the presence of chlorine in seawater has not significant influence on the surface corrosion of CuZn22Al2 with 0.04% As, but it does reduce the resistance to mechanical stress load, so that lower flow rates are recommended than in pure seawater [1236].

Tests in natural seawater on CuZn22Al2 confirmed that in stagnating water chlorine content has practically no influence, but that at flow rates exceeding 3 m/s greater corrosion can be expected [1265].

The presence of sulfides in the seawater can encourage pitting and crevice corrosion in CuZn20Al2 (CW702R, DIN-Mat. No. 2.0460) materials. Crevice corrosion is observed even at content levels of 0.01 ppm sulfide. The concurrent presence of chlorine increases the risk of crevice and pitting corrosion. Ammonia and ammonium salts in the seawater may induce stress corrosion cracking in copper-zinc alloys. In CuZn20Al2, given sufficient tensile stresses, concentrations of 1 ppm ammonia will suffice to cause cracks. Copper-zinc alloys with more than 80% copper develop stress corrosion cracking in exceptional cases only [1236, 1237, 1266–1268].

The two-phase brasses with α+β-structure (36–45% Zn) and the β-brasses (about 50% Cu, 50% Zn) are less corrosion-resistant and more highly susceptible to pitting corrosion and dezincification than the brass types with α-structure in seawater.

The material G-CuZn40Fe (DIN-Mat. No. 2.0590.01, Muntz metal) tends to develop deep dezincification and is used in practical applications with high wall thicknesses only. Muntz metal may show corrosion of 2.1 mm/a (82.68 mpy) in seawater. The alloy improved by addition of 1% tin (marine brass, cf. CuZn38Sn1, CW717R, 2.0530) is less susceptible to dezincification and local corrosion. The corrosion rate is about 0.075 mm/a (2.95 mpy). Both alloys are not recommended for use in seawater without corrosion protection.

Figure 150 lists the corrosion rates for Muntz metal (CuZn40, CW509L, 2.0360), marine brass (CuZn38Sn1) and different α-brass types such as admiralty brass (CuZn28Sn1) CuZn35 and CuZn30, showing their dependence on temperature [1250].

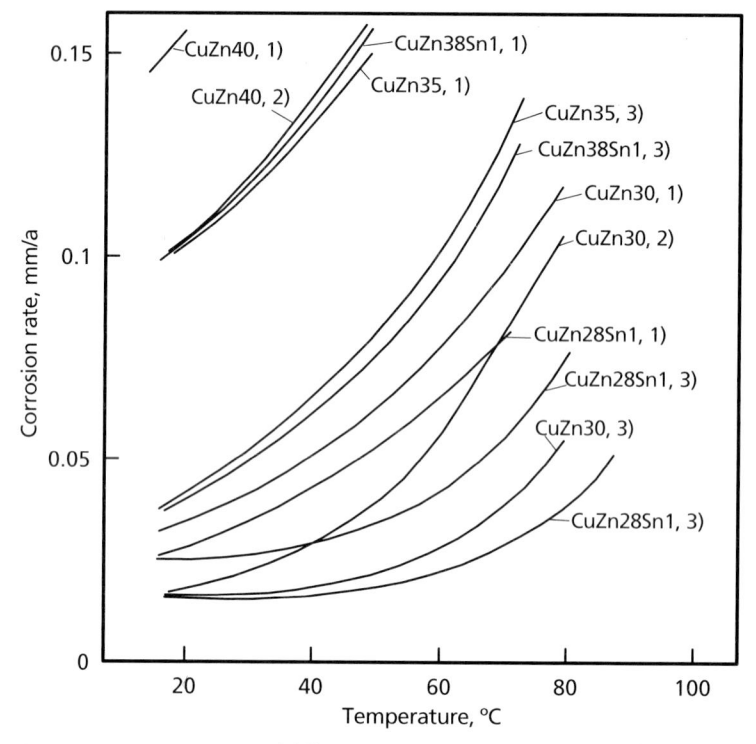

Figure 150: Corrosion rates of different copper materials in seawater as dependent on temperature [1250]
1) = air access, moving
2) = air access, stagnating
3) = no air access, moving

The so-called high-strength brasses with copper contents of 50–60 %, zinc contents of 35–45 % and lesser amounts of manganese, aluminium, iron and nickel, as well as the cast naval brasses, for example G-CuZn34Al2 (CC764S, 2.0596.01), are also used in ship's engine elements, but it is better to use aluminium bronzes instead [1165].

The corrosion rates for the high-strength cast brasses used in larger ships' propellers for seawater flow rates of 8.2 m/s are listed in Table 204 in comparison to standard propeller materials [1175].

The α/β-brasses are not suitable for use in seawater desalinization plants. As shown in Table 205, CuZn alloys can however be effectively protected in seawater circulation systems under desalinization conditions by adding chromate-phosphate inhibitors [1243].

Type	Cu	Zn	Mn	Fe	Al	Ni	Si	Corrosion rate mm/a (mpy)
High-strength brasses cf. DIN EN 1982 [1255] e.g. G-CuZn35Al1, 2.0592.01	58	36	0.7	0.7	0.7			1.800 (70.9)
	52	44	0.7	0.3		2.0		1.850 (72.8)
	58	36	0.4	0.6	1.1	3.0		1.330 (52.4)
	58	34	0.5	0.8	1.5	5.2		1.000 (39.4)
	58	36	1.8	0.8	1.0	3.3		0.780 (30.7)
	57.9	39.8	1.78	0.75	0.7			1.400 (55.1)
Al bronze	83.5		2.5	4.0	10.0			0.900 (35.4)
NiAl bronze	78.5		2.5	4.0	10.0	5.0		0.225 (8.86)
AlMn bronze	74.9		12.0	2.84	8.0	2.2		0.250 (9.84)
NiCuSi	28.7		0.76	2.08		64.7	3.6	0.078 (3.07)

Table 204: Composition (mass%) and corrosion rates of α/β-cast brasses and propeller materials after 60 days in seawater at a flow rate of 8.2 m/s and a temperature of 298 K (25 °C) [1175]

No inhibitor		Inhibited	
Recycle flow 383 K (110 °C) < 25 ppb O_2	Heater exit 394 K (121 °C) 200 ppb O_2	Recycle flow 383 K (110 °C) < 25 ppb O_2	Heater exit 394 K (121 °C) 200 ppb O_2
0.165–0.211 mm/a (6.50–8.31 mpy)	0.483 mm/a (19.0 mpy)	0.005–0.008 mm/a (0.20–0.31 mpy)	0.008 mm/a (0.31 mpy)

Table 205: Corrosion mass loss in CuZn alloys in a seawater desalinization plant with and without addition of a chromate-phosphate inhibitor after 30 days [1243]

In pipes made of CuZnAl alloys in contact with installed parts made of austenitic stainless steels, pronounced corrosion occurred in natural seawater at 303 K (30 °C) after 4 months, whereas no corrosion was observed on pipes in contact with CuNi materials [1269].

Other copper alloys

The copper-silicon alloys CuSi with 1.5 % and 3 % Si (cf. C65100, formerly CuSi2Mn, DIN-Mat. No. 2.1522 and C65500, formerly CuSi3Mn, DIN-Mat. No. 2.1525) show good resistance in seawater up to flow rates of 1.5 m/s and are used as fastening elements for wood, e.g. in cooling towers for seawater or other saltwater [1187, 1270, 1271].

In the tropical waters of the Caribbean off the Panama Canal, the corrosion rates of CuSi bronze with 2.64 % Si are in the 0.006–0.009 mm/a (0.24–0.35 mpy) range and the samples show slight pitting corrosion [1249, 1272]. The CuSi alloys are for the most part resistant against fouling. Figure 151 shows the results of long-term exposure in the Panama Canal Zone [1249].

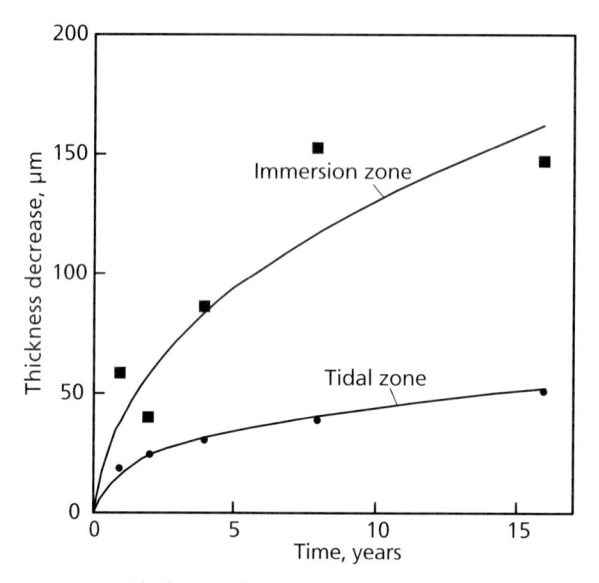

Figure 151: Thickness reduction in samples made of CuSi2.64Fe0.04 in the immersion and tidal zone of the Caribbean [1249]

In seawater-driven impellers made of the alloy CuMnAl with 71.8 % Cu, 16.9 % Mn, 5.8 % Al, 3.6 % Fe and 1.9 % Ni, damage due to corrosion and ultrasonically induced cavitation was observed in the form of microcracking and grain boundary corrosion [1273].

Sodium Chloride

Author: M. B. Rockel / Editor: R. Bender

Copper

Pure copper is less resistant compared with all copper alloys in NaCl solutions. Because of the very low chloride concentrations in tap water (e.g. less than 100 mg/l Cl^-) it is used successfully in household installations, also at elevated temperatures. Deposits of more noble metals or carbon, even in smallest amounts, lead to local corrosion. For example, in the 1980's pipes in household installations failed after several years. It was shown that at the time of making the pipes, the tools brought carbon particles into the surface. These acted as local cathodes and dissolved the surrounding copper matrix in some places to the extent of pipe perforation.

Oxidizing substances such as iron(III) chloride increase the aggressivity of the sodium chloride solutions for copper materials strongly. Sodium dichromate promotes local attack of brass materials.

Numerical values from various sources [1274–1277] for the corrosion of copper and copper alloys can be taken from Table 206.

Material	Concentration of the NaCl, %	Temperature K (°C)	Attack, $g/m^2\ d$	Remarks
Copper				
	3	RT	1.6	test duration 96 h, fully immersed, stationary
	3	RT	2.9	test duration 96 h, fully immersed, agitated with air
	5.8	RT	1.6	fully immersed
	5.8	RT	2.9	fully immersed, air passed through
	5.8	RT	1.7	alternating dip test
	5.8	RT	0.35	spray fog
	3 (pH = 2)	297.5 (24.5)	12	test period 25 days, aerated, acidic

RT: Room temperature

Table 206: Behavior of materials of the copper group with respect to sodium chloride solutions [1274–1277]

Table 206: Continued

Material	Concentration of the NaCl, %	Temperature K (°C)	Attack, g/m² d	Remarks
Cupronickel				
80 % Cu, 20 % Ni	4	RT	5.3	
	8	RT	1.9	
75 % Cu, 25 % Ni	4	RT	1.6	
70 % Cu, 30 % Ni	5.8	RT	0.91	non-aerated, 96 h
	5.8	RT	0.38	aerated, 96 h
	5.8	RT	0.23	alternating dip test
55 % Cu, 45 % Ni	5.8	RT	0.51	aerated, 96 h
	5.8	RT	0.15	non-aerated, 96 h
Copper-aluminium alloys				
90 % Cu, 10 % Al	10	RT	0.41	test duration 125 h, not moving
90 % Cu, 10 % Al	10	RT	0.44	test duration 336 h, not moving, weight increase
95 % Cu, 5 % Al	10	RT	0.07	test duration 336 h, not moving, weight increase
92 % Cu, 81 % Al	20	RT to boiling point	1.32	test duration 145 h, not moving, repeatedly evaporated to dryness
96 % Cu, 4 % Al	20	RT to boiling point	2.26	test duration 145 h, not moving, repeatedly evaporated to dryness
Aluminum bronze 95 % Cu, 5 % Al	sea water	RT	0.82	loss of 5.4 % of the original tensile strength
Aluminum bronze 92 % Cu, 8 % Al	sea water	RT	0.28	loss of 3.6 % of the original tensile strength

RT: Room temperature

Table 206: Behavior of materials of the copper group with respect to sodium chloride solutions [1274–1277]

Table 206: Continued

Material	Concentration of the NaCl, %	Temperature K (°C)	Attack, g/m² d	Remarks
Brass				
70% Cu, 29% Zn, 1% Sn	3	RT	1.1	test duration 96 h, fully immersed, stationary
70% Cu, 29% Zn, 1% Sn	3	RT	1.1	test duration 96 h, fully immersed, agitated with air

RT: Room temperature

Table 206: Behavior of materials of the copper group with respect to sodium chloride solutions [1274–1277]

Copper materials play an important role in the potash industry, in particular as material for preheaters and pumps. According to [1278] the resistance to 363 K (90 °C) hot hard salt solution (129 g magnesium chloride + 115 g sodium chloride + 91 g potassium chloride + 81 g magnesium sulfate + 1,000 g water, in the alternating dip test) is characteristic, also for other liquors with somewhat different composition encountered in the potash industry (see Table 207).

Material	Corrosion in g/m² d	Remarks
Electrolytic copper hard sheet	10	uniform corrosion
Smelted copper over 99.4% copper hard sheet	8	uniform corrosion
Aluminium bronze 96% Cu, 4% Al pipe section	0.8	slight uniform matting
Coinage bronze 91.5% Cu, 8.5% Al hard sheet	0.3	uniform attack, surface dark
Four material bronze 82.9% Cu, 7.5% Al, 4.7% Fe, 4.9% Ni	0.2 to 0.8	uniform attack, iron is partly dissolved preferentially, suitable as casting material for pumps
Aluminium multi-material bronze 92% Cu, 6% Al, 21% Si	0.1	slight uniform attack, suitable for shafts, pipes, fittings, castings
Manganese bronze 85% Cu, 15% Mn rolled sheet	1.4	uniform attack, suitable for valves, fittings, pipes and also castings

Table 207: Behavior of materials of the copper group with respect to hard salt solutions at 363 K (90 °C) [1278]

Table 207: Continued

Material	Corrosion in g/m² d	Remarks
Manganese bronze 0.8 % Mn, 2–3 % Si sheet or pipe	4.4 to 6.5	numerous pore-like holes, unsuitable
Brass 63 % Cu, 37 % Zn pressure brass	1.5	nonuniform attack, zinc depletion, unsuitable
Aluminium brass 76 % Cu, 22 % Zn, 2 % Al hard sheet	0.6 to 0.9	uniform attack
Aluminium brass 70 % Cu, 28 % Zn, 2 % Al	0.6	zinc depletion, unsuitable

Table 207: Behavior of materials of the copper group with respect to hard salt solutions at 363 K (90 °C) [1278]

So called HSM copper with the composition of 2.1–2.6 % Fe and 0.015–0.15 % P as well as 0.05–0.20 % Zn has considerably improved resistance with respect to erosion corrosion in chloride media, compared with pure copper [1279]. Whereas at the beginning of the corrosion tests the behavior of HSM copper hardly differs from that of commercial E-copper (electrolytic copper) or SF copper (DIN Mat. No. 2.0090 with 0.015–0.04 % phosphorus) very much lower corrosion rates are found in saline and alkaline media after longer test times. This is explained by the formation of an iron hydroxide film containing water, which has a low solubility; mechanical damage heals quickly. The formation of this protective film is prevented in acidic solution. HSM copper has excellent resistance to 3.4 % NaCl solution at 313 K (40 °C) and flow velocities up to 3 m/s. As shown in Table 208, the behavior is considerably better than that of SF copper, and the durability of HSM copper comes very close to that of CuNi10Fe1Mn.

In very dilute chloride solutions containing $0.08–1.7 \times 10^{-3}$ mol/l NaCl and under the conditions of positive heat transfer (the metal is hotter than the corrosion medium), copper shows a greater corrosion rate than in the case of equal temperatures of the metal and the medium, cf. Table 209 [1280]. This is explained by the higher rate of oxygen reduction that controls the process.

Alloy	Water flow velocity	Period of uniformly progressing erosion corrosion	Weight loss	Corrosion rate		Deepest local attack after 1 year
					Attacking medium: Water with 3.4% sodium chloride, 4.7 ppm free oxygen, temperature ≈ 313 K (40 °C)	
	m/s		mg/cm^2	g/m^2 d	mm/a (mpy)	mm
SF-Cu	0.61	day 137 to 228	3.9	0.43	0.018 (0.71)	0.150 after
SF-Cu	1.16	day 118 to 265	6.2	0.42	0.018 (0.71)	228 days of progressing corrosion
SF-Cu	1.62	day 60 to 119	6.3	1.07	0.043 (1.69)	see above,
SF-Cu	3.57	day 0 to 119	57.6	4.85	0.198 (7.79)	perforated after 106 days
HSM-Cu	0.76	day 60 to 365	3.4	0.11	0.005 (0.20)	0.038
HSM-Cu	1.10	day 60 to 387	7.0	0.21	0.010 (0.39)	
HSM-Cu	1.53	day 60 to 387	6.2	0.19	0.007 (0.28)	0.025
HSM-Cu	3.42	day 60 to 387	4.3	0.13	0.005 (0.20)	0.030
CuNi10Fe	1.10	day 60 to 387	1.3	0.04	0.0025 (0.10)	0.010
CuNi10Fe	1.59	day 60 to 387	7.8	0.06	0.0025 (0.10)	0.076
CuNi10Fe	3.72	day 60 to 387	1.0	0.03	0.0025 (0.10)	0.023

Table 208: Erosion resistance of HSM copper [1279]

Surface temperature of copper, K (°C)	Corrosion rates, g/m^2 h	
	with heat transfer	in thermal equilibrium
303 (30)	1.50	0.47
313 (40)	1.88	0.54
328 (55)	2.85	0.60
338 (65)	3.43	0.69

Table 209: Corrosion rates of copper in g/m^2 d in 1.7×10^{-3} mol/l NaCl; temperature of medium 296 K (23 °C) [1280]

The mechanism of erosion corrosion of copper and brass in 3 % NaCl solution at 298 K (25 °C) is discussed by [1281]. In all tests an incubation time of 50 h is found. The influence of the flow velocity on the corrosion of Cu and brass, measured as anodic current density, is shown in Figure 152. Three regions of the curve are distinguishable: Up to 2 m/s the dissolution rate depends on the flow velocity, followed by a plateau in which the corrosion rate is independent of the flow velocity up to 12 m/s, and finally as from 12 m/s the corrosion rate increases again considerably with the flow velocity, with an exponential rise as from approximately 20 m/s. Current density/potential curves show that with increasing flow velocity in particular the cathodic currents are increased, to a lesser extent the anodic currents which follow the plotted curve. However, a cathodic reaction controlled by mass transport cannot be assumed; instead it is appropriate to assume a reaction controlled by pore diffusion – because the directly measured very low corrosion rates cannot be explained in any other way.

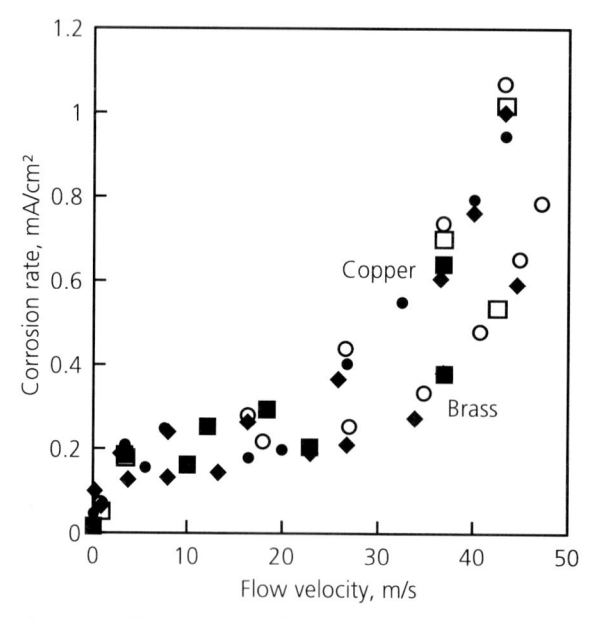

Figure 152: Corrosion rates of copper and brass in 3 % NaCl as a function of the flow velocity at 298 K (25 °C); measured as anodic current density (for Cu $\rightarrow$ Cu^{2+}: 1 mA/cm^2 = 285 g/m^2 d = 11.6 mm/a (457 mpy)) [1281]

Investigations of the durability of copper have been carried out in the pH range from 2 to 10 in a 0.5 M NaCl solution [1282]. Table 210 shows the stationary corrosion potentials obtained at various rotation speeds in an oxygen saturated 0.5 M NaCl solution as a function of the pH-value. In the acidic range the anodic dissolution of copper is largely independent of the pH-value. Current density/potential curves with table relationship are obtained as from –50 mV$_{SCE}$ and a passivating current density at 0 mV. When the passivating current densities are plotted against the square root of the flow velocity, a linear relationship is found, showing that the process is diffusion controlled. A similar relationship is obtained for the anodic corrosion current densities with given potential only slightly above the corrosion potential, see Figure 153. In acidic NaCl solution Cu^+ ions are produced in every case by the anodic dissolution, and depending on their concentration at the surface, these ions either form soluble complexes or are precipitated as CuCl (for Cu → Cu^+: 1 mA/cm^2 = 570 g/m^2 d = 23.2 mm/a (913 mpy)). Current density/potential curves in the alkaline range show above −150 mV a steep increase of the current density; although the CuO layer becomes thermodynamically more stable, and at the same time a protective layer of Cu_2O is formed, strong pitting corrosion takes place at the same time to the extent indicated by the high current densities.

pH-value	Rotation speed, 1/min			
	600	1,000	2,000	3,200
3.0	−270	−275	−278	−280
7.5	−263	−268	−271	−
8.3	−	−268	−270	−
9.2	−188	−190	−191	−193
9.9	−	−271	−130	−132

Table 210: Stationary free corrosion potentials (mV) of a rotating copper electrode in oxygen saturated solution at various pH-values and rotation speeds [1282]

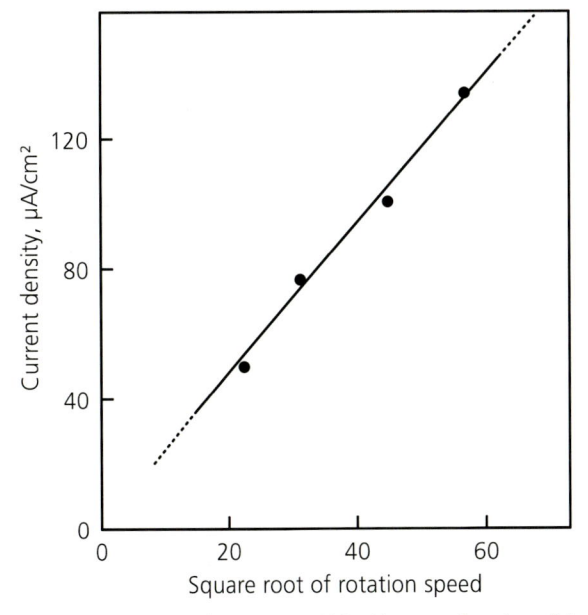

Figure 153: Current densities at −225 mV$_{SCE}$ as a function of the square root of the flow velocity for copper in 0.5 M NaCl [1282]

As belonging to the subject matter of the behavior of copper under anodic polarization, investigations were also carried out in 1 mol/l NaCl solution at pH 2 (addition of HCl) and compared with the corresponding behavior of copper in Na_2SO_4 solution at pH 2 [1283]. One finds regions of the diagram above the free corrosion potential, which lies at −289 mV_{SCE}, in which dissolution of copper to the divalent state takes place. At 0 mV_{SCE} a weak indication of current density decrease is found, but this does not indicate any passivation, because the current densities settle to values > 10 mA/cm^2 when the potential is increased further, corresponding to strong anodic dissolution > 100 mm/a (3,937 mpy).

The copper alloys used in sea water desalination plants are exposed to temperatures greater than 373 K (100 °C). Thus it is of practical interest to clarify the fundamental mechanism of the anodic dissolution of copper in NaCl solution at temperatures up to 448 K (175 °C) [1284]. A strong potential shift towards more negative values is found with increasing temperature, or at a given potential an increase of the anodic dissolution current densities by up to 4 decimal orders of magnitude, which also increase with the flow velocity. The anodic process is diffusion controlled, with $CuCl_2$ as primary corrosion product, and it is independent of the pH-value. Its dependance on the chloride concentration in the concentration range from 0.124 to 1.24 mol/l NaCl is shown in Figure 154 for temperatures in the range from 303 to 374 K (30 to 101 °C).

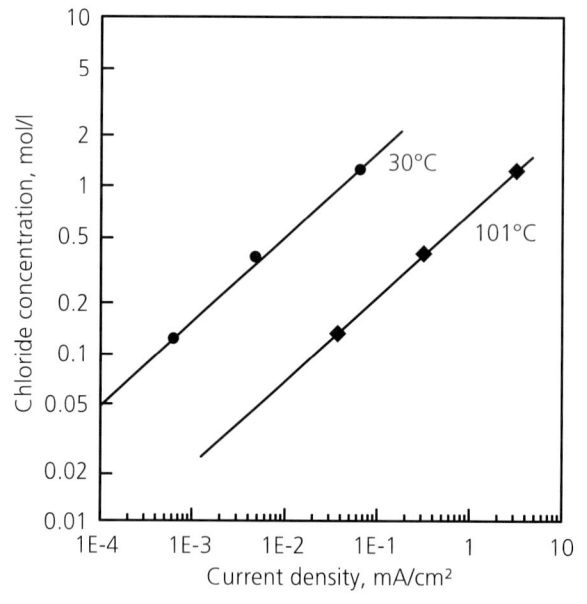

Figure 154: Anodic current densities at the potential −250 mV_{SCE} as function of the chloride concentration at 303 to 374 K (30 and 101 °C) (for Cu → Cu^{2+}: 1 mA/cm^2 = 285 $g/m^2 d$ = 11.6 mm/a (457 mpy)) [1284]

In the context of the corrosion stability of copper alloys utilized in great ocean depths, the influence of the hydrostatic pressure on the behavior of copper in 3.5% NaCl solution and sea water (both adjusted to pH 7.8) at 283 K (10 °C) with an oxygen content of 7 mg/l has been investigated for up to 360 h [1285]. For most metals and alloys a decrease of the corrosion rates with the water pressure has been found under simulating laboratory conditions as well as in field tests, as can be explained by decreasing oxygen content and temperature with increasing ocean depth. However, the results obtained here show that the corrosive attack of copper increases with pressure (approximately 45% increase up to 300 atmospheres). The anodic dissolution is thereby not accelerated, but instead the cathodic reduction.

Inhibition

The inhibition of the corrosion of copper can also be implemented, for example, with specifically applied multi-layers according to the so called Langmuir-Blodgett (LB) technique [1286]. For example, coatings with thicknesses on the molecular level have been produced with N-octadecylbenzidine (NODB) with the addition of 1-docosanol (for stabilizing these layers) with up to 10 monomolecular layers. The copper treated this way was tested with regard to its corrosion resistance in 3.4% NaCl solution, and the effectiveness of the inhibitor was determined. Determinations of the corrosion potential from current density/potential curves and the weight loss were made. Figure 155 shows the effectiveness of the inhibitor in percent as a function of the number of monomolecular layers. Evidently adequate inhibition is already obtained with 10 layers. Values of 92% were still found after 8 weeks, confirming the high stability of this inhibitor. Current density/potential curves show that the cathodic current densities and the corrosion potentials differ only slightly, whereas the anodic curve traces differ significantly, see Figure 156. The individual parameters of the electrochemical measurements are summarized in Table 211. According to this table the corrosion currents decrease to the same extent as the polarization resistance increases, and the effectiveness of the inhibitor increases correspondingly. The results can be summarized as follows:

- The layers applied by the Langmuir-Blodgett technique on the basis of NODB/1-docosanol inhibit the corrosion of copper in 3.4% NaCl more effectively then benzidine.
- The inhibitor effectiveness increases with the number of monomolecular layers.
- Coatings with at least 10 layers are the most effective.
- The corrosion potentials hardly change at all in these processes, indicating that the mechanism is blocking of the anodic dissolution.

However, raster electron microscope images show fine pitting corrosion even on specimens coated with inhibitor, which is much more prominent on copper without inhibitor.

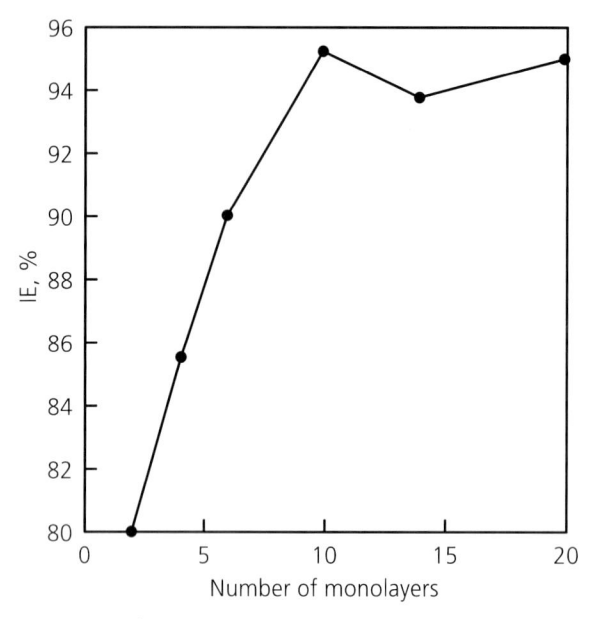

Figure 155: The effectiveness of the inhibitor (IE) NODB/1-docosanol (1:1) on copper as a function of the number of monomolecular layers [1286]

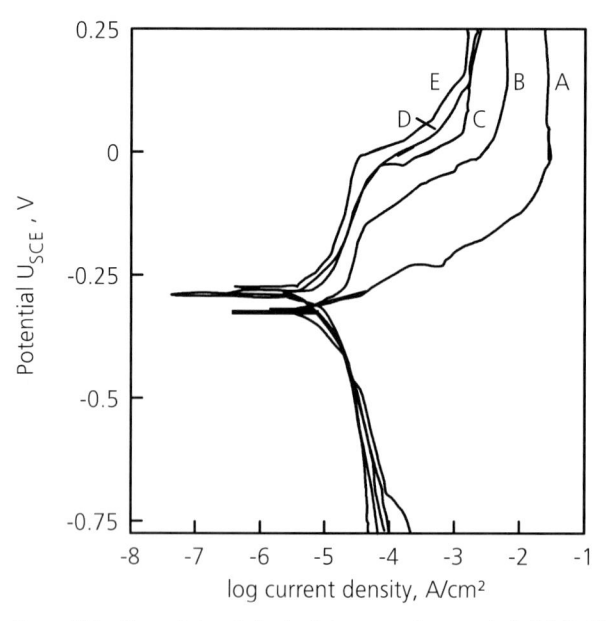

Figure 156: Current density/potential curves of copper in 3.4 % NaCl
A) copper; B) +5 mg/l benzidine; C) 2 monomolecular layers on copper; D) 6 layers E) 10 layers [1286]

Parameter	Corrosion current density A/cm²	Polarization resistance Ω/cm²	Corrosion rate mm/a (mpy)	Inhibitor effectiveness %
Control	6.41×10^{-5}	3.25×10^3	2.550 (100)	–
5 ppm benzidine	1.0841×10^{-5}	5.26×10^3	0.635 (25)	72
2 monolayers	8.78×10^{-6}	8.40×10^3	0.308 (12.1)	86
6 monolayers	6.44×10^{-6}	1.15×10^4	0.226 (8.90)	90
10 monolayers	3.94×10^{-6}	1.54×10^4	0.139 (5.47)	94

Table 211: Electrochemical parameters for the inhibition of copper by LB coatings of NODB/1-docosanol in 3.4% NaCl at 298 K (25 °C) [1286]

The inhibition of copper in 1.5% NaCl solution by piperidine, piperidine dithiocarbamate and Cu(II) complexes is reported by [1287]. As current density/potential curves show, all 3 inhibitors have an influence on the cathodic and in particular the anodic current density curve form at concentrations up to 125 mg/l, as shown for example in Figure 157. Very large inhibiting effects of 92 to 100% are achieved.

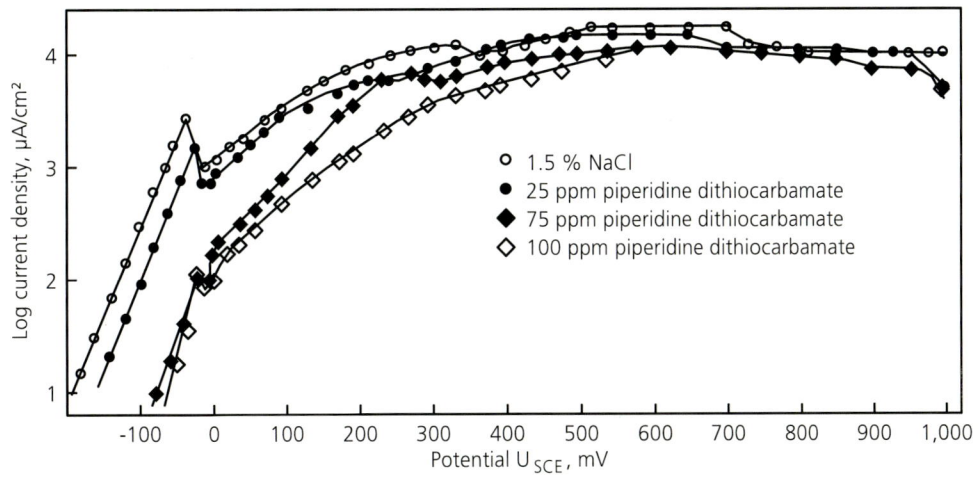

Figure 157: The influence of different concentrations of the inhibitor piperidine dithiocarbamate on the anodic current density/potential curves of copper in 1.5% NaCl at 298 K (25 °C) [1287]

The aforementioned authors have extended their investigations of the influence of inhibitors on the corrosion of copper in 1.5% NaCl solution to diethylamine [1288], using the same test methods. In the results it is found that the three inhibitors diethylamine, diethyl dithiocarbamate and their Cu(II) complexes have a high inhibiting effect of approximately 90% already as from very small quantities of 10 mg/l, increasing to 98–100% at concentrations of 125 mg/l. The inhibiting effect increases with the temperature (298, 308 and 318 K (25, 35 and 45 °C)). The current density/potential curves permit the conclusion that diethylamine is an anodic inhibitor, whereas the other two substances act as mixed inhibitor.

The effectiveness of methyl orange, methyl red and methyl yellow as inhibitors of the corrosion of copper in 5% NaCl solution has been tested specifically in the acidic range, i.e. at pH-values of 3.2 and 5.5 [1289]. Table 212 shows the effectiveness of the inhibitors in percent according to weight loss determinations. The inhibitors are more effective in the acidic range. The current density/potential curves and the corrosion potentials are shifted to more noble potentials; the anodic and the cathodic current densities are decreased by the initiators to the same extent.

Inhibitor	Effectiveness of the inhibition, %	
	pH 5.5	pH 3.2
Methyl yellow	10	73
Methyl red	27	82
Methyl orange	3	34

Table 212: Effectiveness of the inhibition in percent for copper in 5% NaCl according to weight loss determinations [1289]

Heterocyclic compounds on the basis of azole also inhibit the corrosion of copper in chloride solution [1290]. Table 213 summarizes the effectiveness of the inhibitors. A very high effectiveness is found for all substances, with a maximum of 99% for BTA (1,2,3-benzotriazole).

Solution	Corrosion rate $g/m^2\ d$	Inhibitor effect %
0.1 N NaCl	0.63	–
+ 1,2,3-benzotriazole (BTA)	0.01	99
+ 2-benzimidazolethiol (BIE)	0.02	97
+ 2-benzooxazolethiol (BOE)	0.08*)	88
+ benzimidazole (BIA)	0.05	92
+ sulfathiazole (STA)	0.04*)	94
+ 2-benzothiazolethiol (BTE) (saturated)	0.015	98

*) local attack

Table 213: Corrosion rates and effectiveness of inhibitors in percent for copper in 0.1 M NaCl after 10 days exposure at 298 K (25 °C); all inhibitor concentrations 10^{-3} M [1290]

Inhibitors on triazole basis for copper in aerated 3% NaCl solution are reported in [1291]. The following compounds were tested: 1,2,4-triazole (TA); 3-triazole (TTA); diaminotriazole (DAT) and benzotriazole (BTA), in each case in concentrations of 5×10^{-3} mol/l. Polarization measurements were made on a rotating electrode with 1,000 rpm, and weight loss determinations were made for 8 days. Impedance measurements showed an increase of the polarization resistance with decreasing rotation speed. With the exception of TTA, the corrosion current densities (A1) determined from the extrapolated diagram lines gave inhibiting effects up to 99%, confirmed by the weight loss measurements (E2), see Table 214. The inhibiting effect decreases slightly with increasing temperature, but still lies at 94% and 95% respectively for the new inhibitors BTA and DAT. From the current density/potential curves it follows that all inhibitors act anodically, as is also shown by the shift of the free corrosion potential in the positive direction, cf. Table 214. They inhibit the anodic dissolution of Cu by chemical adsorption and formation of a protecting inhibitor layer that blocks the pitting corrosion. The order of decreasing inhibiting effect is BTA > DAT > TA > TTA. The high inhibiting effect of DAT and BTA is retained up to 333 K (60 °C).

Electrolyte	Corrosion potential mV	Corrosion current density $\mu A/cm^2$	Inhibiting effect (E1) %	Inhibiting effect (E2) %
3% NaCl	−224	60	–	–
3% NaCl + TA	−180	0.35	99.4	90
3% NaCl + TTA	−174	12	80	74
3% NaCl + DAT	−288	0.7	99	98
3% NaCl + BTA	−190	0.1	99.8	99

E1: from corrosion current densities
E2: from weight loss measurements
(TA) 1,2,4-triazole; (TTA) 3-triazole; (DAT) diaminotriazole; (BTA) benzotriazole

Table 214: Electrochemical parameters of copper in 3% NaCl solution and with 5×10^{-3} M addition of inhibitors on triazole basis [1291]

The influence of some inhibitors on the corrosion of copper has been tested under heat exchanger conditions in 3.5% NaCl solution [1292]. Relative to isothermal tests at 308 K (35 °C) some inhibitors fail when the temperature difference is 333 K (60 °C), whereas others even there retain higher inhibiting effect, as confirmed by the key data determined from the current density/potential curves, see Table 215. After 100 h test time only 2-mercapto-5-methyl-thiodiazole (2Mc-5Met-TDA) and 2-amino-5-mercapto-thiodiazole (2Am-5Mc-TDA) still showed their good inhibiting effect.

	10^{-3} M		10^{-4} M	
Solution	$1/R_p$	Inhibition, %	$1/R_p$	Inhibition, %
NaCl 3.5 %	9,437.0	–	9,437.0	–
+ 2Mc-5Met-TDA	3.3	99.9	18.1	99.8
+ 2Mc-5Mc-TDA	7.2	99.9	2,964.0	68.6

Table 215: Reciprocal polarization resistance ($1/R_p$) and inhibition in percent of copper under heat exchanger conditions with 333 K (60 °C) temperature difference in 3.5 % NaCl after 1 h [1292]

Salt melt corrosion

The behavior of pure copper in contact with various salt melts on the basis of NaCl-KCl has been determined at high temperatures [1293]. Corrosion rates of 0.6–1.5 × 10^{-4} g/cm²h are found in the temperature range between 973 and 1,173 K (700 and 900 °C), by weight loss determinations as well as from the salt analysis and in agreement with other literature. The measurement of electrochemical potentials gives different results, which are explained by assuming that not the cations, but instead the impurities of the salt melts are the oxidizing substances in the process. Further studies in melts with complexes on tantalum, niobium and hafnium, such as NaCl-KCl-K_2TaF_7, NaCl-KCl-K_2HfF_6 (5 %) as well as NaCl-KCl-K_2NbF_7 (5 %) gave higher corrosion rates, in the case of hafnium by one power of ten, for niobium and tantalum by two powers of ten.

The corrosion potentials of copper in NaCl and KCl melts as well as in a 50/50 mixture of both have been determined in the temperature range of 1,023–1,223 K (750–950 °C) [1294]. As Figure 158 shows, the potential generally become less noble with increasing temperature, whereby the values for NaCl melt lie significantly below those for KCl melts. Because after adding hydrogen into the melt the potentials decreased by a further 0.3–0.4 V, it must be assumed that this removed bound oxygen and thus the surface layers were depassivated. Thus large amounts of oxygen must be built into the surface layer when corrosion takes place, and it was possible to show that the oxygen content in copper (0.05 %) is sufficient for this. Therewith the corrosion potential and thus the salt melt corrosion is controlled by the diffusion of oxygen from the metal into the surface layer.

Copper-aluminium alloys

Copper-aluminium alloys have good durability in NaCl solutions, even in higher concentration. In most cases these alloys are multi-component bronzes with further additive metals such as nickel, iron and manganese. Thus their durability is relatively good not only in neutral, but also in acidic NaCl solutions (utilization in solutions of the paper and pulp industry acidified with sulfuric acid). The stress corrosion cracking behavior, too, is improved. They have high strength and high resistance to

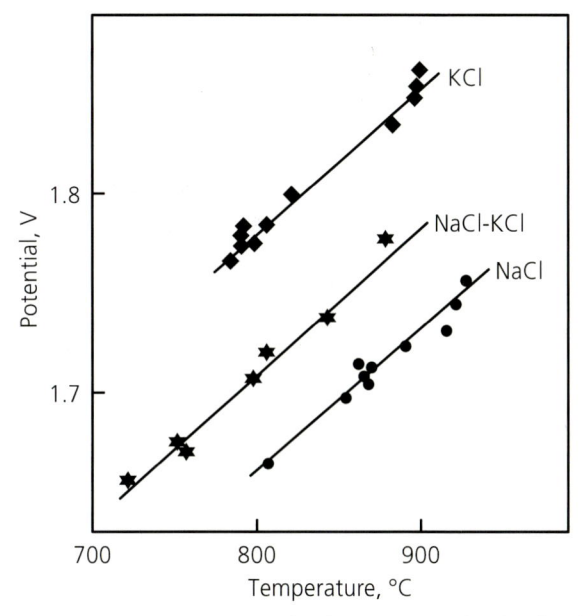

Figure 158: Corrosion potentials of copper in NaCl and KCl melts as well as a 50/50 mixture of both, as a function of the temperature [1294]

erosion, which lies far above that of C-steel. This is the reason for their utilization in ship screws as well as for rotor blades and housings of erosively stressed pumps. For alloys with very high aluminium content (> 10 % Al) possible aluminium depletion must be taken into consideration. It can be avoided by adding the alloying elements nickel and iron, on condition that these are dissolved homogeneously by a heat treatment.

On a copper-aluminium casting alloy (10.6 % Al, 3.2 % Fe, 0.8 % Mn) the influence of the addition of 0.3 and 0.5 % of chromium as well as 1 % of silicon on the corrosion stability in acids as well as NaCl solutions of 2.5 and 10 % has been determined [1295]. The weight losses were determined by dip tests for 48 h at 289 K (16 °C) and 6 h at 353 K (80 °C) in air-saturated solution, supplemented by potentiodynamic (10 V/h) polarization/time and potential/time curves. The weight losses and the corrosion currents determined from the diagram lines of the polarization curves agreed to within 5 %.

A)

Alloy	Temperature K (°C)	5% HCl	10% HCl	2% NaCl	5% NaCl	10% NaCl	5% H_2SO_4	10% H_2SO_4	Sea water
1	289 (16)	5.4	7.3	1.6	2.9	3.3	0.1	0.3	1.5
2	289 (16)	2.9	2.4	2.7	3.8	4.4	0.6	0.7	2.1
1	353 (80)	8.7	13.1	7.7	8.5	10.9	0.6	0.7	2.1
2	353 (80)	3.7	10.9	8.0	9.4	12.9	1.3	0.7	2.4

B)

Alloy	Composition, %	5% HCl	5% H_2SO_4	5% NaCl	Sea water
Aluminium bronze (forged)	Cu 85.4, Al 10.6, Fe 3.1, Mn 0.8	68.7	3.5	0.9	0.2
Brass (cast)	Cu 66.8, Pb 3.8, Sn 0.9, Fe 0.2, Zn balance	76.5	10.2	1.1	0.9
Phosphor bronze (cast)	Cu 88.0, Sn 11.3, Pb 0.3	80.1	9.8	1.1	1.2
Tin bronze (cast)	Cu 85.0, Sn 9.1, Pb 3.6, Zn 2.3	84.7	14.4	1.7	1.2

Table 216: A) Corrosion rates of aluminium bronze in mg/dm² d at 289 and 353 K (16 and 80 °C) Alloy 1: 0.3 % Cr; Alloy 2: 0.5 % Cr, 1 % Si (see the text)
B) Corrosion rates of copper alloys in mg/dm² d at 289 K (16 °C)

The determined corrosion rates are summarized in Table 216 and compared with those of other copper alloys (only at 289 K (16 °C)). Whereas considerably lower corrosion rates are found for the alloys 1 and 2 in the acids, the values are higher in NaCl; this is explained by the short test time, and a strong decrease is expected in tests for a longer period. The alloy variant 2 containing silicon behaves better only in HCl. The results of potential and polarization measurements, which are not reproduced here in detail, are in good agreement with the corrosion rates and support the interpretation that chromium promotes the formation of a more stable protective layer and thus improves the corrosion stability. The influence of the microstructure produced by heat treatment is pronounced. Unfavorable behavior is produced by the β-phase that appears after rapid cooling of the casting alloy. The best behavior is obtained with a microstructure containing chiefly the α-phase, that is obtained by slow cooling from 873 K (600 °C).

The same author compares the corrosion stability between aluminium bronze (6.8 Al, 3.2 Fe, 0.8 Mn) and brass (20.9 Zn, 2 Al, 0.5 Fe, 0.03 As) in chloride solutions as well as acidic media at 298 K (25 °C) by electrochemical measurements [1296]. The results show a better durability of the aluminium bronze in sea water (0.02 mm/a (0.79 mpy)) compared with 0.3 mm/a (11.8 mpy) for brass, HCl and H_2SO_4; this is the reason why it is better than brass in the sea water desalination plant, also under acidic conditions. In contrast thereto, in the NaCl solutions brass shows significantly lower corrosion rates compared with aluminium bronze:

NaCl, mass%	Brass	Aluminium bronze
5	0.016 mm/a (0.63 mpy)	0.178 mm/a (7.01 mpy)
10	0.015 mm/a (0.59 mpy)	0.033 mm/a (1.30 mpy)
15	0.014 mm/a (0.55 mpy)	0.020 mm/a (0.79 mpy)

Aluminium bronze G-CuAl10Ni has been tested in 20% NaCl solution at pH 5.5 and 323 K (50 °C) comparatively with unalloyed and stainless steel with regard to the resistance to erosion corrosion [1297], cf. Table 217. According to the results, the alloy shows very good behavior up to 40 rpm in comparison with carbon steel.

Material designation	Rotation speed, m/s	Measuring time, h	Weight loss, $g/m^2 \, d$
Unalloyed steel Ck 22 (DIN Mat. No. 1.1151)	40	120	1,719
Chrome steel X40Cr13 (DIN Mat. No. 1.4031)	40	240	3
Aluminium bronze G-CuAl10Ni	40	144	29
Austenite-Ferrite type, DIN Mat. No. 1.4462	40	528	1
Chrome steel X40Cr13 DIN Mat. No. 1.4031	70	200	3,725
Austenite-Ferrite type, DIN Mat. No. 1.4462	70	200	1

Table 217: Corrosion rate of materials by erosion corrosion [1297]

Multi-material aluminium bronzes are used for ship screws, pump wheels, heat exchangers and other components chiefly in sea water. Thus many binary multi-component aluminium bronzes and others with addition of iron, nickel and manganese have been investigated with regard to their corrosion stability in 3% sodium chloride solution at room temperature, with special consideration of the casting microstructure states set by various heat treatments [1298]. The composition and heat treatment of the various test alloys, all of which can be classified according to DIN EN 1982 [1299] are summarized in Table 218. In addition to the immersion tests carried out for 800 and 850 h, polarization resistance measurements were carried out at the free corrosion potential (+/– 10 mV), but these measurements permit only qualitative assessments. A tendency for aluminium depletion is immediately evident from the Cu/Al ratio in the solution determined by atomic absorption analysis. The kind and extent of the corrosion can be derived from the stationary corrosion potential. Thus for the specimens in the cast state and with heat treatment at 673 K (400 °C), the corrosion potentials decrease (–70 to –130 mV and –90 to –160 mV$_H$) for increasing aluminium content (10, 11, 13% Al). Aluminium depletion was found on all specimens, documented by re-deposition of copper, combined with aluminium depletion depths up to 1 mm after 800 h exposure in NaCl solution. Among the multi-material alloys, those with high aluminium content or after high heat treatment temperature (χ_2-phase) show negative corrosion potentials and corrode with strong aluminium depletion. But for most of the alloys the corrosion potentials lie above the threshold potential for aluminium depletion (–50 mV) and show excellent durability; this is attributed to the κ-phase which is rich in iron and nickel. The investigation results can be summarized and assessed for practical purposes as follows: Large amounts of χ_2-phases lead in all cases to low corrosion potentials and to poor durability, whereby below a critical potential of −50 mV no protective layer of Cu_2O can be formed any longer, and catastrophic aluminium depletion takes place, with binary aluminium bronzes already as from slight χ_2-precipitations. These can be ignored for the multiphase alloys, because in those the simultaneous presence of the higher potential κ-phase makes the formation of the Cu_2O protective layer possible. The χ_2-phase is (irrespective of the alloy type) less corrosion resistant than Martensite. Conversion of the Martensite to the (α+β)-microstructure with subsequent compensation of the eliquations by diffusion-annealing gives better corrosion resistance.

Test alloys	Composition, mass%					Heat treatment		Cooling mode
	Al	Fe	Ni	Mn	Cu	Annealing temperature K (°C)	Annealing duration h	
1	9.9				balance		cast state	
2	10.8				balance		cast state	
3	12.9				balance		cast state	
4	9.9				balance	693 (420)	80	furnace
5	10.8				balance	693 (420)	80	furnace
6	12.9				balance	693 (420)	80	furnace
7	9.9				balance	1,273 (1,000)	10	ice water
8	12.9				balance	1,273 (1,000)	10	ice water
9	9.9				balance	773 (500*)	4	air
10	12.9				balance	773 (500*)	4	air
11	9.5			0.9	balance		cast state	
12	9.5			0.9	balance	1,273 (1,000)	1/2	furnace; 0.1 °C/min
13	9.5			0.9	balance	873 (600)	1/2	compressed air
14	10.2	4.3	4.9	1.4	balance		cast state	
15	10.2	4.3	4.9	1.4	balance	1,273 (1,000)	1/2	furnace; 0.1 °C/min

* Tempering action after conversion to Martensite (1,223 K (950 °C)/30 min/ice water)

Table 218: Composition and heat treatment of the test alloys [1298]

Table 218: Continued

Test alloys	Composition, mass%					Heat treatment		Cooling mode
	Al	Fe	Ni	Mn	Cu	Annealing temperature K (°C)	Annealing duration h	
16	11.4	5.1	5.1	1.5	balance	1,273 (1,000)	1/2	furnace; 0.1 °C/min
17	13.2	5.6	5.0	1.5	balance	1,273 (1,000)	1/2	furnace; 0.1 °C/min
18	9.9	3.1	5.2	0.2	balance		cast state	
19	9.5	3.1	5.2	0.6	balance	1,273 (1,000)	1/2	furnace; 0.1 °C/min
20	9.7	3.1	5.2	0.6	balance	973 (700)	1/6	air
21	9.8	3.1	5.2	0.6	balance	973 (700)	100	air

* Tempering action after conversion to Martensite (1,223 K (950 °C)/30 min/ice water)

Table 218: Composition and heat treatment of the test alloys [1298]

Copper-aluminium alloys are more susceptible to stress corrosion cracking in NaCl. This was investigated in detail on two alloys containing 9 and 11.8 % Al in 5 % NaCl and 3.5 % NaCl + NaOH solution as well as in various states of the microstructure (heat treatment, dispersion hardened, cold rolled, casting) [1300]. The corrosion potentials all lie around −220 mV$_{SCE}$, the current density/potential curves give current density maxima in the anodic region at about −40 mV. The greatest stress corrosion cracking susceptibility is also found in this potential range for all microstructure variants. Initiated inter-crystalline cracks are then found, as well as aluminium depletion, which at least partly contributes to the stress corrosion cracking susceptibility. Samples containing 11.8 % aluminium show a greater fraction of the γ_2-phase, which is preferentially subject to aluminium depletion, and maximum embrittlement is also found for this alloy variant, whereas specimens containing 9 % aluminium, homogenized after cold forming, show less embrittlement. Corrosion fatigue tests were carried out in 5 % NaCl solution at pH 6.8, in the anodic range at the current density maximum as well as in the strongly cathodic range, see Figure 159. Thereby it was found that the crack propagation rates are very much greater in the anodic than in the cathodic range.

The corrosion fatigue behavior of an aluminium-nickel bronze (9 Al, 5 Ni, 4 Fe; CDA 958) with duplex structure was determined in 0.5 M NaCl solution as a function of the anodic and cathodic polarization as well as the grain size and the frequency [1301], according to the stress/number of load cycles plotted as (σ-N)-diagrams. It is evident from Figure 160 that the load limit decreases by 30 % compared with air at the corrosion potential in the NaCl solution (with N = 10^7). The number of load cycles achieved with a given load decreases with anodic polarization and

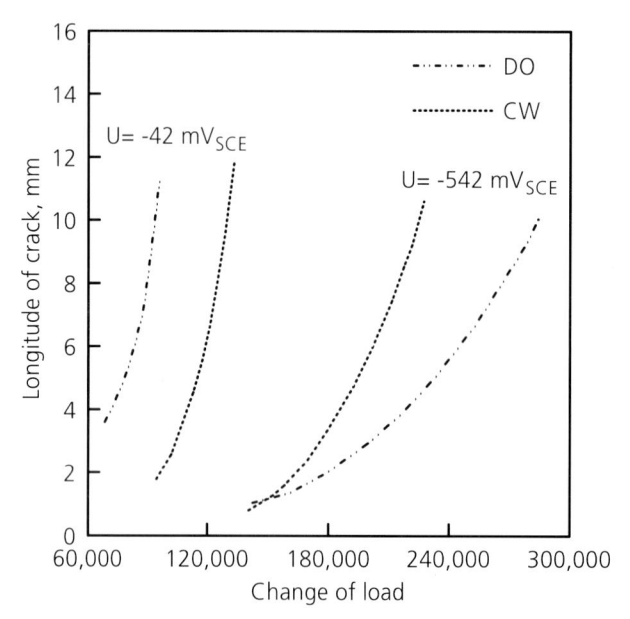

Figure 159: Crack propagation rates as function of the number of load cycles for Cu-9Al in 5% NaCl at pH 6.8; various microstructure states [1300] (DO: recrystallized at 920 K (647 °C) then exposed for 30 min at 523 K (250 °C), CW: cold rolled)

increases with cathodic polarization. The σ-N curve also has a corresponding form for cathodic polarization, lying between the curves in Figure 160, i.e. the corrosion fatigue behavior is significantly improved by cathodic polarization. An influence of the grain size (0.3, 0.5 and 1.5 mm) is not found. However, the curve trace shifts to significantly lower load cycle numbers when the frequency is reduced from 20 to 2 Hz.

As alloying element iron improves the corrosion stability of aluminium bronzes in chloride media, provided that it is dissolved homogeneously. However, because the solubility is limited to about 3.5% iron, alloy variants on the basis of 6.2% Al with up to 11.5% Fe were prepared using the method of rapid quenching, and their corrosion stability in 3.5% NaCl solution was determined electrochemically in comparison with a cast alloy having a corresponding analytical composition [1302]. The improved corrosion stability is shown by the corrosion current densities derived from the diagram lines, Figure 161. As from 3.5% iron the values obtained for rapidly cooled specimens decrease significantly, whereas they increase again slightly for the cast state and lie on a considerably higher level. This increase, as well as the increase found with 11.5% iron (in the rapidly cooled specimen), is explained by the preferential dissolution of iron rich precipitations. The corrosion current densities increase again slightly on heat treatment of the rapidly cooled specimens. For example, after 30 min annealing at 1,073 K (800 °C) of the specimen with 3.6% iron one finds 4.9 $\mu A/cm^2$ (about 0.057 mm/a (2.24 mpy)) for the cast state, 3.6 $\mu A/cm^2$ (about 0.04 mm/a (1.57 mpy)) for the rapidly cooled state and 5.3 $\mu A/cm^2$ (about 0.06 mm/a (2.36 mpy)) for the heat treated state. Evidently the iron rich precipitates produced by the heat treatment at the grain boundaries lead to increased local attack.

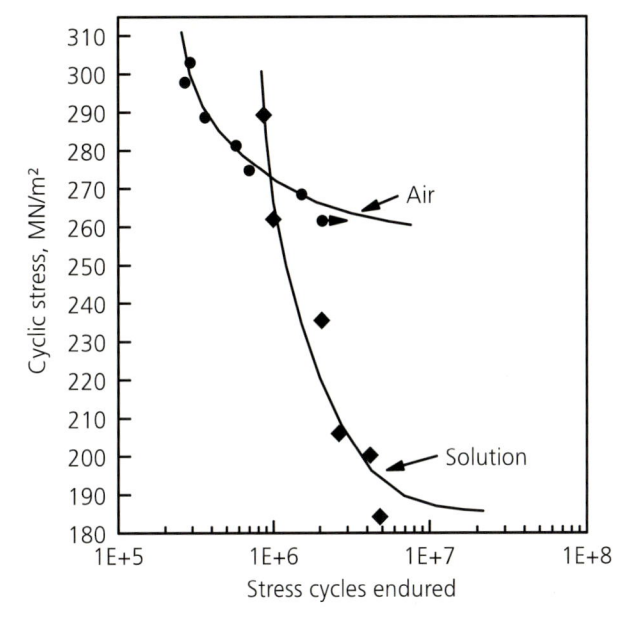

Figure 160: Stress/load cycles diagram for aluminium bronze in aerated 0.5 mol/l NaCl and air at 293 K (20 °C) and 20 Hz [1301]

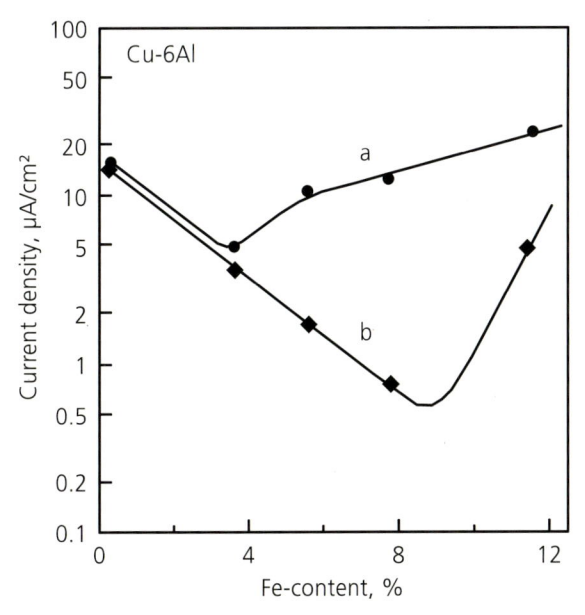

Figure 161: Change of the corrosion current densities with the iron content of 6 aluminium bronzes in 3.5 % NaCl solution a) cast material b) rapidly cooled [1302]

The corrosion resistance of copper-aluminium alloys in chloride solutions is increased by the addition of tin. This was shown by investigations on specimens (7 Al, 2 Sn, 1 Fe, 1 Mn, balance Cu) in 2, 5 and 10% NaCl solution as well as in sea water at 289 and 353 K (16 and 80 °C) [1303]. Table 219 shows that for the chloride media the corrosion rates decrease to approximately one half compared with tin-free CuAl bronzes, also in the two tested acids. The difference between the cast state and the state after annealing is slight; annealing sometimes brings a marginal improvement. The test at 353 K (80 °C) shows a similar positive influence of tin, but here the influence of a heat treatment is more distinct: Rapid or slow cooling from the temperature range of 813–923 K (540–650 °C) gives with about 0.15 mm/a (5.90 mpy) the lowest corrosion rate, whereas it increases by more than a factor of two by cooling from 973–1,223 K (700–950 °C) (0.36–0.381 mm/a (14.2–15 mpy), approximately 0.4 mm/a (15.7 mpy)). Current density/potential curves and measurements of the corrosion potential show that the addition of tin increases the corrosion potential and thus counteracts the lowering of the corrosion potential by the presence of iron and manganese in the alloy.

Condition	5% HCl	10% HCl	2% NaCl	5% NaCl	10% NaCl	5% H$_2$SO$_4$	5% H$_2$SO$_4$	Sea water
Cast state	0.25 (9.84)	0.20 (7.87)	0.09 (3.54)	0.17 (6.69)	0.18 (7.09)	0.06 (2.36)	0.11 (4.33)	0.08 (3.15)
Annealed*	0.21 (8.27)	0.18 (7.09)	0.08 (3.15)	0.15 (5.90)	0.18 (7.09)	0.06 (2.36)	4.72 (0.12)	0.09 (3.54)
Alloy without added tin**	0.51 (20.1)	0.06 (2.36)	0.20 (7.87)	0.30 (11.8)	0.37 (14.6)	0.15 (5.90)	7.87 (0.20)	0.13 (5.11)

* slowly cooled down from 1,223 K (950 °C)
** alloy without addition of tin, as casting

Table 219: Corrosion rates of Cu7Al2Sn (cast state and annealed) at 289 K (16 °C) in mm/a (mpy) [1303]

The potential behavior of CuAl11Ni as well as CuNi3Si has been investigated compared with pure copper in 1 mol/l NaCl with various aeration conditions [1304]. In the presence of oxygen the potential changes produced by **vibration cavitation** lie about 50 mV more negative for all copper materials. In oxygen free solution the corrosion rate is independent of the flow velocity and practically zero. However, after stress by vibration cavitation the surface attack of pure copper is very pronounced, with the appearance strong pit holes evenly distributed over the face end of the specimen. In contrast thereto, the attack of the two bronzes is slight.

The corrosion resistance of the aluminium-manganese bronze, with additions of nickel and iron, used as cast plate in the paper and pulp industry, was tested also in the welded state [1305]. The analysis of the base material was: 7.1 Al, 8.25 Mn, 1.71 Ni, 2.81 Fe, balance Cu; that of the auxiliary welding material was: 10.21 Al, 1.70 Mn, 2.88 Fe, balance Cu; and that of the melted down welding material was:

7.6 Al, 8.37 Mn, 1.61 Ni, 2.41 Fe, 0.66 Si, balance Cu. The weight losses were determined in acidic $Al_2(SO_4)_3$ solution (pH 3.5), in 165 mg/l NaCl + 350 mg/l Na_2SO_4 (pH 3.5) as well as in a pulp solution taken from a paper and pulp production line. Figure 162 shows the corrosion data plotted as a function of time (the numerical values correspond roughly to the corrosion rate in mm/a). It is clearly seen that particularly in the solution containing chloride the welding material (curve 4) shows considerably lower corrosion rates. Similar results are obtained after 700 h of exposure in an original paper and pulp solution with pH-value of 5, at 295–301 K (22–28 °C) and part speeds up to 300 m/min. Corrosion rates of 0.12–0.14 $g/m^2 h$ (equivalent to mm/a) are then found.

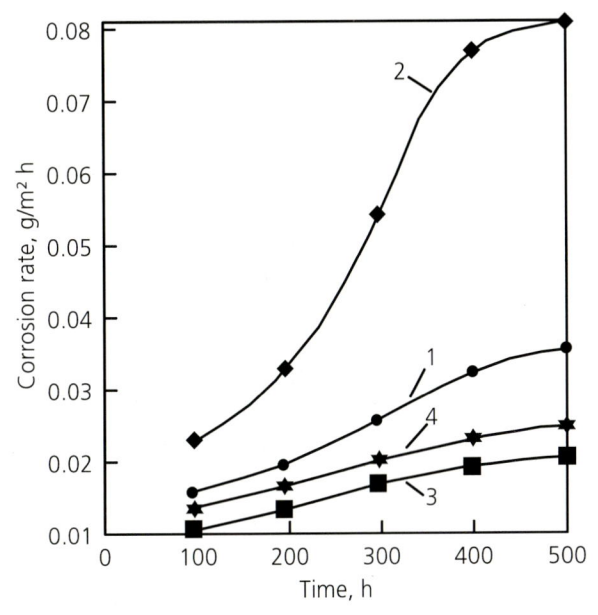

Figure 162: The corrosion rates as a function of time for AlMn bronze; 1 and 2 base material; 3 and 4 welding material; 1 and 3 in $Al_2(SO_4)_3$ solution; 2 and 4 in 165 mg/l NaCl + 350 mg/l Na_2SO_4 [1305]

For inhibition of aluminium bronzes (here: 7 aluminium, 0.04 Fe) in acidic chloride solutions, benzotriazole and thiourea are often used. The synergistic influence of iodide ions in 4 % NaCl solution with a pH-value of approximately 2 (+ 10^{-2} M HCl) at 333 K (60 °C) was tested [1306]. Weight loss determinations and current density/potential curves show that iodide ions alone have no influence on the dissolution behavior of the aluminium bronze. In contrast thereto, iodide ions with a concentration of 100 mg/l potassium iodide improve the inhibiting effect of benzotriazole (to 92 %) and thiourea (to 79 %), when these are present with a concentration of 300 mg/liter.

Copper-nickel alloys

Copper-nickel alloys generally have very good durability in NaCl solutions, as is evident from long standing good experience with their utilization in particular as pipe material in sea water (ships, oil drilling platforms, sea water desalinization) and in heat exchangers (power plant condensers). Prerequisite for the good corrosion resistance is the formation of an optimum protective layer. This is achieved only when the plants and components "are run in properly" before actual operation, i.e. under flowing, well aerated and clean conditions. Thereby pollutants, in particular hydrogen sulfide, must be avoided as far as possible. After damage incidents as consequence of commissioning ships in polluted sea ports, new pipeline systems of ships are now commissioned only on the high sea.

Iron is a specific alloying element (0.7–2% Fe) for copper nickel alloys because it improves the adhesion of the protective layer when incorporated therein. Protective layers must be thin and enriched with iron and nickel (i.e. depleted with respect to copper), and voluminous external layers must be avoided. The higher nickel content in CuNi30Fe (CDA 715, DIN Mat. No. 2.0882, UNS C71500) improves the corrosion stability with respect to CuNi10Fe1Mn (CDA 706, DIN Mat. No. 2.0872, UNS C70600). Stress corrosion cracking, inter-crystalline and galvanic corrosion are hardly known among the copper-nickel alloys. However, they are more susceptible to erosion corrosion, depending strongly at higher temperatures on the pH-value and the oxygen content of the solution. In practice (sea water) utilization of, for example, CuNi10Fe1Mn is restricted to 1.5 m/s and utilization of CuNi30Fe to about 3 m/s. $Fe_2(SO)_4$ is worth considering as possible inhibitor, because Fe^{3+} ions are built into the covering layer and improve the protective effect. The fact must be pointed out that that the behavior of copper-nickel as well as most copper alloys in 3–3.5 % NaCl solutions often differs significantly from the behavior in sea water; this is explained by the presence of other cations and anions in sea water, which there form different surface layers on the alloys. Copper-nickel alloys generally show better behavior in NaCl solutions than in sea water.

An overview of the literature with respect to the corrosion stability of CuNi10-Fe1Mn (former designation CuNi10Fe, DIN Mat. No. 2.0872) in sea water, synthetic sea water and chloride solution in combination with their own measurements is given by [1307]. The goal of the investigations was to define conditions under which firmly adhering layers giving good protection are formed or not formed, or under which they are uniformly or locally destroyed. The investigations were carried out with the rotating disk electrode and on pipes. In general, 4 types of protective layers are found: well adhering and undamaged, well adhering but with pitting corrosion, well adhering but heterogeneous, and poorly adhering and heterogeneous. Table 220 shows the characteristic properties of layers with good and bad adhesion.

	Well adhering and undamaged	Poorly adhering and heterogeneous
Color	yellow, with dark parts	dark brown, green + red patches
Thickness	1–3 µm	1–16 µm
Structure	single layer	multiple layer
Structure of the layer	amorphous	crystalline (+ amorphous)
Determined/ possible compounds	$Fe_2O_3 \times H_2O$, $Ni(OH)_2$	$Cu_2(OH)_2Cl$, $Cu_2(OH)_3CO_3$

Table 220: Characteristic properties of two protective layer types on CuNi10Fe1Mn in chloride solution [1307]

At low flow velocities of 0.1 m/s poorly adhering and heterogeneous layers are obtained. At high flow velocities heterogeneous but well adhering protective layers are obtained. In detail, the variously structured layers are obtained under the following test conditions:

- in NaCl, synthetic and natural sea water
- at various pH-values
- after various pre-treatments (as delivered, chemically cleaned)
- at flow velocities of 0.1–3.5 m/s
- after exposure times of up to 5 months.

The following conclusions can be drawn from the results:

- The protective layers formed in NaCl solution and synthetic sea water are thinner than in natural sea water; all layers contain enriched iron and less nickel.
- The layers obtained at low pH-values of 4.5–6.5 are crystalline, contain more copper and give less protection; the corrosion rates are greater by a factor of 2–3.
- Surface pretreatment with dilute acid leads to the formation of amorphous unattacked protective layers free from pitting corrosion; they contain nickel, iron and copper.
- The higher the flow velocity, the thinner is the protective layer and the lower is its copper and chloride fraction; thin layers are more homogeneous and have better adhesion.
- The protective layers become thinner when their chloride content is less.
- At high flow velocities the exposure duration has no effect on the thickness of the covering layer.

Information regarding the fundamental mechanism for the formation of a protective layer on copper-nickel alloys was obtained by tests made with an alloy containing 9.4 % nickel and 1.7 % iron exposed in air-saturated 3.4 % NaCl solution at room temperature for at most 191 days [1308]. Thereby weight losses of the exposed specimens of various lengths and the weight of the corrosion layers were determined, and in addition thereto the anodic and cathodic polarization curves were recorded before and after the corrosion tests. The corrosion layer is found to consist of a thin very firmly adhering inner layer and a much thicker porous outer layer which can easily be removed by mechanical means. Both layers grow in the course of 1,000 h as a parabolic function of time that also applies for the weight loss, since this is the sum of the metal oxidized in the corrosion layer and the metal which has gone into the corrosive solution, as shown in Figure 163. The strong decrease of the corrosion rate of copper-nickel alloys in the course of time is clearly shown in Figure 164; the corrosion rate decreases by a factor of ten within 100 h and then continues to decrease to very small values. Polarization measurements show that after various exposure times of the specimens, and thus with increasing corrosion layer thickness, the anodic current density/potential curves give significantly lower current densities. A decrease is found with cathodic polarization when the thick corrosion layer is removed mechanically.

As continuation of the work described above, the composition of the protective layers on CuNi10Fe1Mn has been studied in detail by EDAX, X-ray and REM analyses [1309]. Whereas the outer porous protective layer formed after 5 days of exposure has only a small participation in the decreased anodic dissolution process, the high protective effect is produced chiefly by the thinner layer, because due to its low conductance it significantly restricts the cathodic reaction of the oxygen reduction at the transition between the inner and the outer layer.

The influence of the structure of the corrosion layers on the corrosion rate of copper-nickel-iron alloys is reported by [1310]. For this purpose CuNi10Fe1Mn

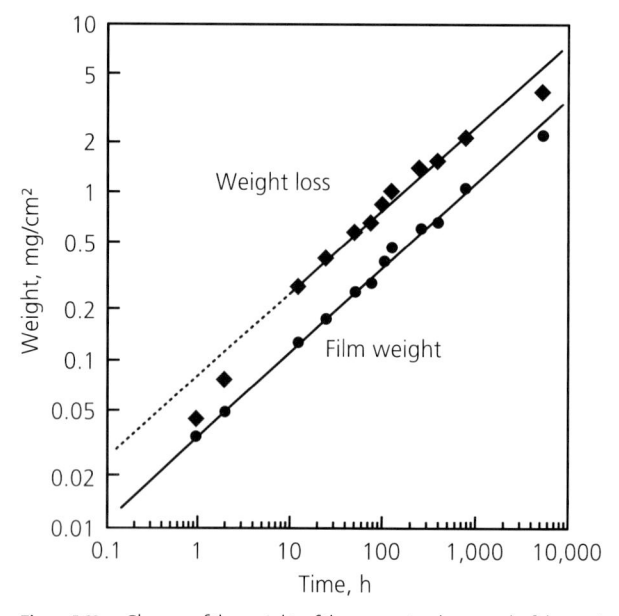

Figure 163: Change of the weight of the corrosion layer and of the entire weight loss of CuNi10Fe1Mn in 3.4 % NaCl solution [1308]

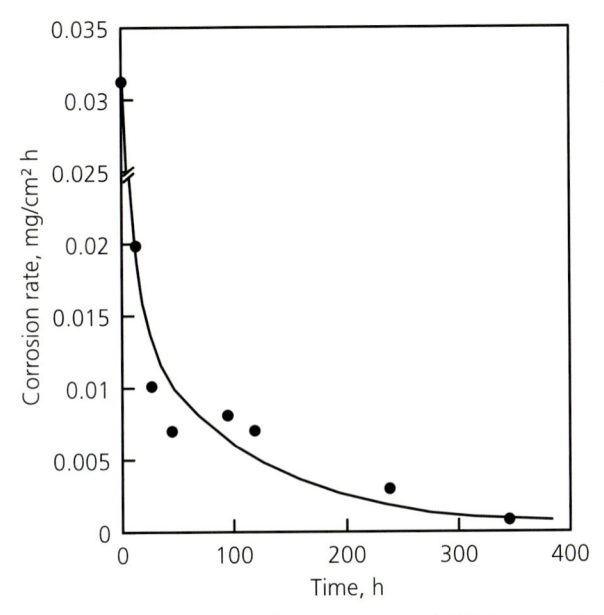

Figure 164: Corrosion rate of CuNi10Fe1Mn in 3.4 % NaCl solution as a function of time [1308]

(DIN Mat. No. 2.0872) and CuNi30Fe (0.41 Fe) were tested comparatively with respect to pure copper in 3.4 % NaCl solution by means of polarization curves and weight loss determinations after 30 and 60 days. The corrosion products were sampled and analyzed after various times. The weight losses plotted as a function of time have the form shown in Figure 165. Table 221 shows the corrosion rates in boiling NaCl solution determined therefrom, the structure of the corrosion layer and its thickness. According thereto CuNi30Fe has the best durability. Whereas anodic current density/potential curves at 298 K (25 °C) indicate no passivation at higher potentials and hardly any differences in form for the 3 materials tested here, a clear differentiation through smaller rises is found for specimens exposed to boiling solution for 30 days, in particular for CuNi30Fe. A similar behavior is found with the cathodic curves too, i.e. closely adjacent cathodic current density increases above the corrosion potential before the exposure, but after exposure for 30 days the hydrogen overvoltages increase significantly and lower current densities are found, indicating strong retardation of the cathodic reaction through layer formation, see Figure 166.

Resistance measurements on the corrosion layer confirm that the conductivity of the formed Cu_2O layer is decreased by incorporation of iron and nickel. Plotting the corrosion rates obtained for the 3 tested materials against the reciprocal value of the sum of the electron and ion resistance gives the linear relationship shown in Figure 167. The high resistance polarization of the CuNi alloys explains their influence on the anodic as well as the cathodic reaction, both of which are retarded, giving the high durability compared with copper. The different character of the protective layer is explained on the basis of a defect structure model.

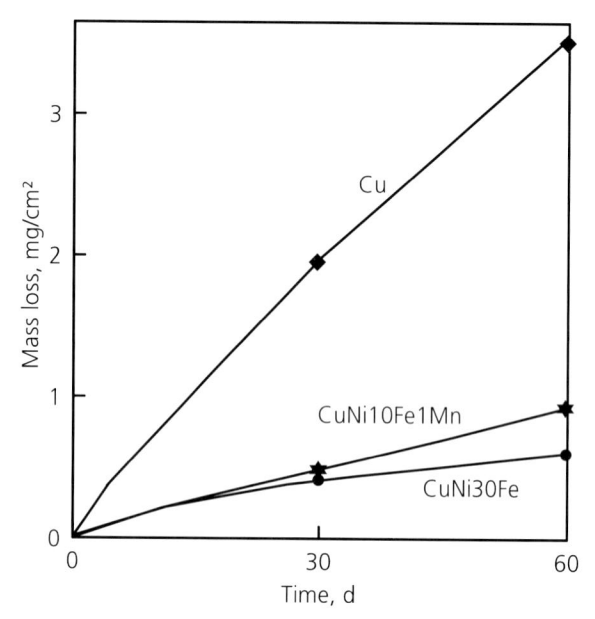

Figure 165: Weight losses as a function of the test time for copper, CuNi10Fe1Mn and CuNi30Fe in boiling 3.4 % NaCl solution [1310]

Alloy	Corrosion rate, g/m² d (approximately mm/a)	Structure of the corrosion layer*	Thickness of the corrosion layer, μm
		Test duration 30 d	
Copper	0.57 (0.02)	Cu_2O	0.35
CuNi10Fe1Mn	0.16 (0.006)	Cu_2O	0.2
CuNi30Fe	0.09 (0.004)	Cu_2O	0.15

* by X-ray diffraction

Table 221: Corrosion rates, structure and thickness of the corrosion layers after exposure for 30 days in boiling 3.4 % NaCl solution [1310]

The statements made above are confirmed in principle by tests made on Cu10Ni material with various iron contents (0, 0.3 and 1.54 % Fe) [1311]. Nickel and iron are very strongly enriched in the corrosion layer, in particular initially and during test periods of up to 30 d. This enrichment appears primarily in the vicinity of the metal oxide boundary layer. Figure 168 shows how the corrosion rates of CuNiFe decrease with increasing iron content in the NaCl solution.

With regard to the partly contradictory results reported in the literature for the corrosion stability of copper materials in natural sea water and in 3–3.5 % NaCl laboratory solutions, mutually comparative investigations have been made in both media [1312]. 4 CuNi alloy variants were prepared in the laboratory with 30, 50

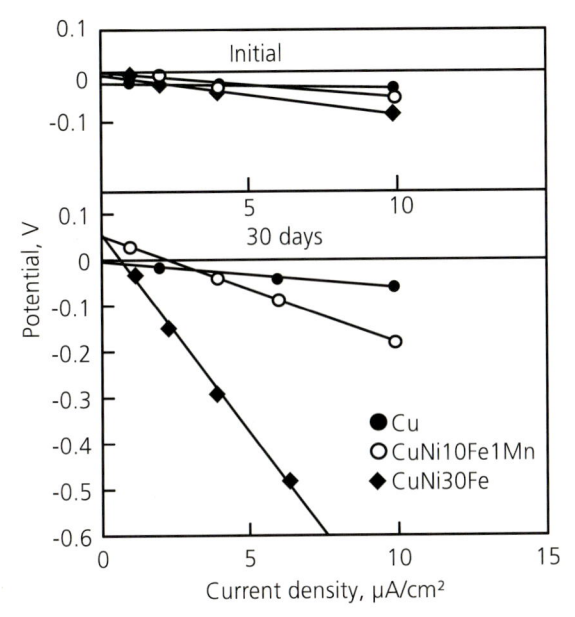

Figure 166: Galvanostatically determined polarization curves for copper, CuNi10Fe1Mn and CuNi30Fe in 3.4 % NaCl at 298 K (25 °C), recorded before and after corrosion in boiling 3.4 % NaCl for 30 days [1310]

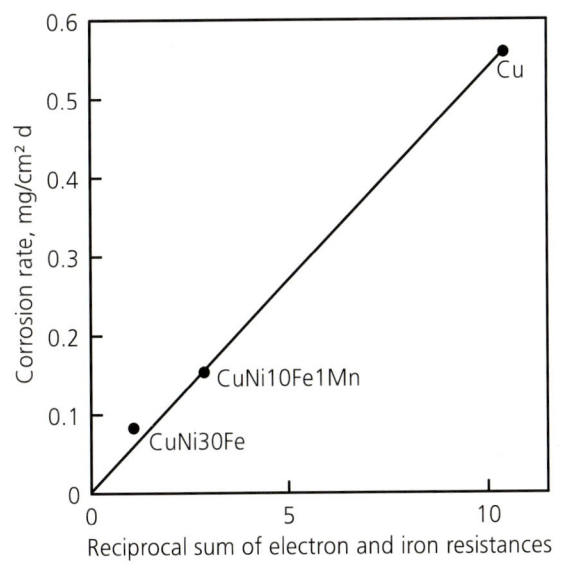

Figure 167: Relationship between the corrosion rate and the reciprocal sum of the electron and ion resistance of the Cu_2O corrosion film of copper, CuNi10Fe1Mn and CuNi30Fe after 30 days in boiling 3.5 % NaCl solution [1310]

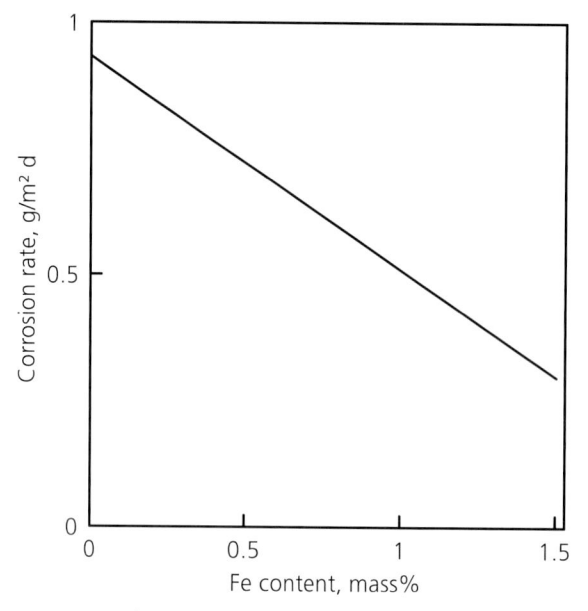

Figure 168: The corrosion rate as a function of the iron content of Cu10Ni in stationary 3.4% NaCl solution [1311]

70 and 90 copper, and the following determinations were made therewith in 3.5% NaCl solution: weight loss, current density/potential curves, analysis of the corrosion products and X-ray analysis of the layer structure. The current density/potential curves in 3.5% NaCl at pH 8.1 differ significantly from those obtained in sea water. Whereas they there lie very close together for all copper-nickel alloys, great differences of the curve forms are found here, see Figure 169. The difference of the corrosion stability in sea water and in NaCl solution become clearer when the weight losses as a function of time are compared, see Table 222. Apart from the CuNi alloy with only 30% copper, much greater dissolution rates are found in the NaCl solution. Calculation of the ratio of copper/nickel in the solution gives values, in particular in the NaCl solution for Cu10Ni and Cu30Ni much greater than 1, indicating strong copper depletion and nickel enrichment. The investigation results on the whole show that copper-nickel alloys have a greater resistance to corrosion in 3.5% NaCl solution than in sea water.

The influence of dissolved ozone on the corrosion stability of Cu30Ni in 0.5 M NaCl solution (pH 5) is reported by [1313]. Already small amounts of ozone of 0.2 mg/l shift the corrosion potential by 100 mV in the anodic direction. Parallel thereto the corrosion current densities decrease by a factor of 3 and the durability increases correspondingly. The addition of ozone produces with Cu30Ni in 0.5 M NaCl solution a shift of the current density/potential curves to more noble potentials. The corrosion potential settles down immediately after immersion without any delay. This effect is based on the formation of a very thin protective layer with a high ratio of O_2 to Cl^- in the layer.

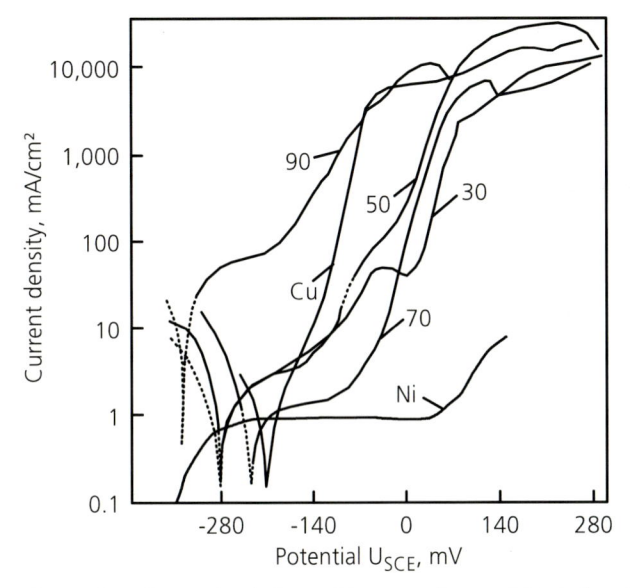

Figure 169: Current density/potential curves of copper/nickel alloys in 3.5 % NaCl at pH 8.1 and 0.5 mV/s (composition of the alloys, % copper) [1312]

Time, h	Alloy composition, % copper							
	90		70		50		30	
	NaCl	Sea water	NaCl	Sea water	NaCl	Sea water	NaCl	Sea water
24	0.13	0.17	0.06	0.31	0.03	0.34	0.02	0.01
60	0.14	0.33	0.08	0.41	0.05	0.46	0.02	0.02
240	0.58	1.20	0.27	0.85	0.14	0.59	0.08	0.02
360	0.99	2.22	0.52	2.58	0.24	0.72	0.08	0.03
720	1.35	2.80	0.66	2.73	0.28	0.98	0.10	0.09

Table 222: Weight loss (mg) depending on the immersion time of copper-nickel alloys in 3.5 % NaCl solution and natural sea water at pH 8.1 [1312]

An overview of the literature on the subject of **erosion corrosion** of copper-nickel alloys in sea water and chloride solutions is given in [1314]. According thereto the partly contradictors statements are based on differing testing techniques, test conditions such as pH-value, temperature, gassing, cavitation, solid materials, surface roughness, surface hardness, specimen dimensions/pipe diameter, test duration, state and composition of the examined materials. Only the most important influencing parameters will be considered briefly below; for more information reference is made to the original papers.

Depending on the iron content of the copper-nickel alloys, the corrosion rate increases with increasing flow velocity, very strongly as from 3–4 m/s with low iron contents, but with iron contents corresponding to the standard only insignificantly up to higher values; this is true for CuNi10Fe1Mn as well as for CuNi30Fe.

Lowered pH-values increase the erosion corrosion considerably, thus in the case of CuNi10Fe1Mn in sea water no protective layer formation is observed at pH 3.6. Nevertheless, the corrosion rate in the deaerated solution (25 ppm O_2) was still low with a value of 0.15 mm/a (5.90 mpy); in aerated solution (5.2 ppm O_2) it was 3 mm/a (118 mpy). At higher temperatures (338 K (65 °C)) a pH change from 8 to 6.5 already produces a significant increase of the erosion corrosion.

In accordance with the kinetics of the protective layer formation, the dependance on the oxygen content of the chloride solution is very distinct. This is shown by the investigation results under simulated conditions of the sea water desalinization at the high temperature of 383 K (110 °C). For example, CuNi10Fe1Mn showed with only 5 ppm O_2 a corrosion rate of 0.064 mm/a (2.52 mpy) with flow velocities of up to 3 m/s; with 75 ppm O_2 the corrosion rate increased significantly to about 0.4 mm/a (15.7 mpy). The values for CuNi30Fe were only insignificantly better.

The test periods have a strong influence in the sense that for short periods very much greater corrosion rates are found. The formation of the protective layer and its growth in thickness takes considerable time. For example, after 183 days the corrosion rate is still found to be 5–7 times greater than after 2,306 days. The other alloying elements apart from copper and nickel have a large influence on the erosion corrosion. The influence of iron on the formation of the protective layer has been investigated comprehensively. This influence is positive for both CuNiFe alloys. For example, increasing iron contents up to 2% in CuNi10Fe1Mn at flow velocities of 4.6 and 9.1 m/s lead to continuous decrease of the corrosion rate from 14 g/m^2 d to 2.5 g/m^2 d (about 0.7 to 0.1 mm/a (27.56 to 3.94 mpy)). Chromium, too, has a significant positive influence already as from 0.4%. This has been taken into consideration in the development of several copper-nickel materials, for example for Cu-30Ni-2.8Cr-0.3Fe-0.7Mn (IN-732 and CA-719) or IN-838 and IN-848, which are used in particular under conditions of high flow velocity. In the same sense the addition of aluminium has a positive influence, evidently for reducing impact corrosion in particular.

Surprisingly the surface hardness has hardly any influence on the erosion corrosion, as tests with soft annealed and cold drawn pipe material have shown. In contrast thereto high surface roughness produces flow turbulence and thus greater susceptibility to erosion.

The addition of $FeSO_4$ has always decreased the corrosion in cooling water plants; evidently the iron behaves similarly as if it were present in the alloy.

Chlorine gas additions to prevent biological and slime production in pipes have negligible influence in reasonable concentrations, but overdose should be avoided. Whereas copper-nickel alloys in (deaerated) sea water can be used at temperatures up to more than 473 K (200 °C), they are subject to strong corrosion in process-concentrated salt solutions, for example in 11% brine.

The influence of sulfide contaminations in 3.5% NaCl solution on the passivating behavior of Cu-30Ni-1.5 Al (UNS C71500) has been investigated by studying cyclic current density/potential curves [1315]. In sulfide-free solution, during the first two oxidation stages found in the course of the current density/potential the passivation layer consists of Cu_2O and $Cu_2(OH)_3Cl$. In stationary solution this passivation layer is thicker and adheres more strongly, thus it is more stable than in agitated solution. In the presence of sulfides (here: 100 mg/l Na_2S) sulfide compounds of nickel and aluminium as well as loosely adhering CuS layers are formed instead of passivation layers. The corrosion rate increases correspondingly.

With regard to the utilization of pipe material in sea water desalinization plants and their stress exposure to high flow velocities, investigations have been carried out in a test loop on 3 CuNi alloys (as well as 2 brass types), whereby apart from the influence of the flow velocity also the impurities such as NH_3 and Na_2S were of interest [1316].

On pipe sections the anodic current density/potential curves as well as the polarization resistances were determined for up to 60 days and at 323 K (50 °C). Figure 170 shows examples of the anodic polarization curves for CuNi30Fe (0.48 Fe, 0.96 Mn) in flowing 3% NaCl solution. With increasing flow velocity the current densities and thus the corrosion rates increase. This is explained by the easier removal of the loose outer CuO layer. The change of the polarization resistance as a function of time leads after 44 days to stationary values, as shown for example with 1 m/s in Figure 171. According to this and corresponding data for 0.5 and 3 m/s, brass (OTS 76: 23.2 Zn, 1.87 Al, 0.037 As) shows the best behavior in pure NaCl solution, followed by CuNi30Fe, which is slightly better than CuNi10Fe1Mn (0.96 Fe; 0.56 Mn). If impurities in the form of sulfide are present in the chloride solution, the alloy CuNi30Fe with higher nickel content shows the best behavior. In contrast thereto, the two copper-nickel alloys behave similarly in the presence of NH_3 in the form of NH_4Cl, but less favorably than brass (OTS 76). For practical purposes it is important to note that the durability of the copper materials can be improved substantially by treatment with fresh water and with oxidizing agents such as oxygen or iron(II) ions (as $FeSO_4$, only 1 ppm suffices), and that new systems should be started this way, because this promotes the formation of a protective layer on the basis of FeOOH, making the materials less sensitive with respect to impurities such as sulfides.

The influence of sulfide impurities on the corrosion stability of CuNiFe alloys as well as brass, bronze and stainless steel in 3.4% NaCl solution is reported by [1317]. For this purpose pipe sections were tested for 60 days at 313 K (40 °C) in a test loop with the addition of 10 mg/l of sulfide as the corresponding amount of 1 M Na_2S solution. This solution was used in one case in the deaerated state (purged with N_2) and in another case with aeration, in alternation with sulfide addition. From the weight changes plotted as a function of time it is possible to estimate the corrosion rates, but the assessment with regard to local corrosion appears to be more important, because condensate pipes generally fail in practice by local corrosion. Table 223 summarizes the results. Whereas CuNi10Fe1Mn (CDA 706, DIN Mat. No. 2.0872, UNS C70600) showed only slight pitting corrosion under all test conditions, CuNi30Fe (CDA 715, DIN Mat. No. 2.0882, UNS C71500) behaved relatively well

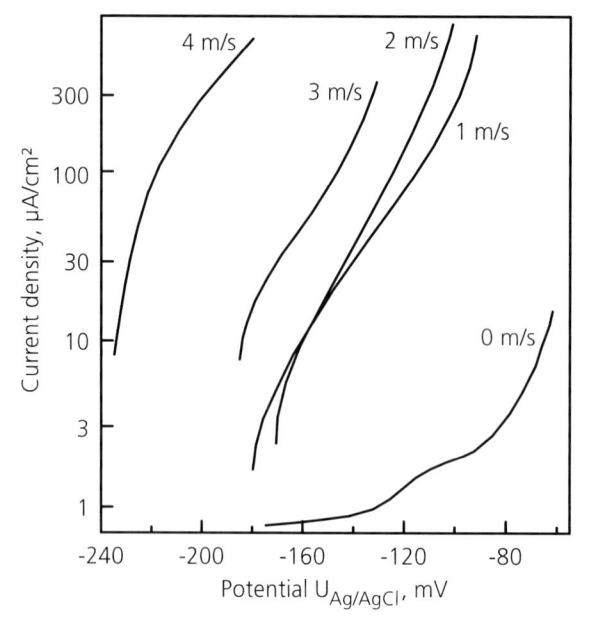

Figure 170: Current density/potential curves of CuNi30Fe with different flow velocities in 3% NaCl at 323 K (50°C) [1316]

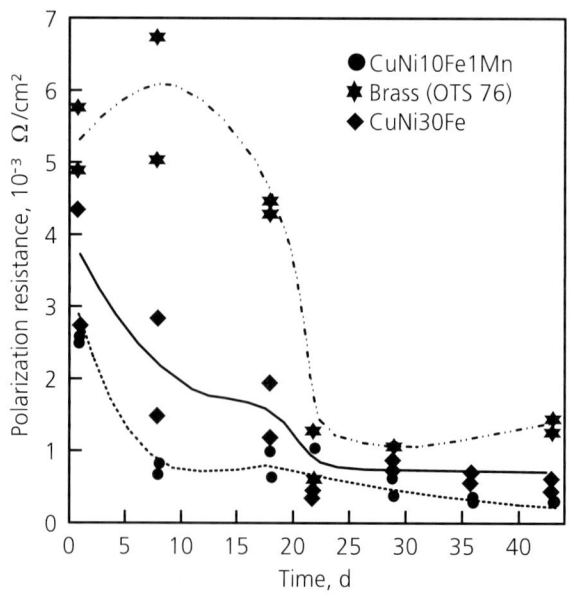

Figure 171: Polarization resistance in 3% NaCl at 1 m/s and 323 K (50°C) for CuNi30Fe, CuNi10Fe1Mn and brass (OTS 76) [1316]

only in the deaerated sulfide solution as well as in aerated 3.4% NaCl solution, but definitely unfavorably in the aerated sulfide solution. Very large weight losses are found according to Figure 172, and they also increase with the flow velocity, together with strong surface roughening. Of all tested materials, CuNi10Fe1Mn showed the best behavior. Regarding the interpretation of the behavior of other materials, see there.

Alloy	Velocity m/s	Aerated + sulfide[1]	Deaerated + sulfide[1]	Brine[1]
Copper-nickel CDA 706, Mat.No. 2.0872, UNS C70600	1.4	lE, lLK	GA	NT
	2.1	lLK	GA	NT
	1.5	NT	NT	GA
	3.3	NT	NT	GA
Copper-nickel CDA 715, DIN Mat. No. 2.0882, UNS C71500	1.4	lLK	GA	NT
	2.1	sE, lLK, lDA	GA	lLK, lE
Admiralty brass, containing arsenic CDA 443 cf. CuZn28Sn1, DIN Mat. No. 2.0470	1.4	sIK; sE	lE	NT
	2.1	sIK; sE	lE	NT
	1.5	NT	NT	GA
	3.3	NT	NT	lE
Aluminium bronze, containing arsenic CDA 687: UNS C68700 cf. CuZn20Al2, DIN Mat. No. 2.0460	1.4	lE, mIK	lE	mE
	2.1	lE, mLK, mIK	NT	mLK, sE
AISI 302	2.1	sLK	lLK	NT
AISI 316	2.1	sSK	NT	NT

1) IK = inter-crystalline corrosion; DA = dealloying; LK = pitting corrosion; E = erosion corrosion; SK = crevice corrosion; GA = general attack with roughening of the surface; NT = not tested
l, m and s = light, moderate and strong

Table 223: Influence of the chloride medium and the flow velocity on the local corrosion of condenser pipe materials [1317]

Copper-nickel alloys with 10% and 30% nickel have been investigated in comparison with pure copper by making impedance measurements in aerated 3.4% NaCl solution for up to 28 days [1318]. The obtained measured curves were interpreted such that for copper only one inner layer of corrosion products produces the durability, whereas for the copper-nickel alloys the better corrosion resistance is produced by an additional outer layer which is also very porous. The corrosion currents determined from the current density/potential curves lie at 2–5 µA/cm² for copper, corresponding to a corrosion rate of 0.05–0.12 mm/a (1.97–4.72 mpy). For CuNi10Fe1Mn the initially high corrosion current densities (30 µA/cm²) had decreased after 672 h

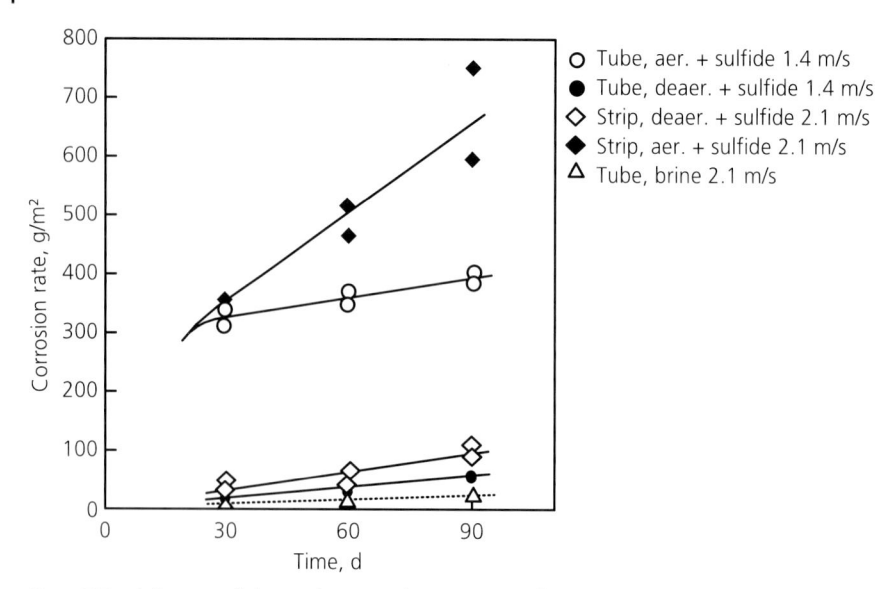

Figure 172: Influence of the medium on the corrosion of CuNi30Fe (CDA 715, DIN Mat. No. 2.0882) [1317]

to only 0.2 µA/cm^2, corresponding to a corrosion rate of < 0.01 mm/a (0.39 mpy), as also found in sea water. CuNi30Fe shows slightly greater values: 0.7 µA/cm^2 corresponding to about 0.01 mm/a (0.39 mpy), which again agrees with experience values in sea water. The higher corrosion resistance of CuNi alloys compared with copper is attributed to the slower cathodic reaction, which is probably reduction of oxygen.

For investigating the corrosion fatigue behavior of a copper-nickel alloy IN-838 (CDA 722) with chromium addition (16.1 Ni, 0.55 Cr, 0.88 Fe, 0.68 Mn) in 0.5 M NaCl solution, the solution-annealed and the precipitation hardened state were tested [1319]. The tests were made with respect to the corrosion potential and with a given anodic current. For the solution-annealed specimen (1,123 K (850 °C), 24 h, water quenched) no susceptibility was found via the corrosion potential, but with anodic polarization corrosion susceptibility was found already with small current density. The fully hardened specimen (923 K (650 °C), 12 h) showed a decrease of the fatigue behavior already at the corrosion potential and even more so with impressed anodic current. Figure 173 shows the behavior of the specimens of the different structures under the various test conditions. The solution-annealed specimen showed hardly any change of the corrosion fatigue behavior in air and in NaCl solution with and without impressed current. In contrast thereto, the precipitation hardened specimen showed a significant decrease of the alternating stress resistance in NaCl compared with air, but on a higher strength level. Trans-crystalline and inter-crystalline attack is found, changing more to the inter-crystalline form in the corrosion medium. The results are described in more detail together with the corrosion fatigue model devised for copper and copper-aluminium alloys.

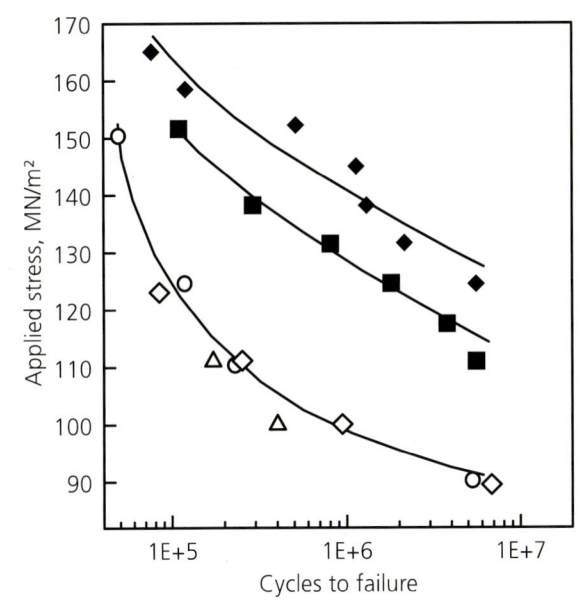

Figure 173: Alternating stress resistance as a function of the number of load cycles for solution-annealed and precipitation hardened CuNiCr in air and in aerated 0.5 M NaCl at the corrosion potential, at 20 Hz [1319]
precipitation hardened: in air (◆), in 0.5 M NaCl (■)
solution-annealed: in air (○), in 0.5 M NaCl (△), in 0.5 M NaCl – 100 µA/cm² (◇)

Benzotriazole is one of the most frequently utilized inhibitors for copper alloys, whose strong interaction with the metal produces a chemisorption polymer layer that resists corrosive ions such as chlorides. From the group of azoles, triazole, aminotriazole and bistriazole have been tested with regard to their inhibiting effect on a Cu25Ni alloy (25.64 Ni, 0.36 Fe, 0.56 Mn) in 3% NaCl solution [1320]. Based on the assumption that these inhibitors influence the cathodic reaction, cathodic current density/potential curves were determined. According to the results and from the determined corrosion current densities, the maximum inhibitor effect of 61% is obtained with a concentration of 5×10^{-3} mol/l triazole. Comparable curves for aminotriazole give maximum inhibitor effect of 92% with the concentration of 10^{-4} mol/l. For bistriazole the values are 91% and 10^{-3} mol/l. All 3 substances effectively inhibit the cathodic process, whereby aminotriazole is evidently the most effective, because the slightest amount already produces a significant decrease of the corrosion current densities (here: from 28 to 2.2 µA/cm² (about 0.05 mm/a (1.97 mpy)), thus a high inhibitor effect is achieved.

Copper-tin alloys (bronzes)

Bronzes show in NaCl solutions good behavior similar to that of the copper-nickel alloys and the copper-zinc alloys (brass) with generally very low corrosion rates. They are only moderately sensitive with regard to pitting corrosion and crevice corrosion.

The influence of an artificially produced patina on two bronzes ((A): 87 Cu, 8 Sn, 5 Pb; (B): 4.76 Sn, 4.58 Pb, 4.35 Zn, 0.88 Ni) on the corrosion stability, for example, with respect to NaCl deposits has been tested in air with 100% relative humidity for 30 days [1321]. The patina was produced by immersion in a 10% solution of K_2S for 3 minutes at 343 K (70 °C). The surfaces were brought into contact with the NaCl solution again at regular intervals of two days; after evaporation of an alcoholic solution NaCl, layers of 0.1, 1, 5 and 10 mg/m² d were produced in this way. All specimens were exposed to the 100% relative humidity atmosphere in a desiccator. Figure 174 shows that no protective effect is achieved with the artificially produced patina, and that, on the contrary, the corrosion rates increase significantly. The fact that the bronze A is slightly inferior is probably due to the lower tin content and the presence of zinc. At any rate the results of the investigation confirm that patina on bronzes decreases the corrosion resistance rather than producing a protective effect.

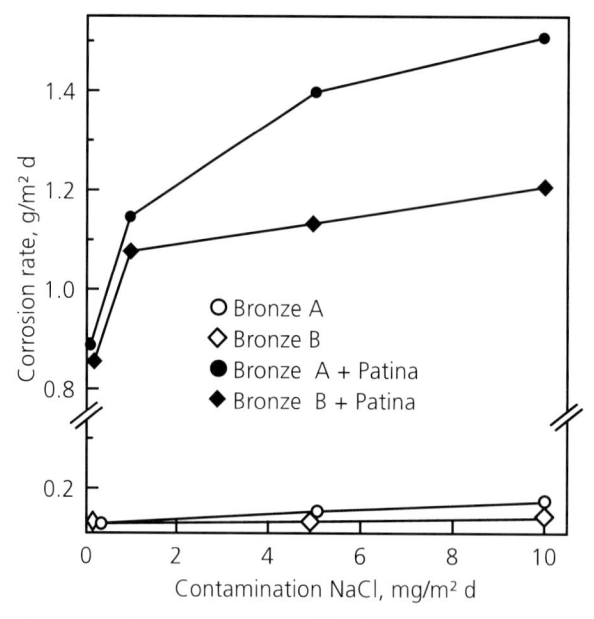

Figure 174: Corrosion rates in NaCl for bronze A and B (cf. the text) with and without patina [1321]

Aluminium bronze shows the best behavior of all copper materials in sulfide contaminated 3.4% NaCl solution at 313 K (40 °C) after the loop test for 90 days test period, i.e. only 10 mg/cm² weight loss with only slight signs of erosion and pitting corrosion (others: up to 50 g/cm² with erosion, pitting and inter-crystalline corrosion) [1317], see also Table 223.

Copper-tin-zinc alloys (red brass (gunmetal))

The copper-tin-zinc alloys commonly utilized as red brass (gunmetal) have high durability, particularly under erosive conditions, confirmed by their behavior with respect to sea water, in which they are used for pump housings, impeller wheels, valves etc. (see also copper-zinc alloys)

Copper-zinc alloys (brass)

The various brass types with 30 to 40 % zinc (alpha brass) have very good durability in NaCl solutions up to about 3 % and are therefore widely utilized in household, industrial and maritime technology. Idle states and contamination with dirt must be avoided, as for the copper-nickel alloys, because otherwise **zinc depletion** (preferential dissolution of the less noble metal zinc) is expected, i.e. local corrosion, possibly with rapid perforation (pipelines). Zinc depletion can be avoided by adding arsenic as alloying element. Sulfide impurities in the chloride solution lead to increased attack in the form of erosion and inter-crystalline corrosion, particularly with simultaneous ventilation. Here brass with the addition of 2–3 % aluminium performs significantly better. The presence of only a few mg/l NH_4 as well as Hg ions in the NaCl solution leads to stress corrosion cracking, which tends to be inhibited with increase of the NaCl concentration.

For the copper-zinc alloys with more than 15 % zinc depletion was formerly regarded as a dangerous type of corrosion [1322]. It takes place preferentially in sea water, river water, tap water and in technical cooling waters (all of which are characterized by the presence of chlorides) and it is also caused by the presence of soldering flux residues (traces of hydrochloric acid). Zinc depletion is promoted by elevated temperatures and higher chloride contents, poor ventilation and, in particular, slow flow velocities, because these make deposits possible. In the actual process copper and zinc initially dissolved together, but the more noble copper deposits again immediately at the corrosion location on the surface, in a very loose spongy form. Of the brass types, in particular those with a heterogeneous α+β-mixed microstructure are susceptible, because the less noble β-mixed crystal is preferentially attacked. A distinction is made between two forms of zinc depletion. In **layer-type dezincification** uniform surface corrosion is found. This form mostly appears in the case of galvanic corrosion in contact with more noble metals, after injuries of the surface and in crevices (screw threads). **Plug-type dezincification** which is confined to small areas is more dangerous, because it progresses in depth only in certain places, and this can soon lead to perforation of thin walled pipes. It always takes place at slow flow velocities and under stationary conditions, then under dirt and solid material deposits. Poor oxygen supply there leads to the appearance of local ventilation galvanic elements. Because the locations underneath cannot passivate and are therefore less noble than their surroundings, brass goes anodically into solution in these places, and the copper is thereafter deposited as porous plug. The process is relatively rapid and depth increase rates of 4.5 mm/a (177 mpy) can be reached. Most instances of damage formerly occurred in condenser pipes in power plant engineering, but now this appears to have been eliminated by utilization of zinc depletion resistant brass types containing arsenic and phosphorus. Two examples of damage incidents are given. On a fixing rod made of brass 70/30, layer zinc depletion took place after two years in a hot water storage tank (363 K (90 °C)) containing 120 mg/l of chloride, from which hot water was drawn only at very long time intervals, so that the conditions were describable as definitely stationary. In a heating system perforations appeared after only 5 months on brass sleeves (70/30) by plug zinc depletion. The liquid mixture utilized in the heating system for heat transport contained apart from distilled water an antifreeze agent whose chloride content of 200 mg/l was too great, and the stipulated mixing ratio of the inhibitors had not been complied with. The plug corrosion took place only in the inside of the pipe, after antifreeze agent had intruded through pitting corrosion. No zinc depletion was found on the outside surfaces, evidently because the high flow velocity there quickly rinsed away the dissolving copper and zinc and prevented a cementation of the copper.

Brass with 76 to 79 % copper, 1.8 to 2.3 % aluminium, 0.02 to 0.06 % arsenic or phosphorus is very resistant to sodium chloride solutions. The arsenic or phosphorus content retards the deposition of copper on the brass surface and thus reduces the danger of plug zinc depletion. Pitting corrosion due to poor ventilation under deposits is not eliminated by the arsenic content, nor is horseshoe-form corrosion produced by strong turbulence [1323].

According to [1324] turbulences arise in a pipe with 19 mm (3/4 inch) diameter with a water flow velocity of 0.7 m (2.2 feet) per second. These turbulences can put gas bubbles on the metal surface and thus induce corrosion by differential ventilation. This fact must also be taken into consideration for pure saturated NaCl solutions. In 3 % NaCl solutions temperature increases of 293 K (20 °C) double the corrosion rate of the special brass (70–72.5 % Cu, 0.9 to 1.3 % Sn, 0.02 to 0.06 % As or P, 27–28.5 % Zn), which at room temperature is about 1 g/m^2 d.

For various types of brass tests have been made with sprayed 20 % NaCl solution at 288 to 300 K (15 to 27 °C), the results of which are shown in Table 224 [1277].

In connection with the development of a measuring technique at higher test temperatures, aluminium brass (with 21 Zn, 2.04 Al, 0.065 As) has been electrochemically investigated at 403 K (130 °C) [1325]. Figure 175 shows current density/potential curves in 0.5 M NaCl solution. A continuous current density increase was found in the solution with pH 5.1, which increased by a factor of 4 with agitation, whereas at pH 8.1 just above the free corrosion potential (for both –375 mV$_{SCE}$) a plateau with 25 µA/cm^2 was found in the range of –350 to –150 mV. Pitting corrosion is found after a steep rise of the current density. Under the same conditions the corrosion current densities were determined as a function of time. For the alloy specified above and another Al-brass (20.6 Zn, 6.9 Ni, 2.2 Al) relatively high initial current densities are found (45 and 10 µA/cm^2 respectively), but after 300 and 600 h respectively they have fallen to very low values of approximately 1 µA/cm^2, corresponding to very low corrosion rates of < 0.02 mm/a (< 0.79 mpy), as has also been confirmed in practice.

Material	Weight loss in mg/cm²					Deepest excavation mm
	28 d	56 d	84 d	168 d	364 d	
Copper	18.1	40.5	54.6	0.0	162.8	0.13
Copper/zinc 97/3	19.7	32.9	47.9	85.9	140.9	0.15
95/5	22.5	35.0	48.7	83.7	140.3	0.15
94/6	23.4	40.5	56.1	88.7	155.0	0.15
90/10	19.2	32.5	45.1	69.4	127.1	0.18
87.5/12.5	20.6	35.5	46.5	65.7	126.6	0.18
85/15	19.7	32.7	40.5	56.8	111.8	0.15
80/20	14.9	31.9	40.5	56.6	110.0	0.10
75/25	13.8	27.6	37.1	50.6	113.6	0.08 (zinc depleted)
70/30	15.2	32.7	43.4	59.2	132.3	0.10 (zinc depleted)
66.7/33.3	13.9	32.1	41.1	62.0	138.4	none, (zinc depleted)
63/37	21.4	43.6	71.6	103.5	214.0	none, (zinc depleted)
60/40	20.0	45.1	86.4	130.8	248.0	none, (zinc depleted)

Table 224: Behavior of brass types in the salt spray test (20 % NaCl) at 288 to 300 K (15 to 27 °C) [1277]

Aluminium brass (2 Al) is adequately suitable for utilization in heat exchanger pipes in sea water desalinization plants, whereas aluminium bronze (5 Al) is utilized for higher chloride concentration solutions, thus also for salt production. Both materials have been tested under the conditions of technical evaporation of industrial sewage waters at 353 K (80 °C) [1326]. The corrosion rates were determined in the original solution (approximately 1.85 g/l chloride and 1.5 g/l sulfate among other substances) and after concentration increase by a factor of 5–50. Figure 176 summarizes the results. In the original solution, the aluminium brass is initially significantly more resistant, but with evaporation concentration by a factor greater than 10 it corrodes continuously and more strongly, whereas the aluminium bronze behaves conversely and finally, with evaporation concentration by a factor of 50, it becomes much more resistant, i.e. by a factor of 6, than the aluminium brass. The behavior of the aluminium bronze follows the decreasing oxygen concentration and is relatively independent of the chloride concentration. Exactly the converse dependancies are found for aluminium brass, i.e. the behavior depends primarily on the chloride content. This is also shown by the comparison with the pure metals copper, zinc and aluminium in Figure 176. Whereas the 3 metals initially, i.e. at low chloride

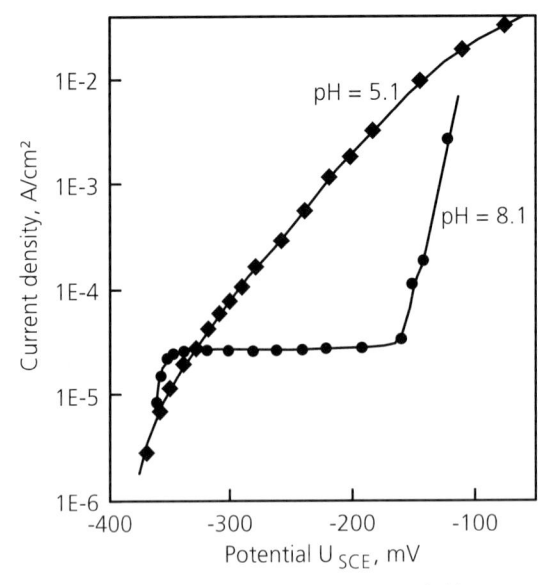

Figure 175: Current density potential curves of Al-brass in 0.5 M NaCl solution at 403 K (130 °C) as well as at pH 5.1 and 8.1 [1325]

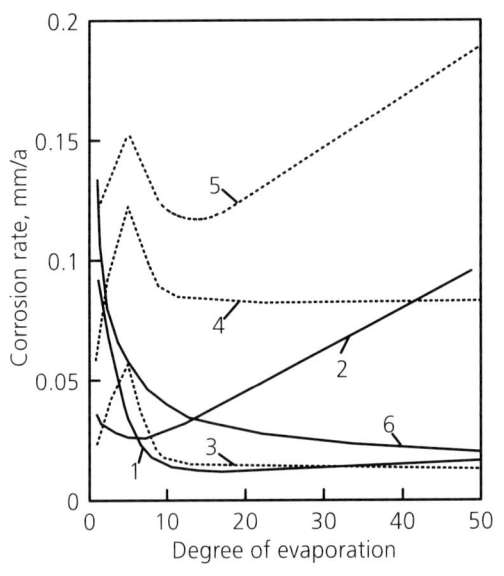

Figure 176: Corrosion rates of brass 70/30 plotted against the degree of evaporation concentration of the original solution, for 1) aluminium bronze, 2) aluminium brass, 3) aluminium, 4) copper, 5) zinc, 6) solubility of oxygen [1326]

concentrations but high oxygen contents, show larger corrosion rates, these decrease at evaporation concentration factors greater than 10, whereas the corrosion rates of zinc increase steeply. This proves that the strong dependance of the aluminium brass on the chloride concentration is unambiguously attributable to the zinc content of this alloy.

With regard to the behavior of condensate pipes made of brass in sulfide contaminated 3.4 % NaCl solution at 313 K (40 °C), cf. [1317] and Table 223. Much higher weight losses are found in sulfide contaminated and aerated NaCl solution in comparison with sulfide-free as well as sulfide containing but deaerated solution, see Figure 177. The brass containing approximately 26 Zn, 1 Sn and 0.06 As showed strong erosion and inter-crystalline corrosion which evidently increase with the flow velocity. Similar behavior as also found for an aluminium brass (2 as well as 3.5 Al), but with significantly lower corrosion rates, so that utilization in sulfide contaminated NaCl solution is possible.

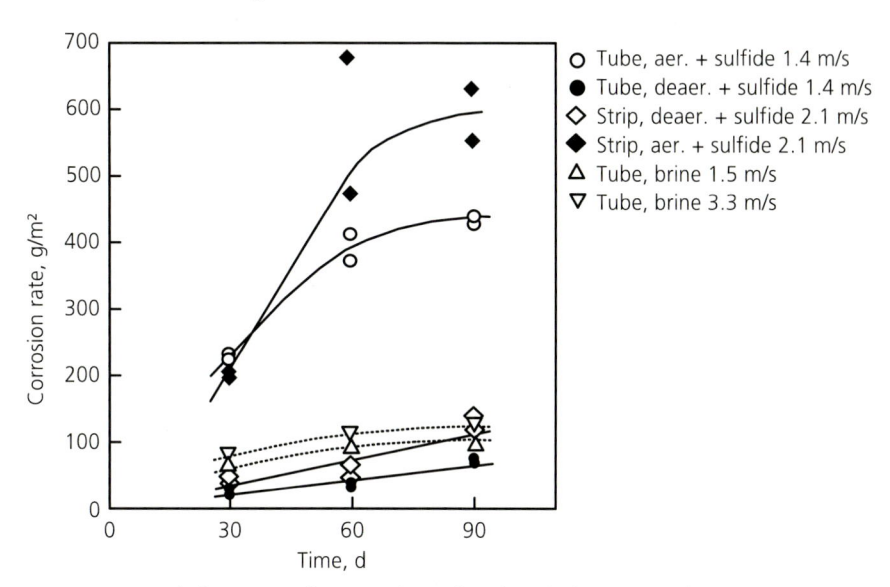

Figure 177: Weight losses as a function of time for admiralty brass (Al-Sn brass) in 3.4 % NaCl + sulfide, aerated and deaerated [1317]

Pitting corrosion and passivation of α-brass (63 Cu, 36.7 Zn) has been examined electrochemically in buffer solutions on the basis of borate and phosphate additionally containing chloride [1327]. The corrosion potentials shift in the borate solution with increasing chloride content (0.01–0.5 mol/l) from approximately –140 mV$_{SCE}$ to approximately –250 mV$_{SCE}$. For the same concentration range of chloride the pitting corrosion potentials for both buffer solutions fall from +300/500 mV$_{SCE}$ down to –200 mV$_{SCE}$, whereby the pitting corrosion susceptibility in borate is higher than in phosphate, cf. Figure 178. In the absence of chlorides, at sufficiently high anodic potentials zinc depletion is observed in both buffer solutions. When chlorides are present in the

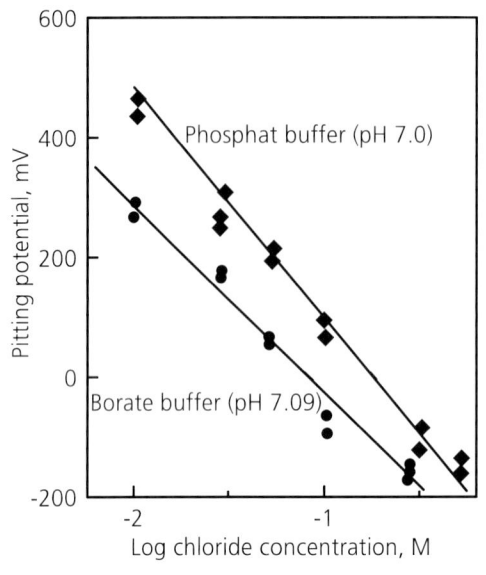

Figure 178: The pitting corrosion potential as a function of the chloride concentration for α-brass in buffer solution [1327]

solution zinc depletion and pitting corrosion take place. Apart from the dependance of the pitting corrosion susceptibility on the buffer solution and on the chloride solution, there is also an influence of the structure of the passivation layer formed before the appearance of pitting corrosion.

With regard to the corrosion stability of brass OTS 70 (28.7 Zn, 1.21 Sn, 0.035 As) and OTS 76 (23.2 Zn, 1.87 Al, 0.037 As) depending on the flow velocity, Figure 179 shows the anodic current density/potential curves in 3% NaCl solution at 323 K (50°C). With increasing flow velocity the current densities and therewith the corrosion rates increase significantly. For the change of the polarization resistance as a function of time and the behavior in the case of contamination with sulfides and ammonium, see under copper-nickel alloys and [1316].

Alpha brass (70/30) was subjected in the annealed state (923 K (650°C)/0.5 h) to stress corrosion cracking tests under constant stress in Mattsson solution (0.02 mol/l copper sulfate + 0.8 mol/l ammonium sulfate) [1328]. Thereby the influence of chlorides on the stress corrosion cracking behavior was investigated. As is shown in Figure 180, the addition of chloride has an inhibiting effect already at low concentrations, as shown by the strongly increased survival times in 0.02 mol/l chloride. The kind of stress corrosion cracking changes from inter-crystalline to trans-crystalline. Higher copper contents of the solution lead to accelerated stress corrosion cracking.

Brass is susceptible to stress corrosion cracking in chloride solutions when mercury is present even in the slightest amounts. Inhibitors can suppress stress corrosion cracking, but, as studies have shown, only in clearly restricted pH ranges of the solution [1329]. The influence of 5 inhibitors was tested by the CERT method and the effectiveness was determined via the increase of the threshold voltage and the

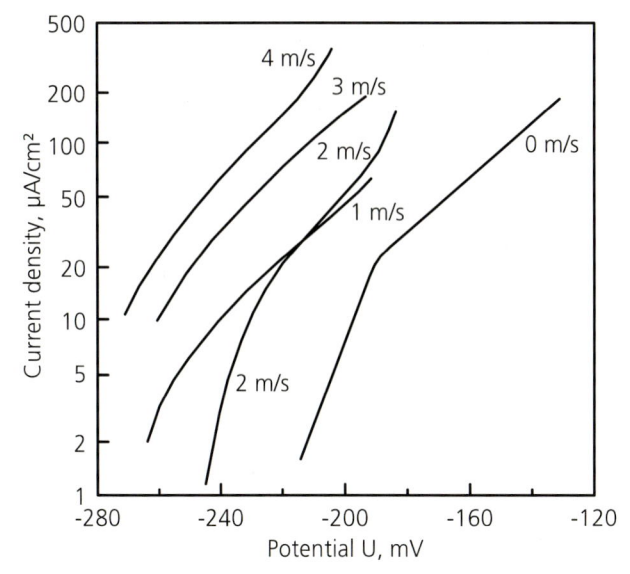

Figure 179: Anodic current density/potential curves of brass OTS 76 at various flow velocities in 3 % NaCl at 323 K (50 °C) [1316]

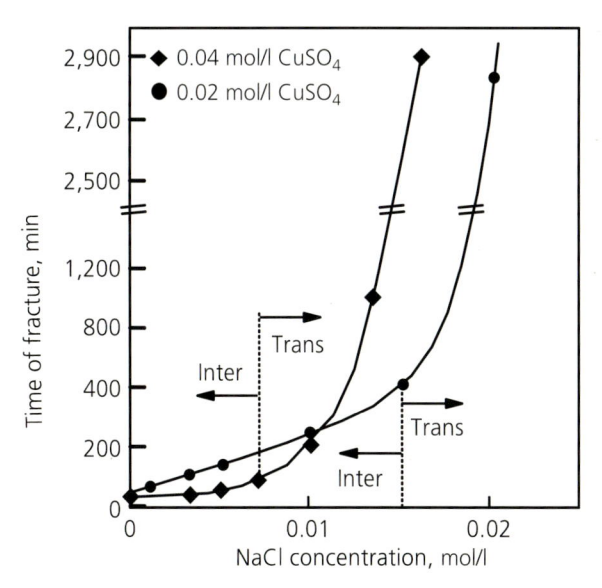

Figure 180: The influence of NaCl additions to the Mattsson solution on the stress corrosion cracking survival time of α-brass [1328]

reduced area in the stress/elongation diagram (relative fracture energy). Already 10^{-4} mol/l $HgCl_2$ makes the values drop to < 10% of the fracture energy and thus indicate extreme susceptibility to stress corrosion cracking. Investigation with the 5 inhibitors gave the pH dependancies shown in Figure 181. On the whole, the test results can be summarized as follows:

- In pure 3.5% NaCl solution stress corrosion cracking is found at no pH-value. It is found only when $HgCl_2$ is added, but extremely small amounts as from 10^{-6} M $HgCl_2$ suffice.
- Stress corrosion cracking of brass (70/30) on 3.5% NaCl containing 10^{-4} M $HgCl_2$ is inhibited by the following substances, in decreasing order of effectiveness: 1,2,3 benzotriazole (BTA) > 1,2,4 triazole > thiourea > indole > urea.
- Each inhibitor has a critical pH-value above which the inhibiting effect is complete and below which the maximum stress and the ductility decrease drastically. These critical pH-values are: 3.8 for BTA; 6.6 for triazole; 8.7 for thiourea; 11.9 for indole and 11.4 for urea.
- The 5 tested inhibitors can be assessed as follows for their efficiency with regard to the covered pH range: BTA covers the greatest pH range and can therefore be used in acidic, neutral and alkaline solutions; triazole and thiourea can be used in neutral and alkaline media, whereas indole and urea can be used only in alkaline solutions.

The zinc depletion of brass can be largely suppressed by adding arsenic as alloying element, provided that α-brass is involved. Arsenic is ineffective for brass with

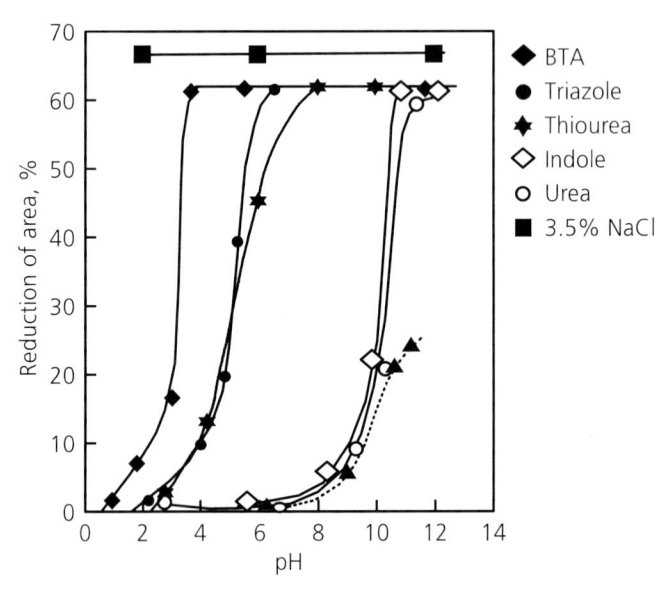

Figure 181: The influence of the pH-value on the relative fracture energy in the stress/elongation diagram of brass (70/30) in 3.5% NaCl + 10^{-4} M $HgCl_2$ (▲) and after the addition of 1,2,3 benzotriazole (BTA), triazole, urea, thiourea and indole, 10^{-3} mol/l each [1329]

α-β-microstructure. Therefore it was the subject matter of an investigation to determine the influence of various inhibitors on the zinc depletion behavior of brass in 3% NaCl solution at 313 K (40°C) by determining weight losses (for 5 days) as well as polarization and impedance measurements [1330]. The following compounds were tested: aminomethyl mercaptotriazole (AMMT), aminoethyl mercaptotriazole (AEMT) and aminopropyl mercaptotriazole (APMT) in concentrations of 50–200 mg/l. The effectiveness of the inhibition, in %, is shown in Table 225. It increases from 60 to 94 % with the concentration of the respective inhibitor, whereby a slight differentiation is found, for example as follows with 200 mg/l: AMPT (94.5%) > AEMT (93.6%) > AMMT (91.5%). The results under flowing conditions are also shown in column 3 of Table 225 (1.5 m/s). It is seen that the effectiveness of the inhibitors is retained. Table 225 further contains the analysis results for the amounts of copper and zinc that went into solution. They unambiguously show the preferential dissolution of zinc, but which decreases with increase of the inhibitor concentration. Thus these substances also inhibit the zinc depletion.

Inhibitor concentration ppm	Inhibitor effect, %		Solution analysis, μg m/l		Inhibition of the dissolution, %	
	Static	Dynamic	Zinc	Copper	of Zinc	of Copper
without inhibitor	–	–	3.362	0.010	–	–
AMMT						
50	60.50		1.068	0.00	68.2	100
100	79.80		0.826	0.00	75.43	100
150	86.50		0.510	0.00	84.80	100
200	91.50	87.40	0.436	0.00	87.03	100
AEMT						
50	68.60		0.920	0.00	72.60	100
100	82.80		0.634	0.00	81.4	100
150	91.50		0.426	0.00	87.30	100
200	93.60	88.80	0.396	0.00	88.22	100
APMT						
50	74.5		0.816	0.00	75.73	100
100	88.7		0.620	0.00	81.55	100
150	92.5		0.442	0.00	86.80	100
200	94.5	90.60	0.35	0.00	89.60	100

AMMT: aminomethyl mercaptotriazole; AEMT: aminoethyl mercaptotriazole; APMT: aminopropyl mercaptotriazole

Table 225: Effectiveness of inhibitors in % for brass (70/30) in 3% NaCl solution at 313 K (40°C) [1330]

The inhibiting effect of benzotriazole on the corrosion of brass in chloride solutions is largely canceled by the presence of sulfides. This is shown by polarization measurements in 0.58 mol/l NaCl solution with addition of 10^{-2} mol/l benzotriazole without and with the presence of 1.25×10^{-4} mol/l sulfide [1331]. The lower anodic and cathodic current densities compared with the inhibitor-free solution after adding the inhibitor are increased again significantly when sulfide is present. The explanation of this is that the BTA complex that forms a protective covering layer is decomposed because the sulfide ions remove copper(I) ions from the protective layer to produce the more stable copper sulfide. This leads to local destruction of the protective layer produced by the inhibitor.

In cooling water systems, to protect brass containing tin and arsenic (admiralty brass: 70 Cu, 29 Zn, 1 Sn, 0.023 As), mixed inhibitors consisting of alkyl amine (10%) and thiazole derivatives (1%) are utilized. Polarization and impedance measurements were made for mixed concentrations of 10 and 50 mg/l of these substances in neutral 0.5 mol/l NaCl solution [1332]. The cathodic current density/potential curves show (cf. Figure 182 a) that, starting from the free corrosion potential at approximately -200 mV$_{SCE}$, there is first of all a linear rise until -700 mV and then a plateau until $-1,200$ mV. The addition of the inhibitor decreases the cathodic current densities by 2 powers of ten, particularly in the vicinity of the free corrosion potential, whereas at higher cathodic potentials the current density approaches the value in the state without inhibitor. In the anodic range the current densities are much lower over a greater potential range, and to a greater extent when the inhibitor concentration is increased (Figure 182 b). From these curve forms it must be concluded that the inhibitor functions anodically. No signs of corrosive attack are found even after 5 h. In a further sequence of tests it was shown that thiazole derivatives alone are less effective and that therefore primarily the alkyl amine produces the inhibiting effect. Thiazole derivatives alone show an inhibiting effect on brass only as from 1 mg/l, whereas in combination with alkyl amine the threshold concentration is 0.5 mg/l.

The dissolution behavior of brass (70/30) has been investigated purely by weight loss determination in deaerated 4% NaCl solution (pH-value around 2) in the temperature range 298–333 K (25–60 °C) [1333]. Figure 183 shows curves of the weight losses as a function of time. These curves have an induction period at all temperatures. This is caused by autocatalysis which requires that a minimum concentration of dissolved copper ions must be present in the corrosive medium before copper goes into solution as $CuCl_2$. The corrosion rates can be calculated from the linear forms of the curves after the induction time and their gradients. The estimated values are approximately 0.04 mm/a (1.57 mpy) at room temperature, approximately 0.06 mm/a (2.36 mpy) at 313 K (40 °C) and 0.1 mm/a (3.94 mpy) at 333 K (60 °C). The weight losses decrease considerably with a commercial inhibitor Armohib® 28, as is confirmed by Figure 184. The inhibiting effect reaches 98% at an inhibitor concentration of 100 mg/l. The influence of anodic and cathodic polarization on the creep behavior of brass has been investigated in 3.5% NaCl solution and in 0.45 M $CH_3COOH + 0.05$ M CH_3COONa [1334]. It is found that when an anodic current is impressed the creep rate increases significantly, returning to the values before the

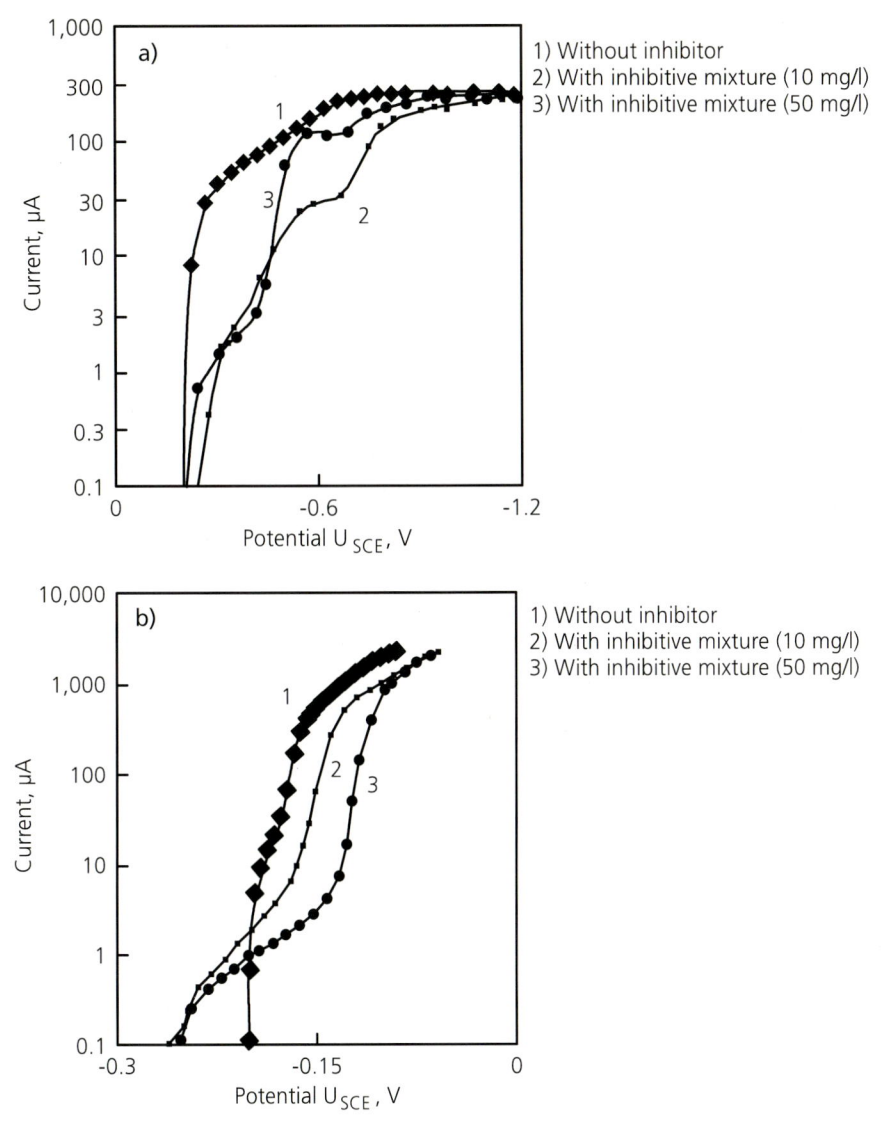

Figure 182: Cathodic (a) and anodic (b) current density/potential curves of brass without inhibitor (1), with mixed inhibitor 10 mg/l (2) and 50 mg/l (3) after 2 h in 0.5 mol/l NaCl. The measuring area was 0.32 cm^2 [1332]
Cathodic range: from the free corrosion potential up to an overvoltage of –1.2 V_{SCE}
Anodic range: from the free corrosion potential up to an overvoltage of 0 V_{SCE}

current was switched on after switching it off again. The creep rate increases linearly with the current density, whereby the rise in chloride solution is considerably greater than in the acetic acid solution, see Figure 185. A similar influence is also found with cathodic polarization, but weaker there by a factor of > 10 with the same current density.

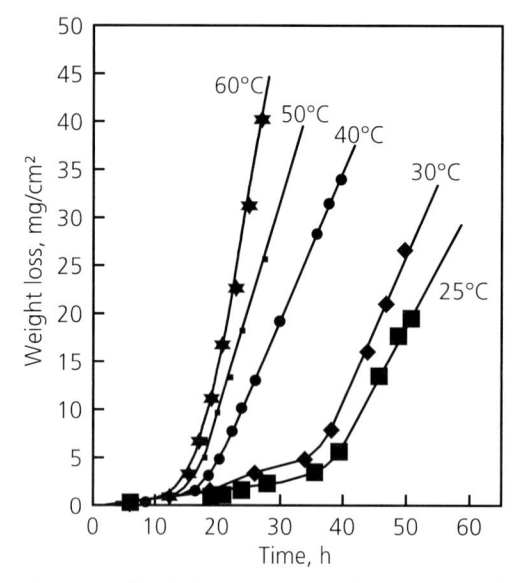

Figure 183: Weight loss/time curves of α-brass in acidified 4% NaCl solution (pH 2) [1333]

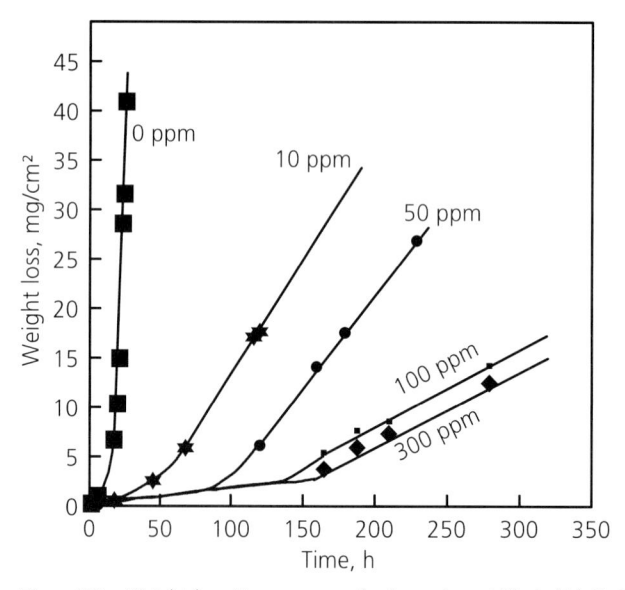

Figure 184: Weight loss/time curves of α-brass in acidified 4% NaCl solution (pH 2) at 333 K (60 °C) after adding various concentrations of the commercial inhibitor Armohib® 28 [1333]

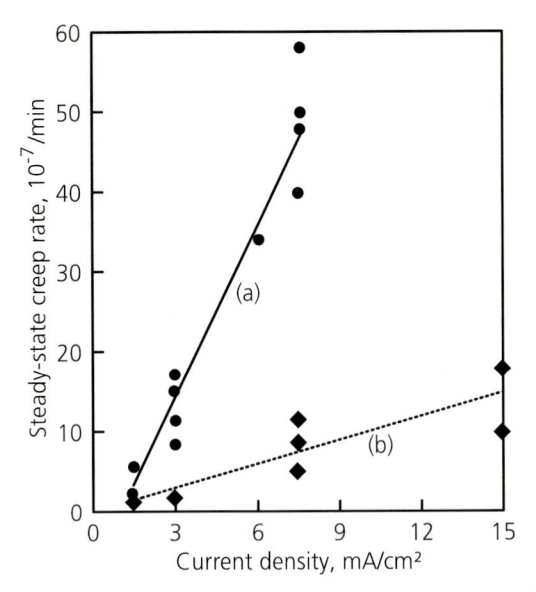

Figure 185: Creep rate of brass at 260 MPa as a function of the current density
a) in 3.5 % NaCl
b) in 0.45 mol/l CH$_3$COOH + 0.05 mol/l CH$_3$COONa [1334]

Many examples of damage incidents of pipes in auxiliary heat exchangers of power plants (not as condenser pipes) are reported by [1335], chiefly on admiralty brass but also on aluminium bronze as well as brass 90/10 and 70/30. Pitting corrosion, zinc depletion, nickel depletion and stress corrosion cracking are the chief forms of damage, caused by the river water used for cooling, which contained varying concentrations of chloride and sometimes undesirable traces of ammonium or nitrates, under conditions of plant shutdown. The damage incidents were documented in pictures, and their causes and damage mechanisms were explained in detail. Instructions were given for selecting suitable alloys as well as regarding the specific requirements already when ordering semifinished products, and also for the initial installation of pipes for heat exchangers, to avoid such damage incidents. CuNi 90-10, CuNi 70-30, admiralty brass, alloy 194 and Al-bronze were recommended as pipe material for auxiliary heat exchangers and are considered to be unproblematic under proper operating conditions. In particular CuNi 90-10 is considered to be the best standard material.

Regarding the subject of zinc depletion, it was shown in an investigation that also α-brass containing the β-phase can be made immune to zinc depletion by adding greater amounts of arsenic [1336]. Several brass types which can be machined, containing 58–63.5 Cu and 1.5–1.75 Pb have been polarized anodically with given current density (5 mA/cm^2) for 3 h in 0.5 mol/l NaCl at 360 K (87 °C). Thereafter the solution was analyzed for copper and zinc, and the zinc depletion factor was determined from the ratio of zinc to copper found in the solution (>> 1 indicates zinc depletion). For 4 of 6 alloy variants zinc depletion factors of 5.7 to 18 were found, indicating strong zinc depletion which was also verified metallographically. For a newly developed casting and forging alloy containing 62.5 copper, 35.5 zinc, 1.8 lead and arsenic increased to 0.17, a value of 1.1 is obtained, which is identical to that of the commercial α-brass (70/30, 0.5 Pb and 0.04 As) that is known to be resistant to zinc depletion. This brass variant was produced as casting and forging material showing good forming behavior and ductility, and very good resistance to uniform corrosion and zinc depletion in field test in process water with high chloride contents (up to 2,000 mg/l).

Investigations of the extent of zinc depletion of α-brass (70/30) and brass 60/40 with α+β-microstructure in NaCl and Na_2SO_4 solutions are reported by [1337]. Specimens were immersed in solutions having various concentrations, and the quantities of copper and zinc which went into solution were determined colorimetrically as a function of time. The results for α-brass are shown in Figure 186, and those for α+β-brass in Figure 187. Initially an increase of the dissolved metals is found with increasing concentration of the sodium chloride solutions (for sodium sulfate the relationship is evidently reversed, and sodium sulfate is less corrosive). Particularly at low NaCl concentrations a zinc to copper ratio is found which for α-brass lies significantly above that of its matrix, indicating that moderate zinc depletion is taking place. Taking into consideration the scale difference, it is seen that for α+β-brass the curves for zinc are considerably steeper, increasing linearly with time, Figure 187. In contrast thereto, the curves for copper become flatter in the course of time. The concentration of zinc in the solution is on the average one order of magnitude greater than that of copper. Compared with the matrix composition of α+β-brass the ratio of Zn:Cu in the solution is greater by a factor of up to 20, i.e. the zinc depletion is very strong.

Investigation of the stress corrosion cracking susceptibility of brass in chloride solution as well as the influence of arsenic and phosphorus was the subject matter of tests on C-ring specimens of pipe material on the basis of 77% copper and 1.8–2.5% aluminium and the rest as zinc, reported in [1338]. The highest susceptibility was found in 1 mol/l NaCl with the addition of CuCl and at pH-values of 2 and 4, in the active range. The course of the cracks is trans-crystalline, but becomes inter-crystalline with higher contents of arsenic and phosphorus, because the material is then also preferentially attacked by inter-crystalline corrosion. With the higher contents of arsenic and phosphorus the greatest stress corrosion cracking susceptibility is also found, whereas, for example, the zinc depletion is inhibited.

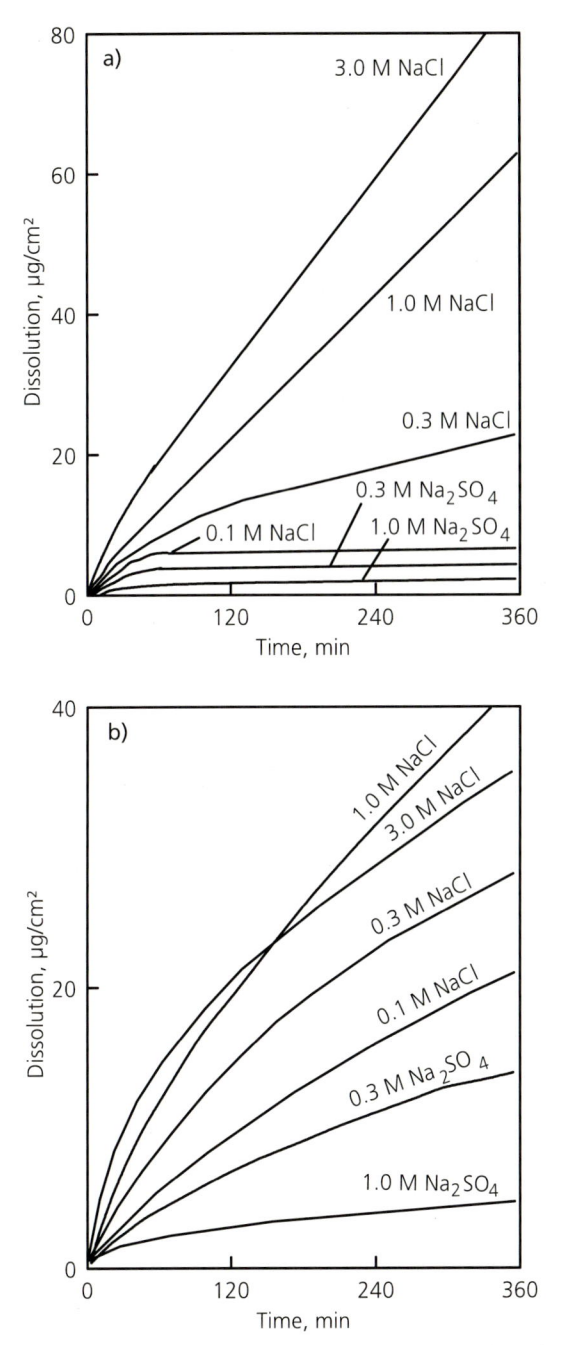

Figure 186: Dissolution of α-brass in sodium chloride and sodium sulfate solutions of various concentrations, a) average values for copper b) average values for zinc [1337]

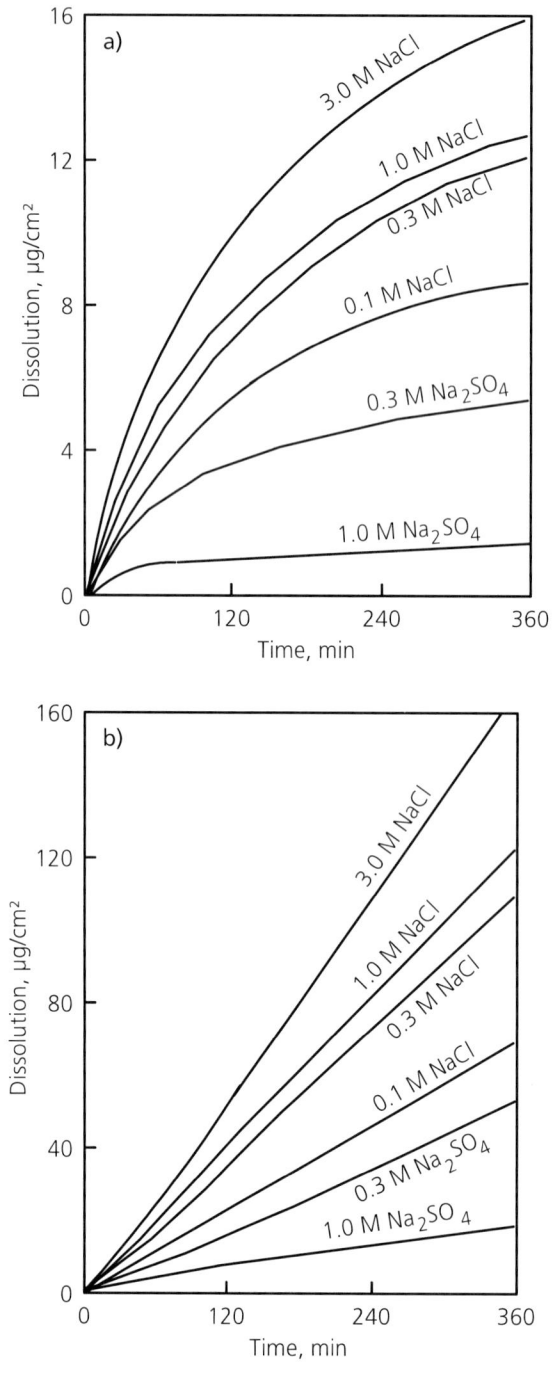

Figure 187: Dissolution of α+β-brass in sodium chloride and sodium sulfate solutions of various concentrations, a) average values for copper b) average values for zinc [1337]

Other copper alloys

Cu-Pd alloys are used as brazing material for copper alloys. Depending on the palladium concentration they show embrittlement after exposure in 3.5% sodium chloride solution, that is strongest with 5% palladium and appears after a very short time [1339]. The inter-crystalline embrittlement is observed, even though neither internal nor external stress loads appear. The effect has been verified in 3.5% NaCl at 323 K (50°C); a violent reaction takes place with all specimens immediately after submersion in the NaCl solution. Whereas additions of palladium up to 0.1% do not produce any embrittlement of specimens in the rolled state, as from 2% palladium first of all embrittlement of the surface regions takes place, and with 5% palladium the embrittlement extends to the center of the specimen. Although copper and palladium have very different galvanic potentials, stress corrosion cracking by anodic dissolution should be ruled out, because no signs of corrosion are found in the fracture microstructure. However, hydrogen embrittlement is a conceivable mechanism. Chlorides promote the diffusion of hydrogen, possibly through the porous structure of the surface layer and via grain boundaries. Irrespective of the embrittlement, weight loss determinations in 3.5% NaCl solution at 323 K (50°C) with 0.1% added palladium show a significant increase of the corrosion rate after 3 days (by a factor of 3), but this falls back to very low values with higher palladium contents.

Copper-manganese alloys with 25 and 50% manganese have been tested in 0.5 mol/l NaCl solution at given potentials in the anodic range with regard to their dissolution behavior [1340]. Both alloys show preferential dissolution of the less noble alloy partner manganese with high dissolution rates similar to the process of zinc depletion of brass and nickel depletion of copper-nickel alloys. Since manganese with −1,180 mV compared with zinc (−763 mV) and nickel (−250 mV) has the least noble equilibrium potential, it is understandable that the preferential dissolution of manganese in copper-manganese alloy takes place so rapidly and becomes significantly less in the order CuZn and CuNi.

The copper manganese alloy Incramute (40–48% Mn; 1.4–2.3% Al) was investigated potentiostatically at −200 mV$_{SCE}$ in 0.3 N NaCl solution with regard to the preferential dissolution of the less noble alloy partner manganese [1341]. A higher dissolution rate is found in the rolling direction than in the transverse direction, because rolling stretches the regions that are rich in manganese. However, heat treatment improves the homogeneity of the phases, reduces the displacement density and increases the grain size. This decreases the dissolution compared with the cold rolled state as delivered. Specimens in the quenched state dissolve more quickly than the furnace cooled homogeneous specimens, because they contain manganese-rich precipitations along the grain boundaries.

Amorphous microstructures evidently have a positive influence on the corrosion stability of copper-bismuth alloys. This was found by electrochemical measurements on an alloy having the composition Bi-35Cu in the amorphous as well as in the crystalline state in 0.1 mol/l NaCl solution [1342]. In acidic chloride solution, first of all for the amorphous state a corrosion potential is found that is less noble by 30 mV compared with pure bismuth, whereas in the heat treated state it lies very close to that of bismuth, so that a less noble behavior of the amorphous alloy would be expected. However, as shown by the current density/potential curves, the anodic dissolution currents, particularly as from a potential of about 500 mV$_H$, are extremely high for the crystalline structure, whereas for the amorphous state with pseudo-fluid structure very low current densities were found, coming close to those of pure bismuth (Figure 188). Although the standard potentials of copper and bismuth lie close together and copper tends to be more positive, the strong passivation of bismuth leads to preferred dissolution of copper in Cu-Bi alloys, so that the surface becomes enriched with bismuth and the alloy passivates like pure bismuth. This passivation is evidently considerably facilitated by the smooth surface of the amorphous state, compared with the crystalline state which has many grain boundaries.

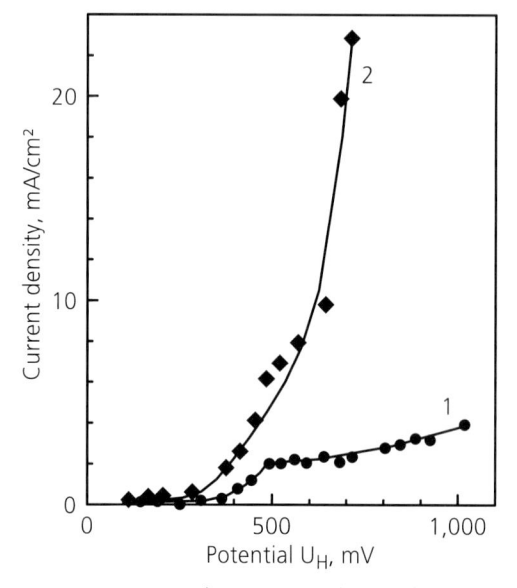

Figure 188: Current density/potential curves for a copper-bismuth alloy in 0.1 mol/l NaCl solution; 1) amorphous, 2) crystalline structure [1342]

Sodium Hydroxide

Author: *P. Drodten* / Editor: *R. Bender*

Copper

Copper is not suitable as a material for evaporators of NaOH solutions because it exhibits a corrosion rate of approximately 5 mm/a (197 mpy) in 50 % NaOH solutions at room temperature. The corrosion resistance of brass is considerably better with corrosion rates of 0.3 to 0.5 mm/a (11.8 to 19.7 mpy) [1343].

Copper pipes in the second evaporating stage in the upper part of a vacuum evaporator for NaOH solutions were attacked when the concentration of the caustic solution exceeded 40 % [1344]. To repair the damage, the copper pipes were replaced by nickel pipes.

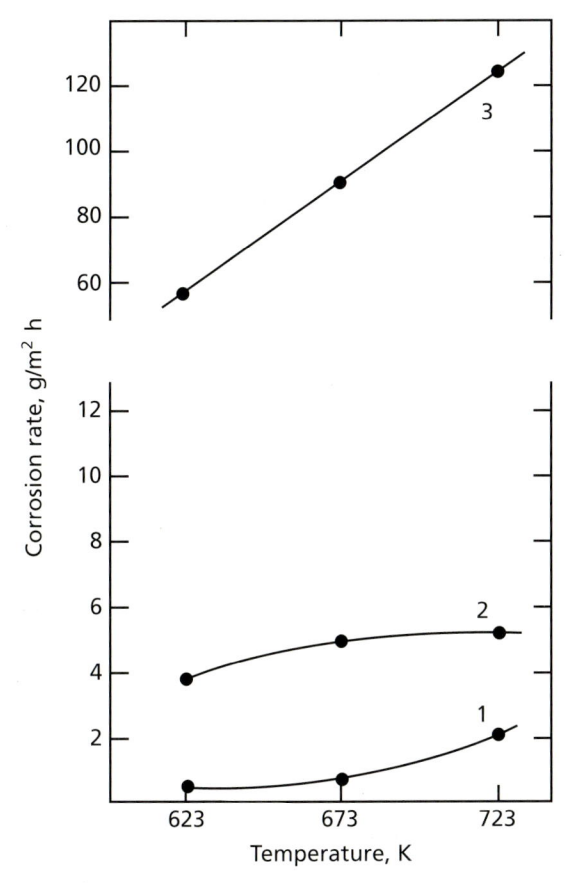

Figure 189: Temperature dependence of the corrosion rate of nickel (1), iron (2) and copper (3) in sodium hydroxide melts [1347]

In an evaluation of evaporators for NaOH solutions, it was found that copper pipes were particularly damaged by cavitation corrosion if these pipes contained increased amounts of sulfur (0.004 % to 0.005 %), while undamaged pipes contained only 0.002 % sulfur. Therefore, the content of sulfur in copper should not exceed 0.0018 % [1345].

The presence of sulfide ions in the NaOH solution considerably increases the corrosion rate of copper. Compared to sulfide-free solutions, even 10 ppm sulfide in very dilute NaOH solution (pH 8) increases the corrosive attack at 303 K (30 °C) by a factor of 6 from 4.8 to 29 g/m^2 h, which corresponds to an increase in the corrosion rate from 4.6 to 28 mm/a (181 to 1,142 mpy). At 358 K (85 °C) it was increased by a factor of 3 from 12 to 34 g/m^2 h, which corresponds to an increase in the corrosion rate from 11 to 33 mm/a (433 to 1,299 mpy). The reaction products are Cu_2O and CuO as well as Cu_2S and CuS [1346].

In sodium hydroxide melts, copper is also considerably more susceptible to corrosion than iron and nickel (Figure 189). At 723 K (450 °C), the corrosion rate of iron is 4.5 g/m^2 h (corresponding to 4.9 mm/a (193 mpy)) and that of nickel is 1.8 g/m^2 h (corresponding to 1.7 mm/a (67 mpy)) while copper is attacked with a corrosion rate of approximately 126 mm/a (4,961 mpy). In potassium hydroxide, the corrosion rate even reaches a value of almost 600 mm/a (23,622 mpy) (Figure 190) [1347].

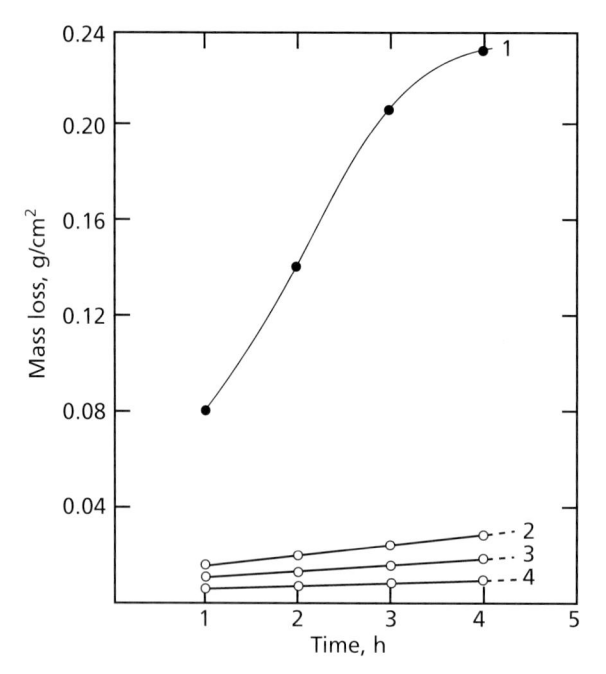

Figure 190: Time dependence of the corrosion of copper in KOH melts (1) at 723 K (450 °C) and in NaOH melts at 723 K (450 °C) (2), 673 K (400 °C) (3) and 523 K (250 °C) (4) [1347]

Copper is cleaned with alkaline solutions, e.g. NaOH solutions with sodium gluconate as an inhibitor [1348].

Copper can be coloured in a defined manner by anodic oxidation in 0.05 mol/l NaOH solutions at 333 K (60 °C) using the operating conditions given below. This produces various thicknesses of the Cu_2O layers [1349].

- alternating current of 50 Hz
- current density: 0.02 A/cm^2
- potential: $U_H = -0.3$ to -0.5 V
- duration: 10 to 60 min

Copper-aluminium alloys

Table 226 gives the corrosion rates for some Cu-Al alloys in NaOH solution [1350].

NaOH	Cu-6Al	Cu-6Al-2Si	Cu-6Al-0.2Mn	Cu-6Al-2Si-0.2Mn
%		Corrosion rate mm/a (mpy)		
20	0.056 (2.20)	0.035 (1.38)	0.043 (1.69)	0.029 (1.14)
40	0.052 (2.05)	0.027 (1.06)	0.028 (1.10)	0.014 (0.55)
60	0.030 (1.18)	0.015 (0.59)	0.018 (0.71)	0.012 (0.47)

Table 226: Corrosion rates (mm/a) of alloyed aluminium bronzes in NaOH solution at 298 K (25 °C) [1350]

According to this, the resistance of the alloy Cu-6Al is noticeably increased by the addition of silicon and manganese. Because of the decrease in the ductility of this alloy caused by these additives, it is not useful to increase their contents above the given values [1350].

In spite of the good corrosion resistance of CuAl alloys, even in concentrated NaOH solutions, stress corrosion cracking can occur under critical conditions (12 % or 25 % cold drawn at 569 to 678 MPa and an extension of 9 % to 25 %, solution annealing at 1,123 K (850 °C) and subsequent quenching). Thus for a Cu-7Al alloy (with and without 1 % Co and 0.25 % Sn) with a drawing rate of 10^6 1/s in dilute NaOH solution (pH 12.3) at 410 MPa, a crack propagation rate of between 10^5 and 10^3 mm/s was observed. The contents of cobalt and/or tin did not have any influence on the cracking behavior [1351].

Copper-nickel alloys

The good corrosion resistance to NaOH solutions is well-known for CuNi alloys, particularly those with nickel contents of more than 20 %, such as e.g. CuNi30 and CuNi40, which may also contain additional amounts of 0.5 to 5.75 % Fe, up to 1.7 % Mn, up to 0.05 % Pb and up to 1 % Zn.

The corrosion rate of alloy CuNi30 in 1 or 2 mol/l NaOH solutions at room temperature is less than 0.005 mm/a (0.20 mpy). As the temperature increases, the corrosion rate increases noticeably; however, the material is still suitable for use in, for example, heat exchangers which are subject to attack by NaOH solutions [1352, 1353].

Copper-nickel alloys with more than 20 % nickel can thus be used as materials for solutions containing sodium hydroxide, even at elevated temperatures.

Copper-tin alloys (bronze)

There is little data in the literature concerning the corrosion behavior of copper-tin alloys in NaOH solutions. The corrosion rate of copper-tin alloys with 8 % Sn in 1 to 2 mol/l NaOH solutions is reported as being below 0.25 mm/a (9.84 mpy) at room temperature and approximately 0.5 mm/a (19.7 mpy) at higher temperatures [1353].

Copper-tin-zinc-alloys (red brass)
Copper-zinc-alloys (brass)

Table 227 gives the corrosion rates for the two brasses CuZn20 (Ms80) and CuZn41 in 1 mol/l NaOH at 308 K (35 °C) with and without additional oxidising agent H_2O_2 [1354].

H_2O_2	Corrosion rate			
	CuZn20		CuZn41	
	$g/m^2 h$	mm/a (mpy)	$g/m^2 h$	mm/a (mpy)
0	0.146	0.15 (5.91)	0.146	0.16 (6.30)
0.5	0.254	0.26 (10.2)	0.246	0.26 (10.2)
1.0	0.250	0.25 (9.84)	0.358	0.38 (14.9)
2.0	0.276	0.28 (11.0)	0.513	0.55 (21.7)
5.0	0.250	0.25 (9.84)	0.379	0.40 (15.7)

Table 227: Dependence of the corrosion rate of two different brasses in 1 mol/l NaOH solutions with different amounts of added H_2O_2 at 308 K (35 °C) and 1 d test period [1354]

According to Table 227, hydrogen peroxide increases the corrosive attack by 1 mol/l NaOH solutions at 308 K (35 °C) on brass (CuZn20 (Ms80) and CuZn41) by the factors 2 and 3.

In another report, corrosion rates of the two brasses CuZn20 (Ms80) and CuZn41 in 1 mol/l NaOH solutions are given as 0.22 and 0.23 mm/a (8.66 and 9.06 mpy), respectively. These rates decreased to 0.042 and 0.061 mm/a (1.65 and 2.40 mpy), respectively, by the addition of 0.25 % sodium diethyldithiocarbamate as an inhibitor [1355].

The following compounds were tested as further inhibitors of the corrosion of brass CuZn37 in 0.2 mol/l, 0.5 mol/l and 1 mol/l NaOH solutions at room temperature: phenol, benzcatechol, resorcin, hydroquinone, pyrogallol, phloroglucine, m-cresol, α-naphthol, Na-rhodizonate, chinalizarin, acriflavine, gallocyanine, glucose, salicyl aldehyde, m-aminophenol, p-aminophenol, hydrazine sulphate, thiourea, acetylacetone, eosin, cupferron, Cupron and tannin [1356, 1357].

Table 228 lists those compounds which decreased the corrosion rate by more than 90 %. Of the investigated compounds, glucose was the most effective: it completely prevented corrosion in 0.2 to 1 mol/l NaOH solutions. This was followed, in order of effectiveness, by phloroglucine, resorcin, chinalizarin, eosine and thiourea. The protective layers formed in the presence of the inhibitors showed a characteristic colouration: reddish yellow for resorcin, bluish for phloroglucine, and reddish black for chinalizarin.

Inhibitor	Corrosion rate mm/a (mpy)	Inhibition %
without inhibitor		
0.2 mol/l NaOH	0.117 (4.61)	
1 mol/l NaOH	0.091 (3.58)	
2% glucose		
in 0.2 mol/l NaOH	0 (0)	100
in 1 mol/l NaOH	0 (0)	100
0.2% phloroglucine		
in 0.2 mol/l NaOH	0.0025 (0.10)	97.9
in 1 mol/l NaOH	0.0025 (0.10)	97.1
1% resorcin		
in 0.2 mol/l NaOH	0.0042 (0.17)	96.4
in 1 mol/l NaOH	0.0059 (0.23)	94.6
0.5% chinalizarin		
in 0.2 mol/l NaOH	0.0091 (0.36)	92.3
in 1 mol/l NaOH	0.0049 (0.19)	94.4
1% thiourea		
in 0.2 mol/l NaOH	0.018 (0.71)	84.4
in 1 mol/l NaOH	0.005 (0.20)	94.3

Table 228: Inhibition of the corrosion of CuZn37 in 0.2 and 1 mol/l NaOH solutions by inhibitors (test period 5 days, room temperature) [1356]

Further compounds that inhibit the corrosion of brasses CuZn30 (Ms70), CuZn37 and CuZn40 (Ms60) in 0.2 mol/l, 0.5 mol/l and 1 mol/l NaOH solutions at room temperature by more than 95 % are furfuryl aldehyde and dithiooxamine [1357].

For electrochemical etching of brass, a solution with 2 to 2.5 mol/l NaOH and 0.1 mol/l ethylene diamine or potassium cyanide are suitable at a potential of $U_H = -0.10$ to -0.21 V and an etching time of 1 to 3 min at 293 K (20 °C) [1358].

For chemical etching, a suitable solution is 2 to 2.5 mol/l NaOH with 1 to 2 vol.% ethylene diamine and 100 g/l $K_3Fe(CN)_6$ at 293 K (20 °C) and an etching time of 1 to 3 min [1358].

Thus, brass can be considered to have good resistance to corrosion in a suitably inhibited sodium hydroxide solution of concentrations up to 1 mol/l. Brasses with high zinc contents (more than 35%) are still resistant in 1 mol/l NaOH solution, even without inhibition.

However, account must be taken of the fact that under certain conditions, stress corrosion cracking has been observed in α-brass in NaOH solution: Figure 191 shows the potential dependence of the crack growth in the α-brass (CuZn30). At a potential of $U_H = -0.6$ V, the crack propagation rate is approximately 3.2 mm/a (126 mpy), and at $U_H = -0.5$ V it is approximately 99 mm/a (3,898 mpy). In this potential range, the cracks are intergranular; however, at higher potentials they are transgranular. The polarity of the potential and the direction of its change allow no clear conclusions to be drawn concerning the possibility of stress corrosion cracking [1359].

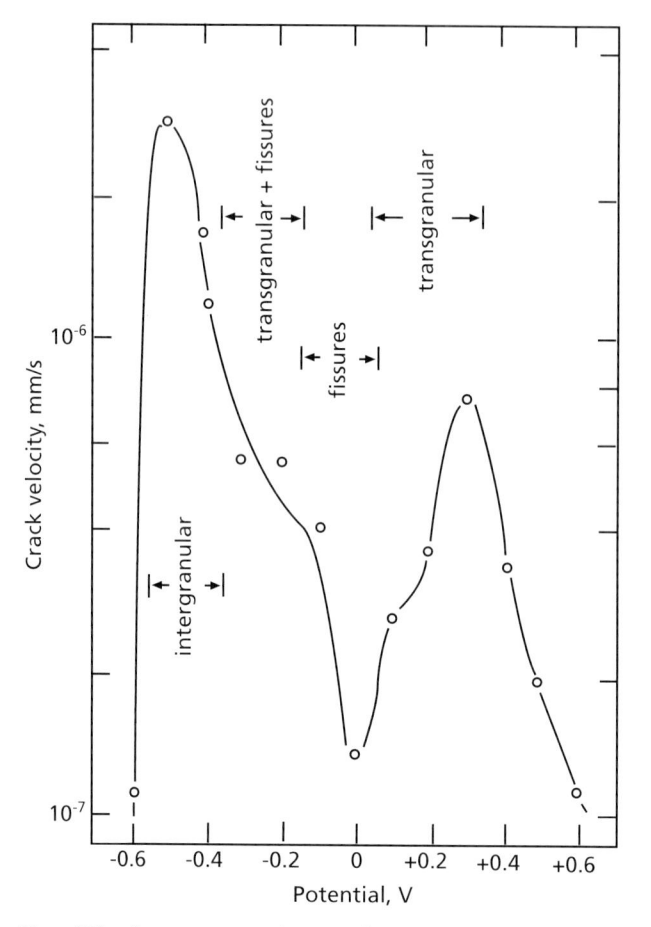

Figure 191: Average propagation rate of stress-corrosion cracks in 70/30 brass in NaOH solution (pH 13) at room temperature as a function of the potential [1359]

Table 229 gives the results of tests on the stress corrosion cracking in NaOH solution at pH 12. In addition, the highest crack propagation rates were observed in the potential range $U_H = -0.6$ to -0.4 V [1359].

Test potential U_H V	No. and type of cracks	Rate of crack propagation mm/s
−0.5	numerous; intergranular	2×10^{-6}
−0.4	numerous; intergranular	1.5×10^{-6}
−0.3	few; short V-shaped	$< 10^{-7}$
+0.2	very short; transgranular	$< 10^{-7}$
+0.4	very short; transgranular	$< 10^{-7}$

Table 229: Results of tests on stress corrosion cracking of brass 70/30 in NaOH solution (pH 12) [1359].

Other copper alloys

Copper-beryllium alloys are used as hardenable spring alloys. The commercially available alloys Beryvac-200® (Cu-2Be) and as the conducting beryllium bronze Beryvac-60® (Cu-CoBe) show a corrosion rate of approximately 0.01 mm/a (0.39 mpy) at 293 K (20 °C) in 50 % NaOH and are thus considered to be corrosion resistant. Even at 373 K (100 °C), the corrosion rate is 0.01 to 0.1 mm/a (0.39 to 3.94 mpy) and they are thus still useable [1360].

Sodium Sulfate

Author: J. Küpper-Feser / Editor: R. Bender

Copper

– Summary –

Aqueous Na_2SO_4 solutions

Pure copper is relatively resistant to neutral aqueous Na_2SO_4 solutions. Because dissolution is regarded as being diffusion-controlled, the flow conditions in the medium can be expected to exert an influence, even if the flow rate is lower than that causing erosion corrosion.

Na_2SO_4 melts and deposits

Cu is not resistant to Na_2SO_4 melts.

– Corrosion in aqueous solutions

Copper with a purity of at least 99.9 % is barely attacked by aqueous sodium sulfate solutions [1361, 1362]. The corrosion rate depends on the degree of aeration, but it is usually less than 0.04 mm/a (1.57 mpy). The corrosion rate is approximately 0.004 mm/a (0.16 mpy) in a 10–20 % sodium sulfate solution at RT. A corrosion rate of 0.06 mm/a (2.36 mpy) was found in an acidic 3.5 % solution (pH 3) at 297 K (24 °C) with intermittent aeration. The test duration was 30 d [1363].

Copper has been used to in heating coils for the dehydration of Glauber's salt [1363].

Copper does not exhibit typical active/passive behavior in sodium sulfate solutions. Continuously increasing current densities are registered as the anodic potential increases [1364].

In a dilute alkaline sodium sulfate solution (1 mol/l Na_2SO_4 + 0.002 mol/l NaOH) a passive film forms on the copper at potentials of –0.3 to 0.0 V_{SCE}. This film consists of an inner layer of Cu_2O and an outer layer of $CuO\cdot Cu(OH)_2$.

At potentials $>0\,V_{SCE}$, the so-called critical potential, copper dissolves as the potential is raised with rapidly increasing current densities, and, at the same time, copper dissolves over large areas thus causing pitting [1365].

In a 0.05–0.1 mol/l sodium sulfate solution, the anodic dissolution rate of copper lies in the potential range from 0.22 to 0.28 V_{SHE}, irrespective of the sulfate concentration. However, as the sulfate concentration increases (>0.1 mol/l), the anodic polarization curve of copper is shifted to overall more negative potentials. This effect is explained by the formation of complexes between copper cations and sulfate anions. In dilute sodium sulfate solutions (<0.1 mol/l) and at current densities up to 1 mA/cm², the dissolution of copper is controlled by the outward diffusion of Cu cations. In concentrated sodium sulfate solutions, the outward diffusion of the copper sulfate complex is the rate-determining step of copper dissolution [1366].

Sodium sulfate can be used to inhibit the corrosion of copper (99.98 %) in nitric acid. Table 230 lists the mass loss rates of copper in 2.0 mol/l nitric acid as a function of the sodium sulfate concentration at 298 K (25 °C). The rates of mass loss were determined calorimetrically from the amount of copper dissolved in the electrolytes. The addition of sodium sulfate shifts the free corrosion potential of copper to more negative potentials [1367].

Na_2SO_4 concentration mol/l	Mass loss rate $g/m^2 h$
0	46.4
5×10^{-3}	38.5
10×10^{-3}	32.0
15×10^{-3}	25.0
20×10^{-3}	19.4
30×10^{-3}	16.7
50×10^{-3}	13.7

Table 230: Influence of the sodium sulfate concentration on the mass loss rate of copper in 2.0 mol/l HNO_3 at 298 K (25 °C) after 1 h [1367]

Hydroxyamino acids were tested as corrosion inhibitors for copper in salt solutions of 0.148 g/l Na_2SO_4, 0.165 g/l NaCl, 0.138 g/l $NaHCO_3$ at 298 K (25 °C) and 313 K (40 °C) for 72 h. They exhibited a similarly good inhibition effect as toxic triazole derivatives and can therefore replace them [1368].

– Corrosion in melts and underneath deposits

Copper is not resistant to melts of sodium sulfate and other alkali metal sulfates. It dissolves in the melt [1369].

Copper-aluminium alloys

– Summary –

Aqueous Na_2SO_4 solutions

CuAl alloys are also resistant to aqueous Na_2SO_4 solutions. The resistance can be significantly increased by alloying with aluminium and tin. There is a considerable decrease in the anodic dissolution current for alloys with Al > 14 atom% and Sn > 3 atom%.

– Corrosion in aqueous solutions

Aluminium bronze with 91 % Cu, 7 % Al, and 2 % Fe (DIN-Mat. No. 2.0932, CuAl8Fe3) is corrosion-resistant in aqueous sodium sulfate solutions. In static sodium sulfate solutions, the corrosion rate at room temperature is approximately 0.005 mm/a (0.20 mpy) [1361]. Aluminium bronze with 8 % aluminium (CuAl8, DIN-Mat. No. 2.0920) corroded at a rate of 0.002 mm/a (0.08 mpy) in a 10 % sodium sulfate solution at 293 to 298 K (20 to 25 °C). Aluminium bronzes have been successfully used for pumps [1363].

The copper-aluminium casting alloys:

CC331G (CuAl10Fe2-C, old 2.0940, G-CuAl10Fe),
CC333G (CuAl10Fe5Ni5-C, old 2.0975, G-CuAl10Ni)
CC332G (CuAl10Ni3Fe2-C, old 2.0970, G-CuAl9Ni), and
2.0962 G-CuAl8Mn (old)

exhibited corrosion rates of < 0.5 mm/a (19.7 mpy) in aqueous sodium sulfate solutions of all concentrations at up to 373 K (100 °C) [1370].

In older production facilities, sodium sulfate was formerly separated from the Glauber's salt melt at 333 K (60 °C) using centrifuges made of aluminium bronze [1363].

The corrosion resistance of copper can be increased by alloying it with aluminium and tin. The corrosion behavior of CuAlSn alloys with Cu (0–20) atom% Al-(0–5) atom% Sn was studied in a 0.1 mol/l sodium sulfate solution in air at room temperature using current density/potential curves (Figure 192). Aluminium contents of > 14–15 atom% and small tin contents (3 atom%) led to a considerable reduction in the anodic dissolution current. Surface analyses confirmed the formation of a protective Al_2O_3 film. The corrosion resistance was dependent on the Al/Sn ratio. Aluminium alone did not lead to a significant improvement in the corrosion behavior. CuAl15 exhibited similar corrosion behavior to pure copper, likewise CuAlSn alloys with < 13 atom% aluminium. Small amounts of tin and an aluminium content > 13 atom% in CuAlSn alloys are decisive for a low anodic dissolution current [1371].

– Corrosion in melts and underneath deposits

Aluminium bronze with 7 % aluminium is only slightly attacked by oil ash deposits (vanadium pentoxide + 10 % sodium sulfate) at 673 K (400 °C). However, at 773 K (500 °C), considerable attack is evident after 100 h [1363].

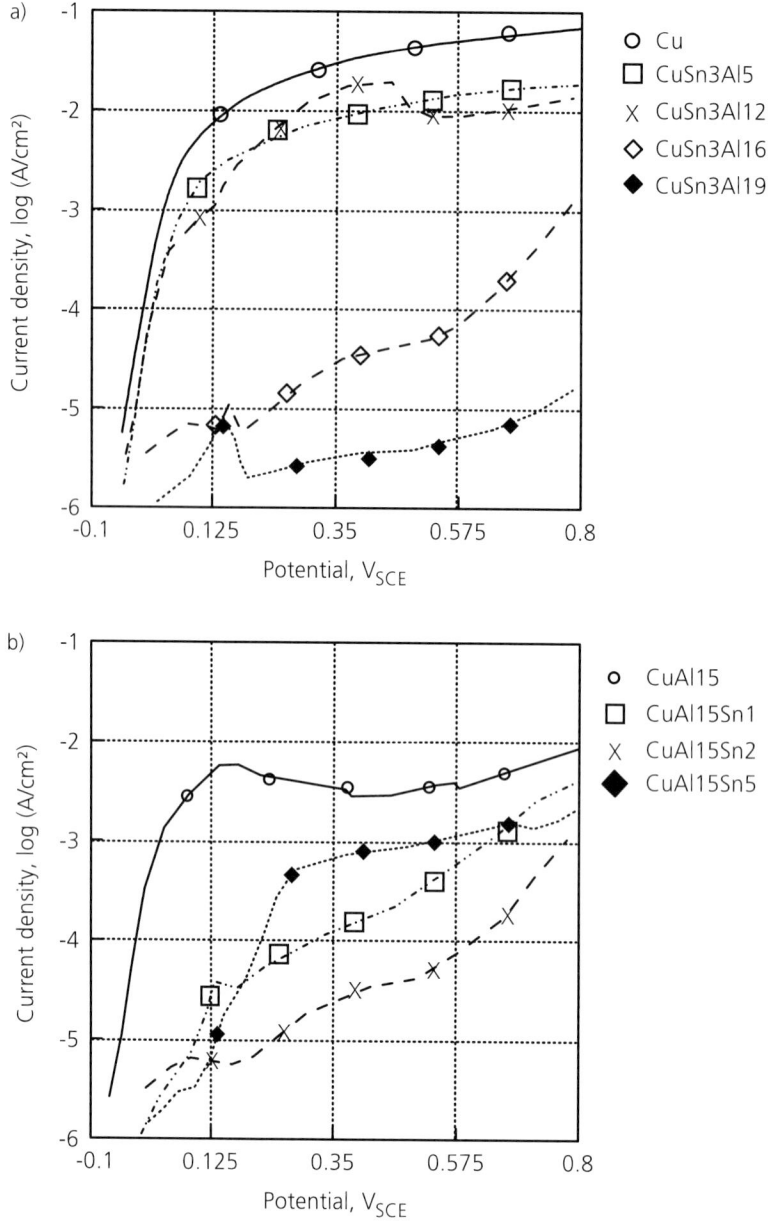

Figure 192: Polarization curves of copper-aluminium-tin alloys (data in atom%) in a 0.1 mol/l sodium sulfate solution at room temperature, scanning rate v = 0.2 mV/s [1371]

Copper-nickel alloys

– Summary –

Aqueous Na_2SO_4 solutions

Copper-nickel alloys have a good resistance to Na_2SO_4 solutions of all concentrations. The presence of chloride ions increases the corrosion rate. Their resistance to erosion corrosion is better than that of CuZn alloys. The possibility of stress corrosion cracking must also be taken into account.

– Corrosion in aqueous solutions

The CuNi alloys Cupronickel 10 (old: 2.0811, CuNi10), Cupronickel 20 (old: 2.0822, CuNi20), and Cupronickel 30 (CuNi30) exhibit good corrosion resistance at RT in sodium sulfate solutions of all concentrations, even at elevated flow rates [1362].
 The alloys

CW352H, CuNi10Fe1Mn (old 2.0872) and
CW354H, CuNi30Mn1Fe (old 2.0882)

are corrosion resistant in saturated sodium sulfate solutions, even at 333 K (60 °C), with a corrosion rate of < 0.12 mm/a (4.72 mpy) [1372].
 At temperatures of up to 373 K (100 °C), the corrosion rate of the copper-nickel casting alloy (CC383H, CuNi30Fe1Mn1NbSi-C, old: DIN-Mat. No. 2.0835, G-CuNi30) is < 0.5 mm/a (19.7 mpy) in sodium sulfate solutions of all concentrations [1370].

– Stress corrosion cracking –

Copper-nickel alloys are used preferentially in heat-exchangers and condensers for seawater and brackish water because of their good thermal conductivity. Because it is known that copper alloys are susceptible to stress corrosion cracking in sodium sulfate solutions at room temperature, CuNi10 alloy CDA 706 was tested accordingly.
 Its stress corrosion cracking behavior was studied in 0.5 mol/l Na_2SO_4 at various pH-values using a tensile test with a low constant strain rate of 10^{-6} 1/s at the free corrosion potential and under polarization. Alloy CDA 706 was resistant to stress corrosion cracking in deaerated 0.5 mol/l Na_2SO_4 solutions in a pH-range from 1.2 to 13 under polarization of approximately –200 mV and +100 mV (with respect to the free corrosion potential) at room temperature. The specimens underwent ductile fracture. At high anodic potentials, the specimens failed due to rapid dissolution of the metal. The good resistance of the CuNi10 alloy CDA 706 in sodium sulfate solutions was attributed to the formation of a protective and ductile oxide film that mainly consisted of Cu_2O. Tensile tests at a strain rate of 10^{-7} 1/s showed that this film only started to tear at an elongation of > 13 % [1373].
 The anodic polarization curves of alloy CDA 706 were measured at room temperature in an argon-purged 0.5 mol/l Na_2SO_4 solution at pH-values from 1.2 to 13. The free corrosion potential dropped as the pH-value increased. No passivation of the

alloy was observed for pH-values of 1.2 to 7. At pH 10, a small passive range can be seen at 150–250 mV$_{SHE}$. In a strongly alkaline sodium sulfate solution at pH 13, the alloy passivates at potentials >60 mV$_{SHE}$ and current densities of <1×10^{-5} A/cm^2 are recorded in the passive range [1373].

– Erosion corrosion –

Erosion corrosion studies were carried out in a ramjet device on, among others, alloy CW352H (CuNi10Fe1Mn) and alloy CW354H (CuNi30Mn1Fe) in chloride-free cooling water (200 mg/l Na$_2$SO$_4$) and chloride-containing cooling water or artificial river water (200 mg/l Na$_2$SO$_4$ + 5 g/l NaCl). The aim was to investigate the influence of the flow rate, temperature, addition of sand (100 mg/l, grain size 0.1–0.2 mm), and air (1–5 vol.%). These CuNi materials are used for heat-exchanger tubes. The investigations and the corrosion rates are summarized in section Copper-zinc alloys. Both CuNi alloys exhibited comparable corrosion rates. At 298 K (25 °C), the erosion corrosion rate was approximately 0.1 mm/a (3.94 mpy) and it increased to approximately 0.25 mm/a (9.84 mpy) when the temperature was raised to 308 K (35 °C). The highest corrosion rates of approximately 0.4 mm/a (15.7 mpy) were obtained at 308 K (35 °C) with added sand.

In chloride-containing cooling water (200 mg/l Na$_2$SO$_4$ + 5 g/l NaCl), the corrosion rates of the CuNi10Fe1Mn and CuNi30Mn1Fe alloys increased by two- to six-fold compared to the pure sodium sulfate solution. The results are summarized in Table 233. Increasing the temperature from 298 K to 308 K (25 to 35 °C) gave an insignificant increase in the erosion corrosion rate of the CuNi30Mn1Fe alloy; however, the rate for the CuNi10Fe1Mn alloy increased by 50 % (Figure 198).

The addition of 5 vol.% air at 298 K (25 °C) and a flow rate of 7.2 m/s increased the corrosion rates of both CuNi alloys by approximately 20 %. An additional temperature increase to 308 K (35 °C) resulted in an insignificant change in the corrosion rate of the CuNi30Mn1Fe1 alloy, whereas the corrosion rate of the CuNi10Fe1Mn alloy rose by approximately 40 %. An increase in the flow rate from 0.9 to 7.2 m/s doubled the corrosion rates of both CuNi alloys at 308 K (35 °C). They have a lower dependence on the flow rate and temperature than CuZn alloys. [1374].

Copper-tin alloys (bronzes)

– Summary –

Aqueous Na$_2$SO$_4$ solutions

CuSn alloys have a good resistance to Na$_2$SO$_4$ solutions. The risk of erosion corrosion must be taken into account if they are to be used for heat-exchanger tubes.

– Corrosion in aqueous solutions

The bronzes CuSn4 and CW453K (CuSn8, old: DIN-Mat. No. 2.1030) are resistant to static aqueous solutions of sodium sulfate at room temperature with a corrosion rate of 0.005 mm/a (0.20 mpy) [1361].

The casting alloys CC480K (CuSn10-C, old G-CuSn10, 2.1050) and CC483K (CuSn12-C, old G-CuSn12, 2.1052) are resistant to moderately resistant to sodium sulfate solutions of all concentrations up to 373 K (100 °C) with corrosion rates of < 0.5 mm/a (< 19.7 mpy) [1370].

Copper-tin-zinc alloys (red brass)

– Summary –

Aqueous Na_2SO_4 solutions

These materials are resistant to moderately resistant in aqueous Na_2SO_4 solutions. Because Cu-base materials are frequently used for heat-exchanger tubes, the associated risk of erosion corrosion must be taken into account. Particularly susceptible areas are the ends of the incoming pipes and sections with a small radius of curvature. In addition to the flow rate, the air and sand content in the medium are important factors influencing erosion corrosion.

– Corrosion in aqueous solutions

The casting alloys G-CuSn10Zn (DIN-Mat. No. 2.1086), CC491K (CuSn5Zn5Pb5-C, old G-CuSn5ZnPb, 2.1096), and CC493K (CuSn7Zn4Pb7-C, old G-CuSn7ZnPb, 2.1090) are resistant to moderately resistant to sodium sulfate solutions of all concentrations up to 373 K (100 °C) with corrosion rates of < 0.5 mm/a (< 19.7 mpy) [1370].

Red brass (85.5 % Cu, 4.8 % Zn, 4.8 % Sn) corroded in a 0.23 % sodium sulfate solution at 473 K (200 °C) with a corrosion rate of 0.004 mm/a (0.16 mpy) [1363].

– Erosion corrosion –

Tin bronze CuSn6ZnAl, which deviates from standard materials because it contains aluminium, is also used for heat-exchanger tubes. Erosion corrosion of such tubes, which often occurs in the inlet zone, was simulated experimentally. The erosion-corrosion rates of CuSn6ZnAl were determined in 200 mg/l Na_2SO_4 solutions and in 200 mg/l Na_2SO_4 + 5 g/l NaCl as a function of the temperature, the flow rate, addition of sand and air (Table 232 and Table 233 as well as Figure 198). A mean corrosion rate of 0.1 mm/a (3.94 mpy) was found in a pure sodium sulfate solution at a flow rate of 7.2 m/s and a temperature of 298 K (25 °C). Raising the temperature to 308 K (35 °C) increased the mean corrosion rate to 0.3 mm/a (11.8 mpy), which is thus somewhat higher than that of the co-tested CuZn and CuNi alloys. The addition of 100 mg/l sand raised the erosion-corrosion rate, whereas 5 vol.% air at 308 K (35 °C) lowered it. In a chloride-containing sodium sulfate solution (200 mg/l Na_2SO_4 + 5 g/l NaCl) with a flow rate of 7.2 m/s, the mean corrosion rate was 0.7 mm/a (27.6 mpy) at 298 K (25 °C) and 0.85 mm/a (33.5 mpy) at 308 K (35 °C). Increasing the flow rate by 0.9 to 7.2 m/s doubled or even tripled the corrosion rates. Injection of air and simultaneously raising the temperature increased the corrosion rates by approximately 100 %. Added sand (100 mg/l, grain size 0.1–0.2 mm) increased erosion corrosion attack by 20 % at 298 K (25 °C). At 308 K (35 °C), the effect of sand was not significantly higher (see Figure 198, as well as Table 232 and Table 233) [1374].

Copper-zinc alloys (brass)

– Summary –

Aqueous Na_2SO_4 solutions

The corrosion behavior of a brass in aqueous Na_2SO_4 solutions is clearly influenced by its chemical composition. Casting alloys, in particular, are no longer resistant at elevated temperatures.

There are further risks due to selective dissolution of zinc along with stress corrosion cracking and erosion corrosion.

The selective dissolution of zinc is dependent not only on the chemical composition of the material, but it is also usually promoted by anodic polarization, which initially causes dissolution of only zinc with a dissolution current density that is essentially independent of the potential. The copper only starts to dissolve above the critical potential.

Stress corrosion cracking can also occur in media that do not contain ammonium and complexing ions. Selective dissolution of zinc plays a decisive role in crack propagation. Zinc progressively dissolves at the crack tip because sliding processes are continuously generating a fresh surface. There is no passivation. Further influencing parameters are the pH-value and sulfate content of the solution.

Erosion corrosion often occurs in heat-exchanger tubes, particularly at the inlet end of the tube and where there is a small radius of curvature. In addition to the flow rate, the air and sand content in the medium as well as the angle of impingement are important factors influencing erosion corrosion.

– Corrosion in aqueous solutions

The copper-zinc alloys
CW501L (CuZn10, old DIN-Mat. No. 2.0230),
CW502L (CuZn15, old DIN-Mat. No. 2.0240), and
CW706R (CuZn28Sn1, old DIN-Mat. No. 2.0470) exhibit very good corrosion resistance to aqueous sodium sulfate solutions at RT. The reported corrosion rates are 0.005 mm/a (0.20 mpy). Alloys with good resistance include

CW505L (CuZn30, old DIN-Mat. No. 2.0265),
CW509L (CuZn40, old DIN-Mat. No. 2.0360),
CW603N (CuZn36Pb3, old DIN-Mat. No. 2.0375),
CW702R (CuZn20Al2, old DIN-Mat. No. 2.0460), and
CW717R (CuZn38Sn1As, old DIN-Mat. No. 2.0530) [1361].

The following casting alloys exhibit corrosion rates below 0.5 mm/a (19.7 mpy) in sodium sulfate solutions at temperatures between 293 and 373 K (20 and 100 °C):

CC750S (CuZn33Pb2-C, old G-CuZn33Pb, DIN-Mat. No. 2.0290),
CC754S (CuZn39Pb1Al-C, old G-CuZn37Pb, DIN-Mat. No. 2.0340),
CC765S (CuZn35Mn2Al1Fe1-C, old G-CuZn35Al1, DIN-Mat. No. 2.0592),
CC764S (CuZn34Mn3Al2Fe1-C, old G-CuZn34Al2, DIN-Mat. No. 2.0596),
CC762S (CuZn25Al5Mn4Fe3-C, old G-CuZn25Al5, DIN-Mat. No. 2.0598), and
CC761S (CuZn16Si4-C, old G-CuZn15Si4, DIN-Mat. No. 2.0492).

They thus have very little resistance, particularly at elevated temperatures [1370].

– Dezincification –

The corrosion behavior of ε-brass Cu14Zn86 (Cu-86 atom% Zn or 86.33 mass% Zn with a hexagonal crystal lattice), γ-brass Cu34Zn66 (Cu-65 atom% Zn or 65.64 mass% Zn with a body-centered cubic lattice) and α-brass Cu69Zn31 (Cu-30 atom% Zn or 30.6 mass% Zn with a face-centered cubic lattice) was studied in acetate-buffered sodium sulfate solutions (1 mol/l Na_2SO_4 + 0.05 mol/l CH_3COONa + 0.05 mol/l CH_3COOH) at pH 5 and 296 K (23 °C). The electrolyte was purged with helium to keep the concentration of dissolved oxygen as low as possible. If anodic polarization is too low, zinc dissolves selectively. Copper also starts to dissolve when the potential exceeds the standard potential of copper. The anodic partial current density/voltage curves for the partial dissolution of zinc and copper were calculated from a chemical analysis of the electrolyte after holding for 1 h and 20 h at a constant anodic potential. Figure 193 shows a plot of the partial current densities (i_{Zn}, i_{Cu}) against the potential for ε-, γ-, and α-brass in a buffered sodium sulfate solution after holding for 20 h. The zinc dissolution curve for the various types of brass shows a potential-independent section of the current density and above a certain potential, the so-called critical potential E_c, a sudden steep increase in the current density that flattens again further on. The curves are shifted to higher potentials as the copper content in the brass increases. The critical potential of ε-brass is reported to be approximately $-950\,mV_{SHE}$, that of γ-brass approximately $0\,mV_{SHE}$, and that of α-brass approximately $+100\,mV_{SHE}$. The dissolution of copper out of γ- and α-brass only starts at approximately $+50\,mV_{SHE}$. The dissolution current densities for ε-brass were only determined up to $-400\,mV_{SHE}$. Up to this potential, only zinc dissolves and no copper. At potentials $E > E_c$, the selective dissolution of zinc from ε-brass (Cu14Zn86) leads to the formation of a porous copper-rich layer on the surface. Copper and zinc dissolve out of γ-brass (Cu34Zn66) at potentials $> +50\,mV_{SHE}$; however, the partial current density for copper dissolution is significantly lower than that of zinc. This means that zinc preferentially dissolves out of γ-brass. Holding tests at potentials $E > E_c$ led to the formation of porous layers on the surface with severe local crack-like attack. At $E > E_c$, copper and zinc dissolve simultaneously out of α-brass (Cu69Zn31) with similarly high current densities. This simultaneous dissolution of both alloying components led to an almost uniform surface corrosion without the formation of a porous layer [1375].

This dezincification potential range is decisively dependent on the composition of the electrolyte solution. Adding 0.2 mol/l sodium chloride to a 0.2 mol/l sodium sulfate solution increased the dissolution rate of the brass over the entire potential range. In this case, the dissolution rates of the mixed salt solution were higher than those obtained by simple addition of the rates measured in solutions to which only one salt had been added. In the potential range of simultaneous dissolution of zinc and copper, a loosely cohesive black film formed on the surface.

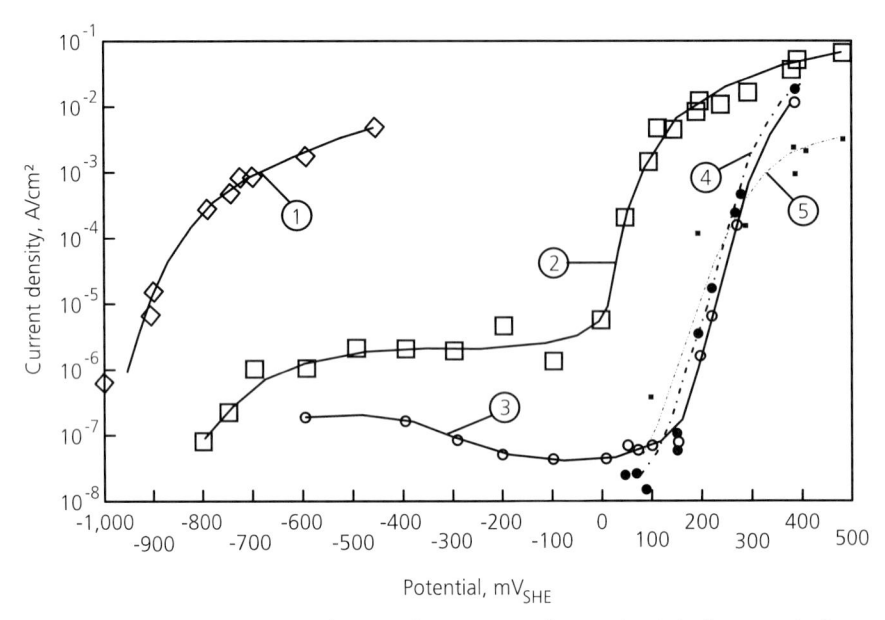

Figure 193: Quasi-stationary partial current density/potential curves (i_{Zn}, i_{Cu}) of ε-, γ-, and α-brass in a buffered (pH 5) sodium sulfate solution at 296 K (23 °C) (solid line: i_{Zn}, dashed line: i_{Cu}) [1375]
curve 1: i_{Zn} of ε-brass (Cu14Zn86)
curve 2: i_{Zn} of γ-brass (Cu34Zn66)
curve 3: i_{Zn} of α-brass (Cu69Zn31)
curve 4: i_{Cu} of α-brass (Cu69Zn31)
curve 5: i_{Cu} of γ-brass (Cu34Zn66)

The simultaneous anodic dissolution of zinc and copper from α-brass was studied in detail in a 1 mol/l sodium sulfate solution acidified with 0.005 mol/l sulfuric acid to pH 2 at 297 K (24 °C). The brass specimens were heat-treated at 873 K (600 °C) (60 h) in a vacuum. The potential of the specimens was adjusted so that it was in the simultaneous steady-state dissolution range (0.20–0.30 V). The potentiostatic holding time was up to 60 min. At a potential of 0.24 V, a steady-state dissolution current of approximately 0.05 mA/cm² was obtained after less than 1 h. Chemical analyses of the electrolyte and analysis of the specimen's surface using soft X-rays (1.5–4 kV) showed that, at the onset of dissolution (<1 h), the less noble zinc was selectively dissolved until the copper had enriched to a layer with a mean thickness of approximately 0.01 µm. Only then did the zinc and copper dissolve simultaneously in a ratio corresponding to the alloy composition [1376].

The dissolution behavior of brass (%, 58.99Cu, 36.94Zn, 1.84Sn, 0.76Fe, 0.04Mn, 1.43Pb) in a nitrogen-purged sodium sulfate solution of various concentrations

(0.066 to 0.5 mol/l) was determined at 303 K (30 °C) with a ring-disk electrode. The dissolution current of brass increased with the sodium sulfate concentration over the entire potential range from −400 mV to +1,300 mV$_{SCE}$. The potential range of selective zinc dissolution was studied in a 0.5 mol/l sodium sulfate solution at −380 mV to −10 mV$_{SCE}$. However, the original brass color appeared underneath the film, thus confirming stoichiometric copper-zinc dissolution [1377].

Brass with zinc contents over 15 % were susceptible to dezincification in sodium sulfate solutions. In this case, more zinc dissolved relative to the Cu:Zn ratio of the alloy. This left porous copper behind. The corrosion of brass 60 Cu60Zn40 (60 mass% Cu, 40 mass% Zn, dual-phase α+β-brass) and brass 70 Cu70Zn30 (70 mass% Cu, 30 mass% Zn, single-phase α-brass) in 0.3 mol/l and 1.0 mol/l sodium sulfate solutions was determined at room temperature using electropolished sheets. The quantities of zinc and copper dissolved in the electrolyte during corrosion were analytically followed over time up to a maximum exposure period of 6 h. The concentration of copper and zinc ions in the solution were determined by calorimetry using a sodium diethyldithiocarbamate solution or dithizone. Figure 195a and Figure 195b show the quantities of dissolved copper and zinc as a function of the corrosion time for brass 60. Figure 196a and Figure 196b show the results for brass 70. Less metal dissolves as the sodium sulfate concentration increases. The dissolution rate of zinc increases linearly over time, whereas that of copper decreases over time. During corrosion of brass 60, 20-times more zinc dissolves than Cu. Dezincification of brass 70 is lower than that of brass 60 [1378].

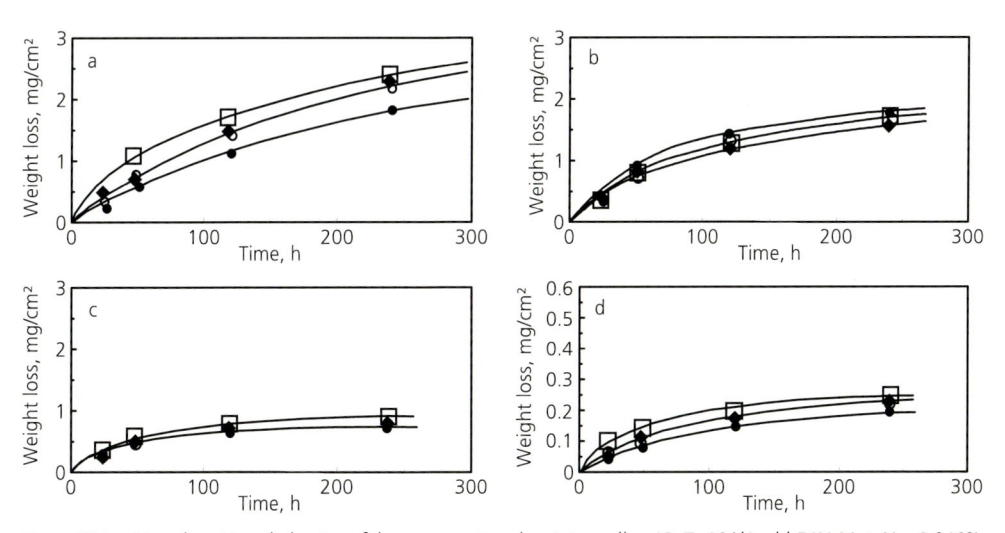

Figure 194: Mass loss/time behavior of the copper-zinc-aluminium alloy (CuZn20Al2, old DIN-Mat. No. 2.0460) in a sodium sulfate solution depending on the sodium chloride concentration and pH-value at 298 K (25 °C) [1379]

- ● 0.1 mol/l Na$_2$SO$_4$
- ○ 0.1 mol/l Na$_2$SO$_4$ + 5·10^{-4} mol/l NaCl
- ◆ 0.1 mol/l Na$_2$SO$_4$ + 5·10^{-3} mol/l NaCl
- □ 0.1 mol/l Na$_2$SO$_4$ + 5·10^{-2} mol/l NaCl

a) pH 3.0
b) pH 3.5
c) pH 4.0
d) pH 7.25

The mass losses of commercial alloy CuZn20Al, used for heat-exchanger tubes, was measured at 298 K (25 °C) in a static 0.1 mol/l sodium sulfate solution depending on the sodium chloride content (from 10^{-4} to 10^{-2} mol/l) and the pH-value (3.0 to 7.25, adjusted with sulfuric acid). The mass losses were determined after 24 h, 48 h, 120 h, and 240 h. The amounts of copper and zinc dissolved in the electrolytes were determined by polarography, and the amount of corrosion products of copper, zinc, and aluminium adhering to the surface of the samples was determined by chemical and X-ray analyses. The results of these studies are shown in Figure 194. The mass losses are high in acidic solutions (pH 3) and decrease with increasing pH.

At a constant pH-value, the mass loss increases with the chloride concentration and exposure time. This is particularly noticeable in acidic sodium sulfate solutions. As the pH-value increases, the influence of the chloride content on the mass loss decreases (Figure 194c and d). Zinc dissolves preferentially during corrosion of alloy CuZn20Al. Dezincification is accelerated by sodium chloride and reaches a maximum value in 0.1 mol/l sodium sulfate + 10^{-3} mol/l sodium chloride solution at pH 3–4. The solid corrosion products on the surface of the sample mainly consist of copper and aluminium. The composition of the corrosion products, determined by X-ray diffraction, depends on the pH-value and the chloride concentration of the sodium sulfate solution. In a pure 0.1 mol/l sodium sulfate solution, a virtually homogeneous surface film of CuO, Cu_2O, and $Cu_3(SO_4)_2 \times 4 H_2O$ forms in a pH range of 3.0 to 7.25 at 298 K (25 °C). In Na_2SO_4 solutions containing chloride, $CuCl$, $[Cu(OH,Cl)]_2 \times 2 H_2O$, $Al_{45}O_{45}(OH)_{45}Cl$, and $Al(OH)_3$ can also be detected in this surface film. In acidic solutions containing chloride, the surface film is very inhomogeneous and thus promotes local corrosive attack as well as dezincification. The susceptibility to stress corrosion cracking is particularly high in acidic solutions containing chloride [1379].

The selective anodic dissolution of zinc from brass with simultaneous tensile loading may cause stress corrosion cracking [1380].

The stress corrosion cracking behavior of α-brass (Cu69Zn31) was studied with a tensile test in an acidic sodium sulfate solution (1 mol/l Na_2SO_4 + 0.005 mol/l H_2SO_4) at pH 2 and under anodic polarization at E = +0.24 V_{SHE}. The electrolyte was purged with helium. The test temperature was 296 K (23 °C). Brass wires with a diameter of 2 mm and a length of 37 mm were placed in an electrochemical measuring cell. The tensile test was started after a constant current density (approximately 10^{-4} A/cm^2 after < 1 h) was obtained at a potential of 0.24 V_{SHE}. The strain rates were 10^{-2}, 10^{-3}, and 10^{-4} 1/min. All specimens exhibited stress corrosion cracking under these conditions. The results demonstrated that stress corrosion cracking of α-brass is possible, even in solutions that do not contain ammonium and complexing ions. The cracks were predominantly intergranular and were accompanied by a few secondary transgranular cracks in the grains. At the preset anodic potential of 0.24 V_{SHE}, there is initial preferential dissolution of zinc from the α-brass. Simultaneous dissolution of zinc and copper takes place after a certain incubation period (< 1 h). The selective dissolution of zinc plays a crucial role in crack propagation.

Zinc progressively dissolves at the crack tip because slip processes are continuously generating a fresh surface [1376].

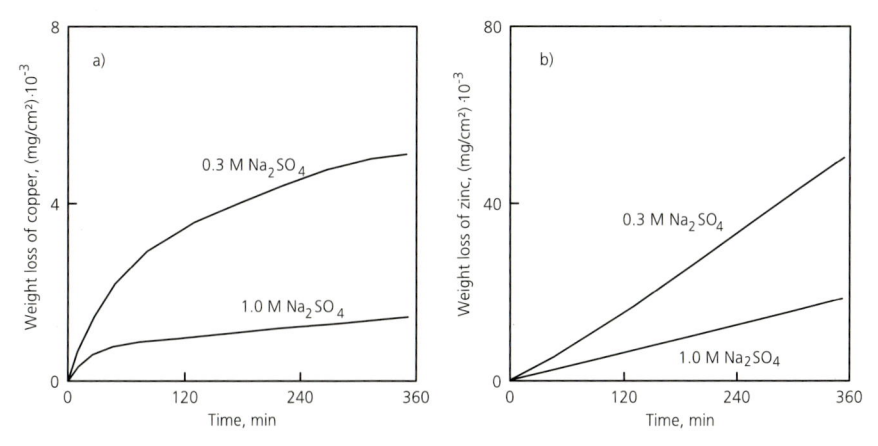

Figure 195: Dissolution of copper and zinc from brass 60 (Cu60Zn40) in sodium sulfate solutions of various concentrations [1378]
a) mean values for copper dissolution
b) mean values for zinc dissolution

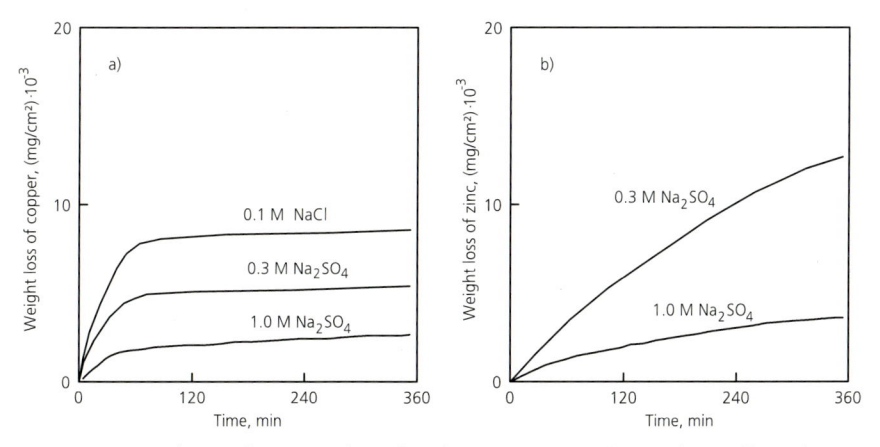

Figure 196: Dissolution of copper and zinc from brass 70 (Cu70Zn30) in sodium sulfate solutions of various concentrations [1378]
a) mean values for copper dissolution
b) mean values for zinc dissolution

The stress corrosion cracking behavior of Admiralty brass CDA 443 (71% Cu, 28% Zn, 1% Sn) in 1 mol/l sodium sulfate solutions was studied depending on the pH-value and potential at room temperature in a slow tensile test. The pH-value of the solution was adjusted to values between 1.3 and 12.7 with sodium hydroxide solution or with sulfuric acid. The specimens were heat-treated at 823 K (550 °C) for 1 h and then left to cool in the furnace. This led to a grain size of 0.05 mm. The strain rate was 1.5×10^{-5} 1/s. Figure 197 shows a plot of the tensile strength and the crack propagation velocity versus the pH-values of the sodium sulfate solution. The tensile strength of the brass in the solution is given relative to the value measured in air. The crack propagation velocity $v = L/t_f$ was calculated from the maximum crack depth, L, and the time-to-fracture, t_f. Admiralty brass CDA 443 is not susceptible to stress corrosion cracking at a corrosion potential of $E_{corr} = 180 \pm 20$ mV$_{SHE}$ between pH 1.3 and pH 10.1. No stress corrosion cracking was observed, even at potentials below 200 mV$_{SHE}$. At potentials between 100–200 mV$_{SHE}$, the tensile strength and the crack propagation velocity are independent of the pH-value. The susceptibility to stress corrosion cracking increases with the potential and passes through a maximum in the pH-range between 4 and 10. Stress corrosion cracking is also influenced by the sulfate concentration. The susceptibility to stress corrosion cracking decreases with the concentration of sulfate ions. For example, no cracking was detected at an SO_4^{2-}-concentration of 10^{-4} mol/l, pH 8, and a potential of 300 mV$_{SHE}$. Admiralty brass CDA 443 does not exhibit stress corrosion cracking at pH-values < 2 and > 12. At the same time, Cu_2O is unstable at pH-values < 2 and > 12. This fact indicates that Cu_2O influences propagation of stress corrosion cracks [1381].

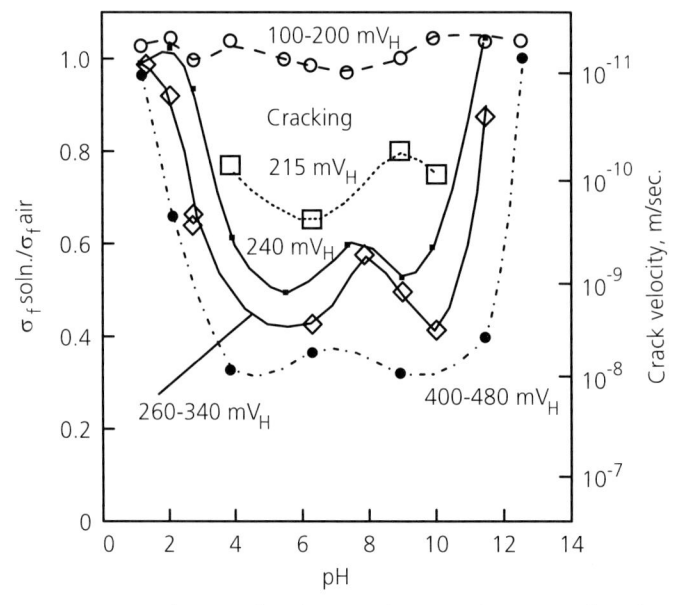

Figure 197: Tensile strength (relative to the strength in air) and crack propagation velocity of Admiralty brass CDA 443, DIN-Mat. No. 2.0470 (71 % Cu, 28 % Zn, 1 % Sn) in a 1 mol/l sodium sulfate solution depending on the pH-value at various potentials and at room temperature [1381]

Commercial brass UNS C27400 (cf. CW508L, CuZn37, old DIN-Mat. No. 2.0321) with 63.3 % Cu, 0.02 % Pb, 0.01 % Fe, remainder Zn exhibits stress corrosion cracking in a nitrogen-purged, alkaline sodium sulfate solution (0.5 mol/l Na_2SO_4 + 0.2 mol/l H_3BO_3 + 0.1 mol/l NaOH) at pH 9.2 in the potential range from 0.30 to

0.40 V_{SHE} at 293 K (20 °C). Transgranular cracks along with serious corrosive attack was observed in a tensile test with a strain rate of 1.7×10^{-5} 1/s. In this potential range (> 0.32 V_{SHE}), the brass dissolves with high current densities. At potentials < 0.29 V_{SHE}, the brass behaves passively with passive current densities of approximately 10^{-5} A/cm².

The passive range in an alkaline sodium sulfate/sodium sulfide solution (0.5 mol/l Na_2SO_4 + 0.05 mol/l Na_2S, pH 12.3) is much wider (from −0.40 to +0.80 V_{SHE}) compared to an alkaline sodium sulfate/borate solution. The current densities in the passive range of 10^{-4} A/cm² are one order of magnitude higher than in the alkaline sodium sulfate/borate solution. In alkaline sodium sulfate/sodium sulfide solutions, brass UNS C27400 is susceptible to stress corrosion cracking in the potential range from +0.40 to +0.80 V_{SHE} (tested in a slow strain test with a strain rate of 1.7×10^{-5} 1/s). These potentials lie within the passive region.

No cracks were detected at potentials < +0.40 V_{SHE}. The free corrosion potential E_{corr} in the sulfide-containing sodium sulfate solution is (-0.43 ± 0.02) V_{SHE}. Anodic polarization to over +0.80 V_{SHE} leads to a considerable increase in the current densities (approximately 10^{-2} A/cm²). This produces a black film on the surface, which was identified as $NaCu_2(SO_4)_2OH \times H_2O$. Pitting corrosion was found underneath the film [1382].

α-β-Brass (Cu-42 mass% Zn) with 76 vol.% β-phase undergoes stress corrosion cracking in 1 mol/l sodium sulfate, pH 7 ± 0.5 on application of a mechanical load. The tests were carried out with notched tensile specimens that were first loaded in air well beyond their yield point (150 MPa) to 270 MPa and then kept at this stress. The sodium sulfate solution was added after 30 min so that the tensile specimen was completely immersed. The corrosion potential was −80 to −120 mV_{SCE}. After polarization of the brass specimen to −20 mV_{SCE}, a crack appeared in the notch within 5 min and the crack propagation velocity increased with the crack length. Crack propagation was arrested by changing the potential of the specimen to −250 mV_{SCE}. After repolarizing to −20 mV_{SCE}, crack propagation restarted. At −20 mV_{SCE} and under a constant load of 270 MPa, the brass exhibited a mean crack propagation velocity of 4×10^{-6} m/s to 11×10^{-6} m/s in the sodium sulfate solution. The crack path followed either the β-β or the α-β-grain boundaries or a transgranular path through the β-phase. Transgranular fracture was only observed in the β-phase. The proportion of transgranular fracture increased with the crack length. When the brass specimen was polarized to −250 mV_{SCE} and then loaded at a constant strain rate of 3.3×10^{-2} 1/s until it fractured, it exhibited a ductile fracture surface. A significantly higher crack propagation velocity of approximately 50×10^{-6} m/s was found when the brass specimen was tested with a constant strain rate of 1.6×10^{-2} 1/s after crack initiation in a sodium sulfate solution at −20 mV_{SCE}. In this case, fracturing occurred after 15 to 25 s and was mainly transgranular. These extremely high crack propagation velocities are attributed to a fracture mechanism with chemisorption of crack-promoting species at the crack tip [1383].

– Erosion corrosion –

Brass, like other copper alloys, are common heat-exchanger materials. In practice, damage frequently occurs in the inlet region and where there is a small radius of curvature because turbulent flow starts there. Erosion is promoted by air, suspended matter, and solids carried in the medium. If the protective oxide layer (Cu_2O) is damaged in these areas and it cannot be replaced quickly enough, material is lost by simultaneous erosion and corrosion. Erosion corrosion damage is rarely observed in areas of lamellar flow, i.e. in straight tubes. The erosion corrosion behavior of a number of materials was studied: brass grades CuZn28Sn1, CuZn20Al2, a nickel-alloyed brass CuZn20Al2+Ni, copper-nickel alloys (CuNi30Mn1Fe1, DIN-Mat. No. 2.0882, CuNi10Fe1Mn, DIN-Mat. No. 2.0872), a non-standard tin bronze that contained aluminium (CuSn6ZnAl), pure titanium, and an austenitic CrNiMo steel (X5CrNiMo17-12-2, DIN-Mat. No. 1.4401). The chemical composition of the copper materials is given in Table 231. Specimens of the various alloys were cut out of commercially available heat exchanger tubes (dimension 23×1 mm) after cutting them in half (diameter 10 mm).

Material/DIN-Mat. No.	Composition, mass%						
	Cu	Sn	As	Zn	Fe	Ni	Al
CuZn28Sn1	balance	1.08	0.026	28.3	0.03	–	–
CuZn20Al2	76.6	–	0.026	balance	0.03	–	2.05
CW354H, CuNi30Mn1Fe/ old 2.0882	balance	–	–	–	0.62	30.6	–
CW352H, CuNi10Fe1Mn/ old 2.0872	balance	–	–	–	1.45	10.2	–
CuSn6ZnAl/–	balance	6.6	–	4.8	–	–	0.83

Table 231: Chemical composition of the tested copper materials [1374]

The samples kept their original curvature. They were only degrased and the oxide skin that had formed during manufacturing was left on the surface. The tests were carried out with a test rig used for studying ramjet erosion corrosion. The samples were secured at a distance of 2 mm from the nozzle opening. The angle of impingement of the liquid jet could be varied from 30° to 90°, although it was normally set to 90°. The effect of entrained air on the erosion corrosion was determined by injecting a controlled amount of air into the feed line to the jets. The effect of entrained sand was determined by adding sand (100 mg/l) with a grain diameter of 0.1–0.2 mm to the solution. The following test solutions were used:

1) 200 mg/l Na_2SO_4 solution
   (conductivity: 300×10^{-6} S/cm), corresponds to chloride-free cooling water
2) 200 mg/l Na_2SO_4 solution + 5 g/l NaCl,
   corresponds to chloride-containing cooling water or river water.

The mass losses were determined by differential weighing after a test time of 4 d. The corrosion rates were calculated taking account of the density.

The alloys were tested at 298 K and 308 K (25 and 35 °C) in a pure sodium sulfate solution containing 3 and 5 vol.% air and at a flow rate of 7.2 m/s. Mass losses were insignificant at lower flow rates. Table 232 gives the corrosion rates of the various types of brass as well as the CuNi and CuSnZn alloys in chloride-free cooling water (200 mg/l Na_2SO_4) with respect to the influence of sand, air, and a higher temperature. The erosion corrosion rate was almost the same for the various materials. The corrosion rate was approximately 0.1 mm/a (3.94 mpy) at 298 K (25 °C) and approximately 0.25 mm/a (9.84 mpy) at 308 K (35 °C). The corrosion rates increased by a factor of 2–3 when the temperature was increased from 298 K to 308 K (25 to 35 °C). The highest values were found for the higher erosive load due to added sand. The addition of 100 mg sand/l to the flowing Na_2SO_4 solution increased the corrosion rates by 20 % to 100 % at a given temperature. An additional temperature increase to 308 K (35 °C) doubled the corrosion rate yet again. In contrast to tests in river water and seawater (chloride-containing sodium sulfate solutions), the addition of 3 % air to a pure sodium sulfate solution did not increase the erosion corrosion rate, but actually improved the resistance e.g. at 308 K (35 °C) and 5 % air. This reduction in the corrosion rate by injecting air was attributed to accelerated passivation of the surface by the higher amount of available oxygen. Because the studied copper alloys are barely susceptible to erosion corrosion and exhibit almost the same corrosion rates in chloride-free cooling water (200 mg/l Na_2SO_4), the brass grades CuZn28Sn1 and CuZn20Al2 are entirely suitable under such conditions [1374].

Material (DIN-Mat. No.)	Temperature K (°C)	Corrosion rate mm/a (mpy)			
		Na_2SO_4 no additives	Na_2SO_4 + sand	Na_2SO_4 +3 % air	Na_2SO_4 +5 % air
CuZn28Sn1 (2.0470)	298 (25)	0.09 ((3.54)	0.12 (4.72)	0.06 (2.36)	0.08 (3.15)
	308 (35)	0.30 (11.8)	0.39 (15.4)	0.27 (10.6)	0.12 (4.72)
CuZn20Al2 (2.0460)	298 (25)	0.07 (2.76)	0.16 (6.30)	0.04 (1.57)	0.07 (2.76)
	308 (35)	0.23 (9.10)	0.36 (14.2)	0.29 (11.4)	0.09 (3.54)
CuNi30Mn1Fe (2.0882)	298 (25)	0.11 (4.33)	0.14 (5.51)	0.10 (3.94)	0.13 (5.12)
	308 (35)	0.26 (10.2)	0.36 (14.2)	0.28 (11.0)	0.16 (6.30)
CuNi10Fe1Mn (2.0872)	298 (25)	0.12 (4.72)	0.16 (6.30)	0.08 (3.15)	0.12 (4.72)
	308 (35)	0.19 (7.48)	0.37 (14.6)	0.27 (10.6)	0.14 (5.51)
CuZn20Al2+Ni (−)	298 (25)	0.09 (3.54)	0.17 (6.70)	0.08 (3.15)	0.10 (3.94)
	308 (35)	0.27 (10.6)	0.33 (13.0)	0.26 (10.2)	0.11 (4.33)
CuSn6ZnAl	298 (25)	0.10 (3.94)	0.20 (7.87)	0.08 (3.15)	0.14 (5.51)
	308 (35)	0.30 (11.8)	0.45 (17.7)	0.30 (11.8)	0.21 (8.26)

Table 232: Corrosion rates of various copper alloys in sodium sulfate solutions (200 mg/l) after addition of sand and air and after increasing the temperature from 298 K to 308 K (25 to 35 °C) at a constant flow rate (7.2 m/s) [1374]

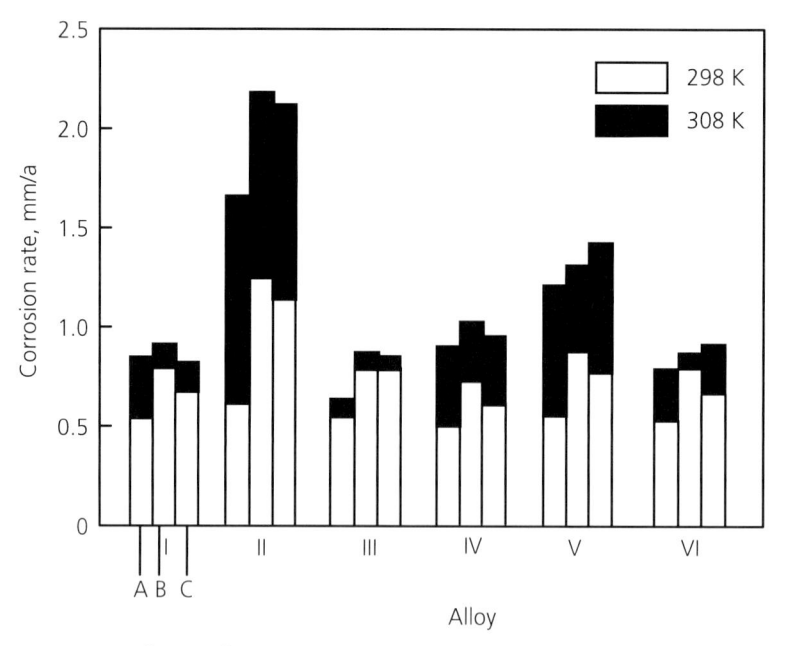

Figure 198: Influence of increasing the temperature from 298 K to 308 K (25 to 35 °C) on the corrosion rate at a flow rate of 7.2 m/s in a sodium sulfate/sodium chloride solution (200 mg/l Na_2SO_4 + 5 g/l NaCl) (A), with added sand (B) and with 5 % air (C) [1374].
I: CuZn28Sn1 (DIN-Mat. No. 2.0470)
II: CuZn20Al2 (DIN-Mat. No. 2.0460)
III: CuNi30Mn1Fe1 (DIN-Mat. No. 2.0882)
IV: CuNi10Fe1Mn (DIN-Mat. No. 2.0872)
V: CuZn20Al2+Ni (–)
VI: CuSn6ZnAl (–)

The various copper alloys and the CrNiMo steel were also tested at temperatures of 298 K (25 °C) and 308 K (35 °C), with and without added sand as well as air injection (3 %, 5 % as well as 1 %) in artificial river water and salty cooling water (200 mg/l Na_2SO_4 + 5 g/l NaCl) using the ramjet erosion corrosion test rig. These tests were carried out with flow rates of 0.9, 1.8, 3.6, and 7.2 m/s. Addition of NaCl to the Na_2SO_4 solution increased the corrosion rates by a factor of 2 to 6. The corrosion rates for the investigated alloys under the various test conditions (temperature, addition of sand or air, flow rate) are given in Table 233. Brass CuZn20Al2 and nickel-alloyed brass CuZn20Al2+Ni exhibited a particularly high susceptibility to erosion corrosion. The copper materials showed a twofold increase, on average, of the corrosion rate on increasing the flow rate from 0.9 to 7.2 m/s at a temperature of 298 K (25 °C). The corrosion rate increased by a factor of 2 to 3 at a temperature of 308 K (35 °C). At a constant flow rate of 7.2 m/s and a temperature of 298 K (25 °C), the corrosion rates in the chloride-containing sodium sulfate solution differed only slightly. In contrast to the chloride-free solution, a temperature increase to 308 K (35 °C) led to lower enhancement factors for the erosion corrosion rate (Figure 198).

CrNiMo steel or pure titanium are suitable for heat exchanger tubes that are subjected to severe erosion corrosion conditions. Pure titanium (DIN-Mat. No. 3.7035) and X5CrNiMo17-12-2 (DIN-Mat. No. 1.4401) did not exhibit any measurable or noticeable mass losses under the test conditions [1374]. The influence of the impingement angle on erosion corrosion was investigated in a chloride-containing sodium sulfate solution at 298 K (25 °C). A decrease in this angle from 90° to 60°, 45° and 30° reduced the corrosion rate under all the conditions described here. The decrease in the corrosion rate on changing the impingement angle from 90° to 30° was almost independent of the materials and amounted to 20–25 % for flow rates of 0.9 and 1.8 m/s and 30–50 % for 3.6 and 7.2 m/s. The corrosion rate of the CuNi10Fe1Mn alloy was least affected by the impingement angle [1374].

Medium	Material	Flow rate, m/s							
		0.9		1.8		3.6		7.2	
		at temperature, K (°C)							
		298 (25)	308 (35)	298 (25)	308 (35)	298 (25)	308 (35)	298 (25)	308 (35)
		Corrosion rate, mm/a (mpy)							
Na_2SO_4 + NaCl	CuZn28Sn1	0.26 (10.2)	0.29 (11.4)	0.32 (12.6)	0.55 (21.6)	0.50 (19.7)	0.66 (26.0)	0.55 (21.6)	0.86 (33.9)
	CuZn20Al2	0.28 (11.0)	0.28 (11.0)	0.31 (12.2)	0.34 (13.4)	0.52 (20.5)	0.51 (20.1)	0.61 (24.0)	1.70 (66.9)
	CuNi30Mn1Fe	0.27 (10.6)	0.27 (10.6)	0.31 (12.2)	0.30 (11.8)	0.47 (18.5)	0.39 (15.4)	0.62 (24.4)	0.56 (22.0)
	CuNi10Fe1Mn	0.28 (11.0)	0.35 (13.8)	0.30 (11.8)	0.46 (18.1)	0.43 (16.9)	0.55 (21.7)	0.51 (20.1)	0.85 (33.5)
	CuZn20Al2+Ni	0.30 (11.8)	0.28 (11.0)	0.30 (11.8)	0.30 (11.8)	0.43 (16.9)	0.41 (16.1)	0.54 (21.3)	0.22 (8.66)
	CuSn6ZnAl	0.30 (11.8)	0.34 (13.4)	0.34 (13.4)	0.48 (18.9)	0.44 (17.3)	0.55 (21.7)	0.52 (20.5)	0.80 (31.5)

Table 233: Corrosion rates of various copper alloys in sodium sulfate/sodium chloride solutions after adding sand and air, depending on the flow rate and the temperature [1374]

Table 233: Continued

Medium	Material	Flow rate, m/s							
		0.9		1.8		3.6		7.2	
		at temperature, K (°C)							
		298 (25)	308 (35)	298 (25)	308 (35)	298 (25)	308 (35)	298 (25)	308 (35)
		Corrosion rate, mm/a (mpy)							
Na$_2$SO$_4$ + NaCl with sand	CuZn28Sn1	0.29 (11.4)	0.48 (18.9)	0.41 (16.1)	0.65 (25.6)	0.62 (24.4)	0.83 (32.7)	0.85 (33.5)	0.88 (34.6)
	CuZn20Al2	0.30 (11.8)	0.40 (15.7)	0.47 (18.5)	0.63 (24.8)	0.81 (31.9)	1.05 (41.3)	1.37 (53.9)	2.13 (83.9)
	CuNi30Mn1Fe	0.31 (12.2)	0.44 (17.3)	0.35 (13.8)	0.54 (21.3)	0.54 (21.3)	0.62 (24.4)	0.80 (31.5)	0.81 (31.9)
	CuNi10Fe1Mn	0.32 (12.6)	0.47 (18.5)	0.42 (16.5)	0.54 (21.3)	0.55 (21.7)	0.73 (28.7)	0.69 (27.2)	0.99 (38.9)
	CuZn20Al2+Ni	0.34 (13.4)	0.45 (17.7)	0.44 (17.3)	0.62 (24.4)	0.56 (22.0)	0.72 (28.3)	0.89 (35.0)	1.29 (50.8)
	CuSn6ZnAl	0.36 (14.2)	0.62 (24.4)	0.48 (18.9)	0.58 (22.8)	0.73 (28.7)	0.73 (28.7)	0.84 (33.1)	0.88 (34.6)
Na$_2$SO$_4$ + NaCl with 1% air	CuZn28Sn1	0.23 (9.1)	0.42 (16.5)	0.33 (13.0)	0.61 (24.0)	0.40 (15.7)	0.72 (28.3)	0.48 (18.9)	0.78 (30.7)
	CuZn20Al2	0.19 (7.48)	0.35 (13.8)	0.27 (10.6)	0.67 (26.4)	0.43 (16.9)	1.03 (40.6)	0.50 (19.7)	1.97 (77.6)
	CuNi30Mn1Fe	0.27 (10.6)	0.32 (12.6)	0.30 (11.8)	0.40 (15.7)	0.47 (18.5)	0.51 (20.1)	0.55 (21.7)	0.62 (24.4)
	CuNi10Fe1Mn	0.28 (11.0)	0.47 (18.5)	0.37 (14.6)	0.53 (20.9)	0.44 (17.3)	0.62 (24.4)	0.51 (20.1)	0.90 (35.4)
	CuZn20Al2+Ni	0.28 (11.0)	0.39 (15.4)	0.32 (12.6)	0.54 (21.3)	0.41 (16.1)	0.59 (23.2)	0.47 (18.5)	1.35 (53.1)
	CuSn6ZnAl	0.29 (11.4)	0.50 (19.7)	0.39 (15.4)	0.59 (23.2)	0.41 (16.4)	0.65 (25.6)	0.47 (18.5)	0.87 (34.3)

Table 233: Corrosion rates of various copper alloys in sodium sulfate/sodium chloride solutions after adding sand and air, depending on the flow rate and the temperature [1374]

Table 233: Continued

Medium	Material	Flow rate, m/s							
		0.9		1.8		3.6		7.2	
		at temperature, K (°C)							
		298 (25)	308 (35)	298 (25)	308 (35)	298 (25)	308 (35)	298 (25)	308 (35)
		Corrosion rate, mm/a (mpy)							
Na$_2$SO$_4$ + NaCl with 3% air	CuZn28Sn1	0.26 (10.2)		0.39 (15.4)		0.44 (17.3)		0.74 (29.1)	
	CuZn20Al2	0.27 (10.6)		0.42 (16.5)		0.56 (22.0)		0.64 (25.2)	
	CuNi30Mn1Fe	0.30 (11.8)		0.33 (13.0)		0.50 (19.7)		0.54 (21.3)	
	CuNi10Fe1Mn	0.31 (12.2)		0.39 (15.4)		0.53 (20.9)		0.56 (22.0)	
	CuZn20Al2+Ni	0.33 (13.0)		0.43 (16.9)		0.54 (21.3)		0.81 (31.9)	
	CuSn6ZnAl	0.33 (13.0)		0.43 (16.9)		0.54 (21.3)		0.81 (31.9)	
Na$_2$SO$_4$ + NaCl with 5% air	CuZn28Sn1	0.29 (11.4)	0.48 (18.9)	0.41 (16.1)	0.66 (26.0)	0.45 (17.7)	0.80 (31.5)	0.80 (31.5)	0.83 (32.7)
	CuZn20Al2	0.28 (11.0)	0.40 (15.7)	0.42 (16.5)	0.71 (28.0)	0.56 (22.0)	1.07 (42.1)	1.18 (46.5)	2.08 (81.9)
	CuNi30Mn1Fe	0.32 (12.6)	0.42 (16.5)	0.37 (14.6)	0.47 (18.5)	0.51 (20.1)	0.55 (21.7)	0.73 (28.7)	0.72 (28.3)
	CuNi10Fe1Mn	0.33 (13.0)	0.51 (20.1)	0.41 (16.1)	0.57 (22.4)	0.53 (20.9)	0.69 (27.2)	0.59 (23.2)	0.96 (37.8)
	CuZn20Al2+Ni	0.32 (12.6)	0.44 (17.3)	0.44 (17.3)	0.55 (21.7)	0.55 (21.7)	0.76 (30.0)	0.89 (35.0)	1.31 (51.6)
	CuSn6ZnAl	0.33 (13.0)	0.52 (20.5)	0.48 (18.9)	0.56 (22.0)	0.48 (18.9)	0.73 (28.7)	0.61 (24.0)	0.88 (34.6)

Table 233: Corrosion rates of various copper alloys in sodium sulfate/sodium chloride solutions after adding sand and air, depending on the flow rate and the temperature [1374]

Other copper alloys

– Corrosion in aqueous solutions

The corrosion rates for casting alloys

> CC761S, CuZn16Si4-C (old G-CuZn15Si4, DIN-Mat. No. 2.0492),
> CC495K, CuSn10Pb10-C (old G-CuPb10Sn, DIN-Mat. No. 2.1176),
> CC496K, CuSn7Pb15-C (old G-CuPb15Sn, DIN-Mat. No. 2.1182),
> CC497K, CuSn5Pb20-C (old G-CuPb20Sn, DIN-Mat. No. 2.1188),
> G-CuPb22Sn, DIN-Mat. No. 2.1166 and
> G-CuPb5Sn, DIN-Mat. No. 2.1170

at temperatures from 293 to 373 K (20 to 100 °C) are reported as being <0.5 mm/a (19.7 mpy) [1370].

They thus have quite good resistance; however, this worsens as the temperature increases and it should therefore be checked.

Copper-gold alloys undergo selective corrosion in acidic sodium sulfate solutions. Copper is the less noble alloying component and it thus dissolves preferentially. The anodic current density/potential curves of Cu68Au32 (Cu-13 atom% Au, corresponding to 31.8 mass% Au) and Cu59Au41 (Cu-18 atom% Au, corresponding to 40.7 mass% Au) were measured at 296 K (23 °C) in 1 mol/l Na_2SO_4 + 0.005 mol/l H_2SO_4 and in 1 mol/l Na_2SO_4 + 0.005 mol/l H_2SO_4 + 0.1 mol/l $CuSO_4$ (Figure 199).

Most of the oxygen in the electrolyte was driven out by purging with helium. The measured points shown in Figure 199 were obtained from electrochemical measurements on the one hand and from an analysis of the amount of dissolved copper on the other. Because these values are in good agreement, the anodic currents can be attributed to the dissolution of copper and the oxygen reduction current is negligible. The partial current density/potential curves show a region in which the current density is independent of the potential. The current density is very low in this region, namely approximately 0.1–1 $\mu A/cm^2$, depending on the alloy composition, and corresponds to the limiting current density of copper dissolution. The resistance range is important for its use e.g. as a dental alloy or as a material for electrical contacts. Increasing the potential even further leads to a high rate of copper dissolution above a critical potential, which is characterized by a very steep curve gradient. As the proportion of gold in the alloy increases, the critical potential is shifted to more noble values, i.e. towards the equilibrium potential of the more noble component. The current density/potential curve of the Cu68Au32 alloy in an acidic sodium sulfate solution containing copper sulfate corresponds to the acidic dissolution behavior of the solution without copper sulfate. However, this does not apply to the higher current densities in the range independent of the potential. The selective dissolution of copper not only leads to Cu depletion in the near-surface zone of the material, but it also causes surface roughening [1384, 1385].

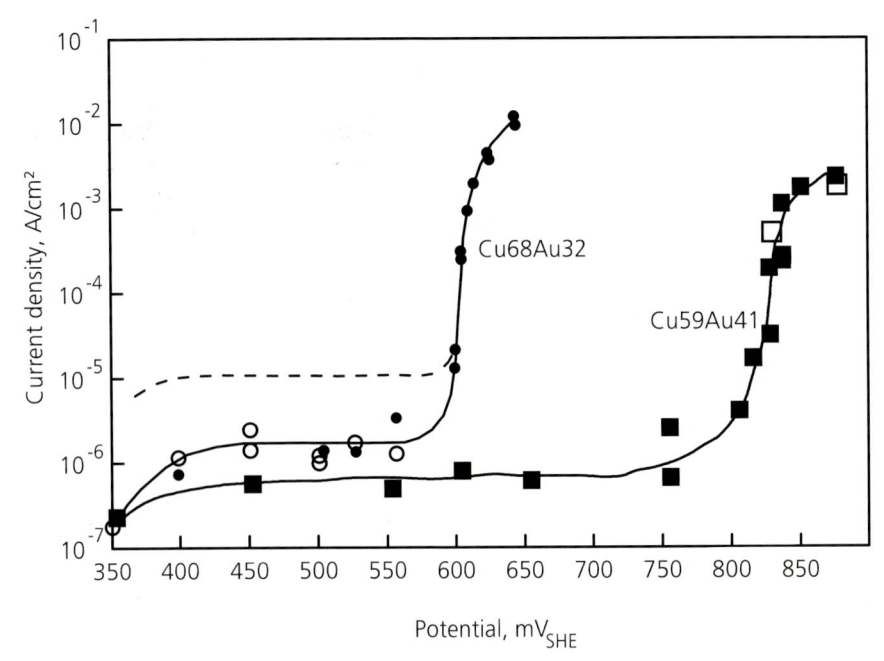

Figure 199: Quasi-stationary partial current density (log i_{Cu})/potential curves for the anodic dissolution of copper from alloys Cu68Au32 (Cu-13 atom% Au, corresponding to 31.8 mass% Au) and Cu59Au41 (Cu-18 atom% Au, corresponding to 40.7 mass% Au) in 1 mol/l Na_2SO_4 + 0.005 mol/l H_2SO_4 at 296 K (23 °C). Dashed line: dissolution current density of copper in 1 mol/l Na_2SO_4 + 0.1 mol/l $CuSO_4$. Open symbols: values from the chemical analysis, filled symbols: measured current densities [1385]

The copper-gold alloy with Cu49Au51 (Cu-25 atom% Au, corresponding to 51 mass% Au) exhibits transgranular stress corrosion cracking in an acidic sodium sulfate solution (1 mol/l Na_2SO_4 + 0.005 mol/l H_2SO_4) in a slow tensile test (strain rate: 10^{-6} 1/s) under anodic polarization of +600 mV (relative to the free corrosion potential) [1386].

Electroplated coatings of amorphous copper-bismuth with 30–35 mass% copper were more resistant to 0.1 mol/l Na_2SO_4 than the same coatings after a heat treatment at 473 K (200 °C), which crystallizes them. The amorphous layers tended to passivate more quickly [1387].

Soil (Underground Corrosion)

Author: G. Elsner / Editor: R. Bender

Copper

Copper and its alloys are among the materials most resistant to the various types of soil (Table 234). For instance, Cu 99.9 or Cu 99.94 exhibit good resistance in various soils, and their susceptibility to pitting corrosion is also low (with the exception of Muntz metal, CuZn40 (cf. CW509L), Table 235). Rather severe corrosion and pitting is to be expected only in soils with high salt contents: sulfides, organic sulfur compounds, ammonia or ammonium salts or with high acidity and poor aeration [1388–1391].

Although copper pipes are no longer laid at all today in the ground, the following must be taken into consideration with underground copper installations: caution is advised in soils containing ash, coal and slag, acidic marsh, peat or loam soils, soil containing sulfides (e.g. when the soil is contaminated by feces) and when there is a possibility of stray currents. In long-term tests (10 years) in such soils corrosion rates of up to 40 µm/a (1.57 mpy), locally as much as 150 µm/a (5.91 mpy), were measured with copper containing oxygen (with and without 0.4 % As). Experience gained with the system formed by copper and water indicate that differences in composition in comparison with grades of copper usual today (E-Cu58 (CW004A) and oxygen-free SF-Cu(CW024A)) do not play an important part [1392–1396].

Copper becomes discolored in the presence of sulfur compounds, whereas in acid-free water in the presence of oxygen a protective black layer of copper oxide is produced which is absent in acid moor water and some groundwaters [1395, 1397].

Metal	Sandy loam	Clay	Marshy soil	Alkaline soil	Sandy loam	Clay	Marshy soil	Alkaline soil
	Material consumption rate $g/m^2\,d$				Deepest pitting mm/a (mpy)			
Aluminium 99.25 Al	0.008	0.029	0.013	0.039	0.053 (2.1)	D	L	D
Duralumin® 94.23 Al; 4.10 Cu	–	0.120	0.012	D	–	D	G	D
AlMn-alloy 97.75 Al; 1.12 Mn	0.032	0.017	0.018	0.026	D	0.036 (1.4)	0.033 (1.3)	0.048 (1.9)

L = local pitting, pits no deeper than 0.15 mm
G = slight corrosion, no pitting
E = dezincification over large areas
D = perforations

Table 234: Corrosion of various metals in different soils [1389]

Table 234: Continued

Metal	Sandy loam	Clay	Marshy soil	Alkaline soil	Sandy loam	Clay	Marshy soil	Alkaline soil
	Material consumption rate g/m² d				Deepest pitting mm/a (mpy)			
Open-hearth steel	0.820	0.470	D	D	D	0.178 (7)	D	D
Cu-steel 0.2 Cu	D	0.454	D	D	D	0.15 (5.9)	D	D
Copper	0.012	0.037	–	0.012	G	L	–	L
Brass Ms60	0.065	0.043	0.008	0.008	G, E	G, E	G, E	G, E
Bronze Sn-Bz 10	0.012	0.038	0.380	0.041	0.028 (1.1)	G	0.028 (1.1)	0.079 (3.1)
Sheet zinc 99.5 Zn	0.290	0.074	0.192	D	0.102 (4)	0.043 (1.7)	0.084 (3.3)	D
Sheet zinc, standard	0.159	0.092	0.375	D	0.132 (5.2)	0.051 (2)	0.094 (3.7)	D
Refined lead 99.99	0.011	0.057	0.052	0.017	0.028 (1.1)	L	0.023 (0.9)	0.028 (1.1)
Lead 99.93 Pb; 0.057 Cu	0.031	0.150	0.064	0.074	0.043 (1.69)	0.064 (2.52)	0.031 (1.22)	0.046 (1.81)

L = local pitting, pits no deeper than 0.15 mm
G = slight corrosion, no pitting
E = dezincification over large areas
D = perforations

Table 234: Corrosion of various metals in different soils [1389]

Soil (Underground Corrosion)

Type of soil	Copper 99.9 Weight loss mg/cm²a	Copper 99.9 Surface appearance*	Copper 99.94 Weight loss mg/cm²a	Copper 99.94 Surface appearance*	Tombac 85.18 Cu/14.80 Zn Weight loss mg/cm²a	Tombac Surface appearance*	Admiralty Brass 71.3 Cu/27.4 Zn/1.3 Sn Weight loss mg/cm²a	Admiralty Surface appearance*	Brass Ms67 with 0.8 Pb Weight loss mg/cm²a	Brass Ms67 Surface appearance*	Brass Ms66 Weight loss mg/cm²a	Brass Ms66 Surface appearance*	Brass Muntz Metal Ms60 Weight loss mg/cm²a	Brass Muntz Surface appearance*	97 Cu, 1 Si, 1.8 Sn Weight loss mg/cm²a	97 Cu, 1 Si, 1.8 Sn Surface appearance*	98 Cu 1.5 Si, 0.2 Mn, 0.1 Fe Weight loss mg/cm²a	98 Cu 1.5 Si Surface appearance*	95.5 Cu, 3.2 Si, 1 Mn, 0.2 Fe Weight loss mg/cm²a	95.5 Cu Surface appearance*	75 Cu, 20 Ni, 5 Zn Weight loss mg/cm²a	75 Cu, 20 Ni Surface appearance*
Acadia clay	6.405	M	6.10	M	5.19	M	4.88	M, d	2.59	M, D	5.19	M, D	10.65	M, D	7.94	P	5.50	M	6.61	M	4.27	d, M
Cecil clay loam	1.89	M	1.86	P	1.83	0.10	2.12	0.15d	3.96	M, d	1.58	0.13d	2.96	P, d	3.66	0.17	2.96	P	2.74	P	1.8	P, d
Hagerstown loam	2.19	P	2.29	P	2.33	0.16	2.41	0.29	3.55	M, d	2.86	0.15	–	0.13, d	4.88	0.19	2.32	P	2.44	0.15	1.86	0.11, d
Lake Charles clay	1.8	M	1.55	M	1.65	M	1.22	M, d	2.44	M, d	1.37	M, d	2.19	d	3.66	0.17	2.29	P	1.52	M	1.02	M, d
Merced building clay	0.11	M	0.189	U	0.165	0.10	0.186	P, d	1.25	M	1.25	M, d	7.33	M, d	9.46	0.31	0.366	M	0.366	d	0.31	d
Muck	2.05	P	2.5	M	1.86	M	2.93	P, d	3.96	M, d	3.35	M, d	3.05	P, d	3.05	0.15	1.65	M	1.71	M	2.26	0.18
Peat	21.6	P	23.2	0.13 S	23.2	0.2	18	0.13	18	P, d	27.2	0.24	27.4	0.2 d	25.6	0.36	20.7	0.17	21.05	P	21.9	0.21
Sharkey clay	3.5	P	1.89	P	3.66	P	5.49	0.21d	3.66	P, d	3.66	0.27d	5.19	P, d	6.10	P	4.27	P	4.27	M	3.96	P, d
Susquehanna clay	2.44	P	2.41	0.10	2.35	0.13	3.35	0.22	4.57	P, d	3.66	P, d	5.19	P, d	5.19	0.25	3.35	P	3.96	0.2	2.79	P, d
Marsh soil (tidal marsh)	19.5	0.11	16.8	M	8.24	P	3.66	M	7.05	M	6.4	P	2.10	M	17.4	0.34	24.4	P	18.6	P	9.75	P
Alkali soil	10.7	0.15	22.6	S	5.19	0.30	3.96	0.33 d	3.65	M	9.15	P, d	64.4	P, D	12.5	0.28	21.9	1.15	10.7	0.33	1.25	P
Alkali soil	7.94	P	27.4	S	8.24	0.13	4.88	0.47	2.2	M, d	21.9	M, d	38.8	M, D	7.64	0.30	17.7	0.11	17.7	P	2.72	P
Mohave sandy loam	1.67	P	4.55	0.16	2.81	P	3.66	M, d	3.96	M, d	5.49	M, d	10.5	P, d	6.10	0.16	3.05	0.24	3.35	P	3.35	M, d
Cinder	47.6	0.44	60	0.55	50.6	0.38	83.9	0.65	119	0.74	250	Z	destroyed	Z	49.15	0.54	83.70	0.59	63.8	0.44	31.42	0.31

M = metal slightly corroded, surface roughened, no pitting
P = pitting corrosion, but not deeper than 0.15 mm
U = no visible corrosion
S = severe surface corrosion. Depth of penetration not measurable, since uniform consumption
D = selective corrosion over large areas
d = selective corrosion in small spots
Z = destroyed by dezincification
* numbers give deepest pitting corrosion in mm per year.

Table 235: Weight loss and type of corrosion of copper and copper alloys in various soils after 2 years [1398]

Despite the wide pH-range of groundwaters, copper laid in the ground is hardly at risk; only if the ground contains particles of carbon, stray currents occur or unsuitable fluxes are used during soldering could a certain hazard arise, but under these conditions protection with bitumenized tapes is possible [1392, 1395–1397, 1399–1401].

Because of its well-known good biocidal properties, copper also exhibits good resistance to microorganisms occurring in the soil [1401].

Corrosion is very much dependent on the gas conditions: in pure oxygen it is more severe than with normal aeration, and with pure nitrogen it is virtually zero [1402].

Copper is virtually not attacked by humic acids; the organic coating found on copper water pipes may therefore consist entirely or partially of humic acids [1403].

Tough-pitch copper, oxygen-free copper, copper containing up to 3% silicon, tin bronze, red brass, and CuZn-alloys containing 15% zinc essentially behave alike. Corrosion is most severe in soils containing ash, rather large quantities of chlorides and sulfides or exhibiting low pH-values [1404, 1405].

Corrosion can be reduced by tin coatings, the effect increasing with the thickness of the coating [1392].

Since copper is cathodic with respect to steel, copper pipes connected, for example, to underground heating oil tanks or piping of steel or cast iron must be well insulated. If pipes of copper-plated steel are involved, the coatings must be perfect, i.e. free from pores, since the steel may otherwise become perforated within a short time [1394–1411]. For these reasons copper should also not be used for grounding underground installations of steel [1412].

When an electrically conducting connection exists between copper and Cr- and CrNi-stainless steels, corrosion conditions are not very clear, since each of the two components may act as anode or cathode. As expected, corrosion of copper is at its most severe in poorly aerated soils containing chlorides, e.g. in the tidal zone [1413].

Grounding electrodes of galvanized steel in local networks are severely corroded if at least one grounding electrode (preferably on transformers) consists of copper, since a cell is then formed (this has been simulated by investigations with model cells in samples of the local soil (Figure 200).

It is therefore recommended that grounding electrodes of one single material, preferably steel, be used for such networks [1412]; occasionally, copper grounding electrodes only are also used [1389].

Since the potential of the reinforcement is more positive than that of ordinary steel, the grounding electrodes for reinforced concrete structures may consist of lead-coated copper [1389].

As grounding electrodes for pipelines laid above ground on supports, insulated conductors of copper are welded on about every 800 meters and connected to grounding electrodes of steel or copper at least 2.5 m deep (or with magnesium or zinc anodes in the soil) [1389].

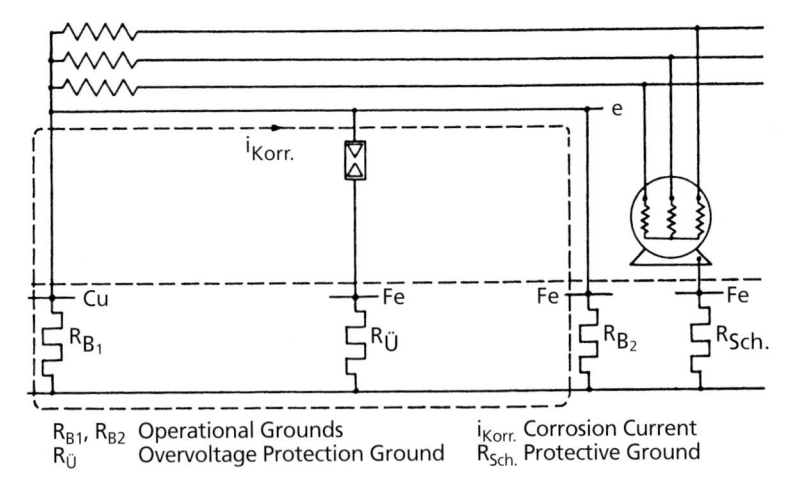

Figure 200: Corrosion current at neutral conductors in power networks with different grounding electrode materials [1412]

R_{B1}, R_{B2} Operational Grounds
$R_{\ddot{U}}$ Overvoltage Protection Ground
$i_{Korr.}$ Corrosion Current
$R_{Sch.}$ Protective Ground

Cables with mineral insulation in which the copper conductor is embedded are sheathed with copper or CuNi-alloy, thus providing both protection against corrosion (moisture, oil, salts) and grounding [1414].

In the case of a system of reinforced concrete structures, cooling water lines and copper grounding electrodes, critical points were cathodically protected, e.g. by zinc anodes in the case of copper-plated iron rods [1415].

Caution is always appropriate with mixed installations: for instance, domestic connections (copper pipes) must not be connected conductively with mains pipes of steel or cast iron, since severe local corrosion may otherwise occur at flaws in the coating of the main pipe [1407, 1410]. This is also true of copper coated lines, whose copper coating must be free from flaws.

In the case of the stainless steels the effect of linking with copper is not critical in practice; only in soils containing chlorides must pitting corrosion of the steel be expected [1413].

Since an electrically conducting connection with steel may lead to corrosion when copper is used as grounding electrode, it is advisable to use grounding electrodes of steel as well [1389].

Tests with soil extracts show that copper-coated cable sheathing of stainless steel exhibits good resistance to soil corrosion; this is true for a wide range of soil resistances (up to 29 000 Ω cm). If the coating is destroyed locally, corrosion of the stainless steel comes to a stop, whereas solid copper cable sheaths would be perforated under these conditions. The steel sheath also remains undamaged under these conditions in acid soils [1416]).

Copper-nickel alloys

CuNi-alloys exhibit good resistance in soils; only if they also contain zinc (e.g. CuNi20Zn5) is corrosion to be expected, and this increases with the zinc content [1391]. In the presence of hydrogen sulfide the copper sulfide produced (which is cathodic with respect to copper) may easily lead to pitting corrosion; hydrogen sulfide concentrations of only 5 ppm are sufficient to cause this [1389].

Copper-tin alloys (bronzes)

As numerous archaeological findings show, tin bronzes exhibit long-term resistance to corrosion in most soils. Corrosion is primarily determined by the carbonic acid content and aeration of the soil. If aeration is poor, a protective patina is also produced in inherently corrosive soils containing carbonic acid, whereas when a plentiful supply of oxygen is present the metal is severely corroded [1389].

Copper-zinc alloys (brass)

The corrosion rate of CuZn-alloys containing more than 27% zinc increases with increasing zinc content and is associated with dezincification. Such alloys should therefore not be laid in the soil [1389, 1404, 1405]. Electrically conducting connections between copper and brass should also be avoided [1410].

In alkaline soils corrosion decreases with increasing zinc content, since at pH 8 to 9.5 in particular there exists a passive range [1389]. It should be noted that Proteus bacteria, for example, may cause stress corrosion cracking of brass because of their ability to form ammonia.

Steam

Author: H. Leyerzapf / Editor: R. Bender

Copper

When copper or copper-containing materials are used in water/steam cycles of high pressure steam power plants, the operating conditions should be adjusted in such a way as to minimize the solubility of copper or copper oxides. Before measures were taken in a power station to reduce copper solubility (fully demineralized feed water; conductivity: less than 0.15 µS/cm; pH range: 6.5 to 7.5; redox potential: +400 mV), relatively large quantities of copper (17 to 35 µg Cu/kg) were found in the steam condensate of the low pressure preheaters, so that copper deposition occurred in the evaporator and on the guide blades and rotor blades of the turbines. Copper-containing layers (predominantly cuprous oxide) up to 1.6 mm in thickness were formed on the turbine blades within three months, causing loss of absorption capacity of the turbines of about 3 % [1417].

Copper-containing deposits can be removed by means of ammoniacal ammonium bicarbonate solutions; turbines need not be opened for this operation [1418].

In comparison with copper alloys such as brass and copper-nickel, copper is less important as a material for steam lines, condenser pipes, gaskets and valves in steam generation plants. The temperature limits for application in dry or wet steam are 533 to 563 K (260 to 290 °C), for arsenic-containing copper about 473 K (200 °C) [1419, 1420]; above these temperatures a noticeable drop in strength is experienced. Oxygen-free copper grades are resistant to pure steam [1421, 1422]. Oxygen-containing copper grades however are sensitive to the action of hydrogen. In these copper grades oxygen is present as cuprous oxide, which reacts with inwardly diffusing hydrogen, forming water, with subsequent embrittlement (so-called "hydrogen disease") [1423].

Stopcocks and steam valves of copper are subject to erosion in the case of steam shocks [1419]. For the same reason impact plates for trapping water drops are recommended when copper condensers are exposed to wet steam [1424].

The ammonia content of steam is of crucial importance when copper and copper alloys are used in steam condensers; copper and copper alloys can be attacked rapidly when the condensate contains more than about 10 mg/kg ammonia in the presence of oxygen. Condensation areas in dead spaces are particularly at risk [1425].

Inhibitors are recommended for protecting copper against corrosion by aggressive impurities in steam. The preferred agents used for reducing the oxygen content are sodium sulfite and hydrazine (hydrazine hydrate). When sodium sulfite is used in high pressure vessels, limitation of the sulfite content to 20 ppm at pressures exceeding 60 bar is recommended, since otherwise sulfurous acid may appear in the steam [1426]. In a Sulzer steam generator with turbine coupling (operating pressure: 265 bar at 873 K (600 °C)), initial malfunctioning, due to deposition of copper oxide in the high pressure turbines, was eliminated by reducing the hydrazine addition from 80 ppb to 35 ppb, addition of hydrazine at the entry to the boiler being reduced restricted to 10 to 15 ppb. As a result no corrosion of the pipes and no deposits in the turbine zone occurred during an operating period of 5 years [1427]. According to another source the permissible hydrazine excess in the vessel is higher (0.2 ppm) an ammonia content of 0.3 to 0.5 ppm being attained in the steam. This ammonia content is the maximum value, beyond which nonferrous metals would be attacked [1428]. Preference is given to injecting the hydrazine into the turbine steam, at 423 to 573 K (150 to 300 °C), as opposed to addition to the feed water preheater. Following hydrazine injection into the turbine steam, the copper content in the condensate was reduced to a fifteenth [1429].

To bind oxygen in steam systems, an activated hydrazine, "Levoxin®", is recommended which, in contrast to the customary hydrazine hydrate, reacts with oxygen even at room temperature [1430].

While the risk of corrosion by oxygen is reduced by addition of hydrazine, it cannot be prevented completely even by up to 300 mg/kg hydrazine. Complete removal of oxygen by chemical reaction is achieved only at hydrazine concentrations exceeding 300 mg/kg (in addition, pH 10 must be maintained). This excludes the use of structural elements of copper and copper alloys. A further problem is the decomposition of excess hydrazine resulting in the formation of ammonia, nitrogen and hydrogen, which may lead to gas accumulation in steam generation plants [1431].

In plants using the so-called "neutral water treatment" copper solubility and subsequent copper deposition is reduced by addition of oxygen, which, however, is restricted to this type of operation [1417].

To prevent corrosion by carbonic acid or CO_2, oxygen or sulfurous acid in return lines for steam condensate, a number of amine-based inhibitors has been developed; among which non-wettable, film-forming primary amines, such as, for example, octadecylamine (or its acetate or phosphate) have been mentioned as particularly effective [1432–1434]. The effectiveness of octadecylamine-based inhibitors is established up to steam pressures of 154 bar [1435].

Other inhibitors mentioned are:

1. Aliphatic long-chain amines, having, for example, 12 to 24 carbon atoms, e.g. 9-aminomethylstearylamine [1436].
2. Aromatic triazoles, for example benzotriazole and substituted 1,3,4-triazoles [1437].
3. Amines volatilized by steam, for example cyclohexylamine and morpholine (tetrahydro-1,4-oxazine), whose action is ascribed to the binding of CO_2 and the alkalization of the steam [1438].

The views as to the suitability of various inhibitors differ. Reference is made e.g. to the decomposition of amine bases, even in the medium pressure range; this phenomenon may even increase the corrosion risk to copper piping e.g. by ammonia and CO_2, formed by the decomposition of morpholine and cyclohexylamine [1439, 1440]. Octadecylamine, which is also approved for sterilization steam, is given a better rating [1441].

Copper-aluminium alloys

Aluminium bronzes and multi-component aluminium bronzes (cast and wrought alloys) are employed in steam generation plants for superheated steam armatures, condenser and preheater pipes, and in sea water evaporators [1442]. For reasons of strength, the maximum approved temperature for copper in contact with steam is 623 to 673 K (350 to 400 °C) [1443]. When exposed to either pure or contaminated high pressure steam, some alloys are susceptible to stress corrosion cracking [1444]. When α-aluminium bronzes containing 5 to 8% aluminium were tested in steam at 450 K (177 °C), fractures of varying morphology occurred under mechanical loading up to 80% of the yield point. Even when the alloys on stress relieved, the risk of stress corrosion cracking in ammonia-containing steam remains [1445]. Pure, air-containing steam (air content 1%), did not cause selective aluminium dissolution at temperatures between 673 and 873 K (400 and 600 °C). Sulfur dioxide (0.1%), however, promotes selective dissolution of aluminium, both from single-phase and two-phase alloys, at 673 K (400 °C) [1446].

The frequent occurrence of stress cracks in single-phase aluminium bronzes in steam is explained by aluminium precipitation at the grain boundaries during the solidification of the melt. Steam attacks these points, the resulting oxide film being destroyed, when mechanical stresses occur [1447]. According to studies by Ampco Metal Inc. (Wisconsin, USA), aluminium segregation and, consequently, the susceptibility to cracking in steam is eliminated, at least to a large extent, by adding about 0.35% of tin or silver as alloying components. An alloy containing (%) 5 to 8 Al, 1.6 to 2.9 Fe, about 0.2 to 0.35 Ag, balance copper, (Ampco grade 8 Alloy) was therefore approved for temperatures up to 533 K (260 °C) and a maximum tensile stress of 700 kg/cm^2 (68.6 MPa) [1447–1449].

While addition of silver or tin produced specific effects, addition of arsenic, beryllium, chromium, nickel, antimony, selenium, silicon or titanium proved to be rather ineffective [1450] or, as did arsenic and antimony, even intensified the susceptibility to intercrystalline corrosion [1451].

Cast multi-component aluminium bronzes, such as the cast nickel-aluminium bronze G-NiAlBz F 60, have been recommended, inter alia, for superheated steam armatures, and the cast manganese-aluminium bronze G-MnAlBz F 42 for turbine blades [1452].

Aluminium bronze failed by stress corrosion cracking in geo-thermal plants when exposed to highly contaminated low pressure steam. This steam contained the following impurities (ppm): 5,400 CO_2, 140 H_2S, 15 NH_3, 0.6 $B(OH)_3$ and 0.03 fluorides [1453].

Copper-nickel alloys

Copper-nickel alloys are frequently used for condensers, inter alia on ships, not in the least because of their resistance to the cooling media, i.e. brackish water and sea water at high flow rates [1454].

Alloys such as CuNi 10 Fe and CuNi30Fe, especially those containing manganese, are suitable materials for multistage distillation units in sea water desalination. An alloy containing 40 % nickel has been recommended for the same application, owing to its even better resistance and greater strength [1455].

Arsenic-Copper	about 477 K (204 °C)
Admiralty Brass (CuZn28Sn)	about 505 K (232 °C)
Copper-Nickel 90 10	about 588 K (315 °C)
Copper-Nickel 80 20	about 644 K (371 °C)
Copper-Nickel 70 30	about 644 K (371 °C)

Table 236: Temperature limits for copper-nickel and some other copper-base materials in high pressure steam [1420]

Tests with copper-nickel alloys in saturated steam at 548 K (275 °C) and in steam at 573 and 773 K (300 and 500 °C), at pressures ranging from atmospheric to about 88 kg/cm^2 (8.6 MPa), with and without addition of oxygen, showed satisfactory behavior only in oxygen-free steam at 573 K (300 °C) and about 88 kg/cm^2. In oxygen-containing steam, however, even in the absence of any other aggressive ingredient and in pure steam without condensate formation, rapid detachment ("exfoliation") of oxide layers occurred [1455, 1456].

Condensers made of copper-nickel alloys (CuNi 10 Fe and CuNi30Fe) are used up to steam inlet temperatures of about 573 K (300 °C), in particular in cases of high erosive and corrosive wear and at cooling water flow rates of about 4 m/s. There is however a risk of corrosion in the presence of aggressive impurities, in the steam, e.g. ammonia, CO_2, oxygen or hydrogen sulfide [1457].

Stress corrosion cracking may occur in ammonia-containing steam at elevated temperatures, for example at 533 K (260 °C), even in the nickel-rich alloy CuNi44 [1458].

While the corrosion of copper-nickel alloys in pure dry steam and in pure wet steam at low flow rates is negligible (less than 0.0025 mm/a (0.10 mpy)), it may reach rates exceeding 1 mm/a (39.4 mpy) in impure steam or at very high flow rates, especially when the direction of flow is changed [1459].

At high steam temperatures (773 and 868 K (500 and 595 °C)), however non-ferrous alloys, with high nickel or copper contents, are more susceptible to corrosion than the steels, developed for these temperatures [1460].

Copper-nickel containing further alloying components are generally suitable for high temperature service: the yield point of annealed CuNi30Fe, is 11 kp/mm^2 (1 MPa) at 673 K (400 °C) and corresponds to that of boiler steel St 35.8 [1461].

Copper-tin alloys (bronzes)
Copper-tin-zinc alloys (red brass)

Copper-tin alloys are employed in steam generation plants for condenser pipes, pressure gauges and shut-off elements (slides and valves) in contact with saturated steam [1462].

The use bronze jets in steam jet vacuum systems is mentioned [1463].

A newly developed copper-nickel-tin bronze, containing (%) 15 Ni, 3 to 5 Sn, balance Cu, is recommended for steam condensers because of its increased resistance to erosion by wet steam [1464].

Red brass (copper-tin-zinc-lead cast alloys DIN EN 1982 [1465]) is used for steam armatures, the temperature limit for red brass, being 498 K (225 °C) [1466]. Bronze gate valves for steam, can be used, according to other information, up to about 561 K (288 °C) [1462].

The mechanical strength of copper-tin alloys restricts their use to medium temperatures: from 473 K (200 °C) onwards, there is a sharp drop in tensile strength, hardness and elongation, which has to be allowed for at elevated temperatures and in welding or soldering [1467]. At 623 K (350 °C), the strength of red brass is only 10 kg/mm^2 (0.98 MPa) [1468]. Structural parts of superheated steam slides (spindle nuts), of red brass, (Rg 5 and Rg 10), proved to be unsuitable in superheated steam at 773 K (500 °C) at a pressure of 128 atmospheres [1469].

A temperature limit of 523 K (250 °C) can therefore be regarded as a guide line for the use of zinc-free bronze [1470].

While the corrosion of CuSn-alloys is negligible (less than 0.0025 mm/a (0.10 mpy)) in pure (dry or wet) steam corrosion rates of approximately 0.9 mm/a (35.4 mpy) were found under aggravated conditions, such as high steam flow rates or impure steam [1471].

Copper-zinc alloys (brass)
Other copper alloys

Brass, and particularly special brass types (DIN EN 1982), are suitable materials for condensers in steam. Cast alloys find application for armatures, high pressure armatures, structural elements of valves and controls, such as seats and cones, and for turbine blades. The temperature limit in steam for special brass (SoMs71) is "about 523 K (250 °C)" [1472].

Brass alloys, especially those having a zinc content of more than 30 %, are susceptible to intercrystalline and even trans-crystalline stress corrosion cracking even at extremely low ammonia concentrations in steam [1473]. Tin-plating on the side facing the steam is proposed as a remedy to the deterioration caused by contaminated steam (ammonia, oxygen) [1474, 1475].

The attack of wet steam on copper-zinc alloys depends on the purity of the steam, on the flow rate and on the number and size of the water drops present in the steam. The corrosion rates therefore fluctuate considerably between 0.0025 mm/a (0.10 mpy) and 1.3 mm/a (51.2 mpy) and may even be higher, in particular as a result of the impact of water drops. The erosion corrosion in such systems may occur in all parts of a steam generation plant: turbine casings, turbine blades, on condenser pipes, valve seats and valve rings. Trapping the water drops from the steam leaving the last blade rim, prior to entry into the condenser, is recommended as a remedy. To mitigate the severe damage caused by the impact of drops on valves, the steam can be channelled in the valves or the water drops can be removed before the steam enters into the valves and shut-off elements [1419].

The reduction in size of the drops by increasing the seed formation rate has been recommended as a remedy for this kind of erosion damage. To reduce the size of the drops, addition of film-forming amines, preferably octadecylamine, to the water/steam cycle in conventional power stations or to the secondary cycle in nuclear power stations, involving pressurized water reactors is proposed [1476].

A type of corrosion typical of brass alloys consists in dezincification on the cooling water side. The causes of this "dezincification" have been described early in the corrosion literature (for example) [1477, 1478]. The use of less susceptible arsenic-containing brass grades or of homogeneous brass containing exclusively a mixed crystals with a zinc content below 30 %, has been recommended as a remedial measure [1478].

A further protective measure against corrosion by water consists in creating protective hydrated iron oxide surface layers [1479], by continuous addition of iron sulfate [1480].

Apart from the copper materials described above, other copper alloys possess only subordinate importance as materials for steam systems. Some low alloy wrought copper alloys, such as CuMn2 (1.5 to 2%, Mn), CuMn5 (4 to 5 % Mn), and the copper-silicon-manganese alloys: CuSi 2 Mn (0.8 to 2.2 Si, 0.3 to 0.7 Mn), and CuSi3Mn (2.7 to 3.6 Si, 0.5 to 1.3 Mn, cf. CW116C). These alloys resemble the American Everdur® grades in composition and behavior [1481]. The behavior of these alloys in steam is similar to that of copper. Above 523 K (250 °C), however, there is a tendency towards stress corrosion cracking [1482].

Material	Composition, wt%	Behavior
Deoxidized Cu (Phosphorized Copper 122®)	99.0 Cu; 0.02 P	A
Commercial Brass (Commercial Bronze 220®)	90.0 Cu; 10.0 Zn	A
Red Brass	85.0 Cu; 15.0 Zn	A
Muntz Metal 280®	60.0 Cu; 40.0 Zn	C
Special Brass (Tobin Bronze 4641)	60.0 Cu; 39.25 Zn; 0.75 Sn	A
Admiralty Brass (Arsenical Admiralty 445®)	71.0 Cu; 27.96 Zn; 1.0 Sn; 0.04 As	A
Multi-Component Aluminium Bronze (Ambraloy 630®)	82.0 Cu; 9.5 Al; 5.0 Ni; 2.5 Fe	A
CuSiMn-Alloy (Everdur 651®)	98.25 Cu; 1.5 Si; 0.25 Mn	B
CuSiMn-Alloy (Everdur 655®)	95.8 Cu; 3.1 Si; 1.1 Mn	B
CuNi 90 10 (Copper Nickel 10%, 706®)	88.35 Cu; 10.0 Ni; 1.25 Fe; 0.40 Mn	A
CuNi 70 30 (Copper Nickel 30%, 715®)	68.9 Cu; 30.5 Ni; 0.6 Mn; 0.5 Fe	A
Silverin (Nickel Silver 18%, 752®)	64.5 Cu; 18.0 Ni; 17.25 Zn; 0.25 Mn	A

A = suitable for most application conditions
B = good corrosion resistance; suitable as replacement for a Group A metal, if some properties, other than corrosion resistance, favor its use
C = satisfactory corrosion resistance

Table 237: Behavior of various copper alloys in steam. (Extract from a corrosion table of Anaconda American Brass Company) [1481]

Sulfonic Acid

Author: K. Hauffe / Editor: R. Bender

Copper

While β-anthraquinonesulfonic acid has a moderate inhibiting effect on the corrosion of copper in a 2.5% sodium chloride solution having pH 4, other sulfonic acids, for example benzenesulfonic acid, 2,3-dihydroxynaphthalene-6-sulfonic acid, 1,2-ethanedisulfonic acid and 1,5-pentanedisulfonic acid, have no inhibiting effect [1483]. The corrosion of copper is even accelerated by the following sulfonic acids: 1-amino-2-hydroxynaphthalene-4-sulfonic acid, 1,5-diamino-4,8-dihydroxyanthraquinone-sulfonic acid, 1,4-dihydroxyanthraquinone-sulfonic acid and p-toluenesulfonic acid.

It is possible to produce a stable surface film on copper and copper alloys in a solution containing 50 to 80% amino-sulfonic acid, 5 to 25% thiourea, 1 to 10% benzotriazole or its derivatives and 0.1 to 10% of an anionic surfactant, the resulting film protecting the material from tarnishing in an atmosphere containing hydrogen sulfide [1484].

To plate oxygen-free copper sheet with silver for integrated circuits, it is immersed for 40 seconds at room temperature in one of the following solutions (g/l): 20 H_2O_2 + 100 H_2NSO_3H + 2.5 $C_{16}H_{33}O(C_2H_4O)_{12}H$ or 35 H_2O_2 + 150 H_2SO_4 + 1 n-$C_3H_7SO_3H$ (propylsulfonic acid) [1485].

Bright, lustrous copper surfaces are obtained by pickling in the following solution at room temperature (g/l): 60 H_2O_2 + 20 H_2SO_4 + 2.5 $C_2H_5SO_3H$ (ethylsulfonic acid) + 2.5 $(C_2H_5O)_2P(O)OH$ [1485].

Copper-zinc alloys (brass)

An aqueous solution used for producing brown surface layers on brass consists of copper carbonate, a coloring agent based on thiosulfate/hydrosulfite, an organic color enhancer and ammonium hydroxide (for adjusting the pH-value to between 10.5 and 12.5). Recommended color enhancers include sulfonic acids [1486].

Sulfur Dioxide

Author: L. Hasenberg / Editor: R. Bender

Copper

Copper is not attacked by dry, anhydrous SO_2 gas at room temperature and up to 373 K (100 °C). Similarly, no corrosive attack occurs in anhydrous, liquid SO_2 [1487, 1488]. Corrosion rates of not more than 0.01 mm/a (0.39 mpy) can be expected in these cases. Corrosion studies in liquefied sulfur dioxide containing 1 % water and 0.21 % oxygen have shown that water alone does not corrode copper in sulfur dioxide – oxygen is also needed. In the above case, the weight loss per unit area after a test duration of 14 days was 0.09 g/m^2 (0.3 mm/a (11.8 mpy)). At a water content of 50 % and 0.21 % oxygen, the weight loss at 298 K (25 °C) after 14 days rose to 3.4 g/m^2 (0.01 mm/a (0.39 mpy)). The corrosion products formed were copper oxides and copper sulfate [1489].

Copper cannot be used in aqueous solutions containing sulfur dioxide, especially if air is present. The corrosion rates expected are in excess of 2.9 mm/a (114 mpy) at room temperature, and rise with the temperature [1487].

Copper cannot be used in sulfite cooking acids in the cellulose and paper industry because of its high corrosion rate, which is about 1 mm/a (39.4 mpy) at room temperature. These solutions contain 4 % SO_2 and 1 % CaO and have a pH of 1.3 [1487].

The corrosion rates caused on copper by dry sulfur dioxide are quoted as not more than 0.5 mm/a (19.7 mpy) up to 422 K (149 °C). Corrosion rates of the same order are also likely in sulfur dioxide with a moisture content of 10 %. At a high moisture content the corrosion rate at 323 K (50 °C) rises to more than 1.27 mm/a (50.0 mpy). Corrosion rates in excess of 1.27 mm/a (50.0 mpy) may occur in "sulfurous acid" at 323 K (50 °C) [1490].

Tables 238 to 240 give a summary of some corrosion rates on copper materials [1491, 1492].

According to these summaries, copper is corroded by pure sulfur dioxide at 292–323 K (19–50 °C) at a rate of less than 0.5 mm/a (19.7 mpy). Sulfur dioxide mixtures lead to corrosion rates in excess of 1 mm/a (39.4 mpy), as do moist SO_2 gas and sulfur dioxide in aqueous solutions [1491].

The corrosion on copper in sulfur dioxide starts at a moisture content of 63 % [1489].

The corrosion rates in "sulfurous acid" at 293 K (20 °C) in all concentrations are quoted as 0.5 to 1 mm/a (19.7 to 39.4 mpy), and at 373 K (100 °C) are quoted as greater than 1 mm/a (39.4 mpy) [1491].

Sulfite waste liquors from cellulose production cause corrosion rates of 0.5–1 mm/a (19.7–39.4 mpy) on copper in the temperature range 293–373 K (20–100 °C) [1491].

Material	Temperature K (°C)	SO$_2$	Corrosion rate mm/a (mpy)	
			Air access	Air exclusion
Copper	293 (20)	gas	0.01 (0.39) static	
	973 (700)	gas		0.06 (2.36) agitated
	1,073 (800)	gas		0.06 (2.36) agitated
	1,173 (900)	gas		0.09 (3.54) agitated
CuZn30 (Ms70)		cannot be used above 573 K (300 °C)		
Brasses generally				
CuAl5	293 (20)	dry gas	< 0.01 static (< 0.39)	
CuAl7	293–1,073 (20–800)	dry gas	if agitated readily suitable	
CuAl10				

Table 238: Corrosion rates of copper materials in gaseous SO$_2$ [1491]

"Sulfurous acid" from a papermill had the following composition: 4.42 % SO$_2$ (2.76 % free SO$_2$ and 1.66 % bound SO$_2$) and 1.46 % CaO [1493].

Table 240 illustrates corrosion studies on copper materials in this aqueous solution of sulfur dioxide [1492].

The results are from a continuous immersion test over 90 days at room temperature. In general, copper materials are scarcely used because of their low corrosion resistance under papermaking conditions. The test results in Table 240 on a copper alloy containing 95 % Cu, 3 % Sn and 2 % Si showing a corrosion rate of only 0.03 mm/a (1.18 mpy) are thus all the more surprising. The absence of oxygen is the reason for the low corrosive attack on this material. The corrosion studies were carried out on specimens in 0.2 l of liquid in closed containers with exclusion of atmospheric oxygen [1493]. In practice, stainless steels, such as, for example, X6CrNiTi18-10 (1.4541, cf. AISI 321) or X6CrNiNb18-10 (1.4550, cf. AISI 347), are used in paper-mills for plant components which are exposed to sulfite liquor. If oxygen were excluded completely, it would also be possible to use copper.

The corrosion of metallic materials in the atmosphere is increased by an increase in pollutants, especially SO$_2$. The critical exposure to SO$_2$ for steel is 6–10 mg/m^2, and about twice this figure for zinc and copper [1494]. Corrosion on copper would, accordingly, be expected at an SO$_2$ exposure of about 20 mg/m^2, although other factors should also be taken into consideration.

A sharp increase in the corrosion rate occurs only at a relative atmospheric humidity of more than 60 % [1495].

Sulfur Dioxide

Medium	Conditions	Temperature K (°C)	Copper	G-CuSn 10-14	G-CuSn 10 Zn	G-CuSn 5-7 Znpb	G-CuPb 25	G-CuPb 5-20 Sn	G-CuAl 9	G-CuAl 10 Fe	G-CuAl 9-10 (Ni)	G-CuAl 9 Mn	G-Cu 65 Zn	G-Cu 60 Zn	G-Cu 55 Zn (Mn)	G-Cu 55 ZnAl 1-2	G-Cu 55 ZnAl 4	G-Cu Zn 16 Si 4	G-CuNi 30
Sulfur	dry	293–323 (20–50)	1	1	1	1	1	1	1	1	1	1	1	3	3	3	3	3	1
Sulfur dioxide	pure	293–323 (20–50)	2	2	2	2	2	2	2	2	2	2	3	3	3	3	3	3	2
Sulfur dioxide	mixtures	293 (20)	4	4	4	4	3	3	4	3	3	4	4	4	4	4	4	4	3
Sulfur dioxide	moist	293 (20)	4	4	4	4	4	4	4	4	4	4	4	4	4	4	4	4	4
Sulfurous acid	all conc.	293 (20)	3	3	3	3	2	2	2	2	2	2	4	4	4	4	4	4	2
Sulfurous acid	all conc.	373 (100)	4	4	4	4	4	4	4	4	4	4	4	4	4	4	4	4	4
Sulfite waste liquors from cellulose production	all conc.	293–373 (20–100)	3	3	3	3	3	3	3	3	3	3	4	4	4	4	4	4	2

corrosion rates: 1: ≤ 0.05 mm/a (≤ 1.97 mpy); 2: ≤ 0.5 mm/a (≤ 19.7 mpy); 3: 0.5 to 1.0 mm/a (19.7 to 39.4 mpy); 4: > 1.0 mm/a (> 39.4 mpy)

Table 239: Corrosion rates of copper materials in sulfur dioxide and "sulfurous acid" [1491]

Material composition, %											Corrosion rate
Cu	Al	Mn	Fe	Sn	Si	Pb	P	Sb	Ni	Zn	mm/a (mpy)
86				14			0.05				no corrosion
95				3	2						0.030 (1.18)
93.5					1.5			5			0.004 (0.16)
93					1			5	1		0.020 (0.79)
85.3	0.2				3.5	10			1		0.001 (0.04)
84.3	0.2	1			3.5	10			1		0.002 (0.08)
85.5	0.5		0.5		2.5				1		no corrosion
76	1				3	20					0.004 (0.16)
84	3				3					10	0.003 (0.12)

Table 240: Corrosion rates of copper materials in "sulfurous acid", analysis in a papermill: 4.4% SO_2 (2.76% free, 1.66% bound SO_2) +1.46% CaO, test duration 90 days at room temperature [1492]

Figure 201 shows the dependence of the weight increase of copper per unit area on the relative atmospheric humidity. This figure also shows that the corrosive attack on copper in an atmosphere containing SO_2 rises suddenly above a relative atmospheric humidity of 60% [1495].

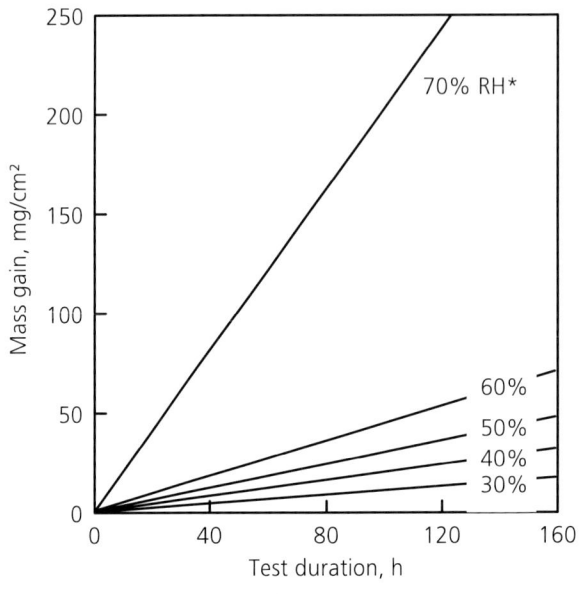

Figure 201: Weight increase of copper in an atmosphere containing SO_2 as a function of the relative humidity at 298 K (25 °C) at a flow rate of 0.36 m/s [1495]
* relative humidity

Of the commodity metals, copper is the noblest. It discolors gradually in air due to oxidation, which leads to a dark coloration on the surface after only a few days, Cu_2O layers in particular being formed. After years of exposure to the open air, a green deposit of basic copper carbonate, sulfate or chloride, named patina, is formed on copper. Corrosion in the open atmosphere is uniform and relatively slight, and even in aggressive atmospheres does not exceed 0.005 mm/a (0.20 mpy) [1496].

These values are confirmed in the industrial atmosphere of Castrop-Rauxel, a city in the Ruhr area in Germany, in long-term studies on copper, in which measured the corrosion rates were 4.9 µm/a (0.19 mpy) in the open atmosphere and 4.2 µm/a (0.17 mpy) under shelter. It can be seen that precipitation, by rinsing of corrosion products on copper, has no influence on the corrosion behavior, as was noted in the case of aluminium [1497].

By increasing the SO_2 content to about 2 mg/m^3 and increasing the airflow rate to 3–5 m/s, the corrosion on copper was also increased to 13 µm/a (0.51 mpy). The corrosion rate on the reverse of the specimens was only 0.3 µm/a (0.12 mpy). In a less aggressive atmosphere, as in Ibbenbüren, a town located about 100 km north of the Ruhr area, Germany, the corrosion rate was 2.7 µm/a (0.11 mpy), as can be seen from Table 241 [1497].

Alloys with a high copper content which are used as conductor materials, should not, where possible, be used in damp atmospheres containing SO_2 because of the high corrosion rates [1498].

Figure 202 contains the results of long-term exposure of copper materials to the aggressive industrial climate of Duisburg, Ruhr area, Germany. After a relatively high initial corrosion in the first 300 days, the corrosion flattens off markedly and slowly levels out to a constant value. The good corrosion resistance of copper in the atmosphere is confirmed by the shape of the curve and the weight increase per unit area of less than 10 mg/cm^2 after 900 days [1499].

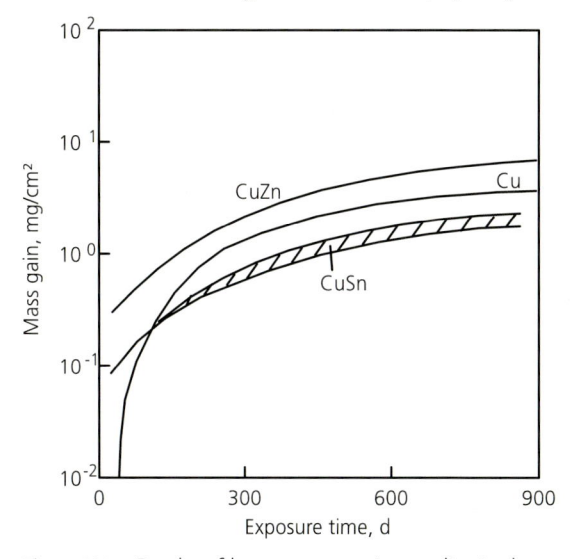

Figure 202: Results of long-term corrosion studies in the aggressive industrial climate of Duisburg, Ruhr area, Germany, on copper materials [1499]

Location	Site of exposure	Corrosion rate, µm/a (mpy)			
		Aluminium	Zinc	Iron	Copper
Castrop-Rauxel	in the open	0.9 (0.04)	5.5 (0.22)	101 (3.98)	4.9 (0.19)
	under shelter	5.2 (0.2)	3.8 (0.15)	103 (4.06)	4.2 (0.17)
	channel before SO_2 injection[1]	5.7 (0.22)	4.0 (0.16)	75 (2.95)	6.8 (0.27)
	channel after SO_2 injection[2]	17* (0.67)	52* (2.05)	142* (5.59)	13* (0.51)
	behind the elements	0.2 (0.008)	1.1 (0.04)	9 (0.35)	0.3 (0.01)
Ibbenbüren	in the open	1.7 (0.07)	7.6 (0.30)	95 (3.74)	2.7 (0.11)
	under shelter	4.6 (0.18)	4.2 (0.17)	101 (3.98)	1.8 (0.07)
	in front of the elements	3.7** (0.15)	4.5** (0.18)	82** (3.23)	2.2** (0.09)
Multiplication factor = r*/r**		4.6	12	1.7	5.9

1) air speed 3–5 m/s
2) air speed 3–5 m/s + 2 mg/m³ SO_2

Table 241: Summary of corrosion results on pure metals in industrial atmospheres [1497, 1500]

Figure 203 shows the severe effect of sulfur dioxide on increasing corrosion in the atmosphere. At a relative atmospheric humidity of about 100 %, copper corrodes after a test duration of 30 days at a weight increase per unit area of less than 0.1 mg/dm² in an SO_2-free atmosphere, and about 3.5 mg/dm² in an atmosphere polluted with 0.01 % SO_2. At a relative humidity of somewhat less than 50 %, the corrosion rate in an atmosphere containing SO_2, as well as in an SO_2-free atmosphere, approaches 0.1 mg/dm² again [1501].

According to [1488], copper is sufficiently resistant in "sulfurous acid". Precise details on the conditions are not given. The corrosion rates in "sulfurous acid" or aqueous solutions of sulfur dioxide increase with the concentration, however. In solutions containing 0.1 and 0.2 g SO_2/100 cm³, corrosion rates at 313 K (40 °C) of 0.002 and 0.005 mm/a (0.08 and 0.20 mpy) respectively are obtained after a test duration of 7 days. Copper is very resistant in these media, but cannot be used in "sulfurous acid" for equipment which comes into contact with foodstuffs [1502].

The corrosion behavior of pure and preoxidized copper in pure SO_2 is investigated in [1503]. Figures 204 and 205 show the weight increase in pure SO_2 under various pressures, and in the temperature range 923–1,203 K (650–930 °C) under an SO_2 pressure of 1.33×10^4 MPa as a function of time.

The maximum corrosion in this temperature range occurs at 1,023 K (750 °C) [1503].

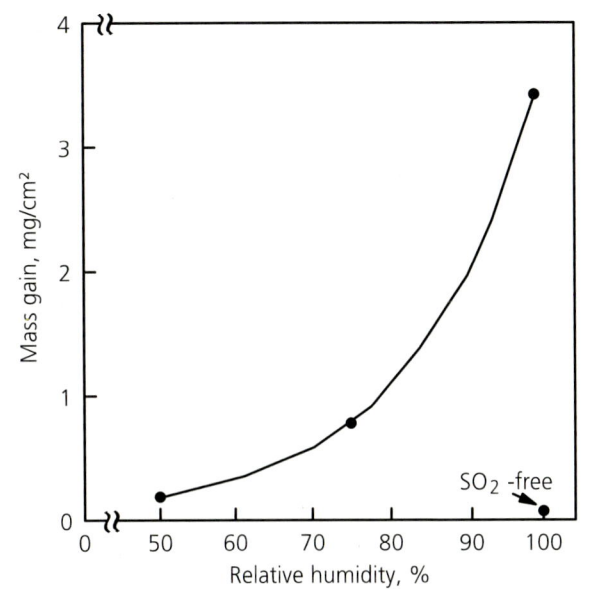

Figure 203: Corrosion of copper in an SO_2-containing (0.01 %) and SO_2-free atmosphere after a test duration of 30 days as a function of the relative atmospheric humidity [1501]

The weight increases linearly with both temperature and pressure. In an SO_2 atmosphere polluted additionally with isobutane, the corrosion behavior of copper is slightly worse, as can be seen from Figure 206 [1509]. No change was found with aluminium.

Corrosion of copper materials in media containing SO_2 is inhibited by triazoles [1504–1508].

Copper-aluminium alloys

The electrochemically noble potential of copper is changed only insignificantly by formation of mixed crystals with aluminium, so that a high corrosion resistance can be expected of copper-aluminium alloys. The corrosion resistance, especially of CuAl alloys containing more than 4 % aluminium, is improved further by the formation of an almost copper-free Al_2O_3 layer. At annealing temperatures above 973 K (700 °C) in dry air, a thin oxide layer of Al_2O_3 is formed. This has similar properties to the layer formed on pure aluminium. Aluminium-containing copper alloys are resistant in media which allow this protective oxide layer to be formed or maintained. The highest corrosion resistance can be expected from the alloys with a minimum aluminium content of 4–10 %.

These alloys are resistant to anhydrous, liquid and dry, gaseous sulfur dioxide. The corrosion rate is not greater than 0.01 mm/a (0.39 mpy) [1487].

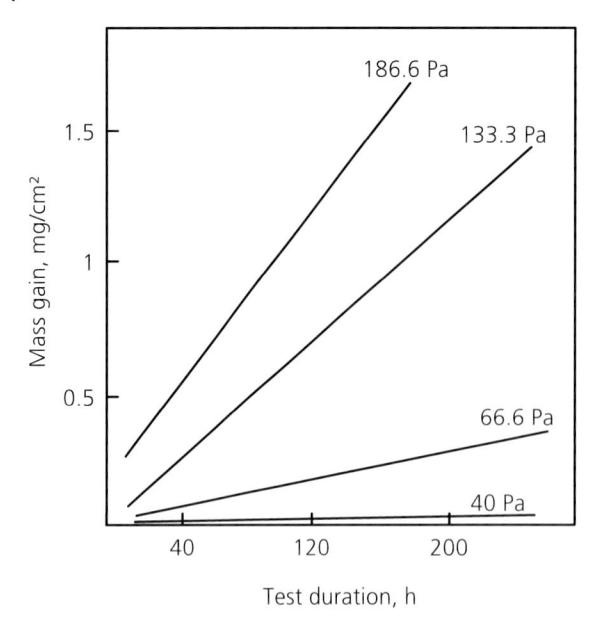

Figure 204: Mass gain of copper at 1,023 K (750 °C) as a function of time in pure SO$_2$ under various pressures [1503]

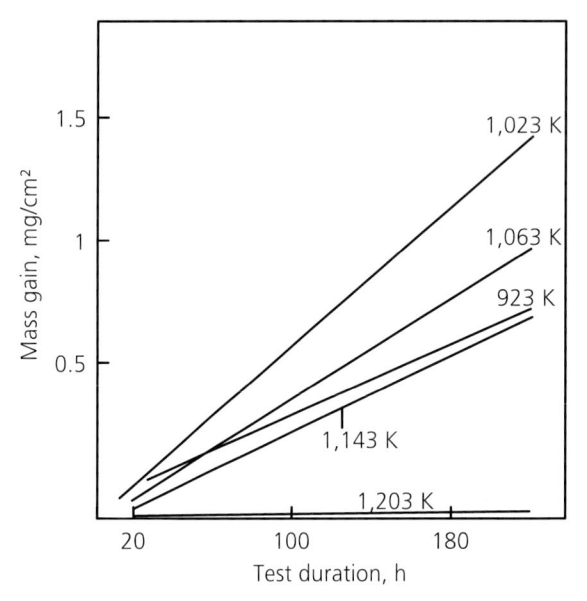

Figure 205: Mass gain of copper in the temperature range 923–1,023 K (650–930 °C) in pure SO$_2$ under a pressure of 1.33×10^4 Pa as a function of time [1503]

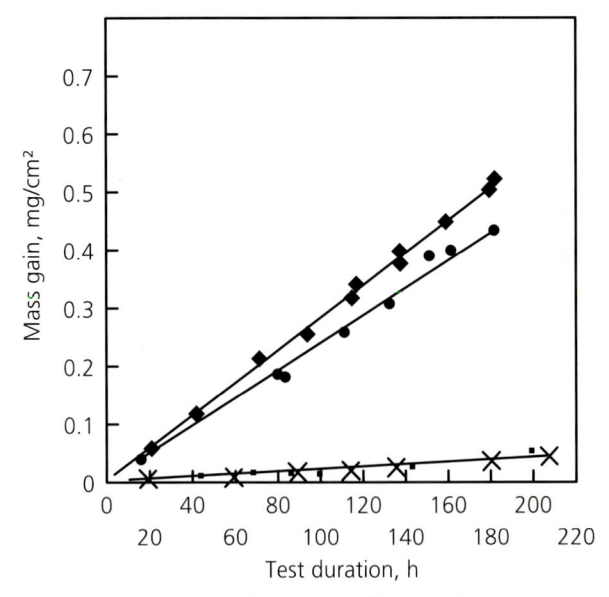

Figure 206: Mass gain of copper as a function of time in various atmospheres at 298 K (25 °C) [1509]
× non-contaminated atmosphere with 78 % rel. humidity
■ as × plus 62.5 ± 12.5 µg/dm^3 isobutane
● as × plus 4.38 ± 0.25 µg/dm^3 SO$_2$
◆ as × plus 62.5 ± 12.5 µg/dm^3 isobutane and 4.38 ± 0.25 µg/dm^3 SO$_2$

The corrosion rate in dry SO$_2$ at 293 K (20 °C) is quoted as being less than 0.01 mm/a (0.39 mpy) for the alloy Cu-5Al. The alloys Cu-7Al and Cu-10Al can be used not only in dry, but also in moist SO$_2$ at 293–1,073 K (20–800 °C) (Table 238) [1491].

Corrosion rates in pure sulfur dioxide are quoted as ≤ 0.5 mm/a (≤ 19.7 mpy) in the temperature range 293–323 K (20–50 °C). In mixtures, they are 1 mm/a (39.4 mpy) or above.

The cast alloys are corroded at 293 K (20 °C) in all concentrations of "sulfurous acid" at a maximum rate of 0.5 mm/a (19.7 mpy), the corrosion rates being more than 1 mm/a (39.4 mpy) under the same conditions at 373 K (100 °C). The corrosion rate in the sulfite liquors in cellulose production is not more than 1 mm/a (39.4 mpy) (Table 239) [1491].

Corrosion studies on copper materials in "sulfurous acid" in a papermill containing 4.42 % SO$_2$ (2.76 % free and 1.66 % bound SO$_2$) and 1.46 % CaO in the absence of oxygen showed corrosion rates of 0.003 mm/a (0.12 mpy) for the alloy Cu-3Al-3Si-10Zn at room temperature. The test duration extended over 90 days. The corrosion rates on all the other copper materials investigated were also low, with two exceptions, as can be seen from Table 240. The reason for the good resistance is the exclusion of oxygen [1492].

In an aggressive atmosphere containing SO$_2$, the corrosion behavior of these alloys is comparable to that of copper, and corrosion rates of 5 µm/a (0.20 mpy) can be expected (compare Table 241 and Figures 201 to 203). The resistance at higher temperatures is better than or equal to that of copper (see Figures 204 and 205).

Copper-nickel alloys

Electrochemically, copper and nickel are two different entities. Copper has a standard potential of +0.345 V and nickel of –0.25 V. Copper/nickel mixed crystals, however, exhibit an excellent chemical resistance, observed on nickel-rich materials in practice. This resistance is based on nickel's resistance due to passivation and its maintenance of this passivation during quite different corrosive attacks.

Copper-nickel alloys are resistant to anhydrous, liquid SO_2 and dry SO_2 gas at low temperatures. The corrosion rate of Cu-30Ni in pure sulfur dioxide gas in the temperature range 293 to 323 K (20 to 50 °C) is quoted as being not more than 0.5 mm/a (19.7 mpy) (Table 239). In moist sulfur dioxide and aqueous solutions, the corrosion rates are in excess of 1 mm/a (39.4 mpy), as is the case in all concentrations of "sulfurous acid" at 373 K (100 °C). A corrosion rate of 0.5 to 1 mm/a (19.7 to 39.4 mpy) can be expected in "sulfurous acid" at 293 K (20 °C) (Table 239) [1491]. Copper-nickel alloys are not recommended for contact with SO_2 (Table 238).

In moist SO_2 gas at 323 K (50 °C), a corrosion rate of more than 1.27 mm/a (50.0 mpy) is quoted [1490]. According to [1487], the corrosion rate up to 373 K (100 °C) on copper alloys containing 10–30 % Ni is up to 0.01 mm/a (0.39 mpy), even in moist sulfur dioxide.

A low-nickel copper alloy containing silicon (0.85–1.1 % Ni, 0.4–0.6 % Si, balance Cu) is corroded at a rate of 0.05 mm/a (1.97 mpy) at 293 K (20 °C) in 3 % "sulfurous acid" [1510].

Copper-tin alloys (bronzes)

The corrosion resistance of copper-tin alloys is comparable to that of copper [1511]. Copper-tin alloys are insensitive to SO_2 and CO_2 vapors and are, therefore, often used in industrial atmospheres and railway tunnels (steam-driven locomotives). The corrosion resistance increases with the tin content, and single-phase alloys should be used in preference in corrosive media.

These alloys also have a very good resistance to sulfite liquors ($Ca(HSO_3)_2$), such as those formed as waste products in the paper industry and in sugar factories (no exact corrosion rates are given) [1511]. The corrosion rate is up to 0.5 mm/a (19.7 mpy) in dry, pure sulfur dioxide, and above 1 mm/a (39.4 mpy) in aqueous and moist SO_2 (Table 239). Corrosion studies are advised before using these materials in media containing SO_2, the exclusion of oxygen leading to a good resistance.

Phosphorus bronzes exhibit a very good resistance in "sulfurous acid" at 293 K (20 °C). They are corroded at a rate of only 0.04 mm/a (1.57 mpy) in static acid solutions containing SO_2. For comparison, the corrosion rates in sulfuric acid are 0.08 mm/a (3.15 mpy) in 1 % acid, 0.15 mm/a (5.91 mpy) in 5 % acid and

0.16 mm/a (6.30 mpy) in 20% acid [1512]. A copper-tin alloy containing 14% Sn and 0.5% Pb showed no material consumption after a test duration of 90 days in a sulfite liquor from the paper industry. The aqueous solution contained 4.42% SO_2 (2.76% free SO_2 and 1.66% bound SO_2) with 1.46% CaO. The bronzes were tested at room temperature in the absence of air in a closed container (Table 240) [1492].

In an aggressive industrial atmosphere, CuSn alloys show the best corrosion behavior of the copper materials, as can be seen from Figure 202. Specimens of copper and Cu-6Sn exposed to this climate showed an increase in thickness due to corrosion products of 15 µm after a period of 800 days, and are, therefore, more resistant than nickel and iron-nickel-cobalt alloys [1499].

Non-restorable bronze sculptures in particular are affected by atmospheric corrosion. Figures of bronze (90% copper, 5% tin and 4% lead), red brass (86% copper, 3% tin, 3% lead, 7% zinc) and brass (80% copper, 3% lead and 16% zinc), for example, surround the Augustusbrunnen in Augsburg (Germany). Puttos of bronze with a lead content of more than 10% stand in the Munich Residence (Germany). In the 19th century, it was recognized that some copper alloys withstand atmospheric corrosion better. Today, the alloys Rg5 (84% Cu, 6% Sn, 7% Zn, 3% Pb) or Rg10 (86% Cu, 10% Sn, 4% Zn) are chiefly used for casting [1513].

Of the substances which pollute the atmosphere, only sulfur dioxide in gaseous form and soot and dust in solid form have a destructive effect on bronzes. Locally increased concentrations of sulfur dioxide were already present in the atmosphere in the 19th century. Measurements in several German towns have shown that it was possible for the sulfur dioxide content to be reduced considerably by the introduction of natural gas and new energy sources instead of coal and oil.

The soot precipitation has also decreased due to conversions and the installation of filters. Road dust, which contains, amongst other things, combustion residues from the exhaust gases of motor vehicles and the rubber worn away from tires, has increased.

Analyses on the corrosion products have shown that only sulfur dioxide from the air or from soot deposits absorbed together with moisture from precipitation attacks bronzes. Apart from a few exceptions, only neutral and basic copper sulfates on a layer of red copper oxide were found as conversion products of the bronzes. By far the most frequent corrosion product on bronzes is basic copper sulfate $CuSO_4 \times 3Cu(OH)_2$, which occurs in nature as brochantite. $CuSO_4 \times 2\,Cu(OH)_2$, or antlerite, is found less frequently. Copper sulfate $CuSO_4 \times 5\,H_2O$ has been found in some cases in the form of blue crystals of millimeter proportions.

An effect of chlorine-containing water on basic copper chlorides was always detectable. Basic copper nitrates have always been produced by synthetic patination. The tin in bronzes is converted into water-soluble stannic acid (H_2SnO_3), which is leached out by the rain. White lead sulfate ($PbSO_4$) occurs as a corrosion product of bronzes containing lead, giving the surface a dirty gray-green appearance.

Compounds like quartz, iron hydrates and gypsum which not originate from bronze also have been detected in patina. Gypsum, which can form crystals several millimeters in size on the dirt deposits on bronzes, is formed by the action of sulfur dioxide on city dust, which contains up to 30% lime.

Microscopic examination of ground cross-sections through the metal, patina and dirt layer reveals a layer of red copper oxide on the metal. Copper oxide and patina are not protective surface layers on the bronze, but a corrosion front penetrating into the metal, the low-tin dendrites of the bronze structure being converted into copper oxide more rapidly than the tin-rich matrix. This corrosion can lead to dissolving of the bronze surface in the form of pitting corrosion. As a rule, the corrosion is uniform. The material loss in such cases was, for example, more than 1 millimeter on a bronze object from the 16th century. From the point of view of corrosion resistance, this value can be considered very good, but is too high for art treasures, because valuable details are lost.

Such corrosion is also to be found, however, on pieces from a more recent period.

Frequent washing and occasional protection using oils and waxes or preferably organic protective coatings, such as, synthetic resin varnishes, is recommended as protection from corrosion by sulfur dioxide [1513].

Copper-tin-zinc alloys (red brass)

These alloys are comparable to the copper-tin alloys and copper in terms of corrosion behavior (Tables 239 and 240).

Their corrosion resistance in dry sulfur dioxide gas is good at low temperatures. Absence of oxygen should be ensured in aqueous solutions (Table 240). A more severe corrosive attack than on copper can be expected in a damp atmosphere containing SO_2, especially on those alloys in which mixing phenomena are found.

Copper-zinc alloys (brass)

Copper-zinc alloys have found a widespread use in industry. The copper content can reach 55%. Zinc is a relatively cheap alloying element, so that as the zinc content of the alloy increases, so does its profitability. On the other hand, its corrosion resistance drops, since zinc is the baser material in this alloy.

Alloys richer in copper should always be used where better corrosion resistance is required.

These materials are resistant to both dry, gaseous and anhydrous, liquid SO_2 – as are other copper materials.

Corrosive attack on brass in liquefied sulfur dioxide increases with water and oxygen content (compare also with copper). In liquefied SO_2 containing 50% water and 0.21% oxygen, the weight loss per unit area at 298 K (25 °C) after a test duration of 14 days was 4.7 g/m^2 (0.014 mm/a). The corrosive attack on the material is uniform and only slightly higher than on copper. The brass turned a blue-green color after this time [1489].

The corrosion resistance of copper-zinc alloys containing 90–60% copper in aqueous sulfur dioxide solutions is similar to that of copper. The corrosion rate extends from 0.01 mm/a (0.39 mpy) to above 3 mm/a (118 mpy) [1487]. A better corrosion resistance occurs at low temperatures in oxygen-free, aqueous sulfur dioxide solutions. A corrosion rate of greater than 1 mm/a (39.4 mpy) is still likely, however, in aqueous sulfur dioxide solutions and moist, liquid SO_2, according to [1491] (Table 239). From this summary, these alloys exhibit the lowest resistance of all copper materials in SO_2 and its aqueous solutions.

The corrosion rate in sulfite liquor from a papermill was only 0.003 mm/a (0.12 mpy) after a test duration of 90 days (Table 240). The alloy consisted of 84% Cu, 10% Zn and 3% each of Al and Si. The sulfite liquor consisted of 4.42% SO_2 in aqueous solution (2.76% free SO_2 and 1.66% bound SO_2) and 1.46% CaO. The tests were carried out at room temperature in static solution in the absence of air and in a closed container [1492]. Copper-zinc alloys exhibit a relatively poor corrosion behavior in an aggressive industrial atmosphere, as can be seen from Figure 202. In this atmosphere, the increase in the thickness of a specimen of Cu30Zn due to the formation of corrosion products was 57 μm after a test duration of 800 days [1499]. This contrasts with 15 μm in the case of copper and copper-tin alloys under the same conditions. Stress corrosion cracking can be caused on brass in an environment containing sulfur dioxide and butane [1514].

Brasses tend to undergo intergranular corrosion, depending on their alloy composition, as a result of grain boundary precipitation. The corrosion rates of CuZnAl alloys at high-temperatures are considerably higher in an atmosphere containing sulfur dioxide than in one free of SO_2. The change in weight in an atmosphere containing O_2 and 1% SO_2 at 1,023 K (750 °C) can be seen in Figure 207. The changes

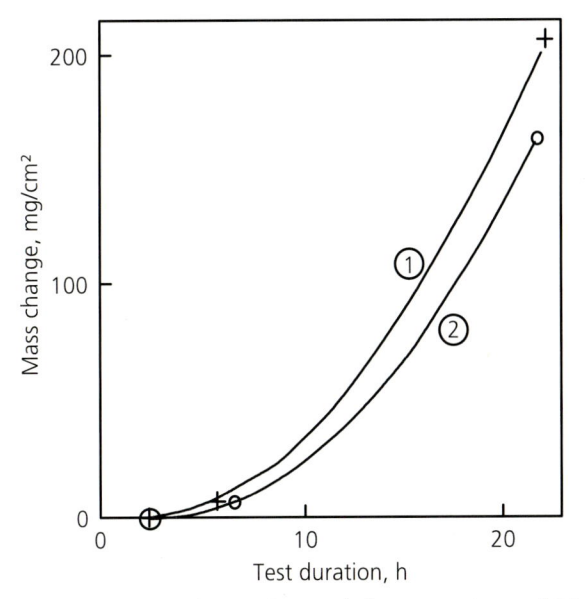

Figure 207: Mass change of CuZnAl alloys in a mixture of SO_2 (p_{SO_2} = 1 × 10^{-3} MPa) and O_2 (p_{O_2} = 9.9 × 10^{-2} MPa) at 1,023 K (750 °C) as a function of time [1515]
① Cu-17Zn-9Al
② Cu-22Zn-7Al

in weight are somewhat lower for alloys with additions of 0.2 at.% each of Ce and Y, but are still higher than in air containing no SO_2 at the same temperature (Figure 208). Weight losses due to the surface layers formed peeling off are observed at test durations longer than 50 h. This peeling occurs after a weight increase per unit area of 0.09 mg/cm^2. The Al_2O_3 formed adheres better to the surface with additions of cerium and yttrium, and does not peel off so quickly. Dense Al_2O_3 layers were not formed on these alloys in gas mixtures containing sulfur, although the thermodynamic conditions necessary for the formation of such layers were present. The surface layers formed could easily be removed with a plastic brush [1515].

As already mentioned in the case of copper, triazoles are effective corrosion inhibitors for copper materials in aqueous solutions of sulfur dioxide.

Naphthotriazole exhibits the best inhibition efficiency with brass in water containing sulfur dioxide [1516].

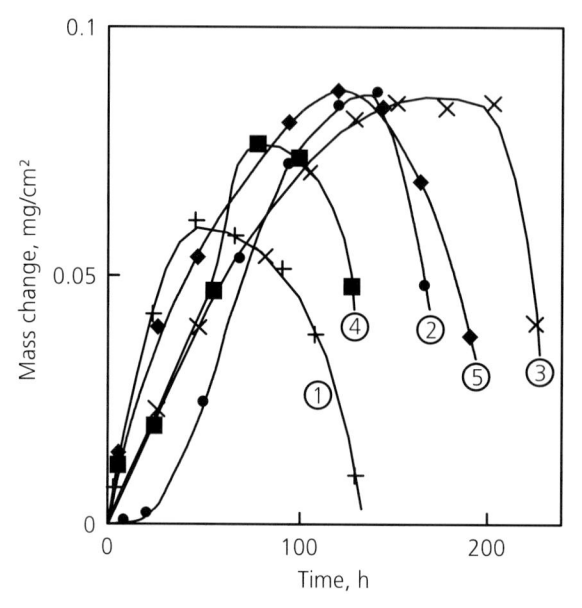

Figure 208: Mass change/time curves during discontinuous oxidation (60 min at 1,023 K (750 °C) and 10 min at room temperature, $p_{O_2} = 2 \times 10^{-2}$ MPa) of CuZnAl alloys in air at 1,023 K (750 °C) [1515]
① Cu-17Zn-9Al
② Cu-17Zn-13Al
③ Cu-22Zn-5Al
④ Cu-22Zn-7Al
⑤ Cu-22Zn-9Al

Other copper alloys

The corrosion behavior of copper alloys with several alloying elements, or alloying elements other than those discussed above, is comparable to that of copper if the contents are low. In special cases, corrosion studies are advisable, especially in moist SO_2 and solutions thereof in the presence of oxygen. Thus, for example, a copper-lead-silicon alloy containing 10% Pb, 3.5% Si, 1% Ni, 0.2% Al with or without 1% Mn and Cu as the balance, shows a corrosion rate of only 0.001 or 0.002 mm/a (0.04 or 0.08 mpy) at room temperature in sulfite liquor (Table 240). The corrosion study was carried out in closed containers in an aqueous solution containing 4.42% SO_2 (2.76% free, 1.66% bound SO_2) and 1.46% CaO over a period of 90 days [1492].

Sulfuric Acid

Author: L. Hasenberg / Editor: R. Bender

Copper

In contrast to high-alloy steels, for example the chromium-nickel steels, copper and copper-based materials do not form passive layers in acid solutions. Copper corrosion is therefore largely determined by transport processes and by the diffusion of dissolved oxygen into the acid as well as the removal of the soluble corrosion products. The latter leads to erosion-corrosion under extreme flow conditions.

A chemical equilibrium exists on a copper surface immersed in an aqueous solution, this equilibrium lying practically completely to the left because of the position of copper in the electromotive series:

$$Cu^0 + H^+ \rightleftharpoons Cu^+ + \tfrac{1}{2} H_2.$$

The reaction in the equation can be forced to proceed to the right on the one hand by

- oxidizing agents, water being formed, or on the other hand by
- complexing substances which bind the copper(I) ion.

It follows that copper is insoluble in non-oxidizing acids without the action of oxidizing agents, as is the case for sulfuric acid in a wide range of temperatures and concentrations. In contrast, copper and its alloys dissolve in oxidizing acids, such as, in concentrated hot sulfuric acid [1517–1532].

Alloy	Composition, %
Type 302	Fe-18Cr-8Ni
Type 302 B	Fe-18Cr-8Ni-2.5Si
Type 304	Fe-18Cr-8Ni-0.08C (max.)
Type 304 L	Fe-18Cr-8Ni-0.03C (max.)
Type 309	Fe-23Cr-12Ni
Type 310	Fe-25Cr-20Ni
Type 314	Fe-25Cr-20Ni-2.5Si
Type 316	Fe-17Cr-12Ni-2.5Mo
Type 317	Fe-19Cr-13Ni-3.5Mo
Type 321	Fe-18Cr-10Ni-Ti5 × C (min.)
Type 329	Fe-27Cr-4Ni-2Mo

Table 242: Composition of the materials mentioned in Figure 209 [1533]

Table 242: Continued

Alloy	Composition, %
Type 330	Fe-16Cr-35Ni
Incoloy® 800	Fe-21Cr-32Ni
Incoloy® 825	21Cr-42Ni-3Mo-2Cu-30Fe-0.9Ti-0.1Al
Inconel® 600	Fe-15Cr-76Ni
Inconel® 601	23Cr-60Ni-1Al-14Fe-0.5Cu
RA 330	Fe-20Cr-35Ni
RA 333	Fe-25Cr-45Ni-3Co-3W-3Mo
Nichrome®	20Cr-80Ni
Carpenter® 20	Fe-20Cr-29Ni-3Cu-2Mo
Carpenter® 20-Cb 3	Fe-20Cr-34Ni-3Cu-2.5Mo-0.6Nb(Cb)
Durimet® 20	Fe-20Cr-29Ni-3Cu-2Mo
Chlorimet® 2	32Mo-62Ni-3Fe
Chlorimet® 3	18Cr-18Mo-60Ni-3Fe
Hastelloy® B	28Mo-61Ni-5Fe-2.5Co-0.3V-1Cr (max.)
Hastelloy® C	15.5Cr-16Mo-54Ni-5Fe-4W-2.5Co-0.4V
Hastelloy® D	81Ni-1Cr-1.5Co-9Si-3Cu-2Fe
Hastelloy® F	46Ni-2.5Co-22Cr-6.5Mo-20Fe-1W-2Nb(Cb)
Hastelloy® G	22Cr-6.5Mo-44Ni-2.5Co-1W-20Fe-2Cu-2Nb(Cb)
Hastelloy® X	22Cr-9Mo-47Ni-1.5Co-0.6W-18.5Fe
Type 430	Fe-16Cr
Armco® 22-13-5	Fe-22Cr-13Ni-5Mo
Corronel® 230	Fe-35Cr-20Ni
50–50	50Cr-50Ni
60–40	60Cr-40Ni
Haveg®	phenolic resin
Karbate®	impregnated graphite

Table 242: Composition of the materials mentioned in Figure 209 [1533]

Hot concentrated sulfuric acid and oleum react with copper to form SO_2:

$2\ H_2SO_4 + Cu^0 \rightarrow CuSO_4 + SO_2 + 2\ H_2O$ [1534].

Figure 209 gives an initial overview of the corrosion resistance of copper in sulfuric acid. In the areas marked and, corrosion rates of less than 0.5 mm/a (19.7 mpy) are achieved in oxygen-free sulfuric acid up to a concentration of 50%. A greater drop in temperature is needed for these rates at between 50 and 60% sulfuric acid.

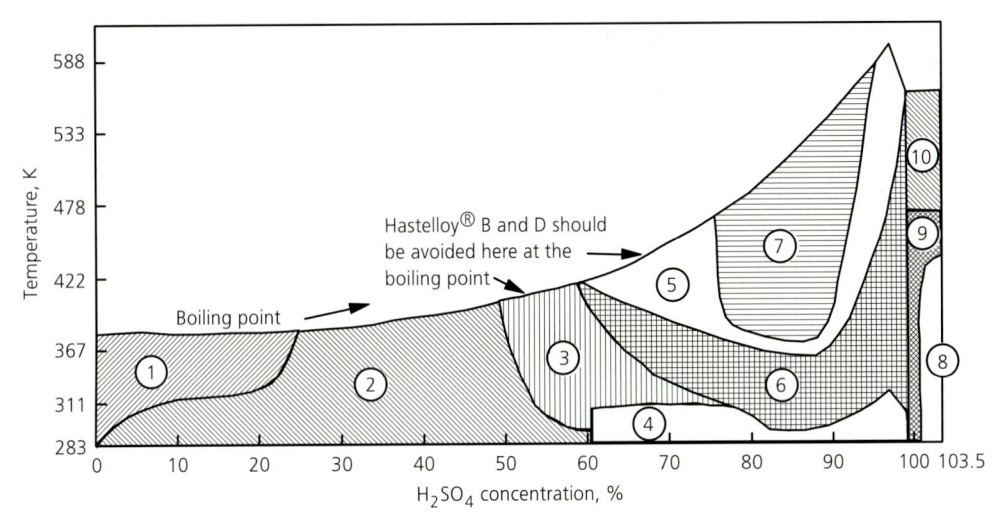

Figure 209: Overview of the areas of use of various materials in sulfuric acid at temperatures up to the boiling point. The corrosion rates in the particular areas are below 0.5 mm/a (19.7 mpy) [1525, 1526, 1533, 1535] (for the materials, see Table 242)

① 10% aluminium bronze (deaerated, Illium®G, glass, Hastelloy®B and D, Durimet®20, Worthite®, lead, copper (O_2-free), Monel® (O_2-free), Haveg®, rubber (up to 350 K (77°C)), Karbate®, Chlorimet®2 (up to 343 K (70°C))

② glass, high-silicon cast iron (14.5% Si), Hastelloy®B and D, Durimet®20 (up to 339 K (66°C)), Worthite® (up to 339 K (66°C)), lead, copper (O_2-free), Monel® (O_2-free), Haveg®, rubber (up to 350 K (77°C)), Karbate®, 10% aluminium bronze (deaerated), Chlorimet®

③ glass, high-silicon cast iron (14.5% Si), Hastelloy®B and D, Durimet®20 (up to 339 K (66°C)), Worthite® (up to 339 K (66°C)), lead, Monel® (air-free), Karbate®, Chlorimet®

④ glass, high-silicon cast iron (14.5% Si), Hastelloy®B and D, Karbate® (up to 96% H_2SO_4), lead, Durimet®, Worthite®, Ni-Resist®, steel

⑤ glass, high-silicon cast iron (14.5% Si), Hastelloy®B and D (0.5 to 1.25 mm/a (19.7 to 49.2 mpy))

⑥ glass, high-silicon cast iron (14.5% Si), Hastelloy®B and D, Karbate® (up to 353 K (80°C) and 96% H_2SO_4), lead (up to 359 K (86°C)), Durimet®20 (up to 339 K (66°C)), Worthite® (up to 339 K (66°C))

⑦ glass, high-silicon cast iron (14.5% Si)

⑧ Hastelloy®C, Illium®G

⑨ 18 8 chromium-nickel steel, Durimet®20, Worthite®

⑩ glass

According to [1518], the corrosion rates for copper in sulfuric acid concentrations up to 50% at 339 K (66°C), and for all concentrations up to 311 K (38°C), are less than 0.51 mm/a (20.1 mpy). As expected, the corrosion rates increase as the temperature increases. Equally high corrosion rates are to be expected in deaerated static H_2SO_4. Corrosion rates of less than 0.05 mm/a (1.97 mpy) are quoted for 20% acid up to 311 K (38 °C), and rates of between 0.51 and 1.27 mm/a (20.1 and 50 mpy) for 60 and 70 % sulfuric acid.

Figure 210 gives a further indication of the corrosion behavior of copper in sulfuric acid. Since no precise data are available, the corrosion rates are to be assumed to be those of solutions with access of air [1536].

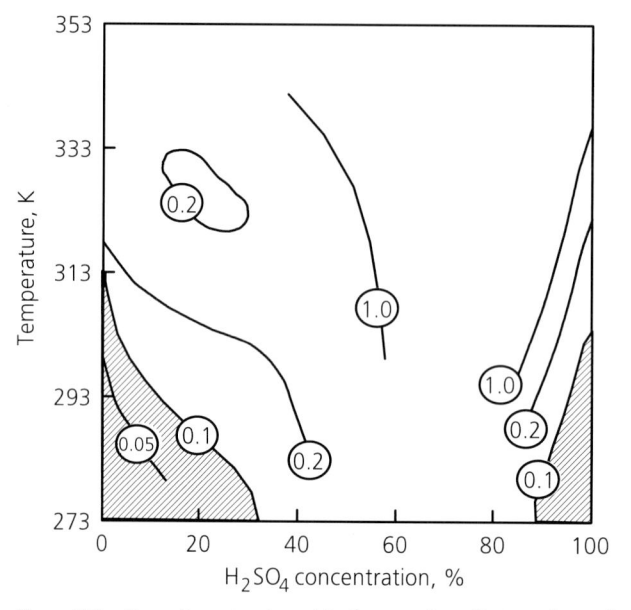

Figure 210: Corrosion rates (mm/a) of copper in sulfuric acid as a function of the concentration and temperature [1536]

Figure 211 shows the corrosion of copper as a function of the sulfuric acid concentration in non-aerated sulfuric acid at 294 ± 1 K (21 ± 1 °C). According to this graph, the corrosion rate of copper in static sulfuric acid in air falls as the concentration increases from 0.15 mm/a (5.91 mpy) in less than 5 % acid, to about 0.01 mm/a (0.39 mpy) in 70 to 80 % acid. The corrosion rate rises again above this concentration. The test solutions were in contact with air, enabling oxygen access. The highest corrosion rates coincide with the maximum oxygen solubility of the medium, whereas the lowest corrosion rates occur at the minimum oxygen solubility.

Figure 212 shows the corrosion of copper in 0.5 % deaerated sulfuric acid above 373 K (100 °C) in comparison with other materials. Since the corrosion rates are considerably more than 0.1 mm/a (3.94 mpy) as the temperature increases, they no longer permit the use of copper under these conditions.

At sulfuric acid concentrations above 70 % and temperatures above 323 K (50 °C) it is no longer possible to use copper even in the absence of atmospheric oxygen. In these cases the sulfate ion has an oxidizing effect on copper [1532].

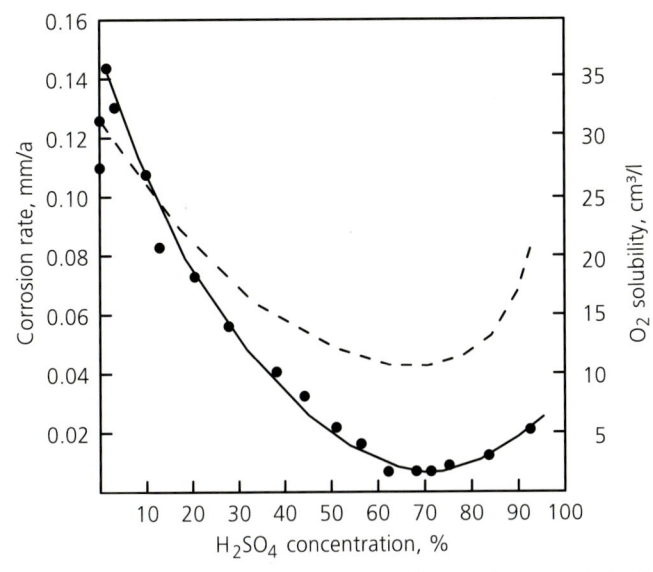

Figure 211: Corrosion of copper in static sulfuric acid in air at 294 ± 1 K (21 ± 1 °C) under the influence of O_2 solubility [1537]
——— Cu
– – – O_2 solubility in H_2SO_4

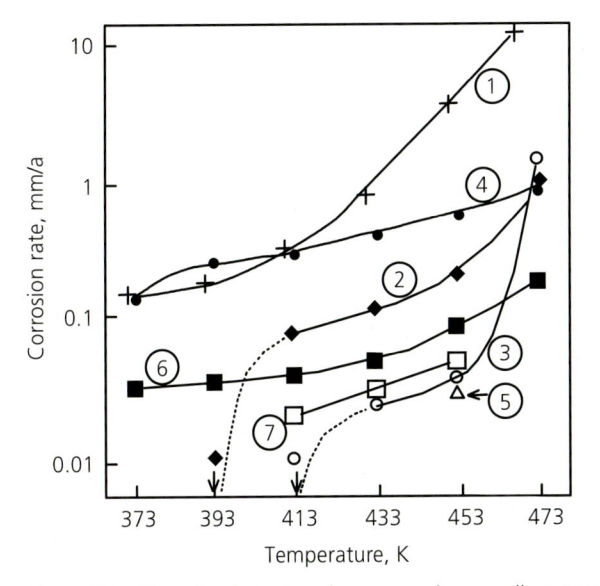

Figure 212: Corrosion behavior of copper and some alloys in sulfuric acid solution containing 0.5 % H_2SO_4 (total absence of O_2) in the temperature range of 373 to 473 K (100 to 200 °C) [1532]
① X5CrNi18-10 (1.4301, cf. AISI 304)
② X2CrNiMo17-12-2 (1.4404, cf. AISI 316 L)
③ Incoloy® 825 (NiCr21Mo)
④ Inconel® 600 (NiCr15Fe)
⑤ Monel® 400 (NiCu 30 Fe)
⑥ electrolytic copper
⑦ Al-bronze

Table 243 shows the corrosion of copper in sulfuric acid solutions, through which various gases were passed, as a function of the temperature. When oxygen is passed through the corrosion rates are considerably greater than those when the solution is gassed with hydrogen. In both cases a rise in the corrosion rate occurs as the temperature increases, this rise being higher at low concentrations than at high concentrations (see Figure 211). Figure 213 shows the influence of air access and agitation on the corrosion of copper at room temperature in 0.14 mol/l sulfuric acid in comparison with other acid solutions (0.27 mol/l). Under these corrosion conditions, more severe attack occurs in sulfuric acid than in the other acids listed, with the exception of HCl.

H_2SO_4 concentration %	Temperature, K (°C)					
	293 (20)		308 (35)		323 (50)	
	O_2	H_2	O_2	H_2	O_2	H_2
	Corrosion rate, mm/a (mpy)					
6	3.76 (148)	0.080 (3.15)	9.65 (380)	0.222 (8.74)	11.0 (433)	0.320 (12.6)
20	3.38 (133)	0.147 (5.79)	5.25 (207)	0.281 (11.1)	2.46 (96.9)	0.25 (9.84)
96.5	1.04 (40.9)	0.142 (5.59)	1.25 (49.2)	0.25 (9.84)	2.14 (84.3)	1.07 (42.1)

Table 243: Corrosion rates of copper in various concentrations of sulfuric acid at various temperatures when oxygen or hydrogen is passed through [1538]

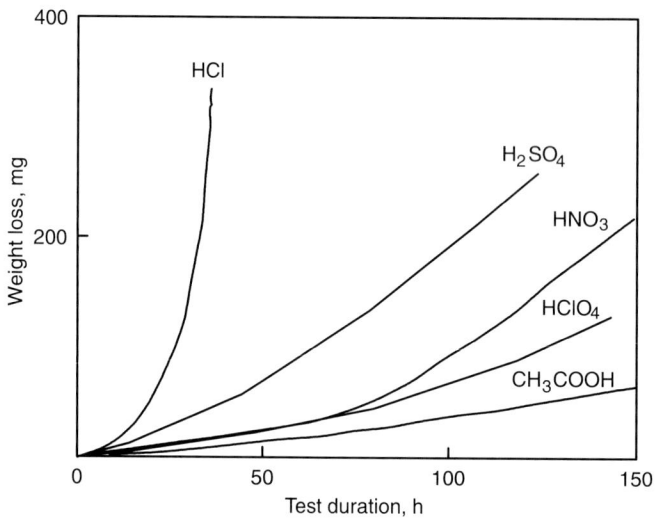

Figure 213: Corrosion of copper in 0.14 mol/l sulfuric acid in comparison with other acid solutions (0.27 mol/l) with air access and agitation of the medium [1539]

As can be seen from Figure 214, the addition of an oxidizing agent intensifies the corrosion many times over. In an agitated medium the rate of corrosion increases as the flow rate increases. For example, in 6 to 30 % air-saturated sulfuric acid, the corrosion rates at a flow rate of 1.3 m/s are more than twice as high as those in static medium [1538].

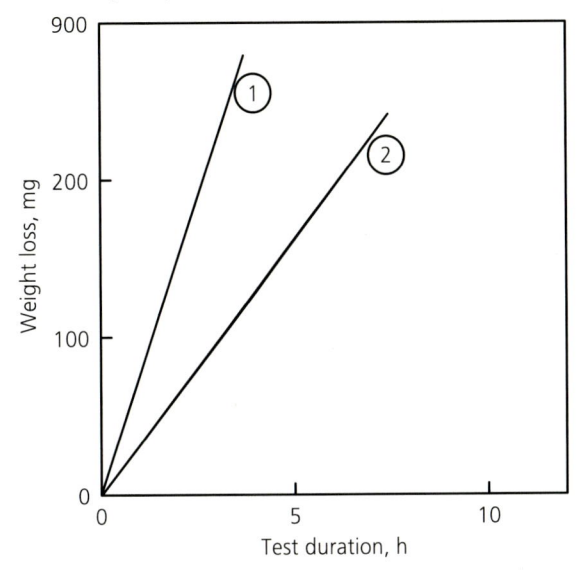

Figure 214: Corrosion of copper in sulfuric acid solutions with added H_2O_2 [1539]
① 0.14 mol/l H_2SO_4 + 0.05 mol/l H_2O_2
② 0.14 mol/l H_2SO_4 + 0.003 mol/l H_2O_2

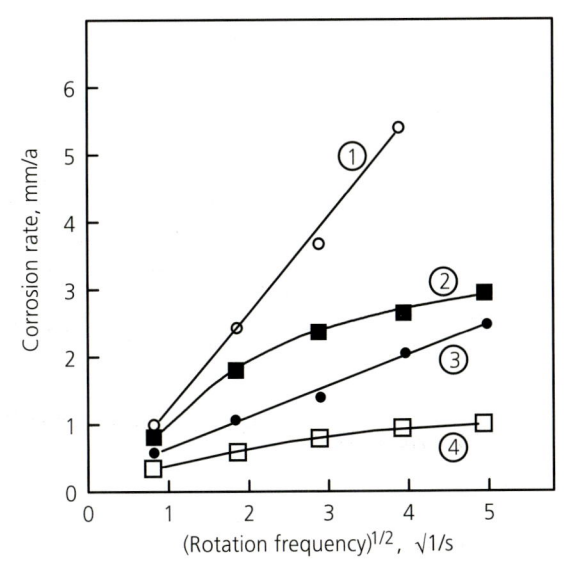

Figure 215: Flow-dependence of the corrosion of copper and aluminium bronze (rotating disks) in 70% sulfuric acid containing air at 323 K (50 °C) without and with a solids content [1532]
① electrolytic copper with Al_2O_3 suspension
② G-CuAl 10 Ni with Al_2O_3 suspension
③ electrolytic copper
④ G-CuAl 10 Ni

Figure 215 shows the influence of the flow rate and the presence of solid particles with an erosive action (corundum) in 70 % sulfuric acid at 323 K (50 °C) on copper and aluminium bronze. As already mentioned, the corrosion rate depends on the flow and increases as the flow rate of the medium and content of solid particles increase. This occurs to a greater degree with copper than with aluminium bronze because copper is softer [1532].

Tables 244 and 245 contain a summary of corrosion rates of copper in various concentrations of sulfuric acid at various temperatures.

Temperature K (°C)	H_2SO_4 concentration %	Corrosion rate, mm/a (mpy)			
		Air access		Air exclusion	
		Agitated	Static	Agitated	Static
293 (20)	1				0.11 (4.33)
	6	3.0–3.9 (118–154)	0.08 (3.15)	0.08 (3.15)	0.04 (1.57)
	10		0.12 (4.72)	0.12 (4.72)	0.1 (3.94)
	20	2.8–3.4 (110–134)	0.27 (10.6)	0.13 (5.12)	0.01 (0.39)
	25			0.12 (4.72)	
	30		0.34 (13.4)		0.0 (0)
	40		0.08 (3.15)		0.01 (0.39)
	45		0.3 (11.8)	0.12 (4.72)	0.01 (0.39)
	50	0.08 (3.15)			
	60	0.12 (4.72)			
	70	2.6 (102)			
	96.5	0.85–1 (33.5–39.4)	0.3 (11.8)	0.13 (5.12)	0.01 (0.39)
308 (35)	6	7.7 (303)	0.1 (3.94)	0.17 (6.69)	0.02 (0.79)
	20	4.2 (165)	0.43 (16.9)	0.23 (9.06)	0.03 (1.18)
	96.5	1.0 (39.4)	0.78 (30.7)	0.2 (7.87)	0.02 (0.79)
313 (40)	10	3.85 (152)			
	25	2.18 (85.8)			
	40	1.31 (51.6)			
	96.5	0.8 (31.5)			

* pit water
** at the waterline in the presence of bichromate

Table 244: Summary of the corrosion rates of copper in sulfuric acid solutions [1531]

Table 244: Continued

Temperature K (°C)	H$_2$SO$_4$ concentration %	Corrosion rate, mm/a (mpy)			
		Air access		Air exclusion	
		Agitated	Static	Agitated	Static
323 (50)	6	8.8 (346)	0.17 (6.69)	0.26 (10.2)	0.08 (3.15)
	20	1.9 (74.8)	0.19 (7.48)	0.2 (7.87)	0.02 (0.79)
	96.5	1.7 (66.9)	0.88 (34.6)	0.8 (31.5)	0.03 (1.18)
333 (60)	10	3.46 (136)			
	25	1.68 (66.1)			
	40	1.36 (53.5)			
	96.5	1.2 (47.2)			
358 (85)	50		0.34 (13.4)		0.016 (0.63)
373 (100)	50		0.75 (29.5)		0.05 (1.97)
395.5 (122.5)	50		2.1 (82.7)		0.1 (3.94)
boiling at 13.33 MPa	30			0.9 (35.4)	
	40			0.9 (35.4)	
	50				1.2 (47.2)
	60			1.6 (63.0)	
	70			0.8 (31.5)	
293 (20)	0.04*		0.05 (1.97)		
	0.14*		0.08 (3.15)		
	0.22*		0.28 (11.0)		
293 (20)	10**		16–32 (630–1,260)		

* pit water
** at the waterline in the presence of bichromate

Table 244: Summary of the corrosion rates of copper in sulfuric acid solutions [1531]

In the case of copper and copper materials, attention should be paid to corrosion at the liquid-gas phase boundary. For example, a corrosion rate of only 0.073 mm/a (2.87 mpy) is caused in 10% H$_2$SO$_4$ at room temperature; whereas at the water line it rises to about 11 mm/a (433 mpy), and in the presence of oxidizing salts, such as Fe(III) compounds and sodium bicarbonate (NaHCO$_3$), from 15 to 22 mm/a (from 591 to 866 mpy) [1539].

The corrosion rates listed in Table 245 were measured above the boiling point of sulfuric acid containing 1 % ethyl alcohol.

H_2SO_4 concentration %	Temperature K (°C)	Corrosion rate mm/a (mpy)
5	493 (220)	0.25 (9.84)
25	493 (220)	17.80 (701)
25	523 (250)	38.00 (1,496)

Table 245: Corrosion behavior of copper in sulfuric acid containing 1 % ethyl alcohol [1521]

The samples were tested in deaerated sulfuric acid over a test duration of one day. Figure 216 shows the course of the corrosion of copper in these solutions over a wide temperature range [1521].

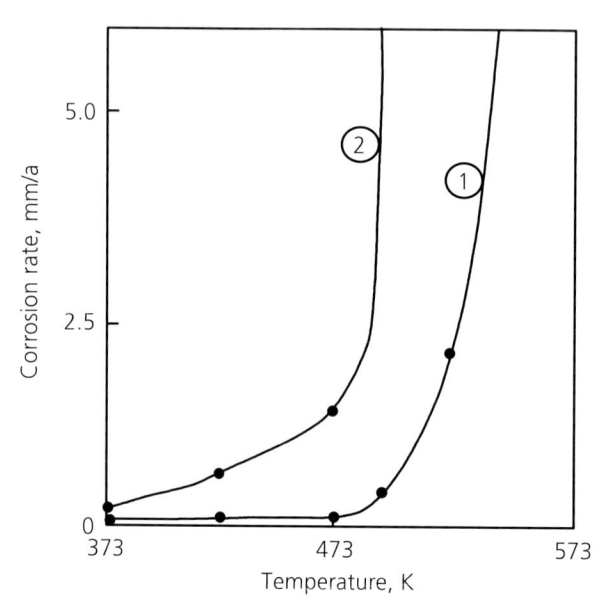

Figure 216: Corrosion of copper in hot deaerated sulfuric acid [1521]
① 5 % H_2SO_4 + 1 % ethyl alcohol
② 25 % H_2SO_4 + 1 % ethyl alcohol

In industrial practice, mineral acids are used as reactants or catalysts in organic solvents. They can therefore be considered as impurities in the solvents and are of importance for the corrosive action of these solvents. They can present an often unheeded risk in the processing industry or to the user. For example, ethanol solutions containing sulfuric acid cause considerable corrosion of copper. In solvents containing acetic acid, sulfuric acid contents have the effect of decreasing corrosion [1540].

Figure 217 gives an overview of the course of corrosion in solvents containing a small amount of sulfuric acid.

The short-term exposure to magnetic fields seems to increase the corrosion on copper, as can be seen from Table 246. The corrosion rates were calculated from the material consumption rates. The corrosion in these studies passes through a maximum at a magnetic field strength of 7,957.7 A/m. This corrosion maximum is reached at a lower magnetic field strength in the case of copper than in the case of steel and nickel and occurs at about the same field strength as that of aluminium [1541]. However, the increases found are not substantially outside the range of variation which occurs in such studies. In addition the corrosion rate increases as the acid concentration increases.

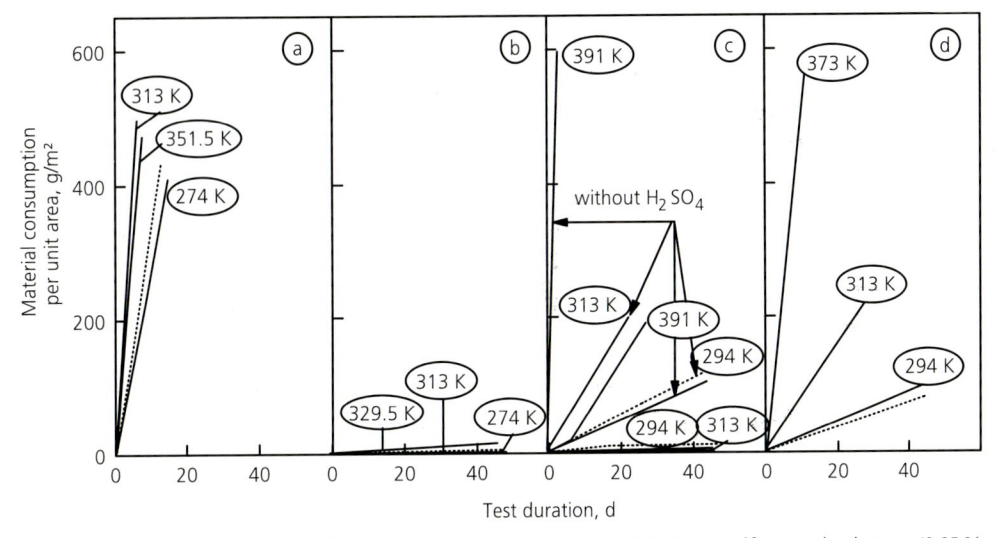

Figure 217: Corrosion-time curves for copper (– electrolytic Cu, ... SF-Cu) in sulfuric acid solutions (0.05 % H_2SO_4) [1540]
ⓐ ethanol (0.1 % H_2O, O_2-saturated, 0.02 % each of acetaldehyde and ethylacetate)
ⓑ acetone (0.2 % H_2O, O_2-saturated)
ⓒ acetic acid (0.5 % H_2O, O_2-saturated)
ⓓ water (O_2-saturated)

Sulfuric acid (1 mol/l) can be used as experimental and comparison solution for evaluation of the resistance of copper materials for supply lines to offset printing machines. The acid is more aggressive than the wetting agents used in these lines, with and without additives. The plot of current density versus potential curves in the medium is a simple and effective method of determining the material/medium compatibility [1542].

Magnetic field strength A/m	H$_2$SO$_4$ concentration, %		
	5	7	15
	Corrosion rate, mm/a (mpy)		
0	0.28 (11.0)	0.56 (22.0)	0.80 (31.5)
5,968.3	0.34 (13.4)	0.70 (27.6)	1.08 (42.5)
7,957.7	0.36 (14.2)	0.76 (29.9)	1.17 (46.1)
28,873.1	0.34 (13.4)	0.71 (28.0)	1.08 (42.5)
31,830.8	0.33 (13.0)	0.68 (26.8)	0.98 (38.6)
39,788.5	0.31 (12.2)	0.66 (26.0)	0.94 (37.0)
47,746.2	0.31 (12.2)	0.64 (25.2)	0.80 (31.5)
55,703.9	0.31 (12.2)	0.61 (24.0)	0.76 (29.9)

Table 246: Corrosion rates of copper (conversion from g/m^{2}h) in flowing sulfuric acid (0.04 m/s) at 293 ± 2 K (20 ± 2 °C) after the short-term action (0.24 min) of a magnetic field of varying strength, test duration 150 h [1541]

– Intergranulare corrosion in sulfuric acid solutions –

Corrosion studies in 50 % sulfuric acid at 373 K (100 °C) have confirmed again that the general corrosion in aerated acid is 10 times greater than that in non-aerated acid. The corrosion rates on non-welded specimens reach values of 0.73 mm/a (28.7 mpy) (non-aerated) and 10.66 mm/a (420 mpy) in the aerated medium after a test duration of 500 h. On welded specimens, the corrosion rates are 1.1 mm/a (43.3 mpy) in a non-aerated medium and 12.6 mm/a (496 mpy) in an aerated medium. After the same test duration in a non-aerated medium, hardly any signs of intergranular corrosion are detectable on non-welded material, although this depends on the purity of the copper. In the heat-affected zone, the penetration rates are between 0.01 and 0.06 mm/a (0.39 and 2.36 mpy) after 500 h. They are of the same order of magnitude if the medium is aerated. Even in the cases with low weight loss the intergranular corrosion is often the cause of failure of copper materials under the action of sulfuric acid [1543].

– Inhibition of corrosion in sulfuric acid –

The corrosion of copper in sulfuric acid can be restricted considerably by inhibitors. In an approximately 0.6 % sulfuric acid solution at room temperature, the activity of the inhibitor increases as the amount added is increased, and in particular from 64 % at 0.12 g/l to 69 % at 0.5 g/l and to 97 % at 1.0 g/l benzotriazole. Quinine and strychnine also prove to be inhibitors of copper corrosion in sulfuric acid solutions. The best inhibitory effect in 0.1 mol/l H$_2$SO$_4$ is achieved by additions of 0.001 mol/l. At 303 K (30 °C) strychnine causes an inhibition of 99 %, compared with 88 % with quinine. In various concentrations of sulfuric acid at various temperatures strychnine is the inhibitor with the better efficiency [1544].

Schiff's bases which are prepared from salicylaldehydes and amines also have an inhibiting effect on copper corrosion. Only aromatic amines have an inhibiting effect here, whereas aliphatic amines stimulate the corrosion of copper in sulfuric acid (5 %). Additions of 0.005 mol/l achieve an inhibition efficiency of more than 90 % in 5 % sulfuric acid at 313 to 353 K (40 to 80 °C). The inhibition efficiency reaches only 60–70 % with additions of 0.001 mol/l. The best effect is achieved by N-2-thiophenylsalicylideneimine [1545].

Additions of thiourea (as inhibitor) in concentrations of up to 500 mg/l to a solution of 170 g/l sulfuric acid and 150 g/l copper sulfate reduce the corrosive attack on copper by one third [1546].

Sulfuric acid itself may reduce copper corrosion in other corrosive media. In boiling glacial acetic acid, through which oxygen was passed to intensify corrosion, the material consumption rate of copper was 1,700 g/m^{2}d (69 mm/a (2,717 mpy)). The corrosion rate was reduced to 3.2 mm/a (126 mpy) by addition of only 0.05 % sulfuric acid [1525].

– Fields of use –

Copper apparatuses were and are used in practice in the saccharification of starch with dilute sulfuric acid at 393 K (120 °C) and 0.4 MPa.

Autoclaves lined with copper withstand mixtures of 20 % sulfuric acid and 1 to 5 % acetic acid during the production of furfurol (2-furaldehyde) from carbohydrates at 393 K (120 °C) [1525].

Copper-aluminium alloys

These materials resemble copper in their corrosion behavior in sulfuric acid solutions (see also section copper). As with copper, oxygen contents in the medium increase the corrosion. Figure 209 gives an overview of the corrosion of aluminium bronze containing 10 % Al in sulfuric acid, although only the corrosion rate of 0.5 mm/a (19.7 mpy), as the application limit, is taken into consideration.

Additions of aluminium to copper considerably improve the corrosion resistance of the latter [1547].

Figure 218 gives a better overview of the resistance of aluminium bronzes (copper-aluminium alloys) in non-aerated sulfuric acid. Figure 219 shows a comparison between the effect of sulfuric acid flushed with nitrogen and that flushed with air on the corrosion behavior of aluminium bronze containing 7 % Al, 2.5 % Fe and balance Cu.

As these graphs show aluminium bronze exhibits a remarkable corrosion resistance in sulfuric acid solutions up to a concentration of 80 % and at temperatures of up to 353 K (80 °C) [1520, 1533, 1548].

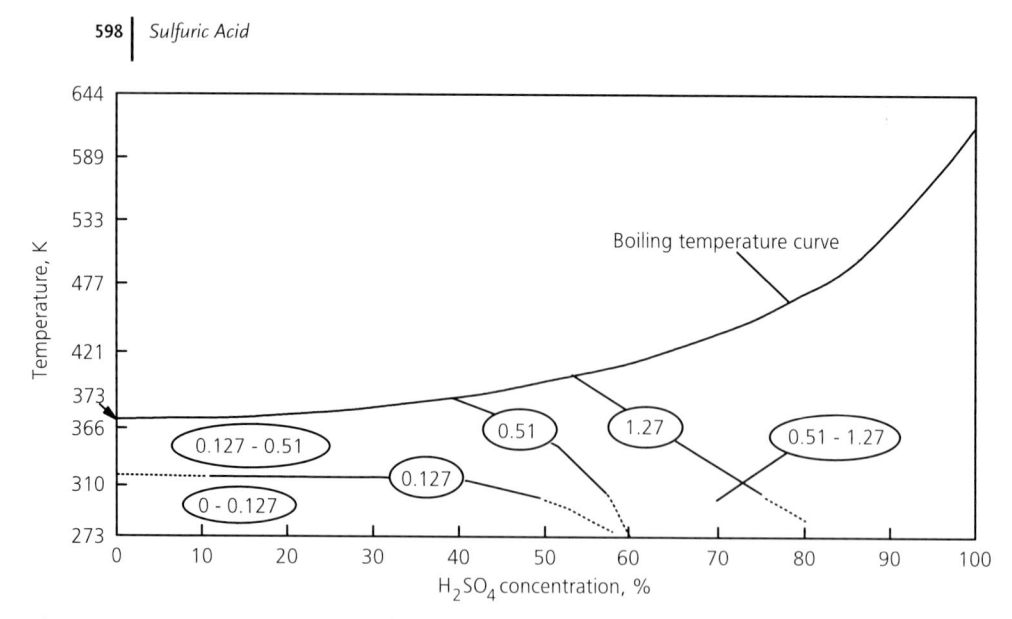

Figure 218: Isocorrosion curves (mm/a) for aluminium bronzes in non-aerated sulfuric acid [1533]

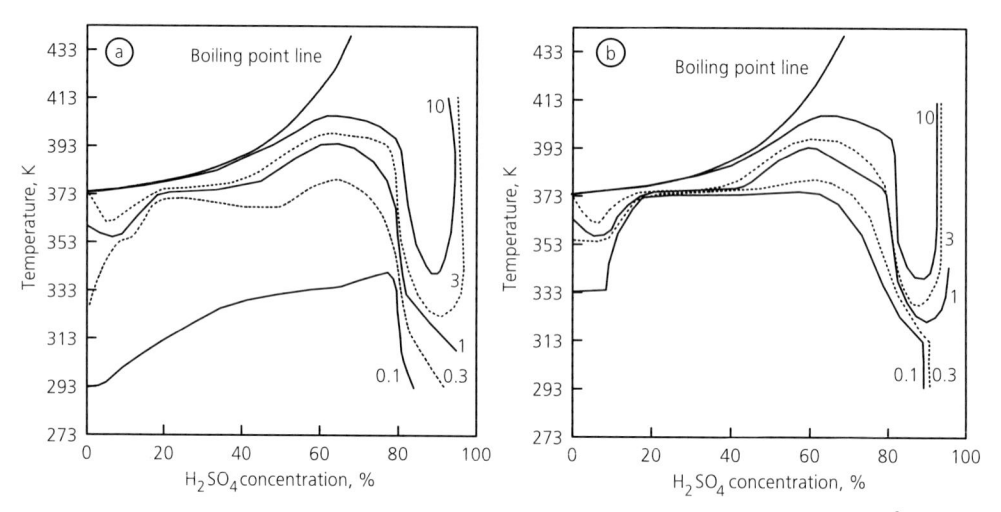

Figure 219: Isocorrosion curves for aluminium bronze (Cu-7Al-2.5Fe) in mm/a (conversion from g/m²h) in sulfuric acid as a function of the concentration and temperature [1548]
ⓐ flushed with air
ⓑ flushed with N$_2$

It is important to mention the fact that the minimum corrosion resistance of aluminium bronze occurs at about 90% sulfuric acid, as can clearly be seen from Figure 219. The action of nitrogen, for example, does not change the region of lowest corrosion resistance. Only the region of the iso-corrosion curve of 0.1 mm/a (3.94 mpy) is shifted significantly towards higher temperatures, especially in the 20 to 40% acid range, whereas the isocorrosion curve at 0.3 mm/a (11.8 mpy) remains practically the same [1548]. Figure 219 shows that on principle markedly increasing corrosion rates are to be expected at temperatures higher than room temperature.

Table 247 contains a summary of the corrosion rates of aluminium bronzes containing 8 and 10 % aluminium in sulfuric acid under various conditions.

As can be seen from the corrosion rates, the corrosive attack is very slight in static acid with exclusion of air. These materials are resistant under the conditions stated. Whereas a considerable increase in corrosion rates must be expected for the alloy CuAl8 (UNS C61000; old 2.0920) in an agitated medium as the concentration of sulfuric acid increases and the temperature rises, this is less noticeable for the alloy CuAl10 (UNS C61000) under the same conditions. CuAl10 therefore has the better resistance and is less sensitive to agitated sulfuric acid solutions than other copper materials [1549].

Table 248 clearly shows this better corrosion behavior in aerated 5 % sulfuric acid in comparison with the behavior of other Cu-alloys. With a corrosion rate of 0.144 mm/a (5.67 mpy) at 289 K (16 °C) aluminium bronze containing 10.6 % Al has a considerably better resistance than the materials compared. Aluminium bronzes are used in contact with sulfuric acid slurries [1549].

Material	Temperature K (°C)	H_2SO_4 concentration %	Corrosion rate, mm/a (mpy)			
			Air access		Air exclusion	
			Agitated	Static	Agitated	Static
CuAl8	293 (20)	0.5	0.12 (4.72)		0.08 (3.15)	
	293 (20)	1	0.1 (3.94)		0.1 (3.94)	
	293 (20)	5	0.1 (3.94)	0.03 (1.18)	0.08 (3.15)	
	293 (20)	10	0.08 (3.15)	0.02 (0.79)		
	293 (20)	25	0.06 (2.36)	0.01 (0.39)		
	293 (20)	40		0.01 (0.39)		
	313 (40)	0.5	0.1 (3.94)	0.04 (1.57)	0.1 (3.94)	0.02 (0.79)
	313 (40)	10	5 (197)			
	313 (40)	25	5.3 (209)			
	323 (50)	10	5 (197)			
	333 (60)	10	5.3 (209)			
	333 (60)	25	5.3 (209)			
	333 (60)	40	2.5 (98.4)			

Table 247: Corrosion rates for copper-aluminium alloys in sulfuric acid solutions under various conditions [1531]

Table 247: Continued

Material	Temperature K (°C)	H$_2$SO$_4$ concentration %	Corrosion rate, mm/a (mpy)			
			Air access		Air exclusion	
			Agitated	Static	Agitated	Static
CuAl10 (CuAl 10Ni)	293 (20)	5	0.046 (1.81)		0.03 (1.18)	0.01 (0.39)
	293 (20)	10			0.03 (1.18)	0.01 (0.39)
	293 (20)	20	0.023 (0.91)			
	293 (20)	35	0.04 (1.57)			
	293 (20)	50	0.002 (0.08)		0.002 (0.08)	
	293 (20)	65	0.002 (0.08)			
	293 (20)	95	0.27 (10.6)		0.01 (0.39)	
	323 (50)	5	0.02 (0.79)			
	323 (50)	20	0.04 (1.57)			
	323 (50)	50	0.002 (0.08)		0.002 (0.08)	
	323 (50)	65	0.002 (0.08)			
	323 (50)	95	1.21 (47.6)			
	373 (100)	5	0.25 (9.84)			
	373 (100)	20	0.17 (6.69)			
	373 (100)	50	0.19 (7.48)			
	373 (100)	65	0.9 (35.4)			
	373 (100)	95	117.6 (4,630)			
	293 (20)	93.8% H$_2$SO$_4$ +1.3% HNO$_3$ +4.9% H$_2$O	0.21 (8.27)			
	323 (50)		3.18 (125)			
	363 (90)		9.83 (387)			
	393 (120)		114.5 (4,508)			

Table 247: Corrosion rates for copper-aluminium alloys in sulfuric acid solutions under various conditions [1531]

Material	Chemical composition, %	Corrosion rate, mm/a (mpy)
Al-bronze (CC331G, CuAl10Fe2-C)	10.6 Al, 3.1 Fe, 0.8 Mn	0.144 (5.67)
Brass (old: G-CuZn28PbSn3)	28.5 Zn, 3.8 Pb, 0.9 Sn	0.435 (17.1)
Bronze (CC482K; CuSn11Pb2-C)	11.3 Sn, 0.5 Pb	0.408 (16.1)
Red brass (old: G-CuSn9Zn2Pb3)	9.1 Sn, 3.6 Pb, 2.3 Zn	0.600 (23.6)

Table 248: Corrosion rates of copper alloys in aerated 5 % sulfuric acid at 289 K (16 °C) [1549]

Alloying additions of silicon and manganese improve the corrosion resistance of aluminium bronzes in sulfuric acid solutions, as electrochemical studies on a Cu-alloy containing 6 % Al, 2 % Si and 0.2 % Mn have shown [1550].

The resistance of several materials in sulfuric acid can also be compared with the aid of Figure 212. In an operating medium containing 0.5 % sulfuric acid aluminium bronze is superior to higher-alloy materials and, as mentioned above, the chromium-nickel steels [1532].

In acid mixtures at room temperature the corrosion rates of aluminium bronzes containing 10 % Al are 0.2 mm/a (7.87 mpy) in solutions of 94 % H_2SO_4 + 1.3 % HNO_3 + 4.7 % H_2O and 0.1 mm/a (3.94 mpy) in solutions of 55 % H_2SO_4 + 5 % SO_3 [1520].

Figure 220 shows the action of acid mixtures of H_2SO_4 and HNO_3. The corrosive attack on aluminium bronze increases rapidly under increasing oxidizing conditions. The material is unusable even at very low nitric acid concentrations [1548].

Aluminium bronzes can also be used in spinning bath solutions (15 % H_2SO_4 + 30 % Na_2SO_4 + 0.3 % $ZnSO_4$). Table 249 gives an overview of the relative corrosion resistance of various metallic materials.

Figure 221 shows the corrosion rate of CuAl10Fe5Ni5-C (CC333G, old G-CuAl10Ni) in 10 % sulfuric acid at 393 K (120 °C) and at various contents of oxygen and chloride ions. This alloy has a good corrosion resistance in hot 10 % sulfuric acid up to chloride concentrations of 5 %, whereas oxygen contents cause an increase in the corrosion rate. At 3 ppm the material can no longer be used, with corrosion rates in excess of 1 mm/a (39.4 mpy).

Aluminium bronzes have advantages over stainless acid-resistant steels under reducing conditions and if low chloride contents are present – conditions in which even the high-alloy steels are depassivated and attacked. Furthermore, construction-related gaps where poor mass transfer occurs have a less adverse effect on copper-based materials in the presence of oxidizing impurities than on steels [1532].

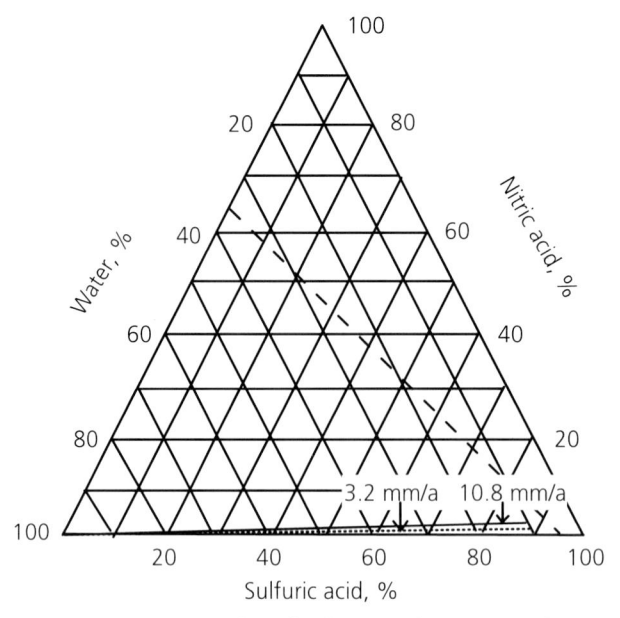

Figure 220: Isocorrosion lines for aluminium bronze (Cu-7Al-2.5Fe) in H_2SO_4-HNO_3 mixtures at room temperature [1548]

Material	Non-pressed	Pressed
Corrix® (Cu-8–10Al-3–4 Fe)	0.22	0.46
Al-bronze	0.50	0.88
Brass	0.34	0.70
P-bronze	0.54	1.00
CrNi-steel (18 8)	1.01	10.20
CrNiMo-steel (18 10 2)	0.09	0.26
Cr-steel (16 % Cr)	95*)	203*)

*) extrapolated

Table 249: Relative corrosive attack (based on phosphorus bronze = 1.00) of spinning bath solutions (15 % H_2SO_4 + 30 % Na_2SO_4 + 0.3 % $ZnSO_4$) on screen fabric made of metallic materials [1519]

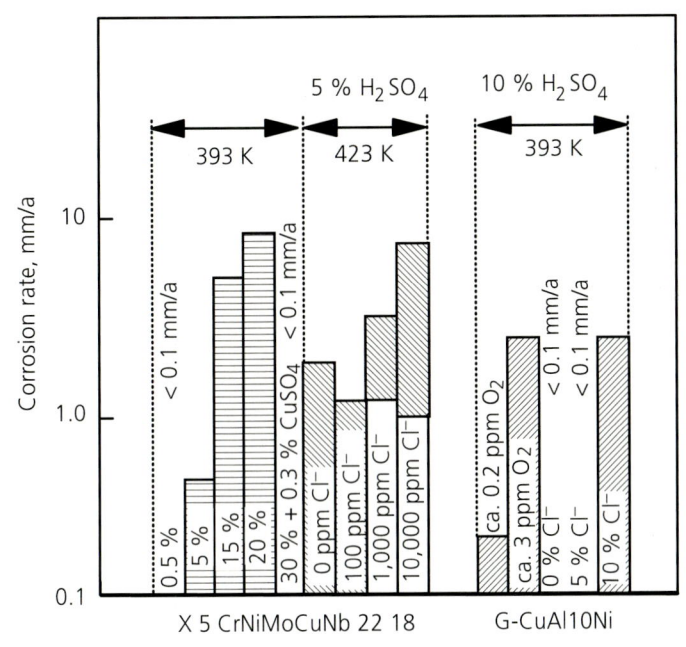

Figure 221: Influence of the sulfuric acid concentration and concentration of chloride impurities on the corrosion behavior of the Cu-alloy CuAl10Fe5Ni5-C (CC333G) and the CrNiMo-steel X5NiCr-MoCuNb22-18 [1532]

Pilot studies on aluminium bronze specimens (CuAl10) in a pickling tank containing a solution of 40 % H_3PO_4, 5 % H_2SO_4, 0.25 % wetting agent and 0.05 % inhibitors with slight agitation and without aeration gave corrosion rates of 2.9 mm/a (114 mpy) over a period of 62 days. The material can be employed only temporarily if required under such conditions [1552].

Figure 215 shows the influence of the flow rate and the presence of solid particles in the corrosion medium 70 % sulfuric acid at 323 K (50 °C) on rotating disks of aluminium bronze CuAl10Fe5Ni5-C (CC333G) in comparison with copper. The dependence of the corrosion rate on the frequency of rotation indicates rate-determining diffusion processes in copper corrosion. The better behavior of aluminium bronze in comparison with copper in the presence of suspended corundum at a high flow rate means that the wear resistance of the alloy under these conditions is a rate-determining factor (the aluminium bronze is harder than pure copper) [1553].

The influence of the flow rate on the corrosion of copper materials in H_2SO_4 is a point to remember [1525, 1554]. Groups of four specimens of an aluminium bronze with the composition 80 % Cu, 10 % Al, 4.5 % Ni, 1 % Mn and Fe were investigated in sulfuric acid, without solid components in the medium, at various temperatures, concentrations and flow rates (test duration 1 week). The results are summarized in Figure 222. According to [1554] the influence of the flow rate on the corrosion rate of nickel-containing aluminium bronze is less than in the case of unalloyed steel, but should not be ignored. The corrosion rate in 65 % sulfuric acid at 338 K (65 °C) and a flow rate of 3 m/s is less than 0.1 mm/a (3.94 mpy). The corrosion rate limit of 1 mm/a (39.4 mpy) is reached at flow rates of 2 to 3 m/s at about 388 K (115 °C), but not before 406 K (133 °C) in static acid. The influence of the flow rate is particularly evident at higher temperatures and rates of 2 to 3 m/s. Under these conditions, shown in Figure 222, uniform corrosion does not always occur. The possibility of the occurrence of intergranular corrosion and pitting is described later.

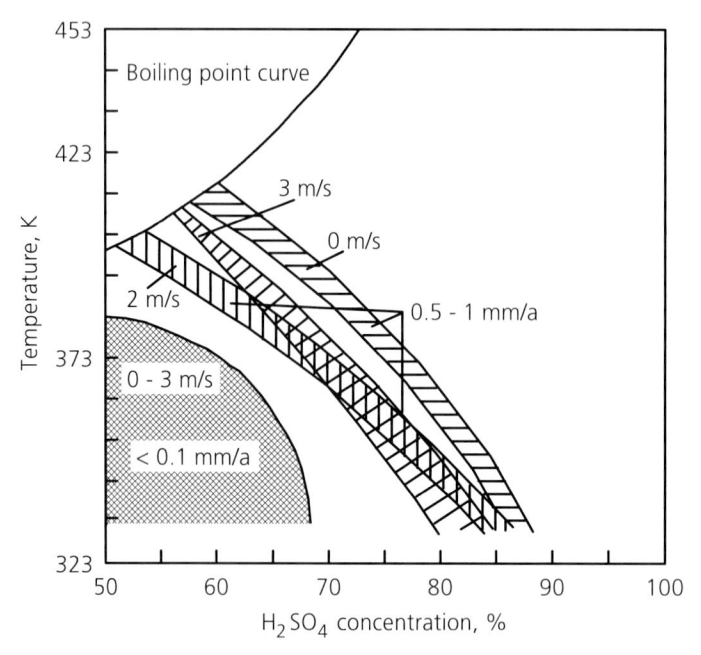

Figure 222: Influence of the flow rate on the corrosion behavior of a CuAl-alloy ((%) 80Cu, 10Al, 4.5Ni, 1Mn, Fe) in sulfuric acid at various temperatures and concentrations [1554]

Figure 223 shows the regions with the same corrosion rate in agitated sulfuric acid with air access for the alloy CuAl10. The low corrosion rates of not more than 0.1 mm/a (3.94 mpy) in the sulfuric acid concentration range of 10 to 80% are a recommendation for the use of aluminium bronzes in this medium at low and slightly elevated temperatures.

The corrosion resistance of aluminium bronzes depends on the alloy composition and their heat treatment [1550, 1555–1557].

Thus according to electrochemical studies alloys with differing aluminium contents exhibit the corrosion rates in sulfuric acid solutions at room temperature listed in Table 250.

Under these conditions the alloy with the higher aluminium content is superior to the brass containing only 2.0% aluminium. The two alloys differ in their corrosion behavior in that the Al-bronze does not show the typical anodic dissolution properties of brass. The formation of a Cu_2O film seems to be the reason for the better resistance [1555].

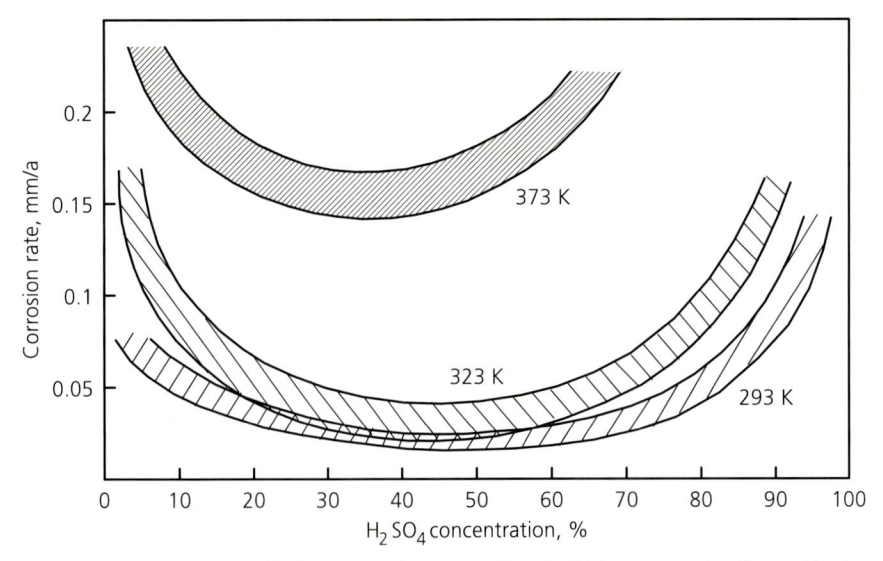

Figure 223: Corrosion rate for the copper-aluminium alloy (CuAl10) in agitated sulfuric acid solution with air access [1554]

Alloy	H_2SO_4 concentration, %		
	5	10	15
	Corrosion rate, mm/a (mpy)		
Cu-6.8Al-3.2Fe-0.8Mn	0.02 (0.79)	0.03 (1.18)	0.04 (1.57)
Cu-20.9Zn-2.0Al10.5Fe-0.03As	0.07 (2.76)	0.18 (7.09)	0.27 (10.63)

Table 250: Corrosion behavior of aluminium bronzes in various concentrations of sulfuric acid at room temperature [1550, 1555–1557]

The dependence of the corrosion behavior of aluminium alloys on the differing aluminium contents and hence on the structure was studied in [1557]. Table 251 shows a summary of the alloys and their structures. The graph of the quasi-steady-state current densities of the alloys in Figure 224 illustrates the better resistance of homogeneous alloys up to a maximum aluminium content of 10% in 0.5 mol/l H_2SO_4. Here the α-phase was found to be conductive to the formation of a protective surface layer within a narrow potential range. The current density of all cubic face centered (cfc) α-alloys is very small and depends only to a small extent on the aluminium content. As soon as the structure becomes multi-phase, as the aluminium content increases or as a result of specific heat treatment, the current density and hence the corrosion increases abruptly [1557].

Studies in pickling solutions containing sulfuric acid produce the same and further results [1556].

Copper-aluminium wrought alloys containing 5 to 10% Al with and without further additions of iron, nickel and manganese were tested in the alternate and continuous immersion test in pickling solutions containing sulfuric acid (20% H_2SO_4 +10% $FeSO_4$) at 313 K (40 °C). In the extruded state corrosive attack of the single-phase α-alloys does not decrease with increasing Al-content, but increases. Small amounts of β'-martensite have little effect on the susceptibility to corrosion, whereas larger amounts have a very adverse effect. The addition of 2% Ni to aluminium bronzes containing 5 to 10% Al does not improve the corrosion resistance. The corrosion resistance is considerably worsened by cold working of the α-alloy [1556].

Alloy	Designation	Heat treatment	Structure
CuAl4	L 1	4 h, 1,223 K (950 °C)/H_2O	α
CuAl6.5	L 2	4 h, 1,223 K (950 °C)/H_2O	α
CuAl7.4	L 3	4 h, 1,223 K (950 °C)/H_2O	α
CuAl9	L 4	72 h, 853 K (580 °C)/H_2O	α
CuAl9	L 5	4 h, 1,223 K (950 °C)/H_2O	α + β'
CuAl10	L 6	72 h, 873 K (600 °C)/H_2O	α
CuAl10	L 7	4 h, 1,223 K (950 °C)/H_2O	α + β'
CuAl10	L 8	2 h, 373 K (100 °C)/H_2O	β' with α-residues
CuAl10.3	L 9	4 h, 1,223 K (950 °C)/H_2O	β' with α-residues
CuAl13	L 10	4 h, 1,223 K (950 °C)/H_2O	γ'_1 with β'_1
CuAl14	L 11	4 h, 1,223 K (950 °C)/H_2O	γ'_1
CuAl16.5	L 12	4 h, 1,223 K (950 °C)/H_2O	heterogeneous, not identified in more detail
CuAl16.5	L 13	4 h, 1,223 K (950 °C) + 48 h, 873 K (600 °C)/H_2O	$\beta'_1 + \gamma_2$

Table 251: Summary of aluminium bronzes for corrosion studies in 0.5 mol/l H_2SO_4 at room temperature (see Figure 224) [1557]

Table 252 gives a summary of the alloys investigated, and Figures 225 and 226 show the corrosion rates under the various test conditions. Heat treatment of the alloys CuAl10Fe and CuAl10Ni with the corresponding structural changes provides no advantages. Uniform corrosion attack of the CuAl5 alloy, preferential intergranular attack of the alloys CuAl8, CuAl9Mn and CuAl10Ni, and dealuminification of CuAl10Fe after heat treatment occur as corrosion phenomena.

Dealuminification is understood as the phenomenon that during corrosive dissolving of the α- and β-phase, and under certain circumstances also the α + χ-structure, copper is redeposited in contrast to aluminium. The materials CuAl5, CuAl8, CuAl9Mn and probably also CuAl8Fe proved to be most suitable for use in pickling solutions containing sulfuric acid [1556].

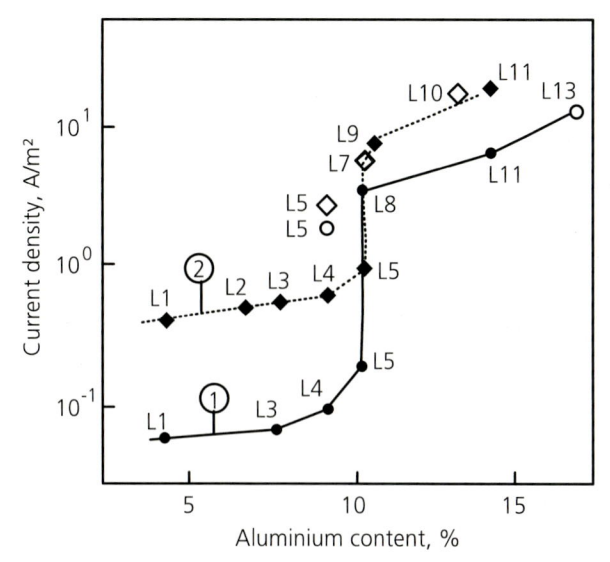

Figure 224: Quasi-steady-state current density of binary CuAl-alloys at $U_H = 0.190$ and 0.245 V as a function of the alloy composition (see Table 251) [1557]
① $U_H = 0.190$ V
② $U_H = 0.245$ V
● ◆ = homogeneous alloys
○ ◇ = heterogeneous alloys

According to [1548], the effect of cold working on the corrosion resistance of the Cu-alloy containing 7% Al and 2.5% Fe is very slight. Figure 227 contains a graph of the experimental results. The results of the study originate from test durations of 72 h and were converted into mm/a. The test duration was shorter than in [1555] (see also Figure 226).

Linings of CuAl-alloys are used, for example, to protect low-alloy steels. As Table 253 shows, protective layers of aluminium bronze CuAl8 applied as linings by plasma welding exhibit a better corrosion behavior in boiling 20% sulfuric acid than the other alloys used.

The pure CuAl8 reference specimen displays a better corrosion resistance than linings on steel because of the local contamination which occurs with the latter during welding. The corrosion rate of only 0.018 mm/a (0.71 mpy) demonstrates that the reference specimen is resistant. The corrosive attack on aluminium bronze CuAl8 in this corrosion solution is linear [1558, 1559].

After a test duration of 150 h at 333 K (60 °C), non-deformed and 60% deformed specimens of the Cu-alloy containing 7% Al and 2.5% Fe exhibited no stress corrosion cracking in 5% H_2SO_4 [1548].

In 5% sulfuric acid, no intergranular corrosion of the above alloy is to be expected, even under various heat treatments [1548]. In this context, see also Figures 224 and 226.

Short designation stating all the main alloying elements	Short designation according to DIN EN 12163	DIN-Material No.	Chemical composition, %								
			Al	Fe	Mn	Ni	Si	Zn	Pb	Sn	Cu
CuAl5	CuAl5	2.0916	4.5	<0.1	<0.1	<0.1	<0.1	<0.1	<0.05	<0.1	bal.*
CuAl5Ni2			4.8	<0.1	<0.1	2.3	<0.1	<0.1	<0.05	<0.1	bal.
CuAl8	CuAl8	2.0920	7.9	0.1	<0.1	<0.1	<0.1	0.05	<0.01	<0.01	bal.
CuAl8Ni2			8.0	0.1	<0.1	2.3	<0.1	0.05	<0.01	<0.01	bal.
CuAl9Mn2FeNi	CuAl9Mn2	2.0960	8.0	0.6	1.9	0.5	<0.1	<0.01	<0.01	0.03	bal.
CuAl10Fe4			10.6	3.8	0.1	0.2	<0.1	0.1	<0.05	0.02	bal.
CuAl10Fe4Mn3Ni	CuAl10Fe3Mn2	CW306G	9.7	3.5	2.6	0.6	0.2	0.2	<0.05	<0.1	bal.
CuAl10Fe4Ni5	CuAl10Ni5Fe4	CW307G	9.1	4.0	0.3	4.9	0.1	0.1	<0.05	<0.1	bal.

* balance Cu

Table 252: Summary of copper-aluminium wrought alloys (aluminium bronzes) studied for their corrosion resistance in pickling solutions containing sulfuric acid (20% H_2SO_4 + 10% $FeSO_4$) at 313 K (40°C) (see Figures 225 and 226) [1556]

Sulfuric Acid

Alloy	Structure (extruded state)	Total decrease in thickness, μm (including dealuminized surface)
Cu-Al5	α-mixed crystals	
Cu-Al5Ni2	α-mixed crystals	
CuAl8 (DIN-Mat. No. 2.0932)	α-mixed crystals	
Cu-Al8Ni2	α-mixed crystals with 1 %* β'-Martensite	
Cu-Al9Mn2FeNi	α-mixed crystals with 23 %* β-Martensite	
Cu-Al10Fe4	85 %* α-, 13.5 %* β'-, 1.5 %* Fe-phase	
Cu-Al10Fe4Mn3Ni	57 %* α-, 31 %* β'-, 10 %* γ$_2$-, 2 %* Fe-phase	
CuAl10Ni5Fe4 (DIN-Mat. No. 2.0966)	α- and κ-phase	

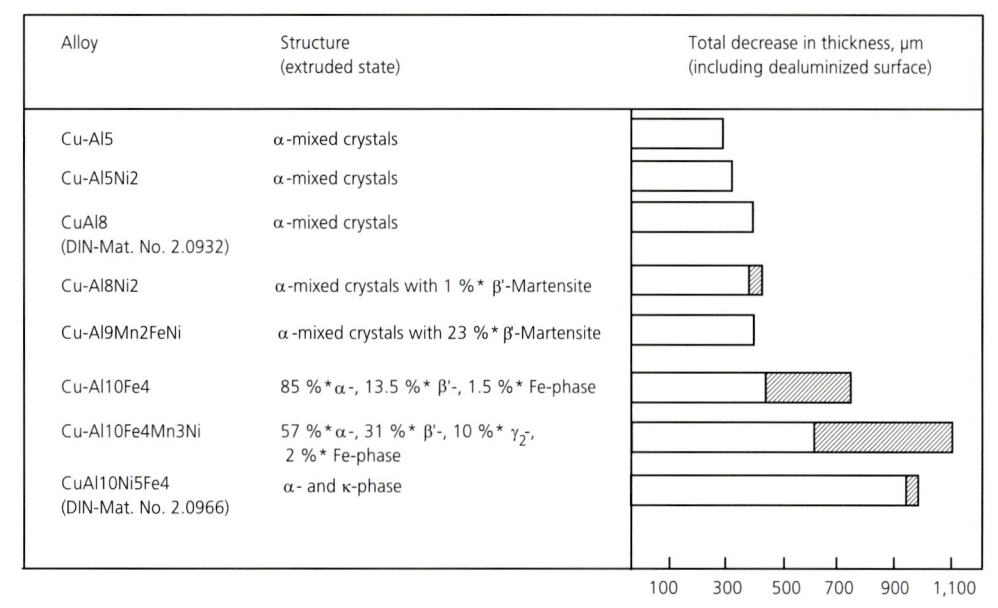

Scale: 100 300 500 700 900 1,100

Figure 225: Decrease in the thickness of copper-aluminium wrought alloys in the extruded state after 30 days of alternate immersion testing in pickling solution containing sulfuric acid (20 % H_2SO_4 + 10 % $FeSO_4$) at 313 K (40 °C) (see Table 252) [1556]
* vol.%
☐ thickness decrease
/ / / dealuminized layer

Alloy	Treatment of extruded rods	Structure	Total thickness-decrease, μm (including dealuminized or by austenite-grain boundary corrosion affected surface layer)
Cu-Al5	–	α-mixed crystals	
Cu-Al5Ni2	–	α-mixed crystals	
	45 % cold deformed	deformed α-mixed crystals	
Cu-Al10Fe4Mn3Ni	–	57 %* α-phase, about 10 %* γ$_2$-phase, about 31 %* β'-martensite, 2 %* Fe-rich phase	
	1 h 1,103 K / H_2O + NaCl	β'-martensite with finely dispersed γ$_2$-phase, little Fe-rich phase	
	1 h 1,103 K / H_2O + NaCl 16 h 773 K / air	α-phase and γ$_2$-phase, a few %* Fe-rich phase	
	16 h 773 K / air	α-phase and γ$_2$-phase, a few %* Fe-rich phase	
CuAl10Ni5Fe4 (DIN-Mat. No. 2.0966)	–	α- and κ-phase	
	30 min 1,243 K / H_2O + NaCl	β'-martensite with finely dispersed κ-phase	
	30 min 1,243 K / H_2O + NaCl + 1 h 873 K	tempered β'-martensite and κ-phase	
	30 min 1,243 K / H_2O + NaCl + 16 h 923 K	α- and κ-phase	
	16 h 923 K	α- and κ-phase	

Scale: 50 100 150 200 250 300 350

Figure 226: Decrease in the thickness of copper-aluminium wrought alloys in the extruded and cold worked state after 50 days of continuous immersion testing in pickling solution containing sulfuric acid (20 % H_2SO_4 + 10 % $FeSO_4$) at 313 K (40 °C) [1556]
* vol.%
☐ decrease in thickness
/ / / dealuminized layer
■ layer with intercrystalline corrosion

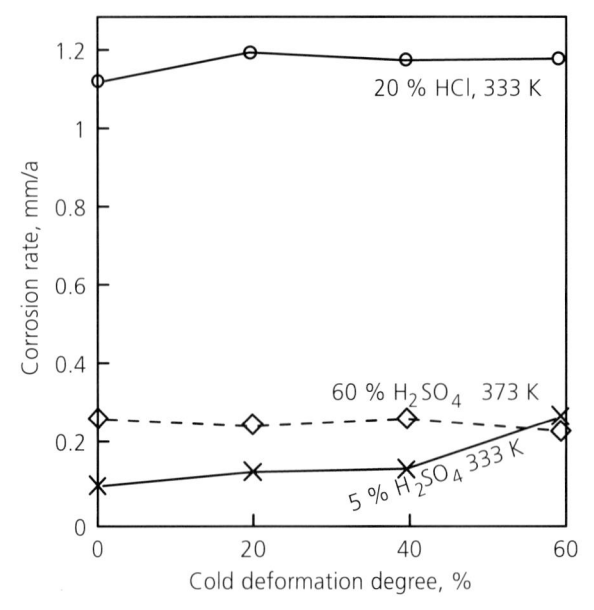

Figure 227: Cold working influence on the corrosion of aluminium bronze (Cu-7Al-2.5Fe) in acid solutions, test duration 72 h [1548]

– Contact (galvanic) corrosion in sulfuric acid solutions –

When chromium-nickel steels come into contact with aluminium bronzes in solutions containing sulfuric acid, a galvanic cell may form as a result of the difference in the electrochemical potentials. This may also occur with other materials and may lead to one of the metals dissolving. The potential difference is increased further by passivation of the chemically resistant steels. The activity of a galvanic cell can be evaluated from current-potential graphs in which the resistance of the external circuit is also taken into account. In the studies, the two metals were introduced into the corresponding sulfuric acid solution and left without external contact until their steady state potentials had been reached. The two metals were then connected conductively via an external resistance (10^6), which was reduced to 1 Ω in each case after equilibrium was established. The surface of the specimens to be investigated was the same in order to eliminate any influence of the surface ratio.

The current density versus potential curves of the galvanic pair Al bronze ((%) Cu-7Al-2.5Fe)/CrNiMoCu-steel 18 22 3 2 in various concentrations of sulfuric acid at 333 K (60 °C) are shown in Figures 228 and 229 for aerated and nitrogen-gassed acid. The corrosion process is controlled cathodically in both cases, which means that the cathode can be polarized whereas the anode retains its potential regardless of the current flow.

The high-alloy steel acts as the cathode in sulfuric acid solutions up to a concentration of 20% and does not dissolve. As the concentration increases, it becomes active, changes to the anode and dissolves [1548]. This behavior is directly related to the oxygen solubility in the solutions which, as Figure 211 shows, decreases as the sulfuric acid concentration increases to 75 %. If the acid is gassed with nitrogen the steel is in the active state regardless of the concentration.

The aluminium bronze described is not substantially influenced by the different types of gassing.

Figure 230 shows the concentration dependence of the steady state potentials of the two materials in sulfuric acid on flushing with air and nitrogen. The activation above a concentration of sulfuric acid of 20 % can be clearly seen in this figure. As the concentration of sulfuric acid increases, the potential difference between the anode and cathode initially decreases, and after the activation of the steel (if it is the anode) increases somewhat again. As Figure 230 shows, the steady state potentials of aluminium bronze change insignificantly. They are about 100 mV more positive with nitrogen flushing than with air flushing.

The points of intersection of the cathodic and anodic polarization curves in Figures 228 and 229 give the cell current densities of the galvanic cell. The current densities are a measure of the corrosion rate if the self-corrosion of the anode is added.

Specimen and location	Material consumption rate per unit area after boiling periods, $g/m^2 h$		Corrosion rate
	20 h	120 h	mm/a (mpy)
NiCu30MnTi (non-annealed) surface, two-layered	1.786	–	1.856 (73.1)
NiTi4 (non-annealed) 2 mm depth, single-layered	4.100	–	4.332 (171)
CuAl8 (non-annealed) surface, single-layered	0.05	–	1.007 (39.6)
CuAl8 (non-annealed) 2 mm depth, single-layered	–	0.948	1.005 (39.5)
CuAl8 (annealed) surface, two-layered	–	0.243	0.273 (10.7)
CuAl8 (annealed) surface, single-layered	–	0.221	0.248 (9.76)
CuAl8 (annealed) 2 mm depth, two-layered	–	0.587	0.659 (25.9)
CuAl8 (annealed) 2 mm depth, single-layered	–	0.100	0.112 (4.41)
CuAl8 reference specimen	–	0.016	0.018 (0.71)

Table 253: Corrosion on copper- and nickel-alloys applied by plasma welding to steel in boiling 20 % H_2SO_4 [1558, 1559]

From the shape of the polarization curves it can be concluded whether corrosion takes place only at the anode or cathode or whether both electrodes are attacked.

Figure 231 shows a comparison between the galvanic short-circuit current densities of the Al-bronze/CrNiMoCu-steel 18 22 3 2 cell and the resistances of the two materials in various concentrations of sulfuric acid at 333 K (60 °C).

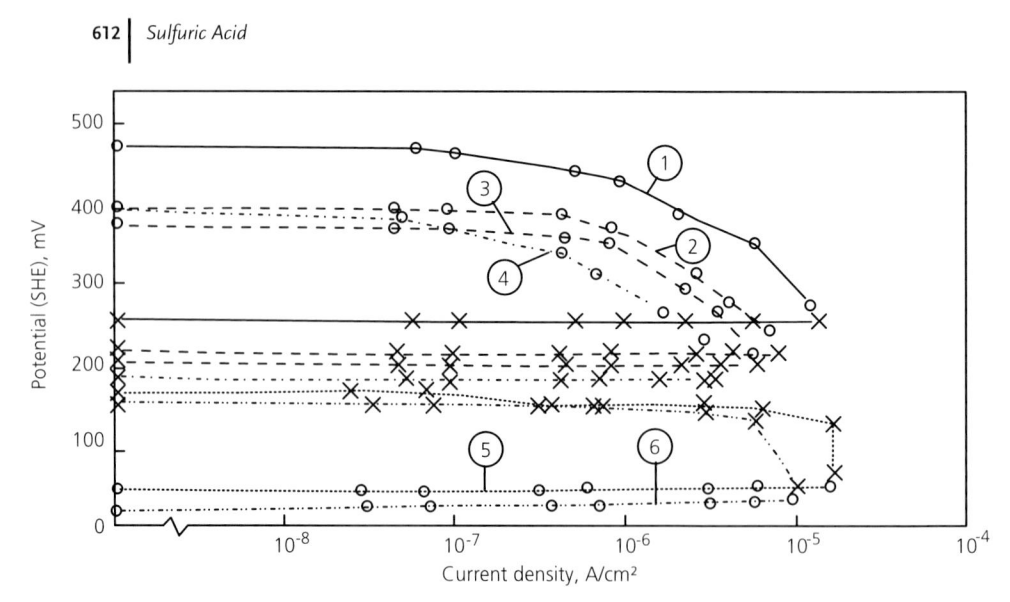

Figure 228: Current density-potential curves of aluminium bronze (Cu-7Al-2.5Fe) (x) and CrNi-MoCu-steel 18 22 3 2 (o) in various concentrations of aerated sulfuric acid at 333 K (60 °C) [1548]
Sulfuric acid concentration:
① 1 %; ② 5 %; ③ 10 %; ④ 20 %; ⑤ 40 %; ⑥ 60 %

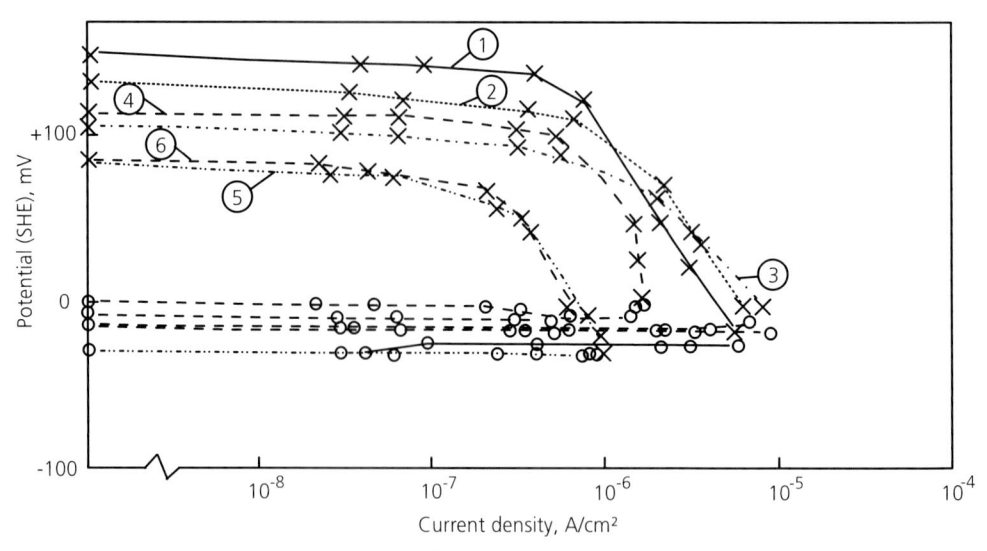

Figure 229: Current density-potential curves of aluminium bronze (Cu-7Al-2.5Fe) (x) and CrNiMoCu-steel 18 22 3 2 (o) in various concentrations of N_2-gassed sulfuric acid at 333 K (60 °C) [1548]
Sulfuric acid concentration:
① 1 %; ② 5 %; ③ 10 %; ④ 20 %; ⑤ 40 %; ⑥ 60 %

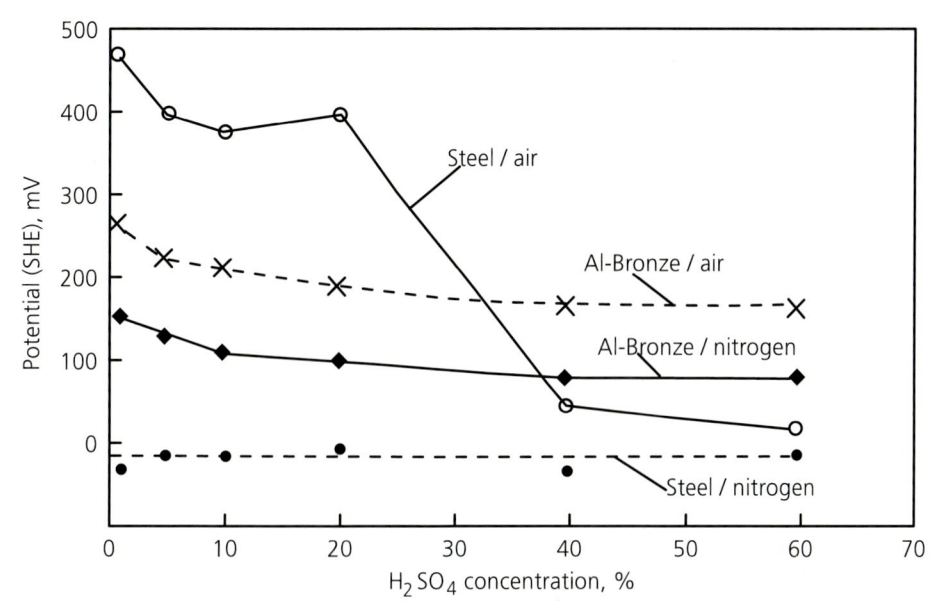

Figure 230: Dependence of the steady state potential of aluminium bronze (Cu-7Al-2.5Fe) and CrNiMoCu-steel 18 22 3 2 in sulfuric acid on the concentration on flushing with nitrogen and air [1548]

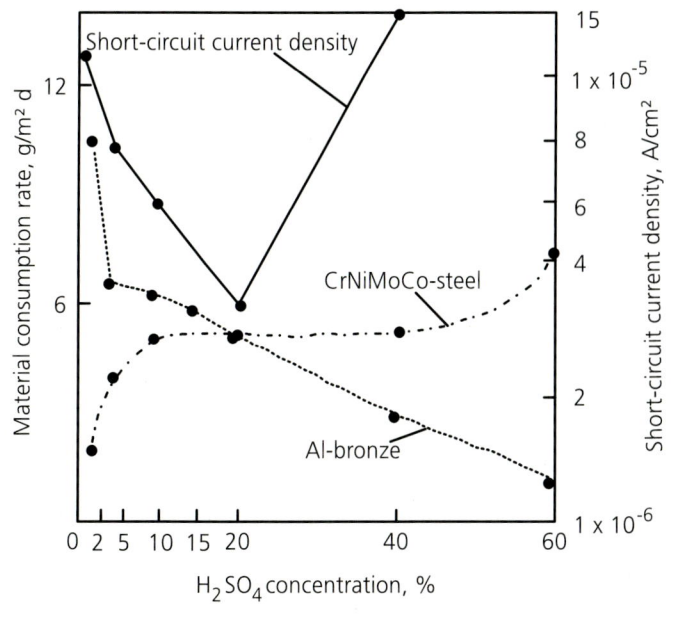

Figure 231: Material consumption rate and galvanic short-circuit current density of aluminium bronze (Cu-7Al-2.5Fe) and CrNiMoCu-steel 18 22 3 2 in aerated sulfuric acid at 333 K (60 °C) [1548]

In the concentration range from 0 to 20% sulfuric acid, in which the Al-bronze is the anode, the short-circuit current densities drop sharply. A change in current direction occurs in the range from 20 to 40% sulfuric acid. The steel becomes the anode in this range [1548].

When designing apparatuses it is important to note this behavior of the material, because under these conditions high corrosion rates may occur on otherwise completely resistant materials.

– Inhibition of corrosion in sulfuric acid –

Schiff's bases which are prepared from salicylaldehydes and amines inhibit corrosion of copper materials only if they have been prepared from aromatic amines. Aliphatic amines have a stimulating effect. The study does not contain further details on the composition of the aluminium bronzes. The best inhibitor efficiency of more than 90% is achieved by addition of 0.005 mol/l N-2-thiophenyl-salicylideneimine [1545].

Copper-nickel alloys

The same considerations as for copper apply to the corrosion behavior of copper-nickel alloys in solutions containing sulfuric acid (see section Copper). According to [1526, 1560], the alloy CuNi30 can be used as the material for pipelines in the concentration range up to 50% H_2SO_4 up to about 333 K (60 °C) (no data on the corrosion of the material).

Corrosion rates of more than 1.22 mm/a (48.0 mpy) are quoted for the alloys CuNi10Fe, CuNi20Fe and CuNi30Fe as screw and nut material in sulfuric acid at room temperature in the concentration range 15 to 70% [1561]. According to [1518], the corrosion resistance of copper-nickel alloys is poorer than that of bronzes and copper. Corrosion rates of between 0.51 and 1.27 mm/a (20.1 and 50 mpy) are quoted for aerated 10% sulfuric acid, but no further data are given in this study. In deaerated 10 to 60% sulfuric acid up to 311 K (38 °C), the corrosion rates are between 0.51 and 1.27 mm/a (20.1 and 50 mpy). The corrosion rates exceed 1.27 mm/a (50 mpy) for the same concentration ranges but higher temperatures.

Figure 209 contains no copper-nickel alloys for use in sulfuric acid, which is certainly a sign that other materials, for example aluminium bronzes, are more suitable since they are included in the list.

In 10% sulfuric acid, the material consumption rate of copper-nickel alloys at 298 K (25 °C) is 1.86 g/m²d (0.08 mm/a (3.15 mpy)). Increased corrosive attack occurred at the phase boundary, that is to say the liquid line [1562].

According to [1520], the corrosion rate on copper-nickel alloys containing 80% Cu, 20% Ni or 55% Cu and 45% Ni in 1 to 60% sulfuric acid at 288 to 293 K (15 to 20 °C) is less than 0.04 mm/a (1.57 mpy).

Nickel silver, a copper alloy containing 56% Cu, 14% Ni and 30% Zn, can no longer be used at 288 K (15 °C) in 5% sulfuric acid, where it has a corrosion rate of 2.3 mm/a (90.6 mpy) [1539].

Material	Temperature	H$_2$SO$_4$ concentration	Corrosion rate, mm/a (mpy)			
			Air access		Air exclusion	
	K (°C)	%	Agitated	Static	Agitated	Static
CuNi10Fe (CuNi10)	293 (20)	5	0.01 (0.39)	0.01 (0.39)	0.01 (0.39)	< 0.01 (< 0.39)
	293 (20)	12.5	0.04 (1.57)	0.04 (1.57)		
	293 (20)	60	0.07 (2.76)	0.05 (1.97)		
CuNi20	293 (20)	5	0.01 (0.39)		0.01 (0.39)	
	293 (20)	60	0.03 (1.18)			
CuNi30Fe (CuNi30)	293 (20)	1	0.07 (2.76)		0.03 (1.18)	0.01 (0.39)
	293 (20)	10	0.1 (3.94)	0.05 (1.97)		
	293 (20)	12.5	0.18 (7.09)			
	293 (20)	20	1 (39.4)			
CuNi30Fe (CuNi30)	373–393 (100–120)	1				0.00 (0)
	373–393 (100–120)	5	0.12 (4.72)			0.003 (0.12)
	373–393 (100–120)	10	0.12 (4.72)			0.03 (1.18)
	373–393 (100–120)	19	0.3 (11.8)			0.01 (0.39)
	373–393 (100–120)	45	16 (630)			0.4 (15.7)
	373–393 (100–120)	50	20.8 (819)			0.52 (20.5)
	373–393 (100–120)	60	48 (1,890)			1.2 (47.2)
	373–393 (100–120)	75	68 (2,677)			1.7 (66.9)
	373–393 (100–120)	96	131 (5,157)			3.3 (130)

Table 254: Corrosion rates for copper-nickel alloys in sulfuric acid solutions [1531]

Corrosion conditions	CuNi10	CuNi30	Cu
	Corrosion rate, mm/a (mpy)		
1% H$_2$SO$_4$, 294 K (21 °C), aerated	0.38 (15.0)	0.40 (15.7)	0.63 (24.8)
4.2% H$_2$SO$_4$, RT, aerated	0.90 (35.4)	0.58 (22.8)	0.70 (27.6)
5% H$_2$SO$_4$, 303 K (30 °C), aerated		0.83 (32.7)	1.0 (39.4)
5% H$_2$SO$_4$, 339–350 K (66–77 °C) aerated, steel pickling bath		2.4 (94.5)	2.5 (98.4)
0.25% H$_2$SO$_4$, 377 K (104 °C) starch digestion		0.83[1] (32.7)	1.9[1] (74.8)
		0.25[2] (9.84)	0.20[2] (7.87)

[1] liquid phase; RT = room temperature
[2] vapor phase

Table 255: Corrosion rates for copper materials in sulfuric acid solutions [1520]

Tables 254 and 255 list the corrosion rates of these alloys. Table 255 allows a comparison with copper within a small concentration range which is rather in favor of the copper-nickel alloys.

In summary, it may be said that these alloys should be used only in cold sulfuric acid solution, and a different material should rather be chosen.

– Inhibitors –

As is also the case with other copper materials (compare section Copper and CuAl-Alloys) corrosive attack on copper-nickel alloys can be inhibited by Schiff's bases. Schiff's bases prepared from salicylaldehydes and aromatic amines act as such inhibitors (aliphatic amines have a corrosion-stimulating effect). The best inhibitor efficiency of 90 % on CuNi10 is achieved by addition of 0.005 mol/l N-2-thiophenylsalicylideneimine to 5 % H_2SO_4 [1545].

Copper-tin alloys (bronzes)

Of the copper alloys containing tin, those having an α-phase of up to 14 % Sn are of particular interest in industry. They are in the form of wrought alloy up to 8 % Sn, and as casting alloy above this figure. Other alloying additions are phosphorus, zinc and lead.

According to [1518], their corrosion resistance corresponds to that of copper. Corrosion rates of less than 0.51 mm/a (20.1 mpy) are quoted up to a sulfuric acid concentration of 50 % at 339 K (66 °C) and at 311 K (38 °C). The corrosion rate in 10 % aerated sulfuric acid at 366 K (93 °C) is between 0.51 and 1.27 mm/a (20.1 and 50 mpy). At the same temperature but at a concentration of between 20 and 40 %, the corrosion rate is less than 0.51 mm/a (20.1 mpy), but then increases to more than 1.27 mm/a (50 mpy) in more highly concentrated acid. Low corrosion rates can be expected in static non-aerated acid (see section Copper). Corrosion rates of less than 0.51 mm/a (20.1 mpy) are usual in such acid conditions up to 366 K (93 °C) in the concentration range up to 50 %, and rates of less than 0.05 mm/a (1.97 mpy) in 20 % acid. In 10 % sulfuric acid up to 450 K (177 °C), the corrosion rates are less than 0.51 mm/a (20.1 mpy), and at higher concentrations and temperatures they exceed this value and indeed may be considerably above it.

According to [1535], tin bronzes can be used in deaerated sulfuric acid up to a concentration of 50 % at 366 K (93 °C). The corrosion resistance of copper-tin alloys increases as the copper content rises [1519, 1520].

Table 256 shows the corrosion rates in 10 % acid as a function of the alloy composition. It confirms the trend mentioned (see also Table 258). The material consumption rate of bronze containing 5 % Sn at room temperature in 10 % sulfuric acid is about 0.6 to 1.8 $g/m^2 d$ (0.03–0.08 mm/a (1.18–3.15 mpy)). The corrosion rate is five times this value at the phase boundary. A considerably higher material consumption rate of 36 $g/m^2 d$ (1.5 mm/a (59.1 mpy)) has been determined for an alloy containing 7 % Sn and 4 % Zn or 10 % Sn and 3 % Sb in vigorously agitated 5 % sulfuric acid at room temperature [1539, 1562].

Chemical composition, %			Corrosion rate mm/a* (mpy)
Cu	Sn	Mn	
96.37	3.42	–	0.08 (3.15)
95.96	0.13	3.49	0.13 (5.12)
94.44	4.94	–	0.17 (6.69)

* calculated from g/m^2 d

Table 256: Corrosion of tin bronzes of various compositions in 10% H_2SO_4 at 288 K (15 °C) [1519]

Table 257 contains a summary of the corrosion rates of various copper-tin alloys in solutions containing sulfuric acid. As can be seen from the example of the CuSn10 alloy, the corrosion rates vary under "identical" conditions, although these are not stated precisely. Merely for this reason it is advisable to perform studies close to the practical conditions of use.

Table 258 contains a comparison of these alloys with other copper materials in sulfuric acid solutions with exclusion of air. As well as the trend of decreasing corrosion resistance as the copper content decreases, the lower resistance of alloys with a high tin content in comparison with other materials, for example lead bronze, is also evident [1520].

The corrosion resistance of a material is not the only decisive factor for choosing materials in chemical plants. Table 259 gives an overview of the suitability of plant components in petroleum refining under the action of sulfuric acid.

As can be concluded from Table 248 and 249, aluminium bronze is not suitable in solutions containing sulfuric acid.

Material	Temperature K (°C)	H_2SO_4 concentration %	Corrosion rate, mm/a (mpy)			
			Air access		Air exclusion	
			Agitated	Static	Agitated	Static
CuSn10	333 (60)	5.3	8.75* (344)			
	298 (25)	10.8	3.26* (128)			
	373 (100)	41.5				0.58* (22.8)
	373 (100)	41.5				3 (118)
	373 (100)	78				3.0* (118)
	373 (100)	78				9 (354)
	373 (100)	95.6				< 200 (< 7,874)

* values from [1531]

Table 257: Corrosion of copper-tin (bronzes) and copper-tin-zinc (red brass) alloys in sulfuric acid solutions [1531, 1549]

Table 257: Continued

Material	Temperature K (°C)	H$_2$SO$_4$ concentration %	Corrosion rate, mm/a (mpy)			
			Air access		Air exclusion	
			Agitated	Static	Agitated	Static
CuSn10Zn	293 (20)	2	0.2 (7.87)		0.1 (3.94)	
	293 (20)	5			0.13 (5.12)	
	293 (20)	20			0.1 (3.94)	
	293 (20)	50			0.002 (0.08)	
	293 (20)	65			0.004 (0.16)	
	293 (20)	95			0.81 (31.9)	
	323 (50)	5			0.36 (14.2)	
	323 (50)	20			0.15 (5.91)	
	323 (50)	50			0.082 (3.23)	
	323 (50)	65			0.077 (3.03)	
	323 (50)	95			2.32 (91.3)	
	373 (100)	5			0.34 (13.4)	
	373 (100)	20			0.31 (12.2)	
	373 (100)	50			0.1 (3.94)	
	373 (100)	65			0.14 (5.51)	
	373 (100)	95			14.8 (583)	
CuSn9Zn6Pb, CuSn9Zn6PbFeSb, CuSn10Zn4PbFeSb	373 (100)	41.5				0.8–3 (31.5–118)
	373 (100)	78				3–9 (118–354)
	373 (100)	95.6				< 200 (< 7,874)

* values from [1531]

Table 257: Corrosion of copper-tin (bronzes) and copper-tin-zinc (red brass) alloys in sulfuric acid solutions [1531, 1549]

Table 257: Continued

Material	Temperature K (°C)	H$_2$SO$_4$ concentration %	Corrosion rate, mm/a (mpy)			
			Air access		Air exclusion	
			Agitated	Static	Agitated	Static
CuSn14, CuSn14Zn, CuSn14ZnPbNiFe, and CuSn14-special alloys	293 (20)	0.5		0.06 (2.36)		
	293 (20)	1	0.1 (3.94)		0.1 (3.94)	
	293 (20)	5	0.23 (9.06)			0.1 (3.94)
	293 (20)	10		0.01 (0.39)		
	293 (20)	25		0.04 (1.57)		
	293 (20)	40		0.02 (0.79)		
	313 (40)	0.5		0.08 (3.15)		
	313 (40)	10		0.1 (3.94)		
	313 (40)	25		3 (118)		
	313 (40)	40		1.3 (51.2)		
	333 (60)	10		1 (39.4)		
	333 (60)	25		3 (118)		
	333 (60)	40		1.35 (53.1)		
	373 (100)	41.5			1.3 (51.2)	0.6 (23.6)
	373 (100)	78			4.5 (177)	3 (118)
	373 (100)	95.5			200 (7,874)	
CuSn8Zn (CuSn10Zn)	293 (20)	93.8% H$_2$SO$_4$ +1.3% HNO$_3$ +4.9% H$_2$O	2.12 (83.5)			
	323 (50)		3.37 (133)			
	393 (120)		77.7 (3,059)			
G-CuSn5Pb, G-CuSn10, G-CuSn10Zn	293 (20)	91.0% H$_2$SO$_4$ +0.8% HNO$_3$ +8.2% H$_2$O	2.06 (81.1)	0.84 (33.1)		
	293 (20)		1.7 (66.9)	0.73 (28.7)		
	293 (20)		1.8 (70.9)	0.75 (29.5)		

* values from [1531]

Table 257: Corrosion of copper-tin (bronzes) and copper-tin-zinc (red brass) alloys in sulfuric acid solutions [1531, 1549]

Material	42% H_2SO_4	78% H_2SO_4
	Corrosion rate, mm/a (mpy)	
Bronze 19–21 Sn	3.0 (118)	9.0 (354)
Bronze 13–15% Sn	1.3 (51.2)	4.5 (177)
Bronze 9–11% Sn	0.60 (23.6)	3.0 (118)
Lead bronze 9–11% Sn, 5–11% Pb	0.60 (23.6)	2.6 (102)
Red brass 8.5–11% Sn, 1–3% Zn, 0–1.5% Pb	0.75 (29.5)	3.0 (118)
Red brass 6–8% Sn, 3–5% Zn, 5–7% Pb	0.90 (35.4)	3.6 (142)
Red brass 4–6% Sn, 4–6% Zn, 4–6% Pb	0.75 (29.5)	3.0 (118)
Red brass 4% Sn, 2% Zn, 1% Pb	0.53 (20.9)	2.3 (90.6)

Table 258: Corrosion of bronzes (CuSn-alloys) and red brass (CuSnZn-alloys) in deaerated sulfuric acid at 373 K (100 °C) [1520]

In aerated 5% sulfuric acid at 289 K (16 °C), aluminium bronze containing 10.6% Al experiences a corrosion rate of 0.144 mm/a (5.67 mpy), whereas tin bronze containing 11.3% of Sn has a corrosion rate of 0.408 mm/a (16.1 mpy). A similar behavior is to be expected in spinning bath solutions.

Copper-tin-zinc alloys (red brass)

These copper materials also behave like copper in sulfuric acid (see section Copper and CuAl-alloys), that is to say the corrosion rate increases with increasing temperature and concentration (in this context see Figure 211), with increasing oxygen content and at the liquid-gaseous phase boundary. Tables 248 and 257 contain a summary of corrosion rates under various conditions for copper-tin-zinc alloys in comparison with other copper materials.

Table 259 contains the corrosion rates of stopcocks and valves in sulfuric acid solutions for a copper-tin-zinc alloy of 8.5 % Sn, 1.54 % Zn, 0.23 % Pb and balance Cu. Deaerated solutions are to be concluded from the low corrosion rates.

H_2SO_4 concentration %	Temperature, K (°C)		
	293 (20)	323 (50)	373 (100)
	Corrosion rate, mm/a (mpy)		
5	0.13 (5.12)	0.36 (14.2)	0.34 (13.4)
20	0.09 (3.54)	0.15 (5.91)	0.32 (12.6)
50	< 0.01 (< 0.39)	0.09 (3.54)	
65	< 0.01 (< 0.39)	0.08 (3.15)	0.14 (5.51)
95	0.81 (31.9)	2.4 (94.5)	15 (591)

Table 259: Corrosion rates on stopcocks and valves made of red brass (8.5 % Sn, 1.54 % Zn, 0.23 % Pb and balance Cu) in sulfuric acid solutions (see also the text and Table 254) [1520]

Copper-tin-zinc alloys behave like bronzes (copper-tin alloys) in sulfuric acid solutions if the Cu-content is at least 90 % or more [1520].

According to [1549], red brass containing more than 85 % Cu still has a good resistance in up to 40 % sulfuric acid hot or cold, and in cold sulfuric acid at a concentration of less than 80 %.

In vigorously agitated 5 % sulfuric acid, an alloy containing 7 % Sn, 4 % Zn and Cu as the balance has a material consumption rate of 36 g/m²d (1.5 mm/a (59.1 mpy)) at room temperature [1539, 1562].

With a relatively high corrosion rate of 0.6 mm/a (23.6 mpy) in aerated 5 % sulfuric acid red brass containing 9.1 % Sn is shown to be a less resistant example of the copper materials (Table 248). Thus in the case of application other materials should be given preference or studies undertaken.

Copper-zinc alloys (brass)

Copper-zinc alloys have better mechanical properties than copper and for this reason are used as a construction material. Alloys containing 10 to 30 % zinc are close to pure copper in their chemical corrosion properties (see also section Copper and CuAl-alloys).

Structural components	Corrosive medium	H$_2$SO$_4$ concentration %	Temperature K (°C)	Durability years	Remarks
Valve (all parts of G-CuSn10)	acid slurry	19 to 90	301–366 (28–93)	1 to 5	corrosion on base and cone
	acid slurry	10 to 30	355–366 (82–93)	1 to 2	erosion and corrosion
Pump (all parts of G-CuSn10)	acid slurry	85	316 (43)	1.5 to 3	general erosion
	recovered acid	78	300 (27)	2	slight repairs
	acid and oil	4 to 5	355 (82)	0.5 to 1.5	corrosion damage

Table 260: Durability of structural components made of the copper-tin alloy G-CuSn10 (CC480K; old: 2.1050.01) in sulfuric acid solutions in the petroleum industry [1549]

The reasons for the local corrosive attack on brass are its different structural development in the various types, intergranular precipitates and stresses.

They are less suitable than copper and bronzes but better than copper-nickel alloys in their corrosion behavior under the action of sulfuric acid [1518]. A homogeneous structure which is free from intergranular precipitates is the most favorable requirement for corrosion resistance. According to the work mentioned, the maximum corrosion rates are 0.51 mm/a (20.1 mpy) up to temperatures of 366 K (93 °C) in the concentration range between 10 and 50 % sulfuric acid. In more highly concentrated sulfuric acid, the corrosion rates are between 0.51 and 1.27 mm/a (20.1 and 50 mpy) at room temperature. At higher temperatures the corrosion rates are more than and in some cases considerably more than 1.27 mm/a (50 mpy). As with all copper materials under such conditions, the corrosion rates are several times higher in acid solutions containing oxygen.

Table 261 contains a summary of corrosion rates for copper-zinc alloys in sulfuric acid solutions under various conditions. As the zinc content increases, the corrosion resistance of these materials in sulfuric acid drops. Aeration and agitation intensify the corrosive attack. The zinc is rapidly dissolved out of alloys containing more than 15 % zinc under the action of sulfuric acid, leading to so-called dezincification. Copper-zinc alloys containing 85 % Cu and 15 % Zn exhibited a very good resistance in the petroleum industry as heating coils for slurry containing sulfuric acid at an acid content of between 30 and 40 % at 355 to 377 K (82 to 104 °C). Pipelines of the same material were destroyed at lower concentrations and elevated temperatures [1563, 1564].

Alloys containing more than 15 % zinc are generally unsuitable for handling acids. Lead-containing alloys do not form a protective sulfate film on the surface of the materials in solutions containing sulfuric acid, and behave like lead-free materials [1565].

According to [1520], brass is unsuitable for handling sulfuric acid at all concentrations.

The corrosion resistance of copper materials is adversely influenced by the zinc content. An alloy (see Table 250) containing (%) Cu-20.9Zn-2.0Al-0.5Fe-0.03As corrodes more in sulfuric acid solutions than an alloy which is richer in aluminium and contains no zinc (electrochemical analysis). At room temperature, the zinc-containing alloy has a corrosion rate of 0.07 mm/a (2.76 mpy) in 5 % H_2SO_4, 0.18 mm/a (7.09 mpy) in 10 % H_2SO_4 and 0.27 mm/a (10.6 mpy) in 15 % H_2SO_4 [1555].

In spinning bath solutions, the relative corrosion resistance of brass is better than, for example, chromium-nickel steels (Table 249). As can be seen from Table 261, its corrosion rates are relatively high in comparison with aluminium bronzes (compare section CuAl-alloys). The silicon-containing alloy exhibits a low corrosion rate at low H_2SO_4 concentrations and low temperatures. Nevertheless, no comparison with other alloys is possible from the table, since the concentration of the acid is only 1 %. This is considerably less suitable than the alloy CuZn40Mn1Pb1 (CW720R; old 2.0580), and not even resistant in 1 % sulfuric acid [1531].

Material	Temperature K (°C)	H_2SO_4 concentration %	Corrosion rate, mm/a (mpy)			
			Air access		Air exclusion	
			Agitated	Static	Agitated	Static
CuZn15 CW502L	293 (20)	3	1 (39.4)	0.2 (7.87)	0.2 (7.87)	
	358 (85)	50		0.3 (11.8)		0.03 (1.18)
	373 (100)	50		0.7 (27.6)		0.2 (7.87)
CuZn16Si4-C CC761S	293 (20)	1	0.13 (5.12)	0.08 (3.15)	0.05 (1.97)	0.02 (0.79)
	323 (50)	1	0.2 (7.87)	0.1 (3.94)		
	353 (80)	1	0.44 (17.3)	0.2 (7.87)		
CuZn30 CW505L	293 (20)	4	2.2 (86.6)		0.5 (19.7)	
	293 (20)	10	< 5 (< 197)			
CuZn36Pb	293 (20)	10			0.5 (19.7)	
	358 (85)	50		0.16 (6.30)		0.03 (1.18)
	373 (100)	50		0.22 (8.66)		0.06 (2.36)
	398 (125)	50		0.8 (31.5)		0.24 (9.45)

Table 261: Corrosion of copper-zinc alloys in sulfuric acid solutions [1531]

Table 261: Continued

Material	Temperature K (°C)	H$_2$SO$_4$ concentration %	Corrosion rate, mm/a (mpy)			
			Air access		Air exclusion	
			Agitated	Static	Agitated	Static
CuZn40 CW509L	293 (20)	0.1	0.8 (31.5)	0.4 (15.7)	0.7 (27.6)	
	293 (20)	10	< 10 (<394)	1.3 (51.2)		
CuZn40MnPb CW720R	293 (20)	1	1.1 (43.3)		0.1 (3.94)	
CuZn44Pb2 CW623N	293 (20)	10		0.1 (3.94)		0.1 (3.94)
CuZn39Pb3 CW614N	293 (20)	25		3 (118)		2 (78.7)
	293 (20)	40		4.2 (165)		2.6 (102)
	313 (40)	10		0.11 (4.33)		
	313 (40)	25		3.64 (143)		
	313 (40)	40		3.63 (143)		
	333 (60)	10		0.13 (5.12)		
	333 (60)	25		1.5 (59.1)		
	333 (60)	40		1.3 (51.2)		

Table 261: Corrosion of copper-zinc alloys in sulfuric acid solutions [1531]

The hydrogen formed during acid corrosion has only an insignificant influence, if any, on the stress corrosion cracking behavior of the brass CuZn30, as confirmed by studies in an electrolyte of approximately 5 % sulfuric acid and 40 mg/l H$_2$SeO$_3$ at room temperature and at a current density of 100 mA/cm^2 (Figure 232) [1566, 1567].

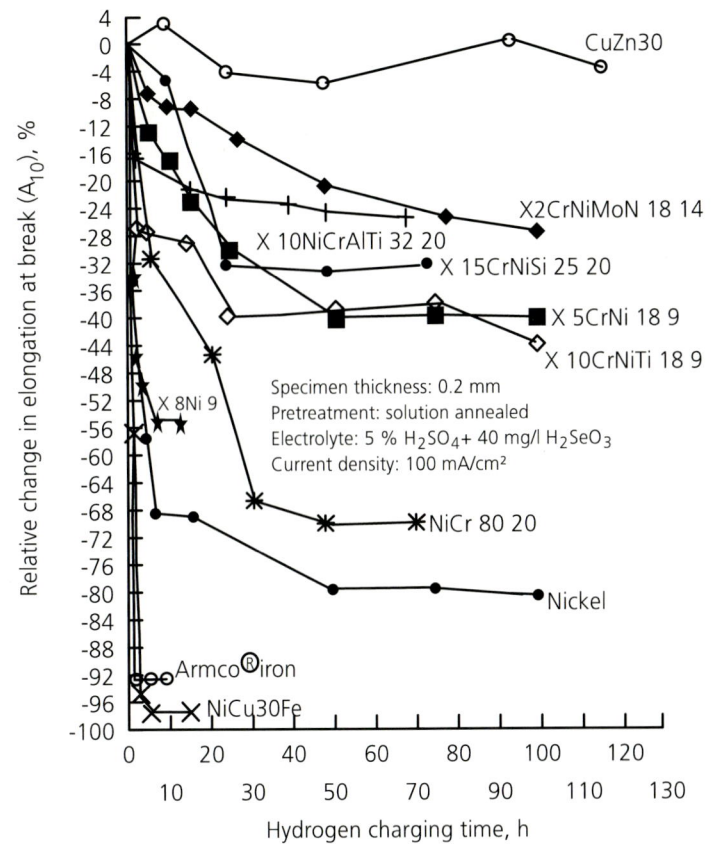

Figure 232: Relative change in the elongation at break as a function of the hydrogen charging time in 5 % sulfuric acid [1566, 1567]

– Inhibitors –

0.12 g/l benzotriazole in 0.5 % sulfuric acid exhibits an inhibitor efficiency of 86 % on brass (CuZn30).

Figure 233 shows the inhibiting effect of picolines on the corrosion of CuZn30 (Ms70) brass in 1 % sulfuric acid. The maximum effect of the inhibitors initially increases with the action time, reaches a peak after a duration of 2 days and then decreases [1568].

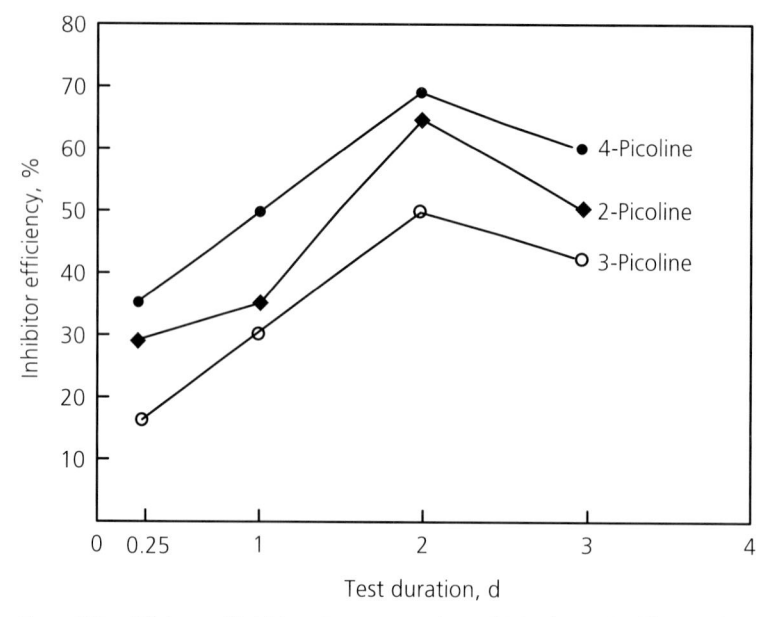

Figure 233: Efficiency of inhibitors in concentrations of 0.1 g/l in 1 % sulfuric acid at 303 K (30 °C) [1568]

Other copper alloys

The corrosion behavior of copper is improved by alloying with silicon. The corrosion resistance in acids increases as the silicon content increases, and the corrosion resistance in sulfuric acid, of course, depends, as with all copper materials (compare in particular sections Copper and CuAl-alloys), on the concentration, aeration and agitation of the acid.

Figure 234 shows the excellent corrosion resistance of silicon bronze in various concentrations of sulfuric acid at 298 and 343 K (25 and 70 °C). Since no further data are available, it is to be assumed that the solutions were deaerated and not agitated. According to this graph, the corrosion rate is less than 0.08 mm/a (3.15 mpy) at 298 K (25 °C) in 10 % sulfuric acid, and still less than 0.2 mm/a (7.87 mpy) at 343 K (70 °C). The corrosion rate in 70 % acid at 343 K (70 °C) is about 0.02 mm/a (0.79 mpy) [1536]. A copper-silicon alloy containing 1 % manganese exhibits the same good corrosion resistance in sulfuric acid at 298 and 343 K (25 and 70 °C) (Table 262).

According to [1559], silicon bronzes are practically resistant in sulfuric acid solutions of up to 70 % at 343 K (70 °C).

The corrosion rate on the alloy Everdur® 1010 (95.8 % Cu-3.1 % Si-1.1 % Mn) was 0.25 mm/a (9.84 mpy) in 5 % sulfuric acid at 493 K (220 °C) with an ethyl alcohol content of 1 %, and the alloy exhibits a significantly better corrosion resistance under these conditions than chromium-nickel steel containing copper at 6.6 mm/a

(260 mpy), or lead at 5.6 mm/a (220 mpy). The test was performed for one day in a deaerated solution [1521].

H_2SO_4 concentration	Temperature, K (°C)	
	298 (25)	343 (70)
	Corrosion rate, mm/a (mpy)	
3	0.07 (2.76)	0.18 (7.09)
10	0.06 (2.36)	0.15 (5.91)
25	0.04 (1.57)	0.09 (3.54)
70	0.02 (0.79)	0.03 (1.18)
90	< 0.10 (< 3.94)	–

Table 262: Corrosion rates for silicon bronze (96 % Cu, 3 % Si, 1 % Mn) in sulfuric acid solutions [1520]

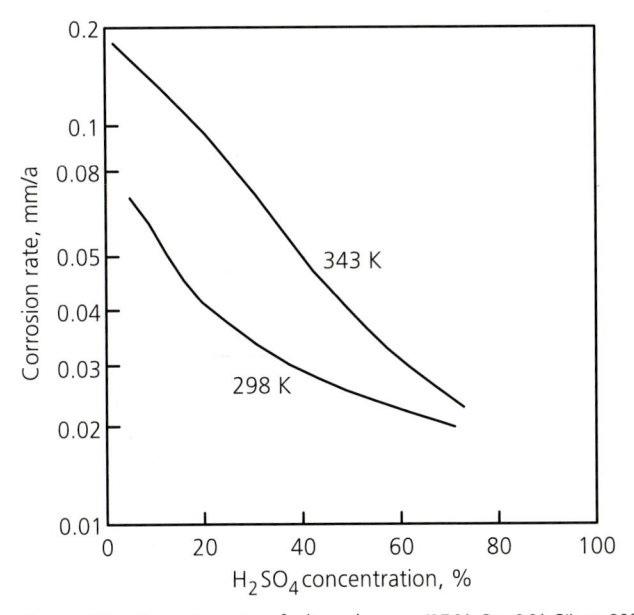

Figure 234: Corrosion rate of silicon bronze (97 % Cu, 3 % Si) at 298 and 343 K (25 and 70 °C) [1536]

Material	Temperature	H$_2$SO$_4$ concentration	Corrosion rate, mm/a (mpy)			
			Air access		Air exclusion	
	K (°C)	%	Agitated	Static	Agitated	Static
CuSi3Mn (CuSi3Zn)	298 (25)	0.5	0.5 (19.7)			
	298 (25)	1		0.06 (2.36)		
	298 (25)	3		0.07 (2.76)		
	298 (25)	10	0.2 (7.87)	0.06 (2.36)		
	298 (25)	25		0.035–0.04 (1.38–1.57)		
	298 (25)	70		0.017–0.02 (0.67–0.79)		
CuSi4	293 (20)	95				0.05 (19.7)
	363 (90)	50				0.06 (2.36)
CuSi3Mn (CuSi3Zn)	343 (70)	3		0.18 (7.09)		
	343 (70)	10		0.07–0.15 (2.76–5.91)		
	343 (70)	25		0.09 (3.54)		
	343 (70)	70		0.01–0.03 (0.39–1.18)		
CuSi3Mn (CuSi3Zn)	468 (195)	0.5		0.42 (16.5)		
	468 (195)	5		destroyed		
CuBe2 (CW101C)	333 (60)	10		0.88 (34.6)		
CuPb8Sn	373 (100)	41.5		3.5 (137)		0.6–0.75 (23.6–29.5)
	373 (100)	78				2.6–3 (102–118)
	373 (100)	95.5				< 100 (< 3,937)

Table 263: Corrosion on copper-silicon, copper-beryllium and copper-lead alloys in sulfuric acid solutions [1531]

Table 263 contains a summary of other alloys based on copper-silicon. In agitated 10% sulfuric acid at 298 K (25 °C), the corrosion rate of CuSi3Mn of 0.06 mm/a (2.36 mpy) in static acid rises to 0.2 mm/a (7.87 mpy). The alloy is destroyed at 468 K (195 °C) in this solution [1531].

Table 264 contains the corrosion rates for spring alloys of beryllium bronze in the hard age-hardened state in various concentrations of sulfuric acid at 293 and 373 K (20 and 100 °C). According to this table, the alloy containing a minor amount of cobalt has the better resistance in 10 % sulfuric acid at 293 K (20 °C) [1569].

H_2SO_4 concentration %	Beryvac® 200 (CuBe2)		Beryvac® 60 (CuCoBe)	
	Temperature, K (°C)			
	293 (20)	373 (100)	293 (20)	373 (100)
	Corrosion rate, mm/a (mpy)			
10	0.1–1.0 (3.94–39.4)	1–10 (39.4–394)	≤ 0.1 (≤ 3.94)	1–10 (39.4–394)
50	≤ 0.1 (≤ 3.94)	0.1–1.0 (3.94–39.4)	≤ 0.1 (≤ 3.94)	0.1–1.0 (3.94–39.4)
96	≤ 0.1 (≤ 3.94)	> 10.0 (> 394)	≤ 0.1 (≤ 3.94)	> 10.0 (> 394)

Table 264: Corrosion rates of spring alloys in the hard age-hardened state in sulfuric acid [1569]

In all cases where the factors with an adverse effect on the corrosion of copper alloys in sulfuric acid can be expected (see also section Copper), in-house corrosion studies are advisable.

Summary of copper materials

The corrosion resistance of copper and copper-based materials in solutions containing sulfuric acid is not based on the formation of surface layers (see section Copper). Oxygen-containing solutions and increasing flow rates of the medium as well as particles contained therein considerably intensify the corrosion. The material suffers particularly severe corrosion at the liquid-solid phase boundary.

The use of copper materials is of interest in the concentration range of 50 to 70 % sulfuric acid, in which the high-alloy chromium-nickel steels are attacked the most.

Materials which are to be recommended from the point of view of corrosion resistance are aluminium and silicon bronzes and copper, roughly in that sequence. In the case of aluminium bronzes, it should be noted that the influence of aeration of the medium is less pronounced than in the case of copper and bronzes or red brass.

Waste Water (Industrial)

Author: E. Heitz, G. Subat / Editor: R. Bender

Copper-nickel alloys

Copper-nickel alloys can fail if the waste waters contain hydrogen sulfide, sulfides, ammonia, or ammonium compounds [1570].

Owing to their good resistance to high concentrations of salt solutions, the alloys CuNi30Fe1Mn (CW354H, 2.0882, CuNi 70/30), and CuNi10Fe1Mn (CW352H, 2.0872, CuNi 90/10) are used in seawater and waste water evaporation plants [1571].

An evaporator plant made of CuNi30Fe1Mn, which processed waste water from the scrubbers of a waste incineration plant, exhibited stress corrosion cracking after only a short operating time. The reason was found to be mercury entrained in the waste (Hg content of the waste water 10.6 mg/l). The plant was able to continue operating without problems after a mercury precipitation step had been installed upstream of the evaporation unit and the damaged sections had been repaired [1572].

In smaller waste water evaporation plants, some of the valves and other fittings are made of partly nickel-plated CuNi alloys [1573]. The evaporator and heat exchangers in seawater desalination plants with vapor compression are frequently made of CuNi10Fe1Mn and CuNi30Fe1Mn [1574].

A comprehensive tabulated compilation of metallic materials, including copper-nickel alloys, used in seawater desalination plants with multistage flash processes is given in [1575]. However, the statements made in [1574] and [1575] only refer to experiences gained with seawater desalination plants. Therefore, the results have only very limited applicability to waste water because the pH-values of seawater are usually neutral to slightly alkaline. Because hot solutions with high chloride contents are produced in the final stage of seawater desalination plants that use evaporation processes, experiences gained with respect to the material's behavior can also be applied to the material behavior in the corresponding plant sections of waste water evaporation plants of flue gas desulfurization plants and waste incineration plants that also have high chloride concentrations, as long as they have comparable pH-values.

Bibliography

[1] Rabald, E.; Bretschneider, H.
DECHEMA-Werkstoff-Tabelle
(DECHEMA Corrosion Data Sheets)
(in German)
Chapters: Aluminium acetate, Ammonium acetate, Lead acetate, Calcium acetate, Potassium acetate, Copper acetate, Sodium acetate
DECHEMA, D-Frankfurt am Main

[2] Nelson, G. A.
Corrosion data survey
NACE, Houston, Tex. (US), (1960) pp. A 1, C 3, F 1, L 1, M 1, N 1, S 1

[3] Nelson, G. A.
Corrosion data survey
NACE, Houston, Tex. (US) (1967) pp. A 4, A 6, B 1, C 1, C 11, C 12, F 1, L 1, L 2, M 1, M 2, M 3, N 1, P 6, S 2, S 4, S 8, S 9, Z 1

[4] Hamner, N. E.
Corrosion data survey
NACE, Houston, Tex. (US) (1974) pp. 8, 14, 44, 66, 110, 146, 160, 208, 212, 214, 236

[5] Escalante, E.; Kruger, J.
Stress corrosion of pure copper
J. Electrochem. Soc. 118 (1971) No. 7; pp. 102–106

[6] Rabald, E.
DECHEMA-Werkstoff-Tabelle Ausg. 1948, pp. 48–50, 60–62, 68–70
(DECHEMA Corrosion Data Sheets (1948) pp. 48–50, 60–62, 68–70 (in German)
DECHEMA, D-Frankfurt am Main

[7] Rabald, E.
Corrosion guide
Elsevier Publishing Co., Amsterdam-London-New York (1968)

[8] Parkins, R. N.; Holroyd, N. J.
Stress corrosion cracking of 70/30 brass in acetate, formate, tartrate, and hydroxide solutions
Corrosion (Houston) 38 (1982) No. 5; pp. 245–255

[9] DECHEMA-Werkstoff-Tabelle "Essigsäure"
(DECHEMA corrosion data sheets "Acetic acid") (in German)
DECHEMA, D-Frankfurt am Main, 1957

[10] Franke, E.
Neue korrosionsbeständige Werkstoffe im Ausland.
IV. Kupfer-Nickel-Legierungen mit geringen Eisengehalten für Kondensatorrohre
(New corrosion-resistant materials in foreign countries. IV. Copper-nickel alloys with small iron contents for condenser tubes) (in German)
Werkst. Korros. 1 (1960) p. 349

[11] Ritter, F.
Korrosionstabellen metallischer Werkstoffe
(Corrosion charts of metallic materials) (in German)
4th ed., p. 88
Springer-Verlag, Wien, 1957

[12] Anuchin, P. I.; Firsov, A. I.; Mikhalyuk, G. F.
The corrosion resistance of various grades of copper in acetic acid solutions
(in Russian)
Gidroliz. Lesokhim. Prom. 18 (1965) 2, p. 12
(C.A. 62 (1965) 15858g)

[13] Le Monnier, E.
in: Kirk-Othmer Encyclopedia of Chemical Technology
2nd ed., vol. 8, p. 386
John Wiley & Sons, Inc., New York-London-Sydney, 1965

[14] Gulyaev, B. N.; Chashchin, A. M.
Replacement of non-ferrous metals in acetic acid rectification columns
(in Russian)
Gidroliz. Lesokhim. Prom. 8 (1955) 3, p. 19

[15] DECHEMA-Werkstoff-Tabelle
"Aliphatische Aldehyde"
(DECHEMA corrosion data sheets "Aliphatic aldehydes") (in German)
DECHEMA, D-Frankfurt am Main, 1978

[16] DECHEMA-Werkstoff-Tabelle
"Peressigsäure und Acetaldehyd-Monoperacetat"
(DECHEMA corrosion data sheets "Peracetic acid and acetaldehyde monoperacetate") (in German)
DECHEMA, D-Frankfurt am Main, 1966

[17] Chemical safety data sheet SD-43
"Acetaldehyde"
Man. Chem. Ass., Washington, 1952

[18] Farkhadov, A. A.; Asatryan, V. G.; Abalyan, N. P.; Belayev, A. I.; Orlovskaya, G. A.
Corrosion of copper and stainless steel in aqueous mixtures of ethylacetate with acetic acid (in Russian)
Zashch. Met. 4 (1968) 3, p. 319

[19] Brockhaus, R.; Förster, G.
in: Ullmanns Encyklopädie der technischen Chemie
(Ullmann's encyclopedia of industrial chemistry) (in German)
4th ed., vol. 11, p. 57
Verlag Chemie, D-Weinheim, 1976

[20] Dillon, C. P.
Role of contaminants in acetic acid corrosion
Mater. Protection 4 (1965) 9, p. 20

[21] Bourrat, J.
The corrosion resistance of stainless steel and alloys in organic acids
Corrosion et Anticorrosion 11 (1963) 4, p. 140

[22] Baker, S.
Materials for chemical plants: copper
Chem. Process Eng. 41 (1960) 11, p. 513

[23] Anonymous
Corrosion resistant materials for the chemical industry
Corros. Prev. Control 9 (1965) 6, p. 35

[24] Rabald, E.
Corrosion Guide, 2nd. ed., p. 3
Elsevier Publishing Company, Amsterdam-London-New York, 1968

[25] Stahl, R.; Kiefer, P.
Korrosion und Korrosionsschutz in der Lebensmittelindustrie
(Corrosion and corrosion protection in the food industry) (in German)
Werkst. Korros. 24 (1973) 6, p. 513

[26] Anuchin, P. I.; Aleeva, T. K.
Corrosion of titanium in wood pulp chemical media (in Rumanian)
Khim. Pererabotka Drevesiny Sb. 26 (1963) p. 5
(C.A. 61 (1964) 14308f)

[27] Konradi, M. V.
Corrosion of copper, aluminium, and iron in solutions of acetic acid and ammonium sulfate (in Russian)
Izv. V.U.Z. Tsvetn. Metall. (1958) 4, p. 165
(C.A. 53 (1959) 7941e)

[28] Heitz, E.
Vergleichende Korrosionsuntersuchungen in Ethanol-, Methanol- und wäßrigen Lösungen
(Comparative corrosion testings in ethanol-, methanol- und aqueous solutions) (in German)
Werkst. Korros. 38 (1987) 12, p. 333

[29] Teeple, H. O.
Corrosion by acetic acid – NACE Report of Task Group T-5A-3
Corrosion 13 (1957) p. 757

[30] Schlain, D.; Kenahan, C. B.; Steele, D. V.
Galvanic corrosion properties of titanium in organic acids
US Bur. Mines, Rep. Invest. No. 5189 (1956) 17 p.

[31] Lositskii, N. T.; Glushko, V. Ya.; Stepanov, S. I.; Vinnichenko, V. V.
Corrosion resistance of copper weld joints produced with an unshielded arc (in Russian)
Khim. Neft. Mashinostr. (1974) 1, p. 18
(C.A. 80 (1974) 136581z)

[32] Johnson, P. R.
Corrosion in the production of acetic acid and its derivatives
Aust. Corr. Eng. 10 (1966) 7, p. 9

[33] Kiss, I.; Versanyi, M.
Testing of the anodic dissolution of copper in anhydrous acetic acid solutions. II. Effect of chlorine ions (in Hungarian)
Magyar Kém. Fol. 80 (1974) 5, p. 225

[34] Robertson, W. D.; Nole, V. F.; Davenport, W. H.; Talboom jr., F. P.
An investigation of chemical variables affecting the corrosion of copper
J. Electrochem. Soc. 105 (1958) 10, p. 569

[35] Anuchin, P. I.
Industrial tests on stainless steels, copper, and titanium in dry-distillation works (in Russian)
Sb. Tr. Tsent. Nauch.-Issled. Proekt. Inst. Lesokhim. Prom. (1966) 17, p. 200
(C.A. 66 (1967) 79019g)

[36] Kiss, L.; Bosquez, A.; Varsanyi, M. L.
Study of the anodic dissolution of copper in nonaqueous acetic acid solutions-IV
Acta Chim. Acad. Sci. Hung. 107 (1981) 1, p. 11

[37] Molodov, A. I.; Yanov, L. A.; Posev, V. V.; Golodnitskaya, D. V.; Maslyuk, G. N.
A study of the discharge-Ionisation of

copper in glacial acetic acid (in Russian)
Elektrokhimiya 15 (1979) 1, p. 122

[38] Patel, N. K.; Patel, L. N.
A réz savas oldódásának korróziós inhibitorai
(Corrosion inhibitors for acid dissolution of copper) (in Hungarian)
Korroz. Figyelö 15 (1975) 6, p. 264

[39] Nelson, G. A.
Corrosion Data Survey 1967 – Corrosion charts A-1/A-2 "Acetic acid"
NACE, Houston (Texas/USA), 1968

[40] Hamner, N. E.
Corrosion Data Survey
5th ed., p. 2, p. 198
NACE, Houston (Texas/USA), March 1974

[41] Kuzyukov, A. N.; Khanzadeev, I. V.
The intercrystalline corrosion of copper in acids (in Russian)
Zashch. Met. 15 (1979) 3, p. 341

[42] Koziukov, A. N.; Khanzadeev, I. V.; Kuziukova, A. N.
Intercrystalline corrosion of stressed copper (in Russian)
Zashch. Met. 21 (1985) 4, p. 586

[43] Evans, L. S.; Morgan, P. E.
Effect of steam temperature on the corrosion of metals under heat-transfer conditions
Br. Corros. J. 2 (1967) p. 150

[44] Evans, L. S.; Morgan, P. E.
Effect of steam temperature on the corrosion of metals under heat-transfer conditions
Br. Corros. J. 2 (1967) p. 150

[45] E.P. 425973 (Du Pont de Nemours) (1933)

[46] DECHEMA-Werkstoff-Tabelle "Essigsäureanhydrid"
(DECHEMA corrosion data sheets "Acetic anhydride") (in German)
DECHEMA, D-Frankfurt am Main, 1957

[47] Le Monnier, E.
in: Kirk-Othmer "Encyclopedia of chemical technology" 2nd ed., vol. 8, p. 405
John Wiley & Sons, Inc., New York-London-Sydney, 1965

[48] Whitman, W. O.; Russel, R. P.
Ind. Eng. Chem. 17 (1925) p. 348

[49] Ford, C. E.
Materials of construction data
Ind. Eng. Chem. 40 (1948) p. 1829

[50] Fuller, T. S. et al.
Am. Soc. Test. Mat. Proceed. 27 (1927) p. 281

[51] DECHEMA-Werkstoff-Tabelle "Essigsäure"
(DECHEMA corrosion data sheets "Acetic acid") (in German)
DECHEMA, D-Frankfurt am Main, 1957

[52] Anonymous
Chemical safety data sheet SD-15 "Acetic anhydride"
Man. Chem. Ass., Washington, 1962

[53] Kühn-Birett
Merkblätter gefährliche Arbeitsstoffe, Blatt E 10 "Essigsäureanhydrid"
(Advisory sheets on hazardous materials handling, paper E 10 "acetic anhydride") (in German)
Verlag Moderne Industrie, Wolfgang Dummer & Co.
D-München, 1977

[54] Rabald, E.
Corrosion guide, 2nd ed., p. 13
Elsevier Publishing Company, Amsterdam-London-New York, 1968

[55] Anonymous
Acetic acid by submerged fermentation
Chem. Eng. 67 (1960) 18, p. 50

[56] Römpps Chemie Lexikon
(Römpp's chemical encyclopedia) (in German)
7th ed., vol. 2, p. 1049
Franckh'sche Verlagshandlung, W. Keller & Co., D-Stuttgart, 1973

[57] Mark, E. M.; La Roux, J. C.
Ind. Eng. Chem. 24 (1932) p. 797

[58] Franke, E.
Das Korrosionsverhalten von Aluminiumbronzen
(The corrosion behavior of aluminium bronzes) (in German)
Werkst. Korros. 2 (1951) p. 298

[59] Flückiger, E.; Zürcher, H.
Nachfermentierung von Emmenthaler Käse in Kupfer- und in Stahlwannen
(Post fermentation of Emmenthaler cheese in copper and steel tubs) (in German)
Schweiz. Milchztg., Wiss. Beil. 59 (1968) p. 3

[60] Yamaguchi, H.
KMC alloy for condenser tubes
(in Japanese)
Kinzoku 26 (1956) p. 782
(C.A. 55 (1961) 313a)

[61] Eisenbrown, Ch. M.; Barbis, P. R.
Corrosion of metals by acetic acid
Chem. Eng. 70 (1963) 9, p. 148

[62] Anonymous
Corrosion by acetic acid and acetic acid mixtures under heat transfer conditions
Mater. Protection 6 (1967) p. 77

[63] Zitter, H.; Kraxner, G.
Korrosionsbeständigkeit von Aluminiumbronzen in Säuren
(Corrosion resistance of aluminium bronzes in acids) (in German)
Werkst. Korros. 14 (1963) 2, p. 80

[64] Franke, E.
Das Korrosionsverhalten von Aluminiumbronzen
(The corrosion behavior of aluminium bronzes) (in German)
Werkst. Korros. 2 (1951) p. 306

[65] Anonymous
Valve corrosion
Mater. Protection 4 (1965) 6, p. 28

[66] Franke, E.
Das Korrosionsverhalten von Aluminiumbronzen
(The corrosion behavior of aluminium bronzes) (in German)
Werkst. Korros. 2 (1951) p. 298

[67] Ried, G.
Übersicht über Werkstoffe für Absperrorgane, die bei angreifenden Medien zu empfehlen sind
(A survey of materials for shut-off units, recommended for use in the presence of corrosive media) (in German)
Werkst. Korros. 15 (1964) 6, p. 468

[68] Oakes, B. D.
Propargylabietylamines as corrosion inhibitors in acid solutions
US Pat. 3 242 094 (Dow Chemical Co.) (22.3.1966)

[69] Shah, R. S.; Desai, C. S.
Corrosion of brass by acetic acid and chlorosubstituted acetic acids
Werkst. Korros. 27 (1976) p. 705

[70] Talati, J. D.; Desai, M. N.; Trivedi, A. M.
Der Einfluß von Wasserstoffsuperoxyd auf die Korrosion von Messing
(The influence of hydrogen peroxide on the corrosion of brass) (in German)
Werkst. Korros. 12 (1961) 7, p. 422

[71] Desai, M. N.; Trivedi, A. K. M.
The influence of hydrogen peroxide on the corrosion of brass by acids
Indian J. Appl. Chem. 21 (1958) p. 137

[72] Anonymous
Copper alloy No. 879 (silicon brass die-casting alloy)
Alloy Dig. (1977) June, Cu-335, 2 p.

[73] Kristal, M. M.; Adugina, N. A.; Krutikov, A. N.; Arest, T. V.
Corrosion resistance of welded bimetal joints of steel-10-copper, steel-10-bronze, and steel-10-brass (in Russian)
Tr. Vses. Nauchn.-Issled. Konstrukt. Inst. Khim. Mashinostr. (1963) 45, p. 66
(C.A. 61 (1964) 14268e)

[74] Patel, M. M.; Patel, N. K.; Vora, J. C.
Thiocresol as a corrosion inhibitor for brass
J. Electrochem. Soc. India 27 (1978) 3, p. 171

[75] Aoyama, T.; Kamioka, M.
Corrosion of nickel-plated brass in wooden boxes (in Japanese)
Corrosion Eng. (Boshoku Gijutsu) 14 (1965) 9, p. 398

[76] Soudan, A.
Corrosion of metals by adhesives (in French)
Galvano-Organo-Trait. Surf. 54 (1986) 569, p. 650

[77] Cermakova, D.; Dohnalova, J.
Corrosion of metals by vapors of organic materials in microclimate (in Czech)
Koroze Ochr. Mater. (1963) 4, p. 5
(C.A. 60 (1964) 10336d)

[78] Loos, W.
Über Korrosionserscheinungen, ausgelöst durch duroplastische Formstoffe. Teil 2: Elektrolytische Korrosion
(Investigations of corrosion caused by thermosetting materials. Part 2: Electrolytic corrosion) (in German)
Z. Werkstofftech. 7 (1976) 3, p. 97

[79] Weisner, E.
Jahresübersicht: Schwermetallguß
(Annual survey: heavy metal cast) (in German)
Gießerei 57 (1970) 4, p. 90

[80] Anonymous
Beryllium und seine Anwendung in der Werkstofftechnik
(Beryllium and its application in material technology) (in German)
Werkst. Korros. 6 (1955) p. 363

[81] Richards, J. T.
Corrosion resistance of beryllium copper
Corrosion 9 (1963) p. 359

[82] White, W. E.; Pupp, C.
Kirk-Othmer Encyclopedia of Chemical Technology, 1st. Ed., Vol. 9 (1966) p. 636
John Wiley & Sons Inc., New York, London, Sydney, 1966

[83] Nelson, G. A.
Corrosion Data Survey (1968) A-3, B-3, B-7, C-8, C-9, T-3
National Ass. Corr. Eng. Houston (Tex.)

[84] DECHEMA-Werkstoff-Tabelle "Mersole, Mersolsulfonsäuren, Mersolate"
DECHEMA corrosion data sheets "mersols, mersolsulfonic acids, mersolates") (in German)
DECHEMA, D-Frankfurt am Main, 1961

[85] Rabald, E.
Corrosion Guide, 2nd Ed. (1968) p. 857
Elsevier Publishing Company, Amsterdam, London, New York, 1968

[86] Eibeck, R. E.
Kirk-Othmer Encyclopedia of Chemical Technology, 2nd Ed., Vol. 9 (1966) p. 676
John Wiley & Sons Inc., New York, London, Sydney, 1966

[87] DECHEMA-Werkstoff-Tabelle "Phosgen"
(DECHEMA corrosion data sheets "phosgene") (in German)
DECHEMA, D-Frankfurt am Main, December 1968

[88] Rasmussen, M. J.; Hopkins jr., H. H.
Preparing plutonium metal via the chloride process
Ind. Eng. Chem. 53 (1961) No. 6; p. 453

[89] DECHEMA-Werkstoff-Tabelle "Fluorcarbonsäurehalogenide"
(DECHEMA corrosion data sheets "fluocarboxylic acid halides") (in German)
DECHEMA, D-Frankfurt am Main, 1957

[90] Thyssen Edelstahlwerke AG, Krefeld
Product Information 1127/2 (Oct. 1976)

[91] DECHEMA-Werkstoff-Tabelle "Acetylchlorid"
(DECHEMA corrosion data sheets "acetylchloride") (in German)
DECHEMA, D-Frankfurt am Main, 1953

[92] DECHEMA-Werkstoff-Tabelle "Buttersäurechlorid (Butyrylchlorid)"
(DECHEMA corrosion data sheets "butyryl chloride" (in German)
DECHEMA, D-Frankfurt am Main, 2007

[93] Bachmann, P.
The (metal) corrosion in the chemical industry
Chimia 10 (1956) Aug.; p. 189

[94] DECHEMA-Werkstoff-Tabelle "Benzoylchlorid"
(DECHEMA corrosion data sheets "benzoyl chloride")
DECHEMA, D-Frankfurt am Main, 1955

[95] Ried, G.
Übersicht über Werkstoffe für Absperrorgane, die bei angreifenden Medien zu empfehlen sind
(Survey of valve materials for corrosive media) (in German)
Werkstoffe u. Korrosion 15 (1964) No. 6; p. 468

[96] Nelson, G. A.
Corrosion Data Survey, 1967 Edition, A-1, A-3, A-4, B-6, B-7, C-4, C-7, C-8, C-13, E-1, F-2, G-1, H-1, M-4, M-7, P-1, P-10
National Ass. Corr. Eng., Houston, Texas

[97] Chemical Safety Data Sheet SD-43 (1952) Acetaldehyde
Man. Chem. Ass., Washington DC, 1952

[98] DECHEMA-Werkstoff-Tabelle "Peressigsäure und Acetaldehyd-Monoperacetat"
(DECHEMA Corrosion Data Sheets "Peracetic Acid and Acetaldehyde Monoperacetate") (in German)
DECHEMA, D-Frankfurt am Main, 1966

[99] Farberov, M. I.; Sperenskaya, V. A.
J. Appl. Chem. USSR 28 (1955) p. 205; (Werkstoffe und Korrosion 8 (1957) No. 8/9; p. 484)

[100] Bernhard, E.
Eidg. Techn. Hochschule, Zürich
Intern. Diary Congr. Proc. 16th, Copenhagen (1962) No. 3; p. 141;
(C. A. 65 (1966) 7891f)

[101] Lingnau, E.
Das Verhalten der Werkstoffe gegenüber Formaldehyd
(The Behavior of Materials towards Formaldehyde) (in German)
Werkstoffe und Korrosion 8 (1957) No. 8/9; p. 480

[102] Teeple, H. O.
Ing. e. Chim. 12 (1954) No. 1; p. 3; No. 2; p. 3; (Corrosion 8 (1952) No. 1; p. 14)

[103] Yakhontov, V. D.
Sbornik Statei po Obshchei Khim. Akad. Nauk USSR 2 (1954) p. 1158; (Werkstoffe und Korrosion 8 (1957) No. 8/9; p. 480) (in Russian)

[104] Diem, H.; Hilt, A.
Ullmanns Encyklopädie der technischen Chemie
(Ullmann's Encyclopedia od Industrial Chemistry) 4th Edition, Vol. 11; p. 68
(in German)
Verlag Chemie GmbH, Weinheim/Bergstraße, 1976

[105] Falbe, J.; Payer, W.
Ullmanns Encyklopädie der technischen Chemie
(Ullmann's Encyclopedia of Industrial Chemistry) 4th Edition, Vol. 7; p. 118
(in German)
Verlag Chemie GmbH, Weinheim/Bergstraße, 1974

[106] Burmester, A.
Materialprüfung 14 (1972) No. 4; p. 122
(in German)

[107] DEGUSSA, D-Frankfurt am Main
Product-Information, Silver and Silver Alloys, Edition 1971 (in German)

[108] Falbe, J.; Weber, J.
Ullmanns Encyklopädie der technischen Chemie
(Ullmann's Encyclopedia of Industrial Chemistry) 4th Edition, Vol. 9; p. 42
(in German)
Verlag Chemie GmbH, Weinheim/Bergstraße, 1975

[109] DECHEMA-Werkstoff-Tabelle
"Butyraldehyd"
(DECHEMA Corrosion Data Sheets "Butyraldehyde") (in German)
DECHEMA, D-Frankfurt am Main, 1955

[110] Jira, R.; Smidt, J.
Ullmanns Encyklopädie der technischen Chemie
(Ullmann's Encyclopedia of Industrial Chemistry 4th Edition, Vol. 7; p. 19)
(in German)
Verlag Chemie GmbH, Weinheim/Bergstraße, 1974

[111] Guzenow, V. K.
Gidroliz i. Lesokhim, Prom 8 (1955) No. 3; p. 21; (C. A. 49 (1955) No. 21; 14396g)
(in Russian)

[112] Sleeman, D. G.
Chem. Eng. 75 (1968) No. 1; p. 42

[113] DECHEMA-Werkstoff-Tabelle
Formaldehyd und polymere Formaldehyde
(Formaldehyde and Polymer Formaldehydes) (in German)
(DECHEMA Corrosion Data-Sheet)
DECHEMA, D-Frankfurt am Main, 1958

[114] DECHEMA-Werkstoff-Tabelle
"Acetaldehyd"
(DECHEMA Corrosion Data Sheets "Acetaldehyde") (in German)
DECHEMA, D-Frankfurt am Main, 1953

[115] DECHEMA-Werkstoff-Tabelle
(DECHEMA Corrosion Data Sheets)
(in German)
Crotonaldehyde, September 1956
DECHEMA, D-Frankfurt am Main, 1956

[116] DECHEMA-Werkstoffe-Tabelle
"Capronaldehyd"
(DECHEMA Corrosion Data Sheets "Capronaldehyde") (in German)
DECHEMA, D-Frankfurt am Main, 1955

[117] Montecatini Soc. Generale per l'Industria Mineraria e Chimica
Brit. P. 847564, 7.9.1960; (C. A. 55 (1961) No. 8; 7291b)

[118] Sulyok, J.
Werkstoffe und Korrosion 26 (1975) p. 858
(in German)

[119] Cermákova, D.
Informat. Problem Klimaschutz 2 (1963) No. 4; p. 99;
(Werkstoffe und Korrosion 17 (1966) No. 5; 409g)

[120] Cermákova, D.; Dohnalova, J.
Statni Vyzkumny Ustav Orchrany Mater., Prague
Koroze Ochrana Mater. (1963) No. 4; p. 75; (C. A. 60 (1964) 10336d)

[121] Cermákova, D.; Vickova, J.
Tr. Mezhdunar. Kongr. Korroz. Metal., 3rd 1966, (Publ. 1968, 4, p. 507 (in Russian); (C.A. 71 (1969) 127950f)

[122] Anonymous
Corrosion Technology 4 (1957) No. 9; p. 301

[123] Anonymous
Chem. Proc. Eng. 37 (1956) No. 1; p. 34
Werkstoffe und Korrosion 8 (1957) No. 6, p. 355c

[124] Poulin, C. J.; Rice, G.
U.S.A. P. 3398031, 20.8.1968; (C. A. 69 (1968) No. 18; 69246b)

[125] Falbe, J.; Hahn, H. D.
Ullmanns Encyklopädie der technischen Chemie
(Ullmann's Encyclopedia of Industrial Chemistry) 4th Edition, Vol. 9; p. 139

(in German)
Verlag Chemie GmbH, Weinheim/Bergstraße, 1975

[126] Anonymous
Chem. Eng. 58 (1951) No. 1; p. 108

[127] Franke, E.
Werkstoffe und Korrosion 2 (1951) No. 8, p. 306

[128] Miller jr., R.
Chem. Eng. 75 (1968) Jan. 15; p. 182

[129] Heyden Chemical Corporation, New York, 1956
Formaldehyde
(Werkstoffe und Korrosion No. 8 (1957) No. 8/9; p. 480)

[130] Farbwerke Hoechst AG
Brit. P. 739263 vom 26.10.1955; (C. A. 50 (1956) 155761)

[131] Chattopadhyay, S. K.
Proc. Anu. Conv. Sugar
Technol. Ass. India (1972) No. 38; p. 31; (C. A. 78 (1973) 112990g)

[132] Dambal, R. P.; Tirumale L. Rama Char
(Electrochem. Lab., Ind. Inst. Sci., Bangalore, India)
J. Electrochem. Soc. India 19 (1970) No. 1; 12 (in English); (C. A. 73 (1970) 104819x)

[133] Desai, M. N.
Werkstoffe und Korrosion 24 (1973); p. 707 (in German)

[134] Rabald, E.
Werkstoffe und Korrosion (Physikalische Eigenschaften)
(Materials and corrosion (physical properties)) (in German) vol. 1
Verlag von Otto Spamer, Leipzig, 1931

[135] Fabian, H.
Ullmanns Encyklopädie der technischen Chemie
(Ullmann's encyclopedia of industrial chemistry) (in German)
4th ed., vol. 15, p. 491
Verlag Chemie GmbH, D-Weinheim, 1978

[136] DECHEMA-Werkstoff-Tabelle
"Methylamine"
(DECHEMA corrosion data sheets "Methylamines") (in German)
DECHEMA, D-Frankfurt am Main, January 1971

[137] DECHEMA-Werkstoff-Tabelle
"Hexamethylentetramin"
(DECHEMA corrosion data sheets "Hexamethylenetetramine") (in German)
DECHEMA, D-Frankfurt am Main, October 1958

[138] DECHEMA-Werkstoff-Tabelle "Amine und Aminierung"
(DECHEMA corrosion data sheets "Amines and Aminisation") (in German)
DECHEMA, D-Frankfurt am Main, May 1953

[139] DECHEMA-Werkstoff-Tabelle
"Diäthylamin"
(DECHEMA corrosion data sheets "Diethylamine") (in German)
DECHEMA, D-Frankfurt am Main, September 1956

[140] DECHEMA-Werkstoff-Tabelle
"Butylamine"
(DECHEMA corrosion data sheets "Butylamines") (in German)
DECHEMA, D-Frankfurt am Main, January 1955

[141] DECHEMA-Werkstoff-Tabelle
"Amylamine"
(DECHEMA corrosion data sheets "Amylamines") (in German)
DECHEMA, D-Frankfurt am Main, September 1953

[142] DECHEMA-Werkstoff-Tabelle
"Octadecylamin"
(DECHEMA corrosion data sheets "Octadecylamine") (in German)
DECHEMA, D-Frankfurt am Main, December 1966

[143] DECHEMA-Werkstoff-Tabelle
"Benzylamin"
(DECHEMA corrosion data sheets "Benzylamine") (in German)
DECHEMA, D-Frankfurt am Main, May 1954

[144] DECHEMA-Werkstoff-Tabelle
"Äthylendiamin"
(DECHEMA corrosion data sheets "Ethylenediamine") (in German)
DECHEMA, D-Frankfurt am Main, March 1953

[145] DECHEMA-Werkstoff-Tabelle
"Cyclohexylamin"
(DECHEMA corrosion data sheets "Cyclohexylamine")
DECHEMA, D-Frankfurt am Main, September 1956

[146] DECHEMA-Werkstoff-Tabelle
"Diphenylamin"
(DECHEMA corrosion data sheets

"Diphenylamine")
DECHEMA, D-Frankfurt am Main,
November 1957

[147] DECHEMA-Werkstoff-Tabelle
"Chloramine"
(DECHEMA corrosion data sheets
"Chloramines")
DECHEMA, D-Frankfurt am Main,
October 1955

[148] Product Information
Beständigkeitstabelle von Kupfer-
Gußwerkstoffen in verschiedenen Medien
(Tables of resistance of cast copper
materials in various media) (in German)
Deutsches Kupfer-Institut, Berlin-
Düsseldorf, 1970

[149] Hamner, N. E.
Corrosion Data Survey, 5th ed. (1974)
National Association of Corrosion
Engineers, Houston/Texas, 1974

[150] Gräfen, H.
Die Praxis des Korrosionsschutzes
(Practice of corrosion protection)
(in German)
expert verlag, D-Grafenau 1, 1981, p. 86

[151] Fiaud, C.
Etude electrochimique du comportement
du cuivre en solution alcaline et de l'action
inhibitrice de la cyclohexalamine
(Electrochemical studies of the behavior of
copper in alkaline solution and of the
inhibiting efficiency of cyclohexylamine)
(in French)
Corrosion Sci. 14 (1974) 4, p. 261

[152] Gugnina, A. P.
Studies of the protective action of volatile
inhibitors of the ferrous and non-ferrous
metals (in Russian)
Tr. Chelyabinsk. Inst. Mekh. Elektrifik.
Sel'sk. (1978) 146, p. 61

[153] Patel, N. K.; Sampat, S. S.; Vora, J. C.;
Trivedi, A. M.
Aliphatische Amine als Inhibitoren für
Essigsäure
(Aliphatic amines as inhibitors for acetic
acid) (in German)
Werkstoffe und Korrosion 21 (1970) 10, p. 809

[154] Baker, S.
Chem. Process Eng. 41 (1960) No. 11,
p. 513

[155] Product Information (The American Brass
Co., Waterbury, Conn. (USA))
Corrosion resistance of copper and copper
alloys
Part I – Resistance to common corrosives
Chem. Eng. 58 (1951) 1, p. 110

[156] Franke, E.
Korrosionsverhalten von Aluminium-
Bronzen
(Corrosion behavior of Al-bronzes)
(in German)
Werkst. Korros. 2 (1951) p. 302

[157] Fabian, R. J.
Special Report No. 202
Mater. Design Eng. (1963) Jan., p. 106

[158] Product Information
Corrosion resistance of copper and copper
alloys No. 1–53, p. 12
The American Brass Co., Waterbury, Conn.
(USA)

[159] Wallbaum, H. J.
in: Landolt-Börnstein "Technik"
(in German)
2nd ed., vol. 4, part b, p. 880
Springer-Verlag, Berlin, 1965

[160] Heitz, E.; v. Meysenbug, C. M.
Die Korrosion von Eisen, Nickel, Kupfer
und Aluminium in organischen
Lösungsmitteln mit geringem
Mineralsäuregehalt (The corrosion of Fe,
Ni, Cu and Al in organic solvents with a
low content of mineral acid) (in German)
Werkst. Korros. 16 (1965) p. 578

[161] Lee, J. A.
Materials of construction for chemical
process industries,
p. 21
McGraw-Hill Book Co., Inc., New York,
1950

[162] Ried, G.
Übersicht über Werkstoffe für Absperr-
Organe, die bei angreifenden Medien zu
empfehlen sind
(Survey of valve units for corrosive media)
(in German)
Werkst. Korros. 15 (1964) p. 468

[163] Wood, W. L.
BIOS Final Report 75, item 22 (Buna-
Werke, Schkopau)
His Majesty's Stationary Office, London

[164] Anonymous
BIOS Report 27/83 and 28/1
His Majesty's Stationary Office, London

[165] Product Information
ABC der Stahlkorrosion
(ABC of steel corrosion) (in German) July
1966, p. 24, 273
Mannesmann AG, D-Düsseldorf

[166] Boguslauskaite, D. E. S.; Puserauskas, A. P.; Sakalauskas, Z. Yu.
Benzotriazole: Corrosion inhibitor of metals in the production of cellulose-acetate fibers (in Russian)
Khim. Volokna (1977) 6, p. 54

[167] Dies, K.
Kupfer und Kupferlegierungen in der Technik
(Copper and copper alloys in technology) (in German)
Springer Verlag, Berlin-Heidelberg-New York, 1967, p. 171

[168] Fabian, R. J.
Corrosion resistance data
Mater. in Design Eng. 37 (1963) p. 106

[169] Dies, K.
Kupfer und Kupferlegierungen in der Technik
(Copper and copper alloys in technology) (in German)
Springer Verlag, Berlin-Heidelberg-New York, 1967, p. 318

[170] Dies, K.
Kupfer und Kupferlegierungen in der Technik
(Copper and copper alloys in technology) (in German)
Springer Verlag, Berlin-Heidelberg-New York, 1967, p. 558

[171] Rabald, E.
Corrosion Guide
2nd ed. (1968) p. 84, 129
Elsevier Publ. Comp, Amsterdam-London-New York

[172] Fabian, R. J.; Vaccari, J. A.
How materials stand up to corrosion and chemical attack
Mater. Eng. 73 (1971) 2, p. 36

[173] Anonymous
Mater. Eng. (Materials' Selector) 74 (1971) 4, p. 11

[174] Fabian, R. J.; Vaccari, J. A.
How materials stand up to corrosion and chemical attack
Mater. Eng. 73 (1971) 2, p. 36

[175] Fitzgerald-Lee, G.
Corrosion resistance of aluminium bronzes
Corrosion Technol. 6 (1959) 9, p. 263, 284

[176] Sheppard, E.; Hise, D. R.; Gegner, P. J. et al.
Performance of titanium vs. other materials in chemical plant exposures
Corrosion 18 (1962) p. 211t

[177] Vaidyanath, L. R. et al.
Corrosion resistance of copper and copper alloys
Proceeding of a Symposium held at Bombay, 1971, p. 18

[178] Bhanot, D.; Chawla, S. L.
Corrosion resistance of pure copper, its alloys and stainless steel in chlorinated lime slurry and calcium hypochlorite solution
Transactions of the SAEST 13 (1978) 3, p. 227

[179] Gordon, R. C.
Glassy silicate corrosion inhibitor with controlled solution rate
US Patent Office No. 3.770.652 (6.11.1973)

[180] Anonymous
Copper and its alloys – wrought
Materials Selector (1983) ME 2-16-27

[181] Rabald, E.
Corrosion Guide, Second Edition
Elsevier Publishing Company, Amsterdam, London, New York, 1968

[182] Heitz, E.
Grundvorgänge der Korrosion von Metallen in organischen Lösungsmitteln (I)
(Corrosion fundamental phenomena of metals in organic solvents (I)) (in German)
Werkstoffe u. Korrosion 21 (1970) 5, p. 360/7

[183] Berg, F. F.
Korrosionsschaubilder
(Corrosion diagrams) (in German)
VDI-Verlag GmbH, Düsseldorf 1969, 2nd Edition

[184] Anonymous
Beständigkeitstabellen von Kupfer-Gußwerkstoffen in verschiedenen Medien
(Corrosion data sheets on copper cast materials against various chemical agents) (in German)
Deutsches Kupferinstitut, Berlin, Düsseldorf, Juni 1970

[185] Kartashova, K. M.; Sukhotin, A. M.; Bobrova, M. M.; Shukan, N. Ya.
Corrosion of metallic construction materials in fatty acids, higher amines and acetates of higher amines (in Russian)
Tr. Gos. in-t prikl. khimii (1971) 67, p. 159

[186] Rabald, E.
DECHEMA-Werkstoff-Tabelle, Fettsäuren (höhere, ab C_6)
(DECHEMA Corrosion Data Sheets, Fatty Acids) (in German)
DECHEMA, D-Frankfurt am Main, 1957

[187] Anonymous
NACE Publication 5A 180
Corrosion of metals by aliphatic organic acids
Mater. Perform. 19 (1980) 9, p. 65

[188] Rabald, E.
DECHEMA-Werkstoff-Tabelle, n-Caprinsäure
(DECHEMA Corrosion Data Sheets, n-Capric Acid) (in German)
DECHEMA, D-Frankfurt am Main, 1955

[189] Knödler, A.; Lübke, H.; Raub, Ch. J.
Goldplattierung von Brillengestellen
(Gold cladding for spectacle frames) (in German)
Galvanotechnik 70 (1979) 6, p. 506

[190] Rabald, E.
Corrosion Guide "Alcohol, butyl alcohols"
2nd ed., p. 28, 120
Elsevier Publishing Comp., Amsterdam-London-New York, 1968

[191] Grigorev, V. P.; Ekilik, V. V.; Ekilik, G. N.
Relation between dielectric penetration ability of solvents and the corrosion characteristics of certain metals in alcohol-HCl systems (in Russian)
Zashch. Met. 9 (1973) 5, p. 594

[192] Vigdorovich, V. I.; Tsygankova, L. E.; Osipova, N. V.
Corrosion of copper in ethanol solutions of hydrogen chloride
J. Appl. Chem. (UdSSR) 49 (1976) p. 2437

[193] Heitz, E.; Kyriazis, C.
Reactions during corrosion of metals in organic solvents
Ind. Eng. Chem. Proc. Res. Develop. 17 (1978) 1, p. 37

[194] Sretenskaya, G. V.; Talaeva, G. V.
Investigation of the corrosion resistance of metallic materials on the production of triacetate fibers (in Russian)
Khim. Volokna (1970) 3, p. 55

[195] Köser, H. J. K.; Lübke, M.
Zum Korrosionsverhalten von metallischen Werkstoffen in chlorkohlenwasserstoffhaltigen Lösungsmittelgemischen
(Corrosion behavior of metallic materials in solvent mixtures containing chlorinated hydrocarbons) (in German)
Werkst. Korros. 31 (1980) p. 186

[196] Köser, H. J. K.; Lübke, M.
Zum Korrosionsverhalten von metallischen Werkstoffen in chlorkohlenwasserstoffhaltigen Lösungsmittelgemischen
(Corrosion behavior of metallic materials in solvent mixtures containing chlorinated hydrocarbons) (in German)
Werkst. Korros. 31 (1980) 3, p. 186

[197] Abd-El Wahab, F. M.; El Shayeb, H. A.
Effect of some alcohols on the dissolution of Cu, Zn and Cu-Zn alloy in nitric acid
Corros. Prev. Control 32 (1985) 1, p. 9

[198] Anonymous
Beständigkeitstabellen von Kupfer-Gußwerkstoffen in verschiedenen Medien
(Charts of resistance of copper cast materials in various media) (in German)
Deutsches Kupfer-Institut und Gesamtverband Deutsche Metallgießereien (1970) p. 22

[199] Anonymous
Corrosion resistance charts
Du Pont Information Service 1975

[200] Rabald, E.; Bretschneider, H.
DECHEMA-Werkstoff-Tabelle "Aluminiumchlorid"
(DECHEMA Corrosion Data Sheets "Aluminium chloride") (in German)
DECHEMA D-Frankfurt am Main, April 1953

[201] Sorokin, Yu. I.; Tsejtlin, Kh. L.; Babitskaya, S. M.; Merzloukhova, L. V.
Corrosion of metals in molten aluminium chloride and its mixtures with sodium chloride (in Russian)
Zashchita Metallov 5 (1969) No. 5; p. 536–540

[202] Tsejtlin, Kh. L.; Sorokin, Yu. J.; Ryzhkova, Zh. S.
Corrosion of plant and safety technology (in Russian)
Khim. Prom. 13 (1977) No. 3; p. 179–182

[203] Trufanova, A. I.; Lazareva, T. A.; Khlebnikova, S. F.
Delayed corrosion of copper in neutral environments containing ionogenic species of nitroaromatic derivatives (in Russian)
Zashchita Metallov 19 (1983) p. 468

[204] Rozenfeld, I. L.; Persiantseva, V. P.; Reizin, B. L.; Shustova, Z. F.; Gavrish, N. M.
A study of some nitrobenzoate salts of amines as corrosion inhibitors for ferrous and non-ferrous metals (in English)
Korr. Metal i Splavov Sb. (1965) p. 341

[205] Oshe, E. K.; Kryakovskaya, N. Yu.
Influence of the oxide surface non-stoichiometry on the efficiency of metal-protecting inhibitors (in Russian)
Zashchita Metallov 19 (1983) p. 393

[206] Kar, G.; Healy, T. W.; Fuerstenau, D. W.
Effects of alkyl amine surfactants on the corrosion of a rotating copper cylinder
Corrosion Sci. 13 (1973) p. 375

[207] Barbier, J.; Fiaud, C.
Study of acetylenic alcohols as volatile inhibitors of the atmospheric corrosion of copper (in French)
Métaux, Corrosion-Ind. 49 (1974) No. 587/588; p. 271

[208] Singh, D. D. N.; Banerjee, M. K.
Vapor phase corrosion inhibitors – a review
Anti-Corrosion 31 (1984) No. 6; p. 4

[209] Ludosan, E.; Cimpoeru, C.
Corrosion behavior and corrosion control of some metallic materials in mixtures of polyethylene glycols and ethoxylated alcohols (in Roumanian)
Revista de Chim. 33 (1982) No. 1; p. 57

[210] Gintsberg, S. A.
Influence of some inhibitors of atmospheric corrosion on the surface resistance of the metal to be protected (in Russian)
Zashchita Metallov 5 (1969) p. 340

[211] Dvali, T. M.
Corrosion resistance of alloys in reaction mixtures of caprolactam production (in Russian)
Zashchita Metallov 9 (1973) p. 58

[212] Dobre, R.
Relations between the chemical constitution of volatile corrosion inhibitors and their protection efficiency (in Roumanian)
Rev. Chim. 26 (1975) p. 313

[213] Stephenson, E. V.
Protective and insulating coatings on non-ferrous metals
Coating 6 (1973) No. 11; p. 333

[214] Graf, L.; Richter, W.
Initiation of stress corrosion cracking of copper-gold and copper-zinc alloys by alternating loading in comparison with static loads. Part 2
Metall 32 (1978) No. 1; p. 48

[215] Gronskij, R. L.; Maklakova, V. P.
Protection of copper alloys against corrosion during the chemical cleaning of heat exchangers (in Russian)
Vodopodgotovka Vodn. Rezhim. Khimkontrol Parosil. Ustanovkakh (1978) No. 6; p. 69

[216] Desai, M. N.; Shah, Y. C.
Inhibiting the corrosion of brass 63/37 in sodium hydroxide solutions (in German)
Werkstoffe u. Korrosion 21 (1970) No. 10; p. 795

[217] Wendler-Kalsch, E.; Gräfen, H.
Korrosionsschadenkunde
(Corrosion damage lore) (in German), 1st ed.
Springer-Verlag, Berlin, 1998

[218] DIN EN 1982 (12/1998)
Kupfer und Kupferlegierungen-Blockmetalle und Gußstücke
(Copper and copper alloy block metals and cast pieces) (in German)
Beuth Verlag GmbH, 10772 Berlin

[219] DIN V 17900 (3/1999)
Kupfer und Kupferlegierungen – Europäische Werkstoffe – Übersicht über Zusammensetzung und Produkte
(Copper and copper alloys – European materials – overview of composition and products) (in German)
Beuth Verlag GmbH, 10772 Berlin

[220] DIN 1787 (1/1973)
Kupfer-Halbzeug
(Copper-semi-finished product) (in German)
Beuth Verlag GmbH, 10772 Berlin

[221] DIN EN 1412 (12/1995)
Kupfer und Kupferlegierungen – Europäisches Werkstoffnummernsystem
(Copper and copper alloys – European material number system) (in German)
Beuth Verlag GmbH, 10772 Berlin

[222] Wendler-Kalsch, E.
Korrosionsverhalten und Deckschichtbildung auf Kupfer und Kupferlegierungen
(Corrosion behavior and surface layer

formation on copper and copper alloys) (in German)
Z. Werkstofftechnik (1982), pp. 129–137

[223] Anonymous
Copper and its alloys – wrought
Mater. Engng. (Mater. Selector 75) 80 (1974) 4, pp. 96–106

[224] Firmenschrift
Beständigkeitstabellen Kupferwerkstoffe (Resistance tables of copper materials) (in German)
Deutsches Kupfer-Institut, Düsseldorf

[225] Firmenschrift
Corrosion Resistance of Copper and Copper Alloys
The American Brass Company, Waterbury (Connecticut/USA)

[226] DIN 8975-2 (05/1978)
Kälteanlagen, Sicherheitstechnische Anforderungen für Gestaltung, Ausrüstung, Aufstellung und Betreiben, Werkstoffauswahl für Kälteanlagen (Refrigerant plants, safety-relevant requirements for organization, equipment, list and operation, material selection for refrigerant plants) (in German)
Beuth Verlag, Berlin

[227] DIN EN 378 (10/2003)
Kälteanlagen und Wärmepumpen – Sicherheitstechnische und umweltrelevante Anforderungen, Teil 2: Konstruktion, Herstellung, Prüfung, Kennzeichnung und Dokumentation (Refrigerant plants and heat pumps – safety-relevant and environmental-relevant requirements, part 2: Construction, production, examination, marking and documentation) (in German)
Beuth Verlag, Berlin

[228] BGV D 4 (04/1991)
BG-Vorschrift: Kälteanlagen, Wärmepumpen und Kühleinrichtungen (BG-regulation: Refrigerant plants, heat pumps and cooling equipments) (in German)
VGB Verwaltungs-Berufsgenossenschaft

[229] Knabe, M.; Reinhold, S.; Schenk, J.
Kupferwerkstoffe widerstehen Ammoniak (Copper materials resist ammonia) (in German)
Metall 52 (1998) 4, pp. 225–229

[230] Braterskaya, G. N.; Lavrenko, V. A.; Kokhanovskii, S. P.; Korobskii, V. V.
Corrosion resistance of electrocontact materials based on copper in corrosive media
Materials Science (Russia) 29 (1993) 2, pp. 155–158

[231] Klatte, H.
Zum Problem der interkristallinen und Spannungsrißkorrosion an den homogenen Kupfer-Gold- und Kupfer-Zink- und an ausscheidungsfähigen Aluminium-Zink-Magnesium-Mischkristallen (The problem of the intergranular and stress corrosion cracking at the homogeneous copper gold and copper zinc and at precipitationable aluminium zinc magnesium mixed crystals) (in German)
Werkst. Korros. (1956) 7, pp. 545–560

[232] Lissner, O.
Sheet Metal Ind. 30 (1953), pp. 4–55

[233] Yamaguchi, T.; Nishiyama, S.; Tanaka, G.
Stress corrosion cracking of the hard drawn high-conductivity copper wire
Proc. Jap. Congr. Mater. Res. (1971), 14, pp. 140–142
(C.A. 76 (1972), 27 586 b)

[234] von Franqué, O.; Stichel, W.
Korrosion unter Wärmedämmstoffen (Corrosion under heat-insulating materials) (in German)
Werkst. Korros. 37 (1986), pp. 340–343

[235] Corrosion Atlas
Third expanded and revised edition, Elsevier 1997

[236] Kikuchi, Y.; Tohmoto, K.; Ozawa, M.; Kanamura, T.; Sakane, T.
Microbiologically influenced corrosion of copper in groundwater
CORROSION 99, Pp 14, Apr. 1999 Report No. 170
Conference: Corrosion 99, San Antonio, TX, USA, 25–30 April 1999

[237] Kikuchi, Y.; Tohmoto, K.; Ozawa, M.; Kanamura, T.; Sakane, T.
Corrosion behavior of copper and its welds by bacteria in underground water
Transactions of the JWRI (Japan Welding Research Institute)
27 (1998) 2, pp. 53–62

[238] Nair, K. V. K.
Biofouling and biocorrosion in industrial cooling systems
Conference: Workshop on failure analysis, corrosion evaluation and metallography
Bombay, India, 6–10 Jan. 1992
Publ.: Library & Information Services, Bhabha Atomic Research Centre, Trombay, Bombay 400 085, India 1992

[239] Aoyama, H. et al.
The behaviour of ammonia in the air cooling zone of large size utility unit condensers
Amer. Power Conf. Paper
Chicago, 1971, 20–22 A

[240] Gonzales, A.
The influence of the design on the corrosion by ammonia of steam condenser tubes: A case history
Corrosion Protection 9 (1978) 5/6, pp. 5–10

[241] Francis, R.
Hot spot corrosion in condenser tubes: Its causes and prevention
Br. Corrosion Journal 22 (1987) 3, pp. 199–201

[242] Francis, R.
Effect of pollutants on corrosion of copper alloys in sea water. Part 1: Ammonia and chlorine
Br. Corrosion Journal 20 (1985) 4, pp. 167–174

[243] Rajaskharan Nair, K. V.; Namboodhiri, T. K. G.
Influence of sulphate and chloride ions on corrosion of 63-37 brass in aqueous ammonia containing copper
Br. Corrosion. J. 32 (1988) 4, pp. 245–249

[244] Bag, S. K.; Chakrborty, S. B.; Roy, A.; Chaudhuri, S. R.
2-aminobenzimidazole as corrosion inhibitor for 70-30 brass in ammonia
British Corrosion Journal 31 (1996) 3, pp. 207–212

[245] Patel, K. M.; Oza, B. N.
Effect of some polyhydroxy derivates of benzene on corrosion rate of 70/30 brass in ammoniacal solutions
Journal of the Electrochemical Society of India 38 (1989) 1, pp. 46–49

[246] Ochoa, T.; Fernández, B.
Corrosión de la latón de aluminio en condiciones de enfriamiento pro agua de mar contaminada por sulfuro y amonio
Rev. Iber. Corros. y Prot. XVIII (1987) 2–6, pp. 149–152

[247] Kato, H.; Nakai, K.
Evaluation of stress corrosion process by X-ray diffraction technique
VIII International Congress on Experimental Mechanics
Society for Experimental Mechanics, Nashville, Tennessee, USA, 1996, pp. 394–398

[248] Pini, G. C.
Typische Schadensfälle durch Spannungsrißkorrosion und Wasserstoffversprödung
(Typical cases of damage by stress corrosion cracking and hydrogen embrittlement) (in German)
Material und Technik 4 (1976) 2, pp. 98–107

[249] Bertocci, U.; Pugh, E. N.
Modeling of the potential at the tip of a transgranular stress corrosion crack in the alpha-brass-ammonia-system
in: Conference: Corrosion Chemistry within Pits, Crevices and Cracks, National Physical Laboratory, Teddington, Middlesex, UK, 1–3 Oct. 1984
HMSO Books, P.O. Box 276, London SW8 5Dt, UK, 1987, pp. 187–198

[250] Bertocci, U.; Pugh, E. N.
Modeling of electrochemical processes during transgranular stress corrosion cracking of copper-base alloys
Conference: 10th International Congress on Metallic Corrosion, Madras, India, 7–11 Nov. 1987, Vol. V
Trans. Tech. Publications, P.O. Box 10, CH-4711 Aedermannshof, Switzerland, 1989, pp. 219–229

[251] Kaufmann, M. J.; Fink, J. L.
A closer look at the transgranular stress corrosion cracking of Cu-30Zn in cuprous ammonia
Metallurgical Transactions A 18A (1987) 8, pp. 1539–1542

[252] Bertocci, U.
Modeling of crack chemistry in the alpha brass-ammonia system
Conference: Embrittlement by the localized crack environment
Philadelphia, Pa., USA. 4–5 Oct. 1983
The Metallurgical Society/AIME, 420

Commonwealth Dr., Warrendale, Pa. 15086, USA, 1984, pp. 49–58
[253] Namboodhiri, T. K. G.; Tripathi, R. S.
The stress-assisted dezincification of 70/30 brass in ammonia
Corrosion Science 26 (1986) 10, p. 745
[254] Whitaker, W.
Metallurgica 39 (1948) 21, p. 66
[255] Class, I.
Überblick über das Gebiet der Spannungsrißkorrosion.
II. Betrachtungen grundsätzlicher Art zur Frage der Spannungsrißkorrosion
(Overview of the area of the stress corrosion cracking. II. Views of fundamental kind for the question of the stress corrosion cracking) (in German)
Werkstoffe u. Korrosion 16 (1965) p. 288
[256] Stettler, R.
Stress cracking and corrosion cracking of brass
Sheet Metal Ind. 27 (1950) p. 119;
(Werkstoffe u. Korrosion 1 (1950) p. 370)
[257] Eichhorn, K.
Zur Korrosionsbeständigkeit der Kondensatorrohre aus Kupferlegierungen
(The corrosion resistance of the condenser tubes from copper alloys) (in German)
Werkstoffe u. Korrosion 8 (1957) p. 657
[258] Raymann, H.
Spannungsrißkorrosion an Ms 72
(Stress corrosion cracking at Ms 72) (in German)
Praktische Metallographie 8 (1971) 2, p. 125
[259] Matthaes, K.
Spannungskorrosion und Festigkeitstheorie
(Stress corrosion and strength durabilty theory) (in German)
Werkstoffe u. Korrosion 8 (1957) p. 261
[260] Graf, L.
Was spricht gegen das Aufreißen von Deckschichten und gegen eine Versprödung als Ursachen der Spannungsrißkorrosion?
(What speaks against breaking surface layers and against an embrittlement as causes of the stress corrosion cracking?) (in German)
Z. Metallkunde 66 (1975) 12, p. 749

[261] Pinchback, T. R.; Clough, S. P.; Heldt, L. A.
Stress corrosion cracking of alpha brass in a tarnishing ammoniacal environment. Fractography and chemical analysis
Metal. Trans. A, 6A (1975) 8, p. 1479;
(C. A. 84 (1976) 78 181p)
[262] Leidheiser jr., H.; Kissinger, R.
Chemical analysis of the liquid within a propagating stress corrosion crack in 70:30 brass immersed in concentrated ammonia hydroxide
Corrosion (Houston) 28 (1972) 6, p. 218
[263] Buecker, B. J.
Watch out for steam-side corrosion in utility condensers
Materials Performance 31 (1992) 9, pp. 68–70
[264] Fromberg, M.
Spannungsrißkorrosion an Teilen aus Cu-Zn-Legierungen
(Stress corrosion cracking at parts from Cu-Zn-alloys) (in German)
Korrosion 16 (1985) 6, pp. 323–325
[265] Marden, K. M.; Lewis, K. R.
Failure analysis of Admiralty brass condenser tubes
Handbook of case histories in failure analysis, vol. 1, pp. 110–114
ASM International, Materials Park, Ohio, 44073-0002, USA, 1992
[266] Tanner, G. M.
Stress corrosion cracking of a brass tube in a generator air cooler unit
Handbook of case histories in failure analysis, vol. 2, pp. 107–110
ASM International, Materials Park, Ohio, 44073-0002, USA, 1993
[267] Rückert, J.
Spannungsrißkorrosion an Kupferlegierungen
(Stress corrosion cracking at copper alloys) (in German)
Materials and Corrosion 47 (1996) 2, pp. 71–77
[268] Sang, C.; Wang, Y.-B.; Chu, W.-Y.; Hsiao, C.-M.
Stress corrosion cracking of alpha + beta brass in ammonia solution under compressive stress
Scripta metallurgica et materialia 25 (1991) 12, pp. 2751–2756

[269] Newman, R. C.; Sharabi, T.; Sieradzki, K.
Film-induced cleavage of alpha-brass
Scripta Metallurgica 23 (1989) 1, pp. 71–74

[270] Venkataraman, B.; Hebbar, K. R.; Sudhaker Nayak, H. V.
Polarization studies on the effect of cold work on stress corrosion failure of alpha-brass in ammonia solutions
Bull. Electrochem. 3 (5) (1987) Sept.–Oct., pp. 399–403

[271] Slattery, P. W.; Smit, J.; Pugh, E. N.
Use of a load-pulsing technique to determine stress corrosion crack velocity
Conference: Environment-sensitive fracture: Evaluation and comparison of test methods
Gaithersburg, Maryland, USA, 26–28 Apr. 1982
ASTM, 1916 Race St., Philadelphia, Pennsylvania 19103, USA, 1984, pp. 399–411

[272] Ranjan, R.; Buck, O.
On life prediction under stress corrosion conditions below K(ISCC)
Conference: Review of progress in quantitative non-destructive evaluation, vol. 3B, Santa Cruz, California, USA, 7–12 Aug. 1983
Plenum Press, 233 Spring St., New York, NY 10013, USA, 1984, pp. 1251–1258
see also:
Ranjan, R.; Buck, O.
A crack initiation model for stress corrosion cracking of α-brass in nontarnishing ammoniacal solution
Conference: Embrittlement by the localized crack environment
Philadelphia, Pa., USA. 4–5 Oct. 1983
The Metallurgical Society/AIME, 420 Commonwealth Dr., Warrendale, Pa. 15086, USA, 1984, pp. 349–361
and:
Ranjan, R.; Buck, O.
Predicting initiation of crack growth under stress corrosion cracking conditions
Conference: Predictive capabilities in environmentally assisted cracking
Miami Beach, Florida, USA, 17–22 Nov. 1985
American Society of Mechanical Engineers, 345 East 47th St., New York, 10017, USA, 1985, pp. 79–89

[273] Thoden, B.; Pletka, H. D.
Failure analysis of heat exchanger tubes. Conventional corrosion or MIC?
Conference: EUROCORR 99: The European Corrosion Congress event no. 227; "Solution of corrosion problems in advanced technologies"
Aachen, 30.08.–02.09.1999

[274] Rao, T. S.; Nair, K. V. K.
Microbiologically influenced stress corrosion cracking failure of admiralty brass condenser tubes in a nuclear power plant cooled by freshwater
Corrosion Science 40 (1998) 11, pp. 1821–1836

[275] Dobrovolskaya, V. P.
Corrosion of copper in ammonium salt solutions (in Russian)
Khim. Prom. 49 (1973) 10, p. 769

[276] Konradi, M. V.
Corrosion of copper, aluminium and iron in acetic acid/ammonium sulfate solutions (in Russian)
Izv. V.U.Z. Tsvetn. Metall. (1958) 4, p. 165

[277] Kiss, L.; Farkas, J.; Körösi, A.; Mandl, J.
A réz anódos oldódásának vizsgálata ammóniát tartalmazó oldatokban
(Testing of the anodic dissolution of copper in ammonia-containing solutions)
(in Hungarian)
Magyar Kémiai Folyoirat 79 (1973) 3, p. 127

[278] Bogdanov, V. P.; Bolotina, R. S.
Investigation of the corrosion behavior of copper in acidic chloride solutions
(in Russian)
Vsb. Khim. i Khim. Tecknol. Vyp 9, Minsk Vyshejsh. Shkola (1975) p. 89

[279] Karelov, S. V.; Doroshkevich, A. P.; Rybnikow, V. I.; Gryaznukhina, L. M.
The passivation of a copper anode in an ammonium sulfate electrolyte (in Russian)
Izv. V.U.Z. Tsetn. Metall. (1981) 6, p. 65

[280] Sorokin, Yu. I.; Tsejtlin, Kh. L.; Balashova, A. A.
Effect of water vapour and its mixture with carbon dioxide on metal corrosion in ammonia at 500 °C (in Russian)
Zashch. Met. 8 (1972) 4, p. 430

[281] Sedzimir, J.; Bujaska, M.
The corrosion of copper in copper(II)ammonium sulphate solutions: The influence of ammonia concentration, of temperature and of the substitution of

sulphates by carbonates
Corros. Sci. 20 (1980) p. 1029

[282] Shams El Din, A. M.; Abd El Kader, J. M.; Badran, M. M.
Galvanic corrosion in the copper/zinc system. II. Effect of NH_4^+-ion
Brit. Corros. J. 16 (1981) 1, p. 38

[283] Dvali, T. M.
Corrosion resistance of steels in caprolactam production mediums (in Russian)
Zashch. Met. 9 (1973) 1, p. 58

[284] Köhler, W.
Inhibition der Korrosion von Kupfer und Messing in Ammoniumchloridlösungen durch Kolloide
(Inhibition of corrosion of Cu and brass in ammonium chloride solutions by colloids) (in German)
Werkst. Korros. 14 (1963) 1, p. 11

[285] Kazantsev, A. A.; Kuznetsov, V. A.
Anodic dissolution of copper in halide solutions (in Russian)
Elektrokhimiya 19 (1983) p. 92

[286] Krivtsova, G. E.; Shutov, A. A.; Vinogradova, T. S.
The corrosion behavior of copper in a complex-ammonia electrolyte for silver plating (in Russian)
Izv. Vyssh. Uchebn. Zavedenij Khim. i Khim. Tekhnol. 22 (1979) p. 1357

[287] Shumilov, V. I.; Korolev, G. V.; Kucherenko, V. I.; Flerov, V. N.
Anodic oxidation of copper in weakly alkaline ammonium chloride solutions (in Russian)
Izv. Vusov. Khim. i Khim. Tekhnol. 26 (1983) p. 126

[288] Sato, Shi.; Sagisaka, Ki.
Application of copper and its alloys as tubes for town gas service lines (in Japanese)
Sumitomo Keikinzoku Giho 16 (1975) No. 3/4, p. 106

[289] Sato, S.; Nagata, K.; Simono, M.
Corrosion and fouling of various heat exchanger tubes in flowing treated sewage (in Japanese)
J. Japan Copper Brass Res. Assoc. 19 (1980) p. 50

[290] Dobrovolskaya, V. P.; Neznamova, T. G.; Barannik, V. P.
8-Mercaptoquinoline and 8-oxiquinoline used as corrosion inhibitors for steel, cast iron and copper in acid media in the presence of ammonium ions (in Russian)
Izv. Vyssh. Uchebn. Zavedenij, Khim. i. Khim. Tekhn. 9 (1966) 1, p. 144

[291] Walker, R.
Corrosion inhibition of copper by tolyltriazole
Corrosion 32 (1976) 8, p. 339

[292] Walker, R.
Triazole, benzotriazole and naphthotriazole as corrosion inhibitors for copper
Corrosion 31 (1975) 3, p. 97

[293] Lewis, G.
Adsorption isotherm for the copper-benzotriazole system
Brit. Corros. J. 16 (1981) 3, p. 169

[294] Patel, A. B.; Patel, N. K.
Influence of phenols as corrosion inhibitors for corrosion of copper in ammonium chloride solution
Indian J. Technol. 13 (1975) 1, p. 47

[295] Fox, P. G.; Bradley, P. A.
1,2,4-Triazole as a corrosion inhibitor for copper
Corros. Sci. 20 (1980) p. 643

[296] Desai, M. N.; Rana, S. S.; Gandhi, M. H.
Corrosion inhibitors for copper
Anticorros. Methods Mater. 17 (1970) 6, p. 17

[297] Prajapati, S. N.; Bhatt, I. M.; Soni, K. P.; Vora, J. C.
Role of organic heterocyclic compounds as corrosion inhibitors for copper and brass (63/37) in ammonium chloride solution
J. Indian Chem. Soc. 53 (1976) p. 723

[298] Franco, J.; Patel, N. K.
Photocorrosion of copper and brass in aqueous solutions of copper and ferric chlorides
J. Electrochem. Soc. India 27 (1978) 3, p. 161

[299] Franco, J.; Patel, N. K.
Photocorrosion of copper and 63/37 brass in aqueous solution of ammonium persulphate with additives
Fertilizer Technology 14 (1977) 1/2, p. 142

[300] Imashimizu, Y.; Watanabé, J.
Morphology change of dislocation etch pits on the (111) surface of Cu crystals (in Japanese)
J. Jpn. Inst. Met. 64 (1982) 2, p. 126

[301] Product Information (C. Dittmann & Co., D-Karlsruhe) Ein verbessertes Ätzmittel für die Herstellung von gedruckten Schaltungen
(An improved etching agent for the manufacture of edge board contacts) (in German)
Metallpraxis/Oberflächentechnik 21 (1972) p. 96

[302] Yepifanova, V. S.; Prusov, Yu. V.; Flerov, V. N.
Activation of copper surface by ammonium chloride during chemical nickel plating (in Russian)
Zashch. Met. 14 (1978) 6, p. 226

[303] Bogenschütz, A. F.; Büttner, U.; Jostan, J. L.; Marten, A.
Präzisionsätzen von Leiterplatten durch Verwendung von Benzotriazol als Flankenschutzmittel
(Precision etching of printed circuits by the use of benzotriazole as an edge protection agent) (in German)
Galvanotechnik 69 (1978) 11, p. 960

[304] Pilavov, Sh. G.; Egorova, A. I.
The anodic behavior of copper in pyrophosphate electrolytes (in Russian)
Zashch. Met. 13 (1977) 6, p. 734

[305] Pozdeev, P. M.; Konkov, V. P.
Some operational characteristics in the handling of ammonium bicarbonate (in Russian)
Sb. Nauchn. Tr. Chelyabinsk Politekhn. Inst. (1976) No. 180, p. 76

[306] Kawakatsu, I.; Ariga, T.
The corrosion property of a copper-phosphorus brazing alloy (in Japanese)
J. Japan Welding Soc. 42 (1973) 8, p. 734

[307] Klimova, G. V.; Kristal, M. M.; Shetulov, D. I.
Investigations on the corrosion resistance of materials used in the production of methylmethacrylate (in Russian)
Khim. Neft. Mashinostr. (1972) 9, p. 20

[308] Babkina, V. Yu.; Chub, E. G.; Vasileva, I. K.
The corrosion resistance of construction materials in caustic soda production (in Russian)
Khim. Neft. Mashinostr. (1978) 2, p. 22

[309] Morrissey, R. J.
The electrodeposition of palladium from chelated complexes
Platinum Met. Rev. 27 (1983) 1, p. 10

[310] Wendler-Kalsch, E.
Korrosionsverhalten und Deckschichtbildung auf Kupfer und Kupferlegierungen
(Corrosion and film formation of copper and copper alloys)
Z. Werkstofftech. 13 (1982) 4, p. 129

[311] Talbot J.
Aspects scientifiques des relations entre la pollution et la corrosion métallique
(Scientific aspects of the relation between impurities and metallic corrosion) (in French)
Corrosion, Traitements, Protection, Finition 21 (1973) p. 97

[312] Starodubtseva, L. A.; Nefedova, L. I.; Tarasko, D. I.; Voronova, G. L.
Increasing the wear resistance of copper alloys (in Russian)
Izv. V.U.Z. Tsvetn. Metall. (1978) 3, p. 138

[313] Cavalotti, P.
Materiali metallici non ferrosi per scambio termico
(Nonferrous metallic materials for heat exchange) (in Italian)
Metall. Ital. 70 (1978) 7/8, p. 343

[314] Miska, K. H.
Copper-nickel cuts corrosion, enhances electrical efficiency
Mater. Eng. 80 (1974) 7, p. 40

[315] Giuliani, L.; Bombara, G.
Influence of pollution on the corrosion of copper alloys in flowing salt water
Brit. Corros. J. 8 (1973) 1, p. 20

[316] Kollek, H.; Brockmann, W.
Einfache Methoden zur Messung der Adhäsionseigenschaften metallischer Oberflächen (Remissionsphotometrie)
(Simple methods for measuring the adhesive properties of metallic surfaces (reflectance photometry)) (in German)
Farbe + Lacke 85 (1979) 10, p. 820

[317] Katkar, A. J.; Balachandra, J.; Vasu, K. I.
On the stress corrosion cracking of nickel silver in sea water
J. Electrochem. Soc. India 29 (1980) 2, p. 108

[318] Ritter, F.
Korrosionstabellen metallischer Werkstoffe
(Corrosion tables of metallic materials) (in German) 3rd ed., p. 14
Springer Verlag, Wien, 1952

[319] Rabald, E.; Bretschneider, H.
DECHEMA-Werkstoff-Tabelle
"Ammoniumchlorid"
(DECHEMA corrosion data sheets
"ammonium chloride") (in German)
DECHEMA, D-Frankfurt am Main, 1955

[320] Leidheiser, H.; Retamoso, C.
Chemische Untersuchungen zur
Spannungsrißkorrosion von
Marinemessing
(Chemical investigations on stress
corrosion cracking of navy brass)
(in German)
Final Report to Project No. 154 B, 23 pages
Intern. Copper Res. Assoc., New York,
1974

[321] Mattsson, E.
Stress corrosion in brass considered
against the background of potential/pH
diagrams
Electrochim. Acta 3 (1961) p. 279

[322] Lefakis, H.; Rostoker, W.
Stress corrosion crack growth rates of brass
and austenitic stainless steel at low stress
intensity factors
Corrosion 33 (1977) 5, p. 178

[323] Chatterjee, U. K.; Goswami, S. K.;
Sircar, S. C.
Role of chloride in the stress corrosion
cracking of alpha-brass
Trans. Ind. Inst. Met. 33 (1980) p. 76

[324] Bradford, S. A.; Lee, T.
Effect of strain rate on stress corrosion
cracking of brass
Corrosion 34 (1978) 3, p. 96

[325] Kermani, M.; Scully, J. C.
The role of the tarnish film in the stress
corrosion crack propagation process in
alpha-brass in neutral ammoniacal
solutions
Corros. Sci. 19 (1979) p. 111

[326] Procter, R. P. M.; Islam, M.
Fracture of tarnish films on alpha-brass
Corrosion 32 (1976) 7, p. 267

[327] Takano, M.; Teramoto, K.; Nakayama, T.
The effects of crosshead speed and
temperature on the stress corrosion
cracking of Cu-30Zn alloy in ammoniacal
solution
Corros. Sci. 21 (1981) 6, p. 459

[328] Takano, M.; Teramoto, K.; Nakayama, T.
Stress corrosion cracking of 70/30 brass in
SERT (Slow Extrusion Rate Technique)
(in Japanese)
J. Japan Copper Brass Res. Assoc. 19 (1980)
p. 64

[329] Misawa, T.; Murakami, H.
Intergranular stress corrosion cracking of
alpha-brass in an ammoniacal nontarnishing solution
Corros. Eng. (Boshoku Gijutsu) Japan 25
(1976) 8, p. 505

[330] Bhakta, U. C.; Lahiri, A. K.
Stress corrosion cracking of alpha-brass in
ammonium sulphate solution
Br. Corros. J. 8 (1973) p. 182

[331] Newman, R. C.; Burstein, G. T.
The anodic behavior of freshly generated
alpha-brass surfaces
Corrosion Sci. 21 (1981) 2, p. 119

[332] Bhakta, U. C.; Lahiri, A. K.
Stress corrosion cracking of alpha-brass in
ammonium sulphate solution
Br. Corros. J. 8 (1973) p. 182

[333] Weiland H.; Paul, M.; Ihle, S.
Untersuchungen zum Zusammenwirken
einiger Wasserinhaltsstoffe bei der
Korrosion von Kondensatrohren aus
CuZn30 und CuZn21Al2
(Study of the interaction of some water
components in the corrosion of condenser
tubes made of CuZn39 and CuZn21Al2)
(in German)
Korrosion 10 (1979) p. 274

[334] Walker, R.
Triazole, benzotriazole and naphthatriazole
as corrosion inhibitors for brass
Corrosion 32 (1976) 10, p. 414

[335] Ponzano, G. P.; Sangiorgi, R.; Bonora, P. L.
Corrosione sotto tensione di un ottone
alpha + beta. II. Comportamento alla
rottura (in Italian)
Ann. Chimica 65 (1975) p. 609

[336] Yamane, T.; Hirao, K.; Yoshimoto, K. et al.
Weibull's distribution of time to stress
corrosion cracking of Cu-Zn alloys
Z. Metallkd. 74 (1983) 9, p. 603

[337] Mai, Y. W.
Crack propagation resistance of alpha-brass
subjected to prior-exposure in corrosive
environments
Corrosion 32 (1976) 8, p. 336

[338] Zajtsev, V. A.; Lozhkin, L. N.;
Skorechelletti, V. V.; Topunov, P. P.
The effect on corrosion cracking of changes

in electrode potential during the
application of tensile stresses (in Russian)
Zashch. Met. 13 (1977) 3, p. 308

[339] Patel, N. K.; Franco, J.; Patel, I. S.
1,5 Diphenyl carbazide as corrosion
inhibitor for 63/37 brass
Metals and Miner. Rev. 14 (1975) 8, p. 20

[340] Bartoniĉek, R.
Anodische Polarisationskurven von
Messinglegierungen in
Ammoniumchloridlösungen
(Anodic polarization curves of brass alloys
in ammonium chloride solutions)
(in German)
Werkst. Korros. 25 (1974) 4, p. 253

[341] Kingerley, D. G.; Longster, M. J.
Stress corrosion of phosphor-bronze
reinforcing tapes on underground power
cables
Corros. Sci. 14 (1974) p. 165

[342] Anonymous
Anti-corrosion methods and materials. 4:
Forms of corrosion. Part 3: Corrosion
cracking
Anticorros. Methods Mater. 19 (1972) 12, p. 9

[343] Desai, M. N.
Corrosion inhibitors for brasses
Werkst. Korros. 24 (1973) 8, p. 707

[344] Nielsen, K.; Rislund, E.
Comparative study of dezincification tests
Br. Corros. J. 8 (1973) p. 106

[345] Romanski, J.; Bialonczyk, M.
Corrosion properties of MM47 type
manganese brass with low copper content
Prace Inst. Odlewn. 6 (1956) 4, p. 15

[346] Prasad, E.; Panchanathan, N. V.
Chemical descaling and the importance of
correct acid inhibition
Steam Fuel Users J. 28 (1978) 1, p. 18

[347] Takemoto, T.; Hori, S.
The effect of pH, copper content and
ammonia concentration on the stress
corrosion cracking susceptibility on an
aged Cu-4% Ti alloy
Corros. Sci. 18 (1978) p. 323

[348] Chatterjee, U. K.; Sircar, S. C.; Banerjee, T.
Transition in mode of cracking and effect of
cold reduction on stress corrosion cracking
in copper-manganese alloys
Corrosion 26 (1970) 6, p. 141

[349] Hönig, A.; Schlenk, K. W.
Aushärtbare Federlegierungen und ihre
Anwendungen
(Age hardenable spring alloys and their
use) (in German)
Chem.-Tech. 6 (1977) p. 409

[350] Potzschke, M.
Grundlegende korrosionschemische
Eigenschaften von Kupferwerkstoffen
(Fundamental corrosion-chemical
characteristics of copper materials)
(in German)
Schiff und Hafen (1981) 11,
(DKI-Sonderdruck), 1982

[351] Fitzgerald, L. D.
Protection of copper metals from
atmospheric corrosion
in: Atmospheric Factors Affecting the
Corrosion of Engineering Metals, p. 152
ASTM STP 646, 1978

[352] Stockle, B.; Fitz, S.; Mach, M.; Pohlmann,
G.; Snethlage, R.
Die atmospharische Korrosion von Kupfer
und Bronze im Rahmen des UN/ECE-
Expositionsprogrammes Zwischenbericht
nach 4-jähriger Bewitterung (The
atmospheric corrosion of copper and
bronze in the context of the UN/ECE
exposure program. Interim report after
4-years of weathering) (in German)
Werkst. Korros. 44 (1993) 1, p. 48

[353] Costas, L. P.
Atmospheric corrosion of copper alloys
exposed for 15 to 20 years
in: Atmospheric Corrosion of Metals,
p. 106
ASTM STP 767, 1982

[354] Flinn, D. R.; Cramer, S. D.; Carter, J. P.;
Hurwitz, D. M.; Linstrom, P. J.
Environmental effects on metallic
corrosion products formed in short-term
atmospheric exposures
in: Materials Degradation Caused by Acid
Rain
American Chemical Society, 1986, p. 119

[355] Starke, E. A.
Aluminium and aluminium alloys:
Selection
in: Encyclopedia of Materials Science and
Engineering, Vol. 1
Pergamon Press, Oxford, 1986, p. 150

[356] Leidheiser jr., H.
The corrosion of copper, tin and their alloys
John Wiley and Sons, Inc., New York,
London, Sydney, Toronto, p. 3

[357] Baboian, R.; Haynes, G.; Sexton, P.
Atmospheric corrosion of laminar composites consisting of copper on stainless steel
in: Atmospheric Factors Affecting the Corrosion of Engineering Metals, p. 185
ASTM STP 646, 1978

[358] Leidheiser, H.
Corrosion of copper alloys
in: Encyclopedia of Materials Science and Engineering, Vol. 2, p. 895
Pergamon Press, Oxford (GB), 1986

[359] Tulka, J.; Zeman, J.
A modified atmospheric corrosion model. Model of low carbon steel
11th International Corrosion Congress, Florence (Italy), 1990, (Proc. Conf.), Vol. 1, p. 1.149

[360] Cramer, S. D.; McDonald, L. G.; Spence, J. W.
Effects of acidic deposition on the corrosion of zinc and copper
12th International Corrosion Congress, Houston (Tx./USA), September 1993 (Proc. Conf.), Vol. 2, p. 722, NACE, Houston (Tx./USA), 1993

[361] Carter, J. P.; Linstrom, P. J.; Flinn, D. R.; Cramer, S. D.
The effects of sheltering and orientation on the atmospheric corrosion of structural metals
Mater. Performance 26 (1987) 7, p. 25

[362] Espada, L.; Merino, P.; Gonzalez, A.; Sanchez, A.
Atmospheric corrosion in marine environments
10th International Congress on Metallic Corrosion, Madras (India), 1987, (Proc. Conf.), Vol. I, p. 3

[363] Holler, P.; Knotkova, D.
Využiti koroznich zkoušek pro hodnoceni korozni agresivity atmosfery (Use of the corrosion experiments for the evaluation of the aggressiveness of the atmosphere) (in Czechian)
Koroze Ochr. Mater. 31 (1987) 3, p. 49

[364] Rychtera, M.; Honzak, J.; Machova, B.
Chronologicka předpověd' atmosfericke koroze medi (Chronological prediction of the atmospheric corrosion of copper) (in Czech)
Electrotechn. obzor 70 (1981) 9, p. 518

[365] Haynie, F. H.; Upham, J. B.
Correlation between corrosion behavior of steel and atmospheric pollution data
in: Corrosion in Natural Environments, p. 33, ASTM STP 558, 1974
Haynie, F. H.; Spence, J. W.; Upham, J. B.
Effects of air pollutants on weathering steel and galvanized steel: a chamber study
in: Atmospheric Factors Affecting the Corrosion of Engineering Metals, p. 30
ASTM STP 646, 1978

[366] Hakkarainen, T.; Ylasaari, S.
Atmospheric corrosion testing in Finland
Atmospheric Corrosion, Hollywood (Fla./USA), October 1980 (Proc. Conf.), p. 787

[367] Golfer, P.; Knotkova, D.; Cherny, M.; Strekalov, P.; Yegutidze, Z.; Kozhukharov, V.; Zaydel, M.
Atmospheric corrosion of unalloyed steel, zinc, copper and aluminium and its connection with certain characteristics of the environmental atmosphere. Analysis of results of 10-year tests conducted in the European CMEA countries (in Russian)
Zashch. Met. 22 (1986) 5, p. 675

[368] Feliu, S.; Morcillo, M.
Datos de corrosion atmosforica en Espana y su interpretacion
(Data on atmospheric corrosion in Spain and its interpretation) (in Spanish)
Rev. Metal. 14 (1978) 4, p. 198
Feliu, J.; Morcillo, M.
La corrosion de los metales en las atmosferas de la Cuenca del Tajo
(The corrosion of the metals in the atmosphere of the area of the Tajo basin) (in Spanish)
Corros. Prot. 11 (1980) 2, p. 7

[369] Espada, L.; Sanchez, A. M.; Vilar, J. C.
Determinacion de la corrosion atmosferica en amplias zonas geograficas
(Determination of the atmospheric corrosion in expanded geographical areas) (in Spanish)
8th European Congress on Metallic Corrosion,
Paris, 1985 (Proc. Conf.), Vol. 2, 7.1–7.5
Rev. Iber. Corros. y Prot. 17 (1986) 3, p. 220

[370] Knotkova, D.; Gullman, J.; Holler, P.; Kucera, V.
Assessment of corrosivity by short-term atmospheric field tests of technically important metals

International Congress on Metallic Corrosion, Toronto (Canada), 1984, (Proc. Conf.), Vol. 3, p. 198

[371] Kucera, V.; Knotkova, D.; Gullman, J.; Holler, P.
Corrosion of structural metals in atmospheres with different corrosivity at eight years in Sweden and Czechoslovakia
10th International Congress on Metallic Corrosion, Madras (India), 1987, (Proc. Conf.), Vol. I, Sessions 1–4, p. 167

[372] Southwell, C. R.; Bultman, J. D.
Atmospheric corrosion testing in the tropics
Atmospheric Corrosion, Hollywood (Fla./USA), October 1980, (Proc. Conf.), p. 943

[373] Fernandez, A.; Leiro, M.; Rosales, B.; Ayllon, E. S.; Varela, F. E.; Gervasi, C. A.; Vilche, J. R.
Techniques applied to the analysis of the atmospheric corrosion of low carbon steel, zinc, copper and aluminum
12th International Corrosion Congress, Houston (Tx./USA), September 1993 (Proc. Conf.), Vol. 2, p. 574
NACE, Houston (Tx./USA), 1993

[374] Nixon, M.
Monitoring environmental contributions to atmospheric corrosion
in: Progress in the Understanding and Protection of Corrosion, p. 53–65
Institute of Materials, London University Press, Cambridge, 1993

[375] Ismail, A. A.; Khanem, N. A.; Abduk-Azim, A. A.
A corrosion map of Cairo and the coastal area of Egypt
Corros. Prev. Control 32 (1985) 4, p. 75

[376] Sundaram, M.; Mohan, P S.; Ananth, V.
Atmospheric corrosion of metals – long term exposure results at Mandapam Camp
10th International Congress on Metallic Corrosion, Madras (India), 1987, (Proc. Conf.), Vol. I, Sessions 1–4 (Key. Eng. Mater (1988) 1, p. 143)

[377] Mehta, P. R.
Atmospheric corrosion of metals in Saurashtra
Salt Research and Industry 15 (1979) 1, p. 51

[378] Binh, D. T.; Huy, V. D.; Strecalov, P. V.
Atmospheric corrosion of metals and alloys after five years of tests in Vietnam
in: Progress in the Understanding and Protection of Corrosion, p. 126
University Press, Institute of Materials London, Cambridge, 1993

[379] Knotkova, D.
Atmospheric corrosivity classification results of the International Testing Program ISOCORRAG
12th International Corrosion Congress, Houston (Tx./USA), September 1993 (Proc. Conf.), Vol. 2, p. 561, NACE, Houston (Tx./USA), 1993

[380] Kucera, V.; Henriksen, J.; Leygraf, C.; Coote, A. T.; Knotkova, D.; Stockle, B.
Materials damage caused by acidifiying air pollutants, 4-year results from an international exposure programme within UN ECE
Corrosion Control for Low-Cost Reliability 12th International Corrosion Congress, Houston (Tx./USA), September 1993, Vol. 2, p. 494
NACE, Houston (Tx./USA), 1993

[381] Franey, J.
Degradation of copper and copper alloys by atmospheric sulfur
in: Degradation of Metals in the Atmosphere, p. 306, ASTM STP 965, Philadelphia, 1988

[382] Henriksen, O.; Johnson, E.; Hohansen, A.; Sarholt-Kristensen, L.
Effect of boron implantation on the corrosion behavior, microhardness and contact resistance of copper and silver surfaces
Copenhagen Univ. Report Nr. 86-05, 1986

[383] Rice, D. W.; Peterson, P.; Rigby, E. B.; Phipps, P. B. P.; Cappell, R. J.; Tremoureux, R.
Atmospheric corrosion of copper and silver
J. Electrochem. Soc. 128 (1981) 2, p. 275

[384] Granese, S. L.; Fernandez, A.; Rosales, B. M.
Chemical characterization of the corrosion products formed on plain C steel, zinc, copper and aluminium
12th International Corrosion Congress, Houston (Tx./USA), September 1993 (Proc. Conf.), Vol. 2, p. 652
NACE Houston (Tx./USA), 1993

[385] Knotkova, D.; Vlčkova, J.; Rozlivka, L.
Defects of steel structures caused by atmospheric corrosion

12th International Corrosion Congress, Houston (Tx./USA), September 1993 (Proc. Conf.), Vol. 2, p. 734
NACE, Houston (Tx./USA), 1993

[386] Gullman, J.
Effects of sea salt on corrosion attack at 8 years exposure of metals in a small geographical area of the Swedish West Coast
12th International Corrosion Congress, Houston (Tx./USA), September 1993 (Proc. Conf.), Vol. 2, p. 642
NACE, Houston (Tx./USA), 1993

[387] Mikhailovskii, Yu. N.; Dzhincharadze, G. Kh.; Shirokolobov, V. N.; Panchenko, J. M.; Strekalov, P. V.
Structure of the aerochemical complex of the atmosphere at coastal corrosion stations of the Institute of Physical Chemistry of the Academy of Science of the USSR and its influence on the corrosion rates of metals
Prot. Met. (USSR) 18 (1982) 1, p. 33

[388] Mikhailovsky, Y. N.; Strekalov, P. V.
Atmospheric corrosion tests in the USSR
Atmospheric Corrosion, Hollywood (Fla./USA), October 1980, (Proc. Conf.), p. 923

[389] Feliu, S.; Morcillo, M.
Relaciones empiricas entre corrosion y variables meteorologicas y de contaminacion (Empirical relations between atmospheric corrosion and meteorological and contamination parameters) (in Spanish)
Corrosion and the Environment 2, Barcelona (Spain), May 1986 (Proc. Conf.), p. 299

[390] Takazawa, H.
Effect of NO_2 on the atmospheric corrosion of metals (in Japanese)
Boshoku Gijutsu 34 (1985) 11, p. 612

[391] Henriksen, J. F.; Rode, A.
Corrosion rates of various metals in SO_2/NO_2 polluted atmospheres
10th Scandinavian Corrosion Congress, Stockholm (Sweden), 1986 (Proc. Cont.), Vol. 7, p. 39

[392] Ericsson, F; Johansson, L.-G.
The role of NO_2 in the atmospheric corrosion of different metals
10th Scandinavian Corrosion Congress, Stockholm (Sweden), 1986, (Proc. Conf.), Vol. 8, p. 43–48

[393] Byrne, S. C.; Miller, A. C.
Effect of atmospheric pollutant gases on the formation of corrosive condensate on aluminium
Aluminium 58 (1982) 12, p. 709

[394] Eriksson, P.; Johansson, L.-G.; Strandberg, H.
Initial stages of copper corrosion in humid air containing SO_2 and NO_2
J. Electrochem. Soc. 140 (1993) 1, p. 53

[395] Groger, W.; Mayr, M.; Ebel, H.; Vernisch, J.
Surface investigations of hot-dip galvanized steel sheets
Radex Rundsch. (1986) 2/3, p. 75

[396] Graedel, T. E.; Kammlot, G. W.; Franey, J. P.
Carbonyl sulfide: Potential agent of atmospheric sulfur corrosion
Science 212 (1981) Mai, p. 663

[397] Protzer, H.; Haselbach, M.
Einfluss von Bitumen and bitumenhaltigen Baustoffen auf Kupfer (Influence of bitumen and bitumen-containing building materials on copper) (in German)
Metall 37 (1983) 3, p. 257

[398] Izumo, S.; Sueyoshi, H.; Kitamaru, K.; Ohzono, Y.
Corrosion of metals in volcanic atmosphere – effect of volcanic ash (in Japanese)
Boshoku Gijutsu 39 (1990) 5, p. 247

[399] Mikhailovsky, Y. N.
Theoretical and engineering principles of atmospheric corrosion of metals
Atmospheric Corrosion, Hollywood (Fla./USA), October 1980 (Proc. Conf.), p. 85

[400] Morcillo, M.; Feliu, S.
Datos de corrosion atmosforica en Espana. Resultados de seis anos de exposition (Atmospheric corrosion data for Spain. Results of six-year exposition) (in Spanish)
Rev. lb. Corrosion y Protection 14 (1983) Numero extraordinario, p. 89

[401] Jostan, J. L.; Mussinger, W.
Untersuchungen der Korrelation zwischen Laborkurzzeittests and Betriebsatmosphären bei der Korrosionsprüfung galvanisch hergestellter Schutzschichten für die Elektronik. Ergebnisse des Forschungs- und Entwicklungsprogramms Korrosion und Korrosionsschutz (Investigations of

the correlation between laboratory short time tests and operating atmospheres during the corrosion test of galvanically manufactured protective layers for electronics. Results of the research and development program corrosion and corrosion protection) (in German)
Werkst. Korros. 35 (1984) 6, p. 297
Jostan, J. L.; Mussinger, W.; Bogenschütz, A. F.
Langzeitkorrosionsprüfungen an Elektrowerkstoffen, Teil 1: Werkstoffproben und Prüfergebnisse unter aggressivem Industrieklima (Long-term corrosion tests at electrical materials, Part 1: Material samples and inspection results under aggressive industrial climate) (in German)
Metalloberfläche 39 (1985) 2, p. 45

[402] Cerveny, L.
Atmosfericka koroze hořčiku přiskladovani v krytych skladech (Atmospheric corrosion of magnesium with storage in protected rooms) (in Czech)
Hutnicke Listy 43 (1988) 1, p. 40

[403] Fukuda, Y.; Fukushima, T.; Sulaiman, A.; Musalam, I.; Yap, L. C.; Chotimongkol, L.; Judabong, S.; Potjanart, A.; Keowkangwal, O.; Yoshihara, K.; Tosa, M.
Indoor corrosion of copper and silver exposed in Japan and ASEAN countries
J. Electrochem. Soc. 138 (1991) 5, p. 1238

[404] Knotkova, D.; Holler, P.; Vlakova, J.
The evaluation of the microclimate effect of the characterisics of environment and corrosion of metals
8th International Congress on Metallic Corrosion, Mainz, 1981, (Proc. Conf.), Vol. 1, p. 859

[405] Furusawa, T.; Udo, Y.; Yanagida, K.; Nagai, T.; Henmi, Z.
Corrosion of metals for electronic component in mixed gases
J. Jpn. Inst. Met. 50 (1986) 5, p. 467 (Japanese)

[406] Sachsenweger, C.; Hansele, K.; Stimper, K.; Sockel, H. G.
Die Wirkung korrosiver Atmosphären auf Niederspannungsisolierungen (The effect of corrosive atmospheres on low voltage isolations) (in German)
Werkst. Korros. 39 (1988) 6, p. 287

[407] Anonymous
Korrosionsschutz von Kupfer (Corrosion protection of copper) (in German)
Galvanotechnik 76 (1985) 1, p. 40

[408] Eschke, K.-R.; Erwe, H.; Stier, F.
Die Nachwirkungen von Dampfphaseninhibitoren nach Entnahme des metallischen Packgutes und der VCI-enthaltenden Umhüllung (The aftereffects from vapor phase inhibitors to withdrawal of the metallic packaging material and the VCI containing casing) (in German)
Neue Verpackung 40 (1987) 2, p. 66

[409] Zak, E. G.; Zhdanova, E. I.; Laikhter, L. B.
Corrosion and corrosion protection of copper with benztriazole derivatives (in Russian)
Ingibitory Korroz. Met. Mosk. Gos. Pedagog. Institut (1980), p. 105

[410] Poulter, G. A.; Sorensen, N. R.
The effect of organic inhibitors on the atmospheric corrosion and solderability of copper
Sandia National Laboratories, Report DE 88000261, August 1987

[411] Vaidyanath, L. R.; Bhandary, V. S.
Corrosion resistance of copper and copper alloys – a survey
Proceedings of a Symposium, Bombay (India), 1971

[412] Cohen, A.
Copper and copper based alloys
in: Process Industries Corrosion – The Theory and Practice, p. 479
NACE, Houston (Texas/USA), 1986

[413] Kobayashi, H.; Yasumori, A.; Kimura, T.; Hayashi, H.; Goto, S.
Corrosion-resistant copper alloys
European Patent Application 0 263 879, A 1
Int. Cl. C 22 C 9/01 vom 27. 11. 86

[414] Ailor, H. W.
Seven-year exposure at Point Reyes, California
in: Corrosion in Natural Environments, p. 75, ASTM STP 558, 1974

[415] Anonymous
Copper alloy no. C 60800 (Aluminium bronze)
Alloy Dig. (Cu-443) (1982) 8

[416] Mshana, J. S.; Vosikovsky, O.; Sahoo, M.
Corrosion fatigue behaviour of nickel-aluminium bronze alloys
Canadian Metallurgical Quarterly 23 (1984) 1, p. 7

[417] Smith, R. D.
An accelerated atmospheric corrosion test (AACT)
in: Atmospheric Factors Affecting the Corrosion of Engineering Metals, p. 17
ASTM STP 646, 1978

[418] Thompson, D. H.
Atmospheric corrosion of copper alloys
in: Metal Corrosion in the Atmosphere, p. 129 ASTM STP 435, 1969

[419] Castillo, A. P.; Popplewell, J. M.
General, localized, and stress-corrosion resistance of a series of copper alloys in natural atmospheres
in: Atmospheric Corrosion of Metals, p. 60
ASTM STP 767, 1982

[420] Heubner, M.; Rockel, M.; Rudolph, G.
Kupfer-Münzwerkstoffe hoher Korrosions- und Anlaufbeständigkeit (Copper coin materials of high corrosion and tarnish stability) (in German)
Z. Metallkd. 73 (1982) 8, p. 522

[421] Herman, R. S.; Castillo, A. P.
Short-term atmospheric corrosion of various copper-base alloys – two- and four-year results
in: Corrosion in Natural Environments, p. 82, ASTM STP 558, 1974

[422] Sudarshan, T. S.; Louthan jr., M. R.; Place, T. A.; Mabie, H. H.
Hydrogen and humidity effects on fatigue behavior of a 70-30 copper-nickel alloy
J. Materials for Energy Systems 8 (1986) 3, p. 291

[423] Sudarshan, T. S; Louthan jr., M. R.; Place, T. A.
Humidity effects on the fatigue behavior of Cu-Ni
Mater. Sci. Eng. 66 (1984) 2, p. L 27

[424] Mattsson, E.; Holm, R.
Copper and copper alloys
in: Metal Corrosion in the Atmosphere, p. 187
ASTM STP 435, 1969

[425] Mertel, J.
Über die atmosphärische Korrosion des Stahls in Havanna (On the atmospheric corrosion of steel in Havana) (in German)
Korrosion (Dresden) 18 (1987) 4, p. 193

[426] Holm, R.; Mattsson, E.
Atmospheric corrosion tests of copper and copper alloys in Sweden – 16-year results
in: Atmospheric Corrosion of Metals, p. 85
ASTM STP 767, 1982

[427] Subramanian, G.; Ananth, V.; Mohan, P. S.; Balasubramanian, T. M.
Selective removal of tin and phosphorus in phosphor bronze
Bull. Electrochem. 4 (1988) 2, p. 173

[428] Ananth, V.; Subramanian, G.; Mohan, P. S.; Palraj, S.
Atmospheric corrosion of 3004 aluminium alloy, phosphor bronze and α-brass
Bulletin of Electrochemistry 2 (1986) 6, p. 541

[429] Soto, L.; Franey, J. P.; Graedel, T. E.; Kammlott, G. W.
On the corrosion resistance of certain ancient Chinese bronze artifacts
Corros. Sci. 23 (1983) 3, p. 241

[430] Girolette, L.; Pozzi, D.; Roggia, E.
Inspection on the state of conservation of the war memorial in Monza
11th International Corrosion Congress, Florence (Italy), 1990, (Proc. Conf.), Vol. 1, p. 1.157

[431] Ghali, E.; Fiset, M.
Etude sur la protection des monuments en bronze de la ville de Quebec (Study about protection of monuments of bronze in Quebec City) (in French)
Metaux Corros.-Ind. 61 (1986) 729, p. 145

[432] Anonymous
Mueller alloy 3300 (Low-leaded brass)
Alloy Dig. (1981) 11

[433] Andrzejaczek, B. J.
Einige Probleme der Korrosion von Messing (Some problems of the corrosion of brass) (in German)
Korrosion (Dresden) 19 (1988) 2, p. 59

[434] Han, M. K.; Baevers, J. A.; Goins, W.
Stress corrosion cracking failure of brass electrical connectors in an outdoor atmospheric environment
Mater. Performance 26 (1987) 7, p. 20

[435] Richards, J. T. (Firmenschrift)
Corrosion resistance of beryllium copper alloys
Brush Wellman Engineered Materials, Cleveland

[436] Vaidyanath, L. R.; Bhandary, V. S.
Corrosion resistance of copper and copper

alloys – A survey Proc. Symp. "Corrosion resistance of copper and copper alloys" Bombay, 1971, p. 1

[437] Papshev, D. D.; Yakovlev, V. M.
The effect of homopolar liquid media on the microhardness of copper (in Russian)
Fiz. Khim. Mekh. Mater. 16 (1980) 6, p. 98

[438] Shevyakova, G. V.; Shishov, Yu. F.
The electrochemical activity of steel/copper couples in benzene solutions of sulphur (in Russian)
Zashch. Met. 13 (1977) p. 576

[439] Bottenberg, K.
Substitution von Benzol durch elementares Fluor in der Gasphase
(Substitution of benzene by elementary fluorine in the gas phase) (in German)
Chemiker Ztg. 96 (1972) p. 84

[440] Ilin, I. N.; Turlais, D. P.; Grishin, V. A. et al.
The structural changes and chemical changes taking place in a heat-transfer surface in the process of boiling
Therm. Eng. 29 (1982) p. 117

[441] Serpuchenko, E. A.; Kurilenko, O. D.; Anistratenko, G. A.
Corrosion of metallic powders and their protection
Russ. Progress Chem. 48 (1982) 6, p. 59

[442] Kretschmar, H.
Reinigen von Leiterplatten
(Cleaning of printed circuits) (in German)
Galvanotechnik 65 (1974) 1, p. 11

[443] Samoilov, Yu. F.; Petrova, T. I.; Pankratov, S. A.
Effect of various impurities in the condensate on the corrosion of brass (in Russian)
Trudy Moskov. Energ. Inst. (1972) 128, p. 55

[444] Nelson, G. A.
Corrosion Data Survey, Shell Devel. Comp., Emeryville, Calif. (1960)

[445] Rabald, E.
Corrosion Guide, 2nd rev. ed. (1968)
Elsevier Pub. Co., Amsterdam, London, New York

[446] Paatsch, W.
Photoelectric measurements during corrosion and inhibition of copper in aqueous solutions
Ber. Bunsenges. phys. Chem. 81 (1971) p. 645

[447] Mayanna, S. M.; Setty, T. H. V.
Corrosion of copper single crystal planes in acidic bromide solution
Indian J. Chem. 12 (1974) p. 1083

[448] Smolyaninov, S.
The influence of halide ions on the corrosion stability and the electrochemical behavior of copper in sulfuric and in hydrochloric acid (in Russian)
Khim. i Khim. Tekhnol. 7 (1964) p. 588; (C.A. 62 (1965) 4900c)

[449] Dockus, K. F.; Krueger, R. H.; Rush, W. F.
Am. Soc. Heating, Refrig. Air. Cond. Eng. J. (1962) Dec.; p. 67

[450] Bogdanov, V. P.; Kadraliev, M. I.
Depolarizing properties of copper ions in halide solutions (in Russian)
Zashch. Metallov 4 (1968) No. 6; p. 740

[451] Levin, A. I.; Pomosov, A. V. et al.
Corrosion and stabilization of powdered metals
I. Corrosion of powdered copper (in Russian)
Korroziya i Metallo-pokritiya, Sbornik Statei (1953) No. 43, p. 118; (C.A. 50 (1956) 16635a)

[452] Marshakov, I. K.; Bogdanov, V. P.
Effect of the solution composition on the corrosion and electrochemical behavior of brasses (in Russian)
Khim. i Khim. Tekhnol. 12 (1969) p. 273; (C.A. 71 (1969) 35398d)

[453] Gurovich, E. I.
Effect of fused halides on nickel, copper, iron, and some steels (in Russian)
Zhur. Priklad. Khim. 33 (1960) p. 2096; (C.A. 55 (1961) 1346f)

[454] Marshakov, I. K.; Bogdanov, V. P.
Effect of the solution composition on the corrosion and electrochemical behavior of brasses (in Russian)
Khim. i Khim. Tekhnol. 12 (1969) p. 273; (C.A. 71 (1969) 35398d)

[455] Uhlig, H.; Gupta, K.; Liang, W.
Critical potentials for stress corrosion cracking of 63-37 brass in ammoniacal and tartrate solutions
J. Electrochem. Soc. 112 (1975) p. 343

[456] Marshakov, I. K.; Ugar, Y. A.; Vigdorovich, V. I.
Mechanism of corrosion damage of Mg-Cu alloys (in Russian)

Zashch. Metallov 1 (1965) p. 406; (C.A. 63 (1965) 12809b)

[457] Rabald, E.
DECHEMA-Werkstoff-Tabelle "Brom" Mai (1954)
(DECHEMA Corrosion Data Sheets "Bromine" May (1954)
DECHEMA, D-Frankfurt am Main

[458] Boet, E.; Crousier, J. P.
Etude de la bromuration du cuivre massif (Bromation of solid copper) (in French)
Corrosion Sci. 12 (1972) No. 11; p. 817

[459] Franco, J.; Patel, N. K.
Electronic property of photosensitive film formed during corrosion of copper in aqueous and non-aqueous solutions of bromine
J. Appl. Electrochem. 3 (1973) p. 303

[460] Lake, D. E.; Gunkler, A. A.
Moisture: Key to bromine corrosion
Chem. Eng. 67 (1960) No. 22; p. 136

[461] Everhart, J. L.
CuNi alloys with high temperature strength and corrosion resistance
Materials in Design Eng. 47 (1958) No. 5; p. 114

[462] Patel, M. M.; Patel, N. K.; Vora, J. C.
Effect of oxidizing agents on the corrosion of 63/37 brass in local supply water
Australasian Corrosion Eng. (1972) Oct., p. 21

[463] Bay, J.
Löten von Kupferrohren für Reinst- und Medizinalgase
Technica 43 (1994) 4, pp. 66–68

[464] Wendler-Kalsch, E.
Korrosionsverhalten und Deckschichtbildung auf Kupfer und Kupferlegierungen
Zeitschrift für Werkstofftechnik (1982), pp. 129–137

[465] Kristiansen, H.
Corrosion of copper by water of various temperatures and carbon dioxide contents
Werkstoffe und Korrosion 28 (1977) 11, pp. 744–748

[466] Broo, A. E.; Beghult, B.; Hedberg, T.
Copper corrosion in drinking water distribution systems – The influence of water quality
Corrosion Science 36 (1997) 6, pp. 1119–1132

[467] DIN EN 12502-2 (März 2005)
Korrosionsschutz metallischer Werkstoffe
Korrosionswahrscheinlichkeit in Wasserleitungssystemen
Teil 2: Übersicht über Einflussfaktoren für Kupfer und Kupferlegierungen
Beuth Verlag GmbH, 10772 Berlin

[468] Wendler-Kalsch, E.; Gräfen, H.
Korrosionsschadenkunde
Springer-Verlag, Berlin, 1988

[469] Wollrab, O.
Über den Einfluss der Wasserbeschaffenheit auf die Lochkorrosion in Trinkwasserleitungen aus Kupfer
Schadenprisma 18 (1989) 3, pp. 45–48

[470] von Franque, O.; Gerth, D.; Winkler, B.
Ergebnisse von Untersuchungen an Deckschichten in Kupferrohren
Werkstoffe und Korrosion 26 (1975) 4, pp. 255–258

[471] Edwards, M.; Schock, M. R.; Meyer, T. E.
Alkalinity, pH, and copper corrosion by-product release
J. Am. Water Work Assoc. 88 (1996) 3, pp. 81–94

[472] Monroe, E. S.
Effects of CO_2 in steam systems
Chemical Engineering 88 (1981) 6, pp. 209–212

[473] Asrar, N.; Mali, A. U.; Ahmad, S. et al.
Early failure of cupro-nickel condenser tubes in thermal desalination plant
Desalination 116 (1998), pp. 135–144

[474] Werkstoff-Datenblätter
CuSn10-C; CuSn7Zn4Pb7-C; CuSn7Zn2Pb3-C; CuSn5Zn5Pb5-C; CuSn5Pb1
Deutsches Kupferinstitut, Düsseldorf, Stand 2005

[475] DECHEMA-Werkstoff-Tabelle "Methylformiat"
(Corrosion data sheets "Methyl formate") (in German)
DECHEMA, D-Frankfurt am Main, January 1961

[476] DECHEMA-Werkstoff-Tabelle "Äthylformiat"
(Corrosion data sheets "Ethyl formate") (in German)
DECHEMA, D-Frankfurt am Main, March 1953

[477] DECHEMA-Werkstoff-Tabelle
"Acetessigester"
(Corrosion data sheets "Acetoacetic ester")
(in German)
DECHEMA, D-Frankfurt am Main, March 1953

[478] DECHEMA-Werkstoff-Tabelle "Äthylacetat"
(Corrosion data sheets "Ethyl acetate")
(in German)
DECHEMA, D-Frankfurt am Main, March 1953

[479] DECHEMA-Werkstoff-Tabelle
"Äthylbutyrat"
(Corrosion data sheets "Ethyl butyrate")
(in German)
DECHEMA, D-Frankfurt am Main, May 1953

[480] DECHEMA-Werkstoff-Tabelle "Äthyllactat"
(Corrosion data sheets "Ethyl lactate")
(in German)
DECHEMA, D-Frankfurt am Main, May 1953

[481] DECHEMA-Werkstoff-Tabelle
"Äthylstearat"
(Corrosion data sheets "Ethyl stearate")
(in German)
DECHEMA, D-Frankfurt am Main, March 1953

[482] DECHEMA-Werkstoff-Tabelle "Alkydharze"
(Corrosion data sheets "Alkyd resins")
(in German)
DECHEMA, D-Frankfurt am Main, March 1953

[483] DECHEMA-Werkstoff-Tabelle "Amylacetat"
(Corrosion data sheets "Amyl acetate")
(in German)
DECHEMA, D-Frankfurt am Main, September 1953

[484] DECHEMA-Werkstoff-Tabelle "Amyllaurat"
(Corrosion data sheets "Amyl laurate")
(in German)
DECHEMA, D-Frankfurt am Main, September 1953

[485] DECHEMA-Werkstoff-Tabelle
"Amylpropionat"
(Corrosion data sheets "Amyl propionate")
(in German)
DECHEMA, D-Frankfurt am Main, September 1953

[486] DECHEMA-Werkstoff-Tabelle
"Benzylacetat"
(Corrosion data sheets "Benzyl acetate")
(in German)
DECHEMA, D-Frankfurt am Main, May 1954

[487] DECHEMA-Werkstoff-Tabelle
"Benzylbenzoat"
(Corrosion data sheets "Benzyl benzoate")
(in German)
DECHEMA, D-Frankfurt am Main, May 1954

[488] DECHEMA-Werkstoff-Tabelle
"Benzylbutylphthalat"
(Corrosion data sheets "Benzyl butylphthalat") (in German)
DECHEMA, D-Frankfurt am Main, May 1954

[489] DECHEMA-Werkstoff-Tabelle
"Benzylsalicylat"
(Corrosion data sheets "Benzyl salicylate")
(in German)
DECHEMA, D-Frankfurt am Main, May 1954

[490] DECHEMA-Werkstoff-Tabelle
"Bornylacetat"
(Corrosion data sheets "Bornyl acetate")
(in German)
DECHEMA, D-Frankfurt am Main, May 1954

[491] DECHEMA-Werkstoff-Tabelle
"Bornylformiat"
(Corrosion data sheets "Bornyl formate")
(in German)
DECHEMA, D-Frankfurt am Main, May 1954

[492] DECHEMA-Werkstoff-Tabelle "Butylacetat"
(Corrosion data sheets "Butyl acetate")
(in German)
DECHEMA, D-Frankfurt am Main, January 1955

[493] DECHEMA-Werkstoff-Tabelle
"Butylbenzoat"
(Corrosion data sheets "Butyl benzoate")
(in German)
DECHEMA, D-Frankfurt am Main, January 1955

[494] DECHEMA-Werkstoff-Tabelle
"Butylbutyrat"
(Corrosion data sheets "Butyl butyrate")
(in German)
DECHEMA, D-Frankfurt am Main, January 1955

[495] DECHEMA-Werkstoff-Tabelle
"Butylglycolat"
(Corrosion data sheets "Butyl glycolate")
(in German)

DECHEMA, D-Frankfurt am Main, January 1955

[496] DECHEMA-Werkstoff-Tabelle "Butyloxalat"
(Corrosion data sheets "Butyl oxalate")
(in German)
DECHEMA, D-Frankfurt am Main, January 1955

[497] DECHEMA-Werkstoff-Tabelle "Butylphthalate"
(Corrosion data sheets "Butyl phthalate")
(in German)
DECHEMA, D-Frankfurt am Main, January 1955

[498] DECHEMA-Werkstoff-Tabelle "Butylstearat"
(Corrosion data sheets "Butyl stearate")
(in German)
DECHEMA, D-Frankfurt am Main, January 1955

[499] DECHEMA-Werkstoff-Tabelle "Celluloseacetobutyrat"
(Corrosion data sheets "Cellulose acetobutyrate") (in German)
DECHEMA, D-Frankfurt am Main, October 1955

[500] DECHEMA-Werkstoff-Tabelle "Cellulosetripropionat"
(Corrosion data sheets "Cellulose tripropionate") (in German)
DECHEMA, D-Frankfurt am Main, October 1955

[501] DECHEMA-Werkstoff-Tabelle "Chinintartrat"
(Corrosion data sheets "Quinine tartrate") (in German)
DECHEMA, D-Frankfurt am Main, October 1955

[502] DECHEMA-Werkstoff-Tabelle "Chlorameisensäureester"
(Corrosion data sheets "Chloromethanoic esters") (in German)
DECHEMA, D-Frankfurt am Main, October 1955

[503] DECHEMA-Werkstoff-Tabelle "Cyanessigsäureäthylester"
(Corrosion data sheets "Cyanoacetic acid ethylester") (in German)
DECHEMA, D-Frankfurt am Main, September 1956

[504] DECHEMA-Werkstoff-Tabelle "Cyclohexanolester"
(Corrosion data sheets "Cyclohexanol ester") (in German)
DECHEMA, D-Frankfurt am Main, September 1956

[505] DECHEMA-Werkstoff-Tabelle "Cyclohexylaminlaurat"
(Corrosion data sheets "Cyclohexylamine laurate") (in German)
DECHEMA, D-Frankfurt am Main, September 1956

[506] DECHEMA-Werkstoff-Tabelle "Dibutylmethylendithioglycolat"
(Corrosion data sheets "Dibutyl methylene dithioglycolate") (in German)
DECHEMA, D-Frankfurt am Main, September 1956

[507] DECHEMA-Werkstoff-Tabelle "Dibutylphthalat"
(Corrosion data sheets "Dibutyl phthalate") (in German)
DECHEMA, D-Frankfurt am Main, September 1956

[508] DECHEMA-Werkstoff-Tabelle "Dibutylthiodiglycolat"
(Corrosion data sheets "Dibutyl thiodiglycolate") (in German)
DECHEMA, D-Frankfurt am Main, September 1956

[509] DECHEMA-Werkstoff-Tabelle "Diglycolsäuredibutylester"
(Corrosion data sheets "Diglycolic acid dibutylester") (in German)
DECHEMA, D-Frankfurt am Main, November 1957

[510] DECHEMA-Werkstoff-Tabelle "Fette und Wachse"
(Corrosion data sheets "Fats and waxes") (in German)
DECHEMA, D-Frankfurt am Main, November 1957

[511] DECHEMA-Werkstoff-Tabelle "Methylacetat"
(Corrosion data sheets "Methyl acetate") (in German)
DECHEMA, D-Frankfurt am Main, January 1961

[512] DECHEMA-Werkstoff-Tabelle "Peressigsäure und Acetaldehyd-Monoperacetat"
(Corrosion data sheets "Peracetic acid and acetaldehyde monoperacetate") (in German)
DECHEMA, D-Frankfurt am Main, December 1966

[513] Beständigkeitstabellen von Kupfer-
Gußwerkstoffen in verschiedenen Medien
(Resistance tables for cast copper materials
in various media) (in German)
Deutsches Kupfer-Institut, Berlin,
Düsseldorf, June 1970

[514] Product Information
ABC der Stahlkorrosion
(ABC of steel corrosion) (in German),
2nd ed., 1966
Mannesmann AG, D-Düsseldorf

[515] Chashichina, O. A.; Kaimakova, V. P.;
Provorov, A. K.
Materials recommended for equipment in
the production of butyl acetate (in Russian)
Gidroliz. Lesokhim. Promst. (1983) 1, p. 7

[516] Wilkins, R. A.; Jenks, R. H.
in: H. H. Uhlig "The Corrosion
Handbook", p. 65
John Wiley & Sons, Inc., New York, 1958

[517] Team of authors
Werkstoffeinsatz und Korrosionsschutz in
der chemischen Industrie
(Application of materials and corrosion
protection in chemical industry)
(in German), pp. 210
VEB Deutscher Verlag für
Grundstoffindustrie, Leipzig, 1973

[518] Anonymous
Leitfaden – Umgang mit leichtflüchtigen
chlorierten und aromatischen
Kohlenwasserstoffen
(Guidelines – use of volatile chlorinated
and aromatic hydrocarbons) (in German),
p. 12, December 1984
Ministerium für Ernährung,
Landwirtschaft, Umwelt und Forsten
Baden-Württemberg

[519] Dilla, W.; Köser, H. J. K.
Zum Einfluß von Chloracetylchloriden
und Chloressigsäuren auf die Korrosion
metallischer Werkstoffe in Trichlorethylen
und Perchlorethylen
(The influence of chloroacetyl chlorides
and chloroacetic acids on the corrosion of
metals in trichloroethylene and
tetrachloroethylene) (in German)
Werkst. Korros. 35 (1984) 1, p. 16

[520] Köser, H. J. K.; Lübke, M.
Zum Korrosionsverhalten von
metallischen Werkstoffen in
chlorkohlenwasserstoffhaltigen Lösungen
(Corrosion behavior of metallic materials
in solvent mixtures containing chlorinated
hydrocarbons) (in German)
Werkst. Korros. 31 (1980) 3, p. 186

[521] Anonymous
Corrosion Data Survey – Metals Section,
6th ed., p. 54
NACE, Houston (Tx./USA), 1985

[522] Product Information
HYDRA® – Metallschläuche
(HYDRA® – flexible metal tubes)
(in German), pocket book No. 301, 1980
Witzenmann GmbH, Pforzheim

[523] Dunkle, H. H.; Fetter, E. C.
Chemical and heat resistance of gasket
materials
Chem. Eng. (1946) Nov., p. 102

[524] Kirk-Othmer
Encyclopedia of chemical technology,
2nd ed., vol. 5, p. 86, pp. 142
John Wiley & Sons Inc., New York, 1964

[525] LaQue, F. L.; Copson, H. R.
Corrosion resistance of metals and alloys,
2nd edition of the American Chemical
Society Monograph Series
Reinhold Publishing Corp., New York

[526] Ishibe, N.; Jackson, L.
Methylchloroform stabilizer composition
employing an alkynyl sulfide
USA Pat. 4,309,301 (05.01.1982); (Dow
Chemical Comp., Midland, Mich./USA)

[527] Dilla, W.; Köser, H. J. K.
Zur Dehydrochlorierung/Dechlorierung
von Trichlorethylen, Perchlorethylen und
1,1,1-Trichlorethan durch Metallchloride
und oxidativ/hydrolytische Abbauprodukte
(On the dehydrochlorination/
dechlorination of trichlorethylene,
perchlorethylene and 1,1,1-trichloroethane
by metal chlorides and oxidative/hydrolytic
degradation products) (in German)
Werkst. Korros. 34 (1983) 5, p. 241

[528] Bachmann, T.
Untersuchungen zum reduktiven Abbau
von umweltbelastenden chlorierten
Kohlenwasserstoffen
(Investigations on the reductive
decomposition of environmentally harmful
chlorinated hydrocarbons) (in German)
Dissertation, Universität Frankfurt am
Main, 1990

[529] Rabald, E.
Corrosion Guide, 2nd ed., p. 530
Elsevier Publishing Comp., Amsterdam-

London-New York, 1968
[530] DECHEMA-Werkstoff-Tabelle
"Hexachloräthan"
(DECHEMA corrosion data sheets
"Hexachloroethane") (in German)
DECHEMA, D-Frankfurt am Main,
October 1958
[531] Anonymous
Beständigkeitstabellen von Kupfer-
Gußwerkstoffen in verschiedenen Medien
(Resistance tables for cast copper materials
in various media) (in German), pp. 22
DKI Deutsches Kupfer-Institut, Berlin
[532] Vaidyanath, L. R.; Bhandary, V. S.
Corrosion resistance of copper and copper
alloys – a survey, p. 9
Indian Copper Information Centre,
Calcutta
[533] Hamner, N. E.
Corrosion Data Survey, 5th ed., pp. 6, 88,
98, 138
NACE, Houston (Tx./USA)
[534] Köser, J. K.
Korrosion in
halogenkohlenwasserstoffhaltigen
Lösungsmittelgemischen. Ergebnisse des
Forschungs- und Entwicklungsprogramms
Korrosion und Korrosionsschutz
(Corrosion in solvent mixtures containing
halogenated hydrocarbons. Results of the
research and development program
corrosion and corrosion protection)
(in German)
Werkst. Korros. 36 (1985) 1, p. 29
[535] DECHEMA-Werkstoff-Tabelle
"Äthylenchlorid"
(DECHEMA Corrosion data sheets "Ethyl
dichloride") (in German)
DECHEMA, D-Frankfurt am Main, March
1953
[536] Wilkins, R. A.; Jenks, R. H.
in: H. H. Uhlig "The Corrosion
Handbook", p. 92
John Wiley & Sons, Inc., New York, 1958
[537] Product Information
Korrosionsverhalten einiger
gebräuchlicher Werkstoffe
(Corrosion behavior of some materials in
use) (in German), Information MS
75.2.200.0
Elasteflex Metallschläuche, H. Skodock,
Hannover

[538] Product Information
List of the suitability of standard range of
chemical hoses for use with various
conveyants, Information vom 05.01.1973
Compoflex Ltd., NL-Delpht
[539] Product Information
Wieland-Buch – Kupferwerkstoffe
(Wieland book – copper materials)
(in German), p. 181
Wieland-Werke AG, Ulm
[540] Anonymous
Anti-corrosion methods and materials – 4.
Forms of corrosion, Part 3: Corrosion
cracking
Anticorros. Methods Mater. 19 (1972) 12,
p. 9
[541] DECHEMA-Werkstoff-Tabelle
"Chlorwasserstoff"
(DECHEMA corrosion data sheets
"Hydrogen chloride") (in German)
DECHEMA, D-Frankfurt am Main, March
1976
[542] Anonymous
Corrosion resistance of copper and copper
alloys. Part 1: Resistance to common
corrosives
Chem. Eng. (1951) January, p. 108
[543] Anonymous
Leitfaden – Umgang mit leichtflüchtigen
chlorierten und aromatischen
Kohlenwasserstoffen
(Manual of how to use volatile chlorinated
and aromatic hydrocarbons) (in German),
revised edition, pp. 49, December 1989
Ministerium für Ernährung,
Landwirtschaft, Umwelt und Forsten
Baden-Württemberg
[544] Anonymous
Korrosionsbeständigkeitstabellen
moderner Werkstoffe
(Resistance tables of modern materials)
(in German); Special print from
Werkstofftechnik (1970) 1 to (1972) 5, pp. 5
Verlag für Technik und Wirtschaft Meynen
KG, Wiesbaden
[545] H. H. Uhlig
"The Corrosion Handbook", p. 65
John Wiley & Sons, Inc., New York, 1958
[546] Baukloh, A.
Korngrenze und Korrosion bei
Kupferwerkstoffen, Gefüge der Metalle
(Grain boundary and corrosion of copper
materials, structures of the metals)

(in German) (Proc. Conf.) 1981, p. 67
Deutsche Gesellschaft für Metallkunde
e.V., Oberursel

[547] Autorenkollektiv
Werkstoffeinsatz und Korrosionsschutz in der chemischen Industrie
(Material employment and corrosion protection in the chemical industry) (in German)
VEB Deutscher Verlag für Grundstoffindustrie, Leipzig, 1973

[548] Heitz, E.; Deuchler, E.
Korrosion in Halogenkohlenwasserstoffen – Ziele, Abwicklung und Ergebnisse eines Gemeinschaftsprojektes
(Corrosion in halogenated hydrocarbons – purpose, blank and results of a joint project) (in German)
Werkst. Korros. 31 (1980) 3, p. 157

[549] Grefe
Korrosions-Atlas,
(Corrosion atlas) (in German) 3rd ed., 1958, pp. 8–9, 23, 44
D-Frankfurt am Main

[550] Hamner, N. E.
Corrosion Data Survey – Metals Section, 5. ed., p. 245
National Association of Corrosion Engineers, Katy (Tx./USA)

[551] Product Information
Korrosionsverhalten einiger gebräuchlicher Werkstoffe,
(Corrosion behavior of some common materials) (in German) Information MS 75.2.200.0
Elasteflex Metallschläuche, Skodock, Hannover

[552] DECHEMA-Werkstoff-Tabelle
"Methylchlorid"
(DECHEMA corrosion data sheet "Methyl chloride") (in German)
DECHEMA, D-Frankfurt am Main, January 1961

[553] Rabald, E.
Corrosion Guide, 2nd ed., p. 457
Elsevier Publishing Comp., Amsterdam-London-New York, 1968

[554] Product Information
HYDRA® – Metallschläuche
(HYDRA® – metal hoses) (in German), paperback no. 301, 1980
Witzenmann GmbH, Pforzheim

[555] Sretenskaya, G V.; Talaeva, G. V.
Corrosion resistance of the metal materials in the triacetate fibre production sector
Khimicheskie Volokna (1970) 3, p. 55
(English translation in: All-Union Scientific Research Institute for Man-Made Fibre)

[556] Anonymous
Corrosion Data Survey – Metals Section, 6. ed., pp. 36, 37
National Association of Corrosion Engineers, Katy (Tx./USA)

[557] Raudune, D. E.; Girlyavichus, A. A.; Sakalauskas, Z. Yu.
Protection of copper from corrosion in regeneration triacetylcellulose solvent media (in Russian)
Zashch. Met. 22 (1986) 2, p. 289

[558] DECHEMA-Werkstoff-Tabelle
"Methylenchlorid"
(DECHEMA corrosion data sheet "Methylene chloride") (in German)
DECHEMA, D-Frankfurt am Main, October 1961

[559] DECHEMA-Werkstoff-Tabelle
"Chloroform"
(DECHEMA corrosion data sheet "Chloroform") (in German)
DECHEMA, D-Frankfurt am Main, October 1955

[560] Rabald, E.
Corrosion Guide, 2nd ed., pp. 179–182
Elsevier Publishing Comp., Amsterdam-London-New York, 1968

[561] Kröhnke, O.; Masing, G.
Die Korrosion von Nichteisenmetallen und deren Legierungen
(The corrosion of non-ferrous metals and their alloys) (in German), Vol. II, p. 112
Verlag von S. Hirzel, Leipzig, 1938

[562] DECHEMA-Werkstoff-Tabelle
"Tetrachlorkohlenstoff"
(DECHEMA corrosion data sheet "Carbon tetratchloride") (in German)
DECHEMA, D-Frankfurt am Main, December 1969

[563] Hamner, N. E.
Corrosion Data Survey – Metals Section, 5. ed.
National Association of Corrosion Engineers, Katy (Tx./USA)

[564] Kirk-Othmer
Encyclopedia of Chemical Technology,
2nd ed., Vol. 5, p. 131
John Wiley & Sons Inc., New York, 1964

[565] Leontaritis, L.; Horn, E. M.
Zur Korrosion metallischer Werkstoffe in Tetrachlorkohlenstoff
(To the corrosion of metallic materials in carbon tetrachloride) (in German)
Werkst. Korros. 31 (1980) 3, p. 179

[566] Heitz, E.
Corrosion of metals in organic solvents
in: Fontana M. O.; Staehle, R. W. "Advances in Corrosion Science and Technology",
Vol. 4, pp. 225
Plenum Press, New York, London, 1974

[567] Kuppers, J. A.
The reactivity of metals with mixtures of carbon tetrachloride and alcohols
J. Electrochem. Soc. 125 (1978) 1, p. 97

[568] Archer, W. L.; Harter, M. K.
Reactivity of carbon tetrachloride with a series of metals
Corrosion 34 (1978) 5, p. 159

[569] Anonymous
Beständigkeitstabellen von Kupfer-Gußwerkstoffen in verschiedenen Medien.
(Resistance tables of copper cast materials in different media) (in German) pp. 24
DKI Deutsches Kupfer-Institut, Berlin

[570] Product Information
Wieland-Buch. Kupferwerkstoffe – Herstellung, Verarbeitung und Eigenschaften
(Wieland book. Copper materials – production, processing and characteristics) (in German) pp. 181
Wieland-Werke AG, Ulm

[571] Vaidyanath, L. R.; Bhandary, V. S.
Corrosion resistance of copper and copper alloys a survey, p. 1
Indian Copper Information Centre, Calcutta

[572] Franke, E.
Das Korrosionsverhalten von Alumiumbronzen
(The corrosion behavior of alumium bronzes) (in German)
Werkst. Korros. 32 (1981) 8, p. 298

[573] H. H. Uhlig
"The Corrosion Handbook", pp. 66, 76, 92, 101, 109, 157
John Wiley & Sons, Inc., New York, 1958

[574] Product Information
Hüls Kunststoffe – Handbuch, Oktober 1963, pp. 287
(Hüls plastics – manual, October 1963) (in German)
Chemische Werke Hüls AG, Marl

[575] DECHEMA-Werkstoff-Tabelle
"Chlorkohlenwasserstoffe – Chlormethane" (in German)
DECHEMA, D-Frankfurt am Main, 1993

[576] Anonymous
Understanding corrosion
Eng. Mater. Design (1971) September, 5

[577] Ried, G.
Übersicht über Werkstoffe für Absperr-Organe, die bei angreifenden Medien zu empfehlen sind
(Overview of materials for shut-off devices, which are recommended at attacking media) (in German)
Werkst. Korros. 15 (1964) 6

[578] Rabald, E.
Corrosion Guide, 2nd ed., pp. 153
Elsevier Publishing Comp., Amsterdam-London-New York, 1968

[579] Kuzharov, A. S.; Shuravleva, S. A.; Schakurova, I. K.
Untersuchung des Einflusses der Reibung auf die Oxidation von Übergangsmetallen in flüssiger Phase
(Investigation of the influence of the friction on the oxidation of transition metals in liquid phase) (in Russian)
J. Phys. Chem. 55 (1981) 11, p. 2872

[580] Anonymous
Informationsdruck Kupfer-Zinn-, Kupfer-Zinn-Zink- und Kupfer-Blei-Zinn-Gußlegierungen (Guß-Zinnbronze, Rotguß und Guß-Zinn-Bleibronze)
(Information printing copper tin, copper tin zinc and copper lead tin cast alloys (cast-tin bronze, red brass and cast tin lead bronze)) (in German)
Information Nr. i.025, pp. 4, 11, 13
DKI Deutsches Kupfer-Institut, Berlin

[581] Product Information
List of the suitability of standard range of chemical hoses for use with various conveyants
Compoflex Ltd., NL-Delpht

[582] Anonymous
Informationsdruck Kupfer-Zink-Legierungen – Messing und Sondermessing
(Information printing copper zinc alloys – brass and special brass) (in German),
Information Nr. i.5, pp. 2, 12
DKI Deutsches Kupfer-Institut, Berlin

[583] Shiga, S.; Hoshino, M.
A comprehensive investigation of chemical environments for stress corrosion cracking of α-brass: Characteristics and practical meanings of non-ammoniacal activities
Environ. Degrad. Eng. Mater. Aggressive Environ, Blacksburg (Va./USA), September 1981 (Proc. Conf.), p. 201

[584] Anonymous
Anti-corrosion methods and materials – 4. Forms of corrosion, Part 3: Corrosion cracking
Anticorros. Methods Mater. 19 (1972) 12, p. 9

[585] Smolin, V. V.
Selection of structural materials for volatile organic solvents regeneration adsorber (in Russian)
Khim. Prom. (1987) p. 366

[586] Anonymous
Corrosion inhibitor checklist
Hydrocarbon Processing 49 (1970) 8, p. 107

[587] Anonymous
Korrosionsbeständigkeitstabellen moderner Werkstoffe
(Corrosion resistance tables of modern materials) (in German) (Special printing from Werkstofftechnik (1970) 1 bis (1972) 5, pp. 5)
Verlag für Technik und Wirtschaft Meynen KG, Wiesbaden

[588] Nelson, G. A.
Corrosion data survey, ed. 1950, p. C-2
Shell-Development Co., Engineering Department, San Francisco

[589] Bogenschütz, A. F.; Jostan, J. L.; Mussinger, W.
Galvanische Korrosionsschutzschichten für elektronische Anwendungen
(Electroplated corrosion protection coatings for electronic applications) (in German)
Metalloberfläche 34 (1980) No. 3; p. 93; No. 7; p. 261

[590] Hauffe, K.
Reaktionen in und an festen Stoffen
(Reactions in and on solids) (in German)
Springer-Verlag, Berlin (D) 1966, 2nd ed., pp. 122–124, 632–640

[591] Tammann, G.; Köster, W.
Die Geschwindigkeit der Einwirkung von Sauerstoff, Schwefelwasserstoff und Halogenen auf Metalle
(Rate of action of oxygen, hydrogen sulfide and halogenes on metals) (in German)
Z. Anorgan. Allg. Chem. 123 (1922) p. 196

[592] Kohlschütter, V.; Krähenbühl, E.
Zur Morphologie fester Reaktionsschichten an Metallen
(Morphology of solid reaction layers on metals) (in German)
Z. Elektrochem. 29 (1923) p. 570

[593] Rowland, P. R.
Reactivity of different faces of a copper single crystal
Nature 164 (1949) p. 1091

[594] Frommer, L.; Polanyi, M.
Über heterogene Elementarreaktionen. I. Einwirkung von Chlor auf Kupfer
(Heterogeneous elementary reactions. I. Action of chlorine on copper) (in German)
Z. Phys. Chem. (A) 137 (1928) p. 201

[595] Reboul, G.; Luce, L. R.
Influence de la forme géométrique des corps solides sur les actions chimiques qu'ils subissent
(Influence of the geometric shape of solid bodies on their chemical reactions) (in French)
C. R. 172 (1921) p. 917

[596] Luce, L. R.
Etudes de l'influence de la courbure des solides sur les phénomènes chimiques et électrolytiques auxquels ils participent
(Investigation into the influence of the curvature of solid bodies on the chemical and electrolytical phenomena they participate in) (in French)
Ann. Physique (10) 11 (1929) p. 167

[597] Tseitlin, K. L.
Corrosion of metals by chlorine at high temperatures (in Russian)
Zh. Prikl. Khim. 27 (1954) p. 889

[598] Tseitlin, K. L.
Effect of temperature on the corrosion of metals by chlorine (in Russian)
Zh. prikl. Khim. 28 (1955) p. 490

[599] Tseitlin, K. L.
Effect of water vapor on the corrosion of metals by chlorine (in Russian)
Zh. Prikl. Khim 29 (1956) p. 1182

[600] Tseitlin, K. L.
The effect of air on the corrosion of metals by chlorine at high temperatures (in Russian)
Zh. prikl. Khim 29 (1956) p. 229

[601] Harrison, L. G.; Ng, Ching Fai
Reactivity and catalytic activity of copper chlorides. I. Kinetics of the reaction of copper(I) chloride with chlorine
Trans. Faraday Soc. 67 (1971) pp. 1787–1800

[602] Franco, J.; Patel, N. K.
Photo-corrosion of copper in aqueous solutions of chlorine, bromine and iodine
J. Electrochem. Soc. India 23 (1974) p. 31

[603] Santoleri, J. J.
Spray nozzle selections
Chem. Eng. Progr. 70 (1974) No. 9, p. 84

[604] Manning, J. A.; Carleton, S. V.
A study of corrosion rate in an operating desalting plant
Mater. Performance 14 (1975) No. 9, p. 34

[605] O'Neal, jr., C.; Borger, R. N.
The effect of chlorine on the performance of couprous metal corrosion inhibitors
Mater. Performance 16 (1977) No. 11, p. 12

[606] Patel, M. M.; Patel, N. K.; Vora, J. C.
The effect of oxidising agents on the corrosion of 63/37 brass in local supply water
Australian Corr. Eng. (1972) Oct.

[607] Sato, S.; Nagata, K.
Evaluation of various heat exchanger tubes for MFS plants (in Japanese)
Sumitomo Light Metal Techn. Rep. 18 (1977) No. 48, p. 11

[608] Sato, S.
Review on coatings and corrosion
Freund Publishing House Ltd., Tel-Aviv (Israel) 1972, p. 150

[609] Sato, S.; Okawa, M.
(Appearance and controlling factors of the anomalous impingement corrosion as the main cause of condenser tube damage) (in Japanese)
Sumitomo Light Metal Tech. Rep. 17 (1976) No. 1, pp. 17–27

[610] Bailey, G. L.
Copper-nickel-iron alloys in sea water
J. Inst. Metals 77 (1951) p. 243

[611] Stewart, W. C.; LaQue, F. L.
Corrosion resisting characteristics of iron-modified 90:10 cupro-nickel alloy
Corrosion (Houston) 8 (1952) p. 259

[612] Verdini, B.; Gauzzi, F.
Studio della reattività superficiale del rame e delle sue leghe (Cu-8,3 % Al, Cu-3,1 % Si) allo stato ricotto e incrudito
(Investigation into the surface reactivity of copper and its alloys (Cu-8,3Al, Cu-3,1Si) in the annealed and deformed state) (in Italian)
La Metallurgia Ital. 65 (1973) 14

[613] Kemplay, J.
Gate valves
Chem. Process Eng. 47 (1966) No. 7, p. 53

[614] Wagner, C.
Reaktionen mit festen Stoffen (Reactions with solids) (in German)
Z. Elektrochem. 47 (1941) p. 696

[615] Hauffe, K.
Reaktionen in und an festen Stoffen (Reactions in and on solids) (in German)
Springer-Verlag, Berlin 1966; 2nd ed., pp. 122–124; 852–859

[616] Fabian, H.
in: Ullmanns Encyklopädie der technischen Chemie (Ullmann's encyclopedia of industrial chemistry) (in German), 4th ed., vol. 15, p. 491
Verlag Chemie GmbH, D-Weinheim, New York, 1978

[617] Fabian, R. J.; Vaccari, J. A.
How materials stand up to corrosion and chemical attack
Mater. Eng. 73 (1971) 2, p. 36

[618] Product Information
Wieland-Buch. Kupferwerkstoffe – Herstellung, Verarbeitung und Eigenschaften (Wieland book of copper materials – Production, processing and properties) (in German), 4. ed., 1978
Wieland-Werke AG, Ulm, Germany

[619] Product Information
Beständigkeitstabellen von Kupfer-Gußwerkstoffen in verschiedenen Medien (Resistance charts of cast copper materials in various media) (in German), ed. 1970
DKI Deutsches Kupfer-Institut, Düsseldorf, Germany

[620] Bogenschütz, A. F.; Büttner, U.; Lostan, J. L.; Marten, A.
Präzisionsätzen von Leiterplatten durch Verwendung von Benzotriazol als Flankenschutzmittel (Precision etching of printed circuits using benzotriazole as an etchant additive) (in German)
Galvanotechnik 69 (1978) 11, p. 960

[621] Dammer, H.-J.
Die cyanidfreie chemische und elektrolytische Entmetallisierung von Werkstücken und Gestellen (The cyanide-free chemical and electrolytic stripping of work pieces and racks) (in German)
Galvanotechnik 71 (1980) 1, p. 29

[622] McCaul, C.; Geld, I.
Long term corrosion tests in Rondout reservoir, New York
Mater. Performance 17 (1978) 5, p. 27

[623] Suzuki, I.; Ishikawa, Y.; Hisamatsu, Y.
The pitting corrosion of copper tubes in hot water
Corrosion Sci. 23 (1983) 10, p. 1095

[624] Suzuki, I.
The prediction of pit initiation time for copper tubes in hot water from water composition
Corrosion Sci. 24 (1984) 5, p. 429

[625] Vestola, J.; Stieb, H.
Haltbarkeit von Saugwalzmänteln (Durability of suction roll shells) (in German)
Wochenblatt für Papierfabrikation 110 (1982) 4, p. 117

[626] Wallbaum, H. J.
Werkstoffe für Saugwalzmäntel von Papiermaschinen (Materials for suction roll shells of paper machines) (in German)
Das Papier 31 (1977) 10A, p. V 143

[627] Forchhammer, P.
Beispiele für Korrosionsschäden in der Papierindustrie und Hinweise zur Schadenverhütung (Examples of corrosion damage in the paper industry and damage prevention tips) (in German)
Wochenblatt für Papierfabrikation 109 (1981) 22, p. 871

[628] Matzke, W. H.
Werkstoffe in geschlossenen Wasserkreisläufen (Materials in closed water systems) (in German)
Wochenblatt für Papierfabrikation 109 (1981) 16, p. 555

[629] Anonymous
Resistance of nickel and high nickel alloys to corrosion by hydrochloric acid, hydrogen chloride and chlorine
Corrosion Engineering Bulletin CEB-3
The International Nickel Company, Inc., New York (USA)

[630] Bowers, D. F.
Corrosion considerations for alkaline papermaking
Tappi Journal 69 (1986) 1, p. 62

[631] Wintzer, P.
Entwicklung und Trend der Chlordioxidbleiche mit integrierter Chloratelektrolyse für die Zellstoffindustrie (Development and trends in chlorine dioxide bleaching with integrated chlorate electrolysis for the pulp industry) (in German)
Chem.-Ing. Tech. 52 (1980) 5, p. 392

[632] DIN EN 12502-2 (03/2005)
Korrosionsschutz metallischer Werkstoffe – Hinweise zur Abschätzung der Korrosionswahrscheinlichkeit in Wasserverteilungs- und speichersystemen – Teil 2: Einflussfaktoren für Kupfer und Kupferlegierungen
(Protection of metallic materials against corrosion – guidance on the assessment of corrosion likelihood in water distribution and storage systems – Part 2: Influencing factors for copper and copper alloys) (in German)
Beuth Verlag GmbH, Berlin

[633] Elzenga, C. H. J.; v. Franqué, O.
Praxiserfahrungen beim Einsatz von Kupferrohren bei der Trinkwasserinstallation
(Practical experience in the use of copper pipes in drinking water installations) (in German)
Gas- u. Wasserfach, Wasser, Abwasser 127 (1986) p. 209

[634] DIN EN 1057 (06/2010)
Kupfer und Kupferlegierungen – Nahtlose Rundrohre aus Kupfer für Wasser- und Gasleitungen für Sanitärinstallationen und Heizungsanlagen
(Copper and copper alloys – seamless, round copper tubes for water and gas in sanitary and heating applications) (in German)
Beuth Verlag GmbH, Berlin

[635] Güte- und Prüfbestimmungen für das Gütezeichen "Kupferrohr/RAL" (Quality and testing regulations for the quality mark, "Kupferrohr/RAL") (in German)
Gütegemeinschaft Kupferrohr e.V., D-Düsseldorf, October 1987

[636] Campbell, H. S.
Pitting corrosion in copper water pipes caused by films of carbonaceous material produced during manufacture
J. Inst. Met. 77 (1950) 4, p. 345

[637] Elzenga, C. H. J.; Boorsma, H. J.; Nijholt, H.
Corrosion aspects in the Netherlands in the use of copper tubes for the transport of cold drinking water
12th Internat. de la Couverure Plomberie, Den Haag, 1972

[638] v. Franqué, O.
Lochkorrosion Typ I bei Kupferrohren in Kalt- und Warmwassersystemen der Hausinstallation
in: Kruse, C.-L. "Korrosion in Kalt- und Warmwassersystemen der Hausinstallation"
(Type I pitting corrosion on copper pipes in hot and cold water systems in domestic installations; in: Kruse, C.-L. "Corrosion in hot and cold water systems in domestic installations") (in German), p. 233
Deutsche Gesellschaft für Metallkunde, D-Oberursel, 1984

[639] Kruse, C.-L.
Sanitär und Heizungstechnik 10 (1981) p. 946

[640] DIN CEN/TS 13388 (09/2004)
Kupfer und Kupferlegierungen – Übersicht über Zusammensetzung und Produkte
(Copper and copper alloys – Compendium of compositions and products) (in German)
Beuth Verlag GmbH, Berlin

[641] Campbell, H. S.
A natural inhibitor of pitting corrosion of copper in tapwaters
J. Appl. Chem. 4 (1954) December, p. 633

[642] Kruse, C.-L.
ndz, Neue Deliwa Zeitschrift (1975) 9

[643] v. Franqué, O.; Winkler, B.
Feldversuche Trinkwasser, Teil 1: Kupfer
(Field trials with drinking water, Part 1: Copper) (in German)
Werkst. Korros. 35 (1984) 12, p. 575

[644] van Muylder, J.; Pourbaix; M.
Prüfung der Lochfraßanfälligkeit von Kupfer in kaltem Leitungswasser
(Testing the susceptibility of copper to pitting corrosion in cold tapwater) (in French)
CEBELCOR 119 (1972) RT 201, 10 p.

[645] Cornwell, F. J.; Wildsmith, G.; Gilbert, P. T.
Pitting corrosion in copper tubes in cold water service
Br. Corros. J. 8 (1973) Sept., p. 202

[646] Yamauchi, S.; Sato, S.
Corrosion of copper tubes in fresh water
Corros. Eng. 30 (1981) 8, p. 469

[647] Dürrschnabel, W.; Sick, H.
Korrosionsversuche an Kupferrohren mit verschieden behandelter Innenoberfläche
(Corrosion tests on copper pipes with differently treated inner surfaces) (in German)
Werkst. Korros. 33 (1982) 11, p. 619

[648] Landgrebe, H.
Wasserfluß stoppt Korrosionsverdruß
(Water flow stops trouble with corrosion) (in German)
IKZ (1980) 4, p. 86

[649] v. Franqué, O.; Gert, D.; Winkler, B.
Ergebnisse von Untersuchungen an Deckschichten in Kupferrohren
(Results of studies into surface layers in copper tubing) (in German)
Werkst. Korros. 26 (1975) 4, p. 255

[650] v. Franqué, O.; Winkler, B.
Korrosion und Korrosionsschäden an Wasserleitungen aus Kupfer
(Corrosion and corrosion damage on water pipelines made of copper) (in German)
IKZ (1981) 23 and 24; (1982) 1

[651] v. Franqué, O.
IKZ (1968) 18, p. 1933

[652] DVGW Arbeitsblatt GW 2
Verbinden von Kupferrohren für die Gas- und Wasserinstallation von Grundstücken und Gebäuden
(Connecting copper pipes for gas and water installation on property and in buildings) (in German)
Wirtschafts- und Verlagsgesellschaft Gas- und Wasser mbH, D-Bonn, July 1983

[653] DKI Informationsdruck
Kupfer, Lebensmittel, Gesundheit,

Sonderdruck Nr. 1.019
(Copper, Food, Health, Special Print
No. 1.019) (in German)
Deutsches Kupfer-Institut D-Berlin

[654] Müller, J.
Zur Entzinkung von Messing
(Dezincification of brass) (in German)
Korrosion 4 (1973) 4, p. 33

[655] Carmichael, A. J.
Combatting corrosion of copper and brass
in potable water systems
Met. Australas. 14 (1982) 5, p. 12

[656] Herbst, N. F.
Dezincification of brass – a problem solved
Met. Australas. 14 (1982) 5, p. 14

[657] Landgrebe, H.
Wasserfluß stoppt Korrosionsverdruß
(Water flow stops trouble with corrosion)
(in German)
IKZ (1980) 4, p. 86

[658] Turner, M. E. D
The influence of water composition on the
dezincification of duplex brass fittings
Proc. Soc. Water Treatment and
Examination 10 (1961) p. 162

[659] Turner, M. E. D.
Further studies on the influence of water
composition on the dezincification of
duplex brass fittings
Proc. Soc. Water Treatment and
Examination 14 (1965) p. 81

[660] Ladeburg, H.
Untersuchungen von
Entzinkungserscheinungen an Fittings aus
Kupferlegierungen
(Examination of dezincification
phenomena on copper alloy fittings)
(in German)
Metall 20 (1966) 1, p. 33

[661] Fabian, R. J.; Vaccari, J. A.
How materials stand up to corrosion and
chemical attack
Mater. Eng. 73 (1971) 2, p. 36

[662] Franco, J.; Patel, N. K.
Photocorrosion of copper and brass in
aqueous solutions of copper and ferric
chlorides
J. Electrochem. Soc. India 27 (1978) 3,
p. 161

[663] Burrows, W. H.; Turner Lewis jun., C.;
Saire, D. E.; Brooks, R. E.
Kinetics of the copperferric chloride
reaction and the effects of certain
inhibitors
Ind. Eng. Chem. Process Des. Develop. 3
(1964) 2, p. 149

[664] Cahoon, J. R.; Bassim, M. N.; Oman, E. G.
Acoustic emission during corrosion
Canadian Metallurgical Quarterly 25 (1986)
1, p. 73

[665] Ayerest, G. G.
Ferric chloride as an etching material
Trans. Inst. Met. Finish. 44 (1966) 4, p. 176

[666] Molvina, L. I.; Ganzhenko, T. S.;
Kucherenko, V. I.
Etching of copper in a ferric chloride
solution based on a water-acetonitrile
mixture
Izv. V.U.Z. Khim. Khim. Tekhnol. 29 (1986)
2, p. 122 (Russian)

[667] Maurer, R.; Erb, U.; Gleiter, H.
Intercrystalline corrosion: Factors
controlling the enhanced corrosion near
grain boundaries
Mater. Sci. Eng. 63 (1984) 2, p. L13

[668] Walker, R.
Triazole, benzotriazole, and naphtotriazole
as corrosion inhibitors for brass
Corrosion 32 (1976) 10, p. 414

[669] Kot, A. A.; Gronskij, R. K.;
Maklakova, V. P.; Zhivotovskij, E. A.
The effect of "aqueous condensate" on the
corrosion of copper alloys in hydrochloric
acid solution
Elektr. Stantsii (1976) 7, p. 26 (Russian)

[670] Maklakova, V. P.; Gronskij, R. K.
Protection of copper alloys against
corrosion in acid solutions of iron and
copper chlorides
Zashch. Met. 13 (1977) 3, p. 338 (Russian)

[671] Swandby, R. K.
Corrosion charts: Guides to materials
selection
Chem. Eng. 69 (1962) 23, p. 186–201

[672] Miska, K. H.
Copper nickel cuts corrosion, enhances
electrical efficiency
Mater. Eng. 80 (1974) 7, p. 40

[673] Royuela, J. J.
Corrosion in ferric solutions. Evaluation by
D.C. electrochemical methods
10th International Congress on Metallic
Corrosion, Madras (India), November 1987
(Proc. Conf.), Vol. 1, p. 465
Key Eng. Mater., 20–28 (1)2

[674] Anonymous
Nonferrous metals. Nickel and its alloys
Mater. Eng. 74 (1971) 4; Mater. Selector, p. 207

[675] Streicher, M. A.
Development of pitting resistant Fe-Cr-Mo alloys
Corrosion 30 (1974) 3, p. 77–91

[676] Mendoca, J.; Roy, D. L.
Studies on the corrosion behavior of 90/10 Cu-Ni-(Fe) alloy in $FeCl_3$
J. Electrochem. Soc. India 33 (1984) 2, p. 181

[677] Coen-Porisini, F.; Imarisio, G.
The compatibility of containment materials for thermochemical hdyrogen production
1st World Hydrogen Energy Conf., Miami Beach (Fla/USA) 1976 (Proc. Conf.), Vol. 1, p. 7A-3

[678] Nickel, O.
Eigenschaften und neuere Anwendungen nickelhaltiger Werkstoffe
VDI Z. 121 (1979) 7, p. 313

[679] Hoffmann, K.
Über die Korrosion einiger Kupfer-Zink-Legierungen im Meerwasser
Trav. Centre. Etud. Oceanogr. (Paris) 6 (1965) 1-2-3-4, p. 103

[680] Mayer, H.
Verfahren zur Herstellung von Formätzteilen aus Messingblech
DOS 1 812 193 (04.06.1970; Siemens AG, Berlin und München)

[681] Gronskij, R. K.; Zhivotovskij, E. A.; Gasnikova, L. P.
Einfluß von oxidierenden Kationen auf die Korrosion von Messing in verdünnten Salzsäurelösungen
Zashch. Met. 11 (1975) 2, p. 183 (Russian)

[682] Langenegger, E. E.; Callaghan, B. G.
Use of an empirical potential shift technique for predicting dezincification rates of α,β-brasses in chloride media
Corrosion 28 (1972) 7, p. 245

[683] Nielsen, K.; Rislund, E.
Comparative study of dezincification tests
Br. Corros. J. 8 (1973) 3, p. 106

[684] Celis, J. P.; Roos, J. R.; Terwinghe, F.
Corrosion behavior of β and martensitic Al brasses
J. Electrochem. Soc. 130 (1983) 12, p. 2314

[685] Skidmore, K. F.; Schwartzbart, H.
Corrosion and dezincification of brasses in water
J. of Testing and Evaluation 4 (1976) 6, p. 426

[686] Tanabe, Z.
Einfluß des Potential-pH-Verhaltens auf die Spannungsrißkorrosion von 70-30-Messing
Shindo Gijutsu Kenkyukai-Shi 7 (1968) 1, p. 99 (in Japanese)

[687] Graf, L.
Was spricht gegen das Aufreißen von Deckschichten und gegen eine Versprödung als Ursachen der Spannungsrißkorrosion?
Z. Metallkd. 66 (1975) 12, p. 749

[688] Graf, L.; Wittich, W.
Bedeutung und Zustandekommen elektrochemischer Prozess bei der Spannungskorrosion homogener, nicht ausscheidungsfähiger Mischkristalle
Z. Metallkd. 56 (1965) 6, p. 380

[689] Graf, L.; Wieling, N.
Untersuchung der Spannungskorrosionsempfindlichkeit von Kupfer-Nickel-Palladium-Legierungen
Werkst. Korros. 22 (1971) 1, p. 35

[690] Antlinger, K.; Horvath, G.
Kupferlegierung mit hoher Beständigkeit gegen chemische Korrosion
OE Patentschrift 336 902 (21.10.1974; Vereinigte Österreichische Eisen- und Stahlwerke – Alpine Montan Aktiengesellschaft, Wien)

[691] Hackerman, N.; Snavely, E. S.; Fiel, L. D.
The anodic polarization behavior of metals in hydrogen fluoride
Corrosion Sci. 7 (1967) p. 39

[692] Yashina, G. M.; Levin, A. I.
Influence of alternating current density on the rate of corrosion of copper. (in Russian) (Proceedings) Sb. "Tezisy Dokl. VIII Perm. Konf. po Zashchite Met. ot Korrozii", Perm 1974, p. 54

[693] De Vries, G.; Grantham, L. F.
Corrosion of metals in molten LiOH-LiF-H_2O system
Electrochem. Technol. 5 (1967) p. 335

[694] Walker, R.
Triazole, benzotriazole and naphthotriazole as corrosion inhibitors for brass
Corrosion (Houston) 32 (1976) p. 414

[695] Gillardeau, J.; Macheteau, Y.; Plurien, P.; Oudar, J.
Some aspects of the fluorination of copper and iron
Oxid. Metals 2 (1970) p. 319

[696] Kent, R. A.; McDonald, J. D.; Margrave, J. L.
Mass-spectrometric studies at high temperatures. IX. The sublimation pressure of copper(II) fluoride
J. Phys. Chem. 70 (1966) p. 874

[697] Steinmetz, E.; Roth, H.
Freie Standard-Bildungsenthalpien der Fluoride nach der Temperaturfunktion
(Free standard enthalpies of formation of fluorides as a function of temperature)
(in German)
J. Less-Common Metals 16 (1968) p. 295

[698] Hauffe, K.
Reaktionen in und an festen Stoffen
(Reactions in and on solid matters)
(in German)
Springer-Verlag Berlin, 1966, p. 629

[699] Schmidt, H. W.
Handling and use of fluorine and fluorine-oxygen mixtures in rocket systems
NASA Sp-3037, NASA, Washington D.C., 1967

[700] Crabtree, J. M.; Lees, C. S.; Little, K.
The copper fluorides, part I – X-ray and electron microscope examination
J. Inorg. Nucl. Chem. 1 (1955) p. 213

[701] Lukyanychev, Y. A.; Nikolaev, N. S.; Astakhov, I. I.; Lukyanychev, V. I.
A study of the mechanism of copper fluorination at high temperatures
Izvestija Akademia Nauk SSSR 147 (1962) p. 1130

[702] Evans, U. R.
Crack-heal mechanism of the growth of invisible films on metals
Nature 157 (1946) p. 732

[703] Gillardeau, J.; Vincent, L.; Oudar, J.
The existence of a nucleation phenomenon in the reaction of fluorine on copper
C.R. Acad. Sci., Paris, Ser. C 263 (1966) 25, p. 1469

[704] O'Donnell, P. M.; Spakowski, A. E.
The fluorination of copper
J. Electrochem. Soc. 111 (1964) p. 633

[705] Kleinberg, S.; Tompkins, J. F.
The compatibility of various metals with liquid fluorine
ASD-TDR-62-250 Air Force Systems Command, Wright-Patterson Air Force Base, Ohio, 1962

[706] Toy, S. M.; English, W. D.; Crane, W. E.
Studies of galvanic corrosion couples in liquid fluorine
Corrosion 24 (1968) p. 418

[707] Jackson, R. B.
Corrosion of metals and alloys by fluorine
NP-8845 Allied Chem. Corp. Morristown, N.J., 1960

[708] Sterner, C. J.; Singleton, A. H.
The compatibility of various metals and carbon with liquid fluorine
WADD-TR-60-436, Wright Air Development Division, Wright-Patterson Air Force Base, Ohio, 1960

[709] Jackson, J. D.
Corrosion in Cryogenic Liquids
Chem. Eng. Progr. 57 (1961) p. 61

[710] Godwin, T. W.; Lorenzo, C. L.
Ignition of several metals in fluorine
American Rocket Soc. Meeting, New York, November 1958, Paper-No. 740–58

[711] Myers, W. R.; DeLong, W. B.
Fluorine corrosion. High temperature attack on metals by fluorine and hydrogen fluoride. Behavior of insulated steel parts in fluorine cells
Chem. Eng. Progr. 44 (1948) p. 359

[712] Steindler, M. J.; Vogel, R. C.
Corrosion of metals in gaseous fluorine at elevated temperatures
Rep. ANL-5560, Argonne National Laboratory Argonne III., 1957

[713] Ritter, R. L.; Smith, H. A.
The kinetics and mechanism of fluorination of copper oxide. II. The reaction of fluorine with copper(I) oxide.
J. Phys. Chem. 71 (1967) p. 2036

[714] Brown, P. E.; Crabtree, J. M.; Duncan, J. F.
The kinetics of the reactions of elementary fluorine with copper metals
J. Inorg. Nucl. Chem. 1 (1955) p. 202

[715] Cannon, W. A.; Asunmaa, S. K.; English, W. D.; Tiner, N. A.
Passivation reactions of nickel and copper alloys with fluorine
Trans. Met. Soc. AIME 242 (1968) p. 1635

[716] O'Donnell, P. M.; Spakowski, A. E.
Reaction of copper and fluorine from 800 to 1200 F

NASA-TN-D-768, NASA, Washington D.C., 1961

[717] Bottenberg, K.
Substitution von Benzol durch elementares Fluor in der Gasphase
(Substitution of benzene by elemental fluorine in the gas phase) (in German)
Chemiker Ztg. 96 (1972) p. 84

[718] Semerikova, I. A.; Mironova, N. I.; Sukhotin, A. M.
Corrosion of metals in 114B2 freon at 50 °C (in Russian)
Khim. Prom, Moscow, 52 (1976) p. 746

[719] Anonymous
Anti-corrosion methods and materials, 3: Forms of corrosion, part 2: Pitting corrosion, interangular attack and selective leaching
Anti-Corrosion 19 (1972) 10, p. 3

[720] Nikolaev, N. S.; Aliev, D. R.; Ikrami, D. D.
Behavior of copper in hydrofluoric acid (in Russian)
Zashch. Metallov 8 (1972) p. 572

[721] Shiga, Sh.
Aqueous acid pickling solution for metals
DOS 2364162 (11.7.1974)

[722] Greene, N. D. jr.; Ahmed, N.
Filament reinforced metallic composites
Mat. Protection 9 (1970) 3, p. 16

[723] Zotikov, V. S.; Bakhmutova, G. B.; Bocharova, N. A.; Bogdanova, N. O.; Kotchetkov, N. I.; Petukhov, V. V.; Semenyuk, E. Ya.
Corrosion of some non-ferrous metals in hydrogen fluoride media (in Russian)
Khim. Prom, Moscow (1975) p. 773

[724] Landau, R.; Rosen, R.
Industrial handling of fluorine
Ind. Eng. Chem. 39 (1947) p. 281

[725] Degnan, T. F.; Burbridge, H. G.
Materials for receiving, handling and storing hydrofluoric acid
Mat. Prot. Perform. 10 (1971) 3, p. 33

[726] Degnan, T. F.
Materials for handling hydrofluoric, nitric and sulfuric acids
Process Ind. Corrosion (1975) p. 229

[727] Hise, D. R.
Corrosion from halogens, process industries
Corrosion (1975) p. 240

[728] Miska, K. H.
Wrought brass
Mat. Eng. 80 (1974) 5, p. 65

[729] Oda, N.; Morioka, N.
Chemical polishing and deburring of copper alloys
Jap. Patent 7852252 (12.5.1978)

[730] Konstantinova, E. V.; Muravev, L. L.
Time factor in corrosion research (in Russian)
Zashch. Metallov 12 (1976) p. 599

[731] Heitz, E.
Grundvorgänge der Korrosion von Metallen in organischen Lösungsmitteln
(Fundamental phenomena of corrosion of metals in organic solvents) (in German)
Werkstoffe u. Korrosion 21 (1970) Nr. 5, p. 360

[732] Rabald, E.
Wirkung von Sauerstoff auf die Korrosion von Metallen in organischen Säuren
(Oxygen effect on metallic corrosion in organic acids) (in German)
Symp. Nerezavejicich Ocelich, Prague (1961) p. 435

[733] Teeple, H. O.
Corrosion by some organic acids and related compounds
Corrosion (Houston) 8 (1952) No. 1, p. 14

[734] Heitz, E.
Corrosion of metals in organic solvents
Advances in Corrosion Science and Technology 4 (1974) p. 149

[735] Lackey, J. Q.; Degnan, T. F.
Formic acid corrosion of common materials of construction
Mater. Performance 13 (1974) No. 7, p. 13

[736] Schütze, K. G.
Korrosionsverhalten metallischer Werkstoffe in Ameisensäure
(Corrosion behavior of metallic materials in formic acid) (in German)
Werkstoffe und Korrosion 38 (1987) No. 1, p. 41

[737] Donovan, P. D.; Stringer, J.
Corrosion of metals and their protection in atmospheres containing organic acid vapours
Brit. Corrosion J. 6 (1971) p. 132

[738] Lavrenko, V. A.; Koval, A. G.; Lebedev, A. A.; Klimovski, Yu. A.; Lyasnikova, R. N.; Frantsevich, I. N.
Copper, silver, and gold corrosion by

decomposition products of some new organic electrical insulators (in Russian)
Fiz. Mat. Tekh. Nauki (1977) No. 1, p. 65

[739] Lavrenko, V. A.; Koval, A. G.; Gordienko, S. P.; Lugovskaja, E. S.; Fenochka, B. V.; Klimovskii, Yu. A.
The corrosion of gold and silver coated copper by the thermal degradation products of chloroprene and silicone rubber
Corrosion Sci. 18 (1978) p. 809

[740] Anonymous
Flammschutz von Kunststoffen schwierig (Trouble with fire retardants in plastics) (in German)
FAZ Beilage "Natur und Wissenschaft" Nr. 27, 25.2.1987, p. 31

[741] Ison, H. C. K.; Tiller, A. K.
Corrosion and protection at the N.P.L.; II. Corrosion advisory service
Anti-Corrosion 19 (1972) No. 11, p. 11

[742] Swandby, R. K.
Corrosion charts: Guides to materials selection
Chem. Eng. 69 (1962) Nov., p. 186

[743] Ried, G.
Übersicht über Werkstoffe für Absperrorgane, die bei angreifenden Medien zu empfehlen sind (Survey on materials for valves recommended in contact with aggressive media) (in German)
Werkstoffe und Korrosion 15 (1964) p. 468

[744] Pilling, N. B.; Bedworth, R. E.
J. Inst. Metals 29 (1923) p. 529

[745] Wagner, C.; Grünewald, K.
Z. phys. Chem. (B) 40 (1938) p. 455
Hauffe, K.
Oxidation of metals
Plenum Press, New York, 1965, p. 159

[746] Evans, U. R.; Miley, H. A.
Nature 139 (1937) p. 283

[747] Pinnel, M. R.; Tomkins, H. G.; Heath, D. E.
Appl. Surface Sci. 2 (1979) p. 558

[748] Tylecote, R. F.
J. Inst. Metals 78 (1950/51) p. 301; 81 (1952/53) p. 681

[749] Fröhlich, K. W.
Z. Metallkunde 28 (1936) p. 368

[750] Hauffe, K.; Kofstad, P.
Z. Elektrochem. 59 (1955) p. 399

[751] Bardeen, J.; Brattain, W. H.; Shockley, W.
J. Chem. Phys. 14 (1946) p. 714

[752] Castellan, G. W.; Moore, W. J.
J. Chem. Phys. 17 (1949) p. 41
Moore, W. J.; Selikson, B.
J. Chem. Phys. 19 (1951) p. 1539; 20 (1952) p. 927

[753] McKewan, W.; Fassell jr., W. M.
J. Metals 5 (1953) p. 1127

[754] Valensi, G.
in: L'oxidation des métaux, Vol. 1
J. Bénard, Gauthier-Villars, Paris, 1963, p. 247

[755] Paidassi, J.
Acta Met. 6 (1958) p. 216

[756] Garnaud, G.
The formation of a double oxide layer on pure copper
Oxid. Metals 11 (1977) p. 127

[757] Meijering, J. L.; Verheijke, M. L.
Acta Met. 7 (1959) p. 331

[758] Tomlinson, W. J.; Yates, J.
Oxidation of copper in CO_2 at 800–1000 °C
J. Electrochem. Soc. 125 (1978) p. 803

[759] Murgulescu, I. G.; Cismaru, D.
Acad. Republ. Pop. Rom. Stud. Cercet. Chim. 7 (1959) p. 197

[760] Fueki, K.; Wagner jr., J. B.
Oxidation studies on Ni and on a 0.37 w/o CrNi-alloy in CO_2 and in CO_2-CO mixtures
J. Electrochem. Soc. 112 (1965) p. 1079

[761] Hallowes, A. R. C.; Voce, E.
Metallurgia 34 (1946) p. 95

[762] Hudson, O. F.; Herbert, T. M.; Ball, F. E.; Bucknall, E. H.
J. Int. Metals 42 (1929) p. 253

[763] Ipatiev, V. V.; Heltukhim, D. V.; Kleimenova, L. I.
Byull. nauchno-tekh. Inf. po Rezultatem nauchno-issled.
Robot, Leningrad Lesotskh. Akad. (1957 No. 47) (in Russian)

[764] Birks, N.; Tattam, N.
The effect of SO_2 upon the high-temperature oxidation of Cu
Corrosion Sci. 10 (1970) p. 857

[765] Nakai, H.; Sugimoto, T.
J. Jap. Inst. Metals 41 (1977) p. 965 (in Japanese)

[766] Oleinikov, G. N.; Levina, R. A.
Zashch. Metallov 13 (1977) p. 492 (in Russian)

[767] Amosov, N. I.; Zolotarev, G. B.
Izv. Vysch. Uchebn. Zaved. Mashinostr. (1974) 10, p. 32 (in Russian)

[768] Yoda, E.; Siegel, B. M.
J. Appl. Phys. 34 (1963) p. 1512

[769] Hembree, G. G.; Cowley, J. M.; Otooni, M. A.
The oxidation of copper studied by electron scattering techniques
Oxid. Metals 13 (1979) p. 331

[770] Gwathmey, A. T.; Lawless, K. K.
The influence of crystal orientation on the oxidation of metals
in: Surface Chemistry of Metals and Semiconductors
Wiley & Son, New York, 1960, p. 483

[771] Rhodin, T. N.
in: Adv. Catalysis, vol. 5
Academic Press, New York, 1953, p. 39

[772] Bénard, J.
Comp. rend 248 (1959) p. 1658
Grønlund, F.; Bénard, J.
Comp. rend 240 (1955) p. 624
Grønlund, F.
J. chim. Physique 53 (1956) p. 660

[773] Menzel, E.; Stössel, W.
Diskontinuierliche Anlaufschichten auf Kupfer
(Discontinuous staining layers on copper) (in German)
Naturwiss. 41 (1954) p. 302
Menzel, E.; Otter, M.
Nickel-Einkristallkugeln mit reiner, unberührter Oberfläche
(Nickel monocrystal spheres with pure virginal surface) (in German)
Naturwiss. 46 (1959) p. 66
Otter, M.
Z. Naturforsch. 14b (1959) p. 355

[774] Rao, D. B.; Heinemann, K.; Douglass, D. L.
Oxide removal and desorption of oxygen from partly oxidized thin films of copper at low pressures
Oxid. Metals 10 (1976) p. 227

[775] Okawa, M.
Sumitomo Light Metal Techn. Rep. 13 (1977) p. 11

[776] De Wilde, R.
Mod. Dev. Powder Metal 10 (1976) p. 285

[777] Rao, T. H.
Technomet 4 (1976/77) p. 31

[778] Jaenicke, W.; Leistikow, S.
Z. phys. Chem. (NF) 15 (1958) p. 175
Jaenicke, W.; Leistikow, S.; Stadler, A.
Mechanical stresses during the oxidation of copper and their influence on oxidation kinetics. III
J. Electrochem. Soc. 111 (1964) p. 1031

[779] Neumeister, H.; Jaenicke, W.
Z. phys. Chem. (NF) 108 (1977) p. 217

[780] Fröhlich, K. W.
Z. Metallkunde 28 (1936) p. 368

[781] Preston, G. D.; Bircumshaw, L. L.
Phil. Mag. 20 (1935) p. 706

[782] Nishimura, H.
Suiyokwai Shi 9 (1938) p. 655

[783] Hallowes, A. P.; Voce, E.
Metallurgia 34 (1946) pp. 95, 119

[784] Spinedi, P.
Met. Ital. 45 (1953) p. 457

[785] Honjo, G.
J. Phys. Soc., Japan 8 (1953) p. 113

[786] Hessembruch, W.
Metalle und Legierungen für hohe Temperaturen
(Metals and alloys for high temperatures) (in German)
Springer-Verlag, Berlin, 1940

[787] Tylecote, R. F.
J. Inst. Metals 78 (1950/51) p. 259

[788] Rönquist, A.; Fischmeister H.
J. Inst. Metals 89 (1960) p. 65

[789] Hauffe, K.
Oxidation of metals
Plenum Press, New York 1965, p. 167

[790] Price, L. E.; Thomas, G. J.
Metal Ind., London 54 (1939) p. 189

[791] Anonymous
Die Aluminium-Bronzen
(The aluminium bronzes) (in German)
DKI (1958) p. 129

[792] Sanderson, M. D.; Scully, J. C.
Room temperature oxidation of Cu and some Cu-alloys
Corrosion Sci. 10 (1970) p. 55

[793] Bénard, J.
Acta Met. 8 (1960) p. 272

[794] Sanderson M. D.; Scully, J. C.
The selective oxidation behavior of some Cu-alloys
Corrosion Sci. 10 (1970) p. 165
The high-temperature of some oxidation-resistant copper-based alloys
Oxid. Metals 3 (1971) p. 59

[795] Hauffe, K.; Ofulue, E.
Zum Mechanismus der Hochtemperatur-Oxidation von Kupfer-Aluminiumlegierungen
(Mechanism of high-temperature oxidation of copper-aluminium alloys) (in German)
Werkstoffe u. Korrosion 23 (1972) p. 351

[796] Hauffe, K.
Reaktionen in und an festen Stoffen, 2. Aufl.
(Reactions in and on solid matter)
(in German)
Springer-Verlag, Berlin, 1966, p. 756

[797] Meijering, J. L.; Druyvesteijn, M. J.
Philips Res. Rep. 2 (1947) pp. 81, 260
Meijering, J. L.
Z. Elektrochem. Ber. Bunsenges phys. Chem. 63 (1959) p. 824 (in German)

[798] Rhines, F. N.; Johnson, W. A.; Anderson, W. A.
Trans. AIME, Techn. Publ. No. 1368 (1941)
Rhines, F. N.
J. Corrosion 4 (1947) p. 15

[799] Meijering, J. L.
Trans. AIME 218 (1960) p. 968

[800] Wagner, C.
J. Metals 6 (1954) p. 154

[801] Hauffe, K.
Oxidation of metals
Plenum Press, New York, 1965, p. 335
Hauffe, K.
Reaktionen in und an festen Stoffen, 2. Aufl.
(Reactions in and on solid matter 2nd ed.)
(in German)
Springer-Verlag, Berlin, 1966, p. 741

[802] Wood, G. C.
High-temperature oxidation of alloys
Oxid. Metals 2 (1970) p. 11

[803] Miyake, S.
Sci. Papers Inst. Phys. Chem. Res. (Tokyo) 29 (1936) p. 67; 31 (1937) p. 161

[804] Hickman, J. W.; Gulbransen, E. A.
Trans. AIME 180 (1949) p. 534

[805] Sartell, J. A.; Bendel, S.; Johnson, T. L.; Li, C. H.
Trans. ASM 50 (1958) p. 1047

[806] Levin, R. L.; Wagner, J. B.
The role of a displacement reaction in the kinetics of oxidation of alloys
J. Electrochem. Soc. 108 (1961) p. 954

[807] Wagner, C.
J. Electrochem. Soc. 99 (1952) p. 369

[808] Whittle, D. P.; Wood, G. C.
Two-phase scale formation on CuNi-alloys
Corrosion Sci. 8 (1968) p. 295

[809] Castle, J. E.; Nasserian-Riabi, M.
The oxidation of cupro-nickel alloys
I. XPS study of inter-diffusion
Corrosion Sci. 15 (1975) p. 537

[810] Heinemann, K.; Rao, D. B.; Douglass, D. L.
In situ oxidation studies on (001) copper-nickel alloy thin films
Oxid. Metals 11 (1977) p. 321

[811] Castle, J. E.
The oxidation of cupro-nickel alloys
II. The kinetics of diffusion in porous inner layers
Corrosion Sci. 19 (1979) p. 475

[812] Nishida, K.; Narita, T.; Tani, T.; Sasaki, G.
High-temperature sulfidation of Fe-Mn alloys
Oxid. Metals 14 (1980) p. 65

[813] Campbell, W. E.; Thomas, U. B.
Trans. Electrochem. Soc. 91 (1947) p. 623

[814] Zschötge, S.
Kleine Werkstoffkunde der Nichteisenmetalle
(Concise introduction inter non-ferrous metal base materials) (in German)
Deutscher Verlag für Schweißtechnik, Düsseldorf, 1967, p. 58

[815] Hauffe, K.
Reaktionen in und an festen Stoffen
(Reactions in and on solid matter)
(in German)
Springer-Verlag, Berlin, 1955, p. 358

[816] Baukloh, W.; Krysko, W. W. G.
Metallwirtsch. 19 (1940) p. 169 (in German)

[817] Kubaschewski, O.; von Goldbeck, O.
Metalloberfläche (A) 8 (1954) p. 33
(in German)

[818] Landolt-Börnstein "Zahlen und Funktionen", 6. Aufl., Bd. 5
Springer-Verlag, Berlin, 1968, p. 674
(in German)

[819] Dunn, J. S.
J. Inst. Metals 46 (1931) p. 25

[820] Beiswenger, S.
Das Oxidationsverhalten von CuZnAl-Legierungen bei hoher Temperatur
(Oxidation behavior of CuZnAl-alloys at high temperature) (in German)
Werkstoffe u. Korrosion 27 (1976) p. 159

[821] Pfeiffer, H.; Thomas, H.
Zunderfeste Legierungen
(Scaling-resistant alloys) (in German)
Springer-Verlag, Berlin, 1963, p. 254

[822] Pfeiffer, H.; Sommer, G.
Untersuchungen zur Klärung der Wirkungsweise von Zusatzelementen in Heizleiterlegierungen
(Investigations aiming at an understanding of the effect of trace elements in heat-resisting alloys) (in German)
Z. Metallkunde 57 (1966) p. 326

[823] Valverde, N.
Das Oxidationsverhalten von CuZnAl-Legierungen mit Zusätzen von Ce und Y bei 750 °C
(Oxidation behavior of CuZnAl-alloys with additions of Ce and Y at 750 °C)
(in German)
Werkstoffe u. Korrosion 29 (1978) p. 644

[824] Wood, S.; Adamonis, D.; Guha, A.; Soffa, W. A.; Meier, G. H.
Met. Trans. A 6 (1975) p. 1793

[825] Swisher, J. H.; Fuchs, E. O.
J. Inst. Metals 98 (1970) p. 129

[826] Ashby, M. F.; Smith, G. C.
J. Inst. Metals 91 (1962) p. 182

[827] Kapteijn, J.; Couperus, S. A.; Meijering, J. L.
Acta Met. 17 (1969) p. 1311

[828] Tomlinson, W. J.; Yates, J.
High-temperature oxidation kinetics of CuSi-alloys containing up to 4.75 wt. % Si in p_{O_2} = 0.01 atm. and pure CO_2
Oxid. Metals 12 (1978) p. 323

[829] Martin, J. W.; Smith, G. C.
J. Inst. Metals 83 (1954) p. 153

[830] Komatsu, N.; Grant, N. J.
Trans. AIME 224 (1962) p. 705

[831] Tomlinson, W. J.; Yates, J.
J. Phys. Chem. Solids 38 (1977) p. 1205

[832] Panda, P. K.; Lahiri, A. K.; Banerjee, T.
Morphology of internal oxidation in Cu-20Mn alloy
Corrosion 28 (1972) p. 55

[833] Nanni, P.; Viani, F.; Elliott, P.; Gesmundo, F.
High-temperature oxidation of copper-manganese alloys
Oxid. Metals 13 (1979) p. 181

[834] Maak, F.; Wagner, C.
Mindestgehalte von Legierungsbestandteilen für die Bildung von Oxidschichten hoher Schutzwirkung gegen Oxidation bei höheren Temperaturen
(Minimum alloying additions required for the formation of surface layers yielding high protection efficiency against oxidation at elevated temperatures) (in German)
Werkstoffe u. Korrosion 12 (1961) p. 273
Maak, F.
Untersuchungen über die Oxidation von Kupfer-Beryllium-Legierungen bei erhöhter Temperatur
(Investigation into the oxidation of copper beryllium alloys at elevated temperatures)
(in German)
Z. Metallkunde 52 (1961) p. 538

[835] Wagner, C.
J. Electrochem. Soc. 103 (1956) p. 627

[836] Lichter, B. D.; Wagner, C.
The attack of copper-gold, silver-gold, nickel-copper, and silver-copper alloys by sulfur at elevated temperatures
J. Electrochem. Soc. 107 (1960) p. 168

[837] Pinnel, M. R.; Tompkins, H. G.; Heath, D. E.
Oxidation kinetics of copper from gold alloy solution at 50–150 °C
J. Electrochem. Soc. 126 (1979) p. 1798

[838] Swaroop, B.; Wagner jr., J. B.
Oxidation of copper and copper-gold alloys in CO_2 at 1000 °C I. Linear kinetics
J. Electrochem. Soc. 114 (1967) p. 685

[839] Thomas, D. E.
J. Metals 3 (1951) p. 926

[840] Hauffe, K.
Oxidation of metals
Plenum Press, New York, 1965, p. 315

[841] Wagner, C.
Internal oxidation of Cu-Pd and Cu-Pt alloys
Corrosion Sci. 8 (1968) p. 889

[842] Khedr, M. G. A.; Gaamoune, B.
Corrosion behavior of Cu in oxidizing and non oxidizing acid mixtures
Metaux – Corrosion Industrie 60 (1985) 713, pp. 1–8

[843] Ahmad, Z.
Effect of tin addition on the corrosion resistance of aluminium bronze
Anti-Corrosion 24 (1977) 1, pp. 8–12

[844] Singh, R. N.; Tiwari, S. K.; Singh, W. R.
Effects of Ta, La and Nd additions on the corrosion behavior of aluminium bronze in mineral acids
Journal of Applied Electrochemistry 22 (1992) 12, pp. 1175–1179

[845] Singh, R. N.; Verma, N.; Tiwari, S. K.; Singh, W. R.
Effect of some rare earth additives on corrosion of aluminium bronze in hydrochloric acid solution
Indian Journal of Chemical Technology 1 (1994) 2, pp. 103–107

[846] Singh, R. N.; Tiwari, S. K.; Verma, N.
The influence of minor additions of Ta and Nd on the corrosion behavior of propeller bronze in mineral acids
Corros. Prev. Control 35 (1988) 2, pp. 43–48

[847] Saber, T. M. H.; Tag El Din, A. M. K.;
Dibutyl thiourea as corrosion inhibitor for acid washing of multistage flash distillation plant
Br. Corros. J. 27 (1992) 2, pp. 139–143

[848] Yakovleva, L. A.; Vakulenko, L. I.; Vdovenko, I. D.; Lisogor, A. I.; Kalinyuk, N. N.; Novitskaya, G. N.
Corrosion disintegration of α-brass in hydrochloric and sulfamic acid solutions
Ukrainskii Khimicheskii Zhurnal 53 (1987) 7, pp. 709–711

[849] Anonymous
High temperature corrosion data. A compilation by technical Unit Committee T-5B on high temperature corrosion
NACE Rep. Comm. T-5B
Corrosion 11 (1955) p. 241t

[850] Product Information
Resistance of nickel and high nickel alloys to corrosion by hydrochloric acid, hydrogen chloride and chlorine
10 M 3-62 3911 A 279, 31 p.
International Nickel Co., Inc., New York

[851] Maude, A. H.
Trans. Amer. Inst. Chem. Eng. 38 (1942) p. 865

[852] Heitz, E.; v. Meysenbug, C. M.
Die Korrosion von Eisen, Nickel, Kupfer und Aluminium in organischen Lösungsmitteln mit geringem Mineralsäuregehalt
(The corrosion of iron, nickel, copper, and aluminium in organic solvents containing small amounts of mineral acid)
(in German)
Werkst. Korros. 16 (1965) p. 578

[853] Rice, D. W. et al.
Atmospheric corrosion of copper and silver
J. Electrochem. Soc. 128 (1981) 1, p. 275

[854] Bulow, C. L.
in: H. H. Uhlig "The Corrosion Handbook"
J. Wiley & Sons, Inc., New York, 1948/1958, p. 94

[855] Ried, G.
Übersicht über Werkstoffe für Absperrorgane, die bei angreifenden Medien zu empfehlen sind
(A survey of materials for blockage units, recommended for use in the presence of corrosive media) (in German)
Werkst. Korros. 15 (1964) p. 468

[856] Furusawa, T.; Udo, Y.; Yanagida, K.; Nagai, T.; Henmi, Z.
Corrosion of metals for electronic components in mixed gases (in Japanese)
J. Jpn. Inst. Met. 50 (1986) 5, p. 467

[857] Degnan, T. F.
Materials of construction for hydrochloric acid and hydrogen chloride
NACE, Houston, Tex. (USA) "Process Industries Corrosion – the Theory and Practice", 1986, p. 265

[858] Product Information
Carpenter® stainless No. 20 Cb-3, 1965, p. 13
The Carpenter Steel Co., Reading (Pa./USA)

[859] Product Information
Corrosion data survey, 1960, p. H 3
Shell Development Co., Emeryville (Cal./USA)

[860] Lee, J. A.
Materials of construction for chemical process industries
McGraw-Hill Book Co., Inc., New York, 1950

[861] Pastonesi, G.
Atti Congr. Corr. 1953
Metall. Ital. 46 (1954) Special ed., p. 37

[862] Park, H. S.; Day, D. E.
Corrosion of bare Ag, Ni/Ag and Cu/Ag films on glass by wet HCl vapor
Sol. Energy Mater. 13 (1986) 5, p. 367

[863] DECHEMA-Werkstoff-TABELLE
"Natriumhypochlorit"

(Sodium hypochlorite) (in German)
DECHEMA, D-Frankfurt am Main,
October 1962

[864] DECHEMA-WERKSTOFF-TABELLE
"Calciumhypochlorit und Chlorkalk"
(Calcium hypochlorite and chlorinated lime) (in German)
DECHEMA, D-Frankfurt am Main,
January 1955

[865] DECHEMA-WERKSTOFF-TABELLE
"Kaliumhypochlorit"
(Potassium hypochlorite) (in German)
DECHEMA, D-Frankfurt am Main,
October 1958

[866] Rabald, E.
Corrosion Guide, 2nd revised edition
Elsevier Publishing Co., Amsterdam-London-New York, 1968

[867] Fabian, R. J.; Vaccari, J. A.
How materials stand up to corrosion and chemical attack
Mater. Eng. 73 (1971) 2, p. 36

[868] Anonymous
Corrosion Data Survey – Metals Section,
6th ed., 1985
NACE, Houston (Tx./USA)

[869] Nelson, J. K.
Materials of construction for alkalies and hypochlorites
Process Industries Corrosion, NACE,
Houston (Tx./USA), 1986, p. 297

[870] Information
Beständigkeitstabellen von Kupfer-Gußwerkstoffen in verschiedenen Medien
(Resistance table of copper cast materials in different media) (in German)
June 1970
Deutsches Kupfer-Institut, Berlin;
Gesamtverband Deutscher Metallgießereien, Düsseldorf

[871] Drovyannikov, N. P.; Pimenova, S. I.
The corrosion resistance of materials in bleaching solutions and in commercial nekal detergent
Ekspluat. Mod. Remont. Oborud.
Neftepererab. Neftekhim. Prom-st.
(Moskau) (1979) 3, p. 8 (in Russian)

[872] Bhanot, D.; Chawla, S. L.
Corrosion resistance of pure copper, its alloys, and stainless steel in chlorinated lime slurry and calcium hypochlorite solution
Trans. Soc. Adv. Electrochem. Sci. Technol.
13 (1978) 3, July–Sept., p. 227

[873] Sick, H.
Kupferwerkstoffe für Verdampfer in Wasser/Wasserwärmepumpen
(Copper alloys for vaporisers in water/water heat pumps) (in German)
Heizung-Lüftung/Klimatechnik-Haustechnik (1979) 3; (DKI offprint)
VDI-Verlag GmbH, Düsseldorf

[874] Stichel, W.
Korrosion und Korrosionsschutz von Metallen in Schwimmhallen
(Corrosion and corrosion protection of metal in indoor swimming pools)
(in German)
Research report no. 126, BAM-file number 1.3/11467
Bundesanstalt für Materialprüfung (BAM),
Berlin

[875] Rabald. E.
Corrosion Guide, 2nd revised edition
Elsevier Publishing Co., Amsterdam-London-New-York, 1968

[876] Anonymous
Corrosion Data Survey-Metals Section, 6th ed., 1985
NACE, Houston (Tx./USA)

[877] DECHEMA-Werkstofftabelle
"Calciumhypochlorit und Chlorkalk"
(Calcium hypochlorite and chlorinated lime) (in German)
DECHEMA, D-Frankfurt am Main,
January 1955

[878] Franke, E.
Das Korrosionsverhalten von Aluminiumbronzen
(The corrosion behavior of aluminium bronzes) (in German)
Werkst. Korros. 2 (1951) 8, p. 298

[879] DECHEMA-Werkstofftabelle
"Natriumhypochlorit"
(Sodium hypochlorite) (in German)
DECHEMA, D-Frankfurt am Main,
October 1962

[880] Gräfen, H.; Baeckmann, W. G.; Föhl, J.;
Herbsleb, G.; Huppartz, W.; Kuron, D.;
Rother, H.-J.; Rüdinger, K.
Die Praxis des Korrosionsschutzes
(The practice of corrosion protection)
(in German)
Kontakt & Studium, vol. 64
expert verlag, Grafenau, 1981

[881] Franke, E.
Das Korrosionsverhalten von Aluminiumbronzen
(The corrosion behavior of aluminium bronzes) (in German)
Werkst. Korros. 2 (1951) 8, p. 298

[882] Carr, R. P.; Holden, M. J.
The performance of tolyltriazole in the presence of sodium hypochlorite under simulated field conditions
Corrosion '83, Anaheim (Calif./USA), April 1983 (Proc. Conf.), Paper no. 283, 6 p.
NACE, Houston (Tx./USA)

[883] Miller, B.
Split-ring disk study of the anodic processes at a copper electrode in alkaline solution
J. Electrochem. Soc. 116 (1969) p. 1675

[884] Shoesmith, D. W.; Rummery, T. E.; Owen, D.; Lee, W.
Anodic oxidation of copper in alkaline solutions. I. Nucleation and growth of cupric hydroxide films
J. Electrochem. Soc. 123 (1976) p. 790

[885] Shoesmith, D. W.; Rummery, T. E.; Owen, D.; Lee, W.
Oxidation of copper in alkaline solutions. II. The open-circuit potential behavior of electrochemically formed cupric hydroxide films
Electrochim. Acta 22 (1977) p. 1403

[886] Fletcher, S.; Barradas, R. G.; Porter, J. D.
The anodic oxidation of copper amalgam and polycrystalline copper electrode in LiOH solution
J. Electrochem. Soc. 125 (1978) p. 1960

[887] Oshe, E. K.; Rozenfeld, I. L.
Photo-electric polarization method of study non-stoichiometric surface oxides on metal electrodes (in Russian)
Zashchita Metallov 5 (1969) p. 524

[888] Janz, G. J.; Tomkins, R. P. T.
Corrosion in molten salts: An annotated bibliography
Corrosion 35 (1979) p. 485

[889] Gurovich, E. I.
Reaktionen von geschmolzenem Lithium-, Natrium und Kaliumhydroxid mit Nickel, Kupfer, Eisen und Stahl
(Reactions of molten lithium, sodium and potassium hydroxide with nickel, copper, iron and steel) (in Russian)
Zhur. Priklad. Khim. 32 (1968) p. 817

[890] de Vries, G.; Grantham, L. F.
Corrosion of metals in the molten LiOH-LiF-H_2O system
Electrochem. Technol. 5 (1967) p. 335

[891] Anonymous
Kupfer – Vorkommen, Gewinnung, Eigenschaften, Verarbeitung, Verwendung
(Copper – occurrence, recovery, properties, processing use) (in German), DKI Information No. 1.004, p. 312
Deutsches Kupfer-Institut, D-Berlin

[892] Anonymous
Schweißen von Kupfer
(The welding of copper) (in German), DKI Information No. i.11, p. 2
Deutsches Kupfer-Institut, D-Berlin

[893] Wallbaum, H. J.
Über das Korrosionsverhalten von Werkstoffen auf Kupferbasis
(The corrosion behavior of copper-based materials) (in German), DKI Information No. s. 139, p. 5
Deutsches Kupfer-Institut, D-Berlin

[894] Huppatz, W.
in: Die Praxis des Korrosionsschutzes, Kontakt und Studium
(Corrosion protection in practice; dealings with and study of) (in German), vol. 64, p. 86
expert verlag, D-Grafenau, 1981

[895] Hamner, N. E.
Corrosion Data Survey, 5th ed., p. 120
National Association of Corrosion Engineers, Houston, Tx., 1974

[896] Product Information
Corrosion resistance charts, Information No. Ch.5.1, p. 281
Du Pont de Nemours International S.A., CH-1211 Genf

[897] Anonymous
Materials of construction, Part II
Chem. Eng. 61 (1954) Nov., p. 214

[898] Product Information
Gebräuchliche Werkstoffe für Elasteflex®-Metallbälge
(Common materials used for Elasteflex® metal bellows) (in German), Table MB 001, p. 11
Hans Skodock, D-Hannover

[899] Product Information
Hydra-Taschenbuch
(Hydra pocketbook) (in German), No. 301, p. 461

Witzenmann GmbH,
Metallschlauchfabrik, D-Pforzheim

[900] Grefe
Korrosions-Atlas
(Corrosion atlas) (in German), p. 22
D-Frankfurt-Höchst, 1950

[901] Cornils, B.; Falbe, J.; Frohning, C. D.; Höver, H.; Teggers, H.
in: Winnacker-Küchler "Chemische Technologie – Organische Technologie"
(Chemical technology – organic technology) (in German), vol. 5, pp. 583
Carl Hanser Verlag, München-Wien, 1981

[902] Giesen, J.; Hanisch, H.
in: Winnacker-Weingaertner "Chemische Technologie – Organische Technologie I"
(Chemical technology – organic technology I) (in German) p. 426
Carl Hanser Verlag, D-München, 1952

[903] Palm, A.
in: Winnacker-Küchler "Chemische Technologie – Organische Technologie I"
(Chemical technology – organic technology I) (in German), 3rd ed., vol. 3, p. 363
Carl Hanser Verlag, D-München, 1971

[904] Kirk-Othmer
Encyclopedia of Chemical Technology, 2nd. ed., vol. 13, p. 379
John Wiley & Sons., Inc., New York, 1967

[905] Bernthsen, A.
A Textbook of Organic Chemistry, p. 749
Blackie & Son Limited, London, 1946

[906] van Rossum, O.
Werkstoff-Fragen bei den Hochdrucksynthesen
(Material problems in high-pressure syntheses) (in German)
Chem.-Ing.-Tech. 25 (1953) 8/9, p. 481

[907] Marscher, F.; Möller, F.-W.; Gelbke, H. P.
in: Ullmanns Encyklopädie der technischen Chemie
(Ullmann's encyclopedia of industrial chemistry) (in German), 4th ed., vol. 16, p. 621
Verlag Chemie, D-Weinheim, 1976

[908] Köcher, R.
Nichteisenmetalle im Apparate- und Anlagenbau
(Non-ferrous metals in construction of equipment and plant) (in German)
Chem.-Ing.-Tech. 55 (1983) 10, p. 752

[909] Dechema-Werkstoff-Tabelle "Essigsäure"
(Corrosion data sheets "Acetic acid")
(in German)
DECHEMA, D-Frankfurt am Main, November 1981

[910] Heitz, E.
Corrosion of metals in organic solvents
Adv. in Corrosion Science and Technology 4 (1974) p. 226

[911] Dettmeier, U.; Grosskinsky, O.-A.; Mack, K.-E.; Wirtz, R.
in: Winnacker-Küchler "Chemische Technologie – Organische Technologie II"
(Chemical technology – organic technology II) (in German), 4th ed., vol. 6, p. 62
Carl Hanser Verlag, München-Wien, 1982

[912] Bernthsen, A.
A Textbook of Organic Chemistry, p. 756
Blackie & Son Limited, London, 1946

[913] Rabald, E.
Corrosion Guide, 2nd revised edition, p. 449
Elsevier Publishing Co., Amsterdam-London-New York, 1968

[914] Dechema-Werkstoff-Tabelle "Aliphatische Aldehyde"
(Corrosion data sheets "Aliphatic Aldehydes") (in German)
DECHEMA, D-Frankfurt am Main, November 1978

[915] Lingnau, E.
Das Verhalten der Werkstoffe gegenüber Formaldehyd
(Behavior of materials towards formaldehyde) (in German)
Werkst. Korros. 8 (1957) 8/9, p. 482

[916] Sharifulina, I. I.; Tsygankova, L. E.; Vigdorovich, V. I.
Corrosion of copper in methanol/HCl (in Russian)
Priklad. Khim. 50 (1977) 10, p. 2274

[917] D'Ans, J.; Lax, E.
Taschenbuch für Chemiker und Physiker
(Paperback for chemists and physicists) (in German), 2nd ed., p. 1326
Springer-Verlag, Berlin, Göttingen-Heidelberg, 1949

[918] Lechner-Knoblauch, U.; Heitz, E.
Corrosion of zinc, copper and iron in contaminated non-aqueous alcohols
Electrochim. Acta 32 (1987) 6, p. 901

[919] Heitz, E.
Vergleichende Korrosionsuntersuchungen in Ethanol-, Methanol- und wäßrigen Lösungen

[919] (Comparative corrosion studies in ethanol, methanol and aqueous solutions) (in German)
Werkst. Korros. 38 (1987) 6, p. 333

[920] Sick, H.
Kupfer für Verdampfer in Wasser/Wasser-Wärmepumpen
(The use of copper for evaporators in reverse cycle heating systems) (in German), DKI Information No. s. 172, p. 4
Deutsches Kupfer-Institut, D-Berlin

[921] Branzoi, V.; Sternberg, S.
Polarization behavior of copper in $CH_3OH + H_2SO_4$ solution (in Rumanian)
Rev. Roum. Chim. 18 (1983) 3, pp. 39, 200

[922] Mansfeld, F.
Passivity and pitting of Al, Ni, Ti, and stainless steel in $CH_3OH + H_2SO_4$
J. Electrochem. Soc. 120 (1973) 2, p. 188

[923] Abd El Wahab, F. M.; El Shayeb, H. A.
Effect of some alcohols on the dissolution of Cu, Zn and Cu-Zn alloy in nitric acid
Corros. Prev. Control 32 (1985) 1, p. 9

[924] Kuppers, J. R.
The reactivity of metals with mixtures of carbon tetrachloride and alcohols
J. Electrochem. Soc. 125 (1978) 1, p. 97

[925] Köser, H. J. K.
Korrosion in halogenkohlenwasserstoffhaltigen Lösungsmittelgemischen
(Corrosion in solvent mixtures containing halogenated hydrocarbons) (in German)
Werkst. Korros. 36 (1985) 1, p. 30

[926] Rabinovits, E.; Herrmann, B.; Levin, R.; Rosenthal, I.
The corrosion effect of solutions of lasing dyes on metals
Corrosion 38 (1982) 9, p. 510

[927] Heitz, E.
Korrosion und Korrosionsschutz in Systemen mit neuen Energieträgern und Brennstoffen
(Corrosion and corrosion protection in systems utilizing new energy sources and fuels) (in German)
Chem.-Ing.-Tech. 58 (1986) 5, p. 357

[928] Foulkes, F. R.; Kalia, R. K.; Kirk, D. W.
A preliminary study of the corrosiveness of methyl fuel/gasoline blends on materials used in automotive fuel systems
Can. J. Chem. Eng. 58 (1980) 10, p. 654

[929] Poteat, L. E.
Automotive materials compatibility with methanol fuel blends
4th Inter. Symp. (Proc. Conf.), University of Miami, Febr. 1978, p. 707

[930] Dechema-Werkstoff-Tabelle "Kohlensäure"
(Corrosion data sheets "Carbonic acid")
(in German)
DECHEMA, D-Frankfurt am Main, March 1987

[931] Benninghoff, H.
Mechanische, chemische und elektrolytische Oberflächenvorbehandlung von Kupfer und Kupferlegierungen
(Mechanical, chemical and electrolytic pretreatment of the surface of copper and copper alloys) (in German), DKI Information No. s. 152, p. 12
Deutsches Kupfer-Institut, D-Berlin

[932] Product Information
Wieland-Buch "Kupferwerkstoffe"
(Wieland book "copper materials")
(in German) p. 176
Wieland-Werke AG, D-Ulm

[933] Gräfen, H.; v. Baeckmann, W. G.; Föhl, J.; Herbsleb, G.; Huppatz, W.; Kuron, D.; Rother, H.-J.; Rüdinger, K.
Die Praxis des Korrosionsschutzes, Kontakt und Studium
(Corrosion protection in practice; dealings with and study of) (in German), vol. 64, p. 102
expert verlag, D-Grafenau, 1981

[934] Anonymous
Löten von Kupfer und Kupferlegierungen
(Soldering copper and its alloys)
(in German), DKI Information No. i.003, p. 11
Deutsches Kupfer-Institut, D-Berlin

[935] Zeiger, H.
Kupferwerkstoffe
(Copper materials) (in German), DKI Information No. s. 188, p. 78
Deutsches Kupfer-Institut, D-Berlin

[936] Anonymous
Ursachen für das Auftreten galvanischer Elemente
(Causes of the formation of a galvanic cell) (in German), p. 14
lecture, held at Fa. Witzenmann, Metallschlauchfabrik, D-Pforzheim

[937] Product Information
Wieland-Buch "Kupferwerkstoffe"

(Wieland book "copper materials")
(in German) p. 186
Wieland-Werke AG, D-Ulm

[938] Anonymous
Corrosion resistance of copper and copper alloys
Chem. Eng. 58 (1951) 1, p. 111

[939] Vaidyanath, L. R.; Bhandary, V. S.
Corrosion resistance of copper and copper alloys – A survey
Proceedings of a Symposium, Bombay, 1971, pp. 1, 3, 17

[940] Rabald, E.
Corrosion Guide, 2nd revised edition, p. 149
Elsevier Publishing Co., Amsterdam-London-New York, 1968

[941] Anonymous
Kupfer-Nickel-Legierungen, Eigenschaften, Bearbeitung, Anwendung
(Copper-nickel alloys – properties, processing, use) (in German), DKI Information No. i.014, p. 2
Deutsches Kupfer-Institut, D-Berlin

[942] Anonymous
Kupfer-Nickel-Zink-Legierungen, Neusilber
(CuNiZn alloys, nickel silver) (in German), DKI Information No. i.13, p. 2
Deutsches Kupfer-Institut, D-Berlin

[943] Vaidyanath, L. R.; Bhandary, V. S.
Corrosion resistance of copper and copper alloys – A survey
Proceedings of a Symposium, Bombay, 1971, p. 17

[944] Product Information
Wieland-Buch "Kupferwerkstoffe"
(Wieland book "copper materials") (in German) p. 81
Wieland-Werke AG, D-Ulm

[945] Product Information
Corrosion resistance reference table
Anaconda B.V., Amsterdam (NL)

[946] Anonymous
Korrosionsbeständigkeitstabellen moderner Werkstoffe
(Corrosion resistant charts of modern materials) (in German), Special print in Werkstofftechnik, 1/1970 to 5/1972
Verlag für Technik und Wirtschaft Meynen KG, D-Wiesbaden

[947] Butler, G.; Mercer, A. D.
Corrosion of some aluminium casting alloys and cast iron in uninhibited alcohol/water coolants
Br. Corros. J. 12 (1977) 3, p. 169

[948] Lingnau, E.
Das Verhalten der Werkstoffe gegenüber Formaldehyd
(Behavior of materials towards formaldehyde) (in German)
Werkst. Korros. 8 (1957) 8/9, p. 482

[949] Anonymous
Kupfer-Zinn-, Kupfer-Zinn-Zink- und Kupfer-Blei-Zinn-Gußlegierungen
(Guß-Zinnbronze, Rotguß und Guß-Zinn-Bleibronze)
(CuSn, CuSnZn and CuPbSn cast alloys (cast tin bronze, red brass and cast tin lead bronze)) (in German), DKI Information No. i.025, p. 4
Deutsches Kupfer-Institut, D-Berlin

[950] Ried, G.
Übersicht über Werkstoffe für Absperr-Organe, die bei angreifenden Medien zu empfehlen sind
(A survey of materials for shut-off units, recommended for use in the presence of corrosive media) (in German)
Werkst. Korros. 15 (1964) 6, p. 468

[951] Remler, R. F.
Metals for use in handling organic solvents
Chemical and Metallurgical Eng. 30 (1924) 13, p. 511

[952] Product Information
Conveyants list (5.1.1973)
Compoflex Company Ltd.

[953] Anonymous
Kupfer-Zinn-, Kupfer-Zinn-Zink- und Kupfer-Blei-Zinn-Gußlegierungen
(Guß-Zinnbronze, Rotguß und Guß-Zinn-Bleibronze)
(CuSn, CuSnZn and CuPbSn cast alloys (cast tin bronze, red brass and cast tin lead bronze)) (in German), DKI Information No. i.025, p. 13
Deutsches Kupfer-Institut, D-Berlin

[954] Sleeman, D. G.
Silver-catalyst process obtains high-strength formaldehyde solutions
Chem. Eng. 75 (1968) 1, p. 42

[955] Stinson, S.
Chemical methods restore, protect rare statues
Chem. and Eng. News (1981) Dec. 14, p. 27

[956] Anonymous

Kupfer-Zinn-Knetlegierungen
(Zinnbronze)
(CuSn wrought alloys (tin bronze))
(in German), DKI Information No. i.15,
p. 2
Deutsches Kupfer-Institut, D-Berlin

[957] Kleinau, M.
Schwermetall-Schleuder- und -Strangguß –
Technische und wirtschaftliche
Möglichkeiten
(Heavy metal spin casts and continuous
casts – technical and economic
possibilities) (in German), DKI
Sonderdruck No. s. 165, p. 2
Deutsches Kupfer-Institut, D-Berlin

[958] Gräfen, H.; v. Baeckmann, W. G.; Föhl, J.;
Herbsleb, G.; Huppatz, W.; Kuron, D.;
Rother, H.-J.; Rüdinger, K.
Die Praxis des Korrosionsschutzes,
Kontakt und Studium
(Corrosion protection in practice; dealings
with and study of) (in German), vol. 64,
p. 98
expert verlag, D-Grafenau, 1981

[959] Anonymous
Kupfer-Zink-Legierungen, Messing und
Sondermessing
(Copper-zinc alloys, brass and high-
strength brass) (in German), DKI
Information No. i.5, p. 2
Deutsches Kupfer-Institut, D-Berlin

[960] Anonymous
Schweißen von Kupferlegierungen
(Welding copper alloys) (in German), DKI
Information No. i. 12, p. 8
Deutsches Kupfer-Institut, D-Berlin

[961] Product Information
Wieland-Buch "Kupferwerkstoffe"
(Wieland book "copper materials")
(in German) p. 171
Wieland-Werke AG, D-Ulm

[962] Anonymous
Kupfer-Werkstoffe
(Copper materials) (in German), DKI
Information No. s. 188, p. 77
Deutsches Kupfer-Institut, D-Berlin

[963] Anonymous
Ursachen für das Auftreten galvanischer
Elemente
(Causes of the formation of a galvanic cell)
(in German), p. 18
lecture, held at bei Fa. Witzenmann,
Metallschlauchfabrik, D-Pforzheim

[964] Anonymous
Schrauben und Muttern aus Kupfer-Zink-
Legierungen
(Screws and nuts made of CuZn alloys)
(in German), DKI Information No. i.17,
p. 4
Deutsches Kupfer-Institut, D-Berlin

[965] Product Information
Fluid compatability of metal hoses, TIFT
78/12
T.I. Flexible Tubes Ltd., Delph Oldham

[966] Giesen, J.; Hanisch, H.
in: Winnacker-Weingaertner,"Chemische
Technologie – Organische Technologie I"
(Chemical technology – organic technology
I) (in German) p. 451
Carl Hanser Verlag, D-München, 1952

[967] Frömberg, M.
Entzinkung von Dichtungsringen aus
Messing für Rektifikationskolonnen
(Dezincification of brass packing rings on
rectifying columns) (in German)
Korrosion, Dresden 16 (1985) 2, p. 109

[968] Luklinska, Z. B.; Castle, J. E.
Microstructural study of initial corrosion
product of aluminium-brass alloy after
exposure to natural seawater
Corros. Sci. 23 (1983) 11, p. 1163

[969] Marsden, T. B.
Einsatz von Kupfer- und
Kupferlegierungsdrähten für
nichtelektrische Anwendungen
(The use of copper and copper alloy wires
for non-electrical purposes) (in German),
DKI Information No. s. 168
Deutsches Kupfer-Institut, D-Berlin

[970] Foulkes, F. R.; Kalia, R. K.; Kirk, D. W.
A preliminary study of the corrosiveness of
methyl fuel/gasoline blends on materials
used in automotive fuel systems
Can. J. Chem. Eng. 58 (1980) 10, p. 654

[971] Titchener, A. L.
Compatibility of vehicle materials with
unconventional fuels
Metals and Energy (Proc. Conf.), Auckland
(N.Z.), May 1980
Australasian Institute of Metals, Victoria
(Austr.), p. 122

[972] Anonymous
Niedriglegierte Kupferwerkstoffe –
Eigenschaften, Verarbeitung, Verwendung
(Low-alloy copper materials – properties,
processing, use)

(in German), DKI Information No. i.8, p. 2
Deutsches Kupfer-Institut, D-Berlin

[973] Palm, A.
in: Winnacker-Küchler, "Chemische Technologie – Organische Technologie I" (Chemical technology – organic technology I) (in German), 3rd ed., vol. 3, p. 387
Carl Hanser Verlag, D-München, 1971

[974] Giesen, J.; Hanisch, H.
in: Winnacker-Weingaertner, "Chemische Technologie – Organische Technologie I" (Chemical technology – organic technology I) (in German) pp. 428, 452
Carl Hanser Verlag, D-München, 1952

[975] Rabald, E.
Corrosion Guide, 2nd revised edition, p. 450
Elsevier Publishing Co., Amsterdam-London-New York, 1968

[976] Anonymous
Niedriglegierte Kupferwerkstoffe – Eigenschaften, Verarbeitung, Verwendung (Low-alloy copper materials – properties, processing, use)
(in German), DKI Information No. i.8, p. 13
Deutsches Kupfer-Institut, D-Berlin

[977] McIlhone, P. G. H.; Lichti, K. A.; Wilson, P. I.
Materials problems associated with the use of alternative liquid motor fuels in N.Z.
Met. Australas 10 (1982) Nov./Dec. 14, p. 16

[978] Patent DAS 1 621 142
Anodisches Beizen von Legierungen (Anodic pickling of alloys) (in German)
Pechiney Co. de Produits Chimique et Electrometallurgique, Paris (20.1.1972)

[979] Straschill, M.
Beizen von Kupfer und Kupferlegierungen (Pickling of copper and copper alloys) (in German)
Metallwarenind.-Galvanotechnik (1954), pp. 164–168

[980] Company publication
Beständigkeitstabellen Kupferwerkstoffe, 1996
(Resistance charts copper materials) (in German)
Deutsches Kupfer-Institut, D-Düsseldorf

[981] Khedr, M. G. A.; Gaamoune, B.
Corrosion behavior of Cu in oxidizing and non-oxidizing acid mixtures
Metaux Corrosion-Industrie 713 (1985) 1, pp. 1–8

[982] Patel, K. M.; Oza, B. N.
Effect of Urea in Corrosion of Brass (70/30) in Mixed Acid ($HNO_3 + H_2SO_4$)
J. Electrochem. Soc. India 38 (1989) 3, pp. 183–188

[983] Chen, N. G.; Kiryukha, A. S.; Chen, L. N.; Sokolyan, L. N.; Panfilova, Z. V.
Inhibitors of acid corrosion of metals EK and EK-1 (in Russian)
Zashch. Met. 9 (1973) 2, p. 211

[984] Antropov, L. I.; Donchenko, M. I.; Saenko, T. V.
Dissolution of copper in dilute nitric acid (in Russian)
Zashch. Met. 14 (1978) 6, p. 657

[985] Könneke, D.; Lacmann, R.
The dissolution of copper single crystal spheres
J. Crystal Growth 46 (1979) p. 15

[986] Joshi, K. M.; Kamat, P. V.
Dissolution of metals in magnetized water at various pH
Indian J. Appl. Chem. 33 (1970) p. 376

[987] Donchenko, M. I.; Sayenko, T. V.
Inhibition of copper corrosion by dimethylolthiourea in dilute nitric acid (in Russian)
Zashch. Met. 15 (1979) 1, p. 96

[988] Desai, M. N.; Rana, S. S.
Inhibition of the corrosion of the copper in nitric acid
Anticorros. Methods Mater. 20 (1973) 2, p. 8

[989] Beccaria, A. M.; Mor, E. D.
Inhibitive effect of tannic acid on the corrosion of copper in acid solutions
Br. Corros. J. 11 (1976) 3, p. 156

[990] Beccaria, A. M.; Mor, E. D.
Inhibiting effect of some monosaccharides on the corrosion of copper in nitric acid
Br. Corros. J. 13 (1978) 4, p. 186

[991] Walker, R.
Benzotriazole as a corrosion inhibitor for immersed copper
Corrosion 29 (1973) 7, p. 290

[992] Schmitt, G.
Application of inhibitors for acid media
Report prepared for the European Federation of Corrosion working party on inhibitors
Brit. Corros. J. 19 (1984) p. 165

[993] Desai, M. N.; Rana, S. S.
Inhibition of the corrosion of copper in nitric acid. Aniline and other substituted aromatic amines
Anticorros. Methods Mater. 20 (1973) 6, p. 16

[994] Shams El Din, A. M.; El Hosary, A. A.; Saleh, R. M.; Abd El Kader, J. M.
Peculiarities in the behaviour of thiourea as corrosion-inhibitor
Werkst. Korros. 28 (1977) 1, p. 26

[995] Uzlyuk, M. V.; Zimina, V. M.; Fedorov, Yu. V.; Pinus, A. M.
A corrosion inhibitor for copper and brass in nitric acid solutions (in Russian)
Vopr. Khim. i Khim. Technol. 62 (1981) p. 17

[996] Blahnik, R.; Kosobud, J.
The corrosion resistance of wrapped and soldered elec. connections in corrosive atmospheres (in Czech)
Koroze Ochr. Mater. 21 (1977) 1, p. 3

[997] Winkley, D. C.; Grand, A. F.
Hydrogen peroxide pickling of copper rod
Wire J. 6 (1973) p. 64

[998] Martschewska, M.; Penew, P.; Janatschkowa, I.; Spassowa, E.
Oxidation von Kupfer durch Wechselstrom mit technischer Frequenz. I. In wäßrigen Natriumnitrat und Natriumkarbonat-Lösungen
(Oxidation of copper by alternating current with a technical frequency. I. In aqueous solutions of sodium nitrate and sodium carbonate) (in German)
Metalloberfläche 27 (1973) 11, p. 416

[999] Anonymous
Pretreatment of copper surfaces before coating with aluminium
UdSSR Pat. 1802058; (Galvanotechnik 70 (1979) 7, p. 685)

[1000] Vivet, F.
Décapage des cuivreux procédé non polluant au peroxyde d'hydrogène
(Copper pickling, nonpolluting hydrogen peroxide process) (in French)
Galvano-Organo 45 (1976) p. 875

[1001] Afendik, K. F.; Krichmar, S. I.
Model of a viscous electrolyte for electropolishing of copper
Sov. Electrochem. 6 (1970) p. 122

[1002] Costas, L. P.
Some chemical aspects of the mercuric nitrate test used for copper alloys
Corrosion 22 (1966) 3, p. 74

[1003] Miska, K. H.
Wrought brass
Mater. Eng. 80 (1974) 5, p. 65

[1004] Fouda, A. S.; Mohamed, A. K.
Substituted phenols as corrosion inhibitors for copper in nitric acid
Werkst. Korros. 39 (1988) 1, p. 23

[1005] Vakulenko, L. I.; Kozlovskaya, N. A.; Shedenko, L. I.; Vdovenko, I. D.
The effect of polyvinyl alcohol on the chemical polishing of copper in electrolytes based on orthophosphoric and nitric acids
Sov. Progress Chem. 47 (1981) 11, p. 81

[1006] Ahmad, Z.; Ghafelehbashi, M.; Nategh, S.
Corrosion resistance of high manganese copper aluminium alloys in sea water
Anticorros. Methods Mater. 21 (1974) 8, p. 13

[1007] Zhuravlev, B. L.; Nazmutdinova, A. S.; Dresvyannikov, A. F.
Coulometric control of the corrosion stability of copper-nickel-chrome coatings (in Russian)
Zashch. Met. 20 (1984) 2, p. 319

[1008] Dubinin, G. N.; Sokolov, V. S.
Improvement in the scaling and corrosion resistance of copper-based alloys
(in Russian)
Izv. V. U. Z. Tsvet. Metall. (1974) 5, p. 133

[1009] Patel, M. M.; Patel, N. K.; Vora, J. C.
Azoles as corrosion inhibitors for 63/37 brass in acidic media
Metals & Minerals Review (1975) 8, p. 4

[1010] Patel, M. M.; Patel, N. K.; Vora, J. C.
Azoles as corrosion inhibitors for 63/37 brass in acidic media
Vishwakarma 16 (1975) 3, p. 4

[1011] Miska, K. H.
Wrought brass
Mater. Eng. 80 (1974) 5, p. 65

[1012] Walker, R.
Triazole, benzotriazole, and naphthotriazole as corrosion inhibitors for brass
Corrosion 32 (1976) 10, p. 414

[1013] Dinnappa, R. K.; Mayanna, S. M.
Haloacetic acids as corrosion inhibitors for brass in nitric acid
Corrosion 38 (1982) 10, p. 525

[1014] Dinnappa, R. K.; Mayanna, S. M.
Effect of thioureas on corrosion of brass in nitric acid
Trans. SAEST 19 (1984) 2, p. 93

[1015] Desai, M. N.; Joshi, J. S.
Alkyl anilines as corrosion inhibitors for 63/37 brass in nitric acid
J. Indian Chem. Soc. 52 (1975) p. 884

[1016] Uzlyuk, M. V.; Zimina, V. M.; Fedorov, Yu. V.; Pinus, A. M.
A corrosion inhibitor for copper and brass in nitric acid solutions (in Russian)
Vopr. Khim. i Khim. Technol. 62 (1981) p. 17

[1017] Soni, K. P.; Bhatt, I. M.
Corrosion and inhibition of copper, brass and aluminium in nitric acid, sulphuric acid and trichloroacetic acid
J. Electrochem. Soc. India 34 (1985) 1, p. 76

[1018] Desai, M. N.; Thakar, B. C.; Shah, D. K.; Gandhi, M. H.
Cathodic protection of 70/30 brass in 2.0 M nitric acid in the presence of organic corrosion inhibitors
Br. Corros. J. 10 (1975) 1, p. 41

[1019] Desai, M. N.
Corrosion inhibitors for brasses
Werkst. Korros. 24 (1973) 8, p. 707

[1020] Srivastava, R. D.; Mukerjee, R. C.; Tripathi, D.
Bath characteristics and corrosion nature of electrodeposited brass
Indian J. Technol. 15 (1977) p. 446

[1021] Patel, N. K.; Patel, B. B.
Influence of toluidines and acridine derivatives on corrosion of 63/37 brass in nitric acid solution
Werkst. Korros. 26 (1975) 2, p. 126

[1022] Desai, M. N.; Joshi, J. S.
p-Substituted aromatic amines as corrosion inhibitors for 63/37 brass in nitric acid
J. Indian Chem. Soc. 52 (1975) Sept., p. 878

[1023] Desai, M. N.; Patel, B. M.; Thakker, B. C.
Meta substituted aromatic amines as corrosion inhibitors for 70/30 brass in nitric acid
J. Indian Chem. Soc. 52 (1975) June, p. 554

[1024] Desai, M. N.; Thaker, B. C.; Patel, B. M.
Heterocyclic amines as corrosion inhibitors for 70/30 brass in nitric acid
J. Electrochem. Soc. India 24 (1975) 4, p. 184

[1025] Desai, M. N.; Shah, V. K.
Aromatic amines as corrosion inhibitors for 70/30 brass in nitric acid
Corros. Sci. 12 (1972) p. 725

[1026] Desai, M. N.; Shah, Y. C.
Thioureas as corrosion inhibitors for 63/67 brass in nitric acid
Trans. SAEST 6 (1971) 3, p. 81

[1027] Boron, K.; Dabrowiecki, K.
Corrosion resistance of brass pipes with varying inner surface coarseness (in Polish)
Rudy Met. Niezelaz. 27 (1982) 1, p. 36

[1028] Leonard, R. B.
Corrosion of metals by aliphatic organic acids
Mater. Performance 19 (1980) 9, p. 65

[1029] Mai, Y. W.
Crack propagation resistance of alpha-brass subjected to prior-exposure in corrosive environments
Corrosion 32 (1976) 8, p. 336

[1030] Dönnges, E.
Ausmaß und Rhythmus der Auflösungsgeschwindigkeit von Metallhalbzeug in angreifenden Agenzien in Abhängigkeit vom Grade der Kaltverformung
(Extent and rhythm of the dissolution rate of semifinished metal products in corrosive media according to the degree of cold working) (in German)
Metall 33 (1979) 12, p. 1269

[1031] Graf, L.; Ata, H. O. K.
Untersuchung der Spannungskorrosionsempfindlichkeit von Ein- und Vielkristallen aus Messing und Kupfer-Gold-Legierungen
(Investigation of stress corrosion sensitivity of single and polycrystalline specimens of brass and copper-gold alloys) (in German)
Z. Metallkde. 64 (1973) 5, p. 366

[1032] Wawra, H.
Verarbeitung von Messingblechen unter dem Gesichtspunkt hoher Spannungskorrosionsbeständigkeit, Teil I
(Processing brass sheets with the aim of achieving high stress corrosion cracking resistance, Part I) (in German)
Blech 19 (1972) 11, p. 591

[1033] Wawra, H.
Verarbeitung von Messingblechen under dem Gesichtspunkt hoher Spannungskorrosionsbeständigkeit, Teil II

(Processing brass sheets with the aim of achieving high stress corrosion cracking resistance, Part II) (in German)
Blech 19 (1972) 12, p. 631

[1034] Peapell, P. N.; Walkley, E. C.
Acoustic emission during stress corrosion cracking
Metall. Mater. Technol. 11 (1979) 2, p. 95

[1035] Altgeld, W.; Läser, L.
Chemisches Glänzen von Kupfer-Zink-Legierungen
(Chemical polishing of copper-zinc alloys) (in German)
Metalloberfläche 36 (1982) 4, p. 177

[1036] Luklinska, Z. B.; Castle, J. E.
Microstructural study of initial corrosion product of aluminium-brass alloy after exposure to natural seawater
Corros. Sci. 23 (1983) 11, p. 1163

[1037] Srivastava, R. D.; Mukerjee, R. C.; Agarwal, A. K.
Corrosion of electroplated Cu-Cd alloys and its inhibition
Corros. Sci. 19 (1979) 1, p. 27

[1038] Srivastava, R. D.; Agarwal, A. K.
Corrosion inhibition of copper-cadmium alloy electroplates in nitric acid solutions
Anticorros. Methods Mater. 24 (1977) 11, p. 6

[1039] Naka, M.; Hashimoto, K.; Masumoto, T.
Corrosion behavior of amorphous and crystalline $Cu_{50}Ti_{50}$ and $Cu_{50}Zr_{50}$ alloys
J. Non-Crystalline Solids 30 (1978) p. 29

[1040] Mushagi, A.; Rayevskaya, M. V.; Loboda, T. P.; Bodak, O. I.
Phase equilibria and corrosion properties of alloys of the Pd-Y-Cu system
Russ. Metall. (1983) 1, p. 169

[1041] Rabald, E.
Corrosion Guide, 2nd revised edition, p. 556
Elsevier Publishing Company, Amsterdam-London-New York, 1968

[1042] McDowell, D. W.
Handling phosphoric acids and phosphate fertilizers-II
Chem. Eng. 82 (1975) 18, p. 121

[1043] Anonymous
Corrosion Data Survey, Metals Section, 6th ed., p. 95
NACE, Houston (Texas/USA), 1985

[1044] Product Information
Du Pont Information Service, Corrosion Resistance Charts, April 1975, p. 296
Du Pont de Nemours International S.A., CH-1211 Geneva

[1045] Fabian, R. J.; Vaccari, J. A.
How materials stand up to corrosion and chemical attack
Mat. Eng. 73 (1971) 2, p. 36

[1046] Product Information
Wieland-Buch. Kupferwerkstoffe. Herstellung, Verarbeitung und Eigenschaften
(Wieland book "copper materials". Production, processing and properties) (in German), 4th ed., 1978, p. 179
Wieland-Werke AG Metallwerke, D-Ulm

[1047] Vaidyanath, L. R.; Shandary, V. S.
Corrosion resistance of copper and copper alloys – a survey
Proceedings of a Symposium, Bombay 1971, p. 1

[1048] Product Information
Beständigkeitstabellen von Kupfer-Gußwerkstoffen in verschiedenen Medien
(Corrosion resistance charts of copper cast alloys in different media) (in German), June 1970, p. 14
DKI Deutsches Kupfer-Institut, D-Berlin, GDM Gesamtverband Deutscher Metallgießereien, D-Düsseldorf

[1049] Berg, F. F.
Korrosionsschaubilder
(Corrosion Diagrams) (in German), 2nd ed., p. 2
VDI-Verlag GmbH, D-Düsseldorf, 1969

[1050] Swandby, R. K.
Corrosion charts: Guides to materials selection
Chem. Eng. (New York) 69 (1962) November, p. 1

[1051] Johnson, C. E.; Mullen, J. L.; Lashmore, D. S.
Corrosion tests of electrodeposited coatings in boiling 100 percent phosphoric acid
Mechanical Properties. Performance and Failure Modes of Coatings (Proc. Conf.), Gaithersburg (USA), May 1983, p. 124
Cambridge University Press, Cambridge (GB)

[1052] Kuzyukov, A. N.; Khanzadeev, I. V.
The intercrystalline corrosion of copper in acids (in Russian)
Zashch. Met. 15 (1979) 3, p. 341

[1053] Marcus L.; Ahrens, R. R.
Chemical resistance of solid materials to concentrated phosphoric acid
Am. Ceram. Soc. Bull. 60 (1981) 4, p. 490

[1054] Itou, K.; Suzuki, H.; Iida, T.; Yamada, T.
Corrosion behavior of metals – copper group and platinum group – in condensed phosphoric acid at higher temperature (in Japanese)
Nagoya Kogyo Daigaku Gakohu 25 (1973) 20, p. 377

[1055] DECHEMA-WERKSTOFF-TABELLE "Phosphorsäure"
(DECHEMA-WERKSTOFF-TABELLE "Phosphoric Acid") (in German)
DECHEMA e. V., D-Frankfurt am Main, December 1966

[1056] Anonymous
Copper Alloy No. C60800 (Cu-443)
Alloy Dig. (1982) August, p. 2

[1057] Franke, E.
Das Korrosionsverhalten von Aluminiumbronzen
(The corrosion behavior of aluminium bronzes) (in German)
Werkst. Korros. 2 (1951) 8, p. 298

[1058] Ahmad, Z.
Improvement of the mechanical properties and corrosion behavior of Cu-Al 6.0 alloy by Si and Mn addition
Werkst. Korros. 30 (1979) 6, p. 433

[1059] Katz, W.
Gießen für die Chemie. Phosphor und Phosphorsäure
(Casting in chemistry. Phosphorus and phosphoric acid) (in German), p. 58
VDI-Verlag GmbH, D-Düsseldorf, 1976

[1060] Product Information
Korrosionbeständigkeit nickelhaltiger Werkstoffe gegenüber Phosphorsäure und Phosphaten
(Corrosion resistance of nickel containing materials against phosphoric acid and phosphates) (in German), 1st ed., No. 61, March 1970, p. 1
International Nickel Deutschland GmbH, D-Düsseldorf

[1061] Product Information
Corrosion resistance of nickel – containing alloys in phosphoric acid, p. 1
The International Nickel Company, Inc., New York (USA)

[1062] Anonymous
Ampcoloy® 521 (Corrosion-resistant cast copper-nickel alloy), (Cu-408)
Alloy Dig. (1980) December, p. 2

[1063] Anonymous
Fertigungs- und Standardlieferprogramm M 3-M 56. Schrauben, Muttern, Zubehör nach DIN, ISO und EURONORM
(Production- and standard delivery programs M 3-M 56. Screws, nuts, fittings according to DIN, ISO and EURONORM) (in German)
Werkstoffe und ihre Veredlung 2 (1980) 5, p. 250

[1064] Desai, M. N.
Corrosion inhibitors for brasses
Werkst. Korros. 24 (1973) 8, p. 707

[1065] Walker, R.
Triazole, benzotriazole, and naphthotriazole as corrosion inhibitors for brass
Corrosion 32 (1976) 10, p. 414

[1066] Machu, W.; Fouad, M. G.
Über die galvanische Korrosion von Gußeisen und Messing Ms 60 in Säuren bei Ab- und Anwesenheit von Inhibitoren II
(On the galvanic corrosion of cast iron and brass MS 60 in acid during absence and presence of inhibitors II) (in German)
Werkst. Korros. 9 (1958) 11, p. 699

[1067] Haseke, H.; Köcher, R.
CuMn2 – Herstellung, Eigenschaften, schweißtechnische Verarbeitung und Anwendung
(CuMn2 – Production, properties, welding processing and use) (in German)
Metall 26 (1972) 4, p. 333

[1068] Hönig, A.; Schlenk, K. W.
Aushärtbare Federlegierungen und ihre Anwendungen
(Curable spring alloys and their use) (in German)
Chemie-Technik 6 (1977) 10, p. 409

[1069] Dechema-Werkstoff-Tabelle "Glycol und Diäthylenglycol"
(Dechema corrosion data sheets "Glycol and Diethyleneglycol") (in German)
DECHEMA, D-Frankfurt am Main, October 1958

[1070] Product Information
Corrosion resistance charts
DuPont de Nemours, Geneva, Apr. 1975

[1071] Thompson, P. F., Lorking, K. F.
Some aspects of the corrosion processes of iron copper and aluminium in ethylene glycol coolant fluids
Corrosion 13 (1957) p. 531

[1072] Beavers, J. A. et al.
Determination of interaction between different waters and chemical antifreeze additives. (Task of solar collector studies for solar heating and cooling applications)
Final Technical Progr. Report (1980) Apr., 75 p.

[1073] Beavers, J. A.; Diegle, R. B.
The effect of degradation of glycols on corrosion of metals used in nonconcentrating solar collectors
Corrosion Conf., Toronto, Apr. 1981, NACE, Houston, paper 24

[1074] Vigdorovich, V. I.; Pchelnikov, I. T.; Cygankova, L. E.
Zashch. Metallov 8 (1972) 4, p. 464 (in Russian)

[1075] Merkulov, A. V.; Naumov, Yu. I.; Flerov, V. N.
Anodic processes on copper in 5 N sodium hydroxide with additions of polyhydric alcohols (in Russian)
Izv. V. U. Z. Khim. Khim. Technol. 20 (1977) 1, p. 104

[1076] Bigeon, M. J.
Corrosion et Anti-Corrosion 4 (1956) 4, p. 139

[1077] Hirth, F. W.; Speckhardt, H.
Beeinflussung von Korrosion und Kavitations-Korrosion in Kühlmitteln durch Inhibitoren
(Inhibition of corrosion and cavitation-corrosion in coolants) (in German)
Z. Werkstofftechnik 8 (1977) 10, p. 321

[1078] White, C. M.; Ivancic, R. E.
Mixtures of dibasic alkali metal arsenates and alkali tetraborates as corrosion inhibitors for aqueous glycols
USA Pat. 2721183 (18. 10. 1955) (Genesee Research Corp.)

[1079] Corrosion inhibitors for aqueous engine coolants
Brit. Pat. 961409 (24. 6. 1964) (US Borax & Chemical Corp.)

[1080] Streatfield, E. L.
Corrosion Technology 4 (1957) 7, p. 239

[1081] Glover, T. J.; Collins, H. H.; Parkinson, G. S.
Behavior of antifreeze solutions in a simulated aluminium alloy engine test rig
Brit. Corrosion J. 2 (1967) p. 209

[1082] Corrosion inhibitor
Brit. Pat. 013707 (22.12.1965) (Monsanto Chemical Ltd.)

[1083] Butler, G.; Mercer, A. D.
Inhibitor formulations for engine coolants
Brit. Corrosion J. 12 (1977) 3, p. 171

[1084] Levy, M.
Ind. Eng. Chem. 50 (1958) 4, p. 657

[1085] Korpics, C. J.
Aromatic triazoles as corrosion inhibitors of copper and copper alloys
Anti-Corrosion 21 (1974) 3, p. 11

[1086] Ozaki, Y. et al.
Corrosion inhibitor for metals
Jap. Pat. 7416020 (19.4.1974) (Yoshitomi Pharmaceutical Industries Ltd.)

[1087] Deigle, R. B.
Study of corrosion in multimetallic systems (Task 2)
Solar collector studies for solar heating and cooling applications
(Final technical congress report)
Battelle Columbus Labs., OH, (April 1980)

[1088] Brown, P. W.
Laboratory simulated service testing of flat plate solar heat transfer liquid containment systems
NBS., Wash., Corrosion 80/102, NACE, Houston/Tx.

[1089] Kuron, D.; Rother, H.-J.; Graefen, H.
Inhibierung von wässrigen und alkoholisch-wässrigen Wärmeträgern
(Inhibition of aqueous and alcoholic-aqueous heat-carriers) (in German)
Werkstoffe u. Korrosion 32 (1981) 10, p. 409

[1090] Smith, E. F.; Castillo, A. P.
Corrosion performance of copper under simulated solar service conditions
Mater. Perform. 21 (1982) 2, p. 39

[1091] Lindner, B.
Die Korrosionsbeständigkeit metallischer Werkstoffe für Automobilbremsleitungen
(The corrosion resistance of metallic materials for breake lines of automobiles) (in German)
Werkstoffe u. Korrosion 23 (1972) p. 187

[1092] Anonymous
Copper-nickel-tubing helps lick corrosion in Volvo brake lines
Nickel Topics 29 (1976) 2, p. 3

[1093] Franke, E.
Das Korrosionsverhalten von Aluminiumbronzen
(The corrosion behavior of aluminium bronzes) (in German)
Werkstoffe u. Korrosion 2 (1951) 8, p. 298

[1094] Kropachev, V. S.; Tolstaya, M. A. et al.
Corrosion wear of ShKh15 steel and BROS-10-10 bronze in aqueous solutions of polyethylene glycol and glycerine
(in Russian)
Zashch. Metallov 16 (1980) 6, p. 721

[1095] Miska, K. H.
Wrought brass
Mater. Eng. 80 (1974) 5, p. 65

[1096] Ritter, F.
Korrosionstabellen metallischer Werkstoffe
(Tables of corrosion resistance for metallic materials) (in German)
Springer-Verlag, Wien, November 1957

[1097] Dechema-Werkstoff-Tabelle "Glycerin"
(Dechema corrosion data sheets "Glycerin") (in German)
DECHEMA, D-Frankfurt am Main, October 1958

[1098] Rabald, E.
Corrosion Guide, 2nd. ed. (1968) p. 337
Elsevier Publishing Company, Amsterdam-London-New York, 1968

[1099] Vigdorovič, V. I.; Cygankova, L. E.; Merkulova, R. I.
Aggressivity of acid glycerol solutions
(in Russian)
Zashch. Metallov 8 (1972) 3, p. 329

[1100] Stahl, R.; Kiefer, P.
Korrosion und Korrosionsschutz in der Lebensmittelindustrie
(Corrosion and corrosion protection in the food industry)
Werkstoffe u. Korrosion 24 (1973) 6, p. 513

[1101] Ried, G.
Übersicht über Werkstoffe für Absperr-Organe, die bei angreifenden Medien zu empfehlen sind
(Survey on materials for shut-off devices to be recommended in contact with aggressive media) (in German)
Werkstoffe u. Korrosion 15 (1964) 6, p. 469, esp. p. 476

[1102] Wyllie, D.; Morgan, A. W.
Prevention of corrosion in glycerol-water hydraulic fluids
J. Appl. Chem. 15 (1965) p. 289

[1103] Afendik, K. F.; Kritschmar, S. I.
Viscous electrolyte for electropolishing copper (in Russian)
Elektrochimija 6 (1970) p. 1

[1104] Axtell, W. G.
Electrolytic polishing
USA Pat. 2645611 (14.7.1953) (Shwayder Brothers Inc.)

[1105] Rabald, E.
Corrosion Guide, 2nd ed. (1968) p. 766
Elsevier Publishing Company, Amsterdam-London-New York, 1968

[1106] Freifeld, M.; Hort, E. V.
in: Kirk-Othmer Encyclopedia of Chemical Technology
2nd ed. 10 (1966) p. 667
John Wiley & Sons, Inc., New York-London-Sydney, 1966

[1107] Dechema-Werkstoff-Tabelle "Butandiole"
(Dechema corrosion data sheets "Butanediols") (in German)
DECHEMA, D-Frankfurt am Main, 1955

[1108] Dechema-Werkstoff-Tabelle "Butin-2-diol-1,4"
(Dechema corrosion data sheets "Butine-2-diol-1.4") (in German)
DECHEMA, D-Frankfurt am Main, 1955

[1109] White, C. M.; Ivancic, R. E.
Mixtures of dibasic alkali metal arsenates and alkali tetraborates as corrosion inhibitors for aqueous glycols
USA Pat. 2721183 (18.10.1955) (Genesee Research Corp.)

[1110] Hamner, N. E.
Corrosion data survey, 5th ed., p. 146
National Association of Corrosion Engineers, Houston (Texas), 1974

[1111] Butts, A.; Giacobbe, J.
Silver offers resistance to many chemicals
Chemical & Metallurgical Engineering 48 (1941) p. 76

[1112] Rabald, E.
Corrosion Guide, 2nd revised ed., p. 156
Elsevier Publishing Company, Amsterdam-London-New York, 1968

[1113] Maaß, E.; Wiederholt, W.
Prüfung von Metallen und Metallegierungen auf ihre Widerstandsfähigkeit gegen die

Einwirkung von Salzlaugen. 2. Mitteilung: Einwirkung der festen Salze auf Metalle und Legierungen
(The testing of metals and metal alloys for their resistance on exposure to salt solutions. 2nd report: Effect of solid salts on metals and alloys) (in German)
Korrosion und Metallschutz 6 (1930) 11, p. 240

[1114] Product Information
Beständigkeitstabellen von Kupfer-Gußwerkstoffen in verschiedenen Medien
(Charts of chemical resistance of copper-cast materials in various media) (in German)
Deutsches Kupfer-Institut, D-Berlin, Düsseldorf, June 1970

[1115] Rauch, A.; Kolb, H.
Über die Korrosion von Kupfer und hochkupferhaltigen Legierungen durch Salzlösungen unter besonderer Berücksichtigung der Verhältnisse in der Kaliindustrie
(On the corrosion of copper and high-copper alloys by salt solutions with special regard to the conditions in the potash industry) (in German)
Korrosion und Metallschutz 6 (1930) p. 127

[1116] Rauch, A.; Kolb, H.
Über die Korrosion von Kupfer und hochkupferhaltigen Legierungen durch Salzlösungen unter besonderer Berücksichtigung der Verhältnisse in der Kaliindustrie
(On the corrosion of copper and high-copper alloys by salt solutions with special regard to the conditions in the potash industry) (in German)
Korrosion und Metallschutz 6 (1930) p. 151

[1117] Rauch, A.; Kolb, H.
Über die Korrosion von Kupfer und hochkupferhaltigen Legierungen durch Salzlösungen unter besonderer Berücksichtigung der Verhältnisse in der Kaliindustrie
(On the corrosion of copper and high-copper alloys by salt solutions with special regard to the conditions in the potash industry) (in German)
Korrosion und Metallschutz 6 (1930) p. 171

[1118] Rauch, A.; Kolb, H.
Über die Korrosion von Kupfer und hochkupferhaltigen Legierungen durch Salzlösungen unter besonderer Berücksichtigung der Verhältnisse in der Kaliindustrie
(On the corrosion of copper and high-copper alloys by salt solutions with special regard to the conditions in the potash industry) (in German)
Korrosion und Metallschutz 6 (1930) p. 193

[1119] Markovic T.; Jaric, O.
Der Einfluß des Lichtes auf die Löslichkeit von Kupfer in Elektrolyten
(The effect of light on the solubility of copper in electrolytes) (in German)
Werkst. Korros. 6 (1955) 11, p. 535

[1120] Franco, J.; Patel, N. K.
Photocorrosion of copper and brass in aqueous solutions of copper and ferric chlorides
J. Electrochem. Soc. India 17 (1978) 3, p. 161

[1121] Gurovich, E. I.
Reaction of molten lithium, sodium, and potassium chlorides with nickel, copper and some steels (in Russian)
Zh. Prikl. Khim. 27 (1964) p. 425
(C.A. 48 (1954) 13590d)

[1122] Franke, E.
Das Korrosionsverhalten von Aluminiumbronzen
(The corrosion behavior of aluminium bronzes) (in German)
Werkst. Korros. 2 (1951) 8, p. 298

[1123] Leogrande, A.; Jung-König, W.; Wincierz, P.
Temperaturabhängigkeit der mechanischen Eigenschaften, Gefügeausbildung und Stapelfehlerenergie einer arsenhaltigen Aluminiumbronze des Typs CuAl5 (AlBz5)
(Temperature dependence of the mechanical properties, structural constitution and stacking fault energy of an arsenic-containing aluminium bronze of the type CuAl5 (AlBz5)) (in German)
Metall 21 (1967), 2, p. 102

[1124] Weiland, H.; Paul, M.; Beyer, B.
Zum Dauerfestigkeits- und Schwingungsrißkorrosionsverhalten von CuAl5
(On the fatigue and corrosion fatigue behavior of the alloy CuAl5) (in German)
Neue Hütte 26 (1981) 12, p. 464

[1125] Maaß, E.; Wiederholt, W.
Prüfung von Metallen und Metallegierungen auf ihre Widerstandsfähigkeit gegen die Einwirkung von Salzlaugen. 3. Mitteilung: Das Verhalten der Metalle und Legierungen in wäßrigen Salzlaugen
(The testing of metals and metal alloys for their resistance on exposure to salt solutions. 3rd report: The behavior of metals and alloys in aqueous salt solutions) (in German)
Korrosion und Metallschutz 6 (1930) 12, p. 265

[1126] Vaidyanath, L. R.; Bhandary, V. S.
Corrosion resistance of copper and copper alloys – A survey Proceeding of a Symposium, Bombay, 1971, p. 1

[1127] Katz, W.
Die Auflösungsgeschwindigkeit von Kupfer in verschiedenen Salzlösungen und der Mechanismus der Kupferkorrosion
(The dissolution rate of copper in various salt solutions and the mechanism of copper corrosion) (in German)
Werkst. Korros. 1 (1950) 10, p. 393

[1128] Maaß, E.; Wiederholt, W.
Prüfung von Metallen und Metallegierungen auf ihre Widerstandsfähigkeit gegen die Einwirkung von Salzlaugen. 1. Mitteilung: Feststellung des Einflusses der verschiedenen Versuchsbedingungen
(The testing of metals and metal alloys for their resistance on exposure to salt solutions. 1rst report: The influence of the various test conditions) (in German)
Korrosion und Metallschutz 6 (1930) 10, p. 218

[1129] Franco, J.; Patel, N. K.
Photocorrosion of copper and brass in aqueous solutions of copper and ferric chlorides
J. Electrochem. Soc. India 17 (1978) 3, p. 161

[1130] Desai, C. S.; Shah, R. S.
Corrosion of brass in solutions of alkali metal salts of various anions
J. Elektrochem. Soc. India 30 (1981) 4, p. 267

[1131] Anonymous
Alkalis can cause unexpected problems
Canad. Chem. Process 53 (1969) 9, p. 73

[1132] Gurovich, E. I.
Reactions of melted lithium, sodium and potassium hydroxide with nickel, copper, iron and steel (in Russian)
Zh. Prikl. Khim. 32 (1959), p. 817

[1133] Kolenko, Z.
Reinigen von Metalloberflächen mit Natriumgluconat-Lösungen
(in Hungarian)
Korróziós Figyelö 14 (1974) 2, p. 54

[1134] Martschewska, M.; Raitschewski, G.; Petrow, C.
Färben von Kupfer – Elektrochemische Untersuchungen an mit Wechselstrom oxidiertem Kupfer
(Coloring of copper – electro-chemical investigations on with alternating current oxidized copper) (in German)
Metalloberfläche 37 (1983), p. 20

[1135] Ahmad, Z.
Improvement of the Mechanical Properties and Corrosion Behaviour of Cu-6,0Al Alloy by Si and Mn Addition
Werkst. Korros. 30 (1979), p. 433

[1136] Bonora, P. L.; Ponzano, G. P.; Bassol, M.
Anodisches Verhalten von Kupferlegierungen in Alkalilösungen
(Anodic behavior of copper alloys in alkali solutions)
Ann. Chim. ital. 65 (1975), p. 677

[1137] Palmer, J. D.
Alkalis Can Cause Unexpected Problems
Canad. Chem. Processing 53 (1969) 9, p. 7

[1138] Pishchulin, V. N.; Kosintsev, V. I.
Korrosion der Elektroden- und Bauwerkstoffe beim Eindampfen von Natriumhydroxidlösungen in direkt elektrisch beheizten Apparaten
(Corrosion of the electrode and building materials when the evaporation sodium hydroxide solutions in directly electrically heated apparatuses) (in Russian)
Izv. Tomsk. Politekhn. In.-Ta (1976) 275, p. 90

[1139] Shah, R. S.; Trivedi, A. M.
NaDDC as a Corrosion Inhibitor of Brass
Werkst. Korros. 25 (1974), p. 521

[1140] Hönig, A.; Schlenk, K. W.
Aushärtbare Federlegierungen und ihre Anwendungen
(Hardenable spring alloys and their applications) (in German)
Chem.-Techn. 6 (1977), p. 409

[1141] Wendler-Kalsch, E.
Korrosionsverhalten und
Deckschichtbildung auf Kupfer und
Kupferlegierungen
(Corrosion behavior and covering layer
formations on copper and copper alloys)
(in German)
Z. Werkstofftech. 13 (1982) 4, p. 129–137

[1142] Gilbert, P. T.
A review of recent work on corrosion
behavior of copper alloys in sea water
Mater. Performance 21 (1982) 21, p. 47–63

[1143] DIN 1787 (01/1973)
Kupfer – Halbzeug
(Copper – half-finished products)
(in German)
Beuth Verlag GmbH, D-Berlin

[1144] Pötzschke, M.
Schiff und Hafen 33 (1981), p. 92

[1145] DIN 50916-1 (08/1976)
Prüfung von Kupferlegierungen;
Spannungsrißkorrosionsversuch mit
Ammoniak, Prüfung von Rohren, Stangen
und Profilen
(Testing of copper alloys; stress corrosion
cracking test in ammonia, testing of tubes,
rods and profiles) (in German)
Beuth Verlag GmbH, D-Berlin

[1146] DIN 50916-2 (09/1985)
Prüfung von Kupferlegierungen;
Spannungsrißkorrosionsprüfung mit
Ammoniak; Prüfung von Bauteilen
(Testing of copper alloys; stress corrosion
cracking test using ammonia; testing of
components) (in German)
Beuth Verlag GmbH, D-Berlin

[1147] Efird, K. D.
Effect of fluid dynamics on the corrosion of
copper-base alloys in sea water
Corrosion 33 (1977) 1, p. 3

[1148] DIN 81249-2 (11/1997)
Korrosion von Metallen in Seewasser und
Seeatmosphäre; Teil 2: Freie Korrosion in
Seewasser
(Corrosion of metals in sea water and sea
atmosphere; Part 2: Free corrosion in
seawater) (in German)
Beuth Verlag GmbH, D-Berlin

[1149] Turnbull, J. M.
Salt refinery corrosion caused by varying
evaporator conditions
Corrosion 16 (1960) 7, p. 11

[1150] Cohen, A.; George, P. F.
Copper alloys in the desalting environment
Mater. Protection 12 (1973) 2, p. 38

[1151] Mor, E. D.; Beccaria, A. M.
Effects of temperature on the corrosion of
copper in sea water at different hydrostatic
pressures
Werkst. Korros. 30 (1979) 8, p. 554

[1152] Seawater Corrosion Handbook
Noyes Data Corporation, New Jersey
(USA), 1979

[1153] Pircher, H.; Pennenkamp, R.
Clad plates in the energy industry –
properties and applications
Stahl Eisen 102 (1982) 12, pp. 619–624

[1154] Pircher, H.; Ruhland, B.; Sussek, G.
Einsatz CuNi10Fe plattierter Bleche für
Schiffs- und Bootskörper
(Use of CuNi10Fe plated sheetmetals for
hulls) (in German)
Thyssen Tech. Ber. (1985) 3, pp. 1–11

[1155] Drodten, P.; Pircher, H.
Behavior of CuNi10Fe1Mn-clad ship plates
in seawater
Werkst. Korros. 41 (1990) 2, pp. 59–64

[1156] DIN 17665 (12/1983)
Kupfer-Knetlegierungen – Kupfer-
Aluminium-Legierungen
(Aluminiumbronze) – Zusammensetzung
(Wrought copper alloys; copper-aluminium
alloys; (aluminium bronze); composition)
(in German)
Beuth Verlag GmbH, D-Berlin

[1157] BWB WL 2.0957 (09/2005)
Kupfer-Aluminium-Gußlegierung –
G-CuAl8Mn8 – Gußstücke
(Copper-aluminium cast alloys –
G-CuAl8Mn8 – castings) (in German)
Werkstoff-Handbuch der Wehrtechnik

[1158] BWB WL 2.0958 (09/2005)
Kupfer-Aluminium-Knetlegierung –
CuAl8Mn – Stangen und
Freiformschmiedestücke
(Copper-aluminium wrought alloys –
CuAl8Mn – poles and smith hammer
forgings) (in German)
Werkstoff-Handbuch der Wehrtechnik

[1159] DIN EN 1982 (12/1998)
Kupfer und Kupferlegierungen –
Blockmetalle und Gußstücke
(Copper and copper alloys – Ingots and
castings) (in German)
Beuth Verlag GmbH, D-Berlin

[1160] BWB WL 2.0967 (09/2005)
Kupfer-Aluminium-Knetlegierung –
CuAl9Ni7 – Stangen und
Freiformschmiedestücke
(Copper-aluminium wrought alloys –
CuAl9Ni7 – Bars and forgings)
(in German)
Werkstoff-Handbuch der Wehrtechnik

[1161] BWB WL 2.0968 (09/2005)
Kupfer-Aluminium-Gußlegierung –
G-CuAl9Ni7 – Gußstücke
(Copper-aluminium cast alloy –
G-CuAl9Ni7 – castings) (in German)
Werkstoff-Handbuch der Wehrtechnik

[1162] BWB WL 2.1357 (06/1997)
Kupfer-Mangan-Aluminium-Gußlegierung
– G-CuMn9Al6Zn4Ni2Fe2 – Gußstücke
(Copper-manganese-aluminium cast alloy
– G-CuMn9Al6Zn4Ni2Fe2 – castings)
(in German)
Werkstoff-Handbuch der Wehrtechnik

[1163] BWB WL 2.1358 (06/1997)
Kupfer-Mangan-Aluminium-Gußlegierung
– G-CuMn10Zn8Al6Ni2Fe2 – Gußstücke
(Copper-manganese-aluminium cast alloy
– G-CuMn10Zn8Al6Ni2Fe2 – castings)
(in German)
Werkstoff-Handbuch der Wehrtechnik

[1164] Ferrara, R. J.; Caton, T. E.
Review of dealloying of cast aluminum
bronze and nickel-aluminum bronze alloys
in sea water services
Mater. Performance 21 (1982) 2, pp. 30–34

[1165] Kennworthy, L.
Inst. Marine Eng. (1965) June, pp. 149–173

[1166] Upton, B.
Corrosion resistance in sea water of
medium strength aluminium bronzes
Corrosion 19 (1963) 6, pp. 205–209

[1167] Baier, J.
Mehrstoff-Aluminium-Bronzen –
Chemische Beständigkeit und
Verwendung
(Multiple substance-aluminium bronzes –
chemical resistance and use) (in German)
Chem.-Ing.-Tech. 40 (1968) 8, pp. 365–370

[1168] Goldspiel, L.; Kershner, J.; Wacker, G.
Modern Castings 47 (1965) January,
pp. 818–823

[1169] Amaud, D.; Paten, R.; Wigy, S.; Mascre, C.
Dealuminization of copper-aluminium
alloys
Fonderie (1964) 226, pp. 403–430; n. C. A.
62 (1965) 10, 11496c

[1170] Anonymous
Inco Nickel Topics 17 (1964) 6, p. 8

[1171] Ahmad, Z.
Improvement of the mechanical properties
and corrosion behavior of Cu-Al 6.0 alloy
by Si and Mn addition
Werkst. Korros. 30 (1979) 6, p. 433

[1172] Ahmad, Z.
Effect of tin addition on the corrosion
resistance of aluminium bronze
Anticorros. Methods Mater. 24 (1977) 1,
p. 8

[1173] Ahmad, Z.
Corrosion resistance of cast aluminium
bronzes containing chromium and silicon
additions
Br. Corros. J. 11 (1976) 3, p. 149

[1174] Schiffrin, D. J.; De Sanchez, S. R.
The effect of pollutants and bacterial
microfouling on the corrosion of copper-
base alloys in sea water
Corrosion 41 (1985) 1, pp. 31–38

[1175] Vanick, J. S.
Nickel-Aluminium-Bronze für
Schiffsschrauben in arktischen Gewässern
(Nickel-aluminium bronze for ship
propellers in arctic water) (in German)
Schiff und Hafen 8 (1956) 10, p. 843

[1176] Feller, H. G.
Untersuchungen der Kavitationskorrosion
an Pumpenwerkstoffen in Meerwasser
(Tests of the cavitation corrosion in pump
materials in seawater) (in German)
Werkst. Korros. 33 (1982) 5, p. 288

[1177] Moore, R. E.
Materials for water desalting plants.
Part 11: Aluminium and alloys, copper
base alloys
Chem. Eng. 70 (1963) 20, p. 124 and 21,
p. 244

[1178] Swales, G. W.
Corrosion problems of the petroleum
industry
S. C. I. Monograph No. 10, p. 203–204

[1179] Basil, J. L.; Wehlan, J. S.
U. S. Nav. Eng. Exper. Stat., Rep. 91 0027 A,
March 1958

[1180] Bird, B.; Moore, K. L.
Brackish cooling water versus refinery heat
exchangers
Mater. Protection 1 (1962) 10, pp. 71–75

[1181] Heitz, E.
Werkstoff- und Korrosionsprobleme bei der Süßwassergewinnnung
(Material- and corrosion problems in fresh water production) (in German)
Chem.-Ing.-Tech. 41 (1969), p. 98

[1182] Otsu, T.
Sumitomo Light Metal, Tech. Rep. 6 (1965) October, pp. 17–45

[1183] Yamaguchi, H.
KMC alloy for condenser tubes
Kinzoku 26 (1956) pp. 782–784; n. C. A. 55 (1961) 1, 313 a)

[1184] Lambert; Sanekès; Lickeran
L'energie maremotrice. Protection of the ouvrages metalliques contre la corrosion marine
Corrosion Anti-corrosion 13 (1965) 2, pp. 81–92

[1185] Inco Nickel Topics 16 (1963) 8, p. 4

[1186] Cohen, A.; George, P. F.
Copper alloys in the desalting environment – final report
Mater. Performance 13 (1974) 8, p. 26

[1187] White, P. E.
Corros. Prev. Control 13 (1966) April, p. 24

[1188] Walston, K. R.
Materials problems in salt water cooling towers
Mater. Performance 14 (1975) 6, p. 22

[1189] Huber, H.
Wirtschaftlich konstruieren mit Kupfer-Aluminium-Legierungen
(Construct economically with copper-aluminium alloys) (in German)
Werkstatt und Betrieb 118 (1985) 11, p. 114

[1190] Tischner, H.
Vergleichende Korrosionsuntersuchungen an Gußbronzen in künstlichem Meerwasser mit variierenden Ammonium- und Sulfidgehalten unter Berücksichtigung der Strömung
(Comparing corrosion tests of cast bronzes in synthetic seawater with varying ammonium- and sulfide contents under consideration of the current) (in German)
in: Behrens, D.; Rahmel, A. "Ergebnisse des Forschungs- und Entwicklungsprogramms 'Korrosion und Korrosionsschutz'", vol. 4 (1983–1986), pp. 125–129
DECHEMA e. V., D-Frankfurt am Main, 1987

[1191] DIN 17664 (12/1983)
Kupfer-Knetlegierungen – Kupfer-Nickel-Legierungen – Zusammensetzung
(Wrought copper alloys; copper-nickel alloys; composition) (in German)
Beuth Verlag GmbH, D-Berlin

[1192] LaQue, F. L.
J. Am. Soc. of Naval Eng. 33 (1941) February 1, p. 29

[1193] Crousier, J.; Beccaria, A.-M.
Behavior of Cu-Ni alloys in natural sea water and NaCl solution
Werkst. Korros. 41 (1990) 4, p. 185

[1194] Syrett, B. C.; MacDonald, D. D.
The validity of electrochemical methods for measuring corrosion rates of copper-nickel alloys in sea water
Corrosion 33 (1979) 11, p. 505

[1195] Beccaria, A.-M.; Crousier, J.
Dealloying of Cu-Ni alloys in natural sea water
Br. Corros. J. 24 (1989) 1, p. 49

[1196] Fink, F. W.
Corrosion of metals in seawater
U. S. Dep. Interior, Office of Saline Water, Res. and Dev. Progr. Report No. 46, December 1960, p. 19

[1197] Efird, K. D.
Potential-pH diagrams for 90-10 and 70-30 CuNi in sea water
Corrosion 31 (1975) 3, p. 77

[1198] Barbery, J.; Pepin-Donat, C.
Rohre aus Kupferlegierungen für Wärmetauscher – Korrosionsverhalten sandgestrahlter Oberflächen in Meerwasser
(Pipes made of copper alloys for heat exchangers – corrosion behavior of sandblasted surfaces in seawater) (in German)
Metall 36 (1982) 11, p. 1181

[1199] Tracy, A. W.; Hungerford, R. L.
Proc. Amer. Soc. Test. Mater. 45 (1945), p. 391

[1200] LaQue, F. L.; Mason jr., J. F.
15th Mid Year Meeting, Am. Petrol. Inst. Dev. Refining, Cleveland (Ohio/USA) May 1950

[1201] Bailey, G. L.
J. Inst. Met. 79 (1951), p. 243

[1202] Gopius, A. E.; Molchanova, V. P.
Investigation of impact corrosion of cupronickel condenser tubes and elaboration of a more stable alloy

Trudy Gosudarst. Nauchn. Issledovatel. i. Proekt. Inst. po Obrab. Tsvetn. Metal. 18 (1960), p.127; (C. A. 55 (1961) 17, 16359c)

[1203] Stewart, W. C.; LaQue, F. L.
Corrosion 8 (1952) 8, p. 259

[1204] Copson, H.
Effects of velocity on corrosion
Corrosion 16 (1960) 2, pp. 86t–92t

[1205] Kievits, F. J.; Ijsseling, F. P.
Research into the corrosion behavior of CuNi10Fe alloys in seawater
Werkst. Korros. 23 (1972) 12, p. 1084

[1206] Richter, H.; Rockel, B.; Rudolph, G.
Kupferlegierungen und Titan in Meerwasser
(Copper alloys and titanium in seawater) (in German)
Schiff und Hafen 2 (1983), special edition "Seawater-corrosion", pp. 29–34

[1207] Drolenga, L. J. P.; Ijsseling, F. P.; Kolster, B. H.
The influence of alloy composition and microstructure on the corrosion behavior of Cu-Ni alloys in seawater
Werkst. Korros. 34 (1983) 4, p. 167

[1208] Badia, F. A.; Kirby, G. N.; Mikalisin, J. R.
Strengthening of annealed cupronickels by chromium
ASM Trans Quarterly 60 (1967) 3, pp. 395–408

[1209] Product information
Copper-Nickel-Alloy IN-732, Inco-Bulletin 2 C, 10-71, January 1969
International Nickel Co., Inc., New York (N.Y./USA)

[1210] Anderson, D. B.; Badia, F. A.
Spring Meeting ASME Research Committee, Heat Exchanger Tubes, ASME Trans., 1971

[1211] Gopius, A. E.
Alloys for seawater pipelines
Trudy Gosudarst., Nauchn.-Issled., i. Proekt. Inst. Obrab. Tsvetn. Metal. (1960) 18, pp. 163–175; n. C. A. 55 (1961) 20, 19701b)

[1212] Nothing, F. W.
Kupfer-Nickel Legierungen unter 50 % Nickel, Information No. 7, 1958, p. 39
(Copper-nickel alloys under 50 % nickel) (in German)
Nickel-Informationsbüro GmbH, D-Düsseldorf

[1213] Ind. Eng. Chem. 51 (1959) 9, p. 1165

[1214] Metallurgia, Manchr. 72 (1965) September, pp. 114–118

[1215] Schoffs, J.
Centre Belge Inform. Cuivre (1963) 11 p.; n. C. A. 61 (1964) 10, 11706h

[1216] Katz, W.
Fortschritte auf dem Werkstoffgebiet – gezeigt auf der 17. ACHEMA 1973, Teil I: Metallische Werkstoffe
(Progress in the field of materials – shown on the 17. ACHEMA 1973, Part I: Metallic materials) (in German)
Werkst. Korros. 24 (1973) 9, p. 790

[1217] Cohen, A.; Rice, L. V.
Experience with copper alloys in the desalting environment
Mater. Protection Performance 9 (1970) 11, pp. 29–35

[1218] Eichhorn, K.
Kondensatorrohre aus Kupferwerkstoffen. Richtlinien für Werkstoffauswahl und Betriebsbedingungen
(Condenser pipes made of copper materials. Guidelines for the selection of materials and operating conditions) (in German)
Werkst. Korros. 21 (1970) 7, p. 536

[1219] Beccaria, A. M.; Wang, Y. Z.; Poggi, G.
Effect of temperature on localized corrosion of Cu-Ni alloy in sea water
Eurocorr, European Corrosion Meeting (1994) vol. 2, pp. 201–212
Chameleon Press, London

[1220] Ijsseling, F. P.; Drolenga, L. J.; Kolster, B. H.
Corrosion rate as a function of temperature in well aerated sea water
Br. Corros. J. 17 (1982) 4, pp. 162–167

[1221] Rothmann, B.; Ebert, S.; Hoffmann, B.; Böhm, H.
Untersuchungen zur Korrosion und zum Korrosionsschutz von meerwassergekühlten Rohrkondensatoren aus Kupferbasislegierungen – Versuche mit rotierenden Proben
(Tests on the corrosion and corrosion protection of seawater cooled pipe condensers made of copper-based alloys – tests with rotating samples) (in German)
Werkst. Korros. 34 (1983) 12, p. 583

[1222] Todd, B.
Marine Corrosion Conference, Biarritz, June 1969
Sea Horse Institut, International Nickel Comp., contribution No. 3

[1223] Syrett, B. C.
Erosion corrosion of copper-nickel alloys in sea water and other aqueous environments – a literature review
Corrosion 32 (1976) 8, p. 242

[1224] Kirk, W. W.
Paper No. 4
Marine Corrosion Conference, Biarritz, 1968

[1225] Glover, T. J.
Copper-nickel alloy for the construction of ship and boat hulls
Br. Corros. J. 17 (1982) 4, pp.155–158

[1226] Tuthill, H. H.
CA 706. Copper nickel alloy hulls – the copper mariners experience and economics
Soc. Naval Architects & Marine Engineering Conference, New York, 11.–13. November 1976
International Nickel Revue (1971) 1, p. 22
Nickel New. (1971) November, p. 12

[1227] Glover, T. J.; Moreton, B. B.
Corrosion and fouling resistance of cupronickel in marine environments
Ind. Corros. 1 (1983) 3, pp. 11–14, 17

[1228] Gilbert, P. T.
Corrosion resisting properties of 90 10 copper-nickel-iron alloy with particular reference to offshore oil and gas applications
Br. Corros. J. 14 (1979) 1, p. 20

[1229] DIN EN 12451 (10/1999)
Kupfer und Kupferlegierungen – Nahtlose Rundrohre für Wärmeaustauscher
(Copper and copper alloys – seamless, round tubes for heat exchangers) (in German)
Beuth Verlag, D-Berlin

[1230] Southwell, C. R. et al.
Naval Research Lab. Rep. NRL 6592, October 1967, p. 14

[1231] Richter, H.
Kupfer-, Nickel- und Titanlegierungen für Seewasseranwendungen
(Copper-, nickel- and titanium alloys for seawater applications) (in German)
Werkst. Korros. 28 (1977) 10, p. 671

[1232] Tiktak, A.
Zehn Jahre Betrieb des Kernkraftwerkes Borssele
(Ten years' running of the nuclear power station Borssele) (in German)
VGB Kraftwerktechnik 65 (1985) 1, pp. 1–6

[1233] Powell, C. A.
Corrosion and biofouling resistance of copper-nickel in offshore and other marine applications
Eurocorr, European Corrosion Meeting (1994) vol. 2, pp. 118–200
Chameleon Press, London

[1234] Ahmad, Z.; Aleem, B. J. A.
The corrosion performance of 90-10 cupronickel in Persian Gulf water containing ammonia
Desalination 9 (1994) 3, pp. 307–323

[1235] Thomas, E. D.; Lucas, K. E.; Peterson, M. H.; Christian, D. K.
Effects of electrolytic chlorination on marine materials
Mater. Performance 27 (1988) 7, p. 36

[1236] Francis, R.
The effects of chlorine on the properties of films on copper alloys in sea water
Corros. Sci. 26 (1986) 3, p. 205

[1237] Francis, R.
Effect of pollutants on corrosion of copper alloys in sea water. I. Ammonia and chlorine
Br. Corros. J. 20 (1985) 4, pp. 167–175

[1238] Syrett, B. C.; Wing, S. S.
Effect of flow on corrosion of copper-nickel alloys in aerated sea water and in sulfide-polluted sea water
Corrosion 36 (1980) 2, pp. 73–84

[1239] Al-Hajji, J. N.; Reda, M. R
The corrosion of copper-nickel alloys in sulfide-polluted seawater. The effect of sulfide concentration
Corros. Sci. 34 (1993) 1, pp. 163–177

[1240] Syrett, B. C.
Protection of copper alloys from corrosion in sulfide polluted sea water
Mater. Performance 20 (1981) 5, p. 50

[1241] Giuliani, L.; Bombara, G.
Influence of pollution on the corrosion of copper alloys in flowing salt water
Br. Corros. J. 8 (1973) 1, p. 20

[1242] Habib, K.; Husain, A.
Stress corrosion cracking of copper-nickel alloys in sulfide polluted natural seawater

at moderate temperatures
Desalination 97 (1994) 1–3, pp. 29–34

[1243] Oakes, B. D.
Inhibition of corrosion under desalination conditions. 5. Corrosion rates on copper alloys with a chromate-phosphate inhibitor
Mater. Performance 13 (1976) 1, p. 44

[1244] Logan, D. P.; Rey, S. P.
Scale control in multiple stage flash evaporators
Mater. Performance 23 (1986) 6, p. 38

[1245] Hack, H. P.; Gudas, J. P.
Inhibition of sulfide-induced corrosion of copper-nickel alloys with ferrous sulfate
Mater. Performance 18 (1979) 3, p. 25

[1246] Tseng, M. D.; Shih, H. C.
The corrosion behavior of 90Cu-10Ni in seawater
Corrosion Prev. Control 41 (1994) 1, pp. 19–21

[1247] Lee, T. S.
The use of sodium dichromate pretreatment for enhanced marine corrosion resistance of C70600
Corrosion 39 (1983) 9, pp. 371–376

[1248] DIN CEN/TS 13388 (10/2004)
Kupfer und Kupferlegierungen – Übersicht über Zusammensetzung und Produkte
(Copper and copper alloys – Compendium of compositions and products) (in German)
Beuth Verlag GmbH, D-Berlin

[1249] Southwell, C. R. et al.
Naval Research Lab. Rep. NRL 6452, Washington, 28.10.1966

[1250] Product Information
Beständigkeitstabellen von Kupfer-Gußwerkstoffen
(Resistance tables of copper cast materials) (in German) June 1960, p. 56
DKI and GDM, D-Düsseldorf

[1251] Sato, S.
Report, 4. Intern. Congr. Metal Corrosion, Amsterdam (NL), September 1969
Zinn und seine Verwendung (1971) 88, p. 14

[1252] Kupfer (1971) 4, p. 10

[1253] Kupfer (1970) 1, pp. 23–25

[1254] Corros. Prev. Control 7 (1960) September, pp. 35–36

[1255] DIN EN 1982 (12/1998)
Kupfer und Kupferlegierungen – Blockmetalle und Gußstücke
(Copper and copper alloys – Ingots and castings) (in German)
Beuth Verlag GmbH, D-Berlin

[1256] Gusmano, G.; Cigna, R.
Effect of solution heat-treatment on the corrosion behavior of an Al-Ti brass (TIBRAL)
Br. Corros. J. 11 (1976) 2, p. 100

[1257] Mattson, E.; Svensson, L.
Corrosion investigation of valves in flowing sea water
Br. Corrosion J. 7 (1972) 5, p. 200

[1258] Schleithoff, K.; Schmitz, F.
Entzinkung von Aluminiummessing (CuZn20Al)
(Dezincification of aluminium brass (CuZn20Al)) (in German)
Praktische Metallographie 20 (1983) 2, pp. 88–91

[1259] Moore, R. E.
Material for water desalting plants
Chem. Eng. 70 (1963) 21, p. 226

[1260] Product information
Hochleistungswerkstoffe, Information N 5091 87-04
(High-performance materials) (in German)
Vereinigte Deutsche Metallwerke AG, D-Werdohl

[1261] May, T. P.; Holmberg, E. G.; Hinde, J.
Seawater corrosion at atmospheric and elevated temperatures
Dechema Monographie 47 (1962), p. 253

[1262] Haferkamp, H.; Hueper, H.; Louis, H.; Tai, P. T.
Corrosion behavior of copper alloys after short time loading by solid particles in sea water
Corrosion 90: Flow-induced corrosion: Fundamental studies and industry experience

[1263] Beccaria, A. M.; Poggi, G.
Behavior of aluminium brass in sea water at various temperatures
Br. Corros. J. 23 (1988) 2, p. 122

[1264] Luklinska, Z. B.; Castle, J. E.
Microstructural study of initial corrosion product of aluminium brass alloy after exposure to natural seawater
Corros. Sci. 23 (1983) 11, p. 1163

[1265] Habib, K.; Amin, K.
Electrochemical behavior of aluminium brass in natural seawater
Desalination 85 (1992) 3, pp. 275–282

[1266] Böhm, H.
Entwicklung wirksamer Maßnahmen zum Schutz von meer- oder brackwassergekühlten Kondensatoren, deren Rohre aus Kupferbasislegierungen infolge ungenügender eigener Schutzschichtbildung Schaden nehmen können
(Development of effective measurements for the protection of seawater – or brackish water cooled condensers, whose pipes made of copper-based alloys could be damaged because of insufficient own production of protective layers)
(in German)
Werkst. Korros. 33 (1982) 11, p. 622

[1267] Baumann, G.
Zerstörung von Messingrohren durch Spannungskorrosion
(Destruction of brass pipes by stress corrosion cracking) (in German)
Werkst. Korros. 13 (1962), p. 737

[1268] Szabò, S.
Spannungskorrosionsversuche an Kondensatorrohren
(Stress corrosion tests on condenser pipes) (in German)
Werkst. Korros. 14 (1963), p. 165

[1269] Carrasiti, F.; Cigna, R.; Gusmano, G.; Calvarano, M.
Electrochemical behavior of metal inserts galvanically coupled to aluminium brass tubes
Br. Corros. J. 21 (1986) 2, p. 129

[1270] Fabian, R. J.
Mater. Design. Eng. 57 (1963) January, Rep. No. 202

[1271] Finnerty, J. J.
Oil & Gas J. 58 (1960) February, p. 93

[1272] Klement, J. F.; Maersch, R. E.
Metal Progr. 75 (1959) February

[1273] Trethewey, K. R.; Haley, T. J.; Dark, C. C.
Effect of ultrasonically induced cavitation on corrosion behavior of a copper-manganese-aluminium alloy
Br. Corros. J. 23 (1988) 1, p. 55

[1274] Katz, W.
Die Auflösungsgeschwindigkeit von Kupfer in verschiedenen Salzlösungen und der Mechanismus der Kupferkorrosion
Werkst. Korros. 1 (1950) p. 398

[1275] Claus, W.; Herrmann, J.
Metallkunde 30 (1939) p. 58

[1276] Fitzgerald-Lee
Corrosion Resistance of Aluminum Bronzes
Corrosion Technol. 6 (1959) pp. 263–266

[1277] DECHEMA MATERIALS TABLE, Sodium Chloride
DECHEMA e.V., D-Frankfurt am Main, 1962

[1278] Katz, W.
in: Tödt, F.
Korrosion und Korrosionsschutz, 2. Aufl.
W. de Gruyter & Co., D-Berlin, 1961, pp. 251–256, 1090–1100

[1279] Weyrauch, E.
HSM-Cu – ein neuer hochfester Kupferwerkstoff
Metall 27 (1973) 11, pp. 1118–1120

[1280] Karavaeva, A. P. et al.
Corrosion of copper with heat transfer to a dilute chloride solution
Zashchita Metallov 13 (1975) 2, pp. 176–180

[1281] Heitz, E.
On the mechanism of erosion corrosion in liquid media
NACE Corrosion, Proceedings of the 5th International Congress of Metallic Corrosion (1974) pp. 477–479

[1282] Faita, G.; Fiori, G.; Salvadore, D.
Copper behavior in acid and alkaline brines – I. Kinetics of anodic dissolution in 0.5 M NaCl and free-corrosion rates in the presence of oxygen
Corros. Sci. 15 (1975) 6/7, pp. 383–392

[1283] Tromans, D.; Silva, J. C.
Behavior of copper in acidic sulfate solution: comparison with acidic chloride
NACE Corrosion 53 (1997) 3, pp. 171–178

[1284] Bacarella, A. L.; Griess, J. C.
The Anodic Dissolution of Copper in Flowing Sodium Chloride Solutions Between 25 and 175 °C
J. Electrochem. Soc. 120 (1973) 4, pp. 459–465

[1285] Mor, E. D.; Beccaria, A. M.
Effects of Hydrostatic Pressure on the Corrosion of Copper in Seawater
Brit. Corr. J. 13 (1978) 3, pp. 142–146

[1286] Jaiswal, A.; Singh, R. A.; Dubey, R. S.

Inhibition of Copper Corrosion in Aqueous Sodium Chloride Solution by N-Octadecylbenzidine/1-Docosanol Mixed Langmuir-Blodgett Films
NACE Corrosion 57 (2001) 4, pp. 307–312

[1287] Upadhyay, B. N.; Singh, M. M.; Rastogi, R. B.
Inhibition of Copper Corrosion in Aqueous Sodium Chloride Solution by Various Forms of Piperidine Moiety
NACE Corrosion 50 (1994) 8, pp. 620–625

[1288] Singh, M. M.; Rastogi, R. B.; Upadhyay, B. N.
Inhibition of copper corrosion by diethylamine, diethyldithiocarbamate and its Cu(II) complex in NaCl solution
Bulletin of Electrochemistry 12 (1996) 1–2, pp. 26–30

[1289] Abd. El Gulil, R. M.
Inhibition of the corrosion of copper by some organic dyes
Bulletin of Electrochemistry 6 (1990) 12, pp. 913–915

[1290] Trabanelli, F. et al.
Inhibition of copper corrosion in chloride solutions by heterocyclic compounds
Werkst. Korr. 24 (1973) 7, pp. 602–606

[1291] Chaouket, F. et al.
New corrosion inhibitors for copper in 3 % NaCl solution
Proceedings of the 8th European Symposium on Corrosion Inhibitors (8 Seic); Ann. Univ. Ferrara, N.S., Sez. V, Suppl. N. 10. 1995

[1292] Zucchi, F.; Fonsati, M.; Trabanelli, G.
Influence of the heat exchange on the inhibiting efficiency of some heterocyclic derivatives against copper corrosion in 3.5 % NaCl solutions
Corros. Sci. 40 (1998) 11, pp. 1927–1937

[1293] Kuznezov, S. A.; Kuznetsova, S. V.
Copper catalytical self-dissolution in NaCl-KCl melt containing rare refractory metal complexes, Conference EUROCORR '99, No. 227, Aachen, Aug./Sept. 1999
DECHEMA, EUROCORR '99 Proceedings, D-Frankfurt am Main, 1999, p. 142

[1294] Krasilnikova, N. A. et al.
Steady potentials of copper in chloride melts
Zashchita Metallov 15 (1979) 4, pp. 488–490

[1295] Ahmed, Z.
Corrosion Resistance of Cast Aluminum Bronzes Containing Chromium and Silicon Additions
Br. Corr. J. 11 (1976) 3, pp. 149–155

[1296] Ahmad, Z.
A comparative study of the corrosion stability of aluminum bronze and 70-30 brass in acid and neutral solutions
Anti-corrosion 27 (1980) 10, pp. 15–19

[1297] Pini, G. C. et al.
Werkstoffprüfung bei hohen Strömungsgeschwindigkeiten
Werkst. Korr. 27 (1976) pp. 693–697

[1298] Süry, P.; Oswald, H. R.
Zum Einfluß der Gefügeausbildung auf das Korrosionsverhalten von Aluminiumbronzen in Kochsalzlösung
Werkst. Korr. 22 (1971) 2, pp. 135–143

[1299] DIN EN 1982 (12/1998)
Kupfer und Kupferlegierungen – Blockmetalle und Gußstücke
Beuth Verlag GmbH, D-Berlin

[1300] Joseph, G.
Stress Corrosion Cracking and Corrosion Fatigue Susceptibility of alpha Aluminum Bronzes in Sodium Chloride Solutions.
Conference: Copper '90. Refining, Fabrication, Markets. Vasteras, Sweden, Oct. 1990
The Institute of Metals, London, 1990, pp. 477–493

[1301] Collins, P.; Duquette, D. J.
Corrosion fatigue behavior of a duplex aluminum bronze
NACE Corrosion 34 (1978) 4, pp. 119–124

[1302] Collins, L. E.; Mitrovic-Scepanovic, V.
Improved Corrosion Resistance of Rapidly Solidified Fe-Al Bronzes
Materials Science and Engineering 99 (1988) pp. 493–496

[1303] Ahmad, Z.
Effect of Tin Addition on the Corrosion Resistance of Aluminum Bronze
Anti-Corros. Meth. Mat. 24 (1977) 1, pp. 8–12

[1304] Althoff, F. C.; Buhl, H.; Voigt, H.
Zum Potentialverhalten der Oberfläche metallischer Werkstoffe bei milder und starker Schwingungskavitation
Werkst. Korr. 24 (1973) 8, pp. 673–684

[1305] Khasanov, Z. Kh.
Corrosion resistance of bronze
Br.AZhNMts 7-2.5-1.5-9 in various media
Avtom. Svarka (1978) 9, pp. 23–25

[1306] El-Warraky, A. A.; El-Dahan, H. A.
Corrosion inhibition of Al-bronze in
acidified 4% NaCl solution
J. Mater. Sci. 32 (1997) 14, pp. 3693–3700

[1307] Kroughman, J. M.; Ijsseling, F. P.
The corrosion stability of CuNi10Fe in sea water
Fourth International Congress on Marine Corrosion and Fouling (1976) pp. 297–312
Met. A. 7808-72 0227

[1308] Kate, C. et al.
On the Mechanism of Corrosion of Cu-9.4Ni-1.7Fe Alloy in Air Saturated Aqueous NaCl Solution I. Kinetic Investigations
J. Electrochem. Soc. 127 (1980) 9, pp. 1890–1896

[1309] Kato, C. et al.
On the Mechanism of Corrosion of Cu-9.4Ni-1.7 Fe Alloy in Air Saturated Aqueous NaCl Solution II. Composition of the Protective Surface Layer
J. Electrochem. Soc. 12 (1980) 9, pp. 1897–1902

[1310] North, R. F.; Pryor, M. J.
The influence of corrosion product structure on the corrosion rate of Cu-Ni alloys
Corros. Sci. 10 (1970) pp. 297–311

[1311] Popplewell, J. M. Hart, R. J.; Ford, J. A.
The effect of iron on the corrosion characteristics of 90-10 cupronickel in quiescent 3.4% NaCl solution
Corros. Sci. 13 (1973) pp. 295–309

[1312] Crousier, J.; Beccaria, A. M.
Behavior of CuNi alloys in natural sea water and NaCl solution
Werkst. Korr. 41 (1990) pp. 185–189

[1313] Lu, H. H.; Duquette, D. J.
The effect of dissolved ozone on the corrosion stability of Cu-30Ni and Type 304L stainless steel in 0.5 M NaCl solutions
NACE Corrosion 46 (1990) 10, pp. 843–852

[1314] Syrett, B. C.
Erosion-Corrosion of Copper-Nickel Alloys in Sea Water and Other Aqueous Environments – A Literature Review
NACE Corrosion 32 (1976) 6, pp. 242–252

[1315] El-Etre, A. Y.
Passivation characteristics of CuNiAl alloy in pure and sulfide polluted solutions of 3.5% NaCl
Bull. Electrochem. 16 (2000) 4. pp. 171–174

[1316] Giuliani, L.; Bombara, G.
Influence of Pollution on the Corrosion of Copper Alloys in Flowing Salt Water
Brit. Corr. J. 8 (1973) pp. 20–24

[1317] Bates, J. F.; Popplewell, J. M.
Corrosion of Condenser Tubes in Sulfide Contaminated Brine
NACE Corrosion 31 (1975) 8, pp. 269–275

[1318] Hack, H. P.; Pickering, H. W.
AC Impedance Study of Copper and Cu-Ni Alloys in Aerated Salt Water
J. Electrochem. Soc. 138 (1991) 3, pp. 690–695

[1319] Hahn, H. N.; Duquette, D. J.
Effect of Heat Treatment on the Fatigue and Corrosion Fatigue Behavior of a CuNiCr Alloy
Metal. Trans. A. 10A (1979) 10, pp. 1453–1460

[1320] Laachach, A. et al.
Corrosion inhibition of a 70/30 cupronickel in a 3% NaCl medium by various azoles
J. Chim. Phys. 89 (1992) pp. 2011–2027

[1321] Bastidas, J. M.
Corrosion of bronze by acetic and formic acid vapors, sulfur dioxide and sodium chloride particles
Materials and Corrosion 46 (1995) pp. 515–519

[1322] Müller, J.
Zur Entzinkung von Messing
Korrosion, Dresden 4 (1973) 4, pp. 33–39

[1323] Nothing, F. W.
Korrosion, Diskussionstagung Lochfraßkorrosion
Verlag Chemie, Weinheim (1960) pp. 49–64
and
Dechema-Einzelfrage Nr. 220
Werkst. Korros. 8 (1957) p. 48

[1324] Kingerley, D. G.
British Chem. Engng. 6 (1961) 1, pp. 20–25

[1325] Taylor, A. H.; Cocks, F. H.
Electrochemical apparatus for corrosion studies in aqueous environments at high temperature and pressure
Br. Corrosion J. 4 (1969) pp. 287–292

[1326] Konstantinova, E. V.
Comparison of the corrosion stability of aluminum brasses and bronze in chloride solution
Zashchita Metallov 10 (1974) 4, pp. 425–427

[1327] Abd El-Rahman, H. A.
Passivation and pitting corrosion of α-brass (Cu/Zn: 6337) in neutral buffer solutions containing chloride ions
Werkst. Korr. 41 (1990) pp. 635–639

[1328] Chatterjee, U. K. et al.
Role of Chloride in the Stress Corrosion Cracking of Alpha Brass
Trans. Indian Inst. Met. 33 (1980) 1, pp. 76–78

[1329] Shih, H. C.; Tzou et al.
Studies of Inhibitors for Preventing Embrittlement of 70-30 Brass in 3.5 % NaCl Aqueous Solution Containing 10^{-4} M $HgCl_2$
NACE Corrosion 49 (1990) 11, pp. 913–920

[1330] Quraishi, M. A.; Farooqi, I. H.; Saini, P. A.
Inhibition of dezincification of 70-30 brass by aminalkyl mercaptotriazoles
Brit. Corr. J. 35 (2000) 1, pp. 78–80

[1331] Ashour, E. A.; Hagazy, H. S.; Ateya, B. G.
Effects of sulfide ions on the integrity of the protective film of benzotriazole on alpha brass in salt water
J. Electrochem. Soc. 147 (2000) 5, pp. 1767–1769

[1332] Lafont, M. C. et al.
Inhibition effect of long chain amines and thiazole derivates on the corrosion of admiralty brass
Proceedings of the 8th European Symposium on Corrosion Inhibitors (8 SEIC); Ann. Univ. Ferrara, N.S., Sez. V, Suppl. N. 10. 1995

[1333] El-Warraky, A. A.
Dissolution of brass 70/30 (Cu/Zn) and its inhibition during the acid wash in distillers
J. Mater. Sci. 31 (1996) 1, pp. 119–127

[1334] Gu, B.; Chu et al.
The Effect of Anodic Polarization on the Ambient Creep of Brass
Corros. Sci. 36 (1994) 8, pp. 1437–1445

[1335] Reynolds, S. D.; Pement, F. W.
Corrosion Failures of Tubing in Power Plant Auxiliary Heat Exchangers
Materials Performance 13 (1974) 9, pp. 21–28

[1336] Wingrove, A. J.
The development of a forgeable brass resistant to dezincification
Aust. Inst. Metals 15 (1970) 3, pp. 161–166

[1337] Pötzl, R.; Lieser, K. H.
Über die Korrosion von Messing in Natriumchlorid- und Natriumsulfatlösungen
Z. Metallkunde 61 (1970) 7, pp. 525–527

[1338] Torchio, S.
Stress corrosion cracking of aluminum brass in acid chloride solutions
Corros. Sci. 21 (1981) 1, pp. 59–68

[1339] Ballard, D. B. et al.
Intergranular Embrittlement of Copper-Palladium Alloys in Salt Water
NACE Corrosion 28 (1972) 10, pp. 368–372

[1340] Keir, D. S.; Pryor, M. J.
The Dealloying of Copper-Manganese Alloys
J. Electrochem. Soc. 127 (1980) 10, pp. 2138–2144

[1341] Min, U. S.; Li, J. C. M.
The microstructure and dealloying kinetics of a Cu-Mn alloy
J. Mater. Res. 9 (1994) 11, pp. 2878–2883

[1342] Polukarov, Yu.; Grinina, V. V.
Electrochemical behavior of alloys with an amorphous structure
Zashchita Metallov 11 (1974) 3, pp. 304–306

[1343] Pishchulin, V. N.; Kosintsev, V. I.
Corrosion of Electrode and Constructional Materials in the Evaporation of NaOH Solutions in Equipment With Direct Electrical Heating (in Russian)
Izv. Tomsk. Politekh. Inst. (1976) 275, p. 90

[1344] Mrsic, T.
Korrosion der Kupferrohre in einem Vakuumverdampfer bei der Herstellung von NaOH nach dem Kalkverfahren (Corrosion of copper pipes in a vacuum evaporator during the production of NaOH after the lime process)
Hemijska Industrija, Belgrad 29 (1975) 1, p. 27

[1345] Krzysziofowicz, K.; Chromic, H.
Factors affecting the Damages of Copper Pipes in Caustic Soda Evaporators
Ochr. Przed Korozja 20 (1978) p. 200

[1346] Mor, E. D.; Beccaria, A. M.
Influence of Sulfides on the Products of Corrosion of Copper in Polluted Sea Water
Corrosion NACE 30 (1974) p. 354

[1347] Gurovich, E. I.
Reaktionen von geschmolzenem Lithium-, Natrium- und Kaliumhydroxid mit Nickel, Kupfer, Eisen und Stahl
(Reactions of melted lithium, sodium and potassium hydroxide with nickel, copper, iron and steel) (in Russian)
Zh. Prikl. Khim. 32 (1959) p. 817

[1348] Kolenko, Z.
Reinigen von Metalloberflächen mit Natriumgluconat-Lösungen
(Cleaning of metallic surfaces with sodium gluconate solutions) (in Hungarian)
Korróziós Figyelö 14 (1974) 2, p. 54

[1349] Martschewska, M.; Raitschewski, G.; Petrow, C.
Färben von Kupfer – Elektrochemische Untersuchungen an mit Wechselstrom oxidiertem Kupfer
(Colouring of Copper) (in German)
Metalloberfläche 37 (1983) 1, p. 20

[1350] Ahmad, Z.
Improvement of the Mechanical Properties and Corrosion Behaviour of Cu-6.0Al Alloy by Si and Mn Addition
Werkst. Korros. 30 (1979) p. 433

[1351] Jones, R. J.; Murphy, S.
Effect of Cobalt Content and Ageing Treatment on the Stress Corrosion Resistance of 7% Al-Bronze
Brit. Corrosion J. 18 (1983) p. 123

[1352] Bonora, P. L.; Ponzano, G. P.; Bassol, M.
Anodisches Verhalten von Kupferlegierungen in Alkalilösungen
(Anodic behaviour of copper alloys in alkaline solutions) (in Italian)
Ann. Chim. Ital. 65 (1975) p. 677

[1353] Palmer, J. D.
Alkalis Can Cause Unexpected Problems
Canad. Chem. Processing 53 (1969) 9, p. 7

[1354] Shah, R. S.; Desai, C. S.
Effect of Hydrogen Peroxide on Corrosion of Brass by Sodium Hydroxide
J. Electrochem. Soc. India 25 (1976) p. 181

[1355] Shah, R. S.; Trivedi, A. M.
NaDDC as a Corrosion Inhibitor of Brass
Werkst. Korros. 25 (1974) 7, p. 521

[1356] Desai, M. N.; Shah, Y. C.
Inhibierung der Korrosion von Messing 63/37 in Natriumhydroxid-Lösungen
(Inhibition of the corrosion of brass 63/37 in sodium hydroxide solutions) (in German)
Werkst. Korros. 21 (1970) 10, p. 795

[1357] Desai, M. N.
Corrosion Inhibitors for Brasses
Werkst. Korros. 24 (1973) 8, p. 707

[1358] Greene, N. D.; Teterin, G. A.
Development of Brass Etchants by Electrochemical Techniques
Corrosion Sci. 12 (1972) p. 57

[1359] Parkins, R. N.; Holroyd, N. J. H.
Stress Corrosion Cracking of 70/30 Brass in Acetate, Formate, Tartrate and Hydroxide Solutions
Corrosion NACE 38 (1982) p. 245

[1360] Hönig, A.; Schlenk K. W.
Aushärtbare Federlegierungen und ihre Anwendungen
(Hardenable spring alloys and their application) (in German)
Chem.-Techn. 6 (1977), p. 409

[1361] Vaidyanath, L. R.; Bhandary, V. S.
Corrosion resistance of copper and copper alloys – a survey
Proceedings of a Symposium, Bombay (India) 1971, Paper Nr. 1

[1362] Anonymous
Copper and its alloys – wrought
Mater. Eng. 74 (1971) 4, pp. 181–190

[1363] DECHEMA-WERKSTOFF-TABELLE
"Natriumsulfat"
DECHEMA, D-Frankfurt am Main, October 1965

[1364] Mansfeld, F.; Parry, E. P.
Technical note – a quantitative electrochemical test for porosity in Permalloy® plated memory wire
Corrosion 26 (1970) 12, pp. 542–544

[1365] Gonzalez, S.; Souto, R. M.; Salvarezza, R. C.; Ariva, A. J.
A mechanistic approach to the electroformation of anodic layers on copper and their breakdown in Na_2SO_4 containing aqueous solutions
Conference: Progress in the Understanding and Prevention of Corrosion, Barcelona (Spain), July 1993, Vol. II
The Institut of Materials, London, Book No. 556, MA Journal Announcement: 9401, pp. 1,161–1,166

[1366] Altukhov, V. K.; Marshakov, I. K.; Vorontsov, E. S.; Klepinina, T. N.

Anodic behavior of copper in sulfate solutions
Elektrokhimiya 12 (1976), pp. 88–91

[1367] Mansour, H.; El-Wafa, M. H.; Noubi, G. A.
Inhibition of corrosion of copper in nitric acid
Bulletin of Electrochemistry 2 (1986), pp. 103–105

[1368] Shaban, A.; Telegdi, J.; Kalman, E.; Singh, G.
New corrosion inhibitors for copper and brass
Conference: Progress in the Understanding and Prevention of Corrosion, Barcelona (Spain), July 1993, Vol. II
The Institut of Materials, London, Book No. 556, MA Journal Announcement: 9401, pp. 916–919

[1369] Rahmel, A.
Elektrochemische Korrosionsprüfung von metallischen Werkstoffen in Alkalisulfatschmelzen
5. Kongress der europäischen Förderation Korrosion 1973, pp. 431–432

[1370] Beständigkeitstabellen von Kupfer-Gußwerkstoffen in verschiedenen Medien, Ausgabe 1970
DKI Deutsches Kupfer-Institut, D-Berlin

[1371] Wojtas, H.; Virtanen, S.; Boehni, H.
The new stainless Cu-Al-Sn-Alloys
Conference: Progress in the Understanding and Prevention of Corrosion, Barcelona (Spain), July 1993; Vol. II
The Institut of Materials, London, Book No. 556, MA Journal Announcement: 9401, pp. 1,228–1,236

[1372] Firmenschrift
Fertigungs- und Standardlieferprogramm M 3 – M 56. Schrauben, Muttern, Zubehör nach DIN, ISO und EURONORM
Werkstoffe und ihre Veredlung 2 (1980) 5

[1373] Pednekar, S. P.; Agrawal, A. K.; Staehle, R. W.
Stress corrosion cracking of Cu-10 Ni alloy CDA 706 in sulfate solutions of various pHs
Corrosion/79, Atlanta (GA/USA), March 1979 (Proc. Conf.), Nr. 44
NACE, Katy (TX/USA), 1979

[1374] Effertz, P.-H.; Forchhammer, P.
Parameter-Studie zum Erosions-Korrosions-Verhalten von Werkstoffen für Wärmetauscherrohre unter dem Einfluß von chemischen und mechanischen Größen bei Staustrahl-Bedingungen
Der Maschinenschaden 62 (1989), p. 68

[1375] Pickering, H. W.; Byrne, P. J.
Partial currents during anodic dissolution of Cu-Zn alloys at constant potential
J. Electrochem. Soc. 116 (1969), p. 1,492

[1376] Pickering, H. W.; Byrne, P. J.
Stress corrosion of α-brass in an acid sulfate solution
Corrosion 29 (1973) 8, pp. 325–328

[1377] Balakrishnan, K.; Venkatesan, V. K.
Anodic behaviour of brass – a ring-disc study
Werkst. Korros. 29 (1978), pp. 113–122

[1378] Pötzl, R.; Lieser, K. H.
Über die Korrosion von Messing in Natriumchlorid- und Natriumsulfat-lösungen
Z. Metallkd. 61 (1970) 7, pp. 525–527

[1379] Beccaria, A. M.; Mor, E. D.; Poggi, G.; Mazza, F.
A study of the corrosion products of aluminium brass formed in sodium sulfate solution in the presence of chlorides
Corros. Sci. 27 (1987) 4, pp. 363–372

[1380] Reynolds, S. D.; Pement, F. W.
Corrosion failures of tubing in power plant auxiliary heat exchangers
Mater. Performance, September (1974), pp. 21–28

[1381] Kawashima, A.; Agrawal, A. K.; Staehle, R. W.
Stress corrosion cracking of admiralty brass in non-ammoniacal sulfate solutions
J. Electrochem. Soc. 124 (1977) 11, pp. 1,822–1,823

[1382] Alvarez, M. G.; Giordano, M.; Manfredi, C.; Galvele, J. R.
Evaluation of stress corrosion cracking susceptibility of α-brass in non-ammoniac environments
Corrosion 46 (1990) 9, pp. 717–726

[1383] Hintz, M. B.; Blanchard, W. K.; Brindley, P. K.; Heldt, L. A.
Further observations of SCC in alpha-beta brass: Considerations regarding the appearance of crack arrest markings during SCC
Metall. Trans. A 17A (1986), pp. 1,081–1,085

[1384] Kaiser, H.; Kaesche, H.
Mechanismen der selektiven elektrolytischen Korrosion homogener Legierungen
Werkst. Korros. 31 (1980) pp. 347–353

[1385] Pickering, H. W.; Byrne, P. J.
On preferential anodic dissolution of alloys in the low-current region and the nature of the critical potential
J. Electrochem. Soc. 118 (1971) 2, p. 209

[1386] Lichter, B. D.; Cassange, T. B.; Flanagan, W. F.; Pugh, E. N.
Metallographic studies of transgranular stress-corrosion cracking in copper-gold alloys
Corrosion, Failure Analysis, and Metallography, Philadelphia (PA/USA), July 1984 (Proc. Conf.), Microstructural Science
Vol. 13, pp. 361–378

[1387] Polukarov, Yu. M.; Grinina, V. V.
Electrochemical behavior of alloys with an amorphous structure
Zashch. Met. 11 (1974) 3, pp. 304–306 (Russian)

[1388] Harris, J. O.
in: Shreir, L. L.
Corrosion
Newnes Butterworth, London-Boston, 2nd ed., 1976, vol. 1
Metal/Environment Reactions, Part 2, p. 61

[1389] DECHEMA Werkstoftabelle "Boden"
Dechema Corrosion Tables "Soil" (German Ed.),
DECHEMA, D-Frankfurt am Main, 1986

[1390] Branch, H. C.
Mater. Protection and Perform. (1973) p. 9

[1391] Franke
Werkstoffe und Korrosion 2 (1951) No. 7; p. 253 (in German)

[1392] Anonymous
Stahl und Eisen 70 (1950) p. 609 (in German)

[1393] Hildebrand, H.; Schwenk, W.
Untersuchungen zur Korrosion von unlegiertem Stahl im Erdboden mit und ohne Elementbildung
(Investigation of the corrosion of unalloyed steel in soil with and without formation of elements) (in German)
Werkst. Korrosion 29 (1978) No. 2, p. 92

[1394] Gilbert, P. T.
Journal of the Inst. of Metals 73 (1964) pp. 139/74

[1395] v. Franqué, O.
Korrosionsverhalten von Kupfer und Kupferlegierungen im Erdboden
(Corrosion behavior of copper and copper alloys in soil) (in German)
DKI special print, 1979, Deutsches Kupfer Institut, D-Berlin, Knesebeckstr. 96

[1396] v. Franqué, O.
3R-international 18 (1979) No. 8/9; p. 532 (in German)

[1397] Baker, S.
Chem. Process. Eng. 41 (1960) No. 11; p. 513

[1398] Logan, K. H.
J. Res. Natl. Bur. Stand., A, 22 (1939) p. 109

[1399] Anonymous
Copper underground: Its resistance to soil corrosion,
CAD/Publication No. 40, London (1958)

[1400] Carrière, J. E.
Water (s'Gravenhage) 54 (1970) No. 11; p. 638 and No. 12; p. 707 (in Dutch)

[1401] Anonymous
Materials in Design Eng. 60 (1964) No. 12; p. 115

[1402] Guillaume, I.; Grimaudeau, J.; Brisou, J.
Corrosion Science 17 (1977) p. 753

[1403] Dechema Corrosion Tables (German Ed.), "Huminic acids", p. 752, DECHEMA, D-Frankfurt am Main, 1958

[1404] NBS Tech. News. Bull. 42 (1958) No. 9; p. 181

[1405] Lorant, M.
Werkstoffe und Korrosion 20 (1969) No. 2; p. 127 (in German)

[1406] Carrière, J. E.
Water 54 (1970) No. 11; p. 707 (in Dutch)

[1407] Zandveld, B.
Water 38 (1954) p. 27 (in Dutch)

[1408] Camitz, G.
Métaux, Corrosion-Ind. 53 (1978) No. 629; p. 12 (in French)

[1409] Carrière, J. E.
Water 54 (1970) No. 11; p. 707 (in Dutch)

[1410] Zandveld, B.
Water 41 (1957) No. 4; p. 37 (in Dutch)

[1411] Anonymous
Water 40 (1956) No. 8; p. 94 (in Dutch)

[1412] Schwenkhagen, H. F.
Werkstoffe und Korrosion 6 (1955) No. 2;
p. 63 (in German)

[1413] Escalante, E.; Gerhold, W. F.
Mater. Perform. 14 (1975) No. 10; p. 16

[1414] Anonymous
Elect. Times 162 (1973) 4. Jan., p. 27/8,
11. Jan., p. 23

[1415] Prinz, W.
3R-international 17 (1978) No. 7; p. 466
(in German)

[1416] Baboian, R.; Haynes, G. S.
Mater. Perform. 14 (1975) No. 11; p. 16

[1417] Borris, B.
VGB Kraftwerkstech. 54 (1974) No. 5;
pp. 324–332 (in particular pp. 328–331)

[1418] Resch, G.
Kupferablagerungen in
Hochdruckdampfturbinen
(Depositions of copper in high-pressure
turbines)
Mitt. Vereinigung Großkesselbes. (1966)
No. 100; pp. 48–51

[1419] Bulow, C. L.
Corrosion in liquid media, atmosphere,
gases, (in Leach, R. H.
High-temperature corrosion; Silver
The corrosion handbook, edited by H. H.
Uhlig. 1948, 6th reprint, 1958, J. Wiley &
Sons, Inc., New York and Chapman & Hall
Ltd., London, p. 84)

[1420] LaQue, F. L.
Corrosion (Houston) 10 (1954) No. 11;
pp. 391–399

[1421] Anonymous
Mater. Eng. (Materials Selector) 74 (1971)
No. 4; p. 181

[1422] Anonymous
Mater. Eng. (Materials Selector) 80 (1974)
No. 4; p. 96

[1423] Rhines, F. N.
High-temperature corrosion, copper and
copper alloys, (in Leach, R. H.
High-temperature corrosion; Silver
The corrosion handbook, edited by H. H.
Uhlig. 1948, 6th reprint, 1958, J. Wiley &
Sons, Inc., New York and Chapman & Hall
Ltd., London, p. 629)

[1424] Gilbert, P. T.
Chem. and Ind. (1959) No. 28; pp. 888–895

[1425] Hömig, H. E.; Class, G.
Report on Feedwater Congress 1966
(in German)
Werkstoffe u. Korrosion 18 (1967) p. 1021

[1426] Stratfield, E. L.
Bull. Cebedeau (1958) IV., No. 42;
pp. 288–290

[1427] Vyhnalek, H. J.
Proceed. Amer. Power Conf. 27 (1965)
pp. 774–781

[1428] Leicester, J.
Trans. Amer. Soc. Mechan. Engrs 78 (1956)
pp. 273–285

[1429] Bodmer, M.
BBC Rev. 54 (1967) pp. 707–709

[1430] Anonymous
Chemieanl. + Verfahren 7 (1974) p. 76

[1431] Richardson, F. W.
Canad. Chem. Process. 47 (1963) No. 6;
pp. 89–91

[1432] Resch, G.; Zinke, K.
VGB Kraftwerkstech. 55 (1975) No. 12;
pp. 824–827

[1433] Eliashek, J.; Mostecky, J.; Vosta, J.;
Talasek, V.; Singer, P.; Voldrich, K.
CSSR Patent 149254 (15 June 1973)

[1434] Schuck, J. J.; Nathan, C. C.; Metcalf, J. R.
Mater. Protection 12 (1973) No. 10;
pp. 42–47

[1435] Obrecht, M. F.
Heating, Piping, Air Conditioning 31
(1959) No. 11; pp. 121–128

[1436] Silverstein, R. M.
US Patent 3382186 (7 May 1968)

[1437] Korpics, C. J.
Mater. Perform. 13 (1974) No. 2; pp. 36–38

[1438] Anonymous
Chem. Eng. 62 (1955) No. 4; p. 140

[1439] Hömig, H. E.
Mitt. Vereinig. Großkesselbes. (1956)
No. 40; pp. 20–25 (in German)

[1440] Held, H.-D.
TÜV-Mitt. 10 (1969) No. 8; pp. 284–288
(in German)

[1441] Edwards, J. E.
Bull. Cebedeau (1960) No. 2; pp. 88–93

[1442] Hummel, O. H.
KEM (1965) Nov. pp. 15–19

[1443] Smith, A. A.
Corrosion, Prevention Control 10 (1963)
No. 6; pp. 29–33

[1444] Zitter, H.; Kraxner, G.
Werkstoffe u. Korrosion 14 (1963)
pp. 80–88 (in particular p. 84)

[1445] Klement, J. F.; Maersch, R. E.; Tully, P. A.
Corrosion (Houston) 15 (1959) No. 7;
pp. 295t–298t

[1446] Piatti, L.; Grauer, R.
Werkstoffe u. Korrosion 14 (1963)
pp. 551–556 (in particular p. 555)

[1447] Anonymous
Corrosion Technol. 6 (1959) No. 1; pp. 25, 28

[1448] Klement, J. F.
Brit. Patent 849683 (Ampco Metal Inc., 28 Sept. 1960)

[1449] Anonymous
Mater. in Design Eng. 48 (1958) No. 2;
pp. 123–124

[1450] Klement, J. F.; Maersch, R. E.; Tully, P. A.
Corrosion (Houston) 16 (1960) No. 10;
pp. 519t–522t

[1451] Otsu, T.; Sato, Sh.
J. Japan Inst. Metals 25 (1961) pp. 471–475 (in Japanese)

[1452] Anonymous
technica (Bale CH) 19 (1970) No. 9; p. 732

[1453] Marshall, T.; Hugill, A. J.
Corrosion (Houston) 13 (1957) No. 5;
pp. 329t–337t

[1454] Bulow, C. L.; Gleasen, C. A.
H. H. Uhlig, Corrosion Handbook 1948, 6th reprint, 1958, J. Wiley & Sons, Inc., New York and Chapman & Hall Ltd., London
Corrosion in liquid media, atmosphere, gases, copper-nickel-alloys; Special topics in corrosion, condenser corrosion pp. 85, 558–559

[1455] Katz, W.
Werkstoffe u. Korrosion 15 (1064)
pp. 977–987 (in particular p. 978)
(in German)

[1456] Castle, J. E.; Harrison, J. T.; Masterson, H. G.
Proceed. 2nd Intern. Congr. Metallic Corr., New York 1963, publ. 1966, pp. 822–827

[1457] Eichhorn, K.
Werkstoffe u. Korrosion 21 (1970)
pp. 535–553 (in German)

[1458] Heldt, L. A.; Pichback, T. R.; Wilkinson, G. A.
Mater. Perform. 14 (1975) No. 4; pp. 42–45

[1459] Bulow, C. L.
Corrosion in liquid media, atmosphere, gases, copper-nickel-alloys, p. 95 in: H. H. Uhlig, Corrosion Handbook 1948, 6th reprint, 1958, J. Wiley & Sons, Inc., New York and Chapman & Hall Ltd., London

[1460] Hawkins, G. A.; Solberg, H. L.
Special topics in corrosion; Corrosion by high-temperature steam p. 520 in: The corrosion handbook, edited by H. H. Uhlig. 1948, 6th reprint, 1958, J. Wiley & Sons, Inc., New York and Chapman & Hall Ltd., London

[1461] Zschötge, S.
Kleine Werkstoffkunde der Nichteisenmetalle
(Concise materials review of non-ferrous metals) (in German) p. 63; Deutscher Verlag für Schweißtechnik, D-Düsseldorf, (1967)

[1462] Kemplay, J.
Chem. Process Eng. 47 (1966) No. 7;
pp. 53–56

[1463] Jackson, D. H.
Chem. Eng. Progr. 49 (1953) No. 2;
pp. 102–104

[1464] Evans, C. J.
Zinn u. Verwendung (1975) No. 104;
pp. 6–9 (in German)

[1465] DIN EN 1982 (08/2008)
Kupfer und Kupferlegierungen – Blockmetalle und Gussstücke; Deutsche Fassung EN 1982:2008
Beuth Verlag GmbH, D-Berlin

[1466] Ullmanns Enzyklopädie der technischen Chemie
Bd. XI, 3rd ed. (1960) "Kupferlegierungen" (Copper Alloys) p. 226
Verlag Urban & Schwarzenberg, München-Berlin 1960

[1467] Zschötge, S.
Kleine Werkstoffkunde der Nichteisenmetalle
(Concise materials review of non-ferrous metals) (in German) p. 58; Deutscher Verlag für Schweißtechnik, D-Düsseldorf, (1967)

[1468] Römpp, H.
Chemie Lexikon
(Chemistry Encyclopedia) (in German)
6th Edition (1966) Vol. III; p. 5519

[1469] Brauch, G.; Winkelmann, W.
Techn. Informat. Armaturen 4 (1969) No. 1; pp. 16–22 (in German)

[1470] Rabald, E.
Corrosion Guide, 2nd edition (1968)
Elsevier, Publ. Co., Amsterdam

[1471] Bulow, C. L.
Corrosion in liquid media, atmosphere, gases; Copper-tin alloys p. 102 in: The corrosion handbook, edited by H. H. Uhlig, 1948, 6th reprint, 1958, J. Wiley & Sons, Inc., New York and Chapman & Hall Ltd., London

[1472] Ullmanns Enzyklopädie der technischen Chemie
Bd. XI, 3rd. ed. (1960) "Kupferlegierungen" (Copper Alloys) p. 222
Verlag Urban & Schwarzenberg, München-Berlin 1960

[1473] Whitaker, M. W.
Metallurgia 39 (1948) pp. 21, 66

[1474] Austin, J.
Engineer 143 (1927) p. 406

[1475] Bünger, J.
Werkstoffe u. Korrosion 1 (1950) pp. 133–136 (in German)

[1476] Czempik, E.; Langner, A.; Pflugbeil, K.; Schindler, K.,
DOS 2450253 (Inst. f. Energetik, Leipzig, DDR, 22 May 1975)

[1477] Bulow, C. L.
Corrosion in liquid media, atmosphere, gases; Copper-zinc alloys pp. 69–70 in: The corrosion handbook, edited by H. H. Uhlig, 1948, 6th reprint, 1958, J. Wiley & Sons, Inc., New York and Chapman & Hall Ltd., London

[1478] Evans, U. R.
Einführung in die Korrosion der Metalle (An introduction to metallic corrosion, translated by E. Heitz) (1965) pp. 110–115
Verlag Chemie GmbH, D-Weinheim

[1479] Lockhart, A. M.
Corrosion Prevention Control 12 (1965) No. 5; pp. 20–27

[1480] Effertz, P. H.; Fichte, W.; Forchhammer, P.
VGB Kraftwerkstech. 54 (1974) No. 2; pp. 82–93 (in German)

[1481] Anonymous
Corrosion resistance of copper metals publ. by Anaconda Amer. Brass Co., Publication B-36, 8th edition (1965) pp. 12–28, in particular p. 26

[1482] Rabald, E.
Corrosion Guide, p. 779, 2nd edition (1968)
Elsevier Publ. Co., Amsterdam-London-New York

[1483] Horner, L.; Pliefke, E.
Corrosion inhibitors Pt. 27 (1)
Inhibitors of the corrosion of copper: does there exist a structure-efficiency relationship? (in German)
Werkstoffe u. Korrosion 33 (1982) p. 98

[1484] Popplewell, A. F.; Finan, M. A.
Chemical agents (acids) inhibiting the corrosion of copper and its alloys
GB Patent 1198312 (from C.A. 73 (1970) 69321c)

[1485] Shoji, Sh.; Tochigi, N.
Acid aqueous pickling solution for metal treatment (in German)
German DOS 2364162 (21 Dec. 1973)

[1486] Brough, R. W.
Composition and process for the production of brown surface layers on brass
US Patent 3816186

[1487] Rabald, E.
Corrosion guide, 2nd ed.
Elsevier Publishing Company, Amsterdam-London-New York, 1968

[1488] Vaidynath, L. R.; Bhandary, V. S.
Corrosion resistance of copper and copper alloys – A survey
Proceedings of a Symposium, Bombay, 1971

[1489] Bollinger, J.
Über die Korrosion verschiedener Metalle in verflüssigtem Schwefeldioxid
(The corrosion of different metals in liquid sulfur dioxide) (in German)
Schweizer Archiv (1952) 10, p. 321

[1490] Anonymous
Corrosion Data Survey, Section 1: Main Tables, pp. 124, 167
National Asssociation of Corrosion Engineers, Houston, 1985

[1491] Product Information
Beständigkeitstabellen von Kupfer-Gußwerkstoffen in verschiedenen Medien (Corrosion resistance charts of copper cast materials in different media) (in German)
Deutsches Kupfer-Institut, Berlin-Düsseldorf, 1970

[1492] Vaders, E.
Korrosion, insbesondere von Kupfer-Zinklegierungen durch Seewasser und chemische Lösungen, Teil II
(Corrosion, particularly of CuZn-alloys, by seawater and chemical solutions, Part II) (in German)
Metall 16 (1962) 12, p. 1210

[1493] Vaders, E.
Korrosion, insbesondere von Kupfer-Zinklegierungen durch Seewasser und chemische Lösungen, Teil I
(Corrosion, particularly of CuZn-alloys, by seawater and chemical solutions, Part I) (in German)
Metall 16 (1962) 6, p. 555

[1494] Schwabe, K.
Korrosion und Umweltschutz
(Corrosion and pollution control) (in German)
Freiberger Forschungshefte (1976) B 179, p. 7

[1495] Rice, D. W.; Peterson, P.; Rigby, E. B.; Phipps, P. B. P.; Cappell, R. J.; Tremoureux, R.
Atmospheric corrosion of copper and silver
J. Electrochem. Soc. 128 (1981) 2, p. 275

[1496] Laub, H.
Korrosion durch die Atmosphäre
(Corrosion by the atmosphere) (in German)
Galvanotechnik 65 (1974) 3, p. 209

[1497] Süthoff, Th.; Reichel, H.-H.
Vergleichende Korrosionsversuche an Trockenkühlelementen für Trockenkühltürme
(Comparative corrosion tests on dry cooling elements for dry cooling towers) (in German)
VGB Kraftwerkstechnik 65 (1985) 9, p. 835

[1498] Bogenschütz, A. F.; Jostan, J. L.; Mussinger, W.
Galvanische Korrosionsschutzschichten für elektronische Anwendungen, Teil 1
(Electrodeposited corrosion protective coatings for electronic applications, Part 1) (in German)
Metalloberfläche 34 (1980) 2, p. 45

[1499] Jostan, J. L.; Mussinger, W.; Bogenschütz, A. F.
Langzeitkorrosionsprüfungen an Elektrowerkstoffen. Teil 1: Werkstoffproben und Prüfergebnisse unter aggressivem Industrieklima
(Long-term corrosion tests on electric materials. Part 1: Material samples and test results in an aggressive industrial climate) (in German)
Metalloberfläche 39 (1985) 2, p. 45

[1500] Kirsch, H.; Forck, B.
Vergleichende Korrosionsuntersuchungen an Trockenkühlelementen
(Comparative corrosion tests on dry cooling elements) (in German)
Werkst. Korros. 35 (1984) 6, p. 299

[1501] Brown, P. W.; Masters, L. W.
Factors affecting the corrosion of metals in the atmosphere
Atmospheric Corrosion (Proc. Conf.), Hollywood, Fla., Oct. 1980
John Wiley and Sons, Inc., Somerset, N. J., 1982

[1502] Stahl, R.; Kiefer, P.
Korrosion und Korrosionsschutz in der Lebensmittelindustrie
(Corrosion and corrosion protection in the food industry) (in German)
Werkst. Korros. 24 (1973) 6, p. 513

[1503] Toumi, C.; Gillot, B.
Corrosion in SO_2 of pure and preoxidized copper at high temperature
Oxid. Met. 16 (1981) 3/4, p. 221

[1504] Walker, R.
Triazole, benzotriazole and naphthotriazole as corrosion inhibitors for copper
Corrosion 31 (1975) 3, p. 97

[1505] Walker, R.
Corrosion inhibition of copper by tolyltriazole
Corrosion 32 (1976) 8, p. 339

[1506] Ferrari, R.; Marabelli, M.; Serra, M.; Starace, G.
Radiochemical study of the use of benzotriazole for the protection of copper exposed to the attack of sulphur dioxide
Br. Corros. J. 12 (1977) 2, p. 118

[1507] Singh, D. D. N.; Banerjee, M. K.
Vapour phase corrosion inhibitors – a review
Anticorros. Methods Mater. 31 (1984) 6, p. 4

[1508] Korpics, C. J.
Aromatic triazoles inhibit corrosion of copper and copper alloys
Mater. Performance 13 (1974) 2, p. 36

[1509] Arshadi, M. A.; Johnson, J. B.; Wood, G. C.
The influence of an isobutane-SO_2 pollutant system on the earlier stages of the

atmospheric corrosion of metals
Corrosion Sci. 23 (1983) 7, p. 763

[1510] Dies, K.
Kupfer und Kupferlegierungen in der Technik
(Copper and copper alloys in technology) (in German), p. 819
Springer-Verlag, Berlin-Heidelberg-New York, 1967

[1511] Dies, K.
Kupfer und Kupferlegierungen in der Technik
(Copper and copper alloys in technology) (in German), p. 555
Springer-Verlag, Berlin-Heidelberg-New York, 1967

[1512] Anonymous
Ni-Resist® (in German)
Nickel-Berichte 12 (1954) 5, p. 72

[1513] Riederer, J.
Korrosionsschäden an Bronzeplastiken
(Corrosion damage on bronze sculptures) (in German)
Werkst. Korros. 23 (1972) 12, p. 1097

[1514] Meletis, E. I.; Hochman, R. F.
A review of the crystallography of stress corrosion cracking
Corros. Sci. 26 (1986) 1, p. 63

[1515] Valverde, N.
Das Oxidationsverhalten von Cu-Zn-Al-Legierungen mit Zusätzen von Ce and Y bei 750 °C
(Oxidation behavior of Cu-Zn-Al alloys with additions of Ce and Y at 750 °C) (in German)
Werkst. Korros. 29 (1978) 10, p. 644

[1516] Walker, R.
Triazole, benzotriazole, and naphthotriazole as corrosion inhibitors for brass
Corrosion 32 (1976) 10, p. 414

[1517] Hommel, G. et al.
Handbuch der gefährlichen Güter, 4. Auflage: Merkblatt 174 "Rauchende Schwefelsäure"; Merkblatt 183 "Schwefelsäure"; Merkblatt 183a "Batterie-Säure" (Schwefelsäure mit nicht mehr als 51% Säure); Merkblatt 184 "Schwefeltrioxid" (stabilisiert)
(Handbook of dangerous goods, 4th edition; sheet 174 "fuming sulfuric acid"; sheet 183 "sulfuric acid"; sheet 183a "battery acid" (sulfuric acid with not more than 51% acid), sheet 184 "sulfur trioxide" (stabilized)) (in German)
Springer Verlag, Berlin-Heidelberg-New York-London-Paris-Tokyo, 1987

[1518] Corrosion Data Survey, Metals Section, 6th ed., 1985
National Association of Corrosion Engineers, Houston (USA)

[1519] Rabald, E.
in: Ullmanns Encyklopädie der technischen Chemie
(Ullmann's encyclopedia of industrial chemistry) (in German), 3rd ed., vol. 1, p. 920
Urban & Schwarzenberg, München-Berlin, 1960

[1520] Rabald, E.
Corrosion Guide, 2nd revised edition
Elsevier Publishing Company, Amsterdam-London-New York, 1968

[1521] Miller, R. F.; Treseder, R. S.; Wachter, A.
Corrosion by acids at high temperatures
Corrosion 10 (1954) 1, p. 7

[1522] Uhlig, H. H.
Korrosion und Korrosionsschutz
(Corrosion and corrosion protection) (in German)
Akademie-Verlag, D-Berlin, 1970

[1523] Anonymous
Corrosion resistance of materials
Mater. Eng. 74 (1971) 4; (Mater. Selector), p. 7

[1524] Fabian, R. J.; Vaccari, J. A.
How materials stand up to corrosion and chemical attack
Mater. Eng. 73 (1971) 2, p. 36

[1525] Rabald, E.
Einiges über das Verhalten von Werkstoffen gegenüber Schwefelsäure
(Some information about the behavior of materials towards sulfuric acid) (in German)
Werkst. Korros. 7 (1956) 11, p. 652

[1526] Swandby, R. K.
Corrosion charts: Guides to materials selection
Chem. Eng. 69 (1962) Nov. 12, p. 186

[1527] Fabian, H.
in: Ullmanns Encyklopädie der technischen Chemie
(Ullmann's encyclopedia of industrial chemistry) (in German), 4th ed., vol. 15, p. 491
Verlag Chemie, Weinheim-New York, 1978

[1528] Product Information
Western zirconium – Corrosion resistance of zirconium and other metals, 1980
Robert Zapp, D-Düsseldorf

[1529] Gräfen, H. et al.
Die Praxis des Korrosionsschutzes, Kontakt und Studium,
(Corrosion protection in practice; dealings with and study of) (in German), vol. 64
expert verlag, D-Grafenau, 1981

[1530] Team of authors
Werkstoffeinsatz und Korrosionsschutz in der chemischen Industrie
(Material application and corrosion protection in the chemical industry) (in German)
VEB Deutscher Verlag für Grundstoffindustrie, D-Leipzig, 1986

[1531] Anonymous
Beständigkeitstabellen von Kupfer-Gußwerkstoffen in verschiedenen Medien
(Chemical resistance charts of copper cast materials in different media) (in German)
DKI Deutsches Kupfer-Institut, D-Berlin, 1970

[1532] Süry, P.
Schwefelsäurekorrosion metallischer Werkstoffe unter extremen Bedingungen
(Corrosion of metallic materials by sulfuric acid under extreme conditions) (in German)
Chemie-Technik 6 (1977) 10, p. 415

[1533] McDowell, D. W.
Handling sulfuric acid
Chem. Eng. 81 (1974) 24, p. 118

[1534] Sander, U.; Rothe, U.; Gerken, R.
in: Winnacker-Küchler "Chemische Technologie – Anorganische Technologie I"
(Chemical technology – inorganic technology I) (in German), 4th ed., vol. 2, p. 35
Carl Hanser Verlag, München-Wien, 1982

[1535] Product Information
Beständigkeit von nickelhaltigen Legierungen und Stählen gegenüber Schwefelsäure
(Resistance of nickel-containing alloys and steels to sulfuric acid) (in German),
October 1961
Nickel-Informationsbüro GmbH, D-Düsseldorf

[1536] Berg. F. F.
Korrosionsschaubilder
(Corrosion diagrams) (in German)
VDI-Verlag GmbH, D-Düsseldorf, 1969

[1537] Damon, G. H.; Cross, R. C.
Corrosion of copper
Ind. Eng. Chem. 28 (1936) 2, p. 231

[1538] Whitman, W. G.; Russel, R. P.
The acid corrosion of metals, effect of oxygen and velocity
Ind. Eng. Chem. 17 (1925) 4, p. 348

[1539] Tödt, F.
Korrosion und Korrosionsschutz
(Corrosion and corrosion protection) (in German)
Verlag Walter de Gruyter & Co., D-Berlin, 1955

[1540] Heitz, E.; v. Meysenbug, C. M.
Die Korrosion von Eisen, Nickel, Kupfer und Aluminium in organischen Lösungsmitteln mit geringem Mineralsäuregehalt
(The corrosion of iron, nickel, copper and aluminium in organic solvents containing a small amount of mineral acid) (in German)
Werkst. Korros. 16 (1965) 7, p. 578

[1541] Klyuchnikov, N. G.; Verizhskaya, E. V.
Corrosion of certain metals in sulfuric acid solutions passed through a magnetic field
(in Russian)
Zashch. Met. 8 (1972) 6, p. 700

[1542] Gugau, M.; Hirth, F. W.; Speckhardt, H.
Korrosionsschutz bei Offsetdruckmaschinen. Bestimmung der Verträglichkeit von Feuchtmitteln mit galvanisch abgeschiedenen Überzügen durch Stromdichte-Potential-Kurven
(Corrosion protection of offset printing. Determination of the compatibility of wetting agents with electrodeposited coatings by using current density potential curves) (in German)
Metalloberfläche 39 (1985) 5, p. 171

[1543] Kuzyukov, A. N.; Khanzadeev, I. V.
Intercrystalline copper corrosion in acids
(in Russian)
Zashch. Met. 15 (1979) 3, p. 341

[1544] Subramanyam, N. C.; Sheshadri, B. S.; Mayanna, S. M.
Quinine and strychnine as corrosion inhibitors for copper in sulphuric acid
Br. Corros. J. 19 (1984) 4, p. 177

[1545] Omar, I. H.; Zucchi, F.; Trabanelli, G.
Schiff bases as corrosion inhibitors of copper and its alloys in acid media
Surface Coatings and Technology 29 (1986) p. 141

[1546] Kolevatova, V. S.
Effect of thiourea on the corrosion of copper in sulfuric acid solutions (in Russian)
Zh. prikl. Khim. 48 (1975) 10, p. 2216

[1547] Franke, E.
Das Korrosionsverhalten von Aluminiumbronzen
(Corrosion behavior of aluminium bronzes) (in German)
Werkst. Korros. 2 (1951) 8, p. 298

[1548] Zitter, H.; Kraxner, G.
Korrosionsbeständigkeit von Aluminiumbronze in Säuren (Corrosion resistance of aluminium bronze in acids) (in German)
Werkst. Korros. 14 (1963) 2, p. 80

[1549] Katz, W.
Gießen für die Chemie – Das Verhalten metallischer Gußwerkstoffe unter korrosiver Beanspruchung (Casting in chemistry – the behavior of metallic cast materials under corrosive attack) (in German)
VDI-Verlag GmbH, D-Düsseldorf

[1550] Ahmad, Z.
Improvement of the mechanical properties and corrosion behavior of CuAl6.0 alloy by Si and Mn addition
Werkst. Korros. 30 (1979) 6, p. 433

[1551] Barker, W.; Evans, T. E.; Williams, K. J.
Effect of alloying additions on the microstructure, corrosion resistance and mechanical properties of nickel-silicon alloys
Br. Corros. J. 5 (1970) March, p. 78

[1552] Product Information
Korrosionsbeständigkeit nickelhaltiger Werkstoffe gegenüber Phosphorsäure und Phosphaten (Corrosion resistance of nickel-containing materials to phosphoric acid and phosphates) (in German)
Information No. 61, March 1970
International Nickel Deutschland GmbH, D-Düsseldorf

[1553] Rolker, J.
Untersuchungen zur Schwefelsäure-Taupunkt-Bestimmung in Rauchgasen (Investigations for determination of the dew point of sulfuric acid in flue gases) (in German)
VGB Kraftwerkstech. 53 (1973) 5, p. 333

[1554] van der Hoeven, H. W.
Strömungsgeschwindigkeit als besonderer Faktor bei der Schwefelsäurekorrosion (Flow rate, a special factor in sulfuric acid corrosion) (in German)
Werkst. Korros. 6 (1955) 2, p. 57

[1555] Ahmad, Z.
A comparative study of the corrosion behaviour of aluminium bronze and 70-30 brass in acid and neutral solutions
Anticorros. Methods Mater. 27 (1980) 10, p. 15

[1556] Altpeter, E.; Heubner, U.; Rudolph, G.; Weidemann, R.
Die Korrosion von Kupfer-Aluminium-Legierungen in schwefelsaurer Beizlösung (The corrosion of CuAl alloys in a sulfuric acid pickling solution) (in German)
Werkst. Korros. 25 (1974) 6, p. 411

[1557] Langer, R.; Kaiser, H.; Kaesche, H.
Zur Korrosion von binären CuAl-Legierungen in Schwefelsäure (On the corrosion of binary CuAl alloys in sulfuric acid) (in German)
Werkst. Korros. 29 (1978) 6, p. 409

[1558] Trarbach, K. O.
Korrosion von Plattierungen bei Chemieanlagen und in der Meerestechnik (Corrosion of platings in chemical plants and in marine technology) (in German)
Maschinenmarkt 90 (1984) 38, p. 936

[1559] Ruge, J.; Trarbach, K. O.
Plasmaauftragschweißen mit Heißdrahtelektrode von Sonderwerkstoffen (Plasma deposition welding of special materials using a hot wire electrode) (in German)
Schweissen Schneiden 34 (1982) 8, p. 369

[1560] Nickel, O.
Eigenschaften und Anwendungen von Reinnickel, Nickel-Basislegierungen und nickelhaltigen NE-Werkstoffen im Rohrleitungsbau
(Properties and uses of pure nickel, nickel-based alloys and nickel-bearing non-ferrous metals in piping) (in German)
3 R international 14 (1975) 8, p. 419

[1561] Anonymous
Fertigungs- und Standardlieferprogramm M3-M56 Schrauben, Muttern, Zubehör nach DIN, ISO und EURONORM (Manufacturing and standard delivery program M3-M56 screws, nuts and

equipment according to DIN, ISO and EURONORM) (in German)
Werkstoffe und ihre Veredlung 2 (1980) 5, p. 250

[1562] Tödt, F.
Korrosion und Korrosionsschutz
(Corrosion and corrosion protection) (in German), 2nd ed.
Verlag Walter de Gruyter & Co., D-Berlin, 1961

[1563] Dechema-Werkstoff-Tabelle
"Schwefelsäure"
(Corrosion data sheets "sulfuric acid") (in German), Dec. 1971
DECHEMA, D-Frankfurt am Main

[1564] Groth, V. J.; Hafsten, R. J.
Corrosion of refinery equipment by sulfuric acid and sulfuric acid sludges
Corrosion 10 (1954) 11, p. 368

[1565] Miska, K. H.
Wrought brass
Mater. Eng. 80 (1974) 5, p. 65

[1566] Erdmann-Jesnitzer, F.
Untersuchungen zur Wasserstoffversprödung und Spannungsrißkorrosion austenitischer Stähle unter besonderer Berücksichtigung des Einflusses einer thermomechanischen Behandlung (HERF)
(Investigation of the hydrogen embrittlement and stress corrosion cracking of austenitic steels with special consideration of the influence of thermomechanical treatment) (in German)
Werkst. Korros. 33 (1982) 4, p. 221

[1567] Erdmann-Jesnitzer, F.; Wessel, A.
Untersuchungen zur Wasserstoffversprödung einiger kubisch-flächenzentrierter Werkstoffe
(Investigation into the hydrogen embrittlement of some cubic face-centered materials) (in German)
Arch. Eisenhüttenwesen 52 (1981) 2, p. 77

[1568] Pushpa Gupta; Chaudhary, R. S.; Namboodhiri, T. K. G.; Prakash, B.
Effect of pyridine and its derivatives on the corrosion of 70/30 brass in 1 % H_2SO_4 solution
Br. Corros. J. 17 (1982) 4, p. 193

[1569] Hönig. A.; Schlenk, K. W.
Aushärtbare Federlegierungen und ihre Anwendungen
(Heat treatable spring alloys and their uses) (in German)
Chemie-Technik 6 (1977) 10, p. 409

[1570] Amsoneit; N.
Eindampfen problematischer Deponie-Sickerwässer
(Evaporation of problematic disposal site percolating water) (in German)
Wasser, Luft und Betrieb 29 (1985) 10, p. 63–66

[1571] Firmenschrift
Hochleistungswerkstoffe, Werkstoffdaten
(High-performance materials, material data) (in German), print N 53093-08, 1993
Krupp VDM, D-Werdohl

[1572] Heyer, B.
Erfahrungen mit einer Naßwäsche und anschließender Eindampfung der Waschwässer hinter einer Müllverbrennungsanlage
(Disposal of industrial wastes) (in German)
GVC/DECHEMA specialized committee conference "Beseitigung von Industrierückständen" November 1982, Bad Dürkheim
Chem.-Ing. Tech. 55 (1983) 9, pp. 739–742

[1573] Company publication
Schmutzwasser-Eindampfanlagen für geringe Durchsatzmengen PROWADEST 10 E, 20 E und 30 E
(Waste water evaporation plants for small throughput quantities PROWADEST 10 E, 20 E und 30 E) (in German)
KMU-Umweltschutz GmbH, Steinen-Höllstein (1995)

[1574] Rennhack, R.
Verdampfungsverfahren zur Meerwasserentsalzung
(Evaporation methods for the desalination of seawater) (in German)
VDI-Berichte (1976) 255, pp. 23–28

[1575] Eggers, H.
Die Meerwasserentsalzung unter besonderer Berücksichtigung der für Meerwasserentsalzungsanlagen geeigneten Werkstoffe
(Seawater desalination, with special consideration to the materials suitable for seawater desalination plants) (in German)
Stahl Eisen 95 (1975) 5, pp. 194–199

Index of materials

1
1.1151   481
1.4016   123, 235, 237–239
1.4031   481
1.4301   235, 237–239, 378, 589
1.4401   235, 237–239, 371, 378, 544, 547
1.4404   589
1.4449   378
1.4462   481
1.4500   371
1.4512   123
1.4541   123, 419, 570
1.4550   570
1.8962   374
194   515

2
2.0040   244
2.0060   244, 424
2.0065   244, 424
2.0070   244, 424
2.0075   308
2.0076   244, 323
2.0082   308
2.0085   308
2.0090   58–59, 184, 198, 201–202, 244, 271, 307, 316, 323, 424–425, 442, 468; 2.0090† see CW024A
2.0109   308
2.0230   336, 536
2.0240   214, 336, 536; 2.0240† see CW502L
2.0250   211, 336
2.0265   336, 460, 536; 2.0265† see CW505L
2.0290   310, 316, 537
2.0290.01   248, 459
2.0321   302, 337, 459–460, 542
2.0335   214; 2.0335† see CW507L
2.0340   310, 316, 537
2.0340.02   248, 459
2.0340.05   248
2.0360   229, 336, 385, 461, 536; 2.0360† see CW509L
2.0375   536
2.0380   337
2.0401   248, 337, 459
2.0402   248, 459
2.0460   74, 80, 95, 316, 425–426, 431, 442, 447, 458–461, 499
2.0470   74, 80, 316, 431, 435, 442, 447, 459–460, 499; 2.0470† see CW706R
2.0490   459
2.0492   311, 316, 537, 550
2.0492.01   459
2.0510   459
2.0525   95
2.0530   461, 536
2.0540   459
2.0580   623
2.0590.01   461
2.0592   311, 316, 537
2.0596   311, 316, 537
2.0596.01   462
2.0598   311, 316, 537
2.0602   229
2.0606   229
2.0740   229
2.0811   429
2.0815.01   231
2.0835   309, 314, 533
2.0835.01   234
2.0852   314
2.0862   433, 441
2.0872   73–74, 189, 231, 314, 316, 425–426, 428, 432–436, 438, 441–444, 447, 450–453, 455, 488, 491, 497, 499, 533, 544–546, 631; 2.0872† see CW352H
2.0878   314, 433
2.0880   432

Index of materials

2.0882 73–74, 188–189, 205, 231, 314, 425–426, 432–434, 439, 441–444, 448–450, 452–453, 455, 458, 488, 497, 499–500, 533, 544–546, 631; 2.0882† see CW354H
2.0883 73–74, 426, 432, 434, 441–443, 450, 452
2.0918 377, 431, 442, 448; 2.0918† see CW300G
2.0920 136–137, 531, 599; 2.0920† 377
2.0932 228, 531
2.0936 429; 2.0936† see CW306G
2.0940 204, 308, 531
2.0941 228
2.0960 429
2.0962 309, 314, 531
2.0962.01 429
2.0966 429
2.0970 308, 314, 531
2.0970.01 232
2.0975 309, 314, 531
2.0975.01 429, 432
2.0976 228
2.0978 429
2.0980.01 429
2.0981 228
2.1016 457
2.1020 190–191, 333; 2.1020† see CW452K
2.1030 534; 2.1030† see CW453K
2.1050 208, 309, 535
2.1050.01 235, 432, 622
2.1052 309, 535
2.1052.01 236
2.1086 310, 315, 535
2.1086.01 210, 458
2.1090 310, 315, 535
2.1090.01 210, 458
2.1091 230
2.1096 310, 315, 535
2.1096.01 210, 248, 458
2.1097 230
2.1166 312, 317, 550
2.1170 311, 317, 550
2.1176 312, 550
2.1177 230
2.1182 312, 550
2.1188 312, 550
2.1245 342
2.1247 342; 2.1247† see CW101C
2.1248 342
2.1285† see CW104C
2.1292 308
2.1310 442
2.1363 387
2.1491 427, 442
2.1504 432
2.1522 463
2.1525 303, 463
2.1830 353
2.1972 426, 432
2.4360 235, 237–238, 240, 253, 371, 374, 378, 434, 448
2.4374 253
2.4375 371
2.4816 235, 237–239, 378
2.4858 371
2.4882 371
2.4883 371, 378

3
3.7035 547
302 499, 585
302 B 585
304 32, 235, 237–239, 378, 585, 589
304 L 585
309 585
310 585
314 585
316 235, 237–239, 371, 378, 499, 585
316 L 589
317 378, 585
321 123, 570, 585
329 585
330 586
347 570

4
409 123
430 123, 235, 237–239, 586
444 390

6
651 345
652 345

7
716 443
720 443
722 74, 76

8
88Cu-4Sn4Zn 16

9
962 231

Index of materials

a

A 588 374
AA 6061 448
admiralty brass 35, 95, 137, 139, 148, 194, 225, 255, 336, 364, 435, 442, 460–461, 499, 507, 512, 515, 542, 555, 562, 565
admiralty bronze 338
AISI 302 499
AISI 304 32, 235, 237–239, 378, 589
AISI 316 235, 237–239, 371, 378, 499
AISI 316 L 589
AISI 317 378
AISI 321 123, 570
AISI 347 570
AISI 409 123
AISI 430 123, 235, 237–239
AISI 444 390
Al-bronze 15–16, 589
alloy 194 515
alloy 302 585
alloy 302 B 585
alloy 304 585
alloy 304 L 585
alloy 309 585
alloy 310 585
alloy 314 585
alloy 316 585
alloy 317 585
alloy 321 585
alloy 329 585
alloy 330 586
alloy 430 586
alloy 651 345
alloy 652 345
alloy 722 74, 76
Alloy No. 716 443
Alloy No. 720 443
AlMg1SiCu 448
AlMn-alloy 553
Alpaka 353
Al-Sn brass 507
aluminium 32–33, 35, 73, 351–352, 506, 553, 574
aluminium brass 223–224, 366, 442, 468, 504–506
aluminium bronze 3, 15, 28, 41, 72, 105, 164, 187–188, 203–204, 228, 232, 280, 295–297, 299, 313, 330–331, 351, 374–378, 393, 400, 403, 424, 428, 430, 442, 458–459, 466–467, 480, 484–485, 487, 499, 502, 505–506, 515, 523, 531, 561, 587, 597–599, 601–603, 606, 608, 612–613, 617, 623, 629
aluminium bronze 637 205

aluminium-manganese bronzes 486
aluminium multi-material bronze 467
aluminium-nickel bronzes 483
aluminium-nickel cast bronzes 137
Ambrac® 206, 208–209, 211
Ambraloy® 194
Ambraloy® 901 188
Ambraloy® 917 188
Ambraloy® 927 338
Ampco® 41
Ampco® alloys 203, 205
Ampco® aluminium bronze 15
Ampco® aluminiumbronze 34
anti-pollution bronze 456
Arsenical Admirality Brass 211
arsenical aluminium bronzes 404
asbestos 371
austenitic CrNiMo steel 544

b

B-150 296
bell bronze 333
Berylco® 717 261
beryllium bronze 117, 374, 414, 527, 629
beryllium conductor bronze 421
Beryvac® 60 117, 421, 527, 629
Beryvac® 200 117, 421, 527, 629
brass 4, 21–22, 29, 33, 35, 38, 41, 54, 59, 65, 77, 80, 84–85, 89, 95, 108, 113, 115–116, 135, 141, 145, 147, 151, 154, 161, 164, 175, 192, 194, 211, 220–221, 223, 239, 246, 254–255, 257, 261, 268, 273, 285, 302, 306, 315, 317, 335, 337, 343–344, 348, 354, 360, 362, 364–365, 374, 382–383, 385–386, 389–391, 393, 401, 412–414, 420, 459–460, 467, 470, 480, 497–498, 502–504, 507–508, 512, 515, 521, 524, 536, 538, 543–544, 546, 554–555, 558, 563, 567, 579, 601–602, 621, 623
brass 60 541
brass 70 541
brass 70/30 515, 527
brass 90/10 515
α-brass 89–90, 111, 114, 117, 147–148, 211–212, 302, 315, 344, 364–365, 420, 503, 507–508, 510, 514, 516–517, 526, 540
α-β-brass 211–212, 344, 543
α+β-brass 516, 518
β-brass 148, 151, 193, 211
γ-brass 211
brass cast alloys 193
brass casting alloys 360

Index of materials

brass copper-zinc alloys 580
brass OTS 70 508
bronze 16, 18, 28, 34, 38, 41, 54, 73, 77,
    108, 142, 144–145, 154, 161, 174, 181,
    190–191, 207, 224, 234, 253, 273,
    299–300, 305, 314–315, 333, 337, 353,
    359, 379, 391, 394, 410, 420, 456, 497,
    502, 524, 534–535, 554, 558, 563,
    578–579, 601, 616, 620–621, 629
bronze materials 77

c

C12200 391–393, 425, 427, 442
C14200 427, 442
C17200 374, 387
C19400 427, 442
C23000 382, 384–385
C26000 383
C27000 374, 384, 386
C27400 542–543
C28000 229
C42000 230
C44300 85, 384, 431, 442
C52100 374
C60800 376, 431, 442
C61000 599
C61300 431, 442
C61400 228
C64900 303
C65100 463
C65500 5, 303, 463
C67500 229
C68700 425, 431, 442, 499
C70400 441
C70600 231, 425, 434–435, 441–443, 447,
    450, 453–454, 488, 497, 499
C71000 206, 433
C71500 231, 425, 434, 441–443, 448–450,
    453–454, 488, 497, 499
C71640 441–443
C72200 425, 452
C75200 229
C83600 230
C86200 229
C86500 229
C86800 229
C87300 5
C92200 230
C93200 230
C93700 230
C94500 230
C94700 231
C94800 231
C94900 231

C95200 228
C95300 228
C95400 228
C95500 228
C95800 228
C96200 231
CA-230 269
CA-719 438, 496
cadmium 33, 35
carbon steel 402, 481
Carpenter® 20 586
Carpenter® 20-Cb 3 586
cartridge brass 255, 337
cast aluminium bronzes 203
cast brass 195, 364
cast bronzes 125, 209, 333
cast copper material 307–312
casting material 353
cast iron 32, 557
cast tin bronzes 207, 333
CB330G 203
CB331G 228
CB333G 228
CB334G 228
CB491K 230
CB493K 230
CB495K 230
CB764S 229
CB765S 229
CC040A 308
CC140C 308
CC331G 39, 187, 203–204, 308, 396, 531,
    601
CC332G 39, 187, 203, 232, 308, 396, 430,
    531
CC333G 39, 187, 203, 309, 396, 429, 432,
    531, 601, 603
CC334G 429
CC380H 231
CC383H 189–190, 234, 309, 396, 533
CC480K 39, 208, 235, 309, 396, 432, 535,
    622
CC482K 601
CC483K 39, 236, 309, 396, 535
CC491K 39, 192, 210, 248, 310, 396, 458,
    535
CC492K 192
CC493K 39, 192, 210, 310, 396, 458, 535
CC495K 39, 216, 312, 317, 396, 550
CC496K 39, 216, 312, 317, 396, 550
CC497K 39, 216, 312, 317, 396, 550
CC750S 248, 310, 459, 537
CC754S 248, 310, 459, 537

Index of materials

CC761S 52, 177–178, 194, 311, 459, 537, 550, 623
CC762S 311, 537
CC764S 311, 462, 537
CC765S 311, 537
CDA 122 220
CDA 320 256
CDA 360 256
CDA 367 256
CDA 377 256
CDA 443 499, 541
CDA 505 16
CDA 510 16
CDA 521 16
CDA 524 16
CDA 544 16
CDA 642 256
CDA 687 499
CDA 706 220, 488, 497, 499, 533
CDA 715 220, 488, 497, 499–500
CDA 722 500
CDA 836 256
CDA 844 256
CDA 848 256
CDA 879 256
CDA 958 483
Ch18N9T 419
Chlorimet® 2 586–587
Chlorimet® 3 235, 237–238, 240, 586
chromium-aluminium steels 402
chromium-nickel steels 402, 585, 610, 623, 626
Ck 22 481
CM239E 283
CM390H 164, 283
CN-7M 235, 237–239, 371
coinage bronze 467
coin alloys 141
cold-rolled brass 161
Constantan® 280
copper 18, 31, 33, 35, 57, 59, 69–71, 119, 122, 124–125, 127–131, 133–135, 140–141, 167, 171–172, 183–187, 198, 221, 227, 243, 249, 251, 267, 279, 293, 305, 307, 313, 351, 354–358, 369, 371–375, 379, 391, 395, 402, 414, 417–419, 423, 442, 444, 458, 465, 470, 492–493, 500, 506, 521–522, 529–530, 554, 559, 567, 569–570, 573–575, 585, 587, 590, 593, 596, 626, 629
copper alloys 71, 136, 138, 171, 240, 251, 257, 303, 352, 421, 465, 519, 527, 579, 583

copper-aluminium alloys 14, 28, 33, 37, 50, 72, 86, 105, 136, 154, 161, 164, 173, 181, 187, 203, 232, 251, 268, 273, 280–281, 290, 295, 305, 313, 330–331, 352, 358, 375, 403–404, 407–408, 412, 420, 466, 478, 483, 500, 523, 530, 561, 575, 597, 599–600, 605
copper-aluminium bronzes 429
copper-aluminium casting alloys 479
copper-aluminium materials 73
copper base alloys 167
copper based alloys 69
copper-beryllium alloys 24, 152, 164, 195, 216, 421, 527
copper-bismuth alloys 519–520
copper bronze 284
copper-cadmium alloys 366
copper cast materials 353
copper electrode 471
copper-gold alloys 65, 367, 550–551
copper-lead alloys 195, 216
copper-lead-silicon alloys 583
copper-lead-tin alloys 195, 216
copper-manganese alloys 519
copper materials 70, 170, 467, 570–572, 581, 593, 596
copper-nickel alloys 15, 28, 34, 38, 53, 73, 86, 106, 137, 139, 154, 161, 164, 174, 188–190, 205–206, 231, 233, 251–252, 261, 268, 273, 283, 298, 305, 314, 321, 331, 353, 359, 379–380, 409, 420, 488, 499, 523, 533, 544, 558, 562, 578, 614–615, 622, 631
copper-nickel casting alloy 533
copper-nickel-iron alloys 490
copper-nickel-palladium alloys 258
copper-nickel-zinc alloys 353
copper (pure) 45, 71–72, 119, 126, 148, 168, 171, 192, 217, 251, 277, 280, 285, 305, 307, 324, 331, 351, 423, 468, 478, 486, 491, 499
copper-silicon alloys 23, 155, 216, 626
copper-tin alloys 16, 28, 34, 38, 54, 77, 108, 142, 154, 161, 174, 190–191, 207, 209, 234, 253, 273, 284–285, 300, 305, 314, 333, 353, 359, 379, 381, 399, 410–412, 420, 456, 502, 524, 534, 558, 563, 578–580, 616–617, 620–621
copper-tin wrought alloys 192
copper-tin-zinc alloys 18, 29, 34, 38, 54, 108, 145, 164, 192, 210, 236, 246, 253, 273, 284, 300, 306, 315, 334, 353, 382, 411, 420, 458, 503, 524, 535, 563, 580, 620–621

copper-zinc alloys 21, 29, 35, 38, 41, 43,
    54, 59, 65–66, 77, 81, 86, 108–109, 141,
    147–150, 154, 161, 164, 175, 192, 211,
    213, 239, 246, 254, 261, 268–269, 273,
    285–286, 302, 306, 315, 335, 337–338,
    353–354, 360–364, 382–384, 412, 420,
    424, 459, 461, 502–503, 524, 536, 556,
    558, 563, 567, 580, 621–624
copper-zinc-aluminium alloys 80, 539
Corrix® 602
Coronel® 230 586
CR024A 372
CrNiMo steel 73
CrNi-steel 220
CrNiTi-steel 6
Cu-1Pd-1Y 367
Cu-2Be 117, 421, 527
Cu-2Pd-2Y 367
Cu-3Al-3Si-10Zn 577
Cu-3Pd 367
Cu-3Y 367
Cu-5Al 577
Cu-5.3Sn3.9Ti0.32P 269
Cu-6Al 523
Cu-6Al-0.2Mn 523
Cu-6Al-2Si 523
Cu-6Al-2Si-0.2Mn 523
Cu-6Cd 366
Cu-6Sn 579
Cu-6Zn-4Pb-4.7Sn 116
Cu-7Al 523, 577
Cu7Al2Sn 486
Cu-7.1Mn 117
Cu-9Al2Mn 268
Cu-10Al 577
Cu10Ni 494
Cu-10Ni-1.4Fe 154
Cu14Zn86 538
Cu-15.8Mn 117
Cu-15Zn 364
Cu-18Ni-17Zn 154
Cu-18Ni-27Zn 107
Cu-20Cd 366
Cu-20Zn-2Al 364
Cu20Zn2.2Al 223
Cu-22.9Mn 117
Cu-22Zn-2Al-0.03As 366
Cu25Ni alloy 501
Cu-28Zn-1Sn 364
Cu-28Zn-1Sn-P 364
Cu28ZnSn 316
Cu-30Ni 154, 494, 578
Cu-30Ni-0.7Fe 106
Cu-30Zn 109, 360–363, 581

Cu-31Zn-2Pb-0.8Sn-0.7Si-0.5Mn-0.9As 116
Cu-31Zn-2Pb-0.9Sn-0.8Si-0.6Mn-
    0.07As 116
Cu-31Zn-2Pb-08.Sn-1.5Al-0.03As 116
Cu-34.5Zn 113
Cu-34.5Zn-2Pb-1Sn-1Al-0.03As 116
Cu-34Zn-1Sn 108
Cu-34Zn-2Pb-0.05As 116
Cu34Zn66 538
Cu-36.7Zn 114
Cu-36Zn 361–362
Cu-36Zn-2Pb-0.03As 116
Cu-36Zn-2Pb-0.15As 116
Cu-37Zn 360, 362–363
Cu-38Zn 113
Cu-38Zn-1Pb 116
Cu-40Zn 113, 360–364
Cu-40Zn-2Pb 116
Cu49Au51 551
Cu-50Ti 367
Cu-50Zr 367
Cu59Au41 550–551
Cu60Zn40 539, 541
Cu65Ni18Zn17 332
Cu68Au32 550–551
Cu69Zn31 538, 540
Cu70Ni30 332
Cu70Zn30 539, 541
Cu 95.8 136
Cu 99.0 125, 143, 145
Cu 99.5 120–121, 133
Cu 99.9 120, 122–123, 125, 129, 136, 371
Cu 99.95 139
Cu 99.99 138, 430
CuAg0.06 143
CuAl2Si2Fe 429
CuAl2.5Si2.5 281
CuAl3 407
CuAl4 407
CuAl5 137, 280, 404
CuAl5As 374, 377, 431, 442, 448
CuAl5.5 407
CuAl6 429
CuAl6.1 430
CuAl7 407
CuAl7Fe1Mn1 429
CuAl7Fe2 138
CuAl7Ni1Fe0.3Cr0.05 430
CuAl7Si2 137
CuAl7.5Si2 281
CuAl8 14, 136–137, 176, 343, 377, 456,
    531, 599
CuAl8Fe3 531
CuAl8Ni2 138

CuAl9 176
CuAl9Mn2 268, 429
CuAl9Ni4.5Fe4Mn1 137
CuAl9Ni5Fe4Mn1.5 430
CuAl10 181, 456, 599
CuAl10Fe 53, 178
CuAl10Fe2-C 308, 531, 601
CuAl10Fe3 378
CuAl10Fe3Mn2 377, 429, 608
CuAl10Fe5Ni5-C 309, 531, 601, 603
CuAl10Ni 177, 181
CuAl10Ni3Fe2-C 232, 308, 430, 531
CuAl10Ni5Fe4 352, 429, 456, 608
CuAl11Ni 486
CuAl11Ni6Fe5 429
CuAl15 531
CuAl alloys 14, 28, 33, 37, 50, 105, 136, 181, 187, 203, 268, 280–281, 290, 330–331, 358, 403–404, 407–408, 420, 530, 575
CuAlFeNi alloys 136
CuAlNi alloy 352
CuAlSn alloys 531
CuAsP 427
CuAu50Cd7 225
CuBe0.6Co2.5 152
CuBe1.7 342
CuBe1.9Ni0.6 152
CuBe2 117, 180, 225, 342, 374, 387, 628–629
CuBe2Pb 342
CuBe alloys 290, 342, 345, 421
CuCd alloys 366
CuCoBe 117, 421, 527, 629
CuCr1-C 308
CuCr alloys 342
Cu-DHP 58–59, 70, 424
Cu-ETP 424
CuFe2P 427, 442
CuFe alloys 342
CuMn2 387
CuMn alloys 117, 289, 340
CuNi 139
CuNi2 139
CuNi3 283
CuNi3Si 486
CuNi4Mn13 139, 141
CuNi4Sn 145
CuNi4Sn4 141
CuNi4.6 283
CuNi5 433
CuNi5Al5Fe1.2Mn0.8 141
CuNi5Fe 379, 433, 441
CuNi5Sn5 380

CuNi6 139
CuNi9 139, 141
CuNi10 44, 429, 433, 533
CuNi10Fe 205, 316, 343, 391, 488, 562, 614–615
CuNi10Fe0.5 452
CuNi10Fe0.7Mn0.2Zn0.2 447
CuNi10Fe1Mn 70, 73–74, 76, 174, 189, 332, 343, 353, 379, 425–426, 428, 432–436, 438, 441–444, 447–448, 450–456, 468, 488, 490, 492–493, 496–499, 533–534, 544–547, 631
CuNi10Fe1.6Mn 426, 432
CuNi10Fe2Mn0.5 447
CuNi10Fe2Mn1 74
CuNi10Fe (CW352H) 445–446
CuNi10Zn25 139
CuNi12Zn24 141
CuNi12Zn24Mn 143
CuNi14Al3 432
CuNi14.5Al2.5 380
CuNi16Al5.5Fe3.7 380
CuNi17Mn5Al2Fe 432
CuNi18Zn15 139
CuNi18Zn20 141
CuNi18Zn21 225
CuNi20 615
CuNi20Fe 206, 332, 343, 379, 433, 614
CuNi20Zn20Mn 143
CuNi20Zn5 136, 139, 380
CuNi23 139
CuNi25 141, 353
CuNi29Sn1 136
CuNi30 44, 164, 177, 179–180, 283, 305, 343, 420, 433, 523, 614
CuNi30Be1 261, 321
CuNi30Fe 177, 180, 488, 491–493, 496–498, 500, 562, 614–615
CuNi30Fe0.5Mn0.5 447
CuNi30Fe1Mn 188, 631
CuNi30Fe1Mn1NbSi-C 189–190, 234, 309, 533
CuNi30Fe1.4 452
CuNi30Fe2Mn 332
CuNi30Fe2Mn2 70, 73–74, 76, 343, 426, 432, 434, 441–443, 450, 452
CuNi30Mn0.5Zn0.2Fe0.06 447
CuNi30Mn1Fe 73–74, 76, 205, 332, 343, 353, 379–380, 425–426, 432–434, 438–439, 441–444, 448–450, 452–455, 458, 533–534, 544–545, 547
CuNi30Mn1Fe1 534, 544, 546
CuNi40 420, 433, 523
CuNi40Fe 442

CuNi43Mn2   139, 141
CuNi44Mn1   353
CuNi45Mn1   280
CuNi50   283–284, 433
CuNi70/30   298, 515, 631
CuNi90/10   174, 515, 631
CuNi alloys   4, 15, 28, 34, 53, 139–141, 164, 224, 280, 283–284, 286, 299, 331–332, 336, 393, 420, 533, 558, 631
CuNiFe alloys   496–497
Cunifer® 30   189, 205
CuNiPd alloys   257, 259
CuNiZn   139
CuNiZn alloy   331
CuP0.4As0.4   143
CuPb alloys   342
CuPbSn alloys   342
CuPbSn-bronze   353
CuPd alloys   291, 519
Cu-PHC   424
Cupro Nickel®   206, 332, 374, 466
Cupro Nickel® 10   533
Cupro Nickel® 706   189, 205
Cupro Nickel® 710   189, 205
Cupro Nickel® 715   189, 205
Cupro Nickel® 754   189, 205
CuSi1.5   155, 341
CuSi2Mn   388, 463
CuSi2.64Fe0.04   464
CuSi2.9Zn0.5   144, 152
CuSi3.0   341
CuSi3   138, 155
CuSi3Mn   180, 303, 387, 463, 564
CuSi3Mn1   52, 143, 341, 345
CuSi4Mn1   5
CuSi alloys   3, 181, 289, 393–394
CuSiMn alloys   341
CuSn1   142
CuSn1P0.4   143
CuSn1.25   16
CuSn2Zn10   146
CuSn3   142
CuSn3Zn7Pb3   145
CuSn4   177, 457, 534
CuSn4Zn4Pb4   143
CuSn4.5P0.4   142, 144, 152
CuSn5   16, 18, 142, 145, 154, 456
CuSn5Fe   456
CuSn5Ni5   381
CuSn5P0.03   143
CuSn5P0.4   142
CuSn5Pb20-C   312, 550
CuSn5Pb4   145
CuSn5Pb5PbZn5   14

CuSn5Zn5Pb5-C   248, 310, 535
CuSn5.1   430
CuSn6   178–179, 190–191, 333
CuSn6Al2   141
CuSn6Al1Si0.2   430
CuSn6P0.03   143
CuSn6ZnAl   535, 544–547
CuSn7   142
CuSn7P0.3   143
CuSn7Pb15-C   312, 550
CuSn7Zn4   382
CuSn7Zn4Pb6   125
CuSn7Zn4Pb7-C   310, 535
CuSn7ZnPb   145–146
CuSn8   16, 136, 144, 154, 381, 534
CuSn8Al1Si0.2   456
CuSn8P   374
CuSn8Pb   139
CuSn9   142
CuSn9P0.3   143
CuSn10   16, 178, 617
CuSn10-C   235, 535
CuSn10-Cu   236, 309
CuSn10Fe4   14
CuSn10† 381
CuSn10Pb10-C   312, 550
CuSn10Pb15   14
CuSn10Zn   181
CuSn10Zn4   382
CuSn10Zn7Pb1   382
CuSn11Pb2-C   601
CuSn12-C   236, 535
CuSn12-Cu   309
CuSn20   14
CuSn24Zn18   225
CuSn25Pb5   144
CuSn alloys   16, 28, 34, 108, 142, 209, 284–285, 333, 410, 534, 563, 579, 620
CuSnFe   139
CuSnP   353
CuSnPb0.2   141
CuSnPb10   145
CuSnZn   353
CuSnZn-alloys   18, 38, 108, 147, 334, 412, 620
CuSnZn cast alloys   334
CuTi-alloys   289
CuZn   139, 423
CuZn8   143
CuZn9Sn6   382
CuZn10 (Ms90)   65, 138–139, 154, 213–214, 336, 344, 354, 536
CuZn10.5   383
CuZn12Sn1   149

Index of materials | 723

CuZn13 114
CuZn15 136, 143, 147, 149, 154, 181,
   213–214, 336, 344, 354, 382, 384–385,
   536, 623
CuZn16Si4 52, 177–178, 194, 383, 456
CuZn16Si4-C 311, 537, 550, 623
CuZn17Al12 288
CuZn17Al8 288
CuZn17Al9 288
CuZn18Al3 287
CuZn18Al5 288
CuZn20 (Ms80) 65, 138, 141, 147, 211,
   213–214, 286, 336, 343, 420, 524
CuZn20Al 316, 354
CuZn20Al2 74, 76, 80, 423, 425–426, 431,
   442, 447, 458–461, 499, 539, 544–547
CuZn20Al2As 80, 95
CuZn20.5Al2 154
CuZn21 114
CuZn22 287
CuZn22Al2 144, 461
CuZn22Al4 285, 287–288
CuZn22Al6 287–288
CuZn22Al8 287–288
CuZn22Al12 287
CuZn22Fe3Al5Mn4 149
CuZn23Mn3Fe2 383
CuZn25Al5Mn4Fe3-C 311, 537
CuZn27Sn1As0.1 85
CuZn28 143, 354
CuZn28Sn 316–317, 354
CuZn28Sn1 74, 80, 137, 139, 144, 154,
   269, 336, 344, 384, 431, 435, 442, 447,
   459–461, 499, 544–547
CuZn28Sn1As 80
CuZn29Sn1 148–149
CuZn30 (Ms70) 52, 65, 82–83, 85,
   110–112, 136, 138, 141, 148–149,
   150–151, 154, 176, 178–179, 181,
   213–214, 261, 336, 339, 344–345, 383,
   460–461, 525–526, 536, 623
CuZn31 114
CuZn31Si1 459
CuZn32 86
CuZn32FeAl 149
CuZn32.5 383
CuZn33Mn 151
CuZn33Pb2-C 248, 310, 537
CuZn33.5Pb0.5 150
CuZn34 125
CuZn34Mn3Al2Fe1-C 311, 537
CuZn35 461
CuZn35Mn2Al1Fe1-C 311, 537
CuZn35Ni2 459

CuZn35SnPbAlMn 151
CuZn35.5Pb3 154, 213, 336
CuZn36 66, 213–215, 374, 383–384, 386
CuZn36Pb 144
CuZn36Pb3 536
CuZn37 21, 65, 70, 86, 141, 143, 147, 225,
   302, 337, 344, 459–460, 524–525, 542
CuZn37Al1 459
CuZn38 114
CuZn38AlFeNiPbSn 95
CuZn38Sn1 461
CuZn38Sn1As 536
CuZn38SnAl 95
CuZn39Pb1Al-C 248, 310, 537
CuZn39Pb2 337, 344
CuZn39Pb3 70, 248, 337, 344, 459, 624
CuZn39Sn1 383
CuZn39.25Sn0.75 154, 336
CuZn40 (Ms60) 144, 151, 213–214, 336,
   344, 354, 385–386, 461, 525, 536, 553,
   624
CuZn40Mn 354
CuZn40Mn1Pb1 623
CuZn40Pb2 248, 354, 459
CuZn40Pb3 154, 213, 336, 344
CuZn41 420, 524
CuZn42 86, 338
CuZn44Pb2 624
CuZn50 144
CuZnAl 151
CuZnAl6 288
CuZnAl alloys 287, 582
CuZn alloys 21, 59, 141, 147–150, 213,
   269, 286, 302, 337–338, 360–364, 556,
   558
CuZnNi 151
CuZnNi-alloys 34
CuZnSn 139
CuZr-alloys 67, 289, 342
CW004A 244, 424, 553
CW005A 244
CW008A 244
CW020A 244, 424
CW021A 244
CW023A 183, 244, 323
CW024A 58–59, 70, 183–184, 198,
   201–202, 243–244, 271, 323, 374, 424,
   427, 553
CW027A 347
CW100C 342
CW101C 117, 180, 225, 342, 374, 387, 628
CW102C 342
CW104C 387
CW107C 427

CW116C 52, 180, 341, 345, 564
CW300G 374, 377, 448
CW301G 429
CW302G 137
CW303G 14, 228
CW306G 377–378, 429, 608
CW307G 429, 608
CW308G 429
CW350H 141
CW352H 70, 73, 174, 189, 205, 231, 343, 379, 425–426, 428, 433–436, 438, 441, 443–444, 447, 451–453, 455, 533–534, 544, 631
CW353H 70, 73, 343, 426, 441, 450, 452
CW354H 73, 188, 205, 231, 343, 379–380, 425–426, 433, 438–439, 441, 443–444, 448, 452–453, 455, 458, 533–534, 544, 631
CW383H 205
CW409J 229
CW450K 177, 457
CW451K 145, 154
CW452K 178–179, 190–191, 333, 343, 381
CW453K 136, 154, 534
CW459K 374, 381
CW491K 52
CW501L 65, 138, 154, 213, 336, 344, 536
CW502L 136, 147, 154, 181, 213–214, 336, 344, 382, 384–385, 536, 623
CW503L 65, 138, 141, 147, 211, 286, 336, 343, 420
CW505L 52, 65, 82, 110, 112, 136, 138, 148, 154, 176, 178–179, 181, 261, 336, 339, 344–345, 383, 460, 536, 623
CW507L 66, 213–215, 374, 383–384, 386
CW508L 21, 65, 70, 141, 147, 225, 337, 344, 459–460, 542
CW509L 213, 229, 336, 344, 385–386, 461, 536, 553, 624
CW603N 536
CW612N 337, 344
CW614N 70, 248, 337, 624
CW617N 248, 459
CW623N 624
CW702R 80, 95, 423, 425–426, 458–461
CW706R 80, 137, 139, 384, 435, 447, 459–460
CW708R 459
CW710R 459
CW715R 95
CW716R 459
CW717R 461, 536
CW719R 148
CW720R 623–624

d
Duralumin® 553
Durichlor® 234, 237–239
Durimet® 20 235, 237–239, 371, 586–587
Duriron® 234, 237–239, 371

e
E-Cu57 244, 424
E-Cu58 244, 424, 553
electrolytic copper 167, 400–401, 467–468
EN AW-6061 448
EP 371
Everdur® 5, 216
Everdur® 1010 216, 626
Everdur® 1015 5, 216

f
fluorocarbon elastomer 267
four material bronze 467
free-cutting brass 339, 348

g
galvanized steel 556
GC-CuAl9.5Ni 232
G-CnPb15Sn 312
G-Cu 307–308
G-Cu55ZnAl 193
G-Cu55ZnAl1 193–194
G-Cu55ZnAl1-2 211
G-Cu55ZnAl2 194
G-Cu55ZnAl4 211
G-Cu55ZnMn 193
G-Cu55Zn(Mn) 211
G-Cu60Zn 193–194
G-Cu65Zn 193–194
G-CuAl8Mn 309, 314, 429, 531
G-CuAl9 188, 203, 308, 314, 396
G-CuAl9-10 309, 314
G-CuAl9Mn 188, 195, 396
G-CuAl9Ni 39, 396, 430, 531
G-CuAl10Fe 39, 177, 187–188, 195, 203–204, 308, 396, 531
G-CuAl10Ni 39, 177, 396, 429, 432, 481, 531, 601
G-CuAl11Ni 429
G-CuNi30 (CW383H) 39, 189–190, 205, 309, 314, 396, 533
G-CuPb5Sn 311, 550
G-CuPb5-20Sn 317
G-CuPb10Sn 312, 550
G-CuPb15Sn 550
G-CuPb20Sn 312, 550
G-CuPb22Sn 312, 317, 550
G-CuPb25 216

Index of materials

G-CuSn5-7ZnPb   315
G-CuSn5ZnPb   39, 52, 145, 178, 210–211, 248, 310, 396, 458, 535
G-CuSn6ZnPb   381
G-CuSn7ZnPb   39, 210–211, 310, 396, 458, 535
G-CuSn9Zn2Pb3   601
G-CuSn10   145, 208, 309, 432, 535, 622
G-CuSn10-14   315
G-CuSn10Zn   39, 145, 192, 210–211, 236, 310, 315, 381, 396, 458, 535
G-CuSn12   535
G-CuZn15Si4   311, 316, 459, 537, 550
G-CuZn16Si4   194, 211–212
G-CuZn25Al5   311, 316, 537
G-CuZn28PbSn3   601
G-CuZn33Pb   248, 310, 316, 459, 537
G-CuZn34Al2   311, 316, 462, 537
G-CuZn35Al1   311, 316, 463, 537
G-CuZn37Pb   316, 537
G-CuZn40Fe   461
GD-CuZn37Pb   248
GD-Ms60   248
German silver   314, 353
GK-CuZn37Pb   248, 459
GK-Ms60   248
glass   371, 587
G-Ms65   248
gold   135
gunmetal   116, 331, 333, 343–344
Gunmetal®   191
GX7NiCrMoCuNb25-20   235, 237–239, 371

h

hardenable spring alloy   387
Hastelloy® B   371, 378, 586–587
Hastelloy® C   235, 237–238, 240, 371, 378, 586
Hastelloy® D   586
Hastelloy® F   586
Hastelloy® G   586
Hastelloy® X   586
Haveg®   586–587
Haveg® 41   371
Herculoy®   5
high-purity lead   378
high-silicon bronzes   345
high-silicon cast iron   587
high-strength brass   344
HSM copper   468–469

i

impervious graphite   371
IN-732   438–439, 496
IN-838   496, 500
IN-848   496
Incoloy® 800   586
Incoloy® 825   586, 589
Inconel® 600   220, 235, 237–239, 586, 589
Inconel® 601   586
Incramute   519
iron   417–419, 521, 574
iron-nickel-cobalt alloys   579

k

Karbate®   586

l

L-62   113
L-68   257
LaBour R55®   235, 237–239
lead   235, 237–238, 240, 371, 448, 554, 587
lead bronze   230, 620
leaded brass   337
leaded naval brass   337
lead-tin bronzes   230, 284
low-alloy copper alloys   281
LS-59-1   154

m

Maillechort   240
manganese brasses   336
manganese bronzes   61, 208–209, 229, 338, 467
marine brass   461
MgCu-alloys   161–162
mild steel   32
Monel®   5, 104–105, 190, 220, 252, 268, 426, 458, 587
Monel® 400   235, 237–238, 240, 253, 371, 373–374, 434, 589
Monel® 411   252
Monel® 505   252
Monel® B   448
Monel® K-500   371
Ms58   248, 354
Ms60   144, 151, 213–214, 336, 344, 354, 385–386, 461, 525, 536, 553, 624
Ms62   213, 215
Ms70   52, 65, 82–83, 85, 110–112, 136, 138, 141, 148–151, 154, 176, 178–179, 181, 213–214, 257, 261, 336, 339, 344–345, 383, 460–461, 525–526, 536, 623,
Ms72   354, 365
Ms80   65, 138, 141, 147, 211, 213–214, 286, 336, 343, 365, 420, 524
Ms85   343, 354

Ms90   65, 138–139, 154, 213–214, 336, 344, 354, 536
multi-component bronzes   478
muntz metal   164, 229, 337, 461, 553, 555, 565

n
N04400   378
N06600   378
N30107   235, 237–238, 240
naval brass   108, 337, 423, 459–460, 462
Nichrome®   586
nickel   220, 235, 237–239, 402, 417–419, 521
nickel-alloyed brass   546
nickel-plated brass   21
nickel silver   17, 107, 135, 139, 189, 206, 229, 331, 614
nickel-silver alloy 752   205
nickel-tin bronzes   231
NiCr15Fe   235, 237–239, 378, 589
NiCr21Mo   371, 589
NiCu30   433
NiCu30Al   253, 371
NiCu30Fe   235, 237–238, 240, 253, 371, 378, 434, 589
NiCu30Fe1.8   448
NiCu50   433
NiCu-alloy   268
NiMo16Cr   235, 237–238, 240
NiMo16Cr†   371, 378
NiMo30†   371, 378
Ni-O-Nel® Alloy 825   371
noble metal   272

o
OF-Cu   244
OTS 76   497, 508
ounce metal   164
oxidizing agents   497

p
Pb 99.9 Cu   371
Pb 99.985   371
Pb 99.99   371
PB985R   371, 378
PB990R   371, 378
PE   371
phenol resin   371
phosphor bronzes   16, 45–46, 51, 114, 142, 144, 208–209, 344, 480
phosphorus bronzes   333–334, 374, 578, 602
platinum   220

polyethylene   371
polytetrafluoroethylene   267
propeller bronze   298
pure copper   45, 71–72, 119, 126, 148, 168, 171, 192, 217, 251, 277, 280, 285, 305, 307, 324, 331, 351, 423, 468, 478, 486, 491, 499
pure titanium   544, 547
PVC   371
PVDC   371

r
RA 330   586
RA 333   586
red brass   18, 20, 29, 34, 38, 54, 108, 145, 164, 191–192, 210, 236, 246–247, 253, 273, 284, 306, 315, 334, 337, 353, 382, 393, 411, 424, 458, 503, 524, 535, 556, 563, 565, 579–580, 601, 620–621, 629
red brass (gunmetal)   503
red bronze   300, 420
red casting   210
red casting grades   210
rolled alloys   207
rubber   371, 587
rural atmosphere   148

s
SAE 302   499
SAE 304   32, 235, 237–239, 378, 589
SAE 316   235, 237–239, 371, 378, 499
SAE 316 L   589
SAE 317   378
SAE 321   123, 570
SAE 347   570
SAE 409   123
SAE 430   123, 235, 237–239
SAE 444   390
SAE CA-360   339
Saran®   371
SE-Cu   244, 424
SF-copper   183, 185, 307
SF-Cu   58–59, 70, 184–185, 198, 201–202, 243–244, 271, 316, 323, 347, 424–425, 427, 442, 553
silicon bronzes   24–25, 29, 45–46, 55, 152, 216, 273, 303, 374, 387, 626–627, 629
silicon bronzes 651   216
silicon red brass   337
silver   99, 325
smelted copper   467
Sn8Pb6   14
Sn-bronze   18
SnPb-casting bronzes   285

SoMs58   354
SoMs71   354
SoMs76   354
special brass   354
spring alloys   117
stainless steel   32, 267, 299, 351, 497, 557
steel   220, 299, 557
steel bronze   331
Supernickel® 702   189, 205
SW-copper   183
SW-Cu   244, 323

t
tantalum   220, 235, 237–238, 240, 267
Teflon®   267
TIBRAL   459
tin   353
tin brass   230
tin bronzes   45–46, 141, 145, 207–208, 284, 333, 343, 400, 411, 424, 456–457, 480, 556, 558, 616–617
tin-lead bronzes   333
tin-zinc alloys   108
titanium   46, 299, 547
Tobin bronze   208–209, 211, 336–337, 565
tombac   193, 336, 343
tough-pitch copper   556
trumpet brass   337

u
UP   371

v
varnished copper   272
Viton®   267

w
Worthite®   587
wrought copper alloys   307

x
X2CrNiMo17-12-2   589
X3CrNiMo18-12-3   378
X5CrNi18-10   235, 237–239, 378, 589
X5CrNiMo17-12-2   235, 237–239, 371, 378, 544, 547
X6Cr17   235, 237–239
X6CrNiNb18-10   570
X6CrNiTi18-10   570
X40Cr13   481

y
yellow brass   22, 337, 339, 341

z
zinc   33, 35, 448, 506, 574

Subject index

a

AAC 138
AAC-test 151
absorption refrigeration plant 77
acadia clay 555
accelerated atmospheric corrosion test 138
acetaldehyde 31
acetic acid 5–6, 14, 16–17, 48, 55, 177–178, 181
acetic acid (crude) 5
acetic acid (dilute) 14
acetic acid-formic acid mixtures 5
acetic acid media 6
acetic acid vapors 7, 15–16, 20–21
acetic anhydride 9, 15, 18, 23, 25
acetic anhydride vapor 12
acetone vapors 41
acetyl cellulose 198, 207
acetyl chloride 27–29
acetyl fluoride 27
acid 298
acidic rain 119, 135
acid mixtures 293
acid pumps 351
acid vessels 33
acid washing 298–300, 303
acoustic emission 365
acridine (as inhibitor) 362
acrolein 31
AEMT (as inhibitor) 511
aeronautics 153
aircraft carrier 441
aliphatic aldehydes 35
alkali chloride solutions 43
alkali hydroxide melts 417
alkaline arsenates 389
alkaline earth hydroxides 45
alkaline soils 553–554, 558
alkaline solutions 3
alkali soil 555
alkanecarboxylic acids 47–49, 54–55
alkane sulfonyl chlorides 28
alloys for springs 421
allyl thiourea (as inhibitor) 360
Altoona 136–137, 149
aluchromizing 359
aluminium acetate 3
aluminium brass pipes 223
aluminium bronze pipe 73
aluminium chloride 61
aluminium chloride melts 61
aluminium chloride solutions 61
aluminium depletion 479, 481, 483
American east coast 139
American west coast 141
amines 39
amine salt 64
m-aminobenzoic acid (as inhibitor) 362
p-aminobenzoic acid (as inhibitor) 362–363
aminoethyl mercaptotriazole (as inhibitor) 511
aminomethyl mercaptotriazole (as inhibitor) 511
m-aminophenol 66
p-aminophenol 66
aminopropyl mercaptotriazole (as inhibitor) 511
amino-sulfonic acid 567
2-aminothiazole (as inhibitor) 363
aminotriazole (as inhibitor) 501
ammonia 150, 424, 452, 461, 631
ammonia gas 69–70, 72
ammonia synthesis 133
ammonia vapor 81
ammonium acetate solutions 3

Subject index

ammonium bicarbonate solutions 104, 559
ammonium bromide 157, 160–161
ammonium bromide solutions 160
ammonium carbonate solutions 106
ammonium chloride solutions 100, 104–105, 107–108, 112, 114–115
ammonium compounds 515
ammonium ions 80
ammonium oxalate 103
ammonium persulfate solutions 106
ammonium salts 97, 424, 461
ammonium sulfate solutions 97–100, 108–109, 111–113
AMMT (as inhibitor) 511
amylamines 37
anaerobic degradation 95
anhydrous ethylene glycol 389
anhydrous fluorophosphoric acids 27
anhydrous potassium chloride 410
p-anidisine (as inhibitor) 362–363
aniline 37
aniline (as inhibitor) 356
animal farm 86
o-anisidine (as inhibitor) 364
anodic dissolution 540
anodic oxidation 523
Antarctic 126
β-anthraquinonesulfonic acid 567
antifouling properties 428
antifreeze agent 504
antifreeze coolants 390
antique bronze statues 334
APMT (as inhibitor) 511
aqua regia 367
aqueous sulfur dioxide solutions 580
archaeological findings 558
Argentina 123
Armohib® 28 512, 514
ascorbic acid 175, 257
astronautics 153
atmospheric corrosion 579
atmospheric humidity 409
autoclaves 325
automobile engines 390
automobile manufacturer 391
auxiliary heat exchangers 515

b

bacteria 72
bacterial microfouling 429
bending fatigue strength 404
benzene 153
benzenesulfonic acid 567
benzidine (as inhibitor) 356
benzimidazole 102
benzimidazole (as inhibitor) 115, 476
2-benzimidazolethiol (as inhibitor) 476
2-benzooxazolethiol (as inhibitor) 476
2-benzothiazolethiol (as inhibitor) 476
benzotriazole 37, 101–103, 114, 198, 222, 261, 390, 512, 567
1,2,3-benzotriazole 391
1,2,3-benzotriazole (as inhibitor) 476
benzotriazole (as inhibitor) 199, 355, 360, 477, 487, 501, 510, 596, 625
benzotriazole monoethanolamine 390
benzoyl chloride 28–29
benztriazole 102, 135
benztriazole (as inhibitor) 115, 135
benzylamine 37, 39
benzylamine benzoate 65
benzylamine cinnamate 65
benzylamine dihydrocinnamate 65
beverage industry 167
bilge water 459
bimetallic pipes 33
biochemical reaction 373
biocidal properties 556
biocide 317
biofilm 72
biological growth development 496
bistriazole (as inhibitor) 501
bleaching lime 45
bleaching of cellulose 237–240
bleaching process 236
bleaching solutions 232, 234, 236, 240, 313
bleaching tower 236
bleach tower 239–240
boat hulls 448
boiler feed water 77, 85
boiler feed water processing 72
boiling formic acid 272
boiling glycerin 391
borate-buffered solution 63
brackish water 428, 430, 433, 441, 448, 456, 459, 562
brass component 303
brass pipe 86
brass radiators 389
brass sheets 109, 366
brass sleeves 504
brazing material 519
bridge spans 447
bromination 163

bromine   163–164
bromine solutions   163
bromine vapor   163
bromine water   164
bromoacetic acid (as inhibitor)   360–361, 363
bronze cancer   334
bronze monuments   147
bronze spindles   16
Brooklyn   141–142, 149
buffer solution   508
buoys   447
butanol   57, 59
butylamines   37
butyraldehyde   31–32, 35
butyric acid   27, 48–49, 53–55
butyryl chloride   27–29

c
cable industry   275
calcium acetate solutions   3
calcium bromide solutions   157, 161
calcium carbide   5, 14
calcium hydroxide solutions   45
calcium hypochlorite   45
calcium hypochlorite solutions   307, 313–316
caproic alcohol   32
caproic aldehyde   31–32
carbon dioxide   276
carbonic acid content   558
carbon monoxide   276
carbon particles   465
carbon tetrachloride   200, 202, 204–205, 208, 210, 212, 214
carbon tetrachloride (dry)   200, 216
carbon tetrachloride (moist)   216
carbon tetrachloride (water free)   206
carboxylic acid esters   175
carboxylic acids   175
Caribbean   457, 464
carnallite liquor   410, 411
carnallite solutions   395, 413
cartridge sleeve   81
catalysts   175, 324
cathode   302
cathodic corrosion protection   455
cathodic polarization   428
cathodic protection   456
caustic potash solution   403–404
cavitation corrosion   522
CCl4 (dry)   205
CCl4 (moist)   205

CCl4 (water free)   201
CCl_4 200, 208, 210, 212
CCl_4/methanol mixture   202
CCl_4/water mixture   200, 213–214
cecil clay loam   555
cellulose production   569, 577
CERT method   508
chemical etching   525
chemical milling   251
chemical plant construction   69, 167
chemical polishing   357
Chicago   134
chinalizarin   525
chinese bronze mirrors   144
chloramines   37, 39
chloride content   146
chloride immission rate   128
chlorinated hydrocarbons   197, 203
chlorinated lime   45
chlorinated water   217, 220
chlorine   133, 217, 219, 225, 452
chlorine as oxidizing agent   74
chlorine gas additions   496
chlorine stream   219
chloroacetaldehyde   31
chloroacetic acid (as inhibitor)   360–361, 363
chloroaniline   39
m-chloroaniline (as inhibitor)   363
o-chloroaniline (as inhibitor)   356, 362
p-chloroaniline (as inhibitor)   362–363
chlorobenzoyl chloride   28–29
chloroform   199, 206–207, 216
chloroform (dry)   199, 205–207, 210, 216
chloroform (gas)   206–207, 210, 216
chloroform (moist)   199, 205–207, 210, 216
chloroform (water free)   201, 206
chlorosulfonic acid   28–29
cinder   555
clay   553–554
cleaning copper   357
cleaning solution   257
coastal atmosphere   119, 121, 123–124, 126, 128, 132, 134, 137–142, 144, 147–151
coastal power plants   424
coastline   123, 132, 141–142
COBARTEC® TT 100   390
coking plants   154
cold climate   126
cold water pipes   172
columns   181

concentration elements 424
condensate drops 73
condensate pipes 497, 507
condenser pipes 95, 298, 437, 441, 444, 447, 504
condensers 69, 72, 80, 86, 167, 430, 438, 444
condenser tubes 404
conductor plates 153
construction of containers 267
coolants 389
cooling liquid 72
cooling water 72–73, 80, 86, 316, 353, 503
cooling water line 72
cooling water systems 512
Copenhagen 119
copper apparatuses 597
copper as contact material 71
copper bromide 160
copper catalysts 32
copper cladding 197
copper components 174
copper corrosion 170
copper depletion 494
copper grounding 556
copper gutters 119
copper(II) chloride 257
copper(II) chloride solutions 251
copper material 307
copper oxide layers 174
copper oxide powder 266
copper pipe materials 424
copper pipes 41, 72, 167, 170, 325, 521, 556
copper powder 78
copper sulfate 78
copper tanks 187
copper tetramine solutions 65
copper tetrammin complex 70
copper tetrammin ions 71
copper wires 356
copper-zinc-aluminium alloys 74
corrosion in carbon tetrachloride 200, 212
corrosion in chloroform 199, 212
corrosion inhibition 455
corrosion inhibitor 7
corrosion in methyl chloride 197, 211
corrosion in methylene dichloride 198, 212
corrosion resistance 295
cowshed 72
crack formation 85
crevice corrosion 74, 444, 461

Cristobal (Panama) 124, 142, 144
critical pitting potential (in drinking water) 245
cross specimen 87
crotonaldehyde 31, 33–34
trans-crystalline crack course 81
trans-crystalline stress corrosion cracking 81–82, 95
cupric chloride 219
cuprous acetate 3
cuprous chloride 219
current density/potential curve 295–296
current density-voltage curve 352
Cuxhaven 132
cyclohexamine chromate 64
cyclohexylamine 37–38
cyclohexylamine benzoate 65
cyclohexylamine carbonate 64, 66
cyclohexylamine chromate 65
cyclohexylamine cinnamate 65
cyclohexylamine dihydrocinnamate 65
Czech Republic 120–121, 133

d

damp chlorine 239, 314
Daytona Beach (FL, USA) 139, 141–142, 147, 149
dead space 73
dealumination 424, 428
dealuminization 187, 331, 409
dealuminizing 203
decopperization 227
dedusting 133
defect structure model 491
dehydrogenation 32, 35
demineralized feed water 559
depassivator 103
deposits 298
desalination unit 220
dezinc 353
dezincification 22, 193, 236, 246–247, 273, 335–336, 338, 424, **459**–461, 537, 539–540, 553–555, 564, 622
dezincification potential 537
dezincification progresses 336
dezincing 303, 316, 354
diacetyl fluoride 27
diaminotriazole (as inhibitor) 477
dibutylthiourea 298
dibutylthiourea (as inhibitor) 300
dichan 135
dichloroacetic acid (as inhibitor) 360–361, 363

Subject index | 733

dichloroethane 188–189, 191
1,1-dichloroethane 184, 194
1,2-dichloroethane 183–184, 189, 191
dicyclohexylamine nitrate 65
dicyclohexylamine nitrite 65
diethanolamine 39
diethylamine 37
diethylamine (as inhibitor) 475
diethylamine, dry 39
diethyl dithiocarbamate (as inhibitor) 475
diethylenetriamine 39
dimethylamine 39
dimethylol thiourea (as inhibitor) 355
dimethyl phthalate 37
dimethyl thiourea (as inhibitor) 360
diphenylamine 37
1,5-diphenyl carbazide 114
diphenyl thiourea (as inhibitor) 360
disinfectant 227
distillation 6
distillation apparatus 391
distillation of methanol 324, 338
dithioglycolic acid (as inhibitor) 361
dithiooxamine 525
domestic connections 557
domestic installations 71, 86, 227, 243
Dowtherm® SR-1 392
Drewsol® TM 392
drinking water 69, 223, 227, 236, 238, 240–241
drinking water distribution systems 171
drinking water lines 72, 243
drinking water pipes 167
drinking water supply 173
drive propellers 424
dry carbon tetrachloride 208
dry chloroform 203, 205, 212
drying pans 3
dry methylene dichloride 205
Duisburg 132

e

Egypt 123
electrical engineering 119, 127
electrochemical etching 525
electrolyte 358
electronic components 217, 225
electropolishing 293, 295, 357, 366
elevated temperatures 275
embrittlement coefficient 364
embrittlement (inter-crystalline) 519
Emmental cheese 14
eosine 525

equilibrium pressure 282
erosion 507
erosion corrosion 74, 246, 468, 470, 481, 488, **495**, 534, 536, 544–545
erosion corrosion behavior 76
erosion corrosion test rig 546
esterification plants 105, 182
esterification tank 182
etching additive 103
ethanedisulfonic acid 567
ethanol 57–59, 328
ethanol synthesis plant 387
ethoxylated alcohol 64
ethyl alcohol 181
ethylamine 38
ethylenediamine 37, 39
ethylene glycol 389–390
ethylene glycol antifreeze agents 390
ethylene glycol solutions 389
ethylsulfonic acid 567
Europe 125
evaporating stage 521
evaporation (industrial sewage water) 505
evaporators 77, 221, 426, 521–522
evaporator system 298
external oxidation 289

f

feed water 85
feed water processing 85
fermentation 14, 72
fermentation processes 14
ferry ship 74
Finland 121, 141, 147
fire extinguishers 267
fishing fleet 448
fittings 86, 181, 351
fittings industry 151
floating islands 447
flow rates 73, 424
flow turbulence 496
flue gas desulfurization plant 631
fluoride film 264
fluorination 27, 264
fluorination of copper 265–266
fluorine 263, 268
fluorine atmosphere 264
fluorine (high-purity) 265
fluorophosphoric acids 27–28
fluorosulfonic acid 27
Foodfreeze® 35 392
food industry 6, 175
foodstuffs industry 393

formaldehyde 31, 33, 332
formaldehyde coolers 33
formaldehyde manufacturing 33
formaldehyde manufacturing plants 34
formaldehyde solutions 31–33, 35
formaldehyde synthesis 334
formaldehyde vapors 35
formic acid 175, 271, 273
fouling 74, 95, 428, 448, 464
fouling organisms 428
fouling resistance 428, 441
fouling-resistant 434
fractionation column 31
fracture mechanics specimens 87
fresh water pipeline 69, 167
friction bearing material 353
fructose (as inhibitor) 355
furfuryl aldehyde 525
furnace gases 280

g

galactose (as inhibitor) 355
gallic acid (as inhibitor) 356
galvanic anodes 455
galvanic corrosion 391
galvanic element 6, 302
galvanic element formation 73
gaseous chlorine 220
gasoline 329, 339, 341
gasoline/methanol mixtures 341
gas welding 335
glacial acetic acid 5, 7, 13, 15
glucose 525
glucose (as inhibitor) 355
glycerin 389, 391, 393
glycerin solutions 393–394
glycols 389, 391
Göteborg 121
Great Britain 146
grounding 557
grounding electrodes 556–557
groundwaters 553, 556
guanidine carbonate 65
guanidine chromate 65

h

Hagerstown loam 555
halogenated hydrocarbons 211
halogenated organic media 211
Harbor Island 225
harbor water 75, 459
hard salt solutions 467–468
heat-affected zone 596

heat exchanger pipes 72, 74, 80, 505
heat exchangers 5, 69, 72–74, 86, 95, 167, 227, 283, 302, 426, 429–431, 434, 444, 481, 515, 524
heat exchanger tubes 536, 547
heating coils 391
heating system 504
heat pump 69
heat recovery applications 450
heat transfer medium 392–393
heat-treated state 296
heptanol 57
hexachlorobutadiene 202
hexachloroethane 187–188, 190, 192, 195, 202
hexamethyleneamine benzoate 64
hexamethyleneimine-3,5-dinitrobenzoate 63
hexamethyleneimine benzoate 65
hexamethyleneimine-m-nitrobenzoate 63
hexamethyleneimine-o-nitrobenzoate 63, 65
hexamethyleneimine-o-nitrophenolate 63
hexamethylenetetramine 37
hexamethylimine-o-nitrobenzoate solution 64
hexamine (as inhibitor) 356
high temperature water 69
hollander machines 232
hot extraction 302
hot spot corrosion **73**–74,
hot water storage tank 504
hot water system 72
household 503
household installations 465
housings 209
Houston 134
hull plating 450
humic acids 556
hydraulic fluids 391
hydrazine 72, 85
hydrazine hydrate 559
hydrazine sulfate (as inhibitor) 362
hydrochloric acid 133, 176–177
hydrochloric and chromic acid 293, 295
hydrochloric and nitric acid 293, 295
hydrofluoric acid 261, 267–268
hydrogen 82
hydrogen charging 625
hydrogen chloride 305
hydrogen chloride (dry) 305
hydrogen chloride gas 305
hydrogen chloride (moist) 305

hydrogen embrittlement   519
hydrogen fluoride   268
hydrogen sulfide   133, 280, 324, 488, 631
hydrolyzer   64
hydroquinone (as inhibitor)   80

i

ignition temperature   219
immersion zone   436, 448–449
impact corrosion   496
impact zone   76
impellers   430, 458
impeller wheels   503
impingement attack   224
Incracoat®   135
Incralac®   135, 147
indole   102
indole (as inhibitor)   115, 510
industrial atmosphere   64, 119, 121–122, 125, 128, 132, 134, 137, 139, 141–142, 145, 149, 151, 217, 573–574, 578, 581
industrial coastal atmosphere   141, 149
industrially pure phosphoric acid   383
industrial urban atmosphere   138
inhibiting effect   38, 158–159, 280, 404, 567, 597, 625
inhibition efficiency   64, 100–101, 114, 222, 261, 355–357, 360, 363, 366, 385–386, 597
inhibition efficiency (in HCl)   300
inhibition of corrosion in sulfuric acid   596, 614
inhibitor   38, 79–80, 114–116, 158–159, 172, 221, 257, 261, 298, 300, 351, 355–357, 360–364, 366, 385–386, 418, 420, 523–524, 560
inhibitor addition   77
inhibitor efficiency   222–223, 625
inorganic acids   175
intercrystalline corrosion   7, 211
inter-crystalline cracking course   81
inter-crystalline stress corrosion cracking   81
intergranular attack   389
intergranular cracks   109
intergranulare corrosion in sulfuric acid   596
internal oxidation   281–282, 285
iron(II) chloride   251
iron(III) chloride   249, 251, 254
iron(III) chloride solutions   251
iron ions (as inhibitor)   74
isobutyraldehyde   35

isobutyric acid   53
isopropanol   328
isopropylamine   38

k

KCl melts   478
Key West (FL, USA)   123, 136, 149
KOH melt   417, 419
Kure Beach (NC, USA)   123, 139–140, 148, 152

l

lactic acid   179
La Jolla (CA, USA)   123–124, 136, 149
Lake Charles clay   555
Langmuir-Blodgett technique   473
layer-type dezincification   **503**
lead bromide   157, 161
lids   441
liquid methyl chloride   197
lithium arsenite   159
lithium borate   159
lithium chromate   159
lithium hydroxide   158, 261, 319
lithium hydroxide solutions   319, 321
lithium molybdate   158
lithium nitrite   158
lithium perchlorate   159
lithium phosphate   159
lithium silicate   159
lithium silicofluoride   159
lithium thiocyanate   159
lithium tungstate   159
lithium vanadate   159
loam soils   553
local cathodes   465
local pitting   553–554
logarithmic growth law   279
loop samples   366
Los Angeles   134

m

machine oil   70
Madrid   132
maize extract   355
Makrolon®   153
Mandapam Camp (India)   142, 148
mannose (as inhibitor)   355
manufacture of butyraldehyde   32
manufacture of propionic acid   47
maritime technology   503
marshy soil   553–554

Mattsson solution   109–111, 113, 117, 508–509
Mattsson test solution   111
mechanical properties   295
medical technology   167
Mediterranean Sea   444
mercaptobenzimidazole   102
2-mercaptobenzimidazole (as inhibitor)   360
mercaptobenzimidazole (as inhibitor)   115
mercaptobenzothiazole   222, 389–390
2-mercaptobenzothiazole (as inhibitor)   360
2-mercaptobenzoxazole   101
mercaptobenzoxazole   102, 114
mercaptobenzoxazole (as inhibitor)   115
mercaptobenzthiazole   102
mercaptobenzthiazole (as inhibitor)   115
merced building clay   555
metaldehyde   31
methanol   57, 59, 323–324, 328, 337, 343
methanolamine, impure   39
methanol fuel mixtures   330
methanolic solutions   326
methanol synthesis   323–324, 331, 338
methanol vapor   325
methyl acetate production   182
methylamine   38
methylamine, dry   39
methylamines   37
methylamine solutions   37–38
N-methylaniline (as inhibitor)   363
methyl chloride   197, 205, 210
methyl chloride (dry)   197
methylene dichloride   198, 200, 203, 205, 210, 212
methylene dichloride/alcohol mixtures   198, 207
methylene dichloride (dry)   198
methylene dichloride (moist)   205, 212, 216
methylene dichloride (pure)   212, 216
methylene dichloride solutions   198
methyl orange (as inhibitor)   476
methyl red (as inhibitor)   476
methyl yellow (as inhibitor)   476
Mexico City   135
MIC   72
microbiological degradation products   95
microbiologically induced corrosion   72
microbiological processes   72
microorganisms   95, 556
milk   175

mineral insulation   557
mineral waters   174
Miraflores (Panama)   124
mixed acids   351
mixed inhibitors   512
mohave sandy loam   555
moist chloroform   203, 210
moist sulfur dioxide   578
molten lithium hydroxide   321
molten potassium chloride   402
monoacetyl fluoride   27
monocarboxylic acids   47
monochloroethane   188–189, 191–192, 194–195
Monza (Italy)   145
Moscow   121–122, 128
muck   555
multi-material aluminium bronzes   481
Murmansk   128

n
NaCl deposits   502
NaCl melt   478
NaCl solution   430, 433, 461
NaOH melt   419
naphthotriazole   101, 114, 261
1,2-naphthotriazole (as inhibitor)   385
naphthotriazole (as inhibitor)   360, 582
naphthylamine   39
α-naphthylamine (as inhibitor)   356, 364
Newark (NJ, USA)   152
New Haven   141–142
New Jersey   134
New York   122, 134, 136–137, 140, 149
New York drinking water   228–231
nickel acetate   21
nickel depletion   515, 519
nickel enrichment   494
nickel pipes   521
nickel plated pipes   72
nitrates   515
nitric and hydrochloric acid   367
nitric and phosphoric acid   293–294, 357–358
p-nitroaniline (as inhibitor)   361–363
nitrogen dioxide   129–130, 133
nitromethane   37
non-aerated phosphoric acid   373, 375, 381
non-agitated phosphoric acid   379
nonanol   57
North America   125, 133–134
North Carolina   225

Northern Indiana   134
nuclear power stations   564
Nutek® 835   392
nut material   614

o

octadecylamine   37
N-octadecylbenzidine (as inhibitor)   473
offset printing machines   595
offshore structures   432
oil cooler   74, 95
oil drilling platforms   488
oil tanks   556
oleum   586
organic acid   273
organic degradation products   74
organic refrigerants   197
organic thermal insulation materials   71
orthophosphoric acid   103
outdoor exposure   86
oxalic acid   178
oxidation   280
oxidation of copper   279
oxidation rate   275
oxidation rate of copper   277–278
oxide layer   352
oxidizing agents   6, 82, 181, 410, 524
oxidizing atmospheres   275
oxidizing gases   275
oxygen corrosion   351
oxygen partial pressure   282
oxygen pressure   275
ozone   126, 133, 494

p

packing industry   135
Panama Canal   448–449, 457, 464
Panama Canal zone   138
paper industry   236, 353, 478, 486, 578
paper manufacturing   236
papermill   572, 577, 581
paraformaldehyde   31
paraldehyde   31
Paris Green   3
passive current densities   352
patina   502, 558
peat   555
pentachloroethane   187–188, 190, 195
pentanedisulfonic acid   567
pentanol   57, 59
peracetic acid   182
perchloroethylene   58, 202
periods of downtime   174

Persian Gulf   451–452
petroleum industry   622
petroleum refining   617
o-phenetidine (as inhibitor)   356, 364
phenol-formaldehyde   33, 35
phenylamine   37
m-phenylenediamine (as inhibitor)   361
o-phenylenediamine (as inhibitor)   361, 366
phenylhydrazine (as inhibitor)   362–363
phenyl thiourea (as inhibitor)   360, 364
phloroglucin (as inhibitor)   80
phloroglucine   525
Phoenix/AR   122
phosgene   27, 29
phosphoric acid mist   388
phosphoric acid vapor   374
phosphoric and sulfuric acid   378
phosphorus trichloride   27
pickle   295
pickling   293, 567
pickling agents   267, 378
pickling copper   251
pickling process   302
pickling solutions   269, 378, 605
pickling station   133
pickling tank   603
pipe   86
pipelines   424, 447
pipelines conveying sea water   74
pipe material   515
pipe material (for sea water desalinization)   497
pipe material in sea water   488
pipe perforation   465
piperidine-3,5-dinitrobenzoate   63
piperidine (as inhibitor)   475
pitting corrosion   85, 232, 234, 238, 240, 246, 267, 313, 329, 336, 353, 428, 430, 435, 443–444, 448, 453, 461, 464, 555
pitting corrosion (in drinking water)   245
pitting corrosion (in H_3PO_4)   369, 375, 379
pitting corrosion (in hypochlorite solution)   307
pitting corrosion – type 1   172
plant shutdown   515
plated structural element   428
plating material   441
plug-type dezincification   **503**
plug-type zinc depletion   141, 150
plug zinc depletion   504
Point Reyes (CA, USA)   137, 140–141, 146, 148, 152

polarisation curves   352
polishing of copper   357
polyethylene glycol   64
polyethylene glycol ethoxylated alcohol   67
polyglycols   389
polyols   389
potash industry   395, 400, 408, 411, 467
potash salt solution   404
potassium acetate solutions   3
potassium bromide   157
potassium bromide melts   160
potassium bromide solution   157, 159
potassium chloride   395, 397, 414
potassium chloride solutions   395, 397–398, 400–403, 409–414
potassium dichromate   77
potassium fluoride   261
potassium hydroxide   320
potassium hydroxide melts   419
potassium hypochlorite solution   313–315
Pourbaix diagram   417
power plant   72
power plant condensers   488
power plant engineering   504
power plants   515
Prague-Letnany   120–122, 133
preheaters   467
pressure steam power plants   559
Prestone® II   393
process water   69
production of acetic acid   5
production of butyl acetate   182
production of ethyl acetate   181
production of furfurol   597
production of triacetyl cellulose   199
propanol   57–59
propeller   430, 462–463
propeller bronze   298
propionaldehyde   31, 35
propionic acid   5, 14, 47–50, 53, 55
propylene glycol   389, 392
protective layer   423
pulp bleaching   236
pulp industry   478, 486
pulp production line   487
pump bodies   458
pump casings   424, 458
pump components   77
pump housings   503
pump impellers   424
pumps   77, 430, 432, 467
pump wheels   481
pure benzene   153–154
pure chlorine dioxide   227
pure fluorine   263
pure glycerin   391
pure potassium chloride solutions   411
pure water   389
pyrocathechol (as inhibitor)   80
pyrogallol (as inhibitor)   80

q
Quebec (Canada)   145
quinine (as inhibitor)   597
quinoline (as inhibitor)   385–386

r
railway tunnels   578
rain box   138
rainwater   131
rapid vinegar process   14
reaction tank   181
rectification columns   324, 338
rectifying columns   5
refrigerant   69–70
refrigeration engineering   70
refrigeration facilities   70
refrigeration machines   197
refrigeration plants   69
relative humidity   127, 129–130, 133–134, 146, 151
repassivatable pitting   246
resistance to erosion   479
resistance to fouling   74, 428
resorcin   525
resorcinol (as inhibitor)   80
Reynolds' number   452
river water   503
Rodine® 213 Special   116
rolling mill   133
rotor blades   479
rural atmosphere   119, 121–124, 126–127, 137–142, 144, 147, 150–152

s
saccharification   597
salient flanges   441
saline heater   441
saline solutions   448
salt deposit   302
salt melt corrosion   478
salt melts   478
salt production   505
salt solution   77
salt spray test   505
saltwater cooling towers   432

Sandy Hook   136, 149
sandy loam   553–554
Scilly Islands   459
screw material   614
sealing rings   324
season cracking   81
seawater   72, 74, 80, 107, 223, 225, 296,
    298–300, 315, 353, 473, 481, 486, 488,
    495–496, 500, 503
seawater desalinization   455, 488, 496
seawater desalinization plants   174, 300,
    427, 431–432, 441, 448, 450, 455, 458,
    460, 462–463, 472, 480, 497, 505, 631
seawater desalinization unit   220
seawater evaporator plant   434
seawater flash distillation plant   298
seawater heater   221
seawater (natural)   492, 495
seawater pipelines   433, 441
seawater (synthetic)   488
semiconductor   153, 244
sharkey clay   555
sharp bends   441
shell sleeve   81
shipbuilding   447–448, 456
ship propellers   430
ships   423, 488
ship screws   479, 481
shut-off element   379
shut-off valves   181, 209
silver catalysts   32
silver chloride   217
Simulated Solar Service Test   392–393
single crystals   217
$SnCl_2$ (as inhibitor)   385–386
sodium acetate solutions   3–4
sodium bromide   160
sodium bromide solution   161
sodium chloride production   133
sodium chloride solution   161
sodium cinnamate   390
sodium dichromate (as inhibitor)   456
sodium gluconate   418
sodium hydroxide   320, 389
sodium hydroxide melts   522
sodium hypochlorite solution   307,
    313–317
sodium selenite   389
sodium sulfate   63
sodium tellurite   389
soils   553–554, 556, 558
solar collectors   390
solar energy systems   391

solar radiation   132, 163
soldering flux residues   503
solid material deposits   503
sorbitol   389
sorbitol solutions   394
sources of ammonia   77
South Carolina   134
space technology   267
Spain   121, 132
special brass   80
spindles   430
spinning bath solutions   601, 620, 623
sponge balls   224
spring alloys   527, 629
spring water supply   77
standard hydrogen electrode   293
State College (PA, USA)   124, 136–137,
    139–140, 149
stationary conditions   503
steam condenser pipe   85
steam condensers   72, 559
steam generation plants   559, 561
steam generators   154
steel hulls   448
steel pickling bath   615
Stockholm   121, 147, 149
storage hall   133
storage tank   325, 391
storage vessel   31
stress corrosion cracking   65, 70–71,
    73–74, 81–82, 85–87, 91, 95, 107–109,
    113, 117, 137, 151, 211, 232, 331, 353,
    357, 364, 395, 424, 434, 461, 523, 526,
    533, 536, 540, 543, 551, 631
stress corrosion cracking (in hypochlorite
    solution)   315–316
stress cracking   336
strychnine (as inhibitor)   596
stuffing boxes   351
sub-arctic   121
subtropical climate   126
subtropical coastal atmosphere   145
sugar factory   116, 578
sugar manufacturing   353
sulfathiazole (as inhibitor)   476
sulfide impurities   497, 503
sulfide ions   80
sulfides   452, 461, 631
sulfochlorination processes   28–29
sulfonic acids   567
sulfur dioxide   129–130, 133, 277–278, 280
sulfur dioxide concentration   127
sulfuric acid   175

sulfurous acid   569–570, 574, 577–578
sulphamic acid   302
Sulzer steam generator   559
sunlight   197
sunlight irradiation   101
Sun Sol® 60   393
Sweden   120–121, 128, 149
swimming pool hall   313
sylvinite liquor   410, 413
sylvinite solutions   403, 410–411, 413
synthesis gas   324
synthetic drinking water   171
synthetic patina   135
synthetic seawater   425, 453, 461

t

tannic acid (as inhibitor)   356
tannin   390
tantalum fibers   267
tap water   465, 503
tarnish   134
tarnishing   141
technical regulations for steam boilers   69
Tedlar®   135
Tennessee Valley Authority   384
tetrachloroethane   186, 191
tetrafluorodibromomethane   267
textile bleach   307
thermal phosphoric acid   370–371, 373, 382, 384–385
thermometrics   293
N-2-thiophenylsalicylideneimine
   (as inhibitor)   597
thiourea   525, 567
thiourea (as inhibitor)   356, 362–363, 487, 510, 597
tidal water   430
tidal zones   448, 464, 556
toluene   153–154
toluene sulfonyl chlorides   27–29
m-toluidine (as inhibitor)   361
o-toluidine (as inhibitor)   7, 356, 362, 364
p-toluidine (as inhibitor)   361
toluyltriazole (as inhibitor)   135
tolyltriazole   101, 222, 391
tolyltriazole (as inhibitor)   316, 355
transgranular cracks   109
TRD   69
triacetate fibers tests   198
triacetyl cellulose   198
1.2.4-triazole (as inhibitor)   477
3-triazole (as inhibitor)   477
triazole (as inhibitor)   360, 501, 510

triazoles   100–101, 114, 261
tributylamine, all conc.   39
trichloroacetic acid (as inhibitor)   360–361, 363–364
trichloroethane   190, 194
1,1,1-trichloroethane   183, 185, 191, 193
1,1,2-trichloroethane   185, 191
triethanolamine   39, 153, 389–390
triethanolamine phosphate   389
triethylamine   37
triethylamine-2,4-dinitronaphtholate   63
triethylamine, moist   39
triethylamine, pure   39
trifluoroacetyl fluoride   27
trimethylamine   39
tropical climate   138
tropical coastal atmosphere   142, 144
tropic seawater   448
turbine steam   560

u

Ulm   132
ultraviolet radiation   126
underground corrosion   553
underground water   72
urban atmosphere   121, 126–127, 134, 144, 147–150
urban industrial atmosphere   140
urea (as inhibitor)   510
urea-formaldehyde resins   35
USA   121, 144, 149

v

vacuum evaporator   393–394, 434, 521
valves   77, 86, 267, 430, 458, 503
vapor atmosphere   71
VCI   135
VCI 25®   135
VCI packing   135
Venice (Italy)   146
ventilation galvanic elements   503
vibration cavitation   **486**
Vietnam   125
vinegar   14, 18, 23
vinegar bacteria   14
Vlaardingen (Netherlands)   146
volatile corrosion inhibitors   135
volcanic ash   131

w

washers   324
Washington/DC   122
wash solution   298

waste incineration plant   631
waste water evaporation plants   631
water chambers   458
water containers   448
water free carbon tetrachloride   206
water impurities   154
water pipes   243, 556
water pipe systems   **171**
water pressure   473
water/steam cycles   559
water vapor   265
weapons   144
welding technology   167
wet phosphoric acid   385
wetting agents   595

wine vinegar   14, 134
wire drawing works   133
Wladiwostok   128
WOL specimens   87
wooden boxes   21

x
xylitol   389

z
zinc anodes   556
zinc depletion   77, 79, 82, 84–85, 89, 141, 150, 211, 215, **503**–504, 507, 510, 515–516, 519
zinc depletion factor   84